Student Solutions Manual

for

Stewart/ Redlin/Watson's

COLLEGE ALGEBRA

2nd Edition

John A. Banks

Brooks/Cole Publishing Company

I(T)P® An International Thomson Publishing Company

Pacific Grove • Albany • Bonn • Boston • Cincinnati • Detroit • London • Madrid • Melbourne
Mexico City • New York • Paris • San Francisco • Singapore • Tokyo • Toronto • Washington

Assistant Editor: *Elizabeth Barelli Rammel*
Cover Design: *Vernon T. Boes*
Cover Photo: *David Ash, Tony Stone Worldwide*
Editorial Assistant: *Carol Ann Benedict*
Marketing Team: *Patrick Farrant and Margaret Parks*
Production Editor: *Mary Vezilich*
Printing and Binding: *Malloy Lithographing*

For more information, contact:

BROOKS/COLE PUBLISHING COMPANY
511 Forest Lodge Road
Pacific Grove, CA 93950
USA

International Thomson Publishing Europe
Berkshire House 168-173
High Holborn
London WC1V 7AA
England

Thomas Nelson Australia
102 Dodds Street
South Melbourne, 3205
Victoria, Australia

Nelson Canada
1120 Birchmount Road
Scarborough, Ontario
Canada M1K 5G4

International Thomson Editores
Campos Eliseos 385, Piso 7
Col. Polanco
11560 México D. F. México

International Thomson Publishing GmbH
Königswinterer Strasse 418
53227 Bonn
Germany

International Thomson Publishing Asia
221 Henderson Road
#05-10 Henderson Building
Singapore 0315

International Thomson Publishing Japan
Hirakawacho Kyowa Building, 3F
2-2-1 Hirakawacho
Chiyoda-ku, Tokyo 102
Japan

Printed in the United States of America

10 9 8 7 6 5 4

ISBN 0-534-33984-0

Table of Contents

Chapter One

Exercises 1.1

1. $a \cdot b = b \cdot a$

3. $(a - b)(a + b) = a^2 - b^2$

5. $(a + b) + c = a + (b + c)$

7. $(a \cdot b)^2 = a^2 \cdot b^2$

9. The average of two numbers, x and y, is $a = \dfrac{x + y}{2}$.

11. The sum of a number and twice its square is $S = x + 2x^2$ where x is the number.

13. The number of days in w weeks is $d = 7$w.

15. The product of two consecutive integers is $P = n(n + 1)$ where n is the smaller integer.

17. The sum of the squares of two numbers, x and y, is $s = x^2 + y^2$.

19. The time it takes an airplane to travel d miles at r miles per hour is $t = \frac{d}{r}$.

21. The area of a square of side x is $A = x^2$.

23. The volume of a box with square base of side x and height $2x$ is $V = x \cdot x \cdot 2x = 2x^3$.

25. A box of length l, width w, and height h has two sides of area $l \cdot h$, two sides of area $l \cdot w$, and two ends with area $w \cdot h$. So the surface area is $S = 2\,l\,h + 2lw + 2w\,h$.

27. The area of a triangle is $A = \frac{1}{2}$base $\cdot$ height. Since the base is twice the height h we get $A = (\frac{1}{2})(2h)(h) = h^2$.

29. The race track consist of a rectangle of length x and width $2r$ and two semi-circles of radius r. Each semi circle has length πr and two straight runs of length x. So the length is $L = 2x + 2\pi r$.

31. This is the volume of the large ball minus the volume of the inside ball. The volume of a ball (sphere) is $\frac{4}{3}\pi(\text{radius})^3$. Thus the volume is $V = \dfrac{4}{3}\pi R^3 - \dfrac{4}{3}\pi r^3 = \dfrac{4}{3}\pi(R^3 - r^3)$.

33. Since the final exam counts double we must divide the total by 4 (2 exams and 2 finals).

    (a) Average $= \dfrac{79 + 83 + 2(88)}{4} = 84.5$.

    (b) Average $= \dfrac{a + b + 2f}{4}$.

    (c) Average $= \dfrac{79 + 83 + 2f}{4} = 85 \Rightarrow 79 + 83 + 2f = 340 \Rightarrow 2f = 178 \Rightarrow f = 89$.

35. (a) $150\ \dfrac{\text{ft}^2}{\text{min}} \times 30\ \text{min} = 4{,}500\ \text{ft}^2$.

    (b) Area $= 150\ \dfrac{\text{ft}^2}{\text{min}} \times \text{T min} = 150\text{T ft}^2$.

(c) The area of the lawn $= 80$ ft $\times$ 120 ft $= 9{,}600$ ft^2. If T is the time required, then, from part (b), $150T = 9600$, so $T = \dfrac{9600}{150} = 64$ min.

(d) If R is the rate of mowing, then $60R = 9600$, so $R = \dfrac{9600}{60} = 160\ \dfrac{\text{ft}^2}{\text{min}}$.

37. (a) $\text{GPA} = \dfrac{4a + 3b + 2c + 1d + 0f}{a + b + c + d + f} = \dfrac{4a + 3b + 2c + d}{a + b + c + d + f}$.

(b) Using $a = 2 \cdot 3 = 6$, $b = 4$, $c = 3 \cdot 3 = 9$, and $d = f = 0$ in the formula from part (a) we obtain $\text{GPA} = \dfrac{4 \cdot 6 + 3 \cdot 4 + 2 \cdot 9}{6 + 4 + 9} = \dfrac{54}{19} = 2.84$.

Exercises 1.2

1. Commutative Property for addition.
3. Associative Property for addition.
5. Distributive Property.
7. Distributive Property.
9. $3(x+y) = 3x + 3y$
11. $4(2m) = (4 \cdot 2)m = 8m$
13. $\frac{4}{3}(-6y) = \left[\frac{4}{3}(-6)\right]y = -8y$
15. $-\frac{5}{2}(2x - 4y) = -\frac{5}{2}(2x) + \frac{5}{2}(4y) = -5x + 10y$
17. (a) $\frac{4}{13} + \frac{3}{13} = \frac{7}{13}$ (b) $\frac{3}{10} + \frac{7}{15} = \frac{9}{30} + \frac{14}{30} = \frac{23}{30}$
19. (a) $\frac{2}{5} \div \frac{9}{10} = \frac{2}{5} \cdot \frac{10}{9} = \frac{4}{9}$
    (b) $\left(4 \div \frac{1}{2}\right) - \frac{1}{2} = \left(4 \cdot \frac{2}{1}\right) - \frac{1}{2} = 8 - \frac{1}{2} = 7\frac{1}{2}$ or $\frac{15}{2}$.
21. (a) False (b) True
23. (a) True (b) True
25. (a) True (b) True
27. (a) $x > 0$ (b) $t < 4$
    (c) $a \geq \pi$ (d) $-5 < x < \frac{1}{3}$
    (e) $|p - 3| \leq 5$
29. (a) $\mathrm{A} \cup \mathrm{B} = \{1, 2, 3, 4, 5, 6, 8\}$ (b) $\mathrm{A} \cap \mathrm{B} = \{2, 4, 6\}$
31. (a) $\mathrm{A} \cup \mathrm{C} = \{1, 2, 3, 4, 5, 6, 7, 8, 9, 10\}$ (b) $\mathrm{A} \cap \mathrm{C} = \phi$
33. (a) $\mathrm{B} \cup \mathrm{C} = \{x \mid x \leq 5\}$ (b) $\mathrm{B} \cap \mathrm{C} = \{x \mid -1 < x < 4\}$
35. $(-3, 0) = \{x \mid -3 < x < 0\}$

    -3 0
37. $[2, 8) = \{x \mid 2 \leq x < 8\}$

    2 8
39. $[-1, 1] = \{x \mid -1 \leq x \leq 1\}$

    -1 1
41. $[2, \infty) = \{x \mid x \geq 2\}$

    2
43. $(-\infty, -2] = \{x \mid x \leq -2\}$

    -2
45. $x \leq 1 \quad \Leftrightarrow \quad x \in (-\infty, 1]$

    1
47. $1 \leq x \leq 2 \quad \Leftrightarrow \quad x \in [1, 2]$

    1 2
49. $-2 < x \leq 1 \quad \Leftrightarrow \quad x \in (-2, 1]$

    -2 1

51. $x > -1 \Leftrightarrow x \in (-1, \infty)$

53. $(-2, 0) \cup (-1, 1) = (-2, 1)$

55. $[-4, 6] \cap [0, 8) = [0, 6]$

57. $(-\infty, -4) \cup (4, \infty)$

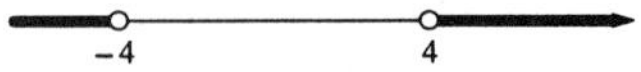

59. $(0, 7) \cap [2, \infty) = [2, 7)$

61. (a) $|100| = 100$ (b) $|-73| = 73$ (c) $|2 - 6| = |-4| = 4$

63. (a) $\left|\sqrt{3} - 3\right| = 3 - \sqrt{3}$ (since $\sqrt{3} < 3$) (b) $\left||-6| - |-4|\right| = |6 - 4| = |2| = 2$

(c) $\dfrac{-1}{|-1|} = \dfrac{-1}{1} = -1$

65. (a) $|17 - 2| = 15$ (b) $|21 - (-3)| = |21 + 3| = |24| = 24$

(c) $\left|-\dfrac{3}{10} - \dfrac{11}{8}\right| = \left|-\dfrac{12}{40} - \dfrac{55}{40}\right| = \left|-\dfrac{67}{40}\right| = \dfrac{67}{40}$

67. (a) Let $x = 0.777\ldots$. So,

$$\begin{aligned} 10x &= 7.7777\ldots \\ x &= 0.7777\ldots \\ \hline 9x &= 7 \end{aligned}$$

Thus $x = \frac{7}{9}$.

(b) Let $x = 0.2888\ldots$. So,

$$\begin{aligned} 100x &= 28.8888\ldots \\ 10x &= 2.8888\ldots \\ \hline 90x &= 26 \end{aligned}$$

Thus $x = \frac{26}{90} = \frac{13}{45}$.

(c) Let $x = 0.575757\ldots$. So,

$$\begin{aligned} 100x &= 57.5757\ldots \\ 1x &= 0.5757\ldots \\ \hline 99x &= 57 \end{aligned}$$

Thus $x = \frac{57}{99} = \frac{19}{33}$.

69. (a) Construct the number $\sqrt{2}$ on the number line transferring the length from the hypothenuse of the right triangle.

(b) Construct the number $\sqrt{2} - 1$ by on the number line by removing the a segment of length 1 from the number constructed in part (a).

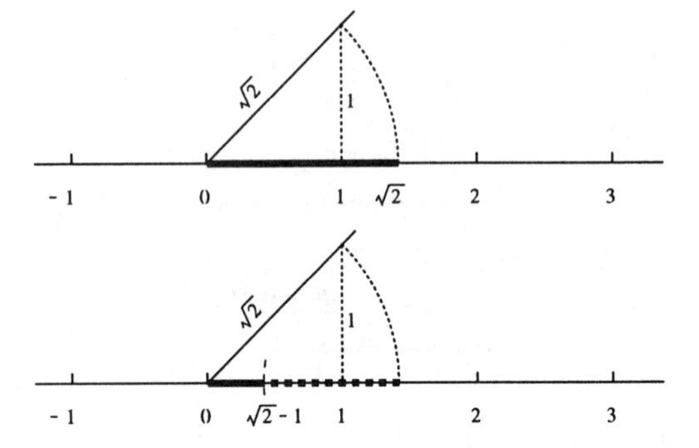

(c) Since $\sqrt{8} = 2\sqrt{2} = \sqrt{2} + \sqrt{2}$, the number $\sqrt{8}$ can be constructed by adding two segments construct in part (a).

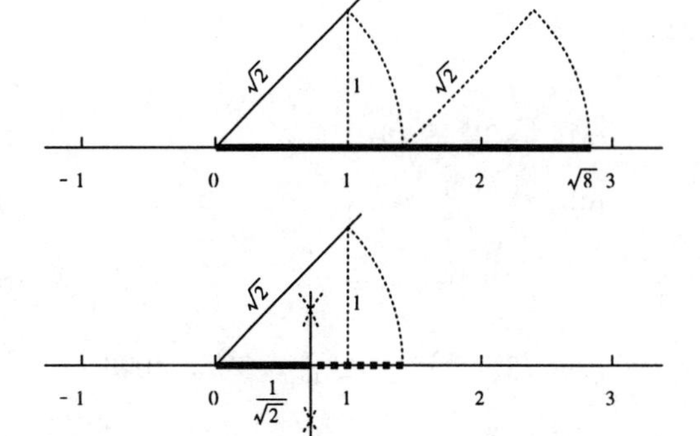

(d) Since $\dfrac{1}{\sqrt{2}} = \dfrac{1}{\sqrt{2}} \cdot \dfrac{\sqrt{2}}{\sqrt{2}} = \dfrac{\sqrt{2}}{2}$, we can construct this number by bisecting the segment constructed in part (a).

71. (a)

x	1	2	10	100	1000
$\dfrac{1}{x}$	1	$\dfrac{1}{2}$	$\dfrac{1}{10}$	$\dfrac{1}{100}$	$\dfrac{1}{1000}$

x	1	.5	.1	.01	.001
$\dfrac{1}{x}$	1	$\dfrac{1}{.5} = 2$	$\dfrac{1}{.1} = 10$	$\dfrac{1}{.01} = 100$	$\dfrac{1}{.001} = 1000$

(b) As x gets large, the fraction $\dfrac{1}{x}$ gets small. Mathematically, we say that $\dfrac{1}{x}$ goes to zero.

(c) As x gets small, the fraction $\dfrac{1}{x}$ gets large. Mathematically, we say that $\dfrac{1}{x}$ goes to infinity.

73. (a) No. Example: π is irrational but $\pi + (-\pi) = 0$ and 0 is not irrational.

(b) No. Example: $\sqrt{2}$ is irrational but $\sqrt{2} \cdot \sqrt{2} = 2$ and 2 is not irrational.

Exercises 1.3

1. (a) $(-2)^4 = 16$ (b) $-2^4 = -16$

   (c) $\pi^0 = 1$

3. (a) $2^{-3}5^4 = \dfrac{5^4}{2^3} = \dfrac{625}{8}$ (b) $\dfrac{10^9}{10^4} = 10^{9-4} = 10^5 = 100,000$

   (c) $\left(2^4 \cdot 2^2\right)^2 = \left(2^6\right)^2 = 2^{12} = 4096$

5. (a) $\sqrt{\dfrac{4}{9}} = \dfrac{\sqrt{4}}{\sqrt{9}} = \dfrac{2}{3}$ (b) $\sqrt[4]{256} = \sqrt[4]{4^4} = 4$

   (c) $\sqrt[6]{\dfrac{1}{64}} = \sqrt[6]{\dfrac{1}{2^6}} = \dfrac{\sqrt[6]{1}}{\sqrt[6]{2^6}} = \dfrac{1}{2}$

7. (a) $\dfrac{\sqrt{72}}{\sqrt{2}} = \sqrt{\dfrac{72}{2}} = \sqrt{36} = 6$ (b) $\dfrac{\sqrt{48}}{\sqrt{3}} = \sqrt{\dfrac{48}{3}} = \sqrt{16} = 4$

   (c) $\sqrt{\dfrac{9}{25}} = \dfrac{\sqrt{9}}{\sqrt{25}} = \dfrac{3}{5}$

9. (a) $\left(\dfrac{4}{9}\right)^{-1/2} = \left(\dfrac{2^2}{3^2}\right)^{-1/2} = \dfrac{2^{-1}}{3^{-1}} = \dfrac{3}{2}$ (b) $\left(-\dfrac{27}{8}\right)^{2/3} = \left(\sqrt[3]{-\dfrac{27}{8}}\right)^2 = \left(-\dfrac{3}{2}\right)^2 = \dfrac{9}{4}$

   (c) $\left(\dfrac{25}{64}\right)^{3/2} = \left(\sqrt{\dfrac{25}{64}}\right)^3 = \left(\dfrac{5}{8}\right)^3 = \dfrac{125}{512}$

11. $\sqrt[3]{108} - \sqrt[3]{32} = 3\sqrt[3]{4} - 2\sqrt[3]{4} = \sqrt[3]{4}$

13. $\sqrt{245} - \sqrt{125} = 7\sqrt{5} - 5\sqrt{5} = 2\sqrt{5}$

15. (a) $128 = 2^7$ (b) $128^2 = \left(2^7\right)^2 = 2^{14}$ (c) $\left(2^9\right)^4 = 2^{36}$

17. (a) $2^8 \cdot 4^{-7} = 2^8 \cdot 2^{-14} = 2^{8-14} = 2^{-6}$ (b) $4\sqrt{2} = 2^2 \cdot 2^{1/2} = 2^{(2+\frac{1}{2})} = 2^{5/2}$

    (c) $\dfrac{1}{\sqrt{2}} = \dfrac{1}{2^{1/2}} = 2^{-1/2}$

19. $t^7 t^{-2} = t^{7-2} = t^5$

21. $\left(12x^2y^4\right)\left(\frac{1}{2}x^5y\right) = \left(12 \cdot \frac{1}{2}\right)x^{2+5}y^{4+1} = 6x^7y^5$

23. $\dfrac{x^9(2x)^4}{x^3} = 2^4 \cdot x^{9+4-3} = 16x^{10}$

25. $b^4\left(\dfrac{1}{3}b^2\right)\left(12b^{-8}\right) = \dfrac{12}{3}b^{4+2-8} = 4b^{-2} = \dfrac{4}{b^2}$

27. $(rs)^3(2s)^{-2}(4r)^4 = r^3s^32^{-2}s^{-2}4^4r^4 = r^3s^32^{-2}s^{-2}2^{2\cdot 4}r^4 = 2^{-2+8}r^{3+4}s^{3-2} = 2^6r^7s = 64r^7s$

29. $\dfrac{(6y^3)^4}{2y^5} = \dfrac{6^4y^{3\cdot 4}}{2y^5} = \dfrac{6^4}{2}\,y^{12-5} = 648y^7$

31. $\dfrac{(x^2y^3)^4(xy^4)^{-3}}{x^2y} = \dfrac{x^8y^{12}x^{-3}y^{-12}}{x^2y} = x^{8-3-2}y^{12-12-1} = x^3y^{-1} = \dfrac{x^3}{y}$

33. $\dfrac{(xy^2z^3)^4}{(x^3y^2z)^3} = \dfrac{x^4y^8z^{12}}{x^9y^6z^3} = x^{4-9}y^{8-6}z^{12-3} = x^{-5}y^2z^9 = \dfrac{y^2z^9}{x^5}$

35. $\left(\dfrac{q^{-1}rs^{-2}}{r^{-5}sq^{-8}}\right)^{-1} = \dfrac{qr^{-1}s^2}{r^5s^{-1}q^8} = q^{1-8}r^{-1-5}s^{2-(-1)} = q^{-7}r^{-6}s^3 = \dfrac{s^3}{q^7r^6}$

37. $x^{2/3}x^{1/5} = x^{(10/15+3/15)} = x^{13/15}$

39. $(4b)^{1/2}(8b^{2/5}) = \sqrt{4}\cdot 8b^{1/2}b^{2/5} = 16b^{(5/10+4/10)} = 16b^{9/10}$

41. $(c^2d^3)^{-1/3} = c^{-2/3}d^{-1} = \dfrac{1}{c^{2/3}d}$

43. $(y^{3/4})^{2/3} = y^{(3/4)\cdot(2/3)} = y^{1/2}$

45. $(2x^4y^{-4/5})^3(8y^2)^{2/3} = 2^3x^{12}y^{-12/5}8^{2/3}y^{4/3} = 2^{3+2}x^{12}y^{(-12/5+4/3)} = \dfrac{32x^{12}}{y^{16/15}}$ (Note that $8^{2/3} = (8^{1/3})^2 = 2^2$).

47. $\left(\dfrac{x^6y}{y^4}\right)^{5/2} = \dfrac{x^{15}y^{5/2}}{y^{10}} = x^{15}y^{5/2-10} = x^{15}y^{-15/2} = \dfrac{x^{15}}{y^{15/2}}$

49. $\left(\dfrac{3a^{-2}}{4b^{-1/3}}\right)^{-1} = \dfrac{3^{-1}a^2}{4^{-1}b^{1/3}} = \dfrac{4a^2}{3b^{1/3}}$

51. $\dfrac{(9st)^{3/2}}{(27s^3t^{-4})^{2/3}} = \dfrac{27s^{3/2}t^{3/2}}{9s^2t^{-8/3}} = 3s^{3/2-2}t^{3/2+8/3} = 3s^{-1/2}t^{25/6} = \dfrac{3t^{25/6}}{s^{1/2}}$

53. $\sqrt[4]{x^4} = |x|$

55. $\sqrt[3]{x^3y} = (x^3)^{1/3}y^{1/3} = x\sqrt[3]{y}$

57. $\sqrt[5]{a^6b^7} = a^{6/5}b^{7/5} = a\cdot a^{1/5}b\cdot b^{2/5} = ab\sqrt[5]{ab^2}$

59. $\sqrt{x^2y^6} = |xy^3|$

61. $\sqrt[3]{\sqrt{64x^6}} = (8|x^3|)^{1/3} = 2|x|$

63. (a) $\dfrac{1}{\sqrt{6}} = \dfrac{1}{\sqrt{6}}\cdot\dfrac{\sqrt{6}}{\sqrt{6}} = \dfrac{\sqrt{6}}{6}$

(b) $\sqrt{\dfrac{x}{3y}} = \dfrac{\sqrt{x}}{\sqrt{3y}} = \dfrac{\sqrt{x}}{\sqrt{3y}}\cdot\dfrac{\sqrt{3y}}{\sqrt{3y}} = \dfrac{\sqrt{3xy}}{3y}$

(c) $\sqrt{\dfrac{3}{20}} = \dfrac{\sqrt{3}}{2\sqrt{5}}\cdot\dfrac{\sqrt{5}}{\sqrt{5}} = \dfrac{\sqrt{3}\sqrt{5}}{10} = \dfrac{\sqrt{15}}{10}$

65. (a) $\dfrac{1}{\sqrt[3]{x}} = \dfrac{1}{\sqrt[3]{x}}\cdot\dfrac{\sqrt[3]{x^2}}{\sqrt[3]{x^2}} = \dfrac{\sqrt[3]{x^2}}{x}$

(b) $\dfrac{1}{\sqrt[5]{x^2}} = \dfrac{1}{\sqrt[5]{x^2}}\cdot\dfrac{\sqrt[5]{x^3}}{\sqrt[5]{x^3}} = \dfrac{\sqrt[5]{x^3}}{x}$

(c) $\dfrac{1}{\sqrt[7]{x^3}} = \dfrac{1}{\sqrt[7]{x^3}}\cdot\dfrac{\sqrt[7]{x^4}}{\sqrt[7]{x^4}} = \dfrac{\sqrt[7]{x^4}}{x}$

67. (a) $69,300,000 = 6.93\times 10^7$

(b) $0.000028536 = 2.8536\times 10^{-5}$

(c) $129{,}540{,}000 = 1.2954\times 10^8$

69. (a) $3.19\times 10^5 = 319{,}000$

(b) $2.670\times 10^{-8} = 0.0000000267$

(c) $7.1 \times 10^{14} = 710{,}000{,}000{,}000{,}000$

71. (a) $5{,}900{,}000{,}000{,}000 \text{ mi} = 5.9 \times 10^{12} \text{ mi}$ (b) $0.0000000000004 \text{ cm} = 4 \times 10^{-13} \text{ cm}$
(c) 33 billion billion molecules $= 33 \times 10^9 \times 10^9 = 3.3 \times 10^{19}$ molecules

73. $(7.2 \times 10^{-9})(1.806 \times 10^{-12}) = 7.2 \times 1.806 \times 10^{-9} \times 10^{-12} \approx 1.3 \times 10^{-20}$

75. $\dfrac{1.295643 \times 10^9}{(3.610 \times 10^{-17})(2.511 \times 10^6)} = \dfrac{1.295643}{3.610 \times 2.511} \times 10^{9+17-6} \approx 1.429 \times 10^{19}$

77. $\dfrac{(0.0000162)(0.01582)}{(594621000)(0.0058)} = \dfrac{(1.62 \times 10^{-5})(1.582 \times 10^{-2})}{(5.94621 \times 10^8)(5.8 \times 10^{-3})} = \dfrac{1.62 \times 1.582}{5.94621 \times 5.8} \times 10^{-5-2-8+3}$
$\approx 7.4 \times 10^{-14}$

79. $9.3 \times 10^6 \text{mi} = 186{,}000 \dfrac{\text{mi}}{\text{sec}} \times t \text{ sec} \quad \Leftrightarrow \quad t = \dfrac{9.3 \times 10^6}{186{,}000} \text{ sec} = 500 \text{ sec} = 8\dfrac{1}{3} \text{ min.}$

81. Since $10^{101} = 10 \cdot 10^{100}$, we have $10^{101} - 10^{100} = 10 \cdot 10^{100} - 1 \cdot 10^{100} = (10-1) \cdot 10^{100} = 9 \times 10^{100}$. But $10^{53} - 10^{10} < 10^{53} < 9 \times 10^{100}$. Therefore the pair 10^{10} and 10^{53} are closer together.

83. (a) Using $v = \dfrac{1}{2}c$ and inserting this into the formula to get $m = \dfrac{m_0}{\sqrt{1 - \frac{v^2}{c^2}}}$. Simplifying we get:

$m = \dfrac{m_0}{\sqrt{1 - \frac{\left(\frac{1}{2}c\right)^2}{c^2}}} = \dfrac{m_0}{\sqrt{1 - \frac{1}{4}}}$. Thus $m = \dfrac{1}{\sqrt{\frac{3}{4}}} m_0 = \dfrac{2}{\sqrt{3}} m_0 = \dfrac{2\sqrt{3}}{3} m_0$. So the rest mass of the spaceship is multiplied by $\frac{2\sqrt{3}}{3} \approx 1.15$. (*Note: There is another way to do this problem. Start with* $c = 3 \times 10^5$ km/sec *and* $v = 1.5 \times 10^5$km/sec. *Substituting with these two values will not alter the result.*)

(b) As the spaceship travels very close to the speed of light, $v \to c$ so the term $\dfrac{v^2}{c^2}$ approaches 1.
So $\sqrt{1 - \dfrac{v^2}{c^2}}$ approaches 0 and we obtain $m = \dfrac{m_0}{\text{very small number}}$ which means $m = (\text{very large number}) m_0$. Hence the mass of the space ship become arbitrary large as its speed approaches the speed of light.

85. (a) Using $f = 0.4$ and substituting $d = 65$, we obtain: $s = \sqrt{30fd} = \sqrt{30 \times 0.4 \times 65} \approx 28$ mi/h.
(b) Using $f = 0.5$ and substituting $s = 50$, we find d. $s = \sqrt{30fd} \quad \Leftrightarrow \quad 50 = \sqrt{30 \cdot (0.5)d}$
$\Leftrightarrow \quad 50 = \sqrt{15d} \quad \Leftrightarrow \quad 2500 = 15d \quad \Leftrightarrow \quad d = \dfrac{500}{3} \approx 167$ feet.

87. Since 1 day $= 86{,}400$sec, 365.25days $= 31{,}557{,}600$sec. Substituting we obtain
$d = \left(\dfrac{6.67 \times 10^{-11} \times 1.99 \times 10^{30}}{4\pi^2}\right)^{1/3} \cdot \left(3.15576 \times 10^7\right)^{2/3} \approx 1.5 \times 10^{11}\text{m} = 1.5 \times 10^8 \text{ km}.$

Exercises 1.4

1. $2(x-1)+4(x+2)=2x-2+4x+8=6x+6$

3. $(2x^2+x+1)+(x^2-3x+5)=3x^2-2x+6$

5. $(x^3+6x^2-4x+7)-(3x^2+2x-4)=x^3+6x^2-4x+7-3x^2-2x+4=$
$x^3+3x^2-6x+11$

7. $2(2-5t)+t^2(t-1)-(t^4-1)=4-10t+t^3-t^2-t^4+1=\ -t^4+t^3-t^2-10t+5$

9. $\sqrt{x}(x-\sqrt{x})=x^{1/2}(x-x^{1/2})=x^{1/2}x-x^{1/2}x^{1/2}=x^{3/2}-x$

11. $\sqrt[3]{y}\,(y^2-1)=y^{1/3}(y^2-1)=y^{1/3}y^2-y^{1/3}=y^{7/3}-y^{1/3}$

13. $(3t-2)(7t-5)=21t^2-15t-14t+10=21t^2-29t+10$

15. $(x+2y)(3x-y)=3x^2-xy+6xy-2y^2=3x^2+5xy-2y^2$

17. $(1-2y$ [illegible] $4y+4y^2$

19. $(2x-5)(x^2-x+1)=2x^3-2x^2+2x-5x^2+5x-5=2x^3-7x^2+7x-5$

21. $x(x-1)(x+2)=(x^2-x)(x+2)=x^3+2x^2-x^2-2x=x^3+x^2-2x$

23. $y^4(6-y)(5+y)=y^4(30+6y-5y-y^2)=y^4(30+y-y^2)=30y^4+y^5-y^6$

25. $(2x^2+3y^2)^2=(2x^2)^2+2(2x^2)(3y^2)+(3y^2)^2=4x^4+12x^2y^2+9y^4$

27. $(x^2-a^2)(x^2+a^2)=(x^2)^2-(a^2)^2=x^4-a^4$ (the difference of squares)

29. $(1+a^3)^3=1+3(a^3)+3(a^3)^2+(a^3)^3=1+3a^3+3a^6+a^9$ (perfect cube)

31. $\left(\sqrt{a}-\frac{1}{b}\right)\left(\sqrt{a}+\frac{1}{b}\right)=(\sqrt{a})^2-\left(\frac{1}{b}\right)^2=a-\frac{1}{b^2}$ (the difference of squares)

33. $(x^2+x-2)(x^3-x+1)=x^5-x^3+x^2+x^4-x^2+x-2x^3+2x-2=$
$x^5+x^4-3x^3+3x-2$

35. $(1+x^{4/3})(1-x^{2/3})=1-x^{2/3}+x^{4/3}-x^{6/3}=1-x^{2/3}+x^{4/3}-x^2$

37. $(1-b)^2(1+b)^2=\Big[(1-b)(1+b)\Big]^2=(1-b^2)^2=1-2b^2+b^4$

39. $(3x^2y+7xy^2)(x^2y^3-2y^2)=3x^4y^4-6x^2y^3+7x^3y^5-14xy^4$

41. $2x+12x^3=2x(1+6x^2)$

43. $6y^4-15y^3=3y^3(2y-5)$

45. $x^2+7x+6=(x+6)(x+1)$

47. $x^2-2x-8=(x-4)(x+2)$

49. $y^2-8y+15=(y-5)(y-3)$

51. $2x^2+5x+3=(2x+3)(x+1)$

53. $9x^2 - 36 = 9(x^2 - 4) = 9(x-2)(x+2)$

55. $6x^2 - 5x - 6 = (3x+2)(2x-3)$

57. $(x-1)(x+2)^2 - (x-1)^2(x+2) = (x-1)(x+2)\left[(x+2)-(x-1)\right] = 3(x-1)(x+2)$

59. $y^4(y+2)^3 + y^5(y+2)^4 = y^4(y+2)^3\left[(1) + y(y+2)\right] = y^4(y+2)^3(y^2+2y+1)$
$= y^4(y+2)^3(y+1)^2$

61. $(a^2-1)b^2 - 4(a^2-1) = (a^2-1)(b^2-4) = (a-1)(a+1)(b-2)(b+2)$

63. $t^3 + 1 = (t+1)(t^2-t+1)$ (the sum of cubes)

65. $4t^2 - 12t + 9 = (2t-3)^2$ (perfect square)

67. $x^3 + 2x^2 + x = x(x^2+2x+1) = x(x+1)^2$

69. $4x^2 + 4xy + y^2 = (2x+y)^2$ (perfect square)

71. $x^4 + 2x^3 - 3x^2 = x^2(x^2+2x-3) = x^2(x-1)(x+3)$

73. $8x^3 - 125 = (2x)^3 - (5)^3 = (2x-5)\left[(2x)^2 + (2x)(5) + (5)^2\right] = (2x-5)(4x^2+10x+25)$ (the difference of cubes)

75. $x^4 + x^2 - 2 = (x^2)^2 + (x^2) - 2 = (x^2+2)(x^2-1) = (x^2+2)(x-1)(x+1)$

77. $y^3 - 3y^2 - 4y + 12 = (y^3 - 3y^2) + (-4y+12) = y^2(y-3) + (-4)(y-3) = (y-3)(y^2-4)$
$= (y-3)(y-2)(y+2)$ (factor by grouping)

79. $2x^3 + 4x^2 + x + 2 = (2x^3 + 4x^2) + (x+2) = 2x^2(x+2) + (1)(x+2) = (x+2)(2x^2+1)$
(factor by grouping)

81. $x^6 - y^6 = (x^3)^2 - (y^3)^2 = (x^3 - y^3)(x^3 + y^3)$ (the difference of squares.)
$= (x-y)(x^2+xy+y^2)(x+y)(x^2-xy+y^2)$ (the difference of cubes and the sum of cubes)

or

$x^6 - y^6 = (x^2)^3 - (y^2)^3 = (x^2-y^2)\left[(x^2)^2 + x^2y^2 + (y^2)^2\right]$ (the difference of cubes)
$= (x-y)(x+y)(x^4 + x^2y^2 + y^4)$ (the difference of squares)
$= (x-y)(x+y)(x^2+xy+y^2)(x^2-xy+y^2)$

However, factoring $x^4 + x^2y^2 + y^4$ into $(x^2+xy+y^2)(x^2-xy+y^2)$ is not an easy task.

83. $x^{5/2} - x^{1/2} = x^{1/2}(x^2-1) = \sqrt{x}(x-1)(x+1)$

85. Start by factoring out the power of x with the smallest exponent, that is, $x^{-3/2}$. So
$x^{-3/2} + 2x^{-1/2} + x^{1/2} = x^{-3/2}(1 + 2x + x^2) = \dfrac{(1+x)^2}{x^{3/2}}$

87. Start by factoring out the power of (x^2+1) with the smallest exponent, that is, $(x^2+1)^{-1/2}$. So
$(x^2+1)^{1/2} + 2(x^2+1)^{-1/2} = (x^2+1)^{-1/2}\left[(x^2+1)+2\right] = \dfrac{x^2+3}{\sqrt{x^2+1}}$

89. Start by factoring $y^2 - 7y + 10$ and then substitute $a^2 + 1$ for y.
$(a^2+1)^2 - 7(a^2+1) + 10 = \left[(a^2+1) - 2\right]\left[(a^2+1) - 5\right] = (a^2-1)(a^2-4) =$
$(a-1)(a+1)(a-2)(a+2)$

91. $x^4 + 3x^2 + 4 = (x^4 + 4x^2 + 4) - x^2 = (x^2+2)^2 - x^2 = \left[(x^2+2) - x\right]\left[(x^2+2) + x\right] =$
$(x^2 - x + 2)(x^2 + x + 2)$

93. $(ac+bd)^2 + (ad-bc)^2 = (a^2c^2 + 2abcd + b^2d^2) + (a^2d^2 - 2abcd + b^2c^2)$
$= a^2c^2 + a^2d^2 + b^2c^2 + b^2d^2$ (rearranging the terms)
$= a^2(c^2+d^2) + b^2(c^2+d^2)$
$= (a^2+b^2)(c^2+d^2)$

95. $4a^2c^2 - (c^2 - b^2 + a^2)^2 = (2ac)^2 - (c^2 - b^2 + a^2)$
$= \left[(2ac) - (c^2 - b^2 + a^2)\right]\left[(2ac) + (c^2 - b^2 + a^2)\right]$ (the difference of squares)
$= (2ac - c^2 + b^2 - a^2)(2ac + c^2 - b^2 + a^2)$
$= \left[b^2 - (c^2 - 2ac + a^2)\right]\left[(c^2 + 2ac + a^2) - b^2\right]$ (regrouping)
$= \left[b^2 - (c-a)^2\right]\left[(c+a)^2 - b^2\right]$ (perfect squares)
$= [b - (c-a)][b + (c-a)][(c+a) - b][(c+a) + b]$ (each factor is a difference of squares)
$= (b - c + a)(b + c - a)(c + a - b)(c + a + b)$

97. When $a > b > 0$, the area of the square with side a minus the area of the square with side b is equal to the sum of the areas of the two rectangles. That is,
$a^2 - b^2 = a(a-b) + b(a-b) = (a-b)(a+b)$.

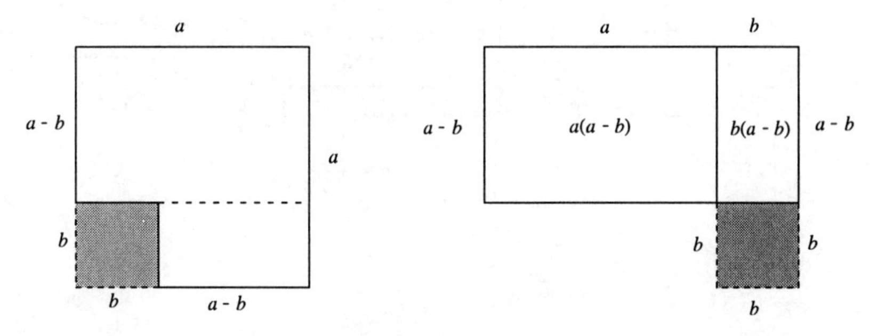

99. The figure on the next page shows how the large $a+b$ by $a+b$ by $a+b$ cube can be separated into one cube whose volume is a^3, three boxes whose volumes are $a \times a \times b$, three boxes whose volumes are $a \times b \times b$, and one cube whose volume is b^3. That is,
$(a+b)^3 = a^3 + 3a^2b + 3ab^2 + b^3$.

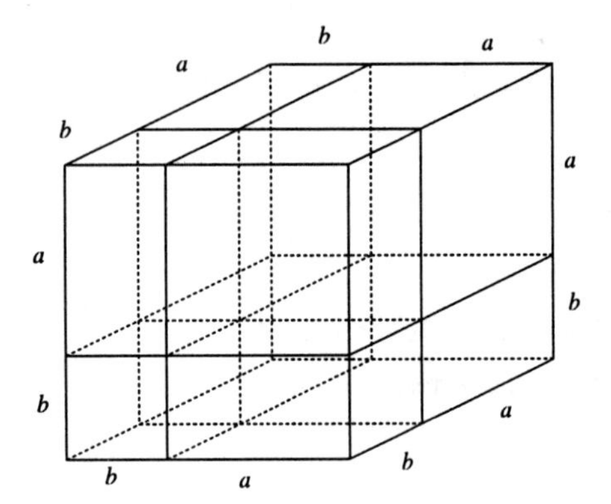
b
a
a
b
a
a
b
b
a
b
b
a

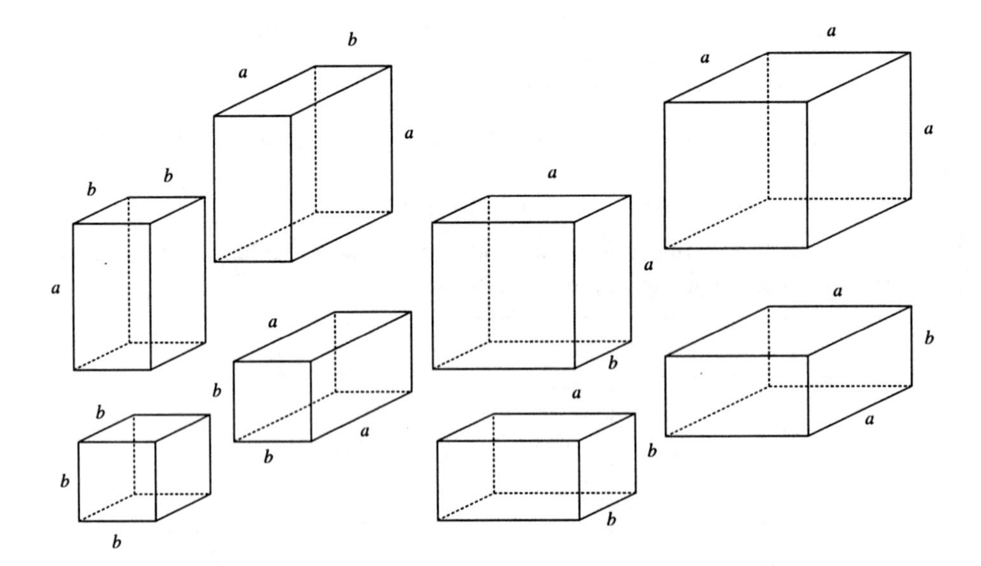
b
a
a
a
a
a
b
b
a
a
a
a
b
a
b
b
b
a
b
a
b
b
b
b
b

Exercises 1.5

1. $$\frac{x^2+3x+2}{x^2+5x+6} = \frac{(x+2)(x+1)}{(x+2)(x+3)} = \frac{x+1}{x+3}$$

3. $$\frac{y-y^2}{y^2-1} = \frac{y(1-y)}{(y-1)(y+1)} = \frac{-y}{y+1}$$

5. $$\frac{2x^3-x^2-6}{2x^2-7x+6} = \frac{x(x^2-x-6)}{(2x-3)(x-2)} = \frac{x(2x+3)(x-2)}{(2x-3)(x-2)} = \frac{x(2x+3)}{2x-3}$$

7. $$\frac{t-3}{t^2+9}\cdot\frac{t+3}{t^2-9} = \frac{(t-3)(t+3)}{(t^2+9)(t-3)(t+3)} = \frac{1}{t^2+9}$$

9. $$\frac{x^2+7x+12}{x^2+3x+2}\cdot\frac{x^2+5x+6}{x^2+6x+9} = \frac{(x+3)(x+4)}{(x+1)(x+2)}\cdot\frac{(x+2)(x+3)}{(x+3)(x+3)} = \frac{x+4}{x+1}$$

11. $$\frac{2x^2+3x+1}{x^2+2x-15}\div\frac{x^2+6x+5}{2x^2-7x+3} = \frac{2x^2+3x+1}{x^2+2x-15}\cdot\frac{2x^2-7x+3}{x^2+6x+5} =$$
$$\frac{(2x+1)(x+1)}{(x-3)(x+5)}\cdot\frac{(2x-1)(x-3)}{(x+1)(x+5)} = \frac{(2x+1)(2x-1)}{(x+5)(x+5)} = \frac{(2x+1)(2x-1)}{(x+5)^2}$$

13. $$\frac{\dfrac{x^3}{x+1}}{\dfrac{x}{x^2+2x+1}} = \frac{x^3}{x+1}\cdot\frac{x^2+2x+1}{x} = \frac{x^3(x+1)(x+1)}{(x+1)x} = x^2(x+1)$$

15. $$\frac{x/y}{z} = \frac{x}{y}\cdot\frac{1}{z} = \frac{x}{yz}$$

17. $$\frac{1}{x+5}+\frac{2}{x-3} = \frac{x-3}{(x+5)(x-3)}+\frac{2(x+5)}{(x+5)(x-3)} = \frac{x-3+2x+10}{(x+5)(x-3)} = \frac{3x+7}{(x+5)(x-3)}$$

19. $$\frac{1}{x+1}-\frac{1}{x+2} = \frac{x+2}{(x+1)(x+2)}+\frac{-(x+1)}{(x+1)(x+2)} = \frac{x+2-x-1}{(x+1)(x+2)} = \frac{1}{(x+1)(x+2)}$$

21. $$\frac{x}{(x+1)^2}+\frac{2}{x+1} = \frac{x}{(x+1)^2}+\frac{2(x+1)}{(x+1)(x+1)} = \frac{x+2x+2}{(x+1)^2} = \frac{3x+2}{(x+1)^2}$$

23. $$u+1+\frac{u}{u+1} = \frac{(u+1)(u+1)}{u+1}+\frac{u}{u+1} = \frac{u^2+2u+1+u}{u+1} = \frac{u^2+3u+1}{u+1}$$

25. $$\frac{1}{x^2}+\frac{1}{x^2+x} = \frac{1}{x^2}+\frac{1}{x(x+1)} = \frac{x+1}{x^2(x+1)}+\frac{x}{x^2(x+1)} = \frac{2x+1}{x^2(x+1)}$$

27. $$\frac{2}{x+3}-\frac{1}{x^2+7x+12} = \frac{2}{x+3}-\frac{1}{(x+3)(x+4)} = \frac{2(x+4)}{(x+3)(x+4)}+\frac{-1}{(x+3)(x+4)}$$
$$= \frac{2x+8-1}{(x+3)(x+4)} = \frac{2x+7}{(x+3)(x+4)}$$

29. $\dfrac{1}{x+3}+\dfrac{1}{x^2-9}=\dfrac{1}{x+3}+\dfrac{1}{(x-3)(x+3)}=\dfrac{x-3}{(x-3)(x+3)}+\dfrac{1}{(x-3)(x+3)}$
$=\dfrac{x-2}{(x-3)(x+3)}$

31. $\dfrac{2}{x}+\dfrac{3}{x-1}-\dfrac{4}{x^2-x}=\dfrac{2}{x}+\dfrac{3}{x-1}-\dfrac{4}{x(x-1)}=\dfrac{2(x-1)}{x(x-1)}+\dfrac{3x}{x(x-1)}+\dfrac{-4}{x(x-1)}$
$=\dfrac{2x-2+3x-4}{x(x-1)}=\dfrac{5x-6}{x(x-1)}$

33. $\dfrac{1}{x^2+3x+2}-\dfrac{1}{x^2-2x-3}=\dfrac{1}{(x+2)(x+1)}-\dfrac{1}{(x-3)(x+1)}$
$=\dfrac{x-3}{(x-3)(x+2)(x+1)}+\dfrac{-(x+2)}{(x-3)(x+2)(x+1)}$
$=\dfrac{x-3-x-2}{(x-3)(x+2)(x+1)}=\dfrac{-5}{(x-3)(x+2)(x+1)}$

35. $\dfrac{\dfrac{x}{y}-\dfrac{y}{x}}{\dfrac{1}{x^2}-\dfrac{1}{y^2}}=\dfrac{\dfrac{x^2-y^2}{xy}}{\dfrac{y^2-x^2}{x^2y^2}}=\dfrac{x^2-y^2}{xy}\cdot\dfrac{x^2y^2}{y^2-x^2}=\dfrac{xy}{-1}=-xy$. An alternative method is to multiply the numerator and denominator by the common denominator of both the numerator and denominator, in this case x^2y^2.

$$\dfrac{\dfrac{x}{y}-\dfrac{y}{x}}{\dfrac{1}{x^2}-\dfrac{1}{y^2}}=\dfrac{\left(\dfrac{x}{y}-\dfrac{y}{x}\right)}{\left(\dfrac{1}{x^2}-\dfrac{1}{y^2}\right)}\cdot\dfrac{x^2y^2}{x^2y^2}=\dfrac{x^3y-xy^3}{y^2-x^2}=\dfrac{xy(x^2-y^2)}{y^2-x^2}=-xy$$

37. $\dfrac{1+\dfrac{1}{c-1}}{1-\dfrac{1}{c-1}}=\dfrac{\dfrac{c-1}{c-1}+\dfrac{1}{c-1}}{\dfrac{c-1}{c-1}+\dfrac{-1}{c-1}}=\dfrac{\dfrac{c}{c-1}}{\dfrac{c-2}{c-1}}=\dfrac{c}{c-1}\cdot\dfrac{c-1}{c-2}=\dfrac{c}{c-2}$. Using the alternative method we obtain:

$$\dfrac{1+\dfrac{1}{c-1}}{1-\dfrac{1}{c-1}}=\dfrac{\left(1+\dfrac{1}{c-1}\right)}{\left(1-\dfrac{1}{c-1}\right)}\cdot\dfrac{c-1}{c-1}=\dfrac{c-1+1}{c-1-1}=\dfrac{c}{c-2}$$

39. $\dfrac{\dfrac{5}{x-1}-\dfrac{2}{x+1}}{\dfrac{x}{x-1}+\dfrac{1}{x+1}}=\dfrac{\dfrac{5(x+1)}{(x-1)(x+1)}+\dfrac{-2(x-1)}{(x-1)(x+1)}}{\dfrac{x(x+1)}{(x-1)(x+1)}+\dfrac{x-1}{(x-1)(x+1)}}=\dfrac{\dfrac{5x+5-2x+2}{(x-1)(x+1)}}{\dfrac{x^2+x+x-1}{(x-1)(x+1)}}$
$=\dfrac{3x+7}{(x-1)(x+1)}\cdot\dfrac{(x-1)(x+1)}{x^2+2x-1}=\dfrac{3x+7}{x^2+2x-1}$

Alternatively,

$$\frac{\dfrac{5}{x-1}-\dfrac{2}{x+1}}{\dfrac{x}{x-1}+\dfrac{1}{x+1}}=\frac{\left(\dfrac{5}{x-1}-\dfrac{2}{x+1}\right)}{\left(\dfrac{x}{x-1}+\dfrac{1}{x+1}\right)}\cdot\frac{(x-1)(x+1)}{(x-1)(x+1)}=\frac{5(x+1)-2(x-1)}{x(x+1)+(x-1)}$$

$$=\frac{5x+5-2x+2}{x^2+x+x-1}=\frac{3x+7}{x^2+2x-1}$$

41. $$\frac{x^{-2}-y^{-2}}{x^{-1}+y^{-1}}=\frac{\dfrac{1}{x^2}-\dfrac{1}{y^2}}{\dfrac{1}{x}+\dfrac{1}{y}}=\frac{\dfrac{y^2}{x^2y^2}-\dfrac{x^2}{x^2y^2}}{\dfrac{y}{xy}+\dfrac{x}{xy}}=\frac{y^2-x^2}{x^2y^2}\cdot\frac{xy}{y+x}=\frac{(y-x)(y+x)xy}{x^2y^2(y+x)}=\frac{y-x}{xy}$$

Alternatively,

$$\frac{x^{-2}-y^{-2}}{x^{-1}+y^{-1}}=\frac{\left(\dfrac{1}{x^2}-\dfrac{1}{y^2}\right)}{\left(\dfrac{1}{x}+\dfrac{1}{y}\right)}\cdot\frac{x^2y^2}{x^2y^2}=\frac{y^2-x^2}{xy^2+x^2y}=\frac{(y-x)(y+x)}{xy(y+x)}=\frac{y-x}{xy}$$

43. $$\frac{\dfrac{1}{a+h}-\dfrac{1}{a}}{h}=\frac{\dfrac{a}{a(a+h)}-\dfrac{a+h}{a(a+h)}}{h}=\frac{\dfrac{a-a-h}{a(a+h)}}{h}=\frac{-h}{a(a+h)}\cdot\frac{1}{h}=\frac{-1}{a(a+h)}$$

45. $$\frac{\dfrac{1-(x+h)}{2+(x+h)}-\dfrac{1-x}{2+x}}{h}=\frac{\dfrac{(2+x)(1-x-h)}{(2+x)(2+x+h)}-\dfrac{(1-x)(2+x+h)}{(2+x)(2+x+h)}}{h}$$

$$=\frac{\dfrac{2-x-x^2-2h-xh}{(2+x)(2+x+h)}-\dfrac{2-x-x^2+h-xh}{(2+x)(2+x+h)}}{h}=\frac{-3h}{(2+x)(2+x+h)}\cdot\frac{1}{h}$$

$$=\frac{-3}{(2+x)(2+x+h)}$$

47. $$\sqrt{1+\left(\frac{x}{\sqrt{1-x^2}}\right)^2}=\sqrt{1+\frac{x^2}{1-x^2}}=\sqrt{\frac{1-x^2}{1-x^2}+\frac{x^2}{1-x^2}}=\sqrt{\frac{1}{1-x^2}}=\frac{1}{\sqrt{1-x^2}}$$

49. $$\frac{2(1+x)^{1/2}-x(1+x)^{-1/2}}{1+x}=\frac{(1+x)^{-1/2}[2(1+x)-x]}{1+x}=\frac{x+2}{(1+x)^{3/2}}$$

51. $$\frac{3(1+x)^{1/3}-x(1+x)^{-2/3}}{(1+x)^{2/3}}=\frac{(1+x)^{-2/3}[3(1+x)-x]}{(1+x)^{2/3}}=\frac{2x+3}{(1+x)^{4/3}}$$

53. $$\frac{2}{3+\sqrt{5}}=\frac{2}{3+\sqrt{5}}\cdot\frac{3-\sqrt{5}}{3-\sqrt{5}}=\frac{2\left(3-\sqrt{5}\right)}{9-5}=\frac{2\left(3-\sqrt{5}\right)}{4}=\frac{3-\sqrt{5}}{2}$$

55. $$\frac{2}{\sqrt{2}+\sqrt{7}}=\frac{2}{\sqrt{2}+\sqrt{7}}\cdot\frac{\sqrt{2}-\sqrt{7}}{\sqrt{2}-\sqrt{7}}=\frac{2\left(\sqrt{2}-\sqrt{7}\right)}{2-7}=\frac{2\left(\sqrt{2}-\sqrt{7}\right)}{-5}=\frac{2\left(\sqrt{7}-\sqrt{2}\right)}{5}$$

57. $\dfrac{1-\sqrt{5}}{3}=\dfrac{1-\sqrt{5}}{3}\cdot\dfrac{1+\sqrt{5}}{1+\sqrt{5}}=\dfrac{1-5}{3\left(1+\sqrt{5}\right)}=\dfrac{-4}{3\left(1+\sqrt{5}\right)}$

59. $\dfrac{\sqrt{r}+\sqrt{2}}{5}=\dfrac{\sqrt{r}+\sqrt{2}}{5}\cdot\dfrac{\sqrt{r}-\sqrt{2}}{\sqrt{r}-\sqrt{2}}=\dfrac{r-2}{5\left(\sqrt{r}-\sqrt{2}\right)}$

61. $\dfrac{\sqrt{x}-\sqrt{x+h}}{h\sqrt{x}\sqrt{x+h}}=\dfrac{\sqrt{x}-\sqrt{x+h}}{h\sqrt{x}\sqrt{x+h}}\cdot\dfrac{\sqrt{x}+\sqrt{x+h}}{\sqrt{x}+\sqrt{x+h}}=\dfrac{x-(x+h)}{h\sqrt{x}\sqrt{x+h}\left(\sqrt{x}+\sqrt{x+h}\right)}$
$=\dfrac{-h}{h\sqrt{x}\sqrt{x+h}\left(\sqrt{x}+\sqrt{x+h}\right)}=\dfrac{-1}{\sqrt{x}\sqrt{x+h}\left(\sqrt{x}+\sqrt{x+h}\right)}$

63. $\sqrt{x^2+x+1}+x=\dfrac{\sqrt{x^2+x+1}+x}{1}\cdot\dfrac{\sqrt{x^2+x+1}-x}{\sqrt{x^2+x+1}-x}=\dfrac{x^2+x+1-x^2}{\sqrt{x^2+x+1}-x}$
$=\dfrac{x+1}{\sqrt{x^2+x+1}-x}$

65. $\dfrac{16+a}{16}=\dfrac{16}{16}+\dfrac{a}{16}=1+\dfrac{a}{16}$. So the statement is true.

67. This statement is false. For example, take $x=2$, then the L.H.S. $=\dfrac{2}{4+x}=\dfrac{2}{4+2}=\dfrac{2}{6}=\dfrac{1}{3}$. While the R.H.S. $=\dfrac{1}{2}+\dfrac{2}{x}=\dfrac{1}{2}+\dfrac{2}{2}=\dfrac{3}{2}$. And $\dfrac{1}{3}\neq\dfrac{3}{2}$.

69. This statement is false. For example, take $x=0$ and $y=1$. Then substituting into the left side we obtain: L.H.S. $=\dfrac{x}{x+y}=\dfrac{0}{0+1}=0$, while the right side yields R.H.S. $=\dfrac{1}{1+y}=\dfrac{1}{1+1}=\dfrac{1}{2}$, and $0\neq\frac{1}{2}$.

71. This statement is true: $\dfrac{-a}{b}=(-a)\left(\dfrac{1}{b}\right)=(-1)(a)\left(\dfrac{1}{b}\right)=(-1)\left(\dfrac{a}{b}\right)=-\dfrac{a}{b}$.

73. This statement is false. For example, take $x=-2$. Then L.H.S. $=\dfrac{x^2+1}{x^2+x-1}$
$=\dfrac{(-2)^2+1}{(-2)^2+(-2)-1}=\dfrac{4+1}{4-2-1}=\dfrac{5}{1}=5$, while R.H.S. $=\dfrac{1}{x-1}=\dfrac{1}{(-2)-1}=\dfrac{1}{-3}=-\dfrac{1}{3}$, and $5\neq-\frac{1}{3}$.

75. (a) $R=\dfrac{1}{\dfrac{1}{R_1}+\dfrac{1}{R_2}}=\dfrac{1}{\dfrac{1}{R_1}+\dfrac{1}{R_2}}\cdot\dfrac{R_1R_2}{R_1R_2}=\dfrac{R_1R_2}{R_2+R_1}$

(b) Substituting $R_1=10$ ohms and $R_2=20$ ohms yields $R=\dfrac{(10)(20)}{(20)+(10)}=\dfrac{200}{30}\approx 6.7$ ohms.

Review Exercises for Chapter 1

1. Commutative Property for addition.

3. Distributive Property.

5. $(-1,3] = \{x \mid -1 < x \leq 3\}$

7. $x > 2 \Leftrightarrow x \in (2,\infty)$

9. $|3 - |-9|| = |3 - 9| = |-6| = 6$

11. $2^{-3} - 3^{-2} = \frac{1}{8} - \frac{1}{9} = \frac{9}{72} - \frac{8}{72} = \frac{1}{72}$

13. $216^{-1/3} = \frac{1}{216^{1/3}} = \frac{1}{\sqrt[3]{216}} = \frac{1}{6}$

15. $\frac{\sqrt{242}}{\sqrt{2}} = \sqrt{\frac{242}{2}} = \sqrt{121} = 11$

17. $2^{1/2}8^{1/2} = \sqrt{2} \cdot \sqrt{8} = \sqrt{16} = 4$

19. $\frac{1}{x^2} = x^{-2}$

21. $x^2 x^m (x^3)^m = x^{2+m+3m} = x^{4m+2}$

23. $x^a x^b x^c = x^{a+b+c}$

25. $x^{c+1}(x^{2c-1})^2 = x^{(c+1)+2(2c-1)} = x^{c+1+4c-2} = x^{5c-1}$

27. $(2x^3y)^2(3x^{-1}y^2) = 4x^6y^2 \cdot 3x^{-1}y^2 = 4 \cdot 3x^{6-1}y^{2+2} = 12x^5y^4$

29. $\frac{x^4(3x)^2}{x^3} = \frac{x^4 \cdot 9x^2}{x^3} = 9x^{4+2-3} = 9x^3$

31. $\sqrt[3]{(x^3y)^2y^4} = \sqrt[3]{x^6y^4y^2} = \sqrt[3]{x^6y^6} = x^2y^2$

33. $\frac{x}{2+\sqrt{x}} = \frac{x}{2+\sqrt{x}} \cdot \frac{2-\sqrt{x}}{2-\sqrt{x}} = \frac{x(2-\sqrt{x})}{4-x}$. Here simplify means rationalize the denominator.

35. $\frac{8r^{1/2}s^{-3}}{2r^{-2}s^4} = 4r^{(1/2)-(-2)}\,s^{-3-4} = 4r^{5/2}s^{-7} = \frac{4r^{5/2}}{s^7}$

37. $78,250,000,000 = 7.825 \times 10^{10}$

39. $\frac{ab}{c} \approx \frac{(0.00000293)(1.582 \times 10^{-14})}{2.8064 \times 10^{12}} = \frac{(2.93 \times 10^{-6})(1.582 \times 10^{-14})}{2.8064 \times 10^{12}}$
$= \frac{2.93 \cdot 1.582}{2.8064} \times 10^{-6-14-12} \approx 1.65 \times 10^{-32}$

41. $12x^2y^4 - 3xy^5 + 9x^3y^2 = 3xy^2(4xy^2 - y^3 + 3x^2)$

43. $x^2 + 3x - 10 = (x+5)(x-2)$

45. $4t^2 - 13t - 12 = (4t+3)(t-4)$

47. $25 - 16t^2 = (5-4t)(5+4t)$

49. $x^6 - 1 = (x^3-1)(x^3+1) = (x-1)(x^2+x+1)(x+1)(x^2-x+1)$

51. $x^{-1/2} - 2x^{1/2} + x^{3/2} = x^{-1/2}(1 - 2x + x^2) = x^{-1/2}(1-x)^2$

53. $4x^3 - 8x^2 + 3x - 6 = 4x^2(x-2) + 3(x-2) = (4x^2+3)(x-2)$

55. $(x^2+2)^{5/2}+2x(x^2+2)^{3/2}+x^2\sqrt{x^2+2}=x^2+2)^{1/2}((x^2+2)^2+2x(x^2+2)+x^2)$
$=\sqrt{x^2+2}(x^4+4x^2+4+2x^3+4x+x^2)=\sqrt{x^2+2}(x^4+2x^3+5x^2+4x+4)$
$=\sqrt{x^2+2}(x^2+x+2)^2$

57. $a^2y-b^2y=y(a^2-b^2)=y(a-b)(a+b)$

59. $(x+1)^2-2(x+1)+1=\left[(x+1)-1\right]^2=x^2$. You can also obtain this result by expanding each term and then simplifying.

61. $(2x+1)(3x-2)-5(4x-1)=6x^2-4x+3x-2-20x+5=6x^2-21x+3$

63. $(2a^2-b)^2=(2a^2)^2-2(2a^2)(b)+(b)^2=4a^4-4a^2b+b^2$

65. $(x-1)(x-2)(x-3)=(x-1)(x^2-5x+6)=x^3-5x^2+6x-x^2+5x-6$
$=x^3-6x^2+11x-6$

67. $\sqrt{x}(\sqrt{x}+1)(2\sqrt{x}-1)=(x+\sqrt{x})(2\sqrt{x}-1)=2x\sqrt{x}-x+2x-\sqrt{x}=2x^{3/2}+x-x^{1/2}$

69. $x^2(x-2)+x(x-2)^2=x^3-2x^2+x(x^2-4x+4)=x^3-2x^2+x^3-4x^2+4x$
$=2x^3-6x^2+4x$

71. $\dfrac{x^2-2x-3}{2x^2+5x+3}=\dfrac{(x-3)(x+1)}{(2x+3)(x+1}=\dfrac{x-3}{2x+3}$

73. $\dfrac{x^2+2x-3}{x^2+8x+16}\cdot\dfrac{3x+12}{x-1}=\dfrac{(x+3)(x-1)}{(x+4)(x+4)}\cdot\dfrac{3(x+4)}{(x-1)}=\dfrac{3(x+3)}{x+4}$

75. $\dfrac{x^2-2x-15}{x^2-6x+5}\div\dfrac{x^2-x-12}{x^2-1}=\dfrac{(x-5)(x+3)}{(x-5)(x-1)}\cdot\dfrac{(x-1)(x+1)}{(x-4)(x+3)}=\dfrac{x+1}{x-4}$

77. $\dfrac{1}{x-1}-\dfrac{x}{x^2+1}=\dfrac{x^2+1}{(x-1)(x^2+1)}-\dfrac{x(x-1)}{(x-1)(x^2+1)}=\dfrac{x^2+1-x^2+x}{(x-1)(x^2+1)}=\dfrac{x+1}{(x-1)(x^2+1)}$

79. $\dfrac{1}{x-1}-\dfrac{2}{x^2-1}=\dfrac{1}{x-1}-\dfrac{2}{(x-1)(x+1)}=\dfrac{x+1}{(x-1)(x+1)}-\dfrac{2}{(x-1)(x+1)}$
$=\dfrac{x+1-2}{(x-1)(x+1)}=\dfrac{x-1}{(x-1)(x+1)}=\dfrac{1}{x+1}$

81. $\dfrac{\frac{1}{x}-\frac{1}{2}}{x-2}=\dfrac{\frac{2}{2x}-\frac{x}{2x}}{x-2}=\dfrac{2-x}{2x}\cdot\dfrac{1}{x-2}=\dfrac{-1(x-2)}{2x}\cdot\dfrac{1}{x-2}=\dfrac{-1}{2x}$

83. $\dfrac{3(x+h)^2-5(x+h)-(3x^2-5x)}{h}=\dfrac{3x^2+6xh+3h^2-5x-5h-3x^2+5x}{h}$
$=\dfrac{6xh+3h^2-5h}{h}=\dfrac{h(6x+3h-5)}{h}=6x+3h-5$

85. (a) Is $\frac{1}{28}x+\frac{1}{34}y\le 15$ when $x=165$ and $y=230$? $\frac{1}{28}(165)+\frac{1}{34}(230)=5.89+6.76$ $=12.65\le 15$, so yes the car can travel 165 city miles and 230 highway mile without running out of gas.

(b) Here we must solve for y when $x = 280$. So $\frac{1}{28}(280) + \frac{1}{34}y \leq 15 \quad \Leftrightarrow \quad 10 + \frac{1}{34}y \leq 15$ $\Leftrightarrow \quad \frac{1}{34}y \leq 5 \quad \Leftrightarrow \quad y \leq 170$. Thus the car can travel up to 170 highway miles without running out of gas.

87. This statement is false. For example, take $x = 1$ and $y = 1$, then L.H.S. $= (x+y)^3 = (1+1)^3$ $= 2^3 = 8$, while the R.H.S. $= x^3 + y^3 = 1^3 + 1^3 = 1 + 1 = 2$, and $8 \neq 2$.

89. This statement is true: $\dfrac{12+y}{y} = \dfrac{12}{y} + \dfrac{y}{y} = \dfrac{12}{y} + 1$.

91. This statement is false. For example, take $a = -1$, then L.H.S. $= \sqrt{a^2} = \sqrt{(-1)^2} = \sqrt{1} = 1$ which does not equal $a = -1$. The true statement is $\sqrt{a^2} = |a|$.

93. This statement is false. For example, take $x = 1$ and $y = 1$, then L.H.S. $= x^3 + y^3 = 1^3 + 1^3$ $= 1 + 1 = 2$, while the R.H.S. $= (x+y)(x^2 + xy + y^2) = (1+1)\left[1^2 + (1)(1) + 1^2\right]$ $= 2(3) = 6$.

95. Substituting we obtain: $\sqrt{1+t^2} = \sqrt{1 + \left[\frac{1}{2}\left(x^3 - \frac{1}{x^3}\right)\right]^2} = \sqrt{1 + \frac{1}{4}\left(x^6 - 2\frac{x^3}{x^3} + \frac{1}{x^6}\right)}$

$= \sqrt{1 + \frac{x^6}{4} - \frac{1}{2} + \frac{1}{4x^6}} = \sqrt{\frac{x^6}{4} + \frac{1}{2} + \frac{1}{4x^6}} = \sqrt{\left(\frac{x^3}{2} + \frac{1}{2x^3}\right)^2}$. Since $x > 0$,

$\sqrt{\left(\frac{x^3}{2} + \frac{1}{2x^3}\right)^2} = \frac{x^3}{2} + \frac{1}{2x^3} = \frac{1}{2}\left(x^3 + \frac{1}{x^3}\right)$.

Chapter 1 Test

1. (a) $[-3, 2]$ (number line: closed dots at -3 and 2) $(4, \infty)$ (number line: open dot at 4)

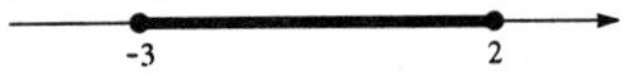

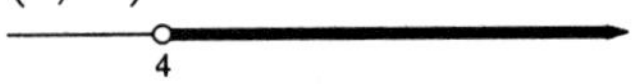

   (b) $x < 5 \Leftrightarrow x \in (-\infty, 5)$ $-2 \le x \le 1 \Leftrightarrow x \in [-2, 1]$

   (c) Distance $= |-22 - 31| = |-53| = 53$

2. (a) $(-3)^4 = 81$ (b) $2^{-4} = \dfrac{1}{2^4} = \dfrac{1}{16}$

   (c) $\dfrac{5^{18}}{5^{12}} = 5^{18-12} = 5^6 = 15{,}625$

3. (a) $\left(\dfrac{2}{3}\right)^{-1} = \dfrac{3}{2}$ (b) $\dfrac{\sqrt{32}}{\sqrt{8}} = \sqrt{\dfrac{32}{8}} = \sqrt{4} = 2$

   (c) $16^{-3/4} = \left(2^4\right)^{-3/4} = 2^{-3} = \dfrac{1}{8}$

4. $\dfrac{(x^2)^a\left(\sqrt{x}\right)^b}{x^{a+b}\,x^{a-b}} = \dfrac{x^{2a}\cdot\left(x^{1/2}\right)^b}{x^{a+b+a-b}} = \dfrac{x^{2a}\cdot x^{b/2}}{x^{2a}} = x^{2a+(b/2)-2a} = x^{b/2}$

5. (a) $\sqrt{200} - \sqrt{8} = 10\sqrt{2} - 2\sqrt{2} = 8\sqrt{2}$

   (b) $(2a^3b^2)\left(3ab^4\right)^3 = 2a^3b^2 \cdot 3^3a^3b^{12} = 54a^6b^{14}$

   (c) $\left(\dfrac{x^2y^{-3}}{y^5}\right)^{-4} = \left(x^2y^{-3-5}\right)^{-4} = \left(x^2y^{-8}\right)^{-4} = x^{2(-4)}y^{-8(-4)} = x^{-8}y^{32} = \dfrac{y^{32}}{x^8}$

   (d) $\left(\dfrac{2x^{1/4}}{y^{1/3}x^{1/6}}\right)^3 = \dfrac{2^3x^{3/4}}{y^{3/3}x^{3/6}} = \dfrac{8x^{3/4}}{yx^{1/2}} = \dfrac{8x^{(3/4)-(1/2)}}{y} = \dfrac{8x^{1/4}}{y} = \dfrac{8\sqrt[4]{x}}{y}$

6. (a) $\dfrac{x^2+3x+2}{x^2-x-2} = \dfrac{(x+1)(x+2)}{(x-2)(x+1)} = \dfrac{x+2}{x-2}$

   (b) $\dfrac{x^2}{x^2-4} - \dfrac{x+1}{x+2} = \dfrac{x^2}{(x-2)(x+2)} - \dfrac{x+1}{x+2} = \dfrac{x^2}{(x-2)(x+2)} + \dfrac{-(x+1)(x-2)}{(x-2)(x+2)}$
   $= \dfrac{x^2-(x^2-x-2)}{(x-2)(x+2)} = \dfrac{x+2}{(x-2)(x+2)} = \dfrac{1}{x-2}$

   (c) $\dfrac{\dfrac{y}{x}-\dfrac{x}{y}}{\dfrac{1}{y}-\dfrac{1}{x}} = \dfrac{\dfrac{y}{x}-\dfrac{x}{y}}{\dfrac{1}{y}-\dfrac{1}{x}}\cdot\dfrac{xy}{xy} = \dfrac{y^2-x^2}{x-y} = \dfrac{(y-x)(y+x)}{x-y} = \dfrac{-(x-y)(y+x)}{x-y} = -(y+x)$

7. (a) $325{,}000{,}000{,}000 = 3.25 \times 10^{11}$ (b) $0.000008931 = 8.931 \times 10^{-6}$

8. (a) $4(3-x) - 3(x+5) = 12 - 4x - 3x - 15 = -7x - 3$

   (b) $(x-5)(2x+3) = 2x^2 + 3x - 10x - 15 = 2x^2 - 7x - 15$

   (c) $\left(\sqrt{x}+\sqrt{y}\right)\left(\sqrt{x}-\sqrt{y}\right) = \left(\sqrt{x}\right)^2 - \left(\sqrt{y}\right)^2 = x - y$

(d) $(3t+4)^2 = (3t)^2 + 2(3t)(4) + (4)^2 = 9t^2 + 24t + 16$

(e) $(2-x^2)^3 = (2)^3 - 3(2)^2(x^2) + 3(2)(x^2)^2 + (x^2)^3 = 8 - 12x^2 + 6x^4 + x^6$

9. (a) $9x^2 - 25 = (3x-5)(3x+5)$

(b) $6x^2 + 7x - 5 = (2x-1)(3x+5)$

(c) $x^3 - 4x^2 - 3x + 12 = x^2(x-4) - 3(x-4) = (x^2-3)(x-4)$

(d) $x^4 + 27x = x(x^3+27) = x(x+3)(x^2-3x+9)$

(e) $3x^{3/2} - 9x^{1/2} + 6x^{-1/2} = 3x^{-1/2}(x^2 - 3x + 2) = 3x^{-1/2}(x-2)(x-1)$

10. $\dfrac{x}{\sqrt{x}-2} = \dfrac{x}{\sqrt{x}-2} \cdot \dfrac{\sqrt{x}+2}{\sqrt{x}+2} = \dfrac{x(\sqrt{x}+2)}{x-4}$

Principles of Problem Solving

1. Let d be the distance traveled to and from work. Let t_1 and t_2 be the times for the trip from home to work and the trip from work to home, respectively. Using $time = \dfrac{distance}{rate}$ we get $t_1 = \dfrac{d}{50}$ and $t_2 = \frac{d}{30}$. Since $average\ speed = \dfrac{distance\ traveled}{total\ time}$ we have $average\ speed = \dfrac{2d}{t_1 + t_2} = \dfrac{2d}{\frac{d}{50} + \frac{d}{30}}$

$\Leftrightarrow$

$$average\ speed = \frac{150(2d)}{150\left(\frac{d}{50} + \frac{d}{30}\right)} = \frac{300d}{3d + 5d} = \frac{300}{8} = 37.5\text{ mi/h}$$

3. We use the formula $d = rt$ (distance = rate $\times$ time). Since the cars each travel at a speed of 40 mi/h, they approach each other at a combined speed of 80 mi/h. (The distance between them decreases at a rate of 80 mi/h.) So the time spent driving till they meet is $t = \dfrac{d}{r} = \dfrac{120}{80} = 1.5$ h. Thus the fly flies at a speed of 100 mi/h for 1.5 hours, and therefore travels a distance of $d = rt = (100)(1.5) = 150$ miles.

5. $\left(\sqrt{3 + 2\sqrt{2}} - \sqrt{3 - 2\sqrt{2}}\right)^2 = 3 + 2\sqrt{2} - 2 \cdot \sqrt{\left(3 + 2\sqrt{2}\right)\left(3 - 2\sqrt{2}\right)} + 3 - 2\sqrt{2}$
$= 6 - 2 \cdot \sqrt{9 - 8} = 6 - 2 = 4$. Therefore, $\sqrt{3 + 2\sqrt{2}} - \sqrt{3 - 2\sqrt{2}} = \sqrt{4} = 2$.

7. By placing two amoebas into the vessel we skip the first simple division which took 3 minutes. Thus when we place two amoebas into the vessel, it will take $60 - 3 = 57$ minutes for the vessel to be full of amoebas.

9. The statement is false. Here is one particular counterexample:

	Player A	Player B
First half	1 hit in 99 at-bats: average $= \frac{1}{99}$	0 hit in 1 at-bat: average $= \frac{0}{1}$
Second half	1 hit in 1 at-bat: average $= \frac{1}{1}$	98 hits in 99 at-bats: average $= \frac{98}{99}$
Entire season	2 hits in 100 at-bats: average $= \frac{2}{100}$	99 hits in 100 at-bats: average $= \frac{99}{100}$

11. Suppose each cup has x spoonfuls of liquid. When one spoonful of cream is added to the coffee, the ratio of cream to liquid is $\frac{1}{x+1}$ and the ratio of coffee to cream is $\frac{x}{x+1}$. Now when a spoonful of the coffee/cream mixture is add to the cup of cream, we are only adding $\frac{x}{x+1}$ of a spoonful of coffee. Thus the ratio of coffee to cream in the cream cup is $\dfrac{\frac{x}{x+1}}{(x-1)+1} = \dfrac{\frac{x}{x+1}}{x} = \dfrac{1}{x+1}$. This is the same as the ratio of cream to coffee (in the coffee cup). So there is the same amount of cream in the coffee cup as there is coffee in the cream cup.

   Another way to see this is to observe that at the end, both cups contain the same amount of liquid. Thus any coffee in the cream cup must be replacing an equal amount of cream that has ended up in the coffee cup.

13. (a) (Baby + Uncle) + Mother = Uncle + Mother = Uncle

    (b) Father · (Grandpa + Aunt) = Father · Grandpa = Father

(c) Commutative: $x + y =$ (older of x and y) $=$ (older of y and x) $= y + x$
$x \cdot y =$ (younger of x and y) $=$ (younger of y and x) $= y \cdot x$
Associative: $(x + y) + z =$ (older of x and y) $+ z =$ older of (older of x and y) and $z =$ older of x and y and z; $x + (y + z) = x +$ (older of y and z) $=$ older of x and (older of y and z) $=$ older of x and y and z. So $(x + y) + z = x + (y + z)$
$(x \cdot y) \cdot z =$ (younger of x and y) $\cdot z =$ younger of (younger of x and y) and $z =$ younger of x and y and z; $x \cdot (y \cdot z) = x \cdot$ (younger of y and z) $=$ younger of x and (younger of y and z) $=$ younger of x and y and z. So $(x \cdot y) \cdot z = x \cdot (y \cdot z)$
Distributive: We consider three cases:
Case (i), x is oldest of x, y, and z: then $x \cdot (y + z) = y + z$ and $x \cdot y + x \cdot z = y + z$, so $x \cdot (y + z) = x \cdot y + x \cdot z$.
Case (ii), y is oldest: then $x \cdot (y + z) = x \cdot y = x$ and $x \cdot y + x \cdot z = x + x \cdot z = x$ (since $x \cdot z$ is no older than x by definition). Thus $x \cdot (y + z) = x \cdot y + x \cdot z$.
Case (iii), z is oldest: then $x \cdot (y + z) = x \cdot z = x$ and $x \cdot y + x \cdot z = x \cdot y + x = x$ (since $x \cdot y$ is no older than x by definition). Thus $x \cdot (y + z) = x \cdot y + x \cdot z$.

(d) Again we consider three cases.
Case (i), x is oldest of x, y, and z: then $x + (y \cdot z) = x$. Also, $x + y = x$ and $x + z = x$, so $(x + y) \cdot (x + z) = x \cdot x = x$. Thus $x + (y \cdot z) = (x + y) \cdot (x + z)$.
Case (ii), y is oldest of x, y, and z: then $y \cdot z = z$, and so $x + (y \cdot z) = x + z$. Also, $x + y = y$, so $(x + y) \cdot (x + z) = y \cdot (x + z) = x + z$. Thus $x + (y \cdot z) = (x + y) \cdot (x + z)$.
The proof in the third case, when z is the oldest, is similar to the case when y is the oldest.

15. Let $r_1 = 8$.
$r_2 = \frac{1}{2}\left(8 + \frac{72}{8}\right) = \frac{1}{2}(8 + 9) = 8.5$
$r_3 = \frac{1}{2}\left(8.5 + \frac{72}{8.5}\right) = \frac{1}{2}(8.5 + 8.471) = 8.485$
$r_4 = \frac{1}{2}\left(8.485 + \frac{72}{8.485}\right) = \frac{1}{2}(8.485 + 8.486) = 8.485$
Thus $\sqrt{72} \approx 8.49$.

Chapter Two

Exercises 2.1

1. (a) $2(4) - 3 \stackrel{?}{=} (4) + 1$
   $8 - 3 \stackrel{?}{=} 5$ Yes, a solution.

   (b) $2\left(\frac{3}{2}\right) - 3 \stackrel{?}{=} \left(\frac{3}{2}\right) + 1$
   $3 - 3 \stackrel{?}{=} \frac{5}{2}$ No, not a solution.

3. (a) $\frac{1}{(-3)} - \frac{1}{(-3)+3} \stackrel{?}{=} \frac{1}{6}$
   The left side is undefined, not a solution.

   (b) $\frac{1}{(3)} - \frac{1}{(3)+3} \stackrel{?}{=} \frac{1}{6}$
   $\frac{1}{3} - \frac{1}{6} \stackrel{?}{=} \frac{1}{6}$ Yes, a solution.

5. (a) $A\left(\frac{By-C}{A}\right) + By + C \stackrel{?}{=} 0$
   $By - C + By + C \stackrel{?}{=} 0$
   $2By \stackrel{?}{=} 0$
   Since this is true for only certain values of B and y, this is not a solution.

   (b) $Ax + B\left(-\frac{Ax+C}{B}\right) + C \stackrel{?}{=} 0$
   $Ax - (Ax + C) + C \stackrel{?}{=} 0$
   $Ax - Ax - C + C \stackrel{?}{=} 0$
   Yes, a solution.

7. Since the equation $x^2 + 16 = 32$ is false when $x = 0$, the equation is not an identity.

9. LHS $= \dfrac{x^2 - 9}{x - 3} = \dfrac{(x+3)(x-3)}{x-3} = x + 3 =$ RHS, so the equation is an identity for $x \neq 3$.

11. LHS $= \dfrac{y + 16}{2} - \dfrac{3}{2}y = \dfrac{y + 16 - 3y}{2} = \dfrac{16 - 2y}{2} = \dfrac{2(8 - y)}{2} = 8 - y =$ RHS, so the equation is an identity.

13. Yes, the equation is linear. $5x - 3 = 2x + 12 \quad \Leftrightarrow \quad 3x = 15 \quad \Leftrightarrow \quad x = 5$.

15. $y^2 - 7y + 4 = 2y^2 + 11 \quad \Leftrightarrow \quad y^2 + 7y + 7 = 0$, so it is not equivalent to a linear equation.

17. Yes, this is equivalent to a linear equation. $\dfrac{x}{x-4} = 1 + \dfrac{3}{x+2} \quad \Leftrightarrow$ (multiplying both sides by LCD) $x(x + 2) = (x - 4)(x + 2) + 3(x - 4) \quad \Leftrightarrow \quad x^2 + 2x = x^2 - 2x - 8 + 3x - 12 \quad \Leftrightarrow$ $x = -20$.

19. $3x - 5 = 7 \quad \Leftrightarrow \quad 3x = 12 \quad \Leftrightarrow \quad x = 4$

21. $x - 3 = 2x + 6 \quad \Leftrightarrow \quad -9 = x$

23. $-7w = 15 - 2w \quad \Leftrightarrow \quad -5w = 15 \quad \Leftrightarrow \quad w = -3$

25. $\frac{1}{2}y - 2 = \frac{1}{3}y \quad \Leftrightarrow \quad 3y - 12 = 2y$ (multiply both sides by the LCD, 6) $\quad \Leftrightarrow \quad y = 12$

27. $2(1 - x) = 3(1 + 2x) + 5 \quad \Leftrightarrow \quad 2 - 2x = 3 + 6x + 5 \quad \Leftrightarrow \quad 2 - 2x = 8 + 6x \quad \Leftrightarrow$ $-6 = 8x \quad \Leftrightarrow \quad x = -\frac{3}{4}$

29. $4\left(y - \frac{1}{2}\right) - y = 6(5 - y) \quad \Leftrightarrow \quad 4y - 2 - y = 30 - 6y \quad \Leftrightarrow \quad 3y - 2 = 30 - 6y \quad \Leftrightarrow$ $9y = 32 \quad \Leftrightarrow \quad y = \frac{32}{9}$

31. $\dfrac{1}{x} = \dfrac{4}{3x} + 1 \quad \Rightarrow \quad 3 = 4 + 3x$ (multiply both sides by the LCD $3x$) $\quad \Leftrightarrow \quad -1 = 3x \quad \Leftrightarrow$ $x = -\frac{1}{3}$

33. $\dfrac{2}{t+6} = \dfrac{3}{t-1} \quad \Rightarrow \quad 2(t-1) = 3(t+6)$ (multiply both sides by the LCD $(t-1)(t+6)$) $\quad \Leftrightarrow$
$2t - 2 = 3t+18 \quad \Leftrightarrow \quad -20 = t$

35. $r - 2[1 - 3(2r+4)] = 61 \quad \Leftrightarrow \quad r - 2(1 - 6r - 12) = 61 \quad \Leftrightarrow \quad r - 2(-6r - 11) = 61 \quad \Leftrightarrow$
$r + 12r + 22 = 61 \quad \Leftrightarrow \quad 13r = 39 \quad \Leftrightarrow \quad r = 3$

37. $\sqrt{3}\,x + \sqrt{12} = \frac{x+5}{\sqrt{3}} \quad \Leftrightarrow \quad 3x + 6 = x + 5$ (multiply both sides by $\sqrt{3}$) $\quad \Leftrightarrow \quad 2x = -1 \quad \Leftrightarrow$
$x = -\frac{1}{2}$

39. $\dfrac{2}{x} - 5 = \dfrac{6}{x} + 4 \quad \Rightarrow \quad 2 - 5x = 6 + 4x \quad \Leftrightarrow \quad -4 = 9x \quad \Leftrightarrow \quad -\dfrac{4}{9} = x$

41. $\dfrac{3}{x+1} - \dfrac{1}{2} = \dfrac{1}{3x+3} \quad \Rightarrow \quad 3(6) - (3x+3) = 2$ (multiply both sides by the LCD $6(x+1)$)
$\Leftrightarrow \quad 18 - 3x - 3 = 2 \quad \Leftrightarrow \quad -3x + 15 = 2 \quad \Leftrightarrow \quad -3x = -13 \quad \Leftrightarrow \quad x = \frac{13}{3}$

43. $\dfrac{2x-7}{2x+4} = \dfrac{2}{3} \quad \Rightarrow \quad (2x-7)3 = 2(2x+4)$ (cross multiply) $\quad \Leftrightarrow \quad 6x - 21 = 4x + 8 \quad \Leftrightarrow$
$2x = 29 \quad \Leftrightarrow \quad x = \frac{29}{2}$

45. $x - \frac{1}{3}x - \frac{1}{2}x - 5 = 0 \quad \Leftrightarrow \quad 6x - 2x - 3x - 30 = 0$ (multiply both sides by the LCD 6) $\quad \Leftrightarrow$
$x = 30$

47. $\dfrac{1}{z} - \dfrac{1}{2z} - \dfrac{1}{5z} = \dfrac{10}{z+1} \quad \Rightarrow \quad 10(z+1) - 5(z+1) - 2(z+1) = 10(10z)$ (multiply both sides by the LCD $10z(z+1)$) $\quad \Leftrightarrow \quad 3(z+1) = 100z \quad \Leftrightarrow \quad 3z + 3 = 100z \quad \Leftrightarrow \quad 3 = 97z \quad \Leftrightarrow$
$\frac{3}{97} = z$

49. $\dfrac{u}{u - \frac{u+1}{2}} = 4 \quad \Rightarrow \quad u = 4\left(u - \dfrac{u+1}{2}\right)$ (cross multiply) $\quad \Leftrightarrow \quad u = 4u - 2u - 2 \quad \Leftrightarrow$
$u = 2u - 2 \quad \Leftrightarrow \quad 2 = u$

51. $\sqrt{x-4} = \sqrt{2x} \quad \Rightarrow \quad x - 4 = 2x \quad \Leftrightarrow \quad -4 = x$. But substituting $x = -4$ into the original equation gives $\sqrt{-8} = \sqrt{-8}$ which is undefined as a real number. Thus there is no solution.

53. $\dfrac{x}{2x-4} - 2 = \dfrac{1}{x-2} \quad \Rightarrow \quad x - 2(2x-4) = 2$ (multiply both sides by the LCD $2(x-2)$) $\quad \Leftrightarrow$
$x - 4x + 8 = 2 \quad \Leftrightarrow \quad -3x = -6 \quad \Leftrightarrow \quad x = 2$. But substituting $x = 2$ into the original equation does not work, since we cannot divide by 0. Thus there is no solution.

55. $\dfrac{3}{x+4} = \dfrac{1}{x} + \dfrac{6x+12}{x^2+4x} \quad \Rightarrow \quad 3(x) = (x+4) + 6x + 12$ (multiply both sides by the LCD $x(x+4)$) $\quad \Leftrightarrow \quad 3x = 7x + 16 \quad \Leftrightarrow \quad -4x = 16 \quad \Leftrightarrow \quad x = -4$. But substituting $x = -4$ into the original equation does not work, since we cannot divide by 0. Thus there is no solution.

57. $2.15x - 4.63 = x + 1.19 \quad \Leftrightarrow \quad 1.15x = 5.82 \quad \Leftrightarrow \quad x = \frac{5.82}{1.19} \approx 5.06$

59. $3.16(x + 4.63) = 4.19(x - 7.24) \quad \Leftrightarrow \quad 3.16x + 14.63 = 4.19x - 30.34 \quad \Leftrightarrow \quad 44.97 = 1.03x$
$\Leftrightarrow \quad x = \frac{44.97}{1.03} \approx 43.66$

61. $PV = nRT \quad \Leftrightarrow \quad R = \dfrac{PV}{nT}$

63. $\dfrac{1}{R} = \dfrac{1}{R_1} + \dfrac{1}{R_2} \quad \Leftrightarrow \quad R_1 R_2 = R R_2 + R R_1$ (multiply both sides by the LCD, $R R_1 R_2$) $\Leftrightarrow$
$R_1 R_2 - R R_1 = R R_2 \quad \Leftrightarrow \quad R_1(R_2 - R) = R R_2 \quad \Leftrightarrow \quad R_1 = \dfrac{R R_2}{R_2 - R}$

65. $\dfrac{ax+b}{cx+d} = 2 \quad \Leftrightarrow \quad ax + b = 2(cx + d) \quad \Leftrightarrow \quad ax + b = 2cx + 2d \quad \Leftrightarrow \quad ax - 2cx = 2d - b$
$\Leftrightarrow \quad (a - 2c)x = 2d - b \quad \Leftrightarrow \quad x = \dfrac{2d - b}{a - 2c}$

67. $a^2x + (a - 1) = (a + 1)x \quad \Leftrightarrow \quad a^2x - (a + 1)x = -(a - 1) \quad \Leftrightarrow \quad (a^2 - (a + 1))x = -a + 1$
$\Leftrightarrow \quad (a^2 - a - 1)x = -a + 1 \quad \Leftrightarrow \quad x = \dfrac{-a + 1}{a^2 - a - 1}$

69. $(4m + a)(6m - a) = (3m + 2a)(8m - 2a) \quad \Leftrightarrow \quad 24m^2 + 2am - a^2 = 24m^2 + 10am - 4a^2$
$\Leftrightarrow \quad -8am = -3a^2 \quad \Leftrightarrow \quad m = \dfrac{3a^2}{8a} = \dfrac{3a}{8}$

71. Substitute $x = 2$ into the equation and solve for k: $3(2) + k - 5 = k(2) - k + 1 \quad \Leftrightarrow$
$6 + k - 5 = 2k - k + 1 \quad \Leftrightarrow \quad 1 + k = k + 1$. Since this last equation is an identity, any value of k will make $x = 2$ a solution.

73. When both sides are divided by $x + 1$, the only solution is eliminated.
$x^2 + 6x + 5 = x^2 - 1 \quad \Leftrightarrow \quad 6x + 6 = 0 \quad \Leftrightarrow \quad 6x = -6 \quad \Leftrightarrow \quad x = -1$

Exercises 2.2

1. If x dollars are invested at 7% simple interest, then each year you will receive $0.07x$ dollars in interest. After two years, this amounts to $2(0.07x) = 0.14x$ dollars.

3. Let n be the first integer of the three consecutive odd integers. Then $n + 2$ is the second consecutive odd integer and $n + 4$ is the third consecutive odd integer. Thus the sum of three consecutive odd integers is $n + (n+2) + (n+4) = 3n + 6$.

5. If w is the width of the rectangle, then the area of the rectangle is $50w$ in^2

7. If s is the initial speed, then $s + 15$ is the speed during the third hour. Thus the distance traveled is $2s + s + 15 = 3s + 15$.

9. If a is the age of the firstborn, then $a - 3$ is the age of the second child, and $(a-3) - 2 = a - 5 =$ age of the third child. Thus the average age of the three children is $\dfrac{a + (a-3) + (a-5)}{3} = \dfrac{3a - 8}{3}$.

11. Let m be the amount invested at $4\frac{1}{2}\%$. Then $12,000 - m$ is the amount invested at 4%. Since *total interest* $=$ (*interest earned at* $4\frac{1}{2}\%$) $+$ (*interest earned* at 4%) we have $525 = 0.045m + 0.04(12,000 - m) \quad\Leftrightarrow\quad 525 = 0.045m + 480 - 0.04m \quad\Leftrightarrow\quad 45 = 0.005m$ $\Leftrightarrow\quad m = \frac{45}{0.005} = 9000$. Thus $\$9,000$ is invested at $4\frac{1}{2}\%$ and $\$12,000 - 9,000 = \$3,000$ is invested at 4%.

13. Let l be the length of the garden. Since *area* $=$ *width* $\cdot$ *length* we obtain the equation: $1125 = 25l \quad\Leftrightarrow\quad l = \frac{1125}{25} = 45$ ft. So the garden is 45 feet long.

15. Let x be the hours the assistant work. Then $2x$ is the hours the plumber worked. Since the *labor charge* $=$ *plumber's labor* $+$ *assistant's labor* we have $4025 = 45(2x) + 25x \quad\Leftrightarrow\quad 4025 = 90x + 25x \quad\Leftrightarrow\quad 4025 = 115x \quad\Leftrightarrow\quad x = \frac{4025}{115} = 35$ Thus the assistant works for 35 hours and the plumber works for $2 \times 35 = 70$ hours.

17. Let p be the number of pennies. Then p is the number of nickels and p is the number of dimes. So the *value of the coins in the purse* $=$ *value of the pennies* $+$ *value of the nickels* $+$ *value of the dimes*. Thus $1.44 = 0.01p + 0.05p + 0.10p \quad\Leftrightarrow\quad 1.44 = 0.16p \quad\Leftrightarrow\quad p = \frac{1.44}{0.16} = 9$. So the purse contains 9 pennies, 9 nickels, and 9 dimes.

19. Let y be the age of the youngest child. Then $2y$ is the age of the oldest child. Since the *average age* $= \dfrac{\textit{sum of the ages}}{4}$, we obtain $10.5 = \dfrac{y + 10 + 11 + 2y}{4} \quad\Leftrightarrow\quad 42 = 3y + 21 \quad\Leftrightarrow$ $3y = 21 \quad\Leftrightarrow\quad y = 7$. So the youngest child is 7 years old.

21. Let x be the number of pounds of $\$3.00$/lb. tea Then $80 - x$ is the number of pounds of $\$2.75$/lb. tea.

	3.00 tea	2.75 tea	mixture
pounds	x	$80 - x$	80
rate (cost per pound)	3.00	2.75	2.90
value	$3.00x$	$2.75(80 - x)$	$2.90(80)$

$3.00x + 2.75(80 - x) = 2.90(80) \quad\Leftrightarrow\quad 3.00x + 220 - 2.75x = 232 \quad\Leftrightarrow\quad 0.25x = 12 \quad\Leftrightarrow$ $x = 48$. The mixture uses 48 pounds of $\$3.00$/lb. tea and $80 - 48 = 32$ pounds of $\$2.75$/lb. tea.

23. Let x be the first consecutive odd integer. Then $x+2$, $x+4$, and $x+6$ are the next consecutive odd integers. So: $x+(x+2)+(x+4)+(x+6)=272 \quad \Leftrightarrow \quad 4x+12=272 \quad \Leftrightarrow \quad 4x=260 \quad \Leftrightarrow \quad x=65$. Thus the consecutive odd integers are 65, 67, 69, and 71.

25. Let x be the width of the strip. Then the length of the mat is $20+2x$ and the width of the mat is $15+2x$. Thus $\textit{perimeter} = 2 \times \textit{length} + 2 \times \textit{width} \quad \Leftrightarrow \quad 102 = 2(20+2x)+2(15+2x) \quad \Leftrightarrow \quad 102 = 40+4x+30+4x \quad \Leftrightarrow \quad 102 = 70+8x \quad \Leftrightarrow \quad 32 = 8x \quad \Leftrightarrow \quad x=4$. Thus the strip of mat is 4 inches wide.

27. Let x be the gross receipts for Friday. Then $2x$ is the gross receipts for Saturday. Using the equation $\textit{average receipts} = \dfrac{\textit{sum of receipts}}{\textit{number of days}}$ we get: $1200 = \dfrac{650+550+300+y+2y}{5} \quad \Leftrightarrow \quad 6000 = 1500+3y \quad \Leftrightarrow \quad 4500 = 3y \quad \Leftrightarrow \quad 1500 = y$. Thus Saturday's receipts were $2y =$ \$3,000.

29. Let t be the time (in seconds) for the sun's light to reach earth. Using the formula $\textit{distance} = \textit{rate} \times \textit{time}$ we get: $1.5 \times 10^{11} = \left(3.0 \times 10^{8}\right)t \quad \Leftrightarrow \quad t = \dfrac{1.5 \times 10^{11}}{3.0 \times 10^{8}} = 0.5 \times 10^{11-8} = 0.5 \times 10^{3} = 500$ seconds or 8 minutes and 20 seconds.

31. Let t be the time in hours after takeoff until the planes pass each other. When the planes pass each other, the total distance they have traveled will equal 2550 km.

	Rate	Time	Distance
K.C. $\rightarrow$ S.F.	800	t	$800t$
S.F. $\rightarrow$ K.C.	900	t	$900t$

$800t+900t = 2550 \quad \Leftrightarrow \quad 1700t = 2550 \quad \Leftrightarrow \quad t = 1.5$. Therefore, they will pass each other 1.5 hours after takeoff.

33. Let d be the distance in miles from Boston to Buffalo.

	Rate	Time	Distance
Boston to Buffalo	50	$\frac{d}{50}$	d
Buffalo to Boston	45	$\frac{d}{45}$	d

Using the equation $\textit{total time} = \begin{pmatrix} \textit{time traveling} \\ \textit{to Buffalo} \end{pmatrix} + \textit{time in Buffalo} + \begin{pmatrix} \textit{time traveling} \\ \textit{to Boston} \end{pmatrix}$ we have: $29 = \dfrac{d}{50} + 10 + \dfrac{d}{45} \quad \Leftrightarrow \quad 19 = \dfrac{d}{50} + \dfrac{d}{45} \quad \Leftrightarrow \quad 19 \cdot 45 \cdot 50 = 45d + 50d \quad \Leftrightarrow \quad 42750 = 95d \quad \Leftrightarrow \quad d = \frac{42750}{95} = 450$. Thus the distance from Boston to Buffalo is 450 miles.

35. Let t be the time in hours that Wendy spent on the train. Then $\frac{11}{2} - t$ is the time in hours that Wendy spent on the bus. Using $\textit{total distance} = \begin{pmatrix} \textit{distance traveled} \\ \textit{by bus} \end{pmatrix} + \begin{pmatrix} \textit{distance traveled} \\ \textit{by train} \end{pmatrix}$ and the table

	Rate	Time	Distance
By train	40	t	$40t$
By bus	60	$\frac{11}{2} - t$	$60\left(\frac{11}{2} - t\right)$

we get the equation $300 = 40t + 60\left(\frac{11}{2} - t\right) \quad \Leftrightarrow \quad 300 = 40t + 330 - 60t \quad \Leftrightarrow \quad -30 = -20t \quad \Leftrightarrow \quad t = \frac{-30}{-20} = 1.5$ hours. So the time spent on the train is $5.5 - 1.5 = 4$ hours.

37. Let r be the speed of the plane from Montreal to Los Angeles. Then $r + 0.20r = 1.20r$ is the speed of the plane from Los Angeles to Montreal.

	Rate	Time	Distance
Montreal to L.A	r	$\frac{2500}{r}$	2500
L.A. to Montreal	$1.2r$	$\frac{2500}{1.2r}$	2500

Using the equation *total time* $= \left(\begin{array}{c}\textit{time traveling}\\ \textit{Montreal to LA}\end{array}\right) + \left(\begin{array}{c}\textit{time traveling}\\ \textit{LA to Montreal}\end{array}\right)$ we have:

$9\frac{1}{6} = \dfrac{2500}{r} + \dfrac{2500}{1.2r} \quad \Leftrightarrow \quad \dfrac{55}{6} = \dfrac{2500}{r} + \dfrac{2500}{1.2r} \quad \Leftrightarrow \quad 55 \cdot 1.2r = 2500 \cdot 6 \cdot 1.2 + 2500 \cdot 6$ $\Leftrightarrow \quad 66r = 18000 + 15000 \quad \Leftrightarrow \quad 66r = 33000 \quad \Leftrightarrow \quad r = \dfrac{33000}{66} = 500$. Thus the plane flew at a speed of 500 mph on the trip from Montreal to Los Angeles.

39. Let x be the amount in mL of 60% acid solution to be used. Then $(300 - x)$ mL of 30% solution would have to be used to yield a total of 300 mL of solution.

	60% acid	30% acid	mixture
mL	x	$300 - x$	300
rate (% acid)	0.60	0.30	0.50
value	$0.60x$	$0.30(300 - x)$	$0.50(300)$

Thus the total amount of pure acid used is $0.60x + 0.30(300 - x) = 0.50(300) \quad \Leftrightarrow$ $0.3x + 90 = 150 \quad \Leftrightarrow \quad x = \frac{60}{0.3} = 200$. So 200 mL of 60% acid solution must be mixed with 100 mL of 30% solution to get 300 mL of 50% acid solution.

41. Let x be the grams of silver added. The weight of the rings is $5 \times 18\text{ g} = 90\text{ g}$

	5 rings	Pure silver	mixture
grams	90	x	$90 + x$
rate (% gold)	0.90	0	0.75
value	$0.90(90)$	$0x$	$0.75(90 + x)$

$0.90(90) + 0x = 0.75(90 + x) \quad \Leftrightarrow \quad 81 = 67.5 + 0.75x \quad \Leftrightarrow \quad 0.75x = 13.5 \quad \Leftrightarrow$ $x = \frac{13.5}{0.75} = 18$. So 18 grams of silver must be added to get the required mixture.

43. Let x be the liters of coolant removed and replaced by water.

	60% antifreeze	60% antifreeze (removed)	water	mixture
liters	3.6	x	x	3.6
rate (% antifreeze)	0.60	0.60	0	0.50
value	$0.60(3.6)$	$-0.60x$	$0x$	$0.50(3.6)$

$0.60(3.6) - 0.60x + 0x = 0.50(3.6) \quad \Leftrightarrow \quad 2.16 - 0.6x = 1.8 \quad \Leftrightarrow \quad -0.6x = -0.36 \quad \Leftrightarrow$ $x = \frac{-0.36}{-0.6} = 0.6$. So 0.6 liters must be removed and replaced by water.

45. Let t be the time in minutes it would take Candy and Tim if they work together. Candy delivers the papers at a rate of $\frac{1}{70}$ of the job per minute while Tim delivers the paper at a rate of $\frac{1}{80}$ of the job per minute. Then the sum of the fractions of the jobs that each can do individually in one minute equals

the fraction of the jobs they can do working together. So we have $\frac{1}{t} = \frac{1}{70} + \frac{1}{80} \Leftrightarrow$ $560 = 8t + 7t \Leftrightarrow 560 = 15t \Leftrightarrow t = 37\frac{1}{3}$ minutes.

47. Let t be the time, in hours, it takes Karen to paint a house alone. Then working together, Karen and Betty can paint a house in $\frac{2}{3}t$ hours. The sum of their individual rates equals their rate working together, so $\frac{1}{t} + \frac{1}{6} = \frac{1}{\frac{2}{3}t} \Leftrightarrow \frac{1}{t} + \frac{1}{6} = \frac{3}{2t} \Leftrightarrow 6 + t = 9 \Leftrightarrow t = 3$. Thus it would take Karen 3 hours to paint a house alone.

49. Let x be the height of the tall tree. Here we use the property that corresponding sides in similar triangles are proportional. The base of the similar triangles starts at eye level of the woodcutter, 5 feet. Thus we obtain the proportion: $\frac{x-5}{15} = \frac{150}{25} \Leftrightarrow 25(x-5) = 15(150) \Leftrightarrow$ $25x - 125 = 2250 \Leftrightarrow 25x = 2375 \Leftrightarrow x = 95$. Thus the tree is 95 feet tall.

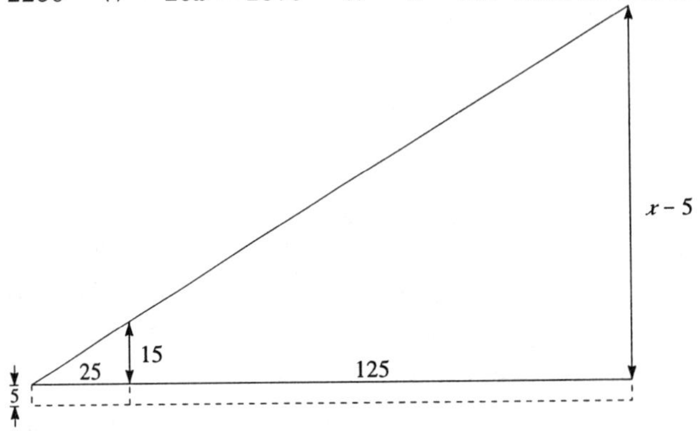

51. Let l be the length of the lot in feet. Then the length of the diagonal is $l + 10$. We apply the Pythagorean Theorem with the hypotenuse the diagonal. $l^2 + 50^2 = (l+10)^2 \Leftrightarrow$ $l^2 + 2500 = l^2 + 20l + 100 \Leftrightarrow 20l = 2400 \Leftrightarrow$ $l = 120$. The length of the lot is 120 feet.

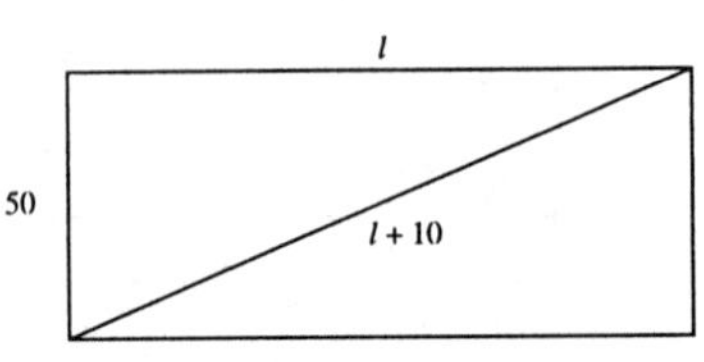

53. Let h be the height in feet of the structure. The structure is composed of a right cylinder with radius 10 and height $\frac{2}{3}h$ and a cone with base radius 10 and height $\frac{1}{3}h$. Using the formulas for the volume of a cylinder and that of a cone we obtain the equation: $1400\pi = \pi(10)^2\left(\frac{2}{3}h\right) + \frac{1}{3}\pi(10)^2\left(\frac{1}{3}h\right)$ $\Leftrightarrow 1400\pi = \frac{200\pi}{3}h + \frac{100\pi}{9}h \Leftrightarrow 126 = 6h + h$ (multiply both sides by $\frac{9}{100\pi}$) $\Leftrightarrow$ $126 = 7h \Leftrightarrow h = 18$. Thus the height of the structure is 18 feet.

55. Let h be the height of the break in feet. Then the portion of the bamboo above the break is $10 - h$. Applying the Pythagorean Theorem we obtain: $h^2 + 3^2 = (10-h)^2 \Leftrightarrow$ $h^2 + 9 = 100 - 20h + h^2 \Leftrightarrow -91 = -20h \Leftrightarrow$ $h = \frac{91}{20} = 4.55$. Thus the break is 4.55 ft above the ground.

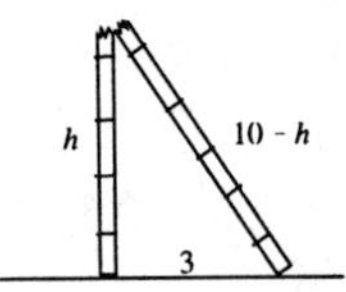

Exercises 2.3

1. $x^2 = 9 \quad \Rightarrow \quad x = \pm 3$

3. $x^2 - 24 = 0 \quad \Leftrightarrow \quad x^2 = 24 \quad \Rightarrow \quad x = \pm\sqrt{24} = \pm 2\sqrt{6}$

5. $8x^2 - 64 = 0 \quad \Leftrightarrow \quad x^2 - 8 = 0 \quad \Leftrightarrow \quad x^2 = 8 \quad \Rightarrow \quad x = \pm\sqrt{8} = \pm 2\sqrt{2}$

7. $x^2 - x - 6 = 0 \quad \Leftrightarrow \quad (x-3)(x+2) = 0 \quad \Leftrightarrow \quad x - 3 = 0$ or $x + 2 = 0 \quad \Leftrightarrow \quad x = 3$ or $x = -2$.

9. $x^2 + 4 = 4x \quad \Leftrightarrow \quad x^2 - 4x + 4 = 0 \quad \Leftrightarrow \quad (x-2)^2 = 0 \quad \Leftrightarrow \quad x - 2 = 0 \quad \Leftrightarrow \quad x = 2$.

11. $2y^2 + 7y + 3 = 0 \quad \Leftrightarrow \quad (2y+1)(y+3) = 0 \quad \Leftrightarrow \quad 2y + 1 = 0$ or $y + 3 = 0 \quad \Leftrightarrow \quad y = -\frac{1}{2}$, or $y = -3$.

13. $6x^2 + 5x = 4 \quad \Leftrightarrow \quad 6x^2 + 5x - 4 = 0 \quad \Leftrightarrow \quad (2x-1)(3x+4) = 0 \quad \Leftrightarrow \quad 2x - 1 = 0$ or $3x + 4 = 0 \quad \Leftrightarrow \quad x = \frac{1}{2}$ or $x = -\frac{4}{3}$.

15. $x^2 + 2x - 2 = 0 \quad \Leftrightarrow \quad x^2 + 2x = 2 \quad \Leftrightarrow \quad x^2 + 2x + 1 = 2 + 1 \quad \Leftrightarrow \quad (x+1)^2 = 3 \quad \Rightarrow \quad x + 1 = \pm\sqrt{3} \quad \Leftrightarrow \quad x = -1 \pm \sqrt{3}$.

17. $x^2 - 6x - 9 = 0 \quad \Leftrightarrow \quad x^2 - 6x = 9 \quad \Leftrightarrow \quad x^2 - 6x + 9 = 9 + 9 \quad \Leftrightarrow \quad (x-3)^2 = 18 \quad \Rightarrow \quad x - 3 = \pm 3\sqrt{2} \quad \Leftrightarrow \quad x = 3 \pm 3\sqrt{2}$.

19. $2x^2 + 8x + 1 = 0 \quad \Leftrightarrow \quad x^2 + 4x + \frac{1}{2} = 0 \quad \Leftrightarrow \quad x^2 + 4x = -\frac{1}{2} \quad \Leftrightarrow \quad x^2 + 4x + 4 = -\frac{1}{2} + 4 \quad \Leftrightarrow \quad (x+2)^2 = \frac{7}{2} \quad \Rightarrow \quad x + 2 = \pm\sqrt{\frac{7}{2}} \quad \Leftrightarrow \quad x = 2 \pm \frac{\sqrt{14}}{2}$.

21. $4x^2 - x = 0 \quad \Leftrightarrow \quad x^2 - \frac{1}{4}x = 0 \quad \Leftrightarrow \quad x^2 - \frac{1}{4}x + \frac{1}{64} = \frac{1}{64} \quad \Leftrightarrow \quad \left(x - \frac{1}{8}\right)^2 = \frac{1}{64} \quad \Rightarrow \quad x - \frac{1}{8} = \pm\frac{1}{8} \quad \Leftrightarrow \quad x = \frac{1}{8} \pm \frac{1}{8}$, so $x = \frac{1}{8} - \frac{1}{8} = 0$ or $x = \frac{1}{8} + \frac{1}{8} = \frac{1}{4}$.

23. $x^2 - 2x - 8 = 0 \quad \Leftrightarrow \quad (x-4)(x+2) = 0 \quad \Rightarrow \quad x - 4 = 0 \quad \Leftrightarrow \quad x = 4$ or $x + 2 = 0 \quad \Leftrightarrow \quad x = -2$

25. $x^2 + 12x - 27 = 0 \quad \Leftrightarrow \quad x^2 + 12x = 27 \quad \Leftrightarrow \quad x^2 + 12x + 36 = 27 + 36 \quad \Leftrightarrow \quad (x+6)^2 = 63 \quad \Rightarrow \quad x + 6 = \pm 3\sqrt{7} \quad \Leftrightarrow \quad x = -6 \pm 3\sqrt{7}$.

27. $3x^2 + 6x - 5 = 0 \quad \Leftrightarrow \quad x^2 + 2x - \frac{5}{3} = 0 \quad \Leftrightarrow \quad x^2 + 2x = \frac{5}{3} \quad \Leftrightarrow \quad x^2 + 2x + 1 = \frac{5}{3} + 1 \quad \Leftrightarrow \quad (x+1)^2 = \frac{8}{3} \quad \Rightarrow \quad x + 1 = \pm\sqrt{\frac{8}{3}} \quad \Leftrightarrow \quad x = -1 \pm \frac{2\sqrt{6}}{3}$.

29. $$2y^2 - y - \frac{1}{2} = 0 \quad \Rightarrow \quad y = \frac{-b \pm \sqrt{b^2 - 4ac}}{2a} = \frac{-(-1) \pm \sqrt{(-1)^2 - 4(2)\left(-\frac{1}{2}\right)}}{2(2)} = \frac{1 \pm \sqrt{1+4}}{4} = \frac{1 \pm \sqrt{5}}{4}$$

31. $4x^2 + 16x - 9 = 0 \quad \Leftrightarrow \quad (2x-1)(2x+9) = 0$ So $2x - 1 = 0 \quad \Leftrightarrow \quad 2x = 1 \quad \Leftrightarrow \quad x = \frac{1}{2}$, or $2x + 9 = 0 \quad \Leftrightarrow \quad 2x = -9 \quad \Leftrightarrow \quad x = -\frac{9}{2}$.

33. $3+5z+z^2=0 \quad \Rightarrow \quad z=\dfrac{-b\pm\sqrt{b^2-4ac}}{2a}=\dfrac{-(5)\pm\sqrt{(5)^2-4(1)(3)}}{2(1)}=\dfrac{-5\pm\sqrt{25-12}}{2}$
$=\dfrac{-5\pm\sqrt{13}}{2}$

35. $x^2-\sqrt{5}x+1=0 \quad \Rightarrow \quad x=\dfrac{-b\pm\sqrt{b^2-4ac}}{2a}=\dfrac{-\left(-\sqrt{5}\right)\pm\sqrt{\left(-\sqrt{5}\right)^2-4(1)(1)}}{2(1)}=$
$\dfrac{\sqrt{5}\pm\sqrt{5-4}}{2}=\dfrac{\sqrt{5}\pm 1}{2}$

37. $\dfrac{x^2}{x+100}=50 \quad \Rightarrow \quad x^2=50(x+100)=50x+5000 \quad \Leftrightarrow \quad x^2-50x-5000=0 \quad \Leftrightarrow$
$(x-100)(x+50)=0$ So $x-100=0 \quad \Leftrightarrow \quad x=100$, or $x+50=0 \quad \Leftrightarrow \quad x=-50$.

39. $\dfrac{x+5}{x-2}=\dfrac{5}{x+2}+\dfrac{28}{x^2-4} \quad \Rightarrow \quad (x+2)(x+5)=5(x-2)+28 \quad \Leftrightarrow$
$x^2+7x+10=5x-10+28 \quad \Leftrightarrow \quad x^2+2x-8=0 \quad \Leftrightarrow \quad (x-2)(x+4)=0 \quad \Leftrightarrow$
$x-2=0$ or $x+4=0 \quad \Leftrightarrow \quad x=2$ or $x=-4$. However, since $x=2$ is inadmissible, so the only solution is $x=-4$.

41. $\dfrac{1}{x-1}-\dfrac{2}{x^2}=0 \quad \Leftrightarrow \quad x^2-2(x-1)=0 \quad \Leftrightarrow \quad x^2-2x+2=0 \quad \Rightarrow$
$x=\dfrac{-(-2)\pm\sqrt{(-2)^2-4(1)(2)}}{2(1)}=\dfrac{2\pm\sqrt{4-8}}{2}=\dfrac{2\pm\sqrt{-4}}{2}$. Since the radicand is negative, there are no real solutions.

43. $ax^2-(2a+1)x+(a+1)=0 \quad \Leftrightarrow \quad [ax-(a+1)](x-1)=0$. So $ax-(a+1)=0 \quad \Leftrightarrow$
$ax=a+1 \quad \Leftrightarrow \quad x=\dfrac{a+1}{a}$, or $x-1=0 \quad \Leftrightarrow \quad x=1$

45. $x^2-0.011x-0.064=0 \quad \Rightarrow \quad x=\dfrac{-(-0.011)\pm\sqrt{(-0.011)^2-4(1)(-0.064)}}{2(1)}$
$=\dfrac{0.011\pm\sqrt{0.000121+0.256}}{2}=\dfrac{0.011\pm\sqrt{0.256121}}{2}\approx\dfrac{0.011\pm 0.506}{2}$

Thus $x\approx\dfrac{0.011+0.506}{2}=0.259$ or $x\approx\dfrac{0.011-0.506}{2}=-0.248$

47. $x^2-2.450x+1.500=0 \quad \Rightarrow \quad x=\dfrac{-(-2.450)\pm\sqrt{(-2.450)^2-4(1)(1.500)}}{2(1)}=$
$\dfrac{2.450\pm\sqrt{6.0025-6}}{2}=\dfrac{2.450\pm\sqrt{0.0025}}{2}=\dfrac{2.450\pm 0.050}{2}$.

Thus $x=\dfrac{2.450+0.050}{2}=1.250$ or $x=\dfrac{2.450-0.050}{2}=1.200$

49. $V=\dfrac{1}{3}\pi r^2 h \quad \Leftrightarrow \quad r^2=\dfrac{3V}{\pi h} \quad \Rightarrow \quad r=\pm\sqrt{\dfrac{3V}{\pi h}}$.

51. $a^2+b^2=c^2 \quad \Leftrightarrow \quad b^2=c^2-a^2 \quad \Rightarrow \quad b=\pm\sqrt{c^2-a^2}$

53. $A = P\left(1 + \frac{i}{100}\right)^2 \quad \Leftrightarrow \quad \frac{A}{P} = \left(1 + \frac{i}{100}\right)^2 \quad \Rightarrow \quad 1 + \frac{i}{100} = \pm\sqrt{\frac{A}{P}} \quad \Leftrightarrow$
$\frac{i}{100} = -1 \pm \sqrt{\frac{A}{P}} \quad \Leftrightarrow \quad i = -100 \pm 100\sqrt{\frac{A}{P}}$

55. Substituting $x = -3$ we have: $(-3)^2 - 2(-3) - k = 0 \quad \Leftrightarrow \quad 9 + 6 - k = 0 \quad \Leftrightarrow \quad -k = -15$ $\Leftrightarrow \quad k = 15$. Substituting $x = 5$ we have: $(5)^2 - 2(5) - k = 0 \quad \Leftrightarrow \quad 25 - 10 - k = 0 \quad \Leftrightarrow$ $-k = -15 \quad \Leftrightarrow \quad k = 15$.

57. $D = b^2 - 4ac = (-6)^2 - 4(1)(1) = 32$. Since D is positive, this equation has two real solutions.

59. $D = b^2 - 4ac = (2.20)^2 - 4(1)(1.21) = 4.84 - 4.84 = 0$. Since $D = 0$, this equation has one real solution.

61. $D = b^2 - 4ac = (r)^2 - 4(1)(-s) = r^2 + 4s$. Since D is positive, this equation has two real solutions.

63. We want to find the values of k that make the discriminant 0. $k^2 - 4(4)(25) = 0 \quad \Leftrightarrow \quad k^2 = 400$ $\Leftrightarrow \quad k = \pm 20$.

65. Let n be one number. Then other number must be $55 - n$, since $n + (55 - n) = 55$. Since the product is 684, we have $(n)(55 - n) = 684 \quad \Leftrightarrow \quad 55n - n^2 = 684 \quad \Leftrightarrow \quad n^2 - 55n + 684 = 0$
$\Rightarrow \quad n = \frac{-(-55) \pm \sqrt{(-55)^2 - 4(1)(684)}}{2(1)} = \frac{55 \pm \sqrt{3025 - 2736}}{2} = \frac{55 \pm \sqrt{289}}{2} = \frac{55 \pm 17}{2}$.
So $n = \frac{55+17}{2} = \frac{72}{2} = 36$ or $n = \frac{55-17}{2} = \frac{38}{2} = 19$. In either case, the two numbers are 19 and 36.

67. Let w be the width of the garden in feet. Then the length is $w + 10$. Thus $875 = w(w + 10) \quad \Leftrightarrow$ $w^2 + 10w - 875 = 0 \quad \Leftrightarrow \quad (w + 35)(w - 25) = 0$. So $w + 35 = 0$ in which case $w = -35$, which is not possible or $w - 25 = 0 \quad \Leftrightarrow \quad w = 25$. Thus the width is 25 feet and the length is 35 feet.

69. Let x be the length of one side of the cardboard, so we start with a piece of cardboard x by x. When 4 inches are removed from each side, the base of the box is thus $(x - 8)$ by $(x - 8)$. Since the volume is 100 in^3, we get: $4(x - 8)^2 = 100 \quad \Leftrightarrow \quad x^2 - 16x + 64 = 25 \quad \Leftrightarrow$ $x^2 - 16x + 39 = 0 \quad \Leftrightarrow \quad (x - 3)(x - 13) = 0$. So $x = 3$ or $x = 13$. But $x = 3$ is not possible, since then the length of the base would be $3 - 8 = -5$, and all lengths must be positive. Thus $x = 13$, and the piece of cardboard is 13 inches by 13 inches.

71. Let w be the width of the lot in feet. The length is $w + 6$. Using the Pythagorean Theorem we have: $w^2 + (w + 6)^2 = (174)^2 \quad \Leftrightarrow \quad w^2 + w^2 + 12w + 36 = 30276 \quad \Leftrightarrow \quad 2w^2 + 12w - 30240 = 0$ $\Leftrightarrow \quad w^2 + 6w - 15120 = 0 \quad \Leftrightarrow \quad (w + 126)(w - 120) = 0$. So $w + 126 = 0$ in which case $w = -126$, which is not possible, or $w - 120 = 0 \quad \Leftrightarrow \quad w = 120$. Thus the width is 120 feet and the length is 126 feet.

73. $v_o = 40$ ft/s
(a) Setting $h = 24$: $24 = -16t^2 + 40t \quad \Leftrightarrow \quad 16t^2 - 40t + 24 = 0 \quad \Leftrightarrow \quad 8(2t - 3)(t - 1) = 0$ $\Leftrightarrow \quad t = 1$ or $t = 1\frac{1}{2}$. Therefore the ball reaches 24 feet in 1 second (on the ascent) and again after $1\frac{1}{2}$seconds (on its descent).

(b) Setting $h = 48$: $48 = -16t^2 + 40t \Leftrightarrow 16t^2 - 40t + 48 = 0 \Leftrightarrow 2t^2 - 5t + 6 = 0$ $\Leftrightarrow t = \frac{5\pm\sqrt{25-48}}{4} = \frac{5\pm\sqrt{-23}}{4}$. However, since the discriminant $D < 0$, there are no real solutions and hence the ball never reaches a height of 48 feet.

(c) The greatest height h is reached only once $\Rightarrow$ $h = -16t^2 + 40t$ has only one solution $16t^2 - 40t + h = 0$ has only one solution $\Leftrightarrow$ $D = (-40)^2 - 4(16)(h) = 0$ $\Leftrightarrow$ $1600 - 64h = 0$ $\Leftrightarrow$ $h = 25$. So the greatest height reached by the ball is 25 feet.

(d) Setting $h = 25$: $25 = -16t^2 + 40t \Leftrightarrow 16t^2 - 40t + 25 = 0 \Leftrightarrow (4t - 5)^2 = 0 \Leftrightarrow$ $t = 1\frac{1}{4}$. Thus the ball reaches the highest point of its path after $1\frac{1}{4}$ seconds.

(e) Setting $h = 0$ (ground level): $0 = -16t^2 + 40t \Leftrightarrow 2t^2 - 5t = 0 \Leftrightarrow t(2t - 5) = 0$ $\Leftrightarrow t = 0$ (start) or $t = 2\frac{1}{2}$. So the ball hits the ground in $2\frac{1}{2}$seconds.

75. (a) The fish population on January 1, 1992 corresponds to $t = 0$: $F = 1000(30 + 17(0) - (0)^2) = 30,000$. To find when the population will again reach this value, we set $F = 30,000$: $30000 = 1000(30 + 17t - t^2) = 30000 + 17000t - 1000t^2 \Leftrightarrow$ $0 = 17000t - 1000t^2 = 1000t(17 - t) \Leftrightarrow t = 0$ or $t = 17$. Thus the fish population will again be the same 17 years later, that is on January 1, 2009.

(b) Setting $F = 0$: $0 = 1000(30 + 17t - t^2) \Leftrightarrow t^2 - 17t - 30 = 0 \Leftrightarrow$ $t = \frac{17\pm\sqrt{289+120}}{-2} = \frac{17\pm\sqrt{409}}{-2} = \frac{17\pm 20.22}{2}$. Thus $t \approx -1.612$ or $t \approx 18.612$. Since $t < 0$ is inadmissible, it follows that the fish in the lake will have died out 18.612 years after January 1, 1992, that is on Aug. 12, 2010.

77. Let x be the rate, in mph, at which the salesman drove between Ajax and Barrington.

Cities	Distance	Rate	Time
Ajax → Barrington	120	x	$\dfrac{120}{x}$
Barrington → Collins	150	$x + 10$	$\dfrac{150}{x+10}$

We have used $Time = \frac{Distance}{Rate}$ to fill in the "Time" column of the table. Since the second part of the trip took 6 min. (or $1/10$ hours) more than the first, we can use the *Time* column to get the equation:
$\dfrac{120}{x} + \dfrac{1}{10} = \dfrac{150}{x+10} \Rightarrow 120(10)(x + 10) + x(x + 10) = 150(10x) \Leftrightarrow$
$1200x + 12000 + x^2 + 10x = 1500x \Leftrightarrow x^2 - 290x + 12000 = 0 \Leftrightarrow$
$x = \dfrac{-(-290) \pm \sqrt{(-290)^2 - 4(1)(12000)}}{2} = \dfrac{290 \pm \sqrt{84100 - 48000}}{2} = \dfrac{290 \pm \sqrt{36100}}{2} =$
$\frac{290 \pm 190}{2} = 145 \pm 95$. Hence, the salesman drove either 50 mi/h or 240 mi/h between Ajax and Barrington. (The first choice seems more likely!)

79. Let r be the rowing rate in km/h of the crew in still water. Then their rate upstream was $r - 3$ km/h and their rate downstream was $r + 3$ km/h.

	Distance	Rate	Time
Upstream	6	$r - 3$	$\dfrac{6}{r-3}$
Downstream	6	$r + 3$	$\dfrac{6}{r+3}$

Since the time to row upstream plus the time to row downstream was 2 hour 40 minutes $= \frac{8}{3}$ hour, we get the equation: $\dfrac{6}{r-3} + \dfrac{6}{r+3} = \dfrac{8}{3} \quad \Leftrightarrow \quad 6(3)(r+3) + 6(3)(r-3) = 8(r-3)(r+3)$ $\Leftrightarrow \quad 18r + 54 + 18r - 54 = 8r^2 - 72 \quad \Leftrightarrow \quad 0 = 8r^2 - 36r - 72 = 4(2r^2 - 9r - 18)$ $= 4(2r+3)(r-6)$ Since $2r + 3 = 0 \quad \Leftrightarrow \quad r = -\frac{3}{2}$ is impossible, the solution is $r - 6 = 0$ $\Leftrightarrow \quad r = 6$. So the rate of the rowing crew in still water is 6 km/h.

81. Let t be the time, in hours it takes Irene to wash all the windows. Then it takes Henry $t + \frac{3}{2}$ hours to wash all the windows, and the sum of the fraction of the job per hour they can do individually equals the fraction of the job they can do together. Since 1 hour 48 minutes $= 1 + \frac{48}{60} = 1 + \frac{4}{5} = \frac{9}{5}$, we have $\dfrac{1}{t} + \dfrac{1}{t + \frac{3}{2}} = \dfrac{1}{\frac{9}{5}} \quad \Leftrightarrow \quad \dfrac{1}{t} + \dfrac{2}{2t+3} = \dfrac{5}{9} \quad \Rightarrow \quad 9(2t+3) + 2(9t) = 5t(2t+3) \quad \Leftrightarrow$ $18t + 27 + 18t = 10t^2 + 15t \quad \Leftrightarrow \quad 10t^2 - 21t - 27 = 0 \quad \Leftrightarrow$ $t = \dfrac{-(-21) \pm \sqrt{(-21)^2 - 4(10)(-27)}}{2(10)} = \dfrac{21 \pm \sqrt{441 + 1080}}{20} = \dfrac{21 \pm 39}{20}$. So $t = \dfrac{21-39}{20} = -\frac{9}{10}$ or $t = \frac{21+39}{20} = 3$. Since $t < 0$ is impossible, all the windows are washed by Irene alone in 3 hours and by Henry alone in $3 + \frac{3}{2} = 4\frac{1}{2}$ hours.

83. Let x be the distance from the center of the earth to the dead spot (in thousand of mile). Now setting $F = 0$: $0 = \dfrac{-K}{x^2} + \dfrac{0.012K}{(239-x)^2} \quad \Leftrightarrow \quad \dfrac{K}{x^2} = \dfrac{0.012K}{(239-x)^2} \quad \Leftrightarrow \quad K(239-x)^2 = 0.012Kx^2$ $\Leftrightarrow \quad 57121 - 478x + x^2 = 0.012x^2 \quad \Leftrightarrow \quad 0.988x^2 - 478x + 57121 = 0$. Using the quadratic formula we obtain:

$$x = \frac{-(-478) \pm \sqrt{(-478)^2 - 4(0.988)(57121)}}{2(0.988)} = \frac{478 \pm \sqrt{228484 - 225742.192}}{1.976} =$$

$\dfrac{478 \pm \sqrt{2741.808}}{1.976} \approx \dfrac{478 \pm 52.362}{1.976} \approx 241.903 \pm 26.499$. So either $x \approx 241.903 + 26.499 \approx 268$ or $x \approx 241.903 - 26.499 \approx 215$. Since 268 is greater than the distance from the earth to the moon, we reject it; thus $x \approx 215$ thousand miles $=$ 215,000 miles.

Exercises 2.4

1. $3-5\,i$: real part 3, imaginary part -5

3. $6\,i$: real part 0, imaginary part 6

5. $\sqrt{2}+\sqrt{-3}=\sqrt{2}+i\sqrt{3}$: real part $\sqrt{2}$, imaginary part $\sqrt{3}$

7. $(4+3\,i)+(5-2\,i)=(4+5)+(3-2)\,i=9+i$

9. $(7-\frac{1}{2}i)+(5+\frac{3}{2}\,i)=(7+5)+(-\frac{1}{2}+\frac{3}{2})\,i=12+i$

11. $(-12+8\,i)-(7+4\,i)=-12+8\,i-7-4\,i=(-12-7)+(8-4)\,i=-19+4\,i$

13. $4(-1+2\,i)=-4+8\,i$

15. $(7-i)(4+2\,i)=28+14\,i-4\,i-2\,i^2=(28+2)+(14-4)\,i=30+10\,i$

17. $(3-4\,i)(5-12\,i)=15-36\,i-20\,i+48i^2=(15-48)+(-36-20)\,i=-33-56\,i$

19. $\dfrac{1}{i}=\dfrac{1}{i}\cdot\dfrac{i}{i}=\dfrac{i}{i^2}=\dfrac{i}{-1}=-i$

21. $\dfrac{2-3\,i}{1-2\,i}=\dfrac{2-3\,i}{1-2\,i}\cdot\dfrac{1+2\,i}{1+2\,i}=\dfrac{2+4\,i-3\,i-6i^2}{1-4\,i^2}=\dfrac{(2+6)+(4-3)\,i}{1+4}=\dfrac{8+i}{5}$ or $\dfrac{8}{5}+\dfrac{1}{5}\,i$

23. $\dfrac{26+39\,i}{2-3\,i}=\dfrac{26+39\,i}{2-3\,i}\cdot\dfrac{2+3\,i}{2+3\,i}=\dfrac{52+78\,i+78\,i+117i^2}{4-9\,i^2}=\dfrac{(52-117)+(78+78)\,i}{4+9}=$
$\dfrac{-65+156\,i}{13}=\dfrac{13(-5+12\,i)}{13}=-5+12\,i$

25. $\dfrac{10\,i}{1-2\,i}=\dfrac{10\,i}{1-2\,i}\cdot\dfrac{1+2\,i}{1+2\,i}=\dfrac{10\,i+20\,i^2}{1-4\,i^2}=\dfrac{-20+10\,i}{1+4}=\dfrac{5(-4+2\,i)}{5}=-4+2\,i$

27. $i^3=i^2\cdot i=-1\cdot i=-i$

29. $i^{100}=\left(i^4\right)^{25}=(1)^{25}=1$

31. $\sqrt{-25}=5\,i$

33. $\sqrt{-3}\,\sqrt{-12}=i\sqrt{3}\cdot 2i\sqrt{3}=6\,i^2=-6$

35. $\left(3-\sqrt{-5}\right)\left(1+\sqrt{-1}\right)=\left(3-i\sqrt{5}\right)(1+i)=3+3\,i-i\sqrt{5}-i^2\sqrt{5}=$
$\left(3+\sqrt{5}\right)+\left(3-\sqrt{5}\right)i$

37. $\dfrac{2+\sqrt{-8}}{1+\sqrt{-2}}=\dfrac{2+2i\sqrt{2}}{1+i\sqrt{2}}=\dfrac{2+2i\sqrt{2}}{1+i\sqrt{2}}\cdot\dfrac{1-i\sqrt{2}}{1-i\sqrt{2}}=\dfrac{2-2i\sqrt{2}+2i\sqrt{2}-4\,i^2}{1-2\,i^2}=$
$\dfrac{(2+4)+(-2\sqrt{2}+2\sqrt{2})\,i}{1+2}=\dfrac{6}{3}=2$

39. $\dfrac{\sqrt{-36}}{\sqrt{-2}\sqrt{-9}}=\dfrac{6\,i}{i\sqrt{2}\cdot 3i}=\dfrac{2}{i\sqrt{2}}\cdot\dfrac{i\sqrt{2}}{i\sqrt{2}}=\dfrac{2i\sqrt{2}}{2\,i^2}=\dfrac{i\sqrt{2}}{-1}=-i\sqrt{2}$

41. $x^2+9=0 \quad \Leftrightarrow \quad x^2=-9 \quad \Rightarrow \quad x=\pm 3i$

43. $x^2-4x+5=0 \quad \Rightarrow \quad x=\dfrac{-(-4)\pm\sqrt{(-4)^2-4(1)(5)}}{2(1)}=\dfrac{4\pm\sqrt{16-20}}{2}=\dfrac{4\pm\sqrt{-4}}{2}=$
$\dfrac{4\pm 2i}{2}=2\pm i$

45. $x^2+x+1=0 \quad \Rightarrow \quad x=\dfrac{-(1)\pm\sqrt{(1)^2-4(1)(1)}}{2(1)}=\dfrac{-1\pm\sqrt{1-4}}{2}=\dfrac{-1\pm\sqrt{-3}}{2}=$
$\dfrac{-1\pm i\sqrt{3}}{2}=-\dfrac{1}{2}\pm\dfrac{i\sqrt{3}}{2}$

47. $2x^2-2x+1=0 \quad \Rightarrow \quad x=\dfrac{-(-2)\pm\sqrt{(-2)^2-4(2)(1)}}{2(2)}=\dfrac{2\pm\sqrt{4-8}}{4}=\dfrac{2\pm\sqrt{-4}}{4}=$
$\dfrac{2\pm 2i}{4}=\dfrac{1}{2}\pm\dfrac{1}{2}i$

49. $t+3+\dfrac{3}{t}=0 \quad \Leftrightarrow \quad t^2+3t+3=0 \quad \Rightarrow \quad t=\dfrac{-(3)\pm\sqrt{(3)^2-4(1)(3)}}{2(1)}=\dfrac{-3\pm\sqrt{9-12}}{2}$
$=\dfrac{-3\pm\sqrt{-3}}{2}=\dfrac{-3\pm i\sqrt{-3}}{2}=-\dfrac{3}{2}\pm\dfrac{i\sqrt{3}}{2}$

51. $x+3-i=2+4i \quad \Leftrightarrow \quad x=(2-3)+(4+1)i=-1+5i$

53. $2x+5+3i=7i \quad \Leftrightarrow \quad 2x=-5-4i \quad \Leftrightarrow \quad x=-\frac{5}{2}-2i$

55. $ix^2-4x+i=0 \quad \Rightarrow \quad x=\dfrac{-(-4)\pm\sqrt{(-4)^2-4(i)(i)}}{2(i)}=\dfrac{4\pm\sqrt{16-4i^2}}{2i}=$
$\dfrac{4\pm\sqrt{16+4}}{2i}=\dfrac{4\pm\sqrt{20}}{2i}=\dfrac{4\pm 2\sqrt{5}}{2i}=\dfrac{2\left(2\pm\sqrt{5}\right)}{2i}\cdot\dfrac{i}{i}=\dfrac{\left(2\pm\sqrt{5}\right)i}{i^2}=-\left(2\pm\sqrt{5}\right)i$

57. $\sqrt{2}x^2+\sqrt{3}x+\sqrt{8}=0 \quad \Rightarrow \quad x=\dfrac{-(\sqrt{3})\pm\sqrt{(\sqrt{3})^2-4\left(\sqrt{2}\right)\left(\sqrt{8}\right)}}{2\left(\sqrt{2}\right)}=\dfrac{-\sqrt{3}\pm\sqrt{3-16}}{2\sqrt{2}}$
$=\dfrac{-\sqrt{3}\pm\sqrt{-13}}{2\sqrt{2}}=\dfrac{-\sqrt{3}\pm i\sqrt{13}}{2\sqrt{2}}=\dfrac{-\sqrt{3}\pm i\sqrt{13}}{2\sqrt{2}}\cdot\dfrac{\sqrt{2}}{\sqrt{2}}=-\dfrac{\sqrt{6}}{4}\pm\dfrac{i\sqrt{26}}{4}$

59. LHS $=\overline{z}+\overline{w}=\overline{(a+bi)}+\overline{(c+di)}=a-bi+c-di=(a+c)+(-b-d)i=$
$(a+c)-(b+d)i$
RHS $=\overline{z+w}=\overline{(a+bi)+(c+di)}=\overline{(a+c)+(b+d)i}=(a+c)-(b+d)i$
Since LHS $=$ RHS this proves the statement.

61. LHS $=(\overline{z})^2=\left(\overline{(a+bi)}\right)^2=(a-bi)^2=a^2-2abi+b^2i^2=(a^2-b^2)-2abi$
RHS $=\overline{z^2}=\overline{(a+bi)^2}=\overline{a^2+2abi+b^2i^2}=\overline{(a^2-b^2)+2abi}=(a^2-b^2)-2abi$
Since LHS $=$ RHS this proves the statement.

63. $z+\overline{z}=(a+bi)+\overline{(a+bi)}=a+bi+a-bi=2a$ which is a real number.

65. $\dfrac{z}{\overline{z}} + \dfrac{\overline{z}}{z} = \dfrac{a+bi}{\overline{a+bi}} + \dfrac{\overline{a+bi}}{a+bi} = \dfrac{a+bi}{a-bi} + \dfrac{a-bi}{a+bi} = \dfrac{a+bi}{a-bi} \cdot \dfrac{a+bi}{a+bi} + \dfrac{a-bi}{a+bi} \cdot \dfrac{a-bi}{a-bi}$
$= \dfrac{a^2+2ab\,i+b^2\,i^2}{a^2-b^2\,i^2} + \dfrac{a^2-2ab\,i+b^2\,i^2}{a^2-b^2\,i^2} = \dfrac{(a^2-b^2)+2ab\,i}{a^2+b^2} + \dfrac{(a^2-b^2)-2ab\,i}{a^2+b^2} = \dfrac{2(a^2-b^2)}{a^2+b^2}$

67. Using the quadratic formula, the solutions to the equation are $x = \dfrac{-b \pm \sqrt{b^2-4ac}}{2a}$. Since both solutions are imaginary, we have $b^2 - 4ac < 0 \quad \Leftrightarrow \quad 4ac - b^2 > 0$, so the solutions are $x = \dfrac{-b}{2a} \pm \dfrac{\sqrt{4ac-b^2}}{2a}\,i$, where $\sqrt{4ac-b^2}$ is a real number. Thus the solutions are complex conjugates of each other.

Exercises 2.5

1. $x^3 = 27 \quad \Leftrightarrow \quad x = 27^{1/3} = 3.$

3. $0 = x^4 - 16 = (x^2+4)(x^2-4) = (x^2+4)(x-2)(x+2)$. $x^2+4=0$ has no real solution, $x-2=0 \quad \Leftrightarrow \quad x=2$, or $x+2=0 \quad \Leftrightarrow \quad x=-2$. Solutions are: ± 2

5. $0 = x^4 - x^3 - 2x^2 = x^2(x^2-x-2) = x^2(x-2)(x+1)$. $x^2=0 \quad \Leftrightarrow \quad x=0$,or $x-2=0$ $\Leftrightarrow \quad x=2$, or $x+1=0 \quad \Leftrightarrow \quad x=-1$. Solutions are: $0,\ 2,\ -1$

7. $0 = x^3 - 5x^2 - 2x + 10 = x^2(x-5) - 2(x-5) = (x-5)(x^2-2)$. So $x-5=0 \quad \Leftrightarrow$ $x=5$, or $x^2-2=0 \quad \Leftrightarrow \quad x^2=2 \quad \Rightarrow \quad x = \pm\sqrt{2}$. Solutions are: $5, \pm\sqrt{2}$

9. $x^3 - x^2 + x - 1 = x^2 + 1 \quad \Leftrightarrow$
$0 = x^3 - 2x^2 + x - 2 = x^2(x-2) + (x-2) = (x-2)(x^2+1)$. Since $x^2+1=0$ has no real solution, the only solution comes from $x-2=0 \quad \Leftrightarrow \quad x=2$.

11. $\sqrt{2x+1}+1 = x \quad \Leftrightarrow \quad \sqrt{2x+1} = x-1 \quad \Rightarrow \quad 2x+1 = (x-1)^2 \quad \Leftrightarrow$
$2x+1 = x^2-2x+1 \quad \Leftrightarrow \quad 0 = x^2-4x = x(x-4)$. Potential solutions are $x=0$ and $x-4$ $\Leftrightarrow \quad x=4$. These are only potential solutions since *squaring* not a reversible operation. We must check each potential solution in the original equation. Checking $x=0$: $\sqrt{2(0)+1}+1 \stackrel{?}{=} (0)$, $\sqrt{1}+1 \stackrel{?}{=} 0$, NO! Checking $x=4$: $\sqrt{2(4)+1}+1 \stackrel{?}{=} (4)$, $\sqrt{9}+1 \stackrel{?}{=} 4$, $3+1 \stackrel{?}{=} 4$, Yes. Only solution is $x=4$.

13. $\sqrt{5-x}+1 = x-2 \quad \Leftrightarrow \quad \sqrt{5-x} = x-3 \quad \Rightarrow \quad 5-x = (x-3)^2 \quad \Leftrightarrow$
$5-x = x^2-6x+9 \quad \Leftrightarrow \quad 0 = x^2-5x+4 = (x-4)(x-1)$. Potential solutions are $x=4$ and $x=1$. We must check each potential solution in the original equation. Checking $x=4$: $\sqrt{5-(4)}+1 \stackrel{?}{=} (4)-2$, $\sqrt{1}+1 \stackrel{?}{=} 4-2, 1+1 \stackrel{?}{=} 2$, Yes. Checking $x=1$: $\sqrt{5-(1)}+1 \stackrel{?}{=} (1)-2$, $\sqrt{4}+1 \stackrel{?}{=} -1$, $2+1 \stackrel{?}{=} -1$, NO! Only solution is $x=4$.

15. $\sqrt{\sqrt{x-5}+x} = 5$. Squaring both sides,$\sqrt{x-5}+x = 25 \quad \Leftrightarrow \quad \sqrt{x-5} = 25-x$.. Squaring both sides, $x-5 = (25-x)^2 \quad \Leftrightarrow \quad x-5 = 625-50x+x^2 \quad \Leftrightarrow \quad 0 = x^2-51x+630$ $= (x-30)(x-21)$. Potential solutions are $x=30$ and $x=21$. We must check each potential solution in the original equation. Checking $x=31$: $\sqrt{\sqrt{(30)-5}+(30)} \stackrel{?}{=} 5$, $\sqrt{\sqrt{(30)-5}+(30)} = \sqrt{\sqrt{25}+30} = \sqrt{35} > 5$, hence not a solution. Checking $x=21$: $\sqrt{\sqrt{(21)-5}+21} \stackrel{?}{=} 5$, $\sqrt{\sqrt{(21)-5}+21} = \sqrt{\sqrt{16}+21} = \sqrt{25} = 5$, hence a solution. Only solution is $x=21$.

17. $2x^4+4x^2+1 = 0$. The LHS is the sum of two nonnegative numbers and a positive number, so $2x^4+4x^2+1 \geq 1 \neq 0$. This equation has no real solution.

19. $0 = (x+5)^2 - 3(x+5) - 10 = [(x+5)-5][(x+5)+2] = x(x+7) \quad \Leftrightarrow \quad x=0$ or $x=-7$. Solutions are: $0, -7$.

21. $4(x+1)^{1/2} - 5(x+1)^{3/2} + (x+1)^{5/2} = 0 \quad \Leftrightarrow \quad \sqrt{x+1}[4-5(x+1)+(x+1)^2] = 0 \quad \Leftrightarrow$
$\sqrt{x+1}(4-5x-5+x^2+2x+1) = 0 \quad \Leftrightarrow \quad \sqrt{x+1}(x^2-3x) = 0 \quad \Leftrightarrow$
$\sqrt{x+1} \cdot x(x-3) = 0 \quad \Leftrightarrow \quad x=-1$ or $x=0$ or $x=3$.

23. Let $u = x^{2/3}$; then $0 = x^{4/3} - 5x^{2/3} + 6 = u^2 - 5u + 6 = (u-3)(u-2) \Leftrightarrow u - 3 = x^{2/3} - 3 = 0 \Leftrightarrow x^{2/3} = 3 \Rightarrow x = 3^{3/2}$, or $u - 2 = x^{2/3} - 2 = 0 \Leftrightarrow x^{2/3} = 2 \Rightarrow x = 2^{3/2}$. Solutions are: $3^{3/2}$, $2^{3/2}$.

25. Let $u = x^{1/6}$. (We choose the exponent $\frac{1}{6}$ because the LCD of 2, 3, and 6 is 6.) Then $x^{1/2} - 3x^{1/3} = 3x^{1/6} - 9 \Leftrightarrow x^{3/6} - 3x^{2/6} = 3x^{1/6} - 9 \Leftrightarrow u^3 - 3u^2 = 3u - 9 \Leftrightarrow 0 = u^3 - 3u^2 - 3u + 9 = u^2(u-3) - 3(u-3) = (u-3)(u^2-3)$. So $u - 3 = 0 \Leftrightarrow x^{1/6} - 3 = 0 \Leftrightarrow x^{1/6} = 3 \Leftrightarrow x = 3^6 = 72$; and $u^2 - 3 = 0 \Leftrightarrow x^{1/3} - 3 = 0 \Leftrightarrow x^{1/3} = 3 \Leftrightarrow x = 3^3 = 27$.

27. $\dfrac{1}{x^3} + \dfrac{4}{x^2} + \dfrac{4}{x} = 0 \Rightarrow 1 + 4x + 4x^2 = 0 \Leftrightarrow (1+2x)^2 = 0 \Leftrightarrow 1 + 2x = 0 \Leftrightarrow 2x = -1 \Leftrightarrow x = -\frac{1}{2}$.

29. $x^2\sqrt{x+3} = (x+3)^{3/2} \Leftrightarrow 0 = x^2\sqrt{x+3} - (x+3)^{3/2} \Leftrightarrow 0 = \sqrt{x+3}[(x^2) - (x+3)] \Leftrightarrow 0 = \sqrt{x+3}(x^2 - x - 3) \Leftrightarrow (x+3)^{1/2} = 0$ or $x^2 - x - 3 = 0 \Leftrightarrow x = -3$ or $x = \frac{1 \pm \sqrt{13}}{2}$.

31. $\sqrt{x + \sqrt{x+2}} = 2$. Squaring both sides, $x + \sqrt{x+2} = 4 \Leftrightarrow \sqrt{x+2} = 4 - x$. Squaring both sides, $x + 2 = (4-x)^2 = 16 - 8x + x^2 \Leftrightarrow 0 = x^2 - 9x + 14 \Leftrightarrow 0 = (x-7)(x-2) \Leftrightarrow x - 7 = 0$ or $x - 2 = 0 \Leftrightarrow x = 7$ or $x = 2$. So $x = 2$ is a solution but $x = 7$ is not, since it does not satisfy the original equation.

33. $x^3 = -8 \Leftrightarrow 0 = x^3 + 8 \Leftrightarrow 0 = (x+2)(x^2 - 2x + 4) \Leftrightarrow x + 2 = 0$ or $x^2 - 2x + 4 = 0 \Leftrightarrow x = -2$ or $x = \dfrac{2 \pm 2i\sqrt{3}}{2} = 1 \pm i\sqrt{3}$

35. $x^3 = 1 \Leftrightarrow x^3 - 1 = 0 \Leftrightarrow (x-1)(x^2 + x + 1) \Leftrightarrow x - 1 = 0$ or $x^2 + x + 1 = 0 \Leftrightarrow x = 1$ or $x = \dfrac{-1 \pm i\sqrt{3}}{2}$

37. $x^3 + x^2 + x = 0 \Leftrightarrow x(x^2 + x + 1) = 0 \Leftrightarrow x = 0$ or $x = \dfrac{-1 \pm i\sqrt{3}}{2}$

39. $x^4 - 6x^2 + 8 = 0 \Leftrightarrow (x^2 - 4)(x^2 - 2) = 0 \Leftrightarrow x = \pm 2$ or $x = \pm\sqrt{2}$.

41. $x^6 - 9x^3 + 8 = 0 \Leftrightarrow (x^3 - 8)(x^3 - 1) = 0 \Leftrightarrow (x-2)(x^2 + 2x + 4)(x-1)(x^2 + x + 1) \Leftrightarrow x = 2$ or $x = \frac{-2 \pm 2i\sqrt{3}}{2} = -1 \pm i\sqrt{3}$ or $x = 1$ or $x = \dfrac{-1 \pm i\sqrt{3}}{2}$

43. $\sqrt{x^2+1} + \dfrac{8}{\sqrt{x^2+1}} = \sqrt{x^2+9}$. Squaring both sides, $(x^2+1) + 16 + \dfrac{64}{x^2+1} = x^2 + 9 \Leftrightarrow \dfrac{64}{x^2+1} = -8 \Leftrightarrow \dfrac{8}{x^2+1} = -1 \Leftrightarrow x^2 + 1 = -8 \Leftrightarrow x^2 = -9 \Leftrightarrow x = \pm 3i$.

45. We have that the volume is 180 ft^3, so $x(x-4)(x+9) = 180 \Leftrightarrow x^3 + 5x^2 - 36x = 180 \Leftrightarrow x^3 + 5x^2 - 36x - 180 = 0 \Leftrightarrow x^2(x+5) - 36(x+5) = 0 \Leftrightarrow (x+5)(x^2 - 36) = 0 \Leftrightarrow (x+5)(x+6)(x-6) = 0 \Rightarrow x = 6$ is the only positive solution. So the box is 2 feet by 6 feet by 15 feet.

47. Let x be the length in miles of the abandoned road to be used. Then the length of the abandoned road not used is $40 - x$ and the length of the new road is $\sqrt{10^2 + (40-x)^2}$ miles by the

Pythagorean Theorem. Since the cost of the road = *cost per mile* × *number of miles*, we have that: $100,000x + 200,000\sqrt{x^2 - 80x + 1700} = 6,800,000 \quad \Leftrightarrow \quad 2\sqrt{x^2 - 80x + 1700} = 68 - x$. Squaring both sides we get: $4x^2 - 320x + 6800 = 4624 - 136x + x^2 \quad \Leftrightarrow$ $3x^2 - 184x + 2176 = 0 \quad \Leftrightarrow \quad x = \frac{184 \pm \sqrt{33856 - 26112}}{6} = \frac{184 \pm 88}{6} \quad \Leftrightarrow \quad x = \frac{136}{3}$ or $x = 16$. Since $45\frac{1}{3}$ is longer than the existing road, 16 miles of the abandoned road should be used. A completely new road would have length $\sqrt{10^2 + 40^2}$, (let $x = 0$), and would cost $\sqrt{1700} \times 200,000 \approx 8.3$ million dollars. So no, it would not be cheaper.

49. Let x be the height of the pile in feet. Then the diameter is $3x$ and the radius is $\frac{3x}{2}$ feet. Since the volume of the cone is 1000 ft^3, we have that: $\frac{\pi}{3}\left(\frac{3x}{2}\right)^2 x = 1000 \quad \Leftrightarrow \quad \frac{3\pi x^3}{4} = 1000 \quad \Leftrightarrow$ $x^3 = \frac{4000}{3\pi} \quad \Leftrightarrow \quad x = \sqrt[3]{\frac{4000}{3\pi}} \approx 7.52$ feet.

51. Let x be the length of the hypotenuse of the triangle, in feet. Then one of the other sides has length $x - 7$ feet, and since the perimeter is 392 feet, the remaining side must have length $392 - x - (x - 7) = 399 - 2x$. From the Pythagorean Theorem, we get
$(x - 7)^2 + (399 - 2x)^2 = x^2 \quad \Leftrightarrow$
$4x^2 - 1610x + 159250 = 0$.
Using the quadratic formula we get

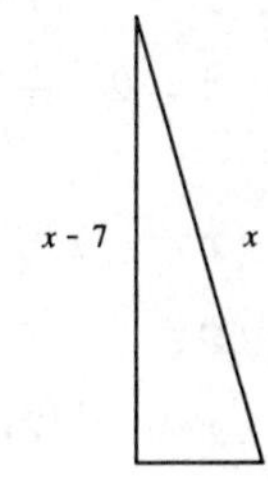

$$x = \frac{1610 \pm \sqrt{1610^2 - 4(4)(159250)}}{2(4)} = \frac{1610 \pm \sqrt{44100}}{8}$$
$= \frac{1610 \pm 210}{8}$ and so $x = 227.5$ or 175.

But if $x = 227.5$, then the side of length $x - 7$ combined with the hypotenuse already exceed the perimeter of 392 feet, and so we must have $x = 175$. Thus the other sides have length $175 - 7 = 168$ and $399 - 2(175) = 49$. The lot has sides of length 49 feet, 168 feet, and 175 feet.

53. $0 = x^4 + 5ax^2 + 4a^2 = (x^2 + a)(x^2 + 4a)$. Since a is positive, $x^2 + a = 0 \quad \Leftrightarrow \quad x^2 = -a \quad \Leftrightarrow$ $x = \pm i\sqrt{a}$. Again, since a is positive, $x^2 + 4a = 0 \quad \Leftrightarrow \quad x^2 = -4a \quad \Leftrightarrow \quad x = \pm 2i\sqrt{a}$. Thus the four solutions are: $\pm i\sqrt{a}$, $\pm 2i\sqrt{a}$.

55. $\sqrt{x + a} + \sqrt{x - a} = \sqrt{2}\sqrt{x + 6}$. Squaring both sides, we have
$x + a + 2(\sqrt{x + a})(\sqrt{x - a}) + x - a = 2(x + 6) \quad \Leftrightarrow$
$2x + 2(\sqrt{x + a})(\sqrt{x - a}) = 2x + 12 \quad \Leftrightarrow \quad 2(\sqrt{x + a})(\sqrt{x - a}) = 12 \quad \Leftrightarrow$
$(\sqrt{x + a})(\sqrt{x - a}) = 6$. Squaring both sides again we have: $(x + a)(x - a) = 36 \quad \Leftrightarrow$
$x^2 - a^2 = 36 \quad \Leftrightarrow \quad x^2 = a^2 + 36 \quad \Leftrightarrow \quad x = \pm\sqrt{a^2 + 36}$. Checking these answers, we see that $x = -\sqrt{a^2 + 36}$ is not a solution (for example, try substituting $a = 8$), but $x = \sqrt{a^2 + 36}$ is a solution.

Exercises 2.6

1. $x=-1$: $(-1)+1 \overset{?}{\geq} 0$. Yes, $0 \geq 0$.
   $x=\frac{1}{2}$: $(\frac{1}{2})+1 \overset{?}{\geq} 0$. Yes.
   $x=2$: $(2)+1 \overset{?}{\geq} 0$. Yes.
   $x=0$: $(0)+1 \overset{?}{\geq} 0$. Yes.
   $x=\sqrt{2}$: $(\sqrt{2})+1 \overset{?}{\geq} 0$. Yes.
   Solutions are: $-1, 0, \frac{1}{2}, \sqrt{2}, 2$

3. $x=-1$: $2(-1)+10 \overset{?}{>} 8$. No, $8 \not> 8$.
   $x=\frac{1}{2}$: $2(\frac{1}{2})+10 \overset{?}{>} 8$. Yes.
   $x=2$: $2(2)+10 \overset{?}{>} 8$. Yes.
   $x=0$: $2(0)+10 \overset{?}{>} 8$. Yes.
   $x=\sqrt{2}$: $2(\sqrt{2})+10 \overset{?}{>} 8$. Yes.
   Solutions are: $0, \frac{1}{2}, \sqrt{2}, 2$

5. $x=-1$: $\dfrac{1}{(-1)} \overset{?}{\leq} \dfrac{1}{2}$. Yes, $-1 \leq \dfrac{1}{2}$.
   $x=\frac{1}{2}$: $\dfrac{1}{(\frac{1}{2})} \overset{?}{\leq} \dfrac{1}{2}$. No.
   $x=2$: $\dfrac{1}{(2)} \overset{?}{\leq} \dfrac{1}{2}$. Yes.
   $x=0$: $\dfrac{1}{(0)} \overset{?}{\leq} \dfrac{1}{2}$. No, the LHS is undefined.
   $x=\sqrt{2}$: $\dfrac{1}{(\sqrt{2})} \overset{?}{\leq} \dfrac{1}{2}$. No.
   Solutions are: $-1, 2$

7. $x=-1$: $(-1)^2+2 \overset{?}{<} 4$. Yes.
   $x=\frac{1}{2}$: $\left(\frac{1}{2}\right)^2+2 \overset{?}{<} 4$. Yes.
   $x=2$: $(2)^2+2 \overset{?}{<} 4$. No.
   $x=0$: $(0)^2+2 \overset{?}{<} 4$. Yes.
   $x=\sqrt{2}$: $(\sqrt{2})^2+2 \overset{?}{<} 4$. No.
   Solutions are: $-1, 0, \frac{1}{2}$

9. $3x \leq 12 \iff x \leq 4$ Interval: $(-\infty, 4]$ Graph: (4)

11. $20 < -4x \iff -5 > x$ Interval: $(-\infty, -5)$ Graph: (-5)

13. $2x-5>3 \iff 2x>8 \iff x>4$ Interval: $(4, \infty)$
    Graph: (4)

15. $7-x \geq 5 \iff -x \geq -2 \iff x \leq 2$ Interval: $(-\infty, 2]$
    Graph: (2)

17. $2x+1<0 \iff 2x<-1 \iff x<-\frac{1}{2}$ Interval: $(-\infty, -\frac{1}{2})$
    Graph: ($-\frac{1}{2}$)

19. $3x+11 \leq 6x+8 \iff 3 \leq 3x \iff 1 \leq x$ Interval: $[1, \infty)$
    Graph: (1)

21. $1-x \leq 2 \iff -x \leq 1 \iff x \geq -1$ Interval: $[-1, \infty)$
    Graph: (-1)

23. $4-3x \leq -(1+8x) \iff 4-3x \leq -1-8x \iff 5x \leq -5 \iff x \leq -1$
    Interval: $(-\infty, -1]$ Graph: (-1)

25. $-1 < 2x - 5 < 7 \quad \Leftrightarrow \quad$ (add 5 to each expression) $4 < 2x < 12 \quad \Leftrightarrow \quad$ (divide each expression by 2) $2 < x < 6$ Interval: $(2, 6)$ Graph:

27. $0 \le 1 - x < 1 \quad \Leftrightarrow \quad$ (add 5 to each expression) $-1 \le -x < 0 \quad \Leftrightarrow \quad$ (multiply each expression by -1, reverse the direction of the inequality) $1 \ge x > 0 \quad \Leftrightarrow \quad$ (expressing in standard form) $0 < x \le 1$

Interval: $(0, 1]$ Graph:

29. $4x < 2x + 1 \le 3x + 2$ Work each part separately.
$4x < 2x + 1 \quad \Leftrightarrow \quad 2x < 1 \quad \Leftrightarrow \quad x < \frac{1}{2}$.
$2x + 1 \le 3x + 2 \quad \Leftrightarrow \quad -1 \le x$.
Putting these conditions together, we have $-1 \le x < \frac{1}{2}$. Interval: $[-1, \frac{1}{2})$
Graph:

31. $1 - x \ge 3 - 2x \ge x - 6$ Work each part separately.
$1 - x \ge 3 - 2x \quad \Leftrightarrow \quad x \ge 2 \quad \Leftrightarrow \quad 2 \le x$.
$3 - 2x \ge x - 6 \quad \Leftrightarrow \quad 9 \ge 3x \quad \Leftrightarrow \quad 3 \ge x \quad \Leftrightarrow \quad x \le 3$
Putting these conditions together, we have $2 \le x \le 3$. Interval: $[2, 3]$
Graph:

33. $\dfrac{2}{3} \ge \dfrac{2x - 3}{12} > \dfrac{1}{6} \quad \Leftrightarrow \quad$ (multiply each expression by 12) $8 \ge 2x - 3 > 2 \quad \Leftrightarrow \quad 11 \ge 2x > 5$
$\Leftrightarrow \quad \frac{11}{2} \ge x > \frac{5}{2} \quad \Leftrightarrow \quad$ (expressing in standard form) $\frac{5}{2} < x \le \frac{11}{2}$. Interval: $\left(\frac{5}{2}, \frac{11}{2}\right]$
Graph:

35. $\dfrac{x - 1}{2} \ge \dfrac{2}{3} \ge 1 - \dfrac{x}{6} \quad \Leftrightarrow \quad$ (multiply each expression by 6) $3(x - 1) \ge 4 \ge 6 - x \quad \Leftrightarrow$
$3x - 3 \ge 4 \ge 6 - x$ Work each part separately.
$3x - 3 \ge 4 \quad \Leftrightarrow \quad 3x \ge 7 \quad \Leftrightarrow \quad x \ge \frac{7}{3} \quad \Leftrightarrow \quad \frac{7}{3} \le x$.
$4 \ge 6 - x \quad \Leftrightarrow \quad -2 \ge -x \quad \Leftrightarrow \quad 2 \le x$
Putting these conditions together, since $2 < \frac{7}{3}$, we have: $\frac{7}{3} \le x$. Interval: $\left[\frac{7}{3}, \infty\right)$
Graph:

37. $\dfrac{1}{x} < 4$. Since $x \ne 0$, we only need to consider two cases: $x > 0$ and $x < 0$.
Case 1: If $x > 0$, then when we multiply both sides of the inequality by x we do not reverse the inequality, and we get: $\dfrac{1}{x} < 4 \quad \Leftrightarrow \quad \dfrac{1}{4} < x$; so $0 < x$ and $\dfrac{1}{4} < x$ yields just $\dfrac{1}{4} < x$.
Case 2: If $x < 0$, then when we multiply both sides of the inequality by x we must reverse the inequality, and we get: $\dfrac{1}{x} < 4 \quad \Leftrightarrow \quad \dfrac{1}{4} > x$; so $x < 0$ and $x < \dfrac{1}{4}$ yields just $x < 0$.
Interval: $(-\infty, 0) \cup (\frac{1}{4}, \infty)$ Graph:

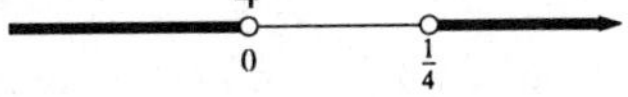

39. $\dfrac{2}{3} \le \dfrac{1}{x - 2} < 1$. The first step is to multiply each expression by $x - 2$. Since we don't know the sign of $x - 2$, we consider two cases: $0 < x - 2$ (or $2 < x$) and $x - 2 < 0$ (or $x < 2$).

Case 1: If $2 < x$, then when we multiply each expression by $x - 2$ we do not reverse the inequality, and we get: $\frac{2}{3} \leq \frac{1}{x-2} < 1 \quad \Leftrightarrow \quad \frac{2(x-2)}{3} \leq 1 < x - 2$. Work each part separately.
$\frac{2(x-2)}{3} \leq 1 \quad \Leftrightarrow \quad 2(x-2) \leq 3 \quad \Leftrightarrow \quad 2x - 4 \leq 3 \quad \Leftrightarrow \quad 2x \leq 7 \quad \Leftrightarrow \quad x \leq \frac{7}{2}$.
$1 < x - 2 \quad \Leftrightarrow \quad 3 < x$. Putting the three conditions together we have: $2 < x$ and $x \leq \frac{7}{2}$ and $3 < x \quad \Rightarrow \quad 3 < x \leq \frac{7}{2}$.
Case 2: If $x < 2$, then when we multiply each expression by $x - 2$ we must reverse the inequality, and we get: $\frac{2}{3} \leq \frac{1}{x-2} < 1 \quad \Leftrightarrow \quad \frac{2(x-2)}{3} \geq 1 > x - 2$. Work each part separately.
$\frac{2(x-2)}{3} \geq 1 \quad \Leftrightarrow \quad 2(x-2) \geq 3 \quad \Leftrightarrow \quad 2x - 4 \geq 3 \quad \Leftrightarrow \quad 2x \geq 7 \quad \Leftrightarrow \quad x \geq \frac{7}{2}$.
$1 > x - 2 \quad \Leftrightarrow \quad 3 > x$. Putting the three conditions together we have: $x < 2$ and $x \geq \frac{7}{2}$ and $3 > x \quad \Rightarrow \quad$ no solution.
Interval: $(3, \frac{7}{2}]$
Graph: 3 $\frac{7}{2}$

41. Inserting the relationship $C = \frac{5}{9}(F - 32)$ we have $5 \leq C \leq 30 \quad \Leftrightarrow \quad 5 \leq \frac{5}{9}(F - 32) \leq 30 \quad \Leftrightarrow$ $9 \leq F - 32 \leq 54 \quad \Leftrightarrow \quad 41 \leq F \leq 86$.

43. (a) Let x be the number of \$3 increases. Then the number of seats sold is $120 - x$. So $P = 200 + 3x \quad \Leftrightarrow \quad 3x = P - 200 \quad \Leftrightarrow \quad x = \frac{1}{3}(P - 200)$. Substituting for x we have that the number of seats sold is $120 - x = 120 - \frac{1}{3}(P - 200) = -\frac{1}{3}P + \frac{560}{3}$.
(b) $90 \leq -\frac{1}{3}P + \frac{560}{3} \leq 115 \quad \Leftrightarrow \quad 270 \leq 360 - P + 200 \leq 345 \quad \Leftrightarrow$ $270 \leq -P + 560 \leq 345 \quad \Leftrightarrow \quad -290 \leq -P \leq -215 \quad \Leftrightarrow \quad 290 \geq P \geq 215$. Putting this into standard order, we have $215 \leq P \leq 290$. So the ticket prices are between \$215 and \$290.

45. We need to solve $6400 \leq 0.35m + 2200 \leq 7100$ for m. So $6400 \leq 0.35m + 2200 \leq 7100 \quad \Leftrightarrow$ $4200 \leq 0.35m \leq 4900 \quad \Leftrightarrow \quad 12000 \leq m \leq 14000$. She plans on driving between $12,000$ and $14,000$ miles.

47. If $A \leq B$, then $0 \leq B - A$. Therefore $0 \leq B - A = (B + C) - (A + C)$ and so $A + C \leq B + C$.

49. If $A \leq B$, then $0 \leq B - A$. If $C < 0$, then $C(B - A) \leq 0$ since this is the product of a negative number and a non-negative number. Thus $CB - CA \leq 0 \quad \Leftrightarrow \quad CB \leq CA \quad \Leftrightarrow \quad CA \geq CB$.

51. $a(bx - c) \geq bc$ (where $a, b, c > 0$) $\quad \Leftrightarrow \quad bx - c \geq \frac{bc}{a} \quad \Leftrightarrow \quad bx \geq \frac{bc}{a} + c \quad \Leftrightarrow$
$x \geq \frac{1}{b}\left(\frac{bc}{a} + c\right) = \frac{c}{a} + \frac{c}{b} \quad \Leftrightarrow \quad x \geq \frac{c}{a} + \frac{c}{b}$.

53. $ax + b < c$ (where $a, b, c < 0$) $\quad \Leftrightarrow \quad ax < c - b \quad \Leftrightarrow \quad x > \frac{c - b}{a}$

55. If $a < b$, then $a + a < a + b$ (adding a to both sides) and $a + b < b + b$ (adding b to both sides). So $a + a < a + b < b + b \quad \Leftrightarrow \quad$ (dividing by 2) $a < \frac{a+b}{2} < b$

57. $\dfrac{a}{b} < \dfrac{c}{d}$ (where $a,\ b,\ c,\ d > 0$) $\Leftrightarrow$ $\dfrac{ad}{b} < c$ $\Leftrightarrow$ $a + \dfrac{ad}{b} < a + c$ $\Leftrightarrow$ $\dfrac{ab}{b} + \dfrac{ad}{b} < a + c$ $\Leftrightarrow$ $\dfrac{a}{b}(b+d) < a + c$ $\Leftrightarrow$ $\dfrac{a}{b} < \dfrac{a+c}{b+d}$, since $b + d > 0$. Similarly, $\dfrac{a}{b} < \dfrac{c}{d}$ $\Leftrightarrow$ $a < \dfrac{bc}{d}$ $\Leftrightarrow$ $a + c < \dfrac{bc}{d} + c$ $\Leftrightarrow$ $a + c < \dfrac{bc}{d} + \dfrac{dc}{d}$ $\Leftrightarrow$ $a + c < \dfrac{c}{d}(b+d)$ $\Leftrightarrow$ $\dfrac{a+c}{b+d} < \dfrac{c}{d}$, since $b + d > 0$. Combining the two inequalities, $\dfrac{a}{b} < \dfrac{a+c}{b+d} < \dfrac{c}{d}$.

Exercises 2.7

1. $(x-2)(x-5) > 0$. The expression on the left of the inequality changes sign where $x = 2$ and where $x = 5$. Thus we must check the intervals in the following table.

Interval	$(-\infty, 2)$	$(2,5)$	$(5,\infty)$
Sign of $x-2$	$-$	$+$	$+$
Sign of $x-5$	$-$	$-$	$+$
Sign of $(x-2)(x-5)$	$+$	$-$	$+$

From the table, the solution set is : $\{x \mid x < 2 \text{ or } 5 < x\}$
Interval: $(-\infty, 2) \cup (5, \infty)$ Graph: 2 5

3. $x^2 - 3x - 18 \le 0 \quad \Leftrightarrow \quad (x+3)(x-6) \le 0$. The expression on the left of the inequality changes sign where $x = 6$ and where $x = -3$. Thus we must check the intervals in the following table.

Interval	$(-\infty, -3)$	$(-3,6)$	$(6,\infty)$
Sign of $x+3$	$-$	$+$	$+$
Sign of $x-6$	$-$	$-$	$+$
Sign of $(x+3)(x-6)$	$+$	$-$	$+$

From the table, the solution set is : $\{x \mid -3 \le x \le 6\}$
Interval: $[-3, 6]$ Graph: -3 6

5. $2x^2 + x \ge 1 \quad \Leftrightarrow \quad 2x^2 + x - 1 \ge 0 \quad \Leftrightarrow \quad (x+1)(2x-1) \ge 0$. The expression on the left of the inequality changes sign where $x = -1$ and where $x = \frac{1}{2}$. Thus we must check the intervals in the following table.

Interval	$(-\infty, -1)$	$(-1, \frac{1}{2})$	$(\frac{1}{2}, \infty)$
Sign of $x+1$	$-$	$+$	$+$
Sign of $2x-1$	$-$	$-$	$+$
Sign of $(x+1)(2x-1)$	$+$	$-$	$+$

From the table, the solution set is : $\{x \mid x \le -1 \text{ or } \frac{1}{2} \le x\}$
Interval: $(-\infty, -1] \cup \left[\frac{1}{2}, \infty\right)$ Graph: -1 $\frac{1}{2}$

7. $3x^2 - 3x < 2x^2 + 4 \quad \Leftrightarrow \quad x^2 - 3x - 4 < 0 \quad \Leftrightarrow \quad (x+1)(x-4) < 0$. The expression on the left of the inequality changes sign where $x = -1$ and where $x = 4$. Thus we must check the intervals in the following table.

Interval	$(-\infty, -1)$	$(-1,4)$	$(4,\infty)$
Sign of $x+1$	$-$	$+$	$+$
Sign of $x-4$	$-$	$-$	$+$
Sign of $(x+1)(x-4)$	$+$	$-$	$+$

From the table, the solution set is : $\{x \mid -1 < x < 4\}$
Interval: $(-1, 4)$ Graph: -1 4

9. $x^2 > 3(x+6) \quad \Leftrightarrow \quad x^2 - 3x - 18 > 0 \quad \Leftrightarrow \quad (x+3)(x-6) > 0$. The expression on the left of the inequality changes sign where $x = 6$ and where $x = -3$. Thus we must check the intervals in the following table.

Interval	$(-\infty,-3)$	$(-3,6)$	$(6,\infty)$
Sign of $x+3$	$-$	$+$	$+$
Sign of $x-6$	$-$	$-$	$+$
Sign of $(x+3)(x-6)$	$+$	$-$	$+$

From the table, the solution set is : $\{x \mid x < -3 \text{ or } 6 < x\}$
Interval: $(-\infty,-3) \cup (6,\infty)$ Graph:

11. $x^2 < 4 \Leftrightarrow x^2 - 4 < 0 \Leftrightarrow (x+2)(x-2) < 0$. The expression on the left of the inequality changes sign where $x = -2$ and where $x = 2$. Thus we must check the intervals in the following table.

Interval	$(-\infty,-2)$	$(-2,2)$	$(2,\infty)$
Sign of $x+2$	$-$	$+$	$+$
Sign of $x-2$	$-$	$-$	$+$
Sign of $(x+2)(x-2)$	$+$	$-$	$+$

From the table, the solution set is : $\{x \mid -2 < x < 2\}$
Interval: $(-2,2)$ Graph:

13. $-2x^2 \le 4 \Leftrightarrow -2x^2 - 4 \le 0 \Leftrightarrow -2(x^2+1) \le 0$. Since $x^2 + 1 > 0$, $-2(x^2+1) \le 0$ for all x. Interval: $(-\infty,\infty)$ Graph:

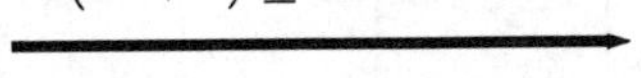

15. $x(x^2-4) \ge 0 \Leftrightarrow x(x+2)(x-2) \ge 0$. The expression on the left of the inequality changes sign where $x = 0$, where $x = -2$ and where $x = 2$. Thus we must check the intervals in the following table.

Interval	$(-\infty,-2)$	$(-2,0)$	$(0,2)$	$(2,\infty)$
Sign of $x+2$	$-$	$+$	$+$	$+$
Sign of x	$-$	$-$	$+$	$+$
Sign of $x-2$	$-$	$-$	$-$	$+$
Sign of $(x+2)(x-2)$	$-$	$+$	$-$	$+$

From the table, the solution set is : $\{x \mid -2 \le x \le 0 \text{ or } 2 \le x\}$
Interval: $[-2,0] \cup [2,\infty)$ Graph:

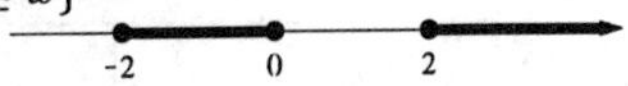

17. $\dfrac{x-3}{x+1} \ge 0$. The expression on the left of the inequality changes sign where $x = -1$ and where $x = 3$. Thus we must check the intervals in the following table.

Interval	$(-\infty,-1)$	$(-1,3)$	$(3,\infty)$
Sign of $x+1$	$-$	$+$	$+$
Sign of $x-3$	$-$	$-$	$+$
Sign of $\dfrac{x-3}{x+1}$	$+$	$-$	$+$

From the table, the solution set is : $\{x \mid x < -1 \text{ or } x \le 3\}$. Since the denominator cannot equal 0 we must have $x \ne -1$.
Interval: $(-\infty,-1) \cup [3,\infty)$ Graph:

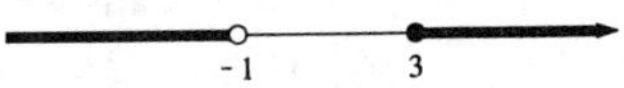

19. $\dfrac{4x}{2x+3} > 2 \Leftrightarrow \dfrac{4x}{2x+3} - 2 > 0 \Leftrightarrow \dfrac{4x}{2x+3} - \dfrac{2(2x+3)}{2x+3} > 0 \Leftrightarrow \dfrac{-6}{2x+3} > 0$. The expression on the left of the inequality changes sign where $x = -\dfrac{3}{2}$. Thus we must check the intervals in the following table.

Interval	$(-\infty, -\frac{3}{2})$	$(-\frac{3}{2}, \infty)$
Sign of -6	$-$	$-$
Sign of $2x+3$	$-$	$+$
Sign of $\dfrac{-6}{2x+3}$	$+$	$-$

From the table, the solution set is : $\left\{x \middle| \ x < -\dfrac{3}{2}\right\}$.

Interval: $\left(-\infty, -\dfrac{3}{2}\right)$ Graph: $-\frac{3}{2}$

21. $\dfrac{2x+1}{x-5} \le 3 \quad \Leftrightarrow \quad \dfrac{2x+1}{x-5} - 3 \le 0 \quad \Leftrightarrow \quad \dfrac{2x+1}{x-5} - \dfrac{3(x-5)}{x-5} \le 0 \quad \Leftrightarrow \quad \dfrac{-x+16}{x-5} \le 0$. The expression on the left of the inequality changes sign where $x = 16$ and where $x = 5$. Thus we must check the intervals in the following table.

Interval	$(-\infty, 5)$	$(5, 16)$	$(16, \infty)$
Sign of $-x+16$	$+$	$+$	$-$
Sign of $x-5$	$-$	$+$	$+$
Sign of $\dfrac{-x+16}{x-5}$	$-$	$+$	$-$

From the table, the solution set is : $\{x \mid x < 5 \text{ or } x \ge 16\}$. Since the denominator cannot equal 0 we must have $x \ne 5$.

Interval: $(-\infty, 5) \cup [16, \infty)$ Graph: 5 16

23. $\dfrac{4}{x} < x \quad \Leftrightarrow \quad \dfrac{4}{x} - x < 0 \quad \Leftrightarrow \quad \dfrac{4}{x} - \dfrac{x \cdot x}{x} < 0 \quad \Leftrightarrow \quad \dfrac{4-x^2}{x} < 0 \quad \Leftrightarrow$ $\dfrac{(2-x)(2+x)}{x} < 0$. The expression on the left of the inequality changes sign where $x = 0$, where $x = -2$, and where $x = 2$. Thus we must check the intervals in the following table.

Interval	$(-\infty, -2)$	$(-2, 0)$	$(0, 2)$	$(2, \infty)$
Sign of $2+x$	$-$	$+$	$+$	$+$
Sign of x	$-$	$-$	$+$	$+$
Sign of $2-x$	$+$	$+$	$+$	$-$
Sign of $\dfrac{(2-x)(2+x)}{x}$	$+$	$-$	$+$	$-$

From the table, the solution set is : $\{x \mid -2 < x < 0 \text{ or } 2 < x\}$

Interval: $(-2, 0) \cup (2, \infty)$ Graph: -2 0 2

25. $\dfrac{x^2-4}{x^2+4} \ge 0 \quad \Leftrightarrow \quad \dfrac{(x+2)(x-2)}{x^2+4} \ge 0$. Since the denominator is always positive, sum of squares, we need only consider the numerator. Thus the expression on the left of the inequality changes sign where $x = -2$ and where $x = 2$. Thus we must check the intervals in the following table.

Interval	$(-\infty, -2)$	$(-2, 2)$	$(2,\infty)$
Sign of $x+2$	$-$	$+$	$+$
Sign of $x-2$	$-$	$-$	$+$
Sign of x^2+4	$+$	$+$	$+$
Sign of $\dfrac{(x+2)(x-2)}{x^2+4}$	$+$	$-$	$+$

From the table, the solution set is : $\{x| \ x \le -2 \text{ or } 2 \le x\}$

Interval: $(-\infty, -2] \cup [2, \infty)$ Graph:

27. $1 + \dfrac{2}{x+1} \le \dfrac{2}{x} \Leftrightarrow 1 + \dfrac{2}{x+1} - \dfrac{2}{x} \le 0 \Leftrightarrow \dfrac{x(x+1)}{x(x+1)} + \dfrac{2x}{x(x+1)} - \dfrac{2(x+1)}{x(x+1)} \le 0 \Leftrightarrow$ $\dfrac{x^2+x+2x-2x-2}{x(x+1)} \le 0 \Leftrightarrow \dfrac{x^2+x-2}{x(x+1)} \le 0 \Leftrightarrow \dfrac{(x+2)(x-1)}{x(x+1)} \le 0$. The expression on the left of the inequality changes sign where $x = -2$, where $x = -1$, where $x = 0$, and where $x = 1$. Thus we must check the intervals in the following table.

Interval	$(-\infty, -2)$	$(-2, -1)$	$(-1, 0)$	$(0, 1)$	$(1, \infty)$
Sign of $x+2$	$-$	$+$	$+$	$+$	$+$
Sign of $x-1$	$-$	$-$	$-$	$-$	$+$
Sign of x	$-$	$-$	$-$	$+$	$+$
Sign of $x+1$	$-$	$-$	$+$	$+$	$+$
Sign of $\dfrac{(x+2)(x-1)}{x(x+1)}$	$+$	$-$	$+$	$-$	$+$

Since $x = -1$ and $x = 0$ yield undefined expressions, we cannot include them in the solution. From the table, the solution set is : $\{x| \ -2 \le x < -1 \text{ or } 0 < x \le 1\}$

Interval: $[-2, -1) \cup (0, 1]$ Graph:

29. $\dfrac{1}{1-x} \le \dfrac{3}{x} \Leftrightarrow \dfrac{1}{1-x} - \dfrac{3}{x} \le 0 \Leftrightarrow \dfrac{x}{x(1-x)} - \dfrac{3(1-x)}{x(1-x)} \le 0 \Leftrightarrow \dfrac{x-3+3x}{x(1-x)} \le 0$

$\Leftrightarrow \dfrac{4x-3}{x(1-x)} \le 0$. The expression on the left of the inequality changes sign where $x = 0$, where $x = \frac{3}{4}$, and where $x = 1$. Thus we must check the intervals in the following table.

Interval	$(-\infty, 0)$	$(0, \frac{3}{4})$	$(\frac{3}{4}, 1)$	$(1, \infty)$
Sign of x	$-$	$+$	$+$	$+$
Sign of $4x-3$	$-$	$-$	$+$	$+$
Sign of $1-x$	$+$	$+$	$+$	$-$
Sign of $\dfrac{4x-3}{x(1-x)}$	$+$	$-$	$+$	$-$

From the table, the solution set is : $\{x| \ 0 < x \le \frac{3}{4} \text{ or } 1 < x\}$

Interval: $\left(0, \frac{3}{4}\right] \cup (1, \infty)$ Graph:

31. $\dfrac{x^2+2x-3}{x^2-7x+6} > 0 \Leftrightarrow \dfrac{(x-1)(x+3)}{(x-1)(x-6)} > 0 \Rightarrow \dfrac{x+3}{x-6} > 0$ (The only exception is at $x = 1$ where $\dfrac{x+3}{x-6}$ is defined and $\dfrac{(x-1)(x+3)}{(x-1)(x-6)}$ is not.) Thus the expression on the left of the inequality changes sign where $x = -3$ and where $x = 6$. Thus we must check the intervals in the following table.

Interval	$(-\infty,-3)$	$(-3,6)$	$(6,\infty)$
Sign of $x+3$	$-$	$+$	$+$
Sign of $x-6$	$-$	$-$	$+$
Sign of $\frac{x+3}{x-6}$	$+$	$-$	$+$

From the table, the solution set is : $\{x \mid x<-3 \text{ or } 6<x\}$.
Interval: $(-\infty,-3)\cup(6,\infty)$ Graph: -3 6

33. $\frac{x-3}{2x+5}\geq 1 \quad\Leftrightarrow\quad \frac{x-3}{2x+5}-1\geq 0 \quad\Leftrightarrow\quad \frac{x-3}{2x+5}-\frac{2x+5}{2x+5}\geq 0 \quad\Leftrightarrow\quad \frac{-x-8}{2x+5}\geq 0$. The expression on the left of the inequality changes sign where $x=-8$ and where $x=-\frac{5}{2}$. Thus we must check the intervals in the following table.

Interval	$(-\infty,-8)$	$(-8,-\frac{5}{2})$	$(-\frac{5}{2},\infty)$
Sign of $-x-8$	$+$	$-$	$-$
Sign of $2x+5$	$-$	$-$	$+$
Sign of $\frac{-x-8}{2x+5}$	$-$	$+$	$-$

From the table, the solution set is : $\{x \mid -8\leq x<-\frac{5}{2}\}$. Since the denominator is zero when $x=-\frac{5}{2}$, we must exclude this value.
Interval: $\left[-8,-\frac{5}{2}\right)$ Graph: -8 $-\frac{5}{2}$

35. $\frac{6}{x-1}-\frac{6}{x}\geq 1 \quad\Leftrightarrow\quad \frac{6}{x-1}-\frac{6}{x}-1\geq 0 \quad\Leftrightarrow\quad \frac{6x}{x(x-1)}-\frac{6(x-1)}{x(x-1)}-\frac{x(x-1)}{x(x-1)}\geq 0 \quad\Leftrightarrow$ $\frac{6x-6x+6-x^2+x}{x(x-1)}\geq 0 \quad\Leftrightarrow\quad \frac{-x^2+x+6}{x(x-1)}\geq 0 \quad\Leftrightarrow\quad \frac{(-x+3)(x+2)}{x(x-1)}\geq 0$. The expression on the left of the inequality changes sign where $x=3$, where $x=-2$, where $x=0$, and where $x=1$. Thus we must check the intervals in the following table.

Interval	$(-\infty,-2)$	$(-2,0)$	$(0,1)$	$(1,3)$	$(3,\infty)$
Sign of $-x+3$	$+$	$+$	$+$	$+$	$-$
Sign of $x+2$	$-$	$+$	$+$	$+$	$+$
Sign of x	$-$	$-$	$+$	$+$	$+$
Sign of $x-1$	$-$	$-$	$-$	$+$	$+$
Sign of $\frac{(-x+3)(x+2)}{x(x-1)}$	$-$	$+$	$-$	$+$	$-$

From the table, the solution set is : $\{x \mid -2\leq x<0 \text{ or } 1<x\leq 3\}$. The points $x=0$ and $x=1$ are excluded from the solution set because they make the denominator zero.
Interval: $[-2,0)\cup(1,3]$ Graph: -2 0 1 3

37. $\frac{x+2}{x+3}<\frac{x-1}{x-2} \quad\Leftrightarrow\quad \frac{x+2}{x+3}-\frac{x-1}{x-2}<0 \quad\Leftrightarrow\quad \frac{(x+2)(x-2)}{(x+3)(x-2)}-\frac{(x-1)(x+3)}{(x-2)(x+3)}<0 \quad\Leftrightarrow$ $\frac{x^2-4-x^2-2x+3}{(x+3)(x-2)}<0 \quad\Leftrightarrow\quad \frac{-2x-1}{(x+3)(x-2)}<0$. The expression on the left of the inequality

changes sign where $x = -\frac{1}{2}$, where $x = -3$, and where $x = 2$. Thus we must check the intervals in the following table.

Interval	$(-\infty, -3)$	$(-3, -\frac{1}{2})$	$(-\frac{1}{2}, 2)$	$(2, \infty)$
Sign of $-2x - 1$	+	+	−	−
Sign of $x + 3$	−	+	+	+
Sign of $x - 2$	−	−	−	+
Sign of $\dfrac{-2x-1}{(x+3)(x-2)}$	+	−	+	−

From the table, the solution set is : $\{x \mid -3 < x < -\frac{1}{2} \text{ or } 2 < x\}$

Interval: $\left(-3, -\frac{1}{2}\right) \cup (2, \infty)$ Graph: -3 $-\frac{1}{2}$ 2

39. $x^4 > x^2 \quad \Leftrightarrow \quad x^4 - x^2 > 0 \quad \Leftrightarrow \quad x^2(x^2 - 1) > 0 \quad \Leftrightarrow \quad x^2(x-1)(x+1) > 0$. The expression on the left of the inequality changes sign where $x = 0$, where $x = 1$, and where $x = -1$. Thus we must check the intervals in the following table..

Interval	$(-\infty, -1)$	$(-1, 0)$	$(0, 1)$	$(1, \infty)$
Sign of x^2	+	+	+	+
Sign of $x - 1$	−	−	−	+
Sign of $x + 1$	−	+	+	+
Sign of $x^2(x-1)(x+1)$	+	−	−	+

From the table, the solution set is : $\{x \mid x < -1 \text{ or } 1 < x\}$

Interval: $(-\infty, -1) \cup (1, \infty)$ Graph: -1 1

41. $128 + 16t - 16t^2 \geq 32 \quad \Leftrightarrow \quad -16t^2 + 16t + 96 \geq 0 \quad \Leftrightarrow \quad -16(t^2 - t - 6) \geq 0 \quad \Leftrightarrow$ $-16(t-3)(t+2) \geq 0$. The expression on the left of the inequality changes sign where $x = -2$, where $t = 3$ and where $t = -2$. However, $t \geq 0$, so the only endpoint is $t = 3$. Thus we check the intervals in the following table.

Interval	$(0, 3)$	$(3, \infty)$
Sign of -16	−	−
Sign of $t - 3$	−	+
Sign of $t + 2$	+	+
Sign of $-16(t-3)(t+2)$	+	−

So $0 \leq t \leq 3$.

43. $\dfrac{600{,}000}{x^2 + 300} < 500 \quad \Leftrightarrow \quad 600{,}000 < 500(x^2 + 300)$ (Note that $x^2 + 300 \geq 300 > 0$, so we can multiply both sides by the denominator and not worry that we might be multiplying both sides by a negative number or by zero.) $1200 < x^2 + 300 \quad \Leftrightarrow \quad 0 < x^2 - 900 \quad \Leftrightarrow$ $0 < (x - 30)(x + 30)$. The possible endpoints are $x = 30$ and $x = -30$. However, since x represents distance, we must have $x \geq 0$.

Interval	$(0, 30)$	$(30, \infty)$
Sign of $x - 30$	−	+
Sign of $x + 30$	+	+
Sign of $(x-30)(x+30)$	−	+

So $x > 30$ and you must stand at least 30 meters from the center of the fire.

45. For $\sqrt{16-9x^2}$ to be defined as a real number we must have $16-9x^2 \geq 0 \Leftrightarrow (4-3x)(4+3x) \geq 0$. So the endpoints are $x=\frac{4}{3}$ and $x=-\frac{4}{3}$.

Interval	$\left(-\infty,-\frac{4}{3}\right)$	$\left(-\frac{4}{3},\frac{4}{3}\right)$	$\left(\frac{4}{3},\infty\right)$
Sign of $4-3x$	$+$	$+$	$-$
Sign of $4+3x$	$-$	$+$	$+$
Sign of $(4-3x)(4+3x)$	$-$	$+$	$-$

Thus $-\frac{4}{3} \leq x \leq \frac{4}{3}$.

47. For $\left(\dfrac{1}{x^2-5x-14}\right)^{1/2}$ to be defined as a real number we must have $x^2-5x-14>0 \Leftrightarrow (x-7)(x+2)>0$. So the endpoints are: $x-7=0 \Leftrightarrow x=7$; $x+2=0 \Leftrightarrow x=-2$.

Interval	$(-\infty,-2)$	$(-2,7)$	$(7,\infty)$
Sign of $x-7$	$-$	$-$	$+$
Sign of $x+2$	$-$	$+$	$+$
Sign of $(x-7)(x+2)$	$+$	$-$	$+$

Thus $x<-2$ or $7<x$; the solution set is $(-\infty,-2)\cup(7,\infty)$.

49. $\dfrac{x^2+(a-b)x-ab}{x+c} \leq 0$ (where $0<a<b<c$) $\Leftrightarrow$ $\dfrac{(x+a)(x-b)}{x+c} \leq 0$. Finding the endpoints we have: $x+a=0 \Leftrightarrow x=-a$; $x-b=0 \Leftrightarrow x=b$; $x+c=0 \Leftrightarrow x=-c$. Endpoints: $-c, -a, b$.

Interval	$(-\infty,-c)$	$(-c,-a)$	$(-a,b)$	(b,∞)
Sign of $x+a$	$-$	$-$	$+$	$+$
Sign of $x-b$	$-$	$-$	$-$	$+$
Sign of $x+c$	$-$	$+$	$+$	$+$
Sign of $\dfrac{(x+a)(x-b)}{x+c}$	$-$	$+$	$-$	$+$

From the table, the solution set is : $\{x \mid -\infty<x<-c \text{ or } -a \leq x \leq b\}$. Note, since $0<a<b<c$ we must have $-c<-b<-a<0$. Also, we exclude the point $x=-c$ because the LHS is undefined at this point. The solution set is $(-\infty,-c)\cup[-a,b]$.

Exercises 2.8

1. $|7 - 3| = |4| = 4$

3. $|50 - 50.1| = |-0.1| = 0.1$

5. Since $a < 4$, $a - 4 < 0$ so $|a - 4| = -(a - 4) = -a + 4 = 4 - a$

7. $|x - 3| = \begin{cases} x - 3 & \text{if } x \geq 3 \\ -(x - 3) & \text{if } x < 3 \end{cases} = \begin{cases} x - 3 & \text{if } x \geq 3 \\ -x + 3 & \text{if } x < 3 \end{cases}$

9. $|3x + 9| = |3(x + 3)| = 3|x + 3|$

11. $\left|\frac{1}{2}x - \frac{5}{2}\right| = \left|\frac{1}{2}(x - 5)\right| = \frac{1}{2}|x - 5|$

13. $|-x^2 - 9| = |-1(x^2 + 9)| = |x^2 + 9| = x^2 + 9$ since $x^2 + 9 > 0$

15. $|4x| = 12 \quad \Leftrightarrow \quad 4x = \pm 12 \quad \Leftrightarrow \quad x = \pm 3$

17. $|x - 2| = 0.05$ is equivalent to $x - 2 = \pm 0.05 \quad \Leftrightarrow \quad x = 2 \pm 0.05 \quad \Leftrightarrow \quad x = 1.95$ or $x = 2.05$.

19. $\left|\dfrac{2x - 1}{x + 1}\right| = 3$ is equivalent to either $\dfrac{2x - 1}{x + 1} = 3 \quad \Leftrightarrow \quad 2x - 1 = 3(x + 1) \quad \Leftrightarrow$ $2x - 1 = 3x + 3 \quad \Leftrightarrow \quad x = -4$; or $\dfrac{2x - 1}{x + 1} = -3 \quad \Leftrightarrow \quad 2x - 1 = -3(x + 1) \quad \Leftrightarrow$ $2x - 1 = -3x - 3 \quad \Leftrightarrow \quad 5x = -2 \quad \Leftrightarrow \quad x = -\frac{2}{5}$. The two solutions are $x = -4$ and $x = -\frac{2}{5}$.

21. $|x - 1| = |3x + 2|$ is equivalent to either $x - 1 = 3x + 2 \quad \Leftrightarrow \quad -2x = 3 \quad \Leftrightarrow \quad x = -\frac{3}{2}$; or $x - 1 = -(3x + 2) \quad \Leftrightarrow \quad x - 1 = -3x - 2 \quad \Leftrightarrow \quad 4x = -1 \quad \Leftrightarrow \quad x = -\frac{1}{4}$. The two solutions are $x = -\frac{3}{2}$ and $x = -\frac{1}{4}$.

23. $|x| < 2 \quad \Leftrightarrow \quad -2 < x < 2$. Solution: $(-2, 2)$

25. $|x - 5| \leq 3 \quad \Leftrightarrow \quad -3 \leq x - 5 \leq 3 \quad \Leftrightarrow \quad 2 \leq x \leq 8$. Solution: $[2, 8]$

27. $|x + 1| \geq 1$ is equivalent to $x + 1 \geq 1 \quad \Leftrightarrow \quad x \geq 0$; or $x + 1 \leq -1 \quad \Leftrightarrow \quad x \leq -2$. Solution: $(-\infty, -2] \cup [0, \infty)$

29. $|x + 5| \geq 2$ is equivalent to $x + 5 \geq 2 \quad \Leftrightarrow \quad x \geq -3$; or $x + 5 \leq -2 \quad \Leftrightarrow \quad x \leq -7$. Solution: $(-\infty, -7] \cup [-3, \infty)$

31. $|2x - 3| \leq 0.4 \quad \Leftrightarrow \quad -0.4 \leq 2x - 3 \leq 0.4 \quad \Leftrightarrow \quad 2.6 \leq 2x \leq 3.4 \quad \Leftrightarrow \quad 1.3 \leq x \leq 1.7$. Solution: $[1.3, 1.7]$

33. $\left|\dfrac{x - 2}{3}\right| < 2 \quad \Leftrightarrow \quad -2 < \dfrac{x - 2}{3} < 2 \quad \Leftrightarrow \quad -6 < x - 2 < 6 \quad \Leftrightarrow \quad -4 < x < 8$. Solution: $(-4, 8)$

35. $|x + 6| < 0.001 \quad \Leftrightarrow \quad -0.001 < x + 6 < 0.001 \quad \Leftrightarrow \quad -6.001 < x < -5.999$. Solution: $(-6.001, -5.999)$

37. $1 \le |x| \le 4$. If $x \ge 0$, then this is equivalent to $1 \le x \le 4$. If $x < 0$, then this is equivalent to $1 \le -x \le 4 \quad \Leftrightarrow \quad -1 \ge x \ge -4 \quad \Leftrightarrow \quad -4 \le x \le -1$. Solution: $[-4, -1] \cup [1, 4]$

39. $\dfrac{1}{|x+7|} > 2 \quad \Leftrightarrow \quad 1 > 2|x+7| \ (x \neq -7) \quad \Leftrightarrow \quad |x+7| < \dfrac{1}{2} \quad \Leftrightarrow \quad -\dfrac{1}{2} < x+7 < \dfrac{1}{2} \quad \Leftrightarrow$ $-\dfrac{15}{2} < x < -\dfrac{13}{2}$ and $x \neq -7$. Solution: $\left(-\dfrac{15}{2}, -7\right) \cup \left(-7, -\dfrac{13}{2}\right)$

41. $|x| > |x-1|$. We solve this exercise by solving the equation $|x| = |x-1|$ and testing the intervals into which the solution(s) divide the real line. $|x| = |x-1| \quad \Rightarrow \quad \left|\dfrac{x-1}{x}\right| = 1 \quad \Leftrightarrow$ $\dfrac{x-1}{x} = -1$ or $\dfrac{x-1}{x} = 1$. Now $\dfrac{x-1}{x} = -1 \quad \Leftrightarrow \quad x - 1 = -x \quad \Leftrightarrow \quad 2x = 1 \quad \Leftrightarrow$ $x = \frac{1}{2}$. And $\dfrac{x-1}{x} = 1 \quad \Leftrightarrow \quad x - 1 = x \quad \Leftrightarrow \quad -1 = 0$, which is impossible. So we check the intervals $\left(-\infty, \frac{1}{2}\right)$ and $\left(\frac{1}{2}, \infty\right)$ by taking a test point from within each interval.

For $\left(-\infty, \frac{1}{2}\right)$ we test $x = 0$. But $|0| \not> |0 - 1|$.

For $\left(\frac{1}{2}, \infty\right)$ we test $x = 1$. And $|1| > |1 - 1|$.

Thus the solution is $\left(\frac{1}{2}, \infty\right)$.

43. $\left|\dfrac{x}{2+x}\right| < 1$. We solve this exercise by solving the equation $\left|\dfrac{x}{2+x}\right| = 1$ and testing the intervals into which the solution(s) divide the real line. $\left|\dfrac{x}{2+x}\right| = 1 \quad \Rightarrow \quad \dfrac{x}{2+x} = \pm 1$. Now, if $\dfrac{x}{2+x} = -1$ then $x = -2 - x \quad \Leftrightarrow \quad x = -1$. And if $\dfrac{x}{2+x} = 1$ then $x = 2 + x \quad \Leftrightarrow \quad 0 = 2$, which is impossible. Since $x = -2$ is not in the domain of the left hand side of the inequality, we must consider the possibility that $x = -2$ is also an endpoint. So we check the intervals $(-\infty, -2)$, $(-2, -1)$ and $(-1, \infty)$ by taking a test point from within each interval.

For $(-\infty, -2)$ we test $x = -3$. But $\left|\dfrac{-3}{2-3}\right| = 3 \not< 1$.

For $(-2, -1)$ we test $x = -\frac{3}{2}$. And $\left|\dfrac{-\frac{3}{2}}{2 - \frac{3}{2}}\right| = 3 \not< 1$.

For $(-1, \infty)$ we test $x = 0$. And $\left|\dfrac{0}{2-0}\right| = 0 < 1$.

Thus the solution is $(-1, \infty)$.

45. $|ab| = \sqrt{(ab)^2} = \sqrt{a^2 b^2} = \sqrt{a^2} \cdot \sqrt{b^2} = |a|\ |b|$. Therefore, $|ab| = |a| \cdot |b|$.

47. $|(x+y) - 5| = |(x-2) + (y-3)| \le |x-2| + |y-3| < 0.01 + 0.04 = 0.05$. Thus $|(x+y) - 5| < 0.05$.

49. $|x - y| = |x + (-y)| \le |x| + |-y| = |x| + |y|$. Thus $|x - y| \le |x| + |y|$.

51. Case 1: $a \ge 0$. LHS $= |a| = a$ while RHS $= \sqrt{a^2} = a$, since $a \ge 0$, and so $|a| = \sqrt{a^2}$.

Case 2: $a < 0$. LHS $= |a| = -a$ and RHS $= \sqrt{a^2} = -a$, since $(-a)^2 = a^2$ and $-a > 0$, and so $|a| = \sqrt{a^2}$. Therefore, $|a| = \sqrt{a^2}$ for any real number a.

Review Exercises for Chapter 2

1. $3x + 12 = 24 \quad \Leftrightarrow \quad 3x = 12 \quad \Leftrightarrow \quad x = 4.$

3. $7x - 6 = 4x + 9 \quad \Leftrightarrow \quad 3x = 15 \quad \Leftrightarrow \quad x = 5.$

5. $\frac{1}{3}x - \frac{1}{2} = 2 \quad \Leftrightarrow \quad 2x - 3 = 12 \quad \Leftrightarrow \quad 2x = 15 \quad \Leftrightarrow \quad x = \frac{15}{2}.$

7. $2(x+3) - 4(x-5) = 8 - 5x \quad \Leftrightarrow \quad 2x + 6 - 4x + 20 = 8 - 5x \quad \Leftrightarrow \quad -2x + 26 = 8 - 5x$ $\Leftrightarrow \quad 3x = -18 \quad \Leftrightarrow \quad x = -6$

9. $\dfrac{x+1}{x-1} = \dfrac{2x-1}{2x+1} \quad \Leftrightarrow \quad (x+1)(2x+1) = (2x-1)(x-1) \quad \Leftrightarrow \quad 2x^2 + 3x + 1 = 2x^2 - 3x + 1$ $\Leftrightarrow \quad 6x = 0 \quad \Leftrightarrow \quad x = 0$

11. $x^2 - 9x + 14 = 0 \quad \Leftrightarrow \quad (x-7)(x-2) = 0 \quad \Leftrightarrow \quad x = 7$ or $x = 2$

13. $2x^2 + x = 1 \quad \Leftrightarrow \quad 2x^2 + x - 1 = 0 \quad \Leftrightarrow \quad (2x-1)(x+1) = 0.$ So either $2x - 1 = 0 \quad \Leftrightarrow$ $2x = 1 \quad \Leftrightarrow \quad x = \frac{1}{2}$; or $x + 1 = 0 \quad \Leftrightarrow \quad x = -1.$

15. $0 = 4x^3 - 25x = x(4x^2 - 25) = x(2x-5)(2x+5) = 0.$ So either $x = 0$; or $2x - 5 = 0 \quad \Leftrightarrow$ $2x = 5 \quad \Leftrightarrow \quad x = \frac{5}{2}$; or $2x + 5 = 0 \quad \Leftrightarrow \quad 2x = -5 \quad \Leftrightarrow \quad x = -\frac{5}{2}$

17. $3x^2 + 4x - 1 = 0 \quad \Rightarrow \quad x = \dfrac{-b \pm \sqrt{b^2 - 4ac}}{2a} = \dfrac{-(4) \pm \sqrt{(4)^2 - 4(3)(-1)}}{2(-3)}$

$$= \frac{-4 \pm \sqrt{16 + 12}}{-6} = \frac{-4 \pm \sqrt{28}}{-6} = \frac{-4 \pm 2\sqrt{7}}{6} = \frac{2(-2 \pm \sqrt{7})}{-6} = \frac{-2 \pm \sqrt{7}}{3}.$$

19. $\dfrac{1}{x} + \dfrac{2}{x-1} = 3 \quad \Leftrightarrow \quad (x-1) + 2(x) = 3(x)(x-1) \quad \Leftrightarrow \quad x - 1 + 2x = 3x^2 - 3x \quad \Leftrightarrow$

$0 = 3x^2 - 6x + 1 \quad \Rightarrow \quad x = \dfrac{-b \pm \sqrt{b^2 - 4ac}}{2a} = \dfrac{-(-6) \pm \sqrt{(-6)^2 - 4(3)(1)}}{2(3)}$

$$= \frac{6 \pm \sqrt{36 - 12}}{6} = \frac{6 \pm \sqrt{24}}{6} = \frac{6 \pm 2\sqrt{6}}{6} = \frac{2(3 \pm \sqrt{6})}{6} = \frac{3 \pm \sqrt{6}}{3}$$

21. $x^4 - 8x^2 - 9 = 0 \quad \Leftrightarrow \quad (x^2 - 9)(x^2 + 1) = 0 \quad \Leftrightarrow \quad (x-3)(x+3)(x^2+1) = 0 \quad \Rightarrow$ $x - 3 = 0 \quad \Leftrightarrow \quad x = 3,$ or $x + 3 = 0 \quad \Leftrightarrow \quad x = -3,$ however $x^2 + 1 = 0$ has no real solution. Solution: $x = \pm 3.$

23. $x^{-1/2} - 2x^{1/2} + x^{3/2} = 0 \quad \Leftrightarrow \quad x^{-1/2}(1 - 2x + x^2) = 0 \quad \Leftrightarrow \quad x^{1/2}(1-x)^2 = 0.$ Since $x^{-1/2} \neq 0$, the only solution comes from $(1-x)^2 = 0 \quad \Leftrightarrow \quad 1 - x = 0 \quad \Leftrightarrow \quad x = 1$

25. $|x - 7| = 4 \quad \Leftrightarrow \quad x - 7 = \pm 4 \quad \Leftrightarrow \quad x = 7 \pm 4$, so $x = 11$ or $x = 3$.

27. Let x be the number of pounds of raisins. Then the number of pounds of nuts is $50 - x$.

	IN		OUT
	raisins	nuts	mixture
pounds	x	$50 - x$	50
rate (cost per pound)	3.20	2.40	2.72
value	$3.20x$	$2.40(50 - x)$	$2.72(50)$

$3.20x + 2.40(50 - x) = 2.72(50) \quad \Leftrightarrow \quad 3.20x + 120 - 2.40x = 136 \quad \Leftrightarrow \quad 0.8x = 16 \quad \Leftrightarrow$ $x = 20$. The mixture uses 20 pounds raisins and $50 - 20 = 30$ pounds of nuts.

29. Let r be the rate the woman runs in mi/h. Then she cycles at $r + 8$ mi/h.

	Rate	Time	Distance
Cycle	$r+8$	$\frac{4}{r+8}$	4
Run	r	$\frac{2.5}{r}$	2.5

Since the total time of the workout is 1 hour we have: $\frac{4}{r+8} + \frac{2.5}{r} = 1$. Multiplying by $2r(r+8)$, we get $4(2r) + 2.5(2)(r+8) = 2r(r+8) \quad \Leftrightarrow \quad 8r + 5r + 40 = 2r^2 + 16r \quad \Leftrightarrow$ $0 = 2r^2 + 3r - 40 \quad \Rightarrow \quad r = \frac{-3 \pm \sqrt{(3)^2 - 4(2)(-40)}}{2(2)} = \frac{-3 \pm \sqrt{9 + 320}}{4} = \frac{-3 \pm \sqrt{329}}{4}$.

Since $r \geq 0$, we reject the negative value. She runs at $r = \frac{-3 + \sqrt{329}}{4} \approx 3.78$ mi/h.

31. Let x be the length of one side in cm, then $28 - x$ is the length of the other side. Using Pythagorean theorem we have: $x^2 + (28 - x)^2 = 20^2 \quad \Leftrightarrow \quad x^2 + 784 - 56x + x^2 = 400 \quad \Leftrightarrow$ $2x^2 - 56x + 384 = 0 \quad \Leftrightarrow \quad 2(x^2 - 28x + 192) = 0 \quad \Leftrightarrow \quad 2(x - 12)(x - 16) = 0$. So $x - 12 = 0 \quad \Leftrightarrow \quad x = 12$ and the other side is $28 - 12 = 16$ or $x - 16 = 0 \quad \Leftrightarrow \quad x = 16$ and the other side is $28 - 16 = 12$. The sides are 12 cm and 16 cm.

33. Let w be width of the pool. Then the length of the pool is $2w$. Thus the volume of the pool is $8(w)(2w) = 8464 \quad \Leftrightarrow \quad 16w^2 = 8464 \quad \Leftrightarrow \quad w^2 = 529 \quad \Rightarrow \quad w = \pm 23$. Since $w > 0$, we reject the negative value. The pool is 23 feet wide, $2(23) = 46$ feet long and 8 feet deep.

35. $(3 - 5i) - (6 + 4i) = 3 - 5i - 6 - 4i = -3 - 9i$

37. $(2 + 7i)(6 - i) = 12 - 2i + 42i - 7i^2 = 12 + 40i + 7 = 19 + 40i$

39. $\frac{2 - 3i}{2 + 3i} = \frac{2 - 3i}{2 + 3i} \cdot \frac{2 - 3i}{2 - 3i} = \frac{4 - 12i + 9i^2}{4 - 9i^2} = \frac{4 - 12i - 9}{4 + 9} = \frac{-5 - 12i}{13} = -\frac{5}{13} - \frac{12}{13}i$

41. $i^{45} = i^{44}i = (i^4)^{11}i = (1)^{11}i = i$

43. $(1 - \sqrt{-3})(2 + \sqrt{-4}) = (1 - \sqrt{3}i)(2 + 2i) = 2 + 2i - 2\sqrt{3}i - 2\sqrt{3}i^2$ $= 2 + (2 - 2\sqrt{3})i + 2\sqrt{3} = (2 + 2\sqrt{3}) + (2 - 2\sqrt{3})i$

45. $x^2 + 16 = 0 \quad \Leftrightarrow \quad x^2 = -16 \quad \Rightarrow \quad x = \pm\sqrt{-16} = \pm 4i$

47. $x^2 + 6x + 10 = 0 \quad \Rightarrow \quad x = \frac{-6 \pm \sqrt{6^2 - 4(1)(10)}}{2(1)} = \frac{-6 \pm \sqrt{36 - 40}}{2} = \frac{-6 \pm \sqrt{-4}}{2}$ $= \frac{-6 \pm 2i}{2} = -3 \pm i$

49. $(1 + 2i)x - 5 = 2x \quad \Leftrightarrow \quad (-1 + 2i)x = 5 \quad \Leftrightarrow \quad x = \frac{5}{-1 + 2i} \quad \Leftrightarrow$ $x = \frac{5}{-1 + 2i} \cdot \frac{-1 - 2i}{-1 - 2i} = \frac{-5 - 10i}{1 - 4i^2} = \frac{-5 - 10i}{5} = -1 - 2i$

51. $x^4 - 256 = 0 \Leftrightarrow (x^2+16)(x^2-16) = 0$. Thus either $x^2 + 16 = 0 \Leftrightarrow x^2 = -16 \Rightarrow x = \pm\sqrt{-16} \Leftrightarrow x = \pm 4i$; or $x^2 - 16 = 0 \Leftrightarrow (x-4)(x+4) = 0 \Rightarrow x - 4 = 0 \Leftrightarrow x = 4$; or $x + 4 = 0 \Leftrightarrow x = -4$. The solutions: $\pm 4i$, ± 4.

53. $x^2 + 4x = (2x+1)^2 \Leftrightarrow x^2 + 4x = 4x^2 + 4x + 1 \Leftrightarrow 0 = 3x^2 + 1 \Leftrightarrow 3x^2 = -1 \Leftrightarrow x^2 = -\frac{1}{3} \Rightarrow x = \pm\sqrt{-\frac{1}{3}} = \pm\frac{\sqrt{3}}{3}i$

55. $3x - 2 > -11 \Leftrightarrow 3x > -9 \Leftrightarrow x > -3$. Interval: $(-3, \infty)$

Graph: -3

57. $-1 < 2x + 5 \le 3 \Leftrightarrow -6 < 2x \le -2 \Leftrightarrow -3 < x \le -1$ Interval: $(-3, -1]$

Graph: -3 -1

59. $x^2 + 4x - 12 > 0 \Leftrightarrow (x-2)(x+6) > 0$. The expression on the left of the inequality changes sign where $x = 2$ and where $x = -6$. Thus we must check the intervals in the following table.

Interval	$(-\infty, -6)$	$(-6, 2)$	$(2, \infty)$
Sign of $x-2$	$-$	$-$	$+$
Sign of $x+6$	$-$	$+$	$+$
Sign of $(x-2)(x+6)$	$+$	$-$	$+$

Interval: $(-\infty, -6) \cup (2, \infty)$ Graph: -6 2

61. $\dfrac{2x+5}{x+1} \le 1 \Leftrightarrow \dfrac{2x+5}{x+1} - 1 \le 0 \Leftrightarrow \dfrac{2x+5}{x+1} - \dfrac{x+1}{x+1} \le 0 \Leftrightarrow \dfrac{x+4}{x+1} \le 0$. The expression on the left of the inequality changes sign where $x = -1$ and where $x = -4$. Thus we must check the intervals in the following table.

Interval	$(-\infty, -4)$	$(-4, -1)$	$(-1, \infty)$
Sign of $x+4$	$-$	$+$	$+$
Sign of $x+1$	$-$	$-$	$+$
Sign of $\dfrac{x+4}{x+1}$	$+$	$-$	$+$

We exclude $x = -1$ since the expression is not defined at this value. Thus the solution is $[-4, -1)$.

Graph: -4 -1

63. $\dfrac{x-4}{x^2-4} \le 0 \Leftrightarrow \dfrac{x-4}{(x-2)(x+2)} \le 0$. The expression on the left of the inequality changes sign where $x = -2$, where $x = 2$, and where $x = 4$. Thus we must check the intervals in the following table.

Interval	$(-\infty, -2)$	$(-2, 2)$	$(2, 4)$	$(4, \infty)$
Sign of $x-4$	$-$	$-$	$-$	$+$
Sign of $x-2$	$-$	$-$	$+$	$+$
Sign of $x+2$	$-$	$+$	$+$	$+$
Sign of $\dfrac{x-4}{(x-2)(x+2)}$	$-$	$+$	$-$	$+$

Since the expression is not defined when $x = \pm 2$, we exclude these values and the intervals are: $(-\infty, -2) \cup (2, 4]$ Graph: -2 2 4

65. $|x-5| \le 3 \quad \Leftrightarrow \quad -3 \le x-5 \le 3 \quad \Leftrightarrow \quad 2 \le x \le 8$. Interval: $[2,\ 8]$.
Graph: 2 8

67. $|2x+1| \ge 1$ is equivalent to $2x+1 \ge 1$or $2x+1 \le -1$.

Case 1: $2x+1 \ge 1 \quad \Leftrightarrow \quad 2x \ge 0 \quad \Leftrightarrow \quad x \ge 0$

Case 2: $2x+1 \le -1 \quad \Leftrightarrow \quad 2x \le -2 \quad \Leftrightarrow \quad x \le -1$.

Interval: $(-\infty,-1] \cup [0,\infty)$ Graph: -1 0

69. (a) For $\sqrt{24-x-3x^2}$ to define a real number we must have $24-x-3x^2 \ge 0 \quad \Leftrightarrow$ $(8-3x)(3+x) \ge 0$. The expression on the left of the inequality changes sign where $8-3x=0 \quad \Leftrightarrow \quad -3x=-8 \quad \Leftrightarrow \quad x=\frac{8}{3}$; or where $x=-3$. Thus we must check the intervals in the following table.

Interval	$(-\infty,-3)$	$\left(-3,\frac{8}{3}\right)$	$\left(\frac{8}{3},\infty\right)$
Sign of $8-3x$	+	+	−
Sign of $3+x$	−	+	+
Sign of $(8-3x)(3+x)$	−	+	−

Interval: $\left[-3,\ \frac{8}{3}\right]$ Graph: -3 $\frac{8}{3}$

(b) For $\dfrac{1}{\sqrt[4]{x-x^4}}$ to define a real number we must have $x-x^4 > 0 \quad \Leftrightarrow \quad x(1-x^3) > 0 \quad \Leftrightarrow$ $x(1-x)(1+x+x^2) > 0$. The expression on the left of the inequality changes sign where $x=0$; or where $x=1$; or where $1+x+x^2=0 \quad \Rightarrow \quad x=\dfrac{-1 \pm \sqrt{1^2-4(1)(1)}}{2(1)}$ $= \frac{1 \pm \sqrt{1-4}}{2}$ which has no real solution. Thus we must check the intervals in the following table.

Interval	$(-\infty,0)$	$(0,\ 1)$	$(1,\infty)$
Sign of x	−	+	+
Sign of $1-x$	+	+	−
Sign of $1+x+x^2$	+	+	+
Sign of $x(1-x)(1+x+x^2)$	−	+	−

Interval: $(0,\ 1)$ Graph: 0 1

71. $|x+y| = |(x-3)+(y+3)| \le |x-3|+|y+3| < \dfrac{k}{2}+\dfrac{k}{2} = k$. Thus $|x+y| < k$.

Chapter 2 Test

1. (a) $2x + 7 = 12 + \frac{5}{2}x \quad \Leftrightarrow \quad 4x + 14 = 24 + 5x \quad \Leftrightarrow \quad x = -10$

   (b) $\dfrac{2x}{x+1} = \dfrac{2x-1}{x} \quad \Leftrightarrow \quad (2x)(x) = (2x-1)(x+1) \;\; (x \neq -1, x \neq 0) \quad \Leftrightarrow$
   $2x^2 = 2x^2 + x - 1 \quad \Leftrightarrow \quad 0 = x - 1 \quad \Leftrightarrow \quad x = 1$

2. Let d = the distance between Ajax and Bixby. Then we have the following table.

	Rate	Time	Distance
Ajax to Bixby	50	$\frac{d}{50}$	d
Bixby to Ajax	60	$\frac{d}{60}$	d

   We use the fact that the total time is $4\frac{2}{5}$ hours to get the equation $\dfrac{d}{50} + \dfrac{d}{60} = \dfrac{22}{5} \quad \Leftrightarrow$
   $6d + 5d = 1320 \quad \Leftrightarrow \quad 11d = 1320 \quad \Leftrightarrow \quad d = 120$. Thus Ajax and Bixby are 120 miles apart.

3. (a) $(6 - 2i) - (7 - \frac{1}{2}i) = 6 - 2i - 7 + \frac{1}{2}i = -1 - \frac{3}{2}i$

   (b) $(1 + i)(3 - 2i) = 3 - 2i + 3i - 2i^2 = 3 + i + 2 = 5 + i$

   (c) $\dfrac{5 + 10i}{3 - 4i} = \dfrac{5 + 10i}{3 - 4i} \cdot \dfrac{3 + 4i}{3 + 4i} = \dfrac{15 + 20i + 30i + 40i^2}{9 - 16i^2} = \dfrac{15 + 50i - 40}{9 + 16} = \dfrac{-25 + 50i}{25}$
   $= -1 + 2i$

   (d) $i^{50} = i^{48} \cdot i^2 = (i^4)^{12} \cdot i^2 = (1)^{12} \cdot (-1) = -1$

   (e) $(2 - \sqrt{-2})(\sqrt{8} + \sqrt{-4}) = (2 - \sqrt{2}i)(2\sqrt{2} + 2i) = 4\sqrt{2} + 4i - 4i - 2\sqrt{2}i^2 =$
   $4\sqrt{2} + 2\sqrt{2} = 6\sqrt{2}$

4. (a) $x^2 - x - 12 = 0 \quad \Leftrightarrow \quad (x - 4)(x + 3) = 0$. So $x = 4$ or $x = -3$.

   (b) $2x^2 + 4x + 3 = 0 \quad \Rightarrow \quad x = \dfrac{-4 \pm \sqrt{4^2 - 4(2)(3)}}{2(2)} = \dfrac{-4 \pm \sqrt{16 - 24}}{4} = \dfrac{-4 \pm \sqrt{-8}}{4}$
   $= \dfrac{-4 \pm 2\sqrt{2}i}{4} = \dfrac{-2 \pm \sqrt{2}i}{2}$

   (c) $\sqrt{3 - \sqrt{x+5}} = 2 \quad \Rightarrow \quad 3 - \sqrt{x+5} = 4 \quad \Leftrightarrow \quad -1 = \sqrt{x+5}$. Squaring both sides again, we get $1 = x + 5 \quad \Leftrightarrow \quad x = -4$. But this does not satisfy the original equation, so there is no solution. (You must always check your final answers if you have squared both sides when solving an equation, since extraneous answers may be introducted.)
   *Some students might stop at the statement $-1 = \sqrt{x+5}$ and recognize that this is impossible, so there can be no solution.

   (d) $x^{1/2} - 3x^{3/2} + 2x^{5/2} = 0 \quad \Leftrightarrow \quad x^{1/2}(1 - 3x + 2x^2) = 0 \quad \Leftrightarrow \quad x^{1/2}(1 - 2x)(1 - x) = 0$.
   So $x^{1/2} = 0 \quad \Leftrightarrow \quad x = 0$; or $x = \frac{1}{2}$ or $x = 1$.

(e) $x^4 + 27x = 0 \Leftrightarrow x(x^3 + 27) = 0 \quad \Leftrightarrow \quad x(x+3)(x^2 - 3x + 9) = 0$. So $x = 0$ or $x = -3$; or $x^2 - 3x + 9 = 0 \Rightarrow x = \dfrac{-(-3) \pm \sqrt{(-3)^2 - 4(1)(9)}}{2(1)}$

$$= \frac{3 \pm \sqrt{9 - 36}}{2} = \frac{3 \pm \sqrt{-27}}{2} = \frac{3 \pm 3\sqrt{3}\,i}{2}$$

5. Let w be the width of the building lot. Then the length of the lot is $w + 70$. So $w^2 + (w + 70)^2 = 130^2 \quad \Leftrightarrow \quad w^2 + w^2 + 140w + 4900 = 16900 \quad \Leftrightarrow$ $2w^2 + 140w - 12000 = 0 \quad \Leftrightarrow \quad 2(w^2 + 70w - 6000) = 0 \quad \Leftrightarrow \quad 2(w + 120)(w - 50) = 0$. So $w = -120$ (which we reject because $w > 0$) or $w = 50$. Thus the lot is 50 by 120.

6. (a) $-1 \le 5 - 2x < 10 \quad \Leftrightarrow \quad -6 \le -2x < 5 \quad \Leftrightarrow \quad 3 \ge x > -\frac{5}{2}$. Expressing in standard form we have: $-\frac{5}{2} < x \le 3$. Interval: $\left(-\frac{5}{2}, 3\right]$ Graph:

(b) $x(x-1)(x-2) > 0$. The expression on the left of the inequality changes sign where $x = 0$, where $x = 1$, and where $x = 2$. Thus we must check the intervals in the following table.

Interval	$(-\infty, 0)$	$(0, 1)$	$(1, 2)$	$(2, \infty)$
Sign of x	$-$	$+$	$+$	$+$
Sign of $x - 1$	$-$	$-$	$+$	$+$
Sign of $x - 2$	$-$	$-$	$-$	$+$
Sign of $x(x-1)(x-2)$	$-$	$+$	$-$	$+$

From the table, the solution set is : $\{x \mid 0 < x < 1 \text{ or } 2 < x\}$

Interval: $(0, 1) \cup (2, \infty)$ Graph:

(c) $|x - 3| < 2$ is equivalent to $-2 < x - 3 < 2 \quad \Leftrightarrow \quad 1 < x < 5$.

Interval: $(1, 5)$ Graph:

(d) $\dfrac{2x+5}{x+1} \le 1 \quad \Leftrightarrow \quad \dfrac{2x+5}{x+1} - 1 \le 0 \quad \Leftrightarrow \quad \dfrac{2x+5}{x+1} - \dfrac{x+1}{x+1} \le 0 \quad \Leftrightarrow \quad \dfrac{x+4}{x+1} \le 0$. The expression on the left of the inequality changes sign where $x = -4$ and where $x = -1$. Thus we must check the intervals in the following table.

Interval	$(-\infty, -4)$	$(-4, -1)$	$(-1, \infty)$
Sign of $x + 4$	$-$	$+$	$+$
Sign of $x + 1$	$-$	$-$	$+$
Sign of $\dfrac{x+4}{x+1}$	$+$	$-$	$+$

Since $x = -1$ makes the expression in the inequality undefined, so we exclude this value.

Interval: $[-4, -1)$ Graph:

7. $5 \le \frac{5}{9}(F - 32) \le 10 \quad \Leftrightarrow \quad 9 \le F - 32 \le 18 \quad \Leftrightarrow \quad 41 \le F \le 50$. Thus the medicine is to be stored at a temperature between $41°F$ to $50°F$.

8. For $\sqrt{4x - 2}$ to be defined as a real number $4x - 2 \ge 0 \quad \Leftrightarrow \quad 4x \ge 2 \quad \Leftrightarrow \quad x \ge \frac{1}{2}$

Focus on Problem Solving

1. We first note that we can obtain 1 gallon by pouring the 13 gallons and all but one of the 7 gallons into the 19 gallon container. Or we can fill the 13 gallon container using the 7 gallon container twice, leaving 1 gallon over.

Action	Result		
	19 gal.	13 gal	7 gal
Start	0	13	7
Fill 19 gal container by adding all 7 gal	19	1	0
Fill 7 gal container from 19 gal container	12	1	7
Add the 7 gal to the 13 gal container	12	8	0
Fill 7 gal container from 19 gal container	5	8	7
Add the 7 gal contents into 13 gal container	5	13	2
Move the contents of 13 gal container to the 19 gal container	18	0	2
Move the 2 gal to the 13 gal container and make 3 gal	18	2	0
Fill 7 gal container from 19 gal container	11	2	7
Add the 7 gal contents into 13 gal container	11	9	0
Fill 7 gal container from 19 gal container	4	9	7
Add the 7 gal contents into 13 gal container	4	13	3
Move the contents of 13 gal container to the 19 gal container	17	0	3
Move the 3 gal to the 13 gal container	17	3	0
Fill 7 gal container from 19 gal container	10	3	7

3. It follows from Property 4 of absolute value that we must solve the following two cases.
Case (i): $5 - |x-1| = 3 \quad \Leftrightarrow \quad 2 = |x-1|$. Again we have two subcases. If $x-1$ is positive then $2 = x-1 \quad \Leftrightarrow \quad x = 3$. If $x-1$ is negative, then $-2 = x-1 \quad \Leftrightarrow \quad x = -1$.
Case (ii): $5 - |x-1| = -3 \quad \Leftrightarrow \quad 8 = |x-1|$. Again we have two subcases. If $x-1$ is positive then $8 = x-1 \quad \Leftrightarrow \quad x = 8$. If $x-1$ is negative, then $-8 = x-1 \quad \Leftrightarrow \quad x = -7$.
Therefore the solutions are $x = -7, -1, 3$, and 8.

5. $|2x-1| = 2x-1$ if $2x-1 \geq 0 \quad \Leftrightarrow \quad x \geq \frac{1}{2}$, and $|2x-1| = -(2x-1)$ if $2x-1 < 0 \quad \Leftrightarrow$ $x < \frac{1}{2}$. $|x+5| = x+5$ if $x+5 \geq 0 \quad \Leftrightarrow \quad x \geq -5$, and $|x+5| = -(x+5)$ if $x+5 < 0 \quad \Leftrightarrow$ $x < -5$. Therefore we consider three cases: $x < -5$, $-5 \leq x < \frac{1}{2}$, and $x \geq \frac{1}{2}$.
Case (i): If $x < -5$, the given equation becomes $-(2x-1) - [-(x+5)] = 3 \quad \Leftrightarrow \quad -x = -3$ $\Leftrightarrow \quad x = 3$, which contradicts $x < -5$.
Case (ii): If $-5 \leq x < \frac{1}{2}$, the given equation becomes $-(2x-1) - (x+5) = 3 \quad \Leftrightarrow \quad -3x = 7$ $\Leftrightarrow \quad x = -\frac{7}{3}$.
Case (iii): If $x \geq \frac{1}{2}$, the given equation becomes $2x - 1 - (x+5) = 3 \quad \Leftrightarrow \quad x - 6 = 3 \quad \Leftrightarrow$ $x = 9$.
So there are two solutions: $x = 9$ and $-\frac{7}{3}$.

7. $|x+1| = x+1$ if $x+1 \geq 0 \quad \Leftrightarrow \quad x \geq -1$, and $|x+1| = -(x+1)$ if $x+1 < 0 \quad \Leftrightarrow$ $x < -1$. And $|x+4| = x+4$ if $x+4 \geq 0 \quad \Leftrightarrow \quad x \geq -4$, and $|x+4| = -(x+4)$ if $x+4 < 0$ $\Leftrightarrow \quad x < -4$. Therefore we consider three cases: $x < -4$, $-4 \leq x < -1$, and $x \geq -1$.

Case (i): If $x < -4$, the given inequality becomes $-(x+1)-(x+4) \le 5 \;\Leftrightarrow\; -2x-5 \le 5 \;\Leftrightarrow\; -2x \le 10 \;\Leftrightarrow\; x \ge -5$.

Case (ii): If $-4 \le x < -1$, the given inequality becomes $-(x+1)+(x+4) \le 5 \;\Leftrightarrow\; 3 \le 5$, which is always true.

Case (iii): If $x \ge -1$, the given inequality is $x+1+x+4 \le 5 \;\Leftrightarrow\; 2x \le 0 \;\Leftrightarrow\; x \le 0$.

Combining the 3 cases, we see that the inequality is satisfied when $-5 \le x \le 0$.

9. $$\begin{array}{r} \text{ABCDE} \\ 4 \\ \hline \text{EDCBA} \end{array}$$

A must be 1 or 2 since 4A consists of just a single digit. Since A is also the last digit of 4E, it must be even and so A = 2. Now E $\geq$ 4A = 8 (since E is the first digit of the product), and since 4E ends in the digit A = 2, E must be 8. So far we have

$$\begin{array}{r} \text{2BCD8} \\ 4 \\ \hline \text{8DCB2} \end{array}$$

Now B must be odd, since it is the last digit of 4D + 3. Also, 4B consists of just a single digit (since there is no "carry" added to the final product $4 \cdot 2 = 8$). Thus B = 1, and so 4D + 3 ends in an 8 $(11 - 3 = 8,\ 21 - 3 = 18$, and so on). The only possibilities are D = 2 or 7, but since 2 is already used, D = 7. Now we have

$$\begin{array}{r} \text{21C78} \\ 4 \\ \hline \text{87C21} \end{array}$$

The digit 7 in the product could only have resulted from the preceding product $(7 = 4 \cdot 1 + \underline{3})$. Thus the product that results in the C must have a 3 as its first digit; that is, 4C + 3 = 30 + C. So 3C = 27 and C = 9. The product is

$$\begin{array}{r} 21978 \\ 4 \\ \hline 87912 \end{array}$$

11. For convenience, let us label the eggs 1, 2, 3, ... , 9. First we weigh $\{1,2,3\}$ against $\{4,5,6\}$.

Case 1: If the first weighing is equal, then the light egg is in $\{7,8,9\}$ and so we weigh 7 against 8. If this second weighing is unequal, then we can tell whether 7 or 8 is lighter. If not, 9 is the light egg.

Case 2: If the first weighing is unequal, consider the lighter group and suppose it is $\{1,2,3\}$. Weigh 1 against 2. If this second weighing is unequal, then we can tell whether 1 or 2 is lighter. If not, 3 is the light egg.

The method is displayed in a *decision* tree below.

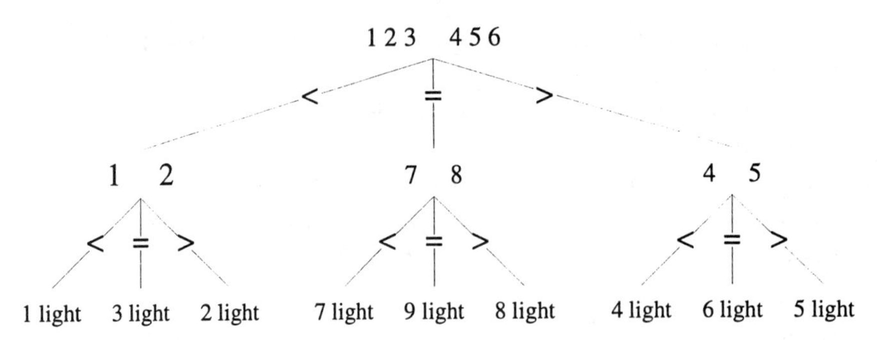

13. Let r be the radius of the earth in feet. Then the circumference (length of the ribbon) is $2\pi r$. When we increase the radius by 1 foot, the new radius is $r+1$ so the new circumference is $2\pi(r+1)$. Thus you need $2\pi(r+1) - 2\pi r = 2\pi$ extra feet of ribbon.

15. Since r_1 and r_2 are roots, $x^2 + px + 8$ factors into $(x - r_1)(x - r_2) = x^2 - (r_1 + r_2)x + 8$. Since the discriminant of the quadratic formula must be positive (we have two distinct real roots) we must have $(r_1 + r_2)^2 - 4(1)(8) > 0 \quad \Leftrightarrow \quad (r_1 + r_2)^2 > 32 \quad \Rightarrow \quad \sqrt{(r_1 + r_2)^2} = |r_1 + r_2| > 4\sqrt{2}$.

Chapter Three

Exercises 3.1

1.

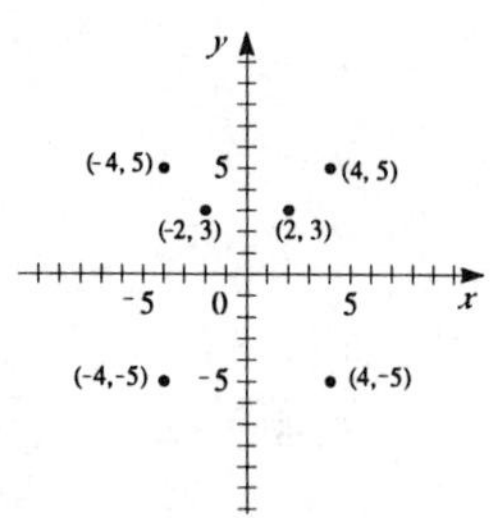

3. $A\,(5,\,1)$ $\quad B\,(1,\,2)$ $\quad C\,(-2,\,6)$ $\quad D\,(-6,\,2)$ $\quad E\,(-4,\,-1)$ $\quad F\,(-2,\,0)$ $\quad G\,(-1,\,-3)$ $\quad H\,(2,\,-2)$

5. $d(A,\,B) = \sqrt{(1-5)^2 + (3-3)^2} = \sqrt{(-4)^2} = 4.$

$d(A,\,C) = \sqrt{(1-1)^2 + (3-(-3))^2} = \sqrt{(6)^2} = 6.$

$d(C,D) = \sqrt{(1-5)^2 + (-3-(-3))^2} = \sqrt{(-4)^2} = 4.$

$d(B,D) = \sqrt{(5-5)^2 + (3-(-3))^2} = \sqrt{(6)^2} = 6.$

The area of the rectangle is $4 \cdot 6 = 24$.

7. (a)

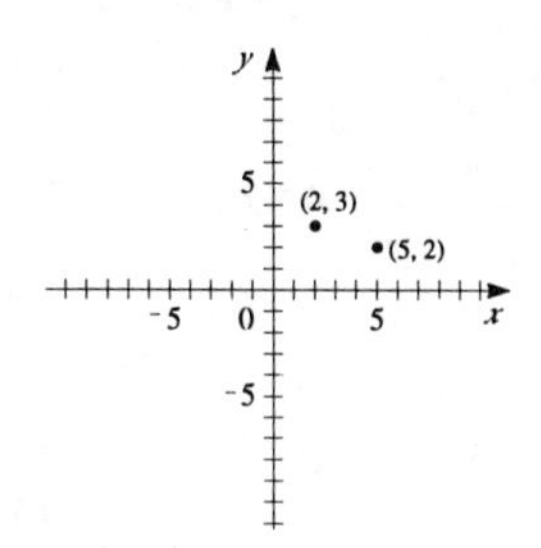

(b) $d = \sqrt{(2-5)^2 + (3-2)^2}$
$= \sqrt{(-3)^2 + (1)^2} = \sqrt{10}$

(c) Midpoint: $\left(\dfrac{2+5}{2}, \dfrac{3+2}{2}\right) = \left(\dfrac{7}{2}, \dfrac{5}{2}\right)$

9. (a)

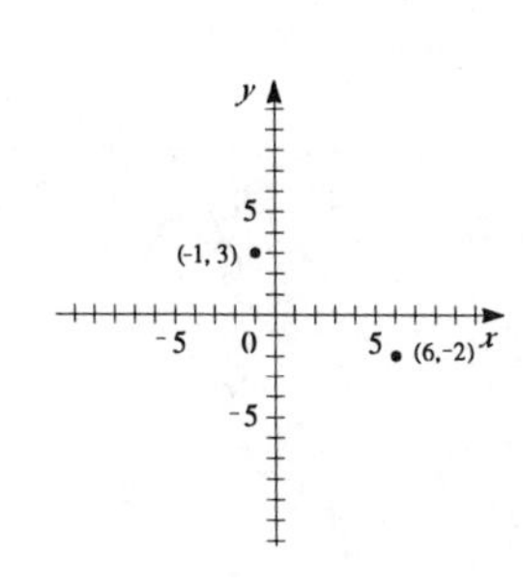

(b) $d = \sqrt{(6-(-1))^2 + (-2-3)^2}$
$= \sqrt{7^2 + (-5)^2} = \sqrt{49+25} = \sqrt{74}$

(c) Midpoint: $\left(\dfrac{6-1}{2}, \dfrac{-2+3}{2}\right) = \left(\dfrac{5}{2}, \dfrac{1}{2}\right)$

11. (a)

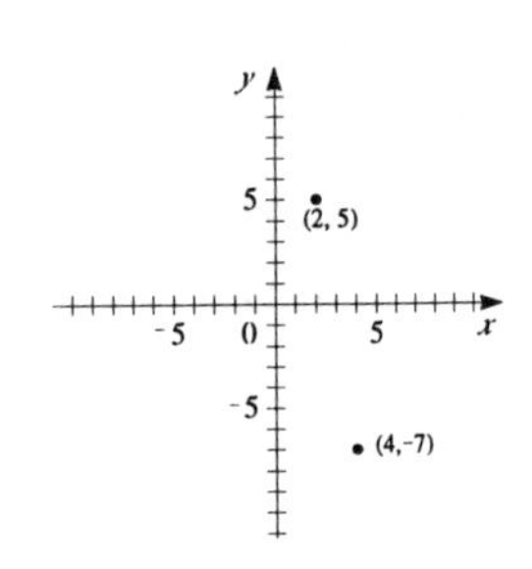

(b) $d = \sqrt{(2-4)^2 + (5-(-7))^2}$
$= \sqrt{(-2)^2 + (12)^2} = \sqrt{4+144}$
$= \sqrt{148} = 2\sqrt{37}$

(c) Midpoint: $\left(\frac{2+4}{2}, \frac{5+(-7)}{2}\right) = (3,-1)$

13. (a)

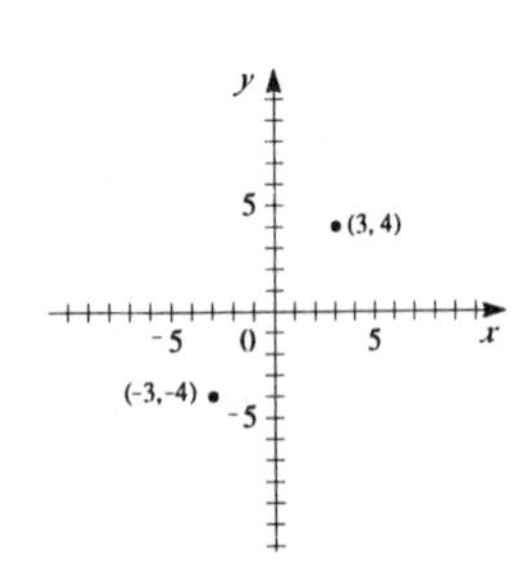

(b) $d = \sqrt{(3-(-3))^2 + (4-(-4))^2}$
$= \sqrt{6^2+8^2} = \sqrt{36+64} = \sqrt{100}$
$= 10$

(c) Midpoint: $\left(\frac{3+(-3)}{2}, \frac{4+(-4)}{2}\right) = (0,0)$

15. From the graph, the quadrilateral $ABCD$ has a pair of parallel sides, so $ABCD$ is a trapezoid. The area $= \left(\frac{b_1+b_2}{2}\right)h$.

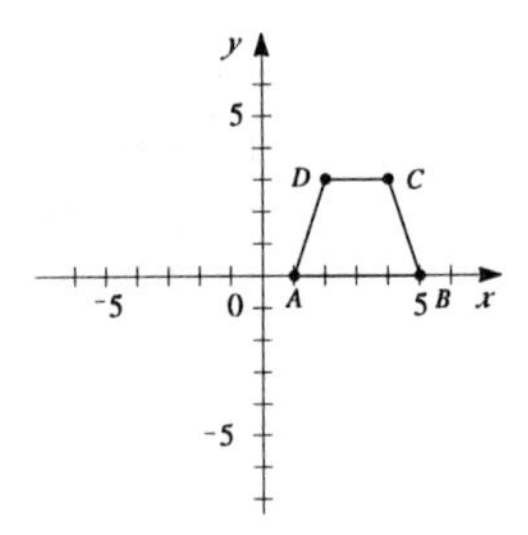

$b_1 = d(A, B) = \sqrt{(1-5)^2 + (0-0)^2} = \sqrt{4^2} = 4.$

$b_2 = d(C, D) = \sqrt{(4-2)^2 + (3-3)^2} = \sqrt{2^2} = 2.$

$h =$ difference in y-coordinate $= |3-0| = 3$. Thus the area of the trapezoid $= \left(\frac{4+2}{2}\right) 3 = 9.$

17.

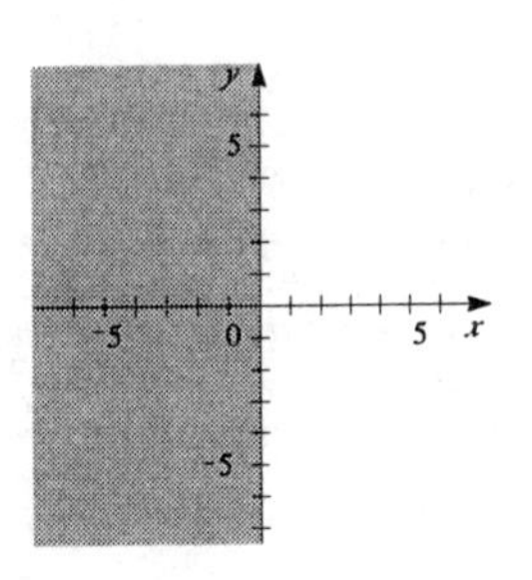

19.

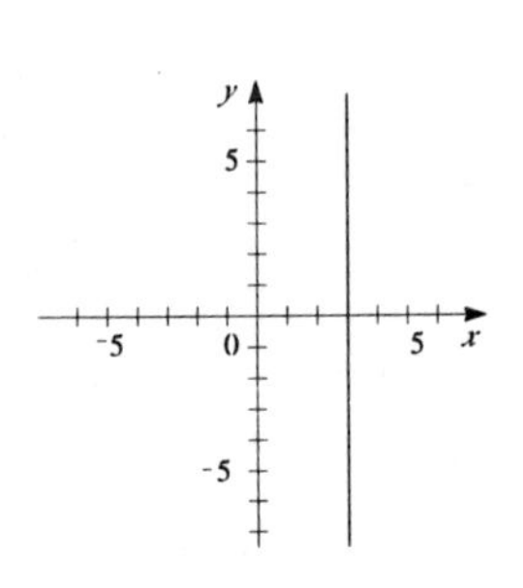

21.

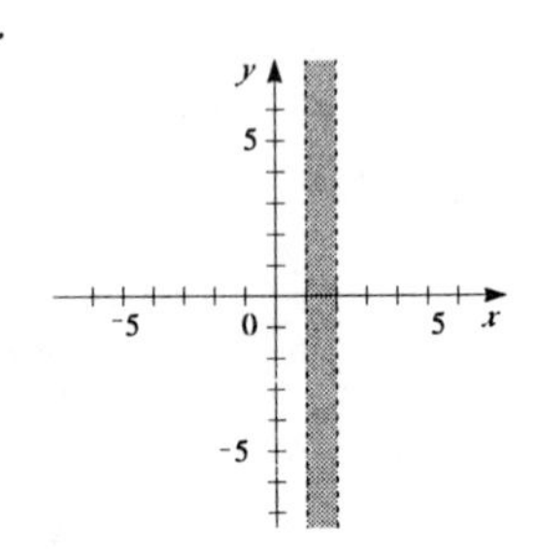

23. $x < 0$ and $y > 0$ or $x > 0$ and $y < 0$.

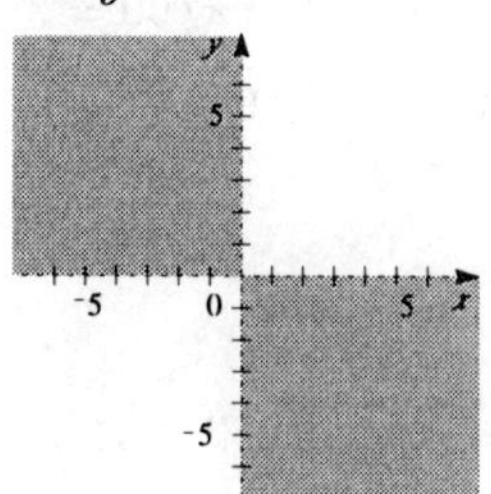

25.

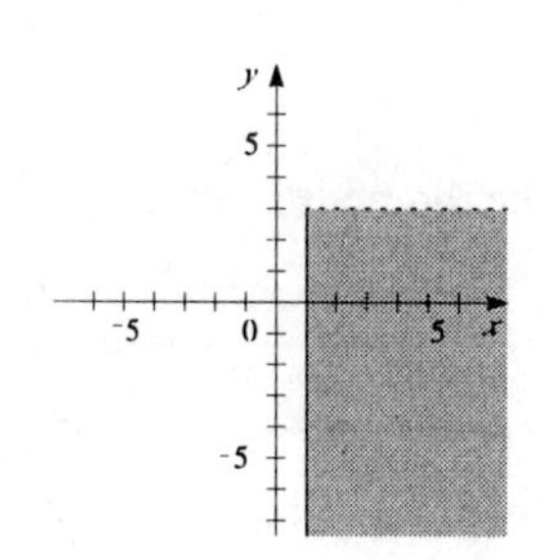

27. 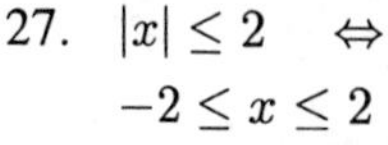
$|x| \le 2 \quad \Leftrightarrow \quad -2 \le x \le 2$

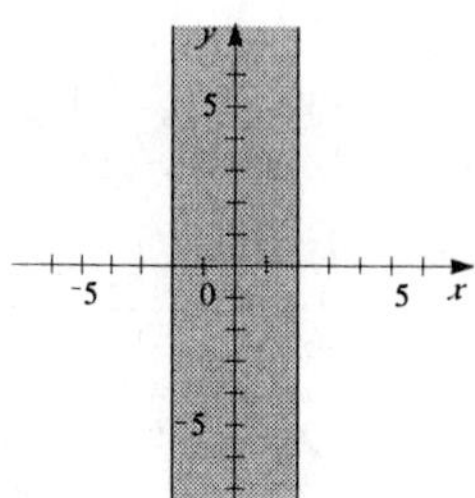

29. $|y - 3| \le 3 \quad \Leftrightarrow \quad -3 \le y - 3 \le 3 \quad \Leftrightarrow \quad 0 \le y \le 6.$

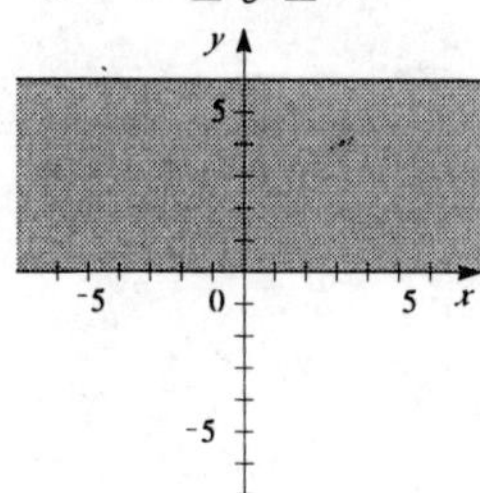

31. $d(0, A) = \sqrt{(6-0)^2 + (7-0)^2} = \sqrt{6^2 + 7^2} = \sqrt{36 + 49} = \sqrt{85}.$
$d(0, B) = \sqrt{(-5-0)^2 + (8-0)^2} = \sqrt{(-5)^2 + 8^2} = \sqrt{25 + 64} = \sqrt{89}.$
Thus point A is closer to the origin.

33. $d(P, R) = \sqrt{(-1-3)^2 + (-1-1)^2} = \sqrt{(-4)^2 + (-2)^2} = \sqrt{16 + 4} = \sqrt{20} = 2\sqrt{5}.$
$d(Q, R) = \sqrt{(-1-(-1))^2 + (-1-3)^2} = \sqrt{0 + (-4)^2} = \sqrt{16} = 4.$ Thus point Q is closer to point R.

35. Since we do not know which pair are isosceles, we find the length of all three sides.
$d(A, B) = \sqrt{(-3-0)^2 + (-1-2)^2} = \sqrt{(-3)^2 + (-3)^2} = \sqrt{9 + 9} = \sqrt{18} = 3\sqrt{2}.$
$d(C, B) = \sqrt{(-3-(-4))^2 + (-1-3)^2} = \sqrt{1^2 + (-4)^2} = \sqrt{1 + 16} = \sqrt{17}.$
$d(A, C) = \sqrt{(0-(-4))^2 + (2-3)^2} = \sqrt{4^2 + (-1)^2} = \sqrt{16 + 1} = \sqrt{17}.$ So sides AC and CB have the same length.

37. (a) Here we have $A = (2, 2)$, $B = (3, -1)$, $C = (-3, -3)$. So
$d(A, B) = \sqrt{(3-2)^2 + (-1-2)^2} = \sqrt{1^2 + (-3)^2} = \sqrt{1 + 9} = \sqrt{10}.$
$d(C, B) = \sqrt{(3-(-3))^2 + (-1-(-3))^2} = \sqrt{6^2 + 2^2} = \sqrt{36 + 4} = \sqrt{40} = 2\sqrt{10}.$
$d(A, C) = \sqrt{(-3-2)^2 + (-3-2)^2} = \sqrt{(-5)^2 + (-5)^2} = \sqrt{25 + 25} = \sqrt{50} = 5\sqrt{2}.$
Since $[d(A, B)]^2 + [d(C, B)]^2 = [d(A, C)]^2$ we conclude that the triangle is a right triangle.

(b) Area of the triangle $= \frac{1}{2} \cdot d(C, B) \cdot d(A, B) = \frac{1}{2} \cdot \sqrt{10} \cdot 2\sqrt{10} = 10.$

39. We show that all sides are the same length (its a rhombus) and then that the diagonals are also equal.
$d(A, B) = \sqrt{(4-(-2))^2 + (6-9)^2} = \sqrt{6^2 + (-3)^2} = \sqrt{36 + 9} = \sqrt{45}.$
$d(B, C) = \sqrt{(1-4)^2 + (0-6)^2} = \sqrt{(-3)^2 + (-6)^2} = \sqrt{9 + 36} = \sqrt{45}.$
$d(C, D) = \sqrt{(-5-1)^2 + (3-0)^2} = \sqrt{(-6)^2 + (-3)^2} = \sqrt{36 + 9} = \sqrt{45}.$

$d(D, A) = \sqrt{(-2-(-5))^2 + (9-3)^2} = \sqrt{3^2 + 6^2} = \sqrt{9+36} = \sqrt{45}$. So the points form a rhombus. $d(A, C) = \sqrt{(1-(-2))^2 + (0-9)^2} = \sqrt{3^2 + (-9)^2} = \sqrt{9+81} = \sqrt{90} = 3\sqrt{10}$ and $d(B, D) = \sqrt{(-5-4)^2 + (3-6)^2} = \sqrt{(-9)^2 + (-3)^2} = \sqrt{81+9} = \sqrt{90} = 3\sqrt{10}$. Since the diagonals are equal, the rhombus is a square.

41. Let $P = (0, y)$ be such a point. Setting the distances equal we have:
$\sqrt{(0-5)^2 + (y-(-5))^2} = \sqrt{(0-1)^2 + (y-1)^2} \quad \Leftrightarrow$
$\sqrt{25 + y^2 + 10y + 25} = \sqrt{1 + y^2 - 2y + 1} \quad \Rightarrow \quad y^2 + 10y + 50 = y^2 - 2y + 2 \quad \Leftrightarrow$
$12y = -48 \quad \Leftrightarrow \quad y = -4$. Thus the point is $P = (0, -4)$. Check:
$\sqrt{(0-5)^2 + (-4-(-5))^2} = \sqrt{(-5)^2 + 1^2} = \sqrt{25+1} = \sqrt{26}$;
$\sqrt{(0-1)^2 + (-4-1)^2} = \sqrt{(-1)^2 + (-5)^2} = \sqrt{25+1} = \sqrt{26}$.

43. We find the midpoint, M of PQ and then the midpoint of PM. $M = \left(\frac{-1+7}{2}, \frac{3+5}{2}\right) = (3, 4)$.
Midpoint of $PM = \left(\frac{-1+3}{2}, \frac{3+4}{2}\right) = \left(1, \frac{7}{2}\right)$.

45. We seek the point $S = (x_1, y_1)$ such that PS is parallel to QR. Thus PS must have the same change in x-coordinate and the same change in y-coordinate as QR. QR's change in x-coordinate $= 4 - 1 = 3$; QR's change in y-coordinate $= 2 - 1 = 1$. Thus $x_1 = -1 + 3 = 2$; $y_1 = -4 + 1 = -3$. So $S = (2, -3)$.

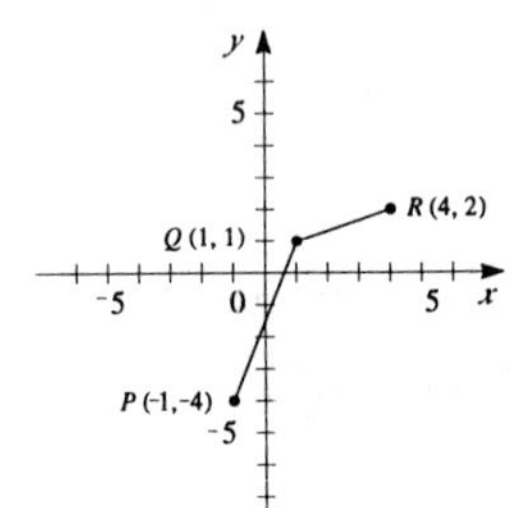

47. (a)

(b) Midpoint of $AC = \left(\frac{-2+7}{2}, \frac{-1+7}{2}\right) = \left(\frac{5}{2}, 3\right)$
Midpoint of $BD = \left(\frac{4+1}{2}, \frac{2+4}{2}\right) = \left(\frac{5}{2}, 3\right)$

(c) Since the they have the same midpoint, we conclude the diagonals bisect each other.

49. (a) The point $(5, 3)$ is moved to $(5+3, 3+2) = (8, 5)$.
(b) The point (a, b) is moved to $(a+3, b+2)$.
(c) $A = (-5, -1)$ so $A' = (-5+3, -1+2) = (-2, 1)$;
$B = (-3, 2)$ so $B' = (-3+3, 2+2) = (0, 4)$;
$C = (2, 1)$ so $C' = (2+3, 1+2) = (5, 3)$.

Exercises 3.2

1. $(0,0)$: $0 \stackrel{?}{=} 2(0)+3 \;\Leftrightarrow\; 0 \stackrel{?}{=} 3$ No. $\left(\frac{1}{2},4\right)$: $4 \stackrel{?}{=} 2\left(\frac{1}{2}\right)+3 \;\Leftrightarrow\; 4 \stackrel{?}{=} 1+3$ Yes.
   $(1,4)$: $4 \stackrel{?}{=} 2(1)+3 \;\Leftrightarrow\; 4 \stackrel{?}{=} 2+3$ No.
   So $\left(\frac{1}{2},4\right)$ is on the graph of this equation.

3. $(0,0)$: $2(0)-0+1 \stackrel{?}{=} 0 \;\Leftrightarrow\; 1 \stackrel{?}{=} 0$ No. $(1,0)$: $2(0)-1+1 \stackrel{?}{=} 0 \;\Leftrightarrow\; -1+1 \stackrel{?}{=} 0$ Yes.
   $(-1,-1)$: $2(-1)-(-1)+1 \stackrel{?}{=} 0 \;\Leftrightarrow\; -2+1+1 \stackrel{?}{=} 0$ Yes.
   So $(1,0)$ and $(-1,-1)$ are points on the graph of this equation.

5. $(0,-2)$: $(0)^2+(0)(-2)+(-2)^2 \stackrel{?}{=} 4 \;\Leftrightarrow\; 0+0+4 \stackrel{?}{=} 4$ Yes.
   $(1,-2)$: $(1)^2+(1)(-2)+(-2)^2 \stackrel{?}{=} 4 \;\Leftrightarrow\; 1-2+4 \stackrel{?}{=} 4$ No.
   $(2,-2)$: $(2)^2+(2)(-2)+(-2)^2 \stackrel{?}{=} 4 \;\Leftrightarrow\; 4-4+4 \stackrel{?}{=} 4$ Yes.
   So $(0,-2)$ and $(2,-2)$ are points on the graph of this equation.

7. To find x-intercepts, set $y=0$: $0=x-3 \;\Leftrightarrow\; x=3$.
   To find y-intercepts, set $x=0$: $y=0-3 \;\Leftrightarrow\; y=-3$.

9. To find x-intercepts, set $y=0$: $0=x^2-9 \;\Leftrightarrow\; x^2=9 \;\Rightarrow\; x=\pm 3$.
   To find y-intercepts, set $x=0$: $y=(0)^2-9 \;\Leftrightarrow\; y=-9$.

11. To find x-intercepts, set $y=0$: $x^2+(0)^2=4 \;\Leftrightarrow\; x^2=4 \;\Rightarrow\; x=\pm 2$.
    To find y-intercepts, set $x=0$: $(0)^2+y^2=4 \;\Leftrightarrow\; y^2=4 \;\Rightarrow\; y=\pm 2$.

13. To find x-intercepts, set $y=0$: $x(0)=5 \;\Leftrightarrow\; 0=5$ which is impossible, no x-intercept.
    Likewise, to find y-intercepts, set $x=0$: $(0)y=5 \;\Leftrightarrow\; 0=5$ which is again impossible, so no y-intercept.

15.

x	y
-4	-4
-2	-2
0	0
1	1
2	2
3	3
4	4

$y=0 \;\Rightarrow\; x=0$. So the x-intercept is 0 and the y-intercept is also 0.
x-axis symmetry: $(-y)=x \;\Leftrightarrow\; -y=x$ which is not the same as $y=x$. Not symmetric.
y-axis symmetry: $y=(-x) \;\Leftrightarrow\; y=-x$ which is not the same as $y=x$. Not symmetric.
Origin symmetry: $(-y)=(-x)$ is the same as $y=x$, so symmetric with respect to the origin.

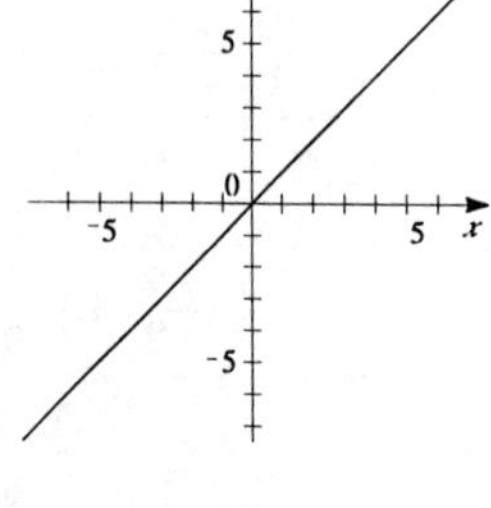

17.

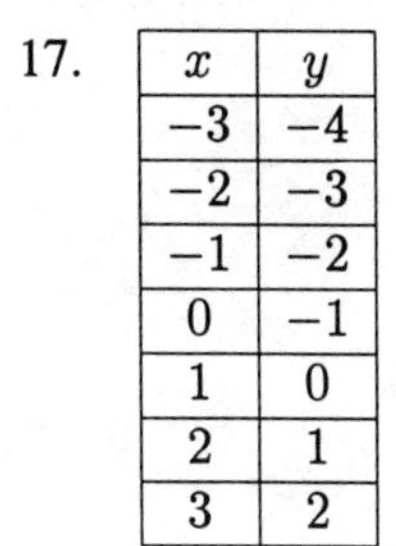

x	y
-3	-4
-2	-3
-1	-2
0	-1
1	0
2	1
3	2

$y=0 \;\Rightarrow\; 0=x-1 \;\Leftrightarrow\; x=1$. The x-intercept is 1. $x=0 \;\Rightarrow\; y=0-1=-1$. The y-intercept is -1.
x-axis symmetry: $(-y)=x-1 \;\Leftrightarrow\; -y=x-1$ which is not the same as $y=x-1$. Not symmetric.
y-axis symmetry: $y=(-x)-1 \;\Leftrightarrow\; y=-x-1$ which is not the same as $y=x-1$. Not symmetric.
Origin symmetry: $(-y)=(-x)-1 \;\Leftrightarrow\; -y=-x-1 \;\Leftrightarrow\; y=x+1$ which is not the same as $y=x-1$. Not symmetric with respect to the origin.

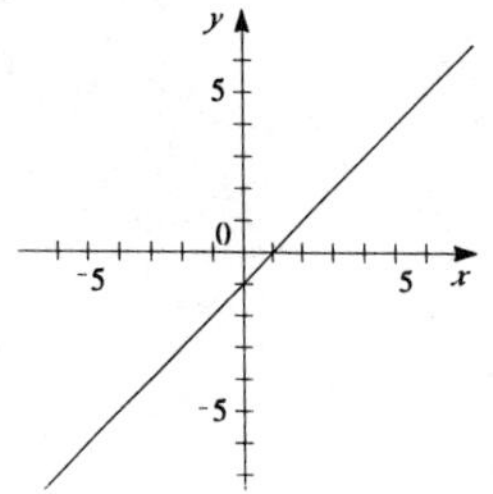

19. Solve for y: $-y = -3x + 5 \quad \Leftrightarrow \quad y = 3x - 5$

x	y
-2	-11
-1	-8
0	-5
1	-2
2	1
3	4
4	7

$y = 0 \quad \Rightarrow \quad 0 = 3x - 5 \quad \Leftrightarrow \quad 3x = 5 \quad \Leftrightarrow$ $x = \frac{5}{3}$. So x-intercept is $\frac{5}{3}$. $x = 0 \quad \Rightarrow$ $y = 3(0) - 5 = -5$. So y-intercept is -5.

x-axis symmetry: $3x - (-y) = 5 \quad \Leftrightarrow \quad 3x + y = 5$ which is not the same as $3x - y = 5$. Not symmetric.

y-axis symmetry: $3(-x) - y = 5 \quad \Leftrightarrow \quad -3x - y = 5$ which is not the same as $3x - y = 5$. Not symmetric.

Origin symmetry: $3(-x) - (-y) = 5 \quad \Leftrightarrow$ $-3x + y = 5$ which is not the same as $3x - y = 5$. Not symmetric with respect to the origin.

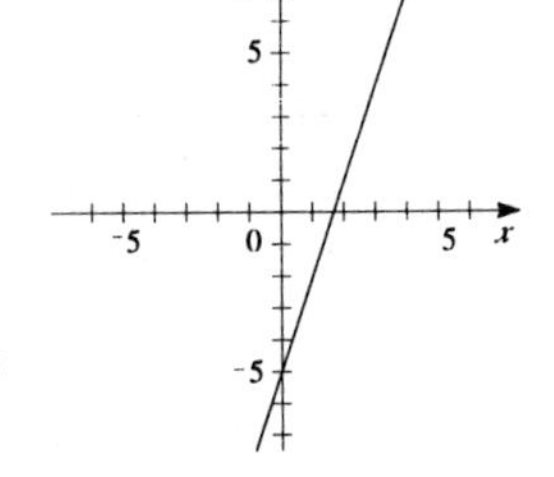

21.

x	y
-3	-8
-2	-3
-1	0
0	1
1	0
2	-3
3	-8

$y = 0 \quad \Rightarrow 0 = 1 - x^2 \quad \Leftrightarrow \quad x^2 = 1 \quad \Rightarrow$ $x = \pm 1$. The x-intercept are 1 and -1. $x = 0$ $\Rightarrow y = 1 - (0)^2 = 1$. The y-intercept is 1.

x-axis symmetry: $(-y) = 1 - x^2 \quad \Leftrightarrow \quad -y = 1 - x^2$ which is not the same as $y = 1 - x^2$. Not symmetric with respect to the x-axis.

y-axis symmetry: $y = 1 - (-x)^2 \quad \Leftrightarrow \quad y = 1 - x^2$. Symmetric with respect to the y-axis.

Origin symmetry: $(-y) = 1 - (-x)^2 \quad \Leftrightarrow$ $-y = 1 - x^2$ which is not the same as $y = 1 - x^2$. Not symmetric with respect to the origin.

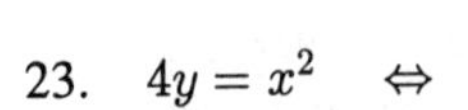

23. $4y = x^2 \quad \Leftrightarrow \quad y = \frac{1}{4}x^2$

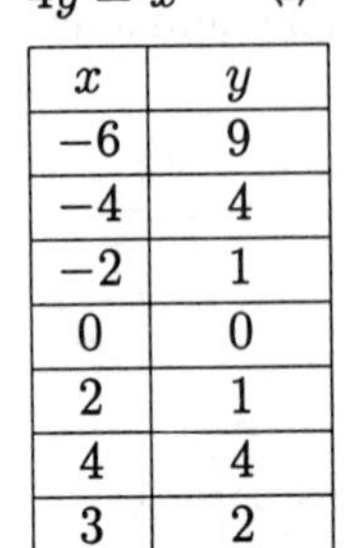

x	y
-6	9
-4	4
-2	1
0	0
2	1
4	4
3	2

$y = 0 \quad \Rightarrow \quad 0 = \frac{1}{4}x^2 \quad \Leftrightarrow \quad x^2 = 0 \quad \Rightarrow \quad x = 0$. The x-intercept is 0. $x = 0 \quad \Rightarrow \quad y = \frac{1}{4}(0)^2 = 0$. The y-intercept is 0.

x-axis symmetry: $(-y) = \frac{1}{4}x^2$ which is not the same as $y = \frac{1}{4}x^2$. Not symmetric with respect to the x-axis.

y-axis symmetry: $y = \frac{1}{4}(-x)^2 \quad \Leftrightarrow \quad y = \frac{1}{4}x^2$. Symmetric with respect to the y-axis.

Origin symmetry: $(-y) = \frac{1}{4}(-x)^2 \quad \Leftrightarrow \quad -y = \frac{1}{4}x^2$ which is not the same as $y = \frac{1}{4}x^2$. Not symmetric with respect to the origin.

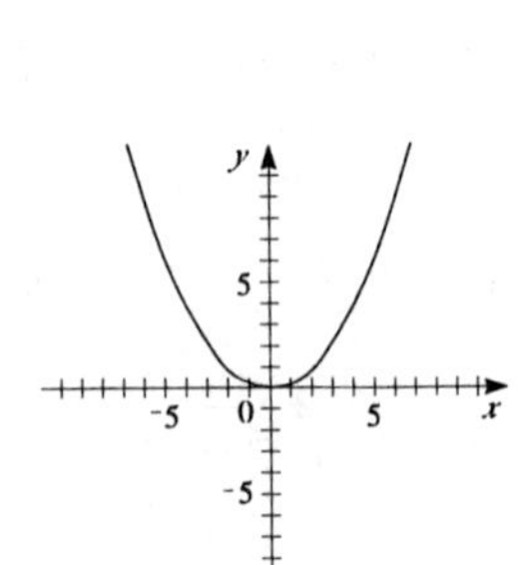

25.

x	y
-4	7
-3	0
-2	-5
-1	-8
0	-9
1	-8
2	-5
3	0
4	7

$y = 0 \;\Rightarrow\; 0 = x^2 - 9 \;\Leftrightarrow\; x^2 = 9 \;\Rightarrow\; x = \pm 3$. The x-intercepts are -3 and 3. $x = 0 \;\Rightarrow\; y = (0)^2 - 9 = -9$. The y-intercept is -9.

x-axis symmetry: $(-y) = x^2 - 9$ which is not the same as $y = x^2 - 9$. Not symmetric with respect to the x-axis.

y-axis symmetry: $y = (-x)^2 - 9 \;\Leftrightarrow\; y = x^2 - 9$. Symmetric with respect to the y-axis.

Origin symmetry: $(-y) = (-x)^2 - 9 \;\Leftrightarrow\; -y = x^2 - 9$ which is not the same as $y = x^2 - 9$. Not symmetric with respect to the origin.

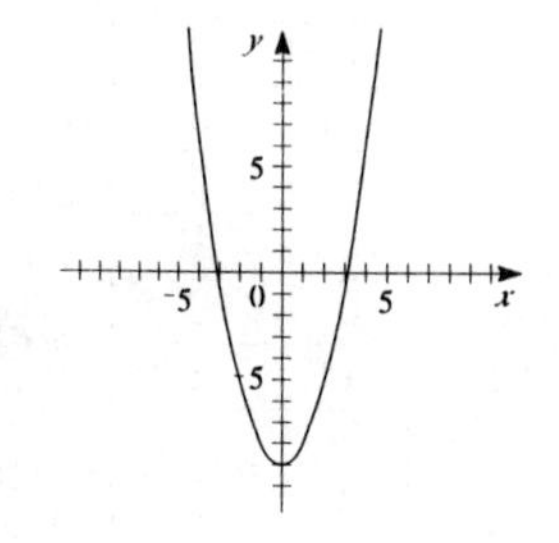

27. $xy = 2 \;\Leftrightarrow\; y = \dfrac{2}{x}$

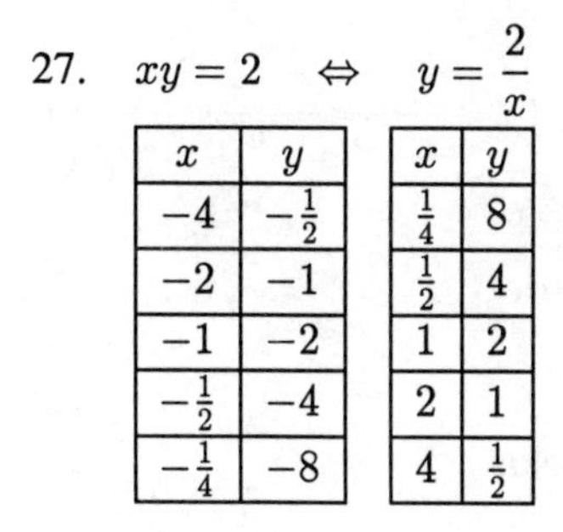

x	y
-4	$-\frac{1}{2}$
-2	-1
-1	-2
$-\frac{1}{2}$	-4
$-\frac{1}{4}$	-8

x	y
$\frac{1}{4}$	8
$\frac{1}{2}$	4
1	2
2	1
4	$\frac{1}{2}$

$y = 0$ or $x = 0 \;\Rightarrow\; 0 = 2$ which is impossible, so this equation has no x-intercept and no y-intercept.

x-axis symmetry: $x(-y) = 2 \;\Leftrightarrow\; -xy = 2$. Which is not the same as $xy = 2$. Not symmetric with respect to the x-axis.

y-axis symmetry: $(-x)y = 2 \;\Leftrightarrow\; -xy = 2$. Which is not the same as $xy = 2$. Not symmetric with respect to the y-axis.

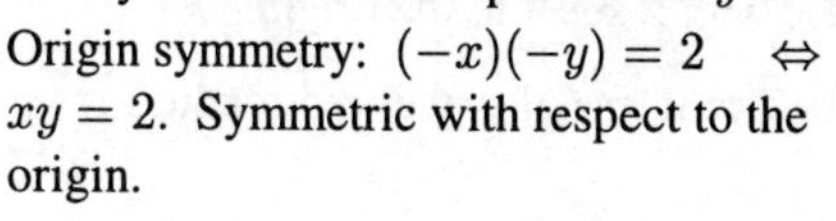

Origin symmetry: $(-x)(-y) = 2 \;\Leftrightarrow\; xy = 2$. Symmetric with respect to the origin.

29.

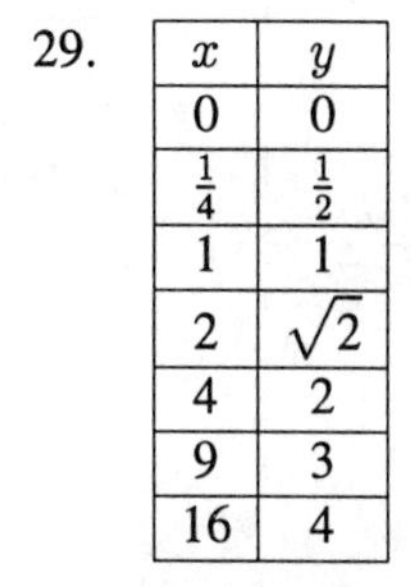

x	y
0	0
$\frac{1}{4}$	$\frac{1}{2}$
1	1
2	$\sqrt{2}$
4	2
9	3
16	4

$y = 0 \;\Rightarrow\; 0 = \sqrt{x} \;\Leftrightarrow\; x = 0$. The x-intercept is 0. $x = 0 \;\Rightarrow\; y = \sqrt{0} = 0$. The y-intercept is 0.

Since we are graphing real numbers and $\sqrt{x}$ is defined to be a non-negative number, the equation is not symmetric with respect to the x-axis nor with respect to the y-axis. The equation is not symmetric with respect to the origin.

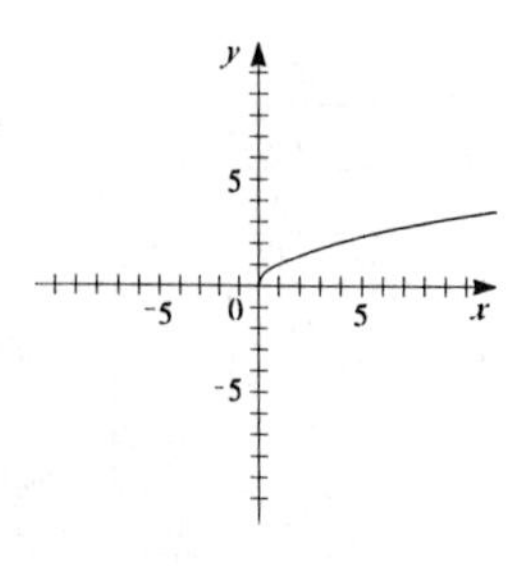

31. Since the radicand (the inside of the square root) cannot be negative, we must have $4 - x^2 \geq 0 \;\Leftrightarrow\; x^2 \leq 4 \;\Leftrightarrow\; |x| \leq 2$.

x	y
-2	0
-1	$\sqrt{3}$
0	4
1	$\sqrt{3}$
2	0

$y = 0 \;\Rightarrow\; 0 = \sqrt{4 - x^2} \;\Leftrightarrow\; 4 - x^2 = 0 \;\Leftrightarrow\; x^2 = 4 \;\Rightarrow\; x = \pm 2$. The x-intercept are -2 and 2. $x = 0 \;\Rightarrow\; y = \sqrt{4 - (0)^2} = \sqrt{4} = 2$. The y-intercept is 2.

Since $y \geq 0$, the graph is not symmetric with respect to the x-axis.

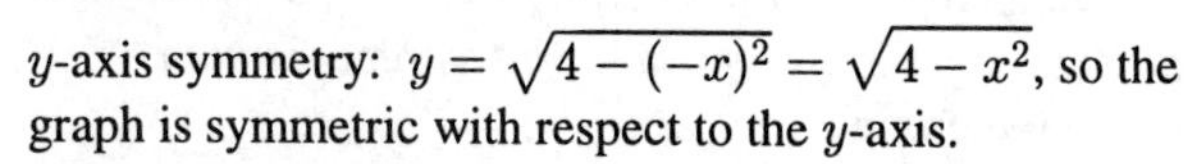

y-axis symmetry: $y = \sqrt{4 - (-x)^2} = \sqrt{4 - x^2}$, so the graph is symmetric with respect to the y-axis.

Also, since $y \geq 0$ the graph is not symmetric with respect to the origin.

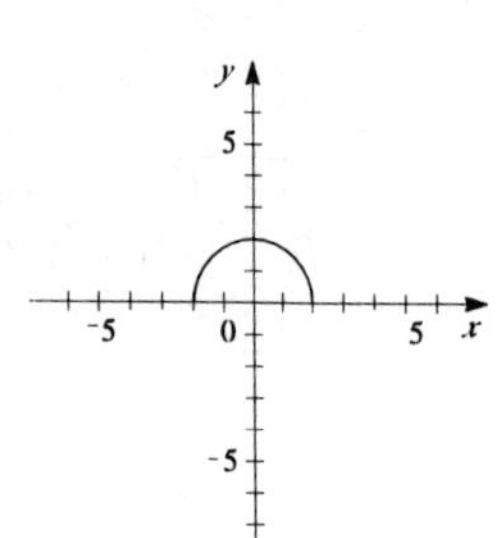

33.

x	y
-3	3
-2	2
-1	1
0	0
1	1
2	2
3	3

$y=0 \quad \Rightarrow \quad 0=|x| \quad \Leftrightarrow \quad x=0$. The x-intercept is 0. $x=0 \quad \Rightarrow \quad y=|0|=0$. The y-intercept is 0.

Since $y \geq 0$, the graph is not symmetric with respect to the x-axis.

y-axis symmetry: $y=|-x|=|x|$. So the graph is symmetric with respect to the y-axis.

Since $y \geq 0$, the graph is not symmetric with respect to the origin.

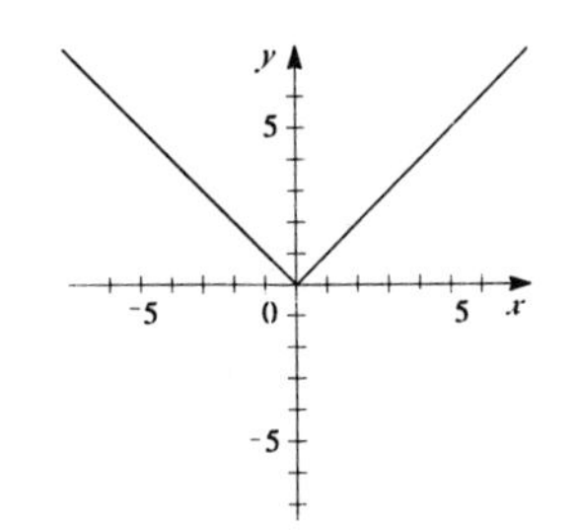

35.

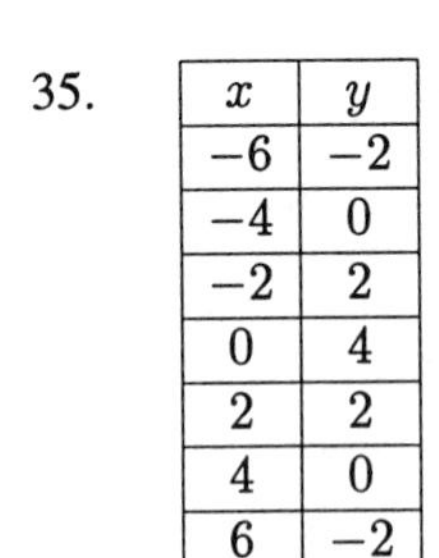

x	y
-6	-2
-4	0
-2	2
0	4
2	2
4	0
6	-2

$y=0 \quad \Rightarrow \quad 0=4-|x| \quad \Leftrightarrow \quad |x|=4$ $\Rightarrow x= \pm 4$. The x-intercept are -4 and 4. $x=0$ $\Rightarrow \quad y=4-|0|=4$. The y-intercept is 4.

x-axis symmetry: $(-y)=4-|x| \quad \Leftrightarrow \quad y=-4+|x|$ which is not the same as $y=4-|x|$. Not symmetric with respect to the x-axis.

y-axis symmetry: $y=4-|-x|=4-|x|$. Symmetric with respect to the y-axis.

Origin symmetry: $(-y)=4-|-x| \quad \Leftrightarrow$ $y=-4+|x|$ which is not the same as $y=4-|x|$. Not symmetric with respect to the origin.

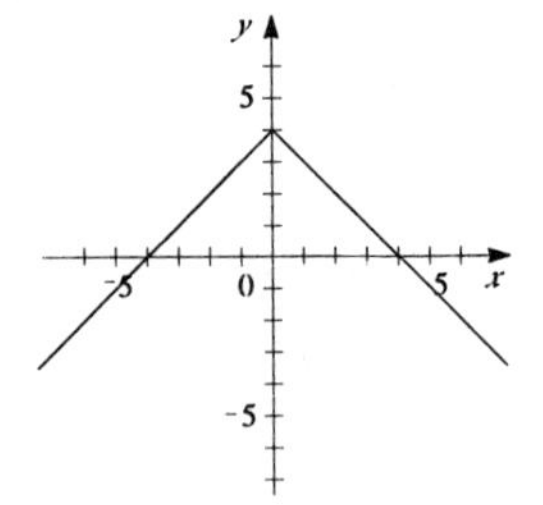

37. Since $x=y^3$ is solved for x in terms of y, we insert values for y and find the corresponding values of x.

x	y
-27	-3
-8	-2
-1	-1
0	0
1	1
8	2
27	3

$y=0 \quad \Rightarrow \quad x=(0)^3=0$. The x-intercept is 0. $x=0$ $\Rightarrow \quad 0=y^3 \quad \Rightarrow \quad y=0$. The y-intercept is 0.

x-axis symmetry: $x=(-y)^3=-y^3$ is not the same as $x=y^3$. Not symmetric with respect to the x-axis.

y-axis symmetry: $(-x)=y^3 \quad \Leftrightarrow \quad x=-y^3$ is not the same as $x=y^3$. Not symmetric with respect to the y-axis.

Origin symmetry: $(-x)=(-y)^3 \quad \Leftrightarrow \quad -x=-y^3$ $\Leftrightarrow \quad x=y^3$. Symmetric with respect to the origin.

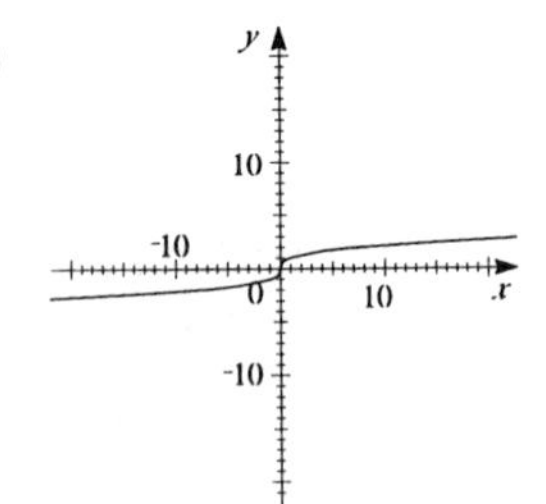

39.

x	y
-3	81
-2	16
-1	1
0	0
1	1
2	16
3	81

$y=0 \quad \Rightarrow \quad 0=x^4 \quad \Rightarrow \quad x=0$. The x-intercept is 0. $x=0 \quad \Rightarrow \quad y=0^4=0$. The y-intercept is 0.

x-axis symmetry: $(-y)=x^4 \quad \Leftrightarrow \quad y=-x^4$ which is not the same as $y=x^4$. Not symmetric with respect to the x-axis.

y-axis symmetry: $y=(-x)^4=x^4$. Symmetric with respect to the y-axis.

Origin symmetry: $(-y)=(-x)^4 \quad \Leftrightarrow \quad -y=x^4$ which is not the same as $y=x^4$. Not symmetric with respect to the origin.

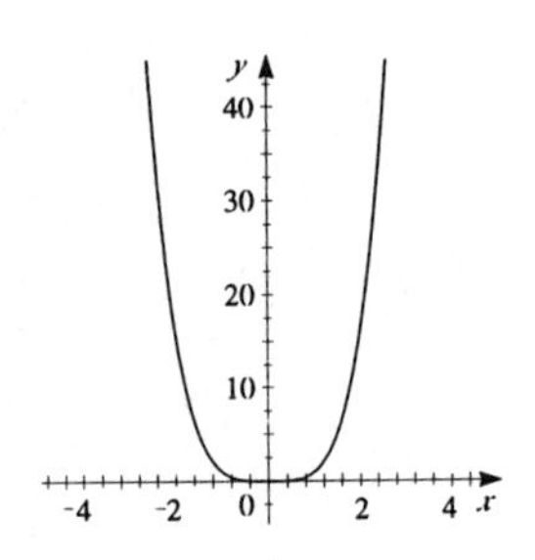

41. x-axis symmetry: $(-y) = x^4 + x^2 \quad \Leftrightarrow \quad y = -x^4 - x^2$ which is not the same as $y = x^4 + x^2$. Not symmetric with respect to the x-axis.
y-axis symmetry: $y = (-x)^4 + (-x)^2 = x^4 + x^2$. Symmetric with respect to the y-axis.
Origin symmetry: $(-y) = (-x)^4 + (-x)^2 \quad \Leftrightarrow \quad -y = x^4 + x^2$ which is not the same as $y = x^4 + x^2$. Not symmetric with respect to the origin.

43. x-axis symmetry: $x^2(-y)^2 + x(-y) = 1 \quad \Leftrightarrow \quad x^2y^2 - xy = 1$ which is not the same as $x^2y^2 + xy = 1$. Not symmetric with respect to the x-axis.
y-axis symmetry: $(-x)^2y^2 + (-x)y = 1 \quad \Leftrightarrow \quad x^2y^2 - xy = 1$ which is not the same as $x^2y^2 + xy = 1$. Not symmetric with respect to the y-axis.
Origin symmetry: $(-x)^2(-y)^2 + (-x)(-y) = 1 \quad \Leftrightarrow \quad x^2y^2 + xy = 1$. Symmetric with respect to the origin.

45. x-axis symmetry: $(-y) = x^3 + 10x \quad \Leftrightarrow \quad y = -x^3 - 10x$ which is not the same as $y = x^3 + 10x$. Not symmetric with respect to the x-axis.
y-axis symmetry: $y = (-x)^3 + 10(-x) \quad \Leftrightarrow \quad y = -x^3 - 10x$ which is not the same as $y = x^3 + 10x$. Not symmetric with respect to the y-axis.
Origin symmetry: $(-y) = (-x)^3 + 10(-x) \quad \Leftrightarrow \quad -y = -x^3 - 10x \quad \Leftrightarrow \quad y = x^3 + 10x$. Symmetric with respect to the origin.

47. Symmetric with respect to the y-axis.

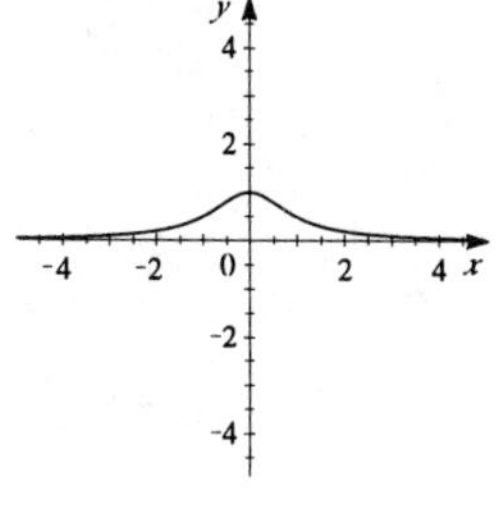

49. Symmetric with respect to the origin.

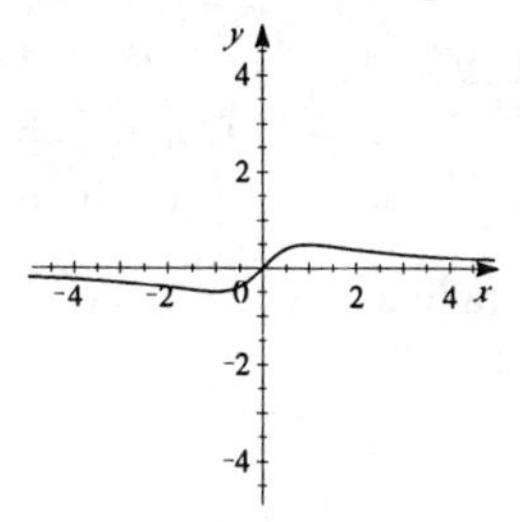

51. Using $h = 2$, $k = -1$, and $r = 3$ we get $(x-2)^2 + (y-(-1))^2 = 3^2 \quad \Leftrightarrow$ $(x-2)^2 + (y+1)^2 = 9$.

53. The equation of a circle centered at the origin is $x^2 + y^2 = r^2$. Using the point $(4, 7)$ we solve for r^2: $(4)^2 + (7)^2 = r^2 \quad \Leftrightarrow \quad 16 + 49 = 65 = r^2$. Thus the equation of the circle is $x^2 + y^2 = 65$.

55. The center is at the midpoint of the line segment: $\left(\frac{-1+5}{2}, \frac{1+5}{2}\right) = (2, 3)$. The radius is one half the diameter so $r = \frac{1}{2}\sqrt{(-1-5)^2 + (1-5)^2} = \frac{1}{2}\sqrt{36+16} = \frac{1}{2}\sqrt{52} = \sqrt{13}$. Thus the equation of the circle is $(x-2)^2 + (y-3)^2 = \left(\sqrt{13}\right)^2$ or $(x-2)^2 + (y-3)^2 = 13$.

57. Since the circle is tangent to the x-axis, it must contain the point $(7, 0)$, so the radius is the change in the y coordinate. That is, $r = |-3-0| = 3$. So the equation of the circle is $(x-7)^2 + (y-(-3))^2 = 3^2$ which is $(x-7)^2 + (y+3)^2 = 9$.

59. From the figure the center of the circle is at $(-2, 2)$. The radius is the change in the y coordinate, so $r = |2-0| = 2$. Thus the equation of the circle is $(x-(-2))^2 + (y-2)^2 = 2^2$ which is $(x+2)^2 + (y-2)^2 = 4$.

61. Complete the square. $x^2+y^2-2x+4y+1=0 \quad\Leftrightarrow\quad x^2-2x+\_\_+y^2+4y+\_\_=-1$
$\Leftrightarrow \quad x^2-2x+\left(\frac{-2}{2}\right)^2+y^2+4y+\left(\frac{4}{2}\right)^2=-1+\left(\frac{-2}{2}\right)^2+\left(\frac{4}{2}\right)^2 \Leftrightarrow$
$x^2-2x+1+y^2+4y+4=-1+1+4 \quad\Leftrightarrow\quad (x-1)^2+(y+2)^2=4.$
Center: $(1,-2)$, radius $=2$

63. $x^2+y^2-4x+10y+13=0 \quad\Leftrightarrow\quad x^2-4x+\_\_+y^2+10y+\_\_=-13 \quad\Leftrightarrow$
$x^2-4x+\left(\frac{-4}{2}\right)^2+y^2+10y+\left(\frac{10}{2}\right)^2=-13+\left(\frac{4}{2}\right)^2+\left(\frac{10}{2}\right)^2 \quad\Leftrightarrow$
$x^2-4x+4+y^2+10y+25=-13+4+25 \quad\Leftrightarrow\quad (x-2)^2+(y+5)^2=16.$
Center: $(2,-5)$, radius $=4$

65. $x^2+y^2+x=0 \quad\Leftrightarrow\quad x^2+x+\_\_+y^2=0 \quad\Leftrightarrow\quad x^2+x+\left(\frac{1}{2}\right)^2+y^2=\left(\frac{1}{2}\right)^2 \quad\Leftrightarrow$
$x^2+x+\frac{1}{4}+y^2=\frac{1}{4} \quad\Leftrightarrow\quad \left(x+\frac{1}{2}\right)^2+y^2=\frac{1}{4}$. Center: $\left(-\frac{1}{2},0\right)$, radius $=\frac{1}{2}$.

67. First divide by 2, then complete the square. $2x^2+2y^2-x+y=1 \quad\Leftrightarrow$
$x^2+y^2-\frac{1}{2}x+\frac{1}{2}y=\frac{1}{2} \quad\Leftrightarrow\quad x^2-\frac{1}{2}x+\_\_+y^2+\frac{1}{2}y+\_\_=\frac{1}{2} \quad\Leftrightarrow$
$x^2-\frac{1}{2}x+\left(\frac{-1/2}{2}\right)^2+y^2+\frac{1}{2}y+\left(\frac{1/2}{2}\right)^2=\frac{1}{2}+\left(\frac{-1/2}{2}\right)^2+\left(\frac{1/2}{2}\right)^2 \quad\Leftrightarrow$
$\left(x-\frac{1}{4}\right)^2+\left(y+\frac{1}{4}\right)^2=\frac{1}{2}+\frac{1}{16}+\frac{1}{16}=\frac{5}{8}.$
Center: $\left(\frac{1}{4},-\frac{1}{4}\right)$, radius $=\sqrt{\frac{5}{8}}$.

69. $x^2+y^2+4x-10y=21 \quad\Leftrightarrow\quad x^2+4x+\_\_+y^2-10y+\_\_=21$
$\Leftrightarrow \quad x^2+4x+\left(\frac{4}{2}\right)^2+y^2-10y+\left(\frac{-10}{2}\right)^2=21+\left(\frac{4}{2}\right)^2+\left(\frac{-10}{2}\right)^2$
$\Leftrightarrow \quad (x+2)^2+(y-5)^2=21+4+25=50.$
Center: $(-2,5)$, radius $=\sqrt{50}=5\sqrt{2}$

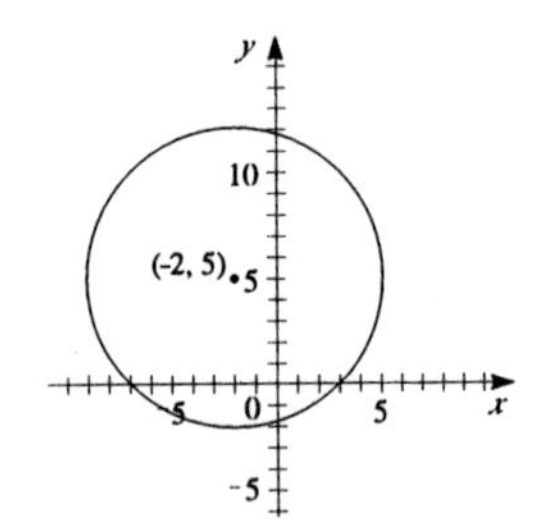

71. $x^2+y^2+6x-12y+45=0 \quad\Leftrightarrow$
$x^2+6x+\_\_+y^2-12y+\_\_=-45 \quad\Leftrightarrow$
$x^2+6x+\left(\frac{6}{2}\right)^2+y^2-12y+\left(\frac{-12}{2}\right)^2=-45+\left(\frac{6}{2}\right)^2+\left(\frac{-12}{2}\right)^2 \quad\Leftrightarrow$
$(x+3)^2+(y-6)^2=-45+9+36=0.$
Center: $(-3,6)$, radius $=0$. This is a degenerate circle whose graph consists only of the point $(-3,6)$.

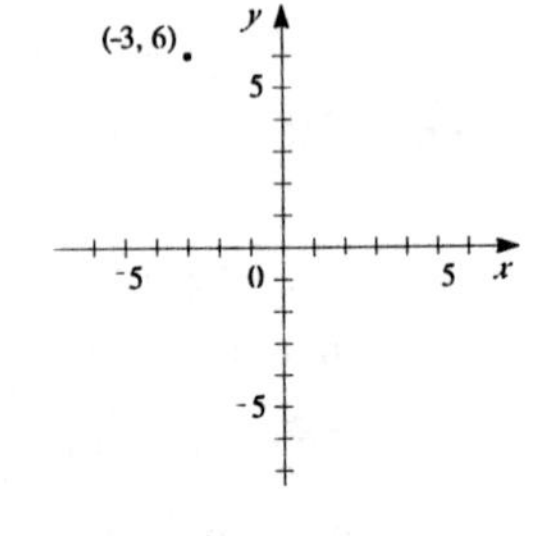

73. $\{(x,y)|\, x^2+y^2\le 1\}$. This is the set of points inside (and on) the circle $x^2+y^2=1$.

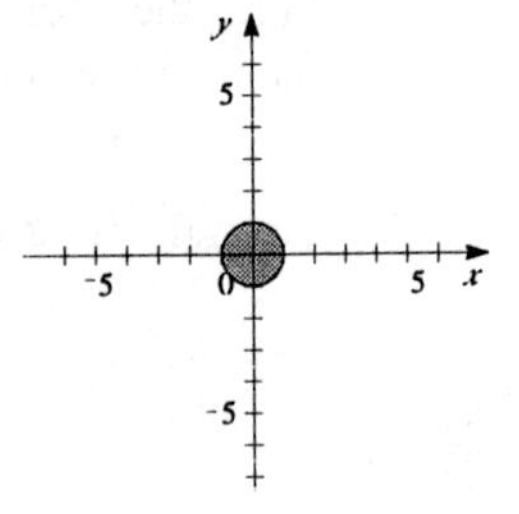

75. $\{(x,y) | 1 \le x^2 + y^2 < 9\}$. This is the set of points outside the circle $x^2 + y^2 = 1$ and inside (but not on) the circle $x^2 + y^2 = 9$.

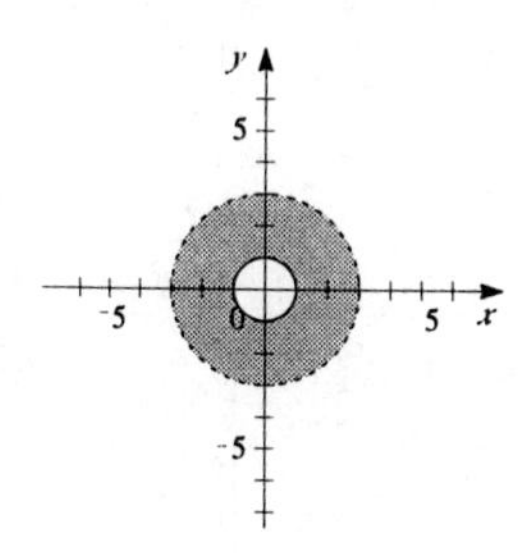

77. $x^2 + y^2 - 4y - 12 = 0 \quad \Leftrightarrow \quad x^2 + y^2 - 4y + \_\_ = 12 \quad \Leftrightarrow$
$x^2 + y^2 - 4y + \left(\frac{-4}{2}\right)^2 = 12 + \left(\frac{-4}{2}\right)^2 \quad \Leftrightarrow \quad x^2 + (y-2)^2 = 16$. Center: $(0,2)$, radius $= 4$. So the circle $x^2 + y^2 = 4$, center: $(0,0)$ and radius $= 2$, sits completely inside the larger circle. Thus the area $= \pi 4^2 - \pi 2^2 = 16\pi - 4\pi = 12\pi$.

79. $x^2 + y^2 + ax + by + c = 0 \quad \Leftrightarrow \quad x^2 + ax + \_\_ + y^2 + by + \_\_ = -c \quad \Leftrightarrow$

$$x^2 + ax + \left(\frac{a}{2}\right)^2 + y^2 + by + \left(\frac{b}{2}\right)^2 = -c + \left(\frac{a}{2}\right)^2 + \left(\frac{b}{2}\right)^2 \quad \Leftrightarrow$$

$\left(x + \frac{a}{2}\right)^2 + \left(y + \frac{b}{2}\right)^2 = -c + \frac{a^2 + b^2}{4}$. Thus this is a circle only when $-c + \frac{a^2 + b^2}{4} > 0$

Center: $\left(-\frac{a}{2}, -\frac{b}{2}\right)$, radius $= \sqrt{-c + \frac{a^2 + b^2}{4}} = \frac{1}{2}\sqrt{a^2 + b^2 - 4ac}$.

Exercises 3.3

1. $y = x^4 + 2$

(a) $[-2, 2]$ by $[-2, 2]$

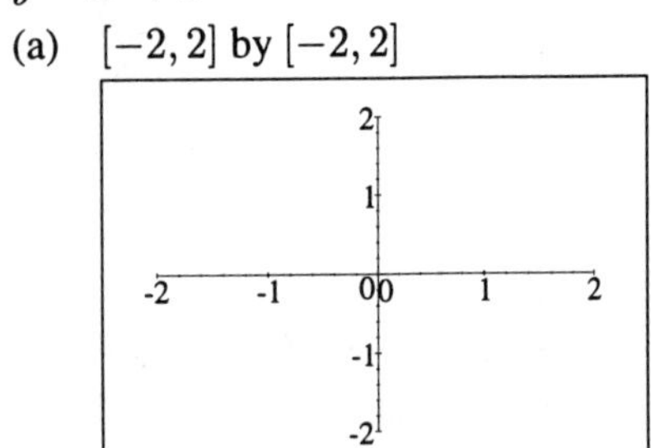

(b) $[0, 4]$ by $[0, 4]$

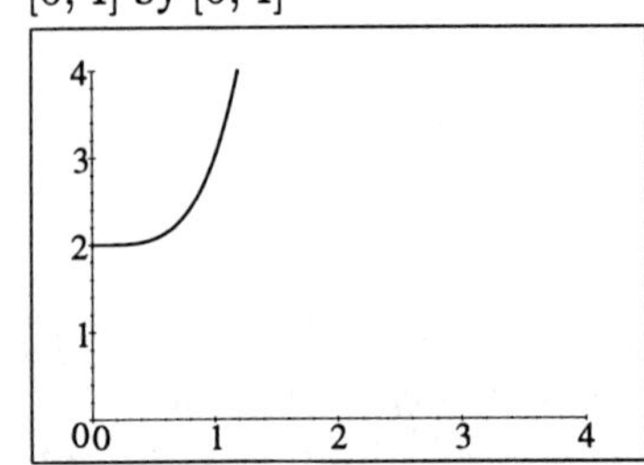

(c) $[-8, 8]$ by $[-4, 40]$

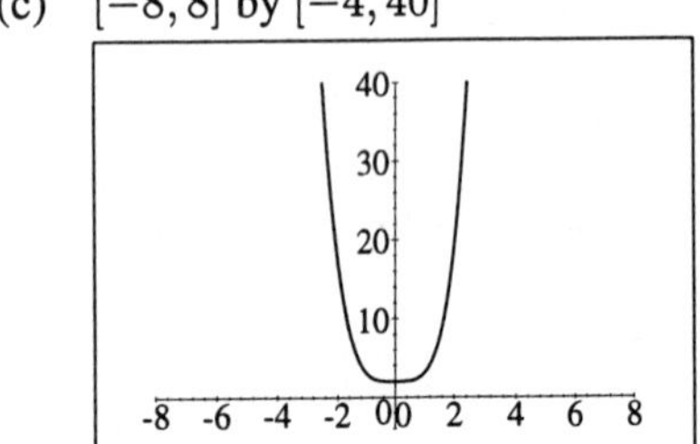

(d) $[-40, 40]$ by $[-80, 800]$

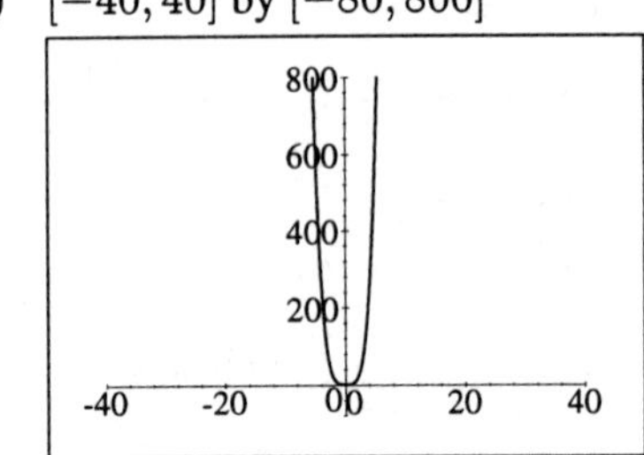

The viewing rectangle in part (c) produces the most appropriate graph of the equation.

3. $y = 100 - x^2$

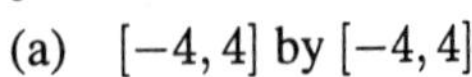
(a) $[-4, 4]$ by $[-4, 4]$

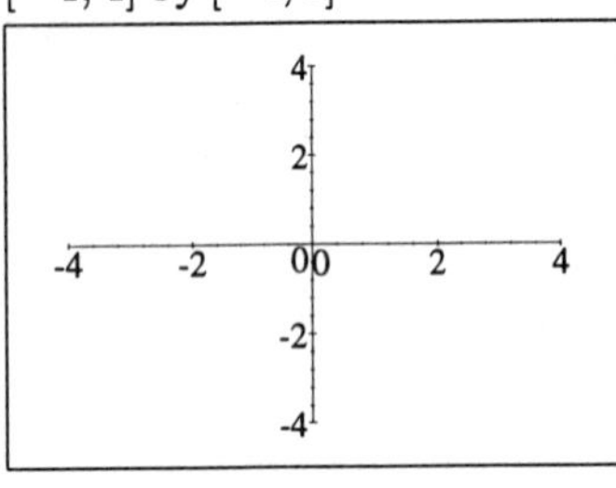

(b) $[-10, 10]$ by $[-10, 10]$

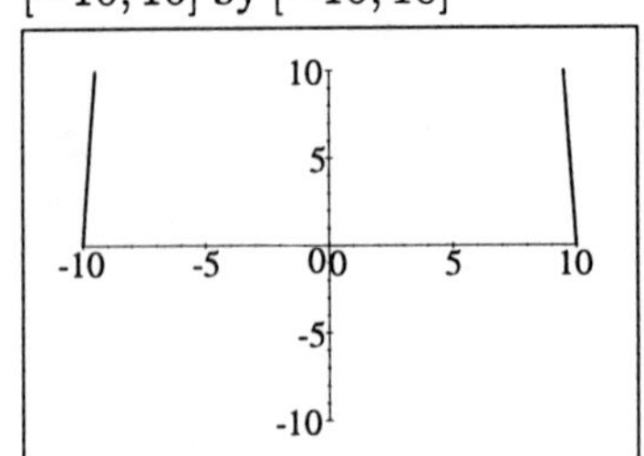

(c) $[-15, 15]$ by $[-30, 110]$

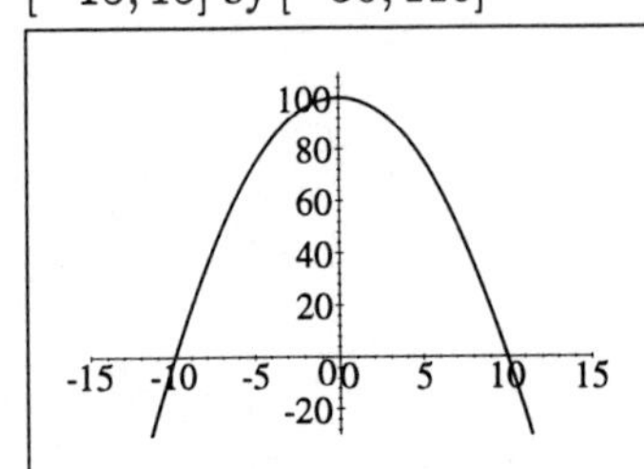

(d) $[-4, 4]$ by $[-30, 110]$

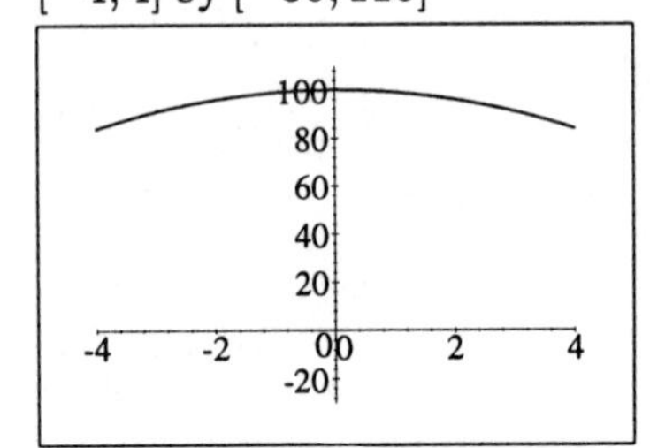

The viewing rectangle in part (c) produces the most appropriate graph of the equation.

5. $y = 10 + 25x - x^3$

(a) $[-4, 4]$ by $[-4, 4]$

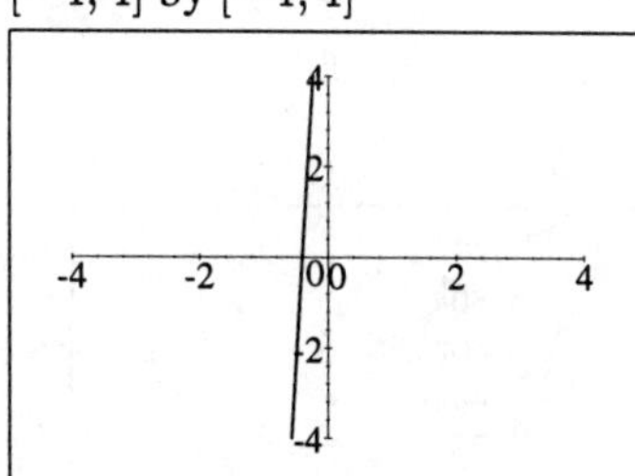

(b) $[-10, 10]$ by $[-10, 10]$

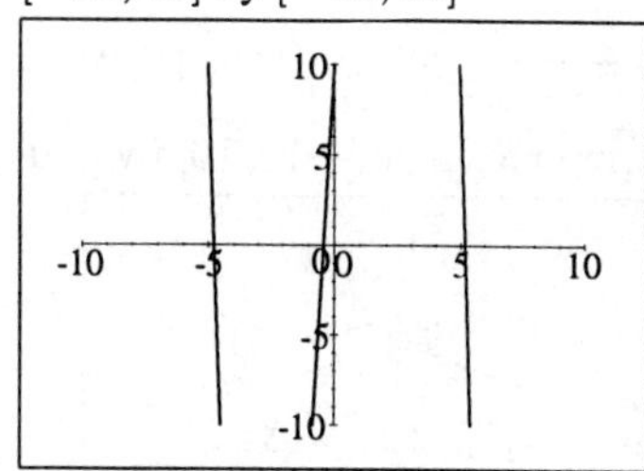

(c) $[-20, 20]$ by $[-100, 100]$

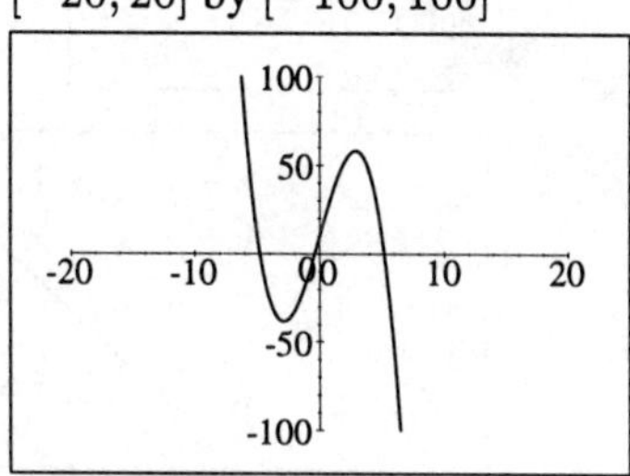

(d) $[-100, 100]$ by $[-200, 200]$

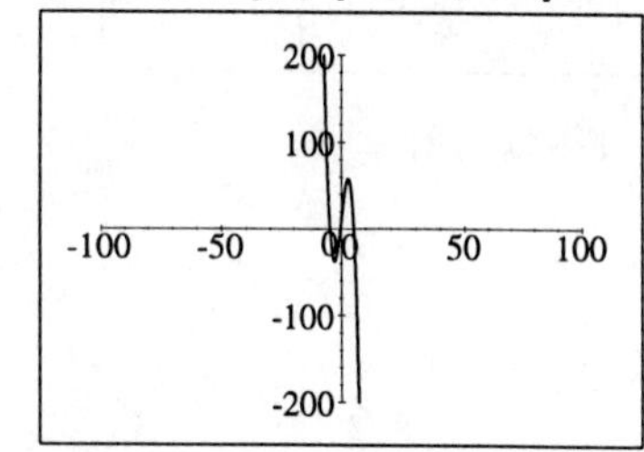

The viewing rectangle in part (c) produces the most appropriate graph of the equation.

7. $y = 100x^2$

$[-2, 2]$ by $[-10, 400]$

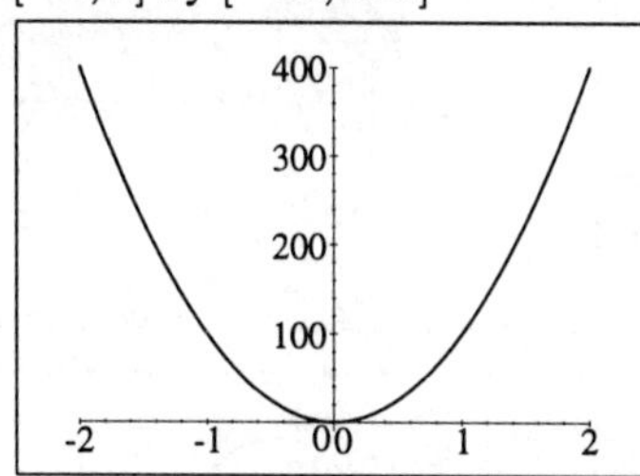

9. $y = 4 + 6x - x^2$

$[-4, 10]$ by $[-10, 20]$

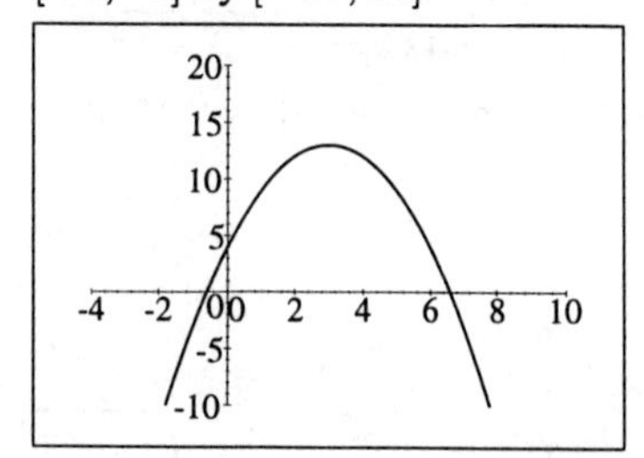

11. $y = \sqrt[4]{256 - x^2}$. We require that $256 - x^2 \geq 0 \quad \Rightarrow \quad -16 \leq x \leq 16$, so we graph $y = \sqrt[4]{256 - x^2}$ in the viewing rectangle $[-20, 20]$ by $[-1, 5]$.

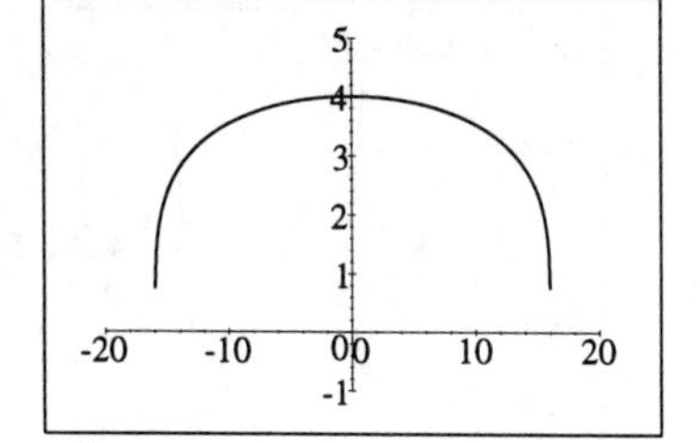

13. $y = 0.01x^3 - x^2 + 5$

$[-50, 150]$ by $[-2000, 2000]$

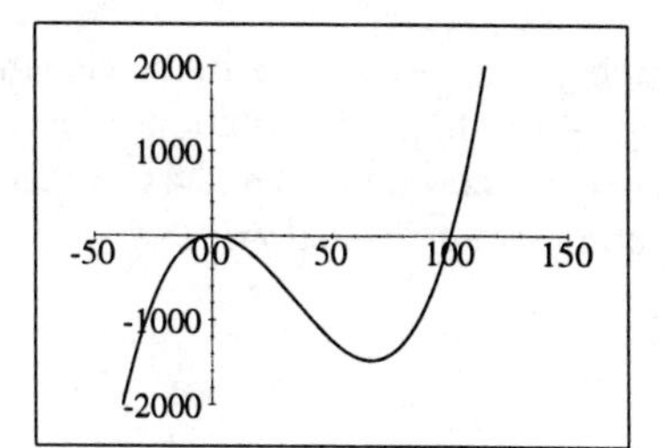

15. $y = \dfrac{1}{x^2+25}$. Since $x^2 + 25 \geq 25$

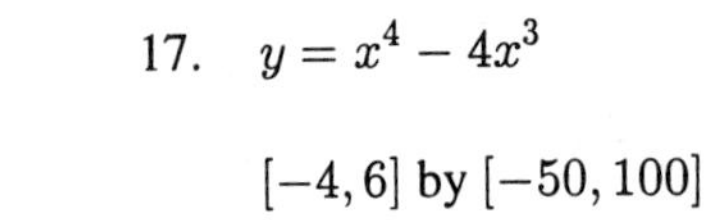

$0 < \dfrac{1}{x^2+25} \leq \dfrac{1}{25}$. Thus we use the viewing rectangle $[-10, 10]$ by $[-0.1, 0.1]$.

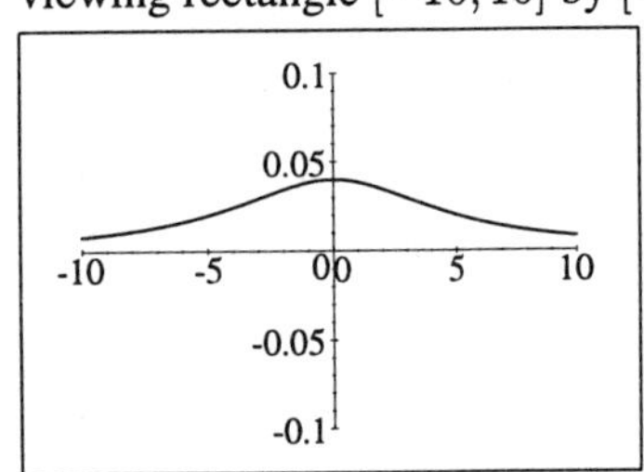

17. $y = x^4 - 4x^3$

$[-4, 6]$ by $[-50, 100]$

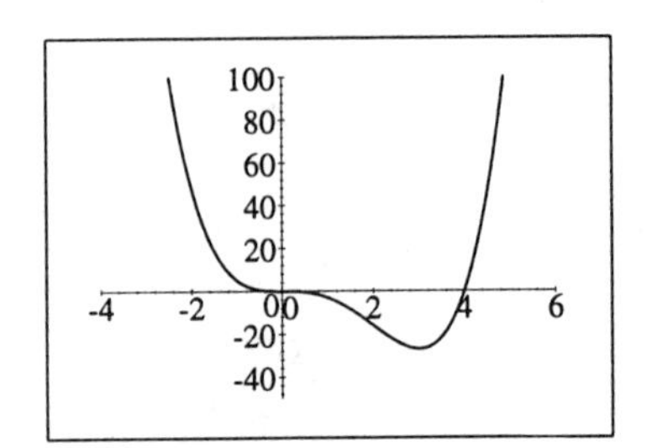

19. $y = 1 + |x - 1|$
$[-3, 5]$ by $[-1, 5]$

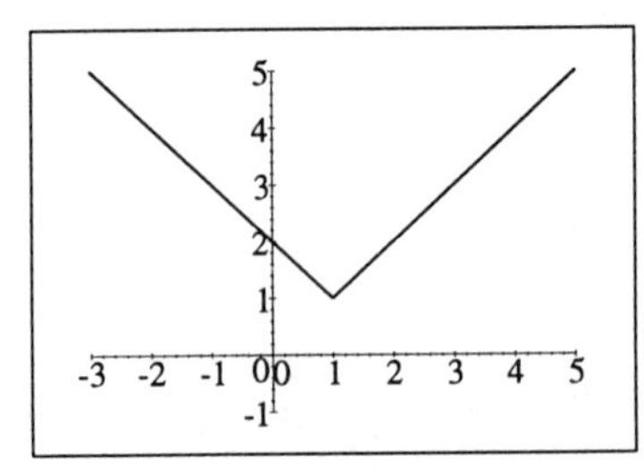

21. $x^2 + y^2 = 9 \quad \Leftrightarrow \quad y^2 = 9 - x^2 \quad \Rightarrow$
$y = \pm\sqrt{9 - x^2}$. So we graph the functions
$y_1 = \sqrt{9 - x^2}$ and $y_2 = -\sqrt{9 - x^2}$.
Viewing rectangle: $[-6, 6]$ by $[-4, 4]$.

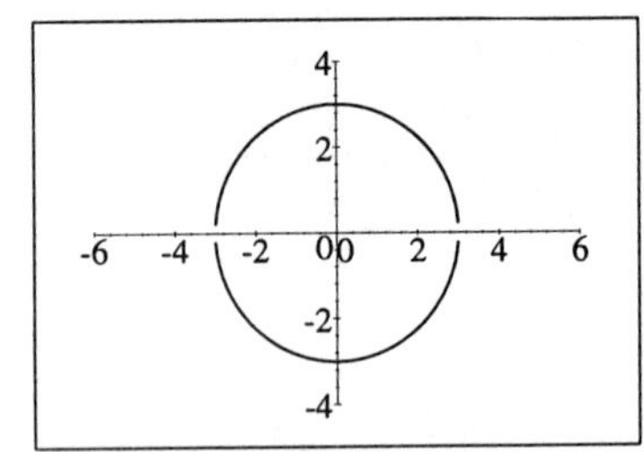

23. $4x^2 + 2y^2 = 1 \quad \Leftrightarrow \quad 2y^2 = 1 - 4x^2 \quad \Leftrightarrow$
$y^2 = \dfrac{1 - 4x^2}{2} \quad \Rightarrow \quad y = \pm\sqrt{\dfrac{1 - 4x^2}{2}}$. So we graph
the functions $y_1 = \sqrt{\dfrac{1 - 4x^2}{2}}$ and $y_2 = -\sqrt{\dfrac{1 - 4x^2}{2}}$.
Viewing rectangle: $[-1.2, 1.2]$ by $[-0.8, 0.8]$.

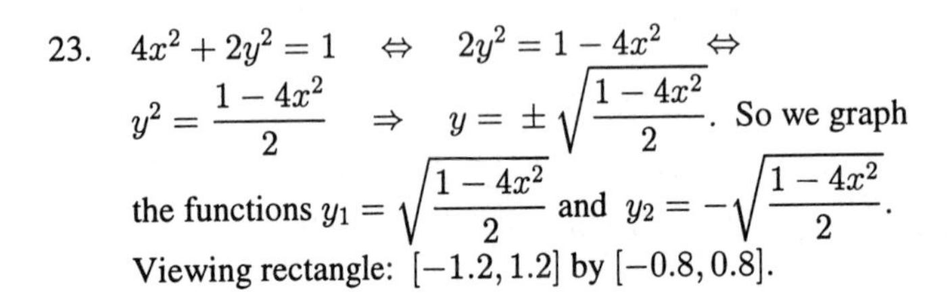

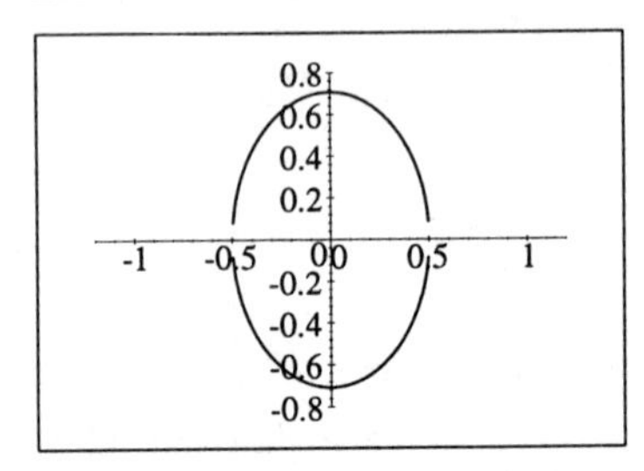

25. We graph $y = x^2 - 7x + 12$ in the viewing rectangle $[2, 5]$ by $[-0.1, 0.1]$. The solutions are: $x = 3.0$ and 4.0. Note: We can check that 3 and 4 are exact solutions by substituting into the equation.

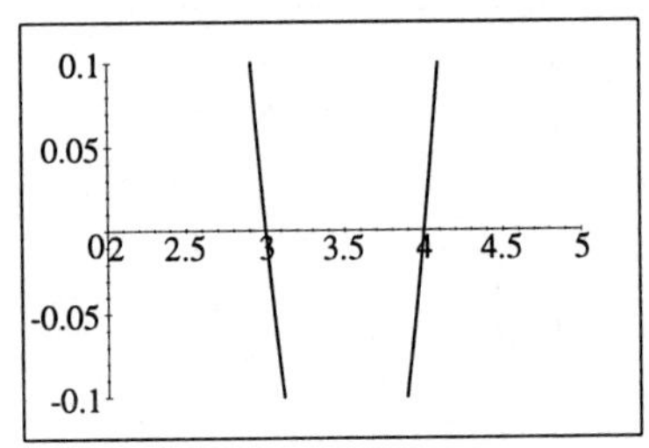

27. We graph $y = x^3 - 6x^2 + 11x - 6$ in the viewing rectangle $[-1, 4]$ by $[-0.1, 0.1]$. The solutions are: $x = 1.0$, 2.0, and 3.0. Note: We can check that 1, 2, and 3 are exact solutions by substituting into the equation.

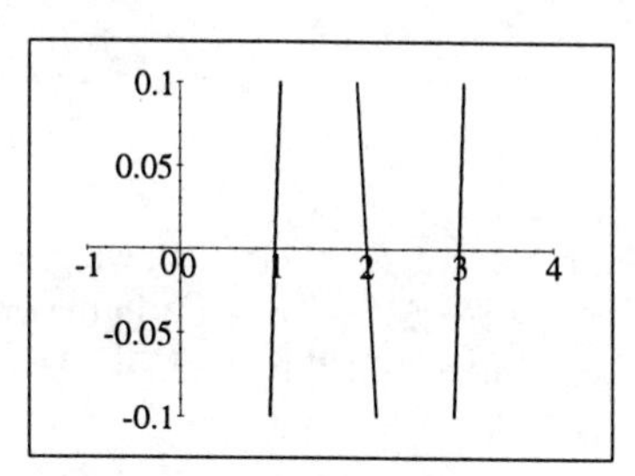

29. We first graph $y = x - \sqrt{x+1}$ in the viewing rectangle $[1.5, 2]$ by $[-0.1, 0.1]$ and find the solution is near 1.6. We isolate the solution in the second viewing rectangle, $[1.6, 1.7]$ by $[-0.01, 0.01]$, as $x \approx 1.62$.

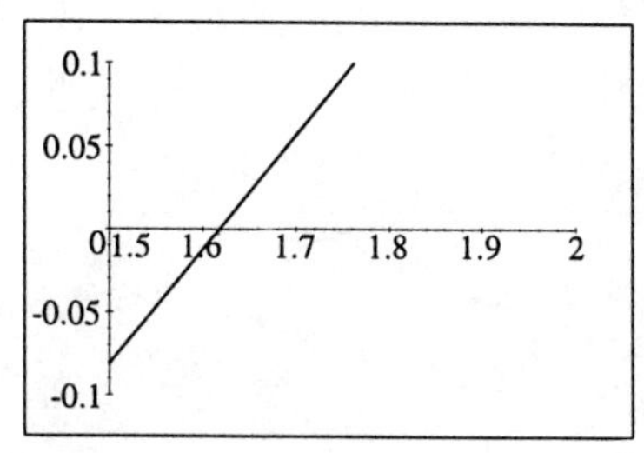

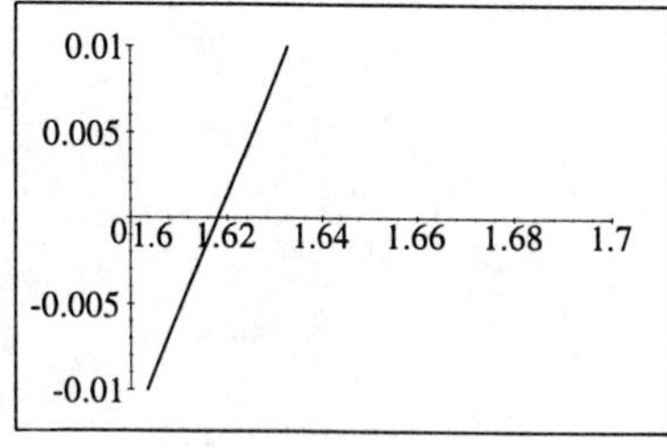

31. We graph $y = x^{1/3} - x$ in the viewing rectangle $[-3, 3]$ by $[-1, 1]$. The solutions are: $x = -1.0, 0.0, 1.0$. We can check that -1, 0, and 1 are exact solutions by substituting into the equation.

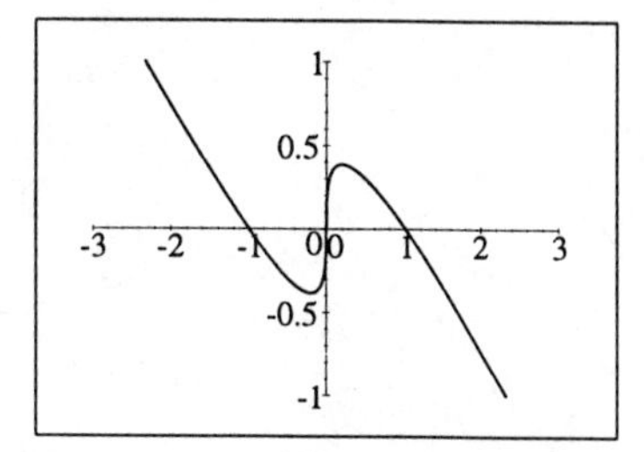

33. We graph $y = x^2 - 3x - 10$ in the viewing rectangle $[-6, 5]$ by $[-10, 2]$. Thus the solution to the inequality is: $[-2.0, 5.0]$.

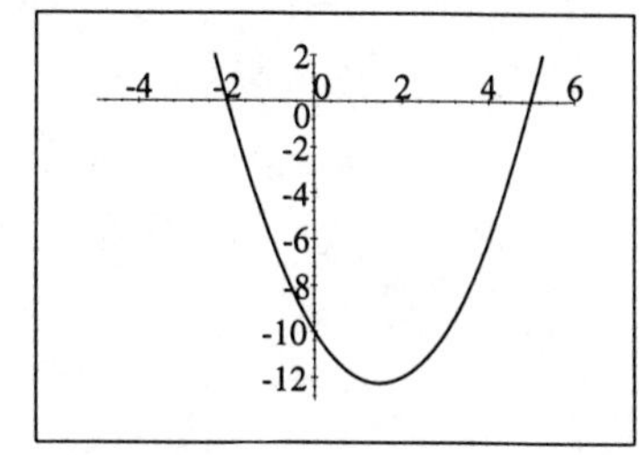

35. Since $x^3 + 11x \leq 6x^2 + 6 \quad \Leftrightarrow$ $x^3 - 6x^2 + 11x - 6 \leq 0$, we graph $y = x^3 - 6x^2 + 11x - 6$ in the viewing rectangle $[0, 5]$ by $[-5, 5]$. The solution set is $(-\infty, 1.0] \cup [2.0, 3.0]$.

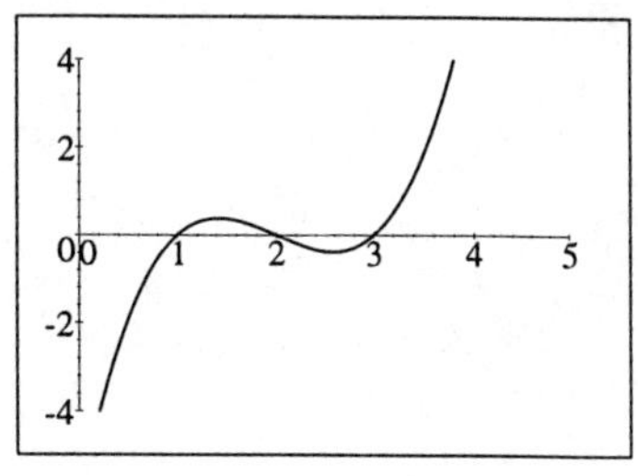

37. Since $x^{1/3} \le x \quad \Leftrightarrow \quad x^{1/3} - x < 0$. In Exercise 31, we graphed $y = x^{1/3} - x$ in the viewing rectangle $[-3, 3]$ by $[-1, 1]$. From this we find the solution set is $(-1.0, 0) \cup (1.0, \infty)$.

39. Since $(x+1)^2 < (x-1)^2 \quad \Leftrightarrow$ $(x+1)^2 - (x-1)^2 < 0$, we graph $y = (x+1)^2 - (x-1)^2$ in the viewing rectangle $[-2, 2]$ by $[-5, 5]$. The solution set is $(-\infty, 0)$.

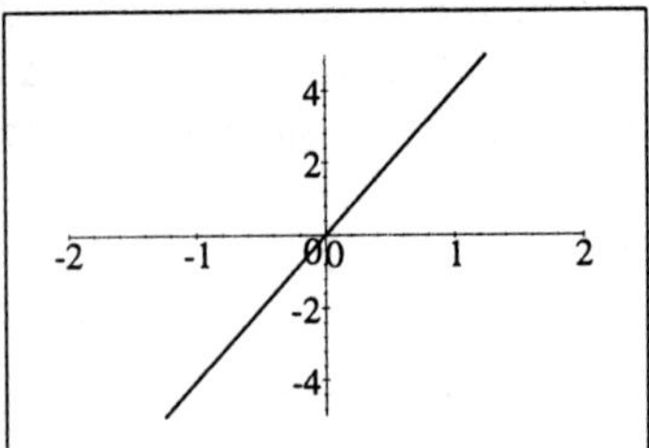

Exercises 3.4

1. $m = \dfrac{y_2 - y_1}{x_2 - x_1} = \dfrac{4 - 12}{2 - 4} = \dfrac{-8}{-2} = 4$

3. $m = \dfrac{y_2 - y_1}{x_2 - x_1} = \dfrac{3 - -6}{-1 - 1} = \dfrac{9}{-2} = -\dfrac{9}{2}$

5. $m = \dfrac{y_2 - y_1}{x_2 - x_1} = \dfrac{2 - 0}{2 - 0} = \dfrac{2}{2} = 1$

7. For ℓ_1 we find two points, $(-1, 2)$ and $(0, 0)$. Thus the slope of ℓ_1 is $m = \dfrac{y_2 - y_1}{x_2 - x_1} = \dfrac{2 - 0}{-1 - 0}$ $= -2$. For ℓ_2 we find two points, $(0, 2)$ and $(2, 3)$. Thus the slope of ℓ_2 is $m = \dfrac{y_2 - y_1}{x_2 - x_1} = \dfrac{3 - 2}{2 - 0}$ $= \frac{1}{2}$. For ℓ_3 we find the points $(2, -2)$ and $(3, 1)$. Thus the slope of ℓ_3 is $m = \dfrac{y_2 - y_1}{x_2 - x_1}$ $= \dfrac{1 - (-2)}{3 - 2} = 3$. For ℓ_4 we find the points $(-2, -1)$ and $(2, -2)$. Thus the slope of ℓ_4 is $m = \dfrac{y_2 - y_1}{x_2 - x_1} = \dfrac{-2 - (-1)}{2 - (-2)} = \dfrac{-1}{4}$.

9. First find two points, $(0, 4)$ and $(4, 0)$. So the slope is $m = \frac{0-4}{4-0} = -1$. Since the y-intercept is 4, the equation of the line is $y = mx + b = -1x + 4$ so $y = -x + 4$, or $x + y - 4 = 0$.

11. We cho0se the two intercepts points, $(0, -3)$ and $(2, 0)$. So the slope is $m = \frac{0-(-3)}{2-0} = \frac{3}{2}$. Since the y-intercept is -3, the equation of the line is $y = mx + b = \frac{3}{2}x - 3$, or $3x - 2y - 6 = 0$.

13. Using the equation $y - y_1 = m(x - x_1)$ we get $y - 3 = 1(x - 2) \quad \Leftrightarrow \quad -x + y = 1 \quad \Leftrightarrow$ $x - y + 1 = 0$.

15. Using the equation $y - y_1 = m(x - x_1)$ we get $y - 7 = \frac{2}{3}(x - 1) \quad \Leftrightarrow \quad 3y - 21 = 2x - 2 \quad \Leftrightarrow$ $-2x + 3y = 19 \quad \Leftrightarrow \quad 2x - 3y + 19 = 0$.

17. First find the slope: $m = \dfrac{y_2 - y_1}{x_2 - x_1} = \dfrac{6 - 1}{1 - 2} = \dfrac{5}{-1} = -5$. Substituting into $y - y_1 = m(x - x_1)$ we get: $y - 6 = -5(x - 1) \quad \Leftrightarrow \quad y - 6 = -5x + 5 \quad \Leftrightarrow \quad 5x + y - 11 = 0$.

19. Using $y = mx + b$ we have $y = 3x + (-2)$ or $3x - y - 2 = 0$.

21. We are given two points $(1, 0)$ and $(0, -3)$, thus the slope is $m = \dfrac{y_2 - y_1}{x_2 - x_1} = \dfrac{-3 - 0}{0 - 1} = \dfrac{-3}{-1} = 3$. Using the y-intercept we have $y = 3x + (-3)$ or $y = 3x - 3$ or $3x - y - 3 = 0$.

23. Since the equation of a horizontal line passing through (a, b) is $y = b$, the equation of a horizontal line passing through $(4, 5)$ is $y = 5$.

25. Since $x + 2y = 6 \quad \Leftrightarrow \quad 2y = -x + 6 \quad \Leftrightarrow \quad y = -\frac{1}{2}x + 3$, the slope of this line is $-\frac{1}{2}$. Thus the line we seek is given by $y - (-6) = -\frac{1}{2}(x - 1) \quad \Leftrightarrow \quad 2y + 12 = -x + 1 \quad \Leftrightarrow$ $x + 2y + 11 = 0$.

27. Any line parallel to $x = 5$ will have undefined slope and be of the form $x = a$. Thus the equation of the line is $x = -1$.

29. First find the slope of $2x+5y+8=0$: $2x+5y+8=0 \quad \Leftrightarrow \quad 5y=-2x-8 \quad \Leftrightarrow$ $y=-\frac{2}{5}x-\frac{8}{5}$. So the slope of the line that is perpendicular to $2x+5y+8=0$ is $m=-\frac{1}{-2/5}=\frac{5}{2}$. The equation of the line we seek is $y-(-2)=\frac{5}{2}(x-(-1)) \quad \Leftrightarrow \quad 2y+4=5x+5 \quad \Leftrightarrow$ $5x-2y+1=0$.

31. First find the slope of the line through $(2,5)$ and $(-2,1)$: $m=\dfrac{1-5}{-2-2}=\dfrac{-4}{-4}=1$. The equation of the line we seek is $y-7=1(x-1) \quad \Leftrightarrow \quad x-y+6=0$.

33. (a)

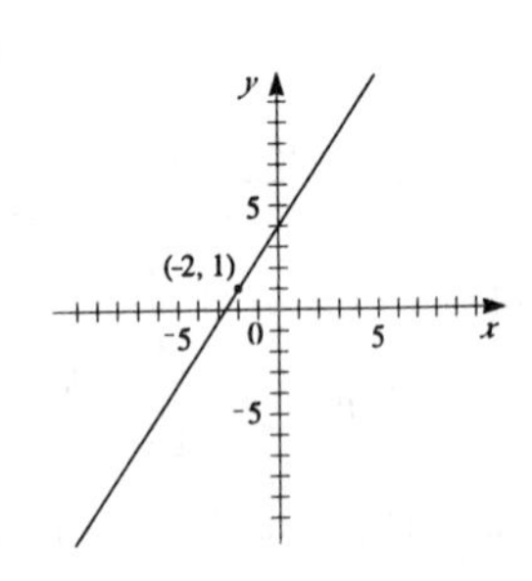

(b) $y-1=\frac{3}{2}(x-(-2)) \quad \Leftrightarrow \quad 2y-2=3(x+2) \quad \Leftrightarrow$
$2y-2=3x+6 \quad \Leftrightarrow \quad 3x-2y+8=0$

35.

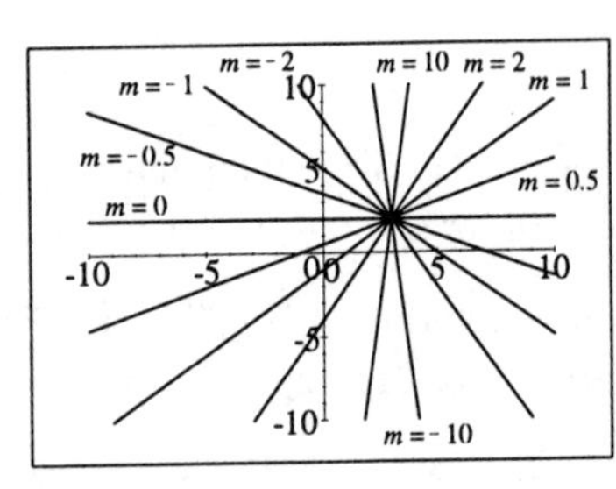

Each of the lines contains the point $(3,2)$ because $y=2+m(x-3) \quad \Leftrightarrow \quad y-2=m(x-3)$.

37. $x+y=3 \quad \Leftrightarrow \quad y=-x+3$. So the slope $=-1$ and the y-intercept $=3$.

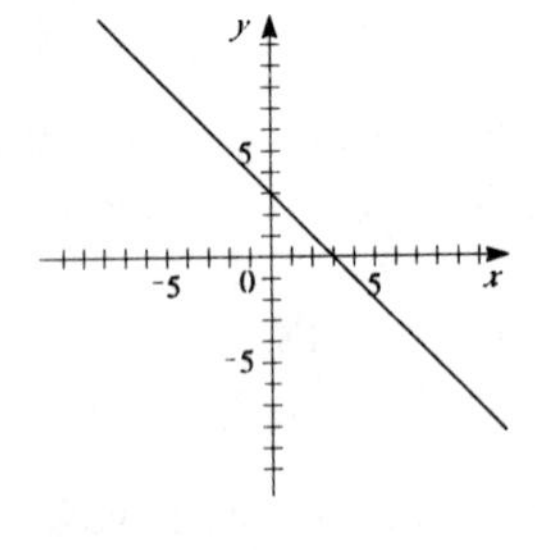

39. $x+3y=0 \quad \Leftrightarrow \quad 3y=-x \quad \Leftrightarrow$ $y=-\frac{1}{3}x$. So the slope $=-\frac{1}{3}$ and the y-intercept $=0$.

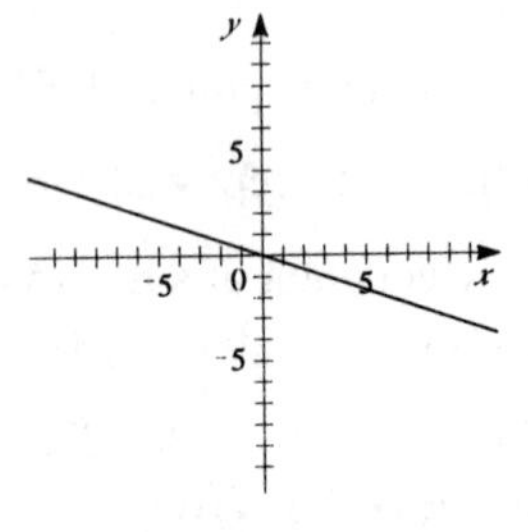

41. $\frac{1}{2}x - \frac{1}{3}y + 1 = 0 \quad \Leftrightarrow \quad -\frac{1}{3}y = -\frac{1}{2}x - 1$
$\Leftrightarrow \quad y = \frac{3}{2}x + 3$. So the slope $= \frac{3}{2}$ and the y-intercept $= 3$.

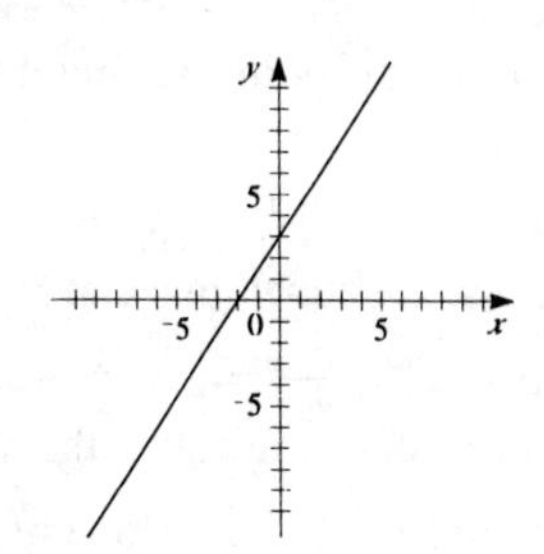

43. $y = 4$ can also be express as $y = 0x + 4$. So the slope $= 0$ and the y-intercept $= 4$.

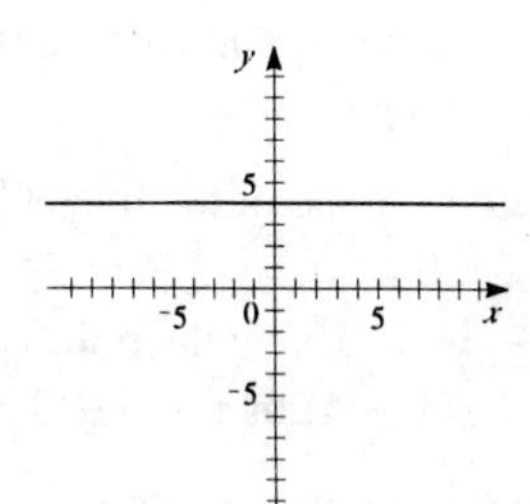

45. $3x - 4y = 12 \quad \Leftrightarrow \quad -4y = -3x + 12$
$\Leftrightarrow \quad y = \frac{3}{4}x - 3$. So the slope $= \frac{3}{4}$ and the y-intercept $= -3$.

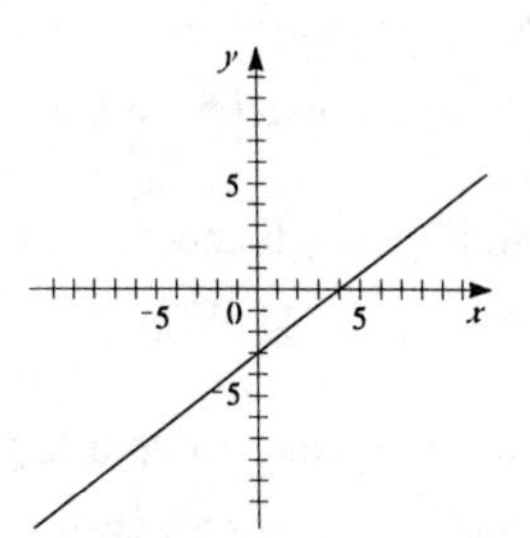

47. $3x + 4y - 1 = 0 \quad \Leftrightarrow \quad 4y = -3x + 1$
$\Leftrightarrow \quad y = -\frac{3}{4}x + \frac{1}{4}$. So the slope $= -\frac{3}{4}$ and the y-intercept $= \frac{1}{4}$.

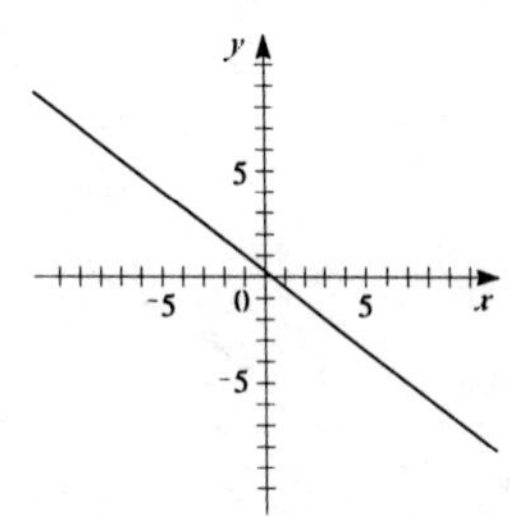

49. We first plot the points to find the pairs of points that determine each side. Next we find the slopes of opposite sides.

Slope of $AB = \frac{4-1}{7-1} = \frac{3}{6} = \frac{1}{2}$. Slope of $DC = \frac{10-7}{5-(-1)} = \frac{3}{6} = \frac{1}{2}$. Since these slope are equal, this pair of sides are parallel.

Slope of $AD = \frac{7-1}{-1-1} = \frac{6}{-2} = -3$. Slope of $BC = \frac{10-4}{5-7} = \frac{6}{-2} = -3$. Since these slope are equal, this pair of sides are parallel.

Hence $ABCD$ is a parallelogram.

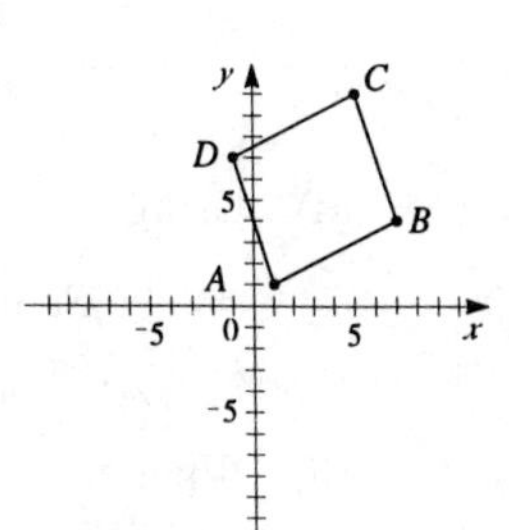

51. We first plot the points to find the pairs of points that determine each side. Next we find the slopes of opposite sides.

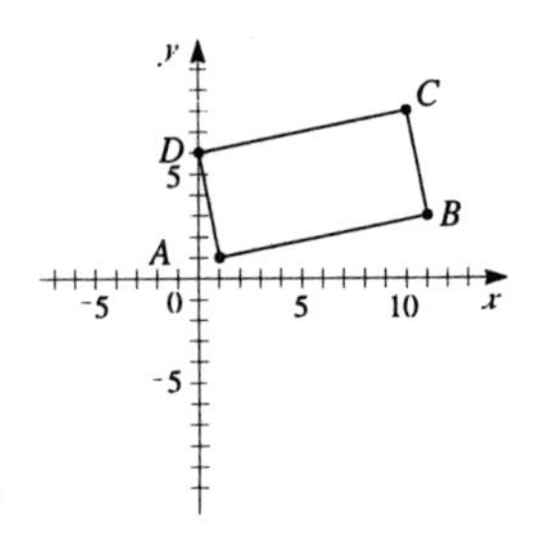

Slope of $AB = \frac{3-1}{11-1} = \frac{2}{10} = \frac{1}{5}$. Slope of $DC = \frac{6-8}{0-10} = \frac{-2}{-10} = \frac{1}{5}$. Since these slope are equal, this pair of sides are parallel.

Slope of $AD = \frac{6-1}{0-1} = \frac{5}{-1} = -5$. Slope of $BC = \frac{3-8}{11-10} = \frac{-5}{1} = -5$. Since these slope are equal, this pair of sides are parallel.

Since (slope of AB) $\times$ (slope of AD) $= \frac{1}{5} \times (-5) = -1$, the first pair of sides are each perpendicular to the second pair of sides. So the sides form a rectangle.

53. We need the slope of the line AB and the midpoint. Midpoint of $AB = \left(\frac{1+7}{2}, \frac{4-2}{2}\right) = (4, 1)$. The slope of AB is $m = \frac{-2-4}{7-1} = \frac{-6}{6} = -1$. The slope of the perpendicular bisector will have slope of $\frac{-1}{m} = \frac{-1}{-1} = 1$. Using the point-slope form, the equation of the perpendicular bisector is $y - 1 = 1(x - 4)$ or $x - y - 3 = 0$.

55. (a) We start with the two points $(a, 0)$ and $(0, b)$. The slope of the line they are on is $\frac{b-0}{0-a} = -\frac{b}{a}$. So the equation of the line containing them is $y = -\frac{b}{a}x + b$ (using the slope-intercept form). Dividing by b (since $b \neq 0$) gives $\frac{y}{b} = -\frac{x}{a} + 1 \quad \Leftrightarrow \quad \frac{x}{a} + \frac{y}{b} = 1$.

(b) Setting $a = 6$ and $b = -8$ we get $\frac{x}{6} + \frac{y}{-8} = 1 \quad \Leftrightarrow \quad 4x - 3y = 24 \quad \Leftrightarrow$ $4x - 3y - 24 = 0$.

57. Let h be the change in your horizontal distance, in feet. Then $-\frac{6}{100} = \frac{-1000}{h} \quad \Leftrightarrow$ $h = \frac{100{,}000}{6} = 16,667$. So the change in your horizontal distance is $16,667$ feet.

59. (a)

(b) The slope is the cost per toaster oven, \$6. The y-intercept, \$3000, is the monthly fixed cost, the cost that is incurred no matter how many toaster ovens are produced.

61. (a) Using n in place of x and t in place of y, we find the slope $= \frac{t_2 - t_1}{n_2 - n_1} = \frac{80 - 70}{168 - 120} = \frac{10}{48}$ $= \frac{5}{24}$. So the linear equations is $t - 80 = \frac{5}{24}(n - 168) \quad \Leftrightarrow \quad t - 80 = \frac{5}{24}n - 35 \quad \Leftrightarrow$ $t = \frac{5}{24}n + 45$.

(b) When $n = 150$, the temperature is approximately given by $t = \frac{5}{24}(150) + 45 = 76.25°\text{F}$ $\approx 76°\text{F}$

63. (a) We are given $\frac{\text{change in pressure}}{\text{10 feet change in depth}} = \frac{4.34}{10} = 0.434$. Using P for pressure and d for depth with the point $P = 15$ when $d = 0$, we have $P - 15 = 0.434(d - 0) \quad \Leftrightarrow$ $P = 0.434d + 15$.

(b) When $P = 100$, then $100 = 0.434d + 15 \Leftrightarrow 0.434d = 85 \Leftrightarrow d = 195.9$ feet. Thus the pressure is 100 lb/in^3 at a depth of approximately 196 feet.

65. (a) Using D in place of x and C in place of y, we find the slope $= \dfrac{C_2 - C_1}{D_2 - D_1} = \dfrac{460 - 380}{800 - 480} = \dfrac{80}{320}$ $= \frac{1}{4}$. So the linear equations is $C - 460 = \frac{1}{4}(D - 800) \Leftrightarrow C - 460 = \frac{1}{4}D - 200 \Leftrightarrow$ $C = \frac{1}{4}D + 260$.

(b) Substituting $D = 1500$ we get $C = \frac{1}{4}(1500) + 260 = 635$. Cost of driving 1500 miles is \$635.

(c)

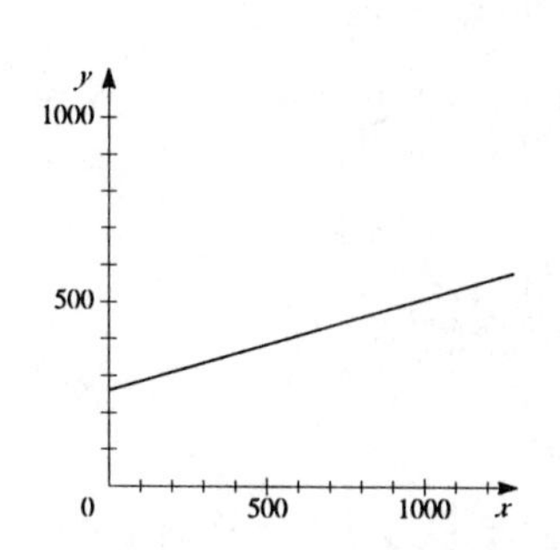

The slope of the line represents the cost per mile, \$0.25.

(d) The y-intercept represents the fixed cost, \$260.

(e) Because you have fixed monthly costs such as insurance and car payments, as well as costs that occur as you drive, such as, gasoline, oil, tires, etc., and the cost of these for each additional mile driven is a constant.

67. (a) Viewing rectangle is $[0, 50]$ by [0,30].

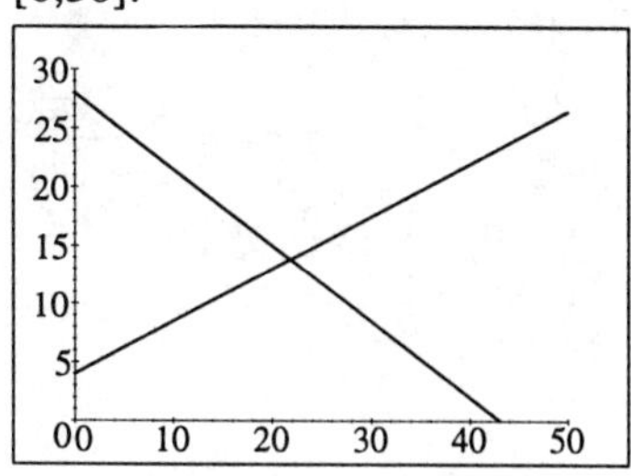

(b) $(p, y) = (21.82, 13.82)$

(c) Setting the two equations equal to each other we have $-0.65p + 28 = 0.45p + 4 \Leftrightarrow 24 = 1.1p \Leftrightarrow$ $p = 21.8181...$ So the price is \$21.82. Substituting we find $y = 0.45(21.82) + 4 = 13.82$.

Exercises 3.5

1. $f(x) = 3x + 1$

3. $f(x) = (x + 2)^2$

5. Divide by 3 then subtract 5

7. Square, multiply by 2, then subtract 3

9. Machine diagram for $f(x) = \sqrt{x}$. Arrow diagram for $f(x) = \sqrt{x}$.

0 square root 0
(input) (output)

1 square root 1

4 square root 2

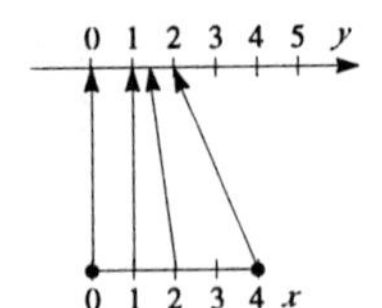

11. $f(1) = 2(1) + 1 = 3; \quad f(-2) = 2(-2) + 1 = -3; \quad f(\frac{1}{2}) = 2(\frac{1}{2}) + 1 = 2;$
$f(\sqrt{5}) = 2(\sqrt{5}) + 1 = 2\sqrt{5} + 1; \quad f(a) = 2(a) + 1 = 2a + 1;$
$f(-a) = 2(-a) + 1 = -2a + 1; \quad f(a + b) = 2(a + b) + 1 = 2a + 2b + 1.$

13. $g(2) = \dfrac{1-(2)}{1+(2)} = \dfrac{-1}{3} = -\dfrac{1}{3}; \quad g(-2) = \dfrac{1-(-2)}{1+(-2)} = \dfrac{3}{-1} = -3; \quad g(\pi) = \dfrac{1-(\pi)}{1+(\pi)} = \dfrac{1-\pi}{1+\pi};$
$g(a) = \dfrac{1-(a)}{1+(a)} = \dfrac{1-a}{1+a}; \quad g(a-1) = \dfrac{1-(a-1)}{1+(a-1)} = \dfrac{1-a+1}{1+a-1} = \dfrac{2-a}{a};$
$g(-a) = \frac{1-(-a)}{1+(-a)} = \frac{1+a}{1-a}$

15. $f(0) = 2(0)^2 + 3(0) - 4 = -4; \quad f(2) = 2(2)^2 + 3(2) - 4 = 8 + 6 - 4 = 10;$
$f(\sqrt{2}) = 2(\sqrt{2})^2 + 3(\sqrt{2}) - 4 = 4 + 3\sqrt{2} - 4 = 3\sqrt{2};$
$f(1 + \sqrt{2}) = 2(1+\sqrt{2})^2 + 3(1 + \sqrt{2}) - 4 = 2(1 + 2\sqrt{2} + 2) + 3 + 3\sqrt{2} - 4$
$= 6 + 3 - 4 + 4\sqrt{2} + 3\sqrt{2} = 5 + 7\sqrt{2}; \quad f(-x) = 2(-x)^2 + 3(-x) - 4 = 2x^2 - 3x - 4;$
$f(x + 1) = 2(x + 1)^2 + 3(x + 1) - 4 = 2x^2 + 4x + 2 + 3x + 3 - 4 = 2x^2 + 7x + 1;$
$2f(x) = 2[2x^2 + 3x - 4] = 4x^2 + 6x - 8; \quad f(2x) = 2(2x)^2 + 3(2x) - 4 = 8x^2 + 6x - 4.$

17. $f(a) = 3(a) + 2 = 3a + 2; \quad f(a) + f(h) = (3a + 2) + (3h + 2) = 3a + 3h + 4;$
$f(a + h) = 3(a + h) + 2 = 3a + 3h + 2; \quad \dfrac{f(a+h) - f(a)}{h} = \dfrac{(3a + 3h + 2) - (3a + 2)}{h}$
$= \dfrac{3a + 3h + 2 - 3a - 2}{h} = \dfrac{3h}{h} = 3.$

19. $f(a) = 5; \quad f(a) + f(h) = 5 + 5 = 10; \quad f(a + h) = 5; \quad \dfrac{f(a+h) - f(a)}{h} = \dfrac{5-5}{h} = 0.$

21. $f(a) = 3 - 5a + 4a^2$; $f(a) + f(h) = (3 - 5a + 4a^2) + (3 - 5h + 4h^2)$
$= 6 - 5a - 5h + 4a^2 + 4h^2$; $f(a+h) = 3 - 5(a+h) + 4(a+h)^2$
$= 3 - 5a - 5h + 4(a^2 + 2ah + h^2) = 3 - 5a - 5h + 4a^2 + 8ah + 4h^2$;
$\frac{f(a+h) - f(a)}{h} = \frac{(3 - 5a - 5h + 4a^2 + 8ah + 4h^2) - (3 - 5a + 4a^2)}{h}$
$= \frac{3 - 5a - 5h + 4a^2 + 8ah + 4h^2 - 3 + 5a - 4a^2}{h} = \frac{-5h + 8ah + 4h^2}{h} = \frac{h(-5 + 8a + 4h)}{h}$
$= -5 + 8a + 4h$

23. $f(2+h) = 8(2+h) - 1 = 15 + 8h$; $f(x+h) = 8(x+h) - 1 = 8x + 8h - 1$;
$\frac{f(x+h) - f(x)}{h} = \frac{(8x + 8h - 1) - (8x - 1)}{h} = \frac{8h}{h} = 8.$

25. $f(2+h) = \frac{1}{2+h}$; $f(x+h) = \frac{1}{x+h}$;
$\frac{f(x+h) - f(x)}{h} = \frac{\frac{1}{x+h} - \frac{1}{x}}{h} = \frac{1}{h}\left(\frac{x}{x(x+h)} - \frac{x+h}{x(x+h)}\right) = \frac{1}{h}\left(\frac{-h}{x(x+h)}\right) = \frac{-1}{x(x+h)}$

27. $f(x) = 2x$. Since there is no restrictions, the domain is the set of real numbers. Also, since $f\left(\frac{c}{2}\right) = 2\left(\frac{c}{2}\right) = c$ the range is the set of real numbers.

29. $f(x) = 2x$. The domain is restricted by the exercise to $[-1, 5]$, so $-1 \le x \le 5 \quad \Leftrightarrow \quad -2 \le 2x \le 10 \quad \Leftrightarrow \quad -2 \le f(x) \le 10 \quad \Leftrightarrow \quad -2 \le y \le 10$. Thus the range is $[-2, 10]$.

31. $f(x) = 6 - 4x$. The domain is restricted by the exercise to $[-2, 3]$, so $-2 \le x \le 3 \quad \Leftrightarrow \quad 8 \ge -4x \ge -12 \quad \Leftrightarrow \quad 14 \ge 6 - 4x \ge -6 \quad \Leftrightarrow \quad -6 \le f(x) \le 14 \quad \Leftrightarrow \quad -6 \le y \le 14$. Thus the range is $[-6, 14]$.

33. $f(x) = 2 - x^2$. There is no restriction on the value of x, so the domain is the set of real numbers. Since $-x^2 \le 0 \quad \Leftrightarrow \quad 2 - x^2 \le 2 \quad \Leftrightarrow \quad f(x) \le 2$, the range is $(-\infty, 2]$.

35. $h(x) = \sqrt{2x - 5}$. Since the square root is defined as a real number only for non-negative numbers, we require that $2x - 5 \ge 0 \quad \Leftrightarrow \quad 2x \ge 5 \quad \Leftrightarrow \quad x \ge \frac{5}{2}$. So the domain is $\{x | x \ge \frac{5}{2}\}$. In interval notation the domain is $[\frac{5}{2}, \infty)$. Since the square root function is non-negative, the range of $h(x)$ is $[0, \infty)$.

37. $F(x) = 3 + \sqrt{1 - x^2}$. We require $1 - x^2 \ge 0 \quad \Leftrightarrow \quad 1 \ge x^2 \quad \Rightarrow \quad -1 \le x \le 1$, thus the domain is $[-1, 1]$. Next, find the range: $-1 \le x \le 1 \quad \Leftrightarrow \quad 0 \le x^2 \le 1 \quad \Leftrightarrow \quad -1 \le -x^2 \le 0 \quad \Leftrightarrow \quad 0 \le 1 - x^2 \le 1 \quad \Leftrightarrow \quad 0 \le \sqrt{1 - x^2} \le 1 \quad \Leftrightarrow \quad 3 \le 3 + \sqrt{1 - x^2} \le 4 \quad \Leftrightarrow \quad 3 \le F(x) \le 4$. Thus the range of $F(x)$ is $[3, 4]$.

39. $f(x) = \frac{1}{x - 3}$. Since the denominator cannot equal 0 we have $x - 3 \ne 0 \quad \Leftrightarrow \quad x \ne 3$. Thus the domain is $\{x | \, x \ne 3\}$. Another way of expressing this domain is $(-\infty, 3) \cup (3, \infty)$.

41. $f(x)=\dfrac{x+2}{x^2-1}$. Since the denominator cannot equal 0 we have $x^2-1\neq 0 \quad\Leftrightarrow\quad x^2\neq 1 \quad\Rightarrow$ $x\neq\pm 1$. Thus the domain is $\{x|\,x\neq\pm 1\}$. Another way of expressing this domain is $(-\infty,-1)\cup(-1,1)\cup(1,\infty)$.

43. $f(x)=\dfrac{x+2}{x^2+1}$. We require $x^2+1\geq 1$, the denominator never is 0, so the domain is the set of real numbers.

45. $f(x)=\sqrt{x-5}$. $x-5\geq 0 \quad\Leftrightarrow\quad x\geq 5$. Thus the domain is $\{x|\,x\geq 5\}$. The domain can also be expressed as $[5,\infty)$.

47. $f(t)=\sqrt[3]{t-1}$. Since the odd root is defined for all real numbers, the domain is the set of real numbers.

49. $g(x)=\dfrac{\sqrt{2+x}}{3-x}$. We require $2+x\geq 0$ and the denominator cannot equal 0. Now $2+x\geq 0$ $\Leftrightarrow\quad x\geq -2$, and $3-x\neq 0 \quad\Leftrightarrow\quad x\neq 3$. Thus the domain is $\{x|\,x\geq -2$ and $x\neq 3\}$, which can be expressed as $[-2,3)\cup(3,\infty)$.

51. $F(x)=\dfrac{x}{\sqrt{x-10}}$. Since the denominator cannot equal 0 and the input to the square root must be non-negative we have $x-10>0 \quad\Leftrightarrow\quad x>10$. Thus the domain is $\{x|\,x>10\}$. In interval notation this is $(10,\infty)$.

53. $G(x)=\sqrt{x}+\sqrt{1-x}$. Since the input to both the square roots must be non-negative, we have $x\geq 0$ and $1-x\geq 0$. Thus the domain is $\{x|\,0\leq x$ and $x\leq 1\}$. The domain in interval notation is $[0,1]$.

55. $f(x)=\sqrt{4x^2-1}$. Since the input to the square root must be non-negative we have $4x^2-1\geq 0$ $\Leftrightarrow\quad 4x^2\geq 1 \quad\Leftrightarrow\quad x^2\geq\frac{1}{4} \quad\Rightarrow\quad x\leq-\frac{1}{2}$ or $x\geq\frac{1}{2}$. Thus the domain is $\{x|\,x\leq-\frac{1}{2}$ or $x\geq\frac{1}{2}\}$. In interval notation the domain is $(-\infty,-\frac{1}{2}]\cup[\frac{1}{2},\infty)$.

57. $g(x)=\sqrt[4]{x^2-6x}$. Since the input to an even root must be non-negative we have $x^2-6x\geq 0$ $\Leftrightarrow\quad x(x-6)\geq 0$. Using the method of Section 2.7 we have

	$(-\infty,0)$	$(0,6)$	$(6,\infty)$
Sign of x	$-$	$+$	$+$
Sign of $x-6$	$-$	$-$	$+$
Sign of $x(x-6)$	$+$	$-$	$+$

Thus the domain is $(-\infty,0]\cup[6,\infty)$.

59. $\phi(x)=\sqrt{\dfrac{x}{\pi-x}}$. Since the input to a square root must be non-negative we have $\dfrac{x}{\pi-x}\geq 0$. Using the method of Section 2.7 we have

	$(-\infty,0)$	$(0,\pi)$	(π,∞)
Sign of x	$-$	$+$	$+$
Sign of $\pi-x$	$+$	$+$	$-$
Sign of $x(\pi-x)$	$-$	$+$	$-$

Also we must have $\pi-x\neq 0$or $x\neq\pi$. Thus the domain is $[0,\pi)$.

Exercises 3.6

1. (a) $f(-1) = 2$; $f(0) = 0$; $f(1) = 2$; $f(3) = 3$.

   (b) Domain: $[-3, 3]$. Range: $[0, 3]$.

   (c) Increasing: $[0, 3]$. Decreasing: $[-3, 0]$.

3. (a) $g(-4) = 3$; $g(-2) = 2$; $g(0) = -2$; $g(2) = 1$; $g(4) = 0$.

   (b) Domain: $[-4, 4]$. Range: $[-2, 3]$.

   (c) Increasing: $[0, 2]$. Decreasing: $[-4, 0]$ and $(2, 4]$.

5. (a) $f(0) = 3 > \frac{1}{2} = g(0)$

   (b) $f(-3) \approx -\frac{3}{2} < 2 = g(-3)$

   (c) $x = -2$ and $x = 2$.

7. The graphs in parts (a) and (c) are graphs of function of x.

9. The given graph is the graph of a function of x. Domain: $[-3, 2]$. Range: $[-2, 2]$.

11. The given graph is not the graph of a function of x.

13. (a) $f(x) = 1 - x$

   (b) Domain: All real numbers.

   (c) Decreasing for all real numbers.

15. (a) $f(x) = x^2 - 4x$

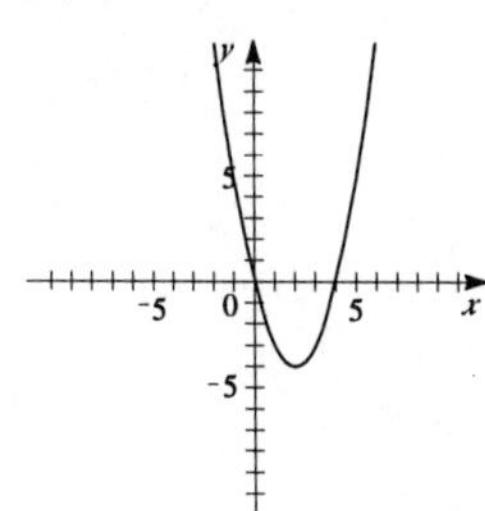

   (b) Domain: All real numbers.

   (c) Increasing: $[2, \infty)$. Decreasing: $(-\infty, 2]$.

17. (a) $f(x) = \sqrt{9 - x}$

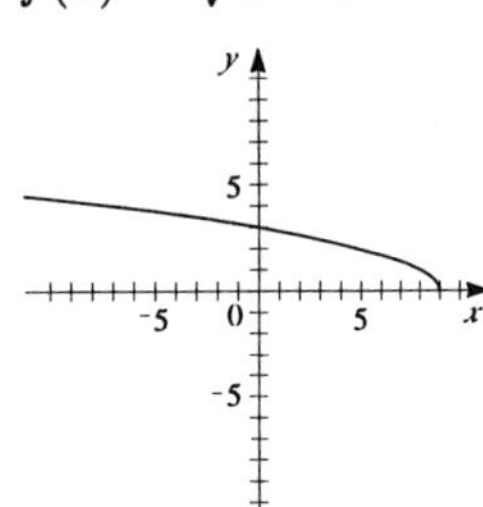

(b) We require $9 - x \geq 0 \quad \Leftrightarrow \quad 9 \geq x$. Domain: $(-\infty, 9]$.

(c) Decreasing: $(-\infty, 9]$.

19. (a) $f(x) = \sqrt{16 - x^2}$

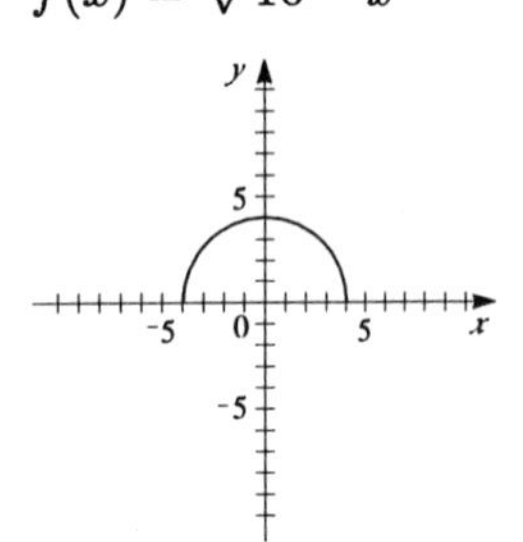

(b) We require $16 - x^2 \geq 0 \quad \Leftrightarrow \quad x^2 \leq 16 \quad \Leftrightarrow \quad |x| \leq 4 \quad \Leftrightarrow \quad -4 \leq x \leq 4$. Therefore the domain is $[-4, 4]$.

(c) Increasing: $[-4, 0]$. Decreasing: $[0, 4]$.

21. $f(x) = 3$

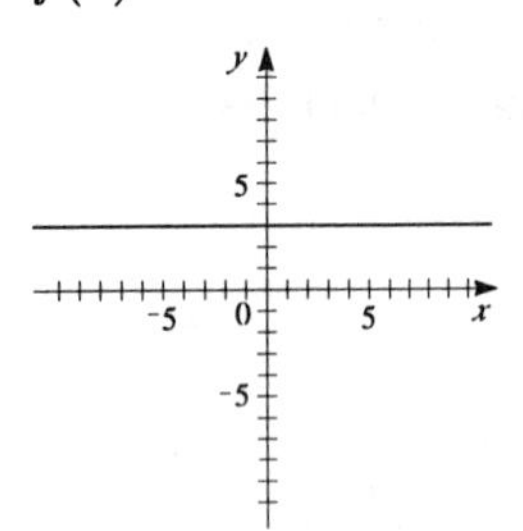

23. $f(x) = 2x + 3$

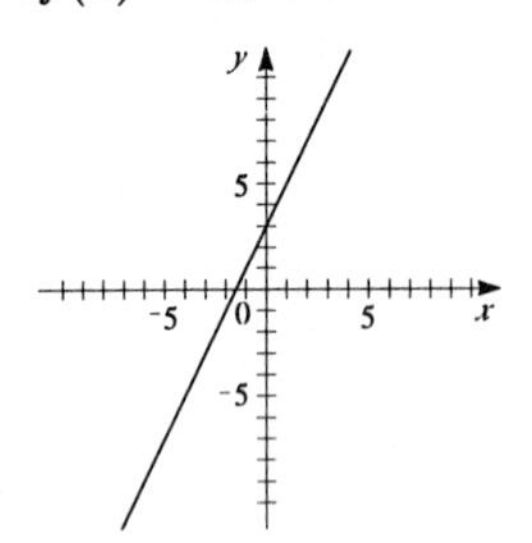

25. $f(x) = -x + 4, \; -1 \leq x \leq 4$

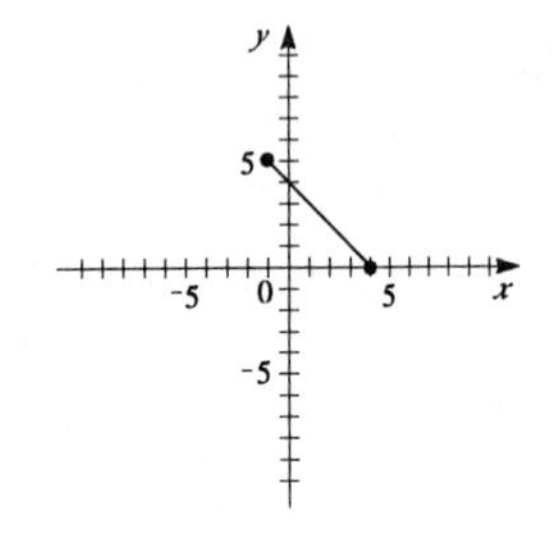

27. $f(x) = -x^2$

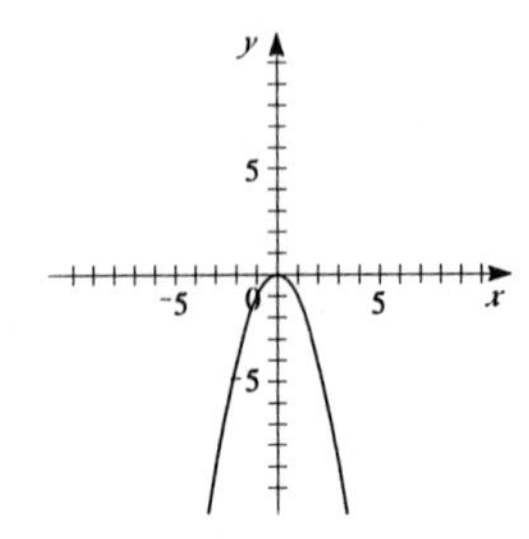

29. $f(x) = x^2 + 2x + 1$

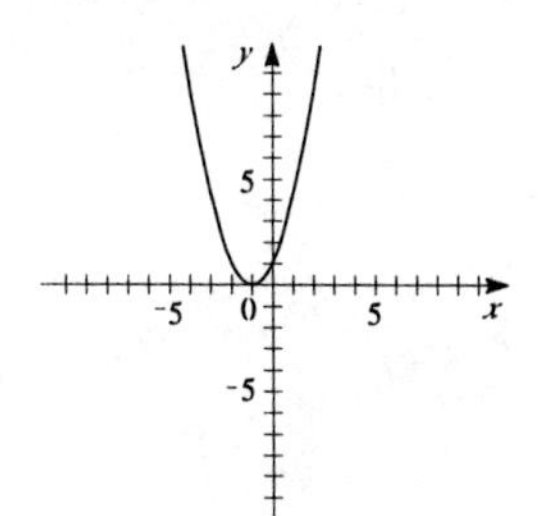

31. $g(x) = x^3 - 8$

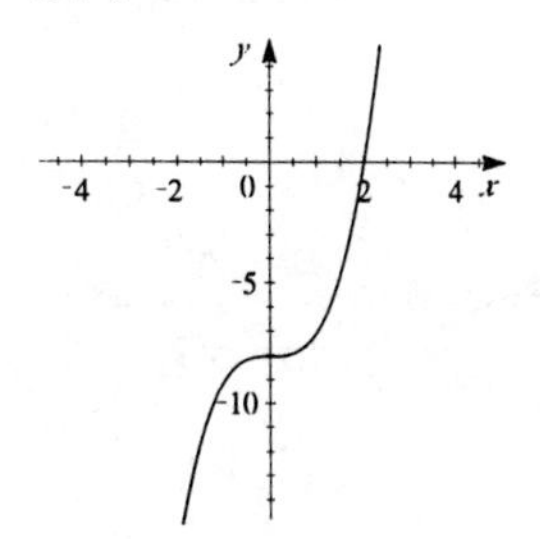

33. $g(x) = \sqrt{-x}$

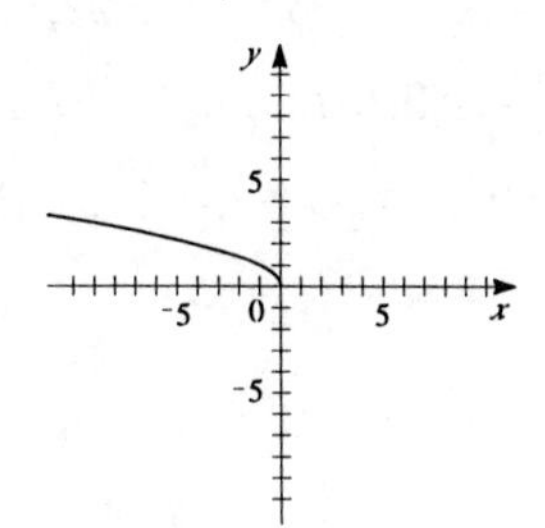

35. $F(x) = \frac{1}{x}$

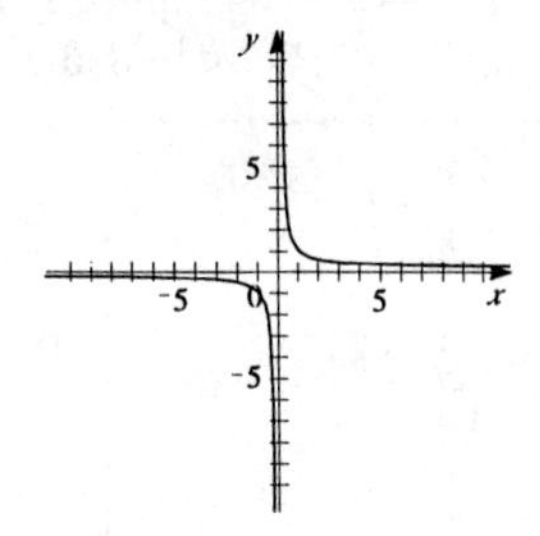

37. $H(x) = |2x|$

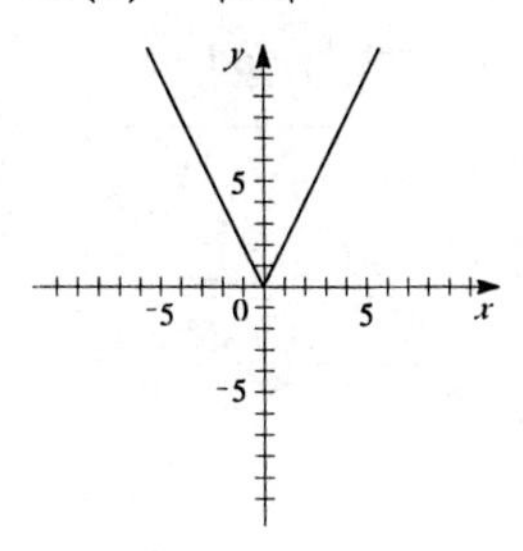

39. $G(x) = |x| + x$

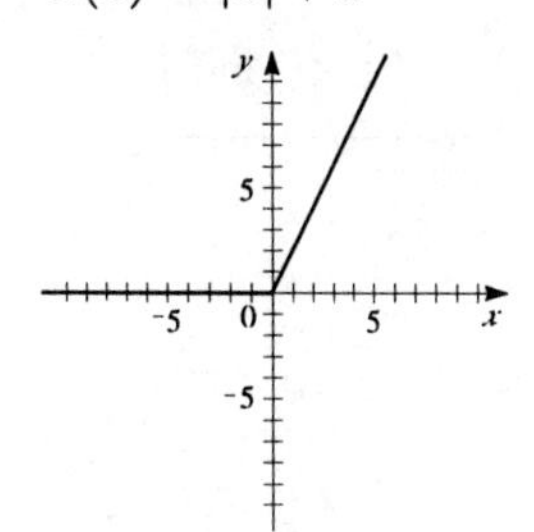

41. $f(x) = |2x - 2|$

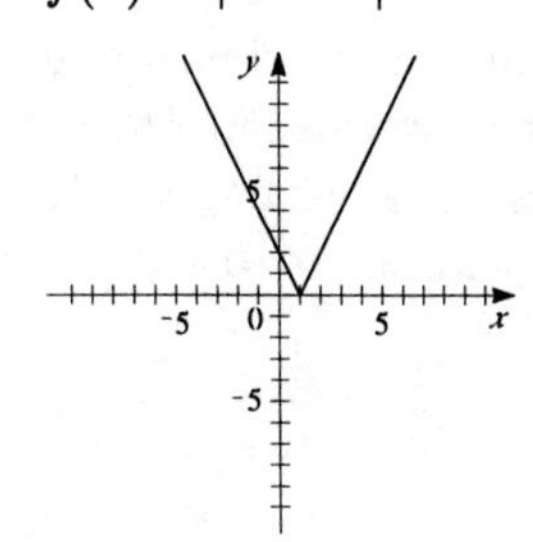

43. $f(x) = \dfrac{x^2 - 1}{x - 1} = \dfrac{(x+1)(x-1)}{x-1}$. Thus $f(x)$ is undefined at $x = 1$. But for $x \neq 1$ we have $f(x) = \dfrac{(x+1)(x-1)}{x-1} = x + 1$.

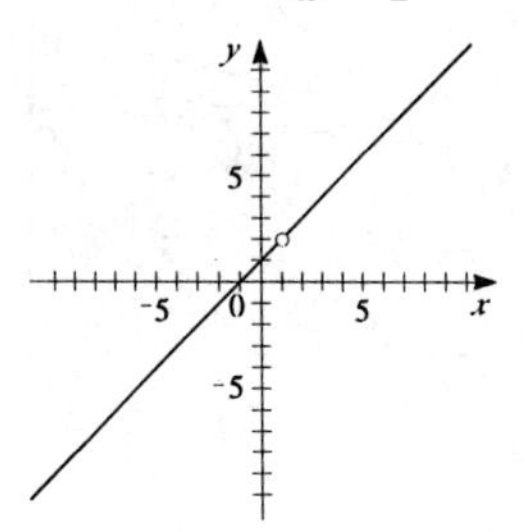

45. (a) $f(x) = x^{2/5}$ graphed in the viewing rectangle $[-10, 10]$ by $[-5, 5]$.

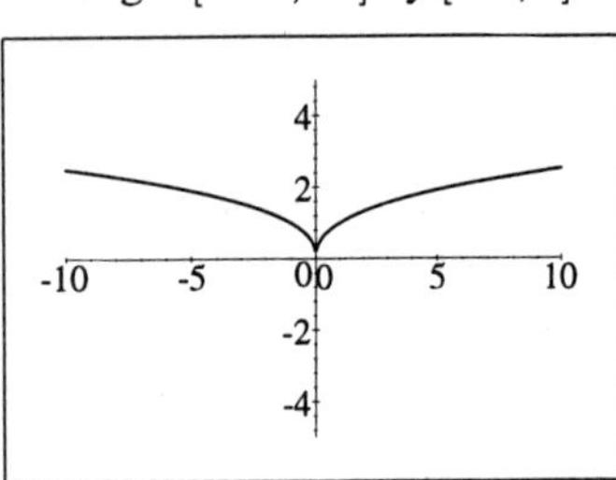

(b) Increasing: $[0, \infty)$.

Decreasing: $(-\infty, 0]$.

47. (a) $f(x) = x^3 + 2x^2 - x - 2$ graphed in the viewing rectangle $[-5, 5]$ by $[-3, 3]$.

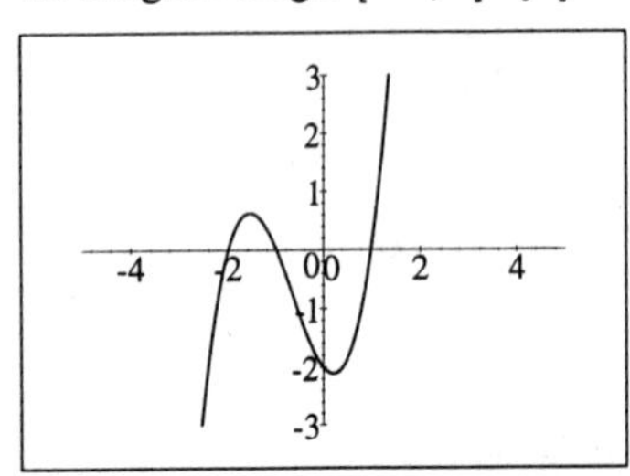

(b) Increasing: $(-\infty, -1.55]$ and $[0.22, \infty)$.

Decreasing: $[-1.55, 0.22]$.

49. (a) $f(x) = x^c$ for $c = \frac{1}{2}, \frac{1}{4}$, and $\frac{1}{6}$.

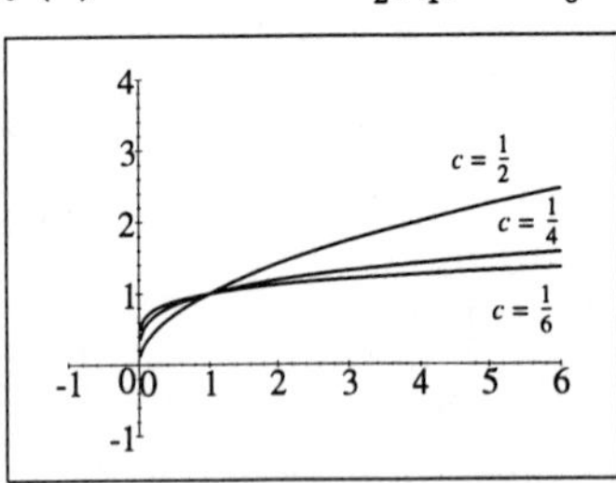

(b) $f(x) = x^c$ for $c = 1, \frac{1}{3}$, and $\frac{1}{5}$.

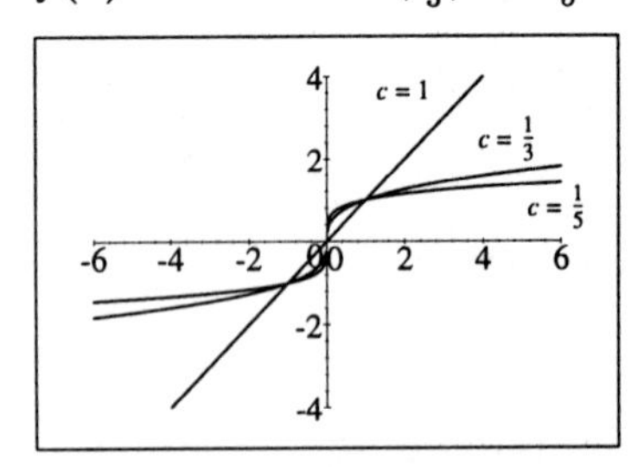

(c) $f(x) = x^c$ for $c = \frac{1}{2}, \frac{1}{3}, \frac{1}{4}$, and $\frac{1}{5}$.

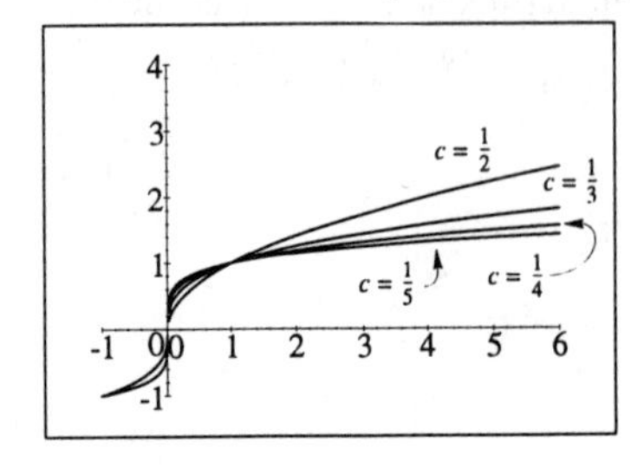

(d) Graphs of even roots are similar to $y = \sqrt{x}$, graphs of odd roots are similar to $y = \sqrt[3]{x}$. As c increases, the graph of $y = \sqrt[c]{x}$ becomes steeper near $x = 0$ and flatter when $x > 1$.

51. (a) $f(x) = x^2 + c$ for $c = 0, 2, 4$ and 6. (b) $f(x) = x^2 + c$ for $c = 0, -2, -4$, and -6.

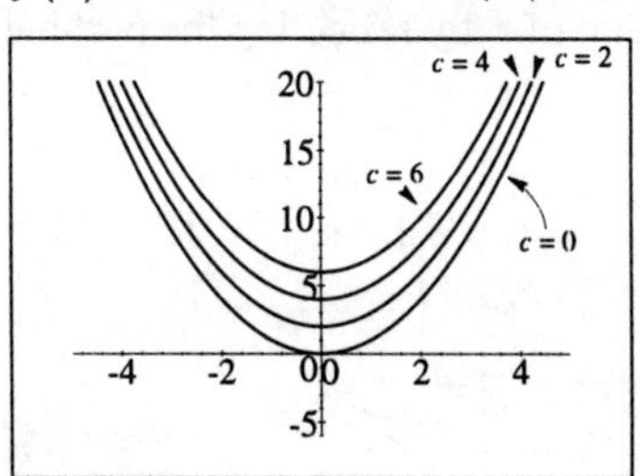

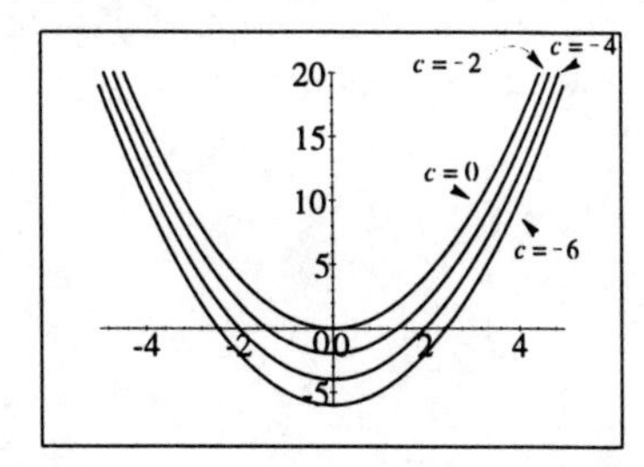

(c) The graphs in part (a) are obtained by shifting the graph of $f(x) = x^2$ upward c units, $c > 0$. While the graphs in part (b) are obtained by shifting the graph of $f(x) = x^2$ downward c units, $c < 0$.

53. (a) $f(x) = (x - c)^3$ for $c = 0, 2, 4$ and 6. (b) $f(x) = (x - c)^3$ for $c = 0, -2, -4$, and -6.

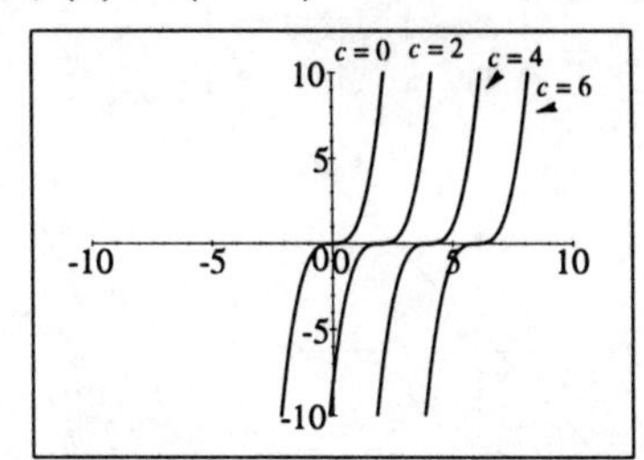

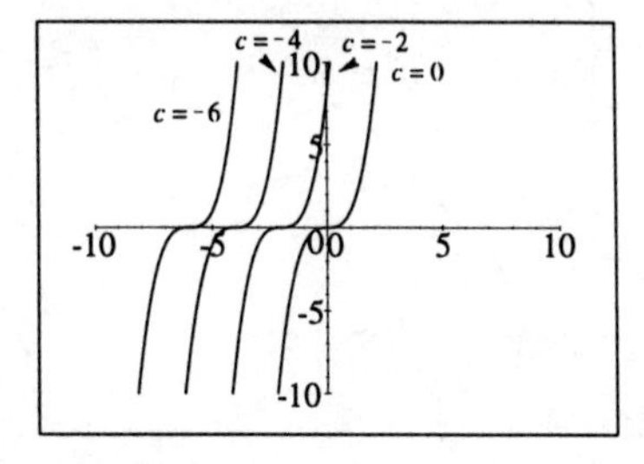

(c) The graph in part (a) are obtained by shifting the graph of $f(x) = x^3$ right c units, $c > 0$. The graphs in part (b) are obtained by shifting the graph of $f(x) = x^3$ left c units, $c < 0$.

55. The viewing rectangle $[0, 120]$ by $[0, 200000]$ is shown. The function $g(x) = \dfrac{x^3}{10}$ starts out lower than the function $f(x) = 10x^2$ but by $x = 100$, $g(x)$ catches $f(x)$ and is eventually larger.

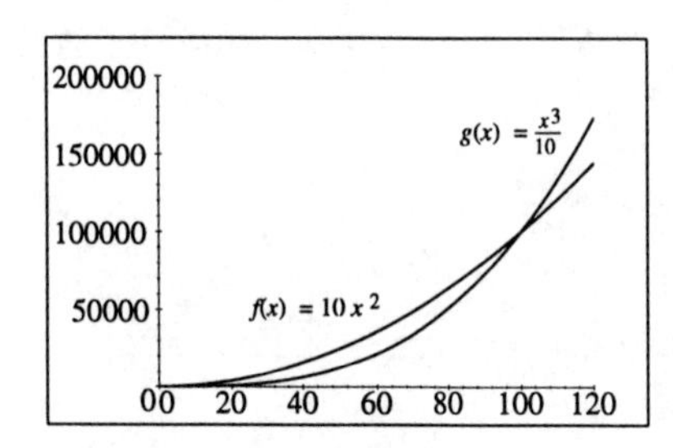

57. (a) The graphs of $f(x) = x^2 + x - 6$ and $g(x) = |x^2 + x - 6|$ are shown in the viewing rectangle $[-10, 10]$ by $[-10, 10]$.

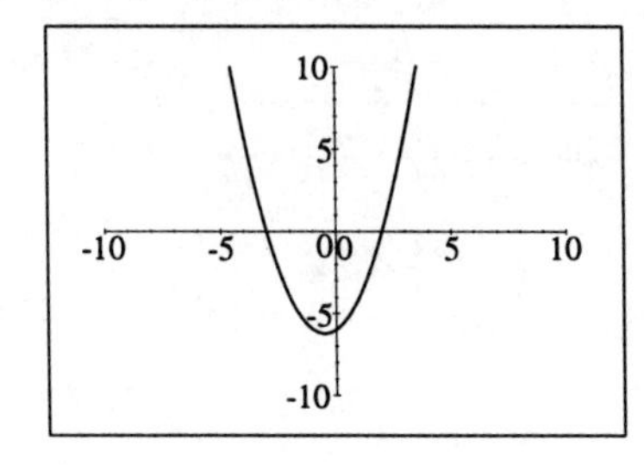

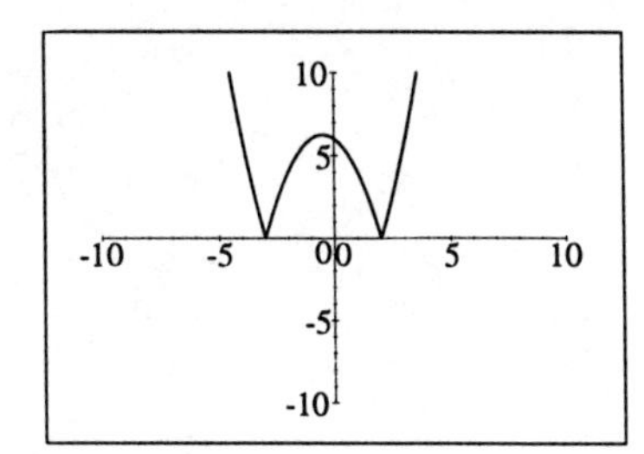

(b) For those values of x where $f(x) \geq 0$, the graphs of f and g coincide, and for those values of x where $f(x) < 0$, the graph of g is obtained from that of f by reflecting the part below the x-axis about the x-axis.

59. $f(x) = \begin{cases} 0 & \text{if } x < 2 \\ 1 & \text{if } x \geq 2 \end{cases}$

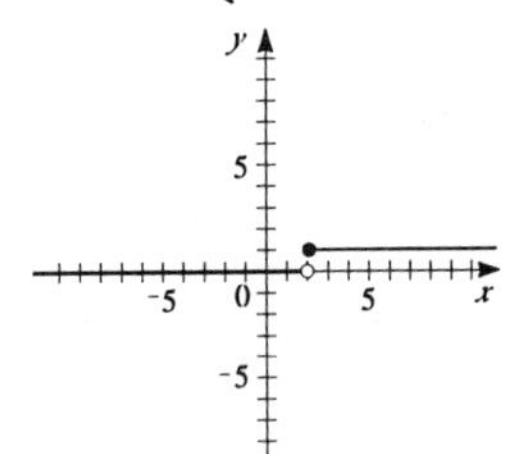

61. $f(x) = \begin{cases} 3 & \text{if } x < 2 \\ x - 1 & \text{if } x \geq 2 \end{cases}$

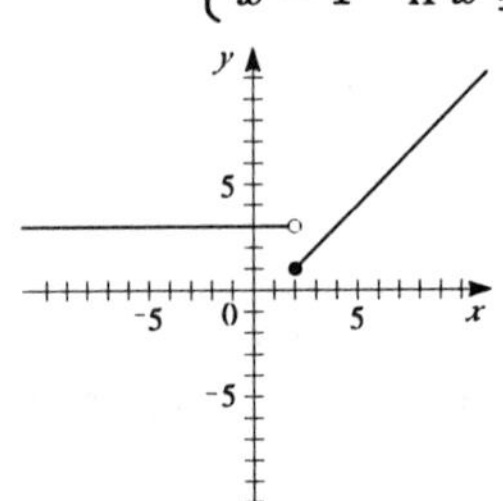

63. $f(x) = \begin{cases} x & \text{if } x \leq 0 \\ x + 1 & \text{if } x > 0 \end{cases}$

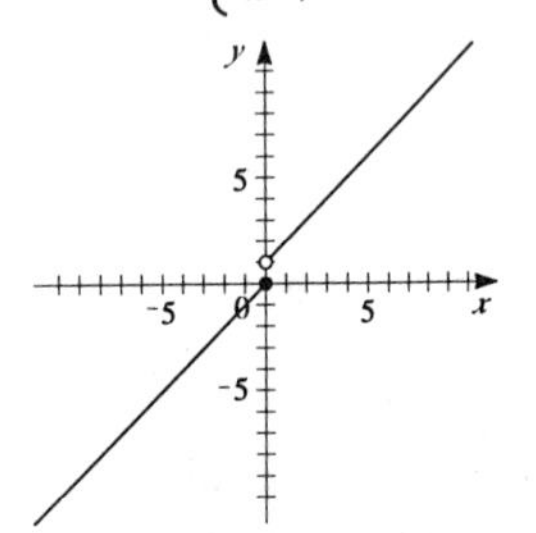

65. $f(x) = \begin{cases} x + 1 & \text{if } x \neq 1 \\ 1 & \text{if } x = 1 \end{cases}$

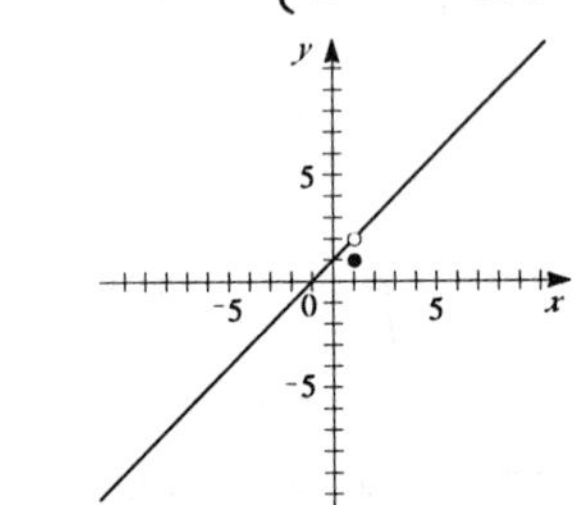

67. $f(x) = \begin{cases} -1 & \text{if } x < -1 \\ 1 & \text{if } -1 \leq x \leq 1 \\ -1 & \text{if } x > 1 \end{cases}$

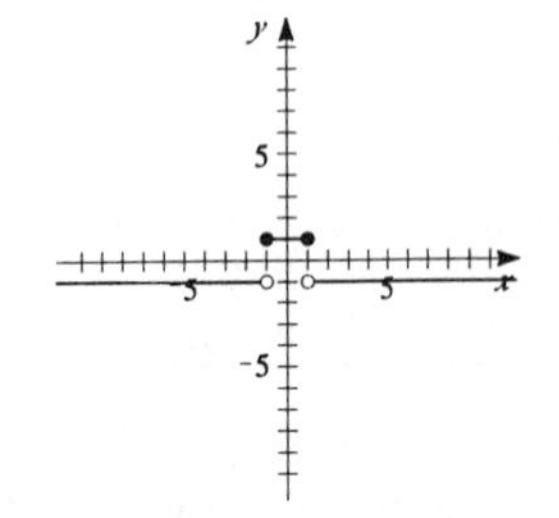

69. $f(x) = \begin{cases} 2 & \text{if } x \leq -1 \\ x^2 & \text{if } x > -1 \end{cases}$

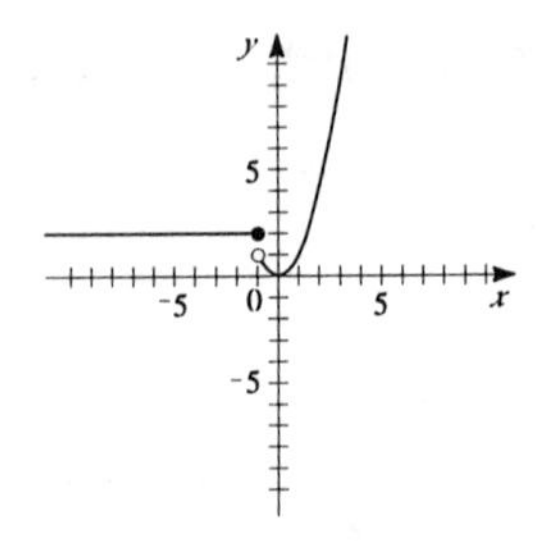

71. $f(x) = \begin{cases} 0 & \text{if } |x| \leq 2 \\ 3 & \text{if } |x| > 2 \end{cases}$

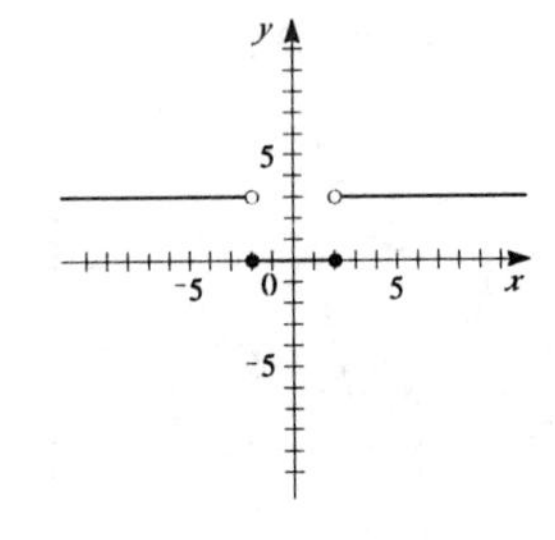

73. $f(x) = \begin{cases} x+2 & \text{if } x \le -1 \\ x^2 & \text{if } x > -1 \end{cases}$

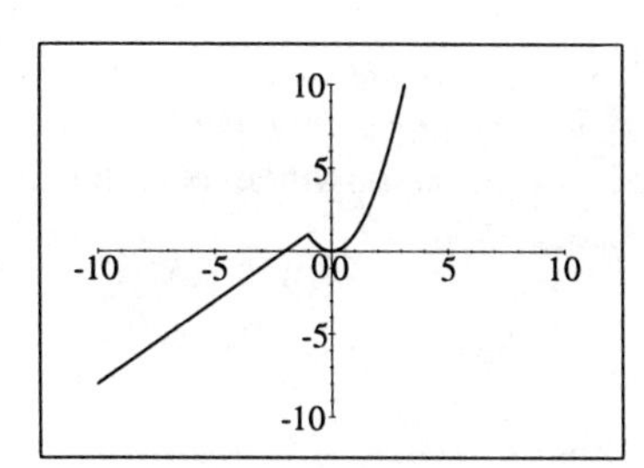

75. $f(x) = \begin{cases} x^3 - 2x + 1 & \text{if } x \le 0 \\ x - x^2 & \text{if } 0 < x < 1 \\ \sqrt[4]{x-1} & \text{if } x > 1 \end{cases}$

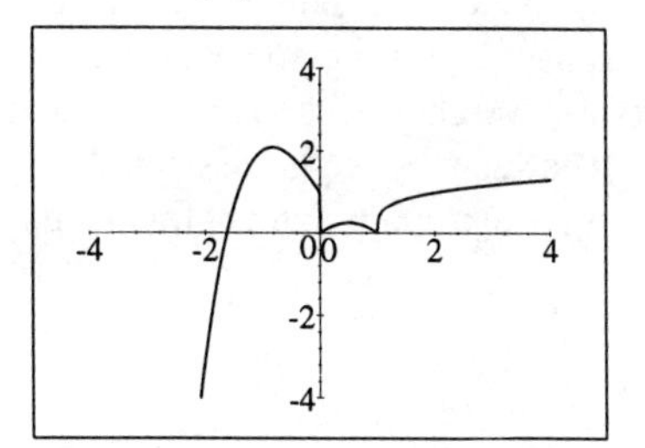

77. $C(x) = \begin{cases} 2.00 & \text{if } 0 < x \le 1 \\ 2.20 & \text{if } 1 < x \le 1.1 \\ 2.40 & \text{if } 1.1 < x \le 1.2 \\ \vdots & \\ 4.00 & \text{if } 1.9 < x < 2 \end{cases}$

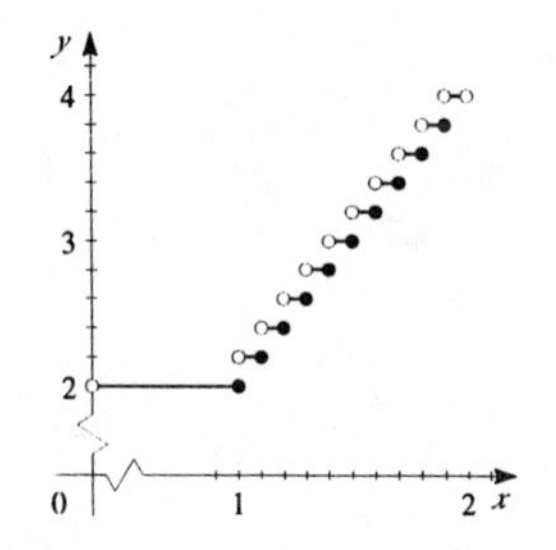

79. The slope of the line segment joining the points $(-2, 1)$ and $(4, -6)$ is $m = \frac{-6-1}{4-(-2)} = -\frac{7}{6}$. Using the point-slope form we have $y - 1 = -\frac{7}{6}(x+2) \quad \Leftrightarrow \quad y = -\frac{7}{6}x - \frac{7}{3} + 1 \quad \Leftrightarrow \quad y = -\frac{7}{6}x - \frac{4}{3}$. Thus the function is $f(x) = -\frac{7}{6}x - \frac{4}{3}$ for $-2 \le x \le 4$.

81. The parabola has equation $x + (y-1)^2 = 0 \quad \Leftrightarrow \quad (y-1)^2 = -x \quad \Rightarrow \quad y - 1 = \pm\sqrt{-x}$ Since we seek the bottom half of the parabola, we choose $y - 1 = -\sqrt{-x} \quad \Leftrightarrow y = 1 - \sqrt{-x}$. So the function is $f(x) = 1 - \sqrt{-x},\ x \le 0$.

Exercises 3.7

1. This person appears to gain weight steadily until the age of 21 when this person's weight gain slows down. At age 31, this person experiences a sudden weight loss, but recovers and by about age 40 this person's weight seems to level off at around 200 pounds. It then appears the person dies at about age 68. The sudden weight loss could be due to a number of reasons, among them is major illness, weight loss program, prison confinement (POW), etc.

3.

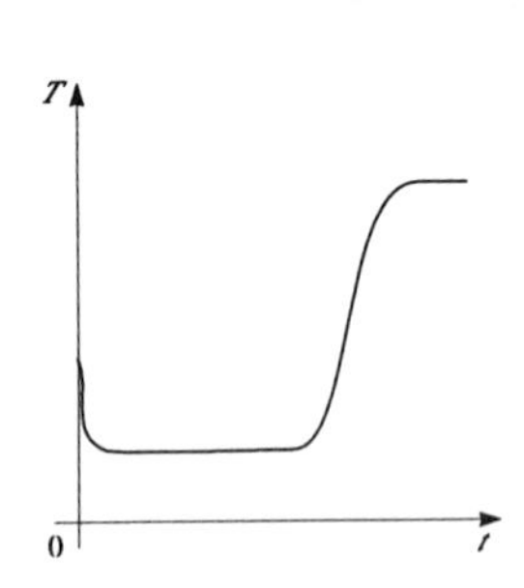

5.

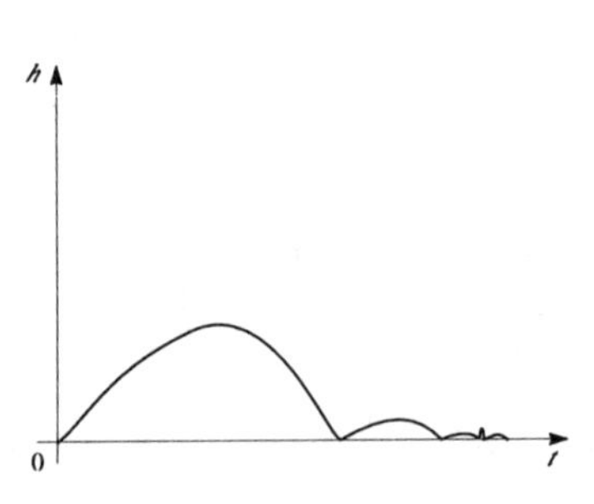

7.

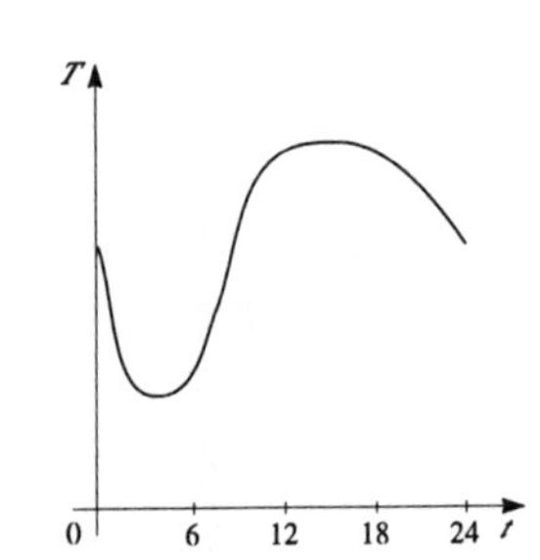

9. Let P be the perimeter of the rectangle and y be the length of the unknown side. Since $P = 2x + 2y$ and the perimeter is 20 we have $2x + 2y = 20 \quad \Leftrightarrow \quad x + y = 10 \quad \Leftrightarrow \quad y = 10 - x$. Since area is $A = xy$, substituting gives $A = x(10 - x) = 10x - x^2$. And since A must be positive, the domain is $0 < x < 10$.

11. Let h be the height of an altitude of the equilateral triangle whose side has length x, see diagram. Thus the area is given by $A = \frac{1}{2}xh$. By the Pythagorean Theorem, $h^2 + \left(\frac{1}{2}x\right)^2 = x^2 \quad \Leftrightarrow \quad h^2 + \frac{x^2}{4} = x^2 \quad \Leftrightarrow \quad h^2 = \frac{3x^2}{4} \quad \Leftrightarrow$ $h = \frac{\sqrt{3}x}{2}$. Substituting into the area of a triangle we get $A = \frac{1}{2}xh = \frac{1}{2}x\left(\frac{\sqrt{3}x}{2}\right) = \frac{\sqrt{3}x^2}{4}$, where $x > 0$.

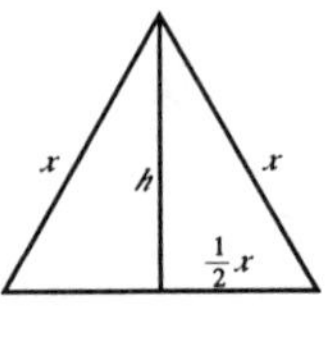

13. We solve the formula for the area of a circle for r. $A = \pi r^2 \quad \Leftrightarrow \quad r^2 = \frac{A}{\pi} \quad \Rightarrow \quad r = \sqrt{\frac{A}{\pi}}$, $A > 0$.

15. Let h be the height of the box in feet. The volume of the box is $V = 12 \quad \Rightarrow \quad x^2h = 12 \quad \Leftrightarrow$ $h = \frac{12}{x^2}$. The surface area, A, of the box is sum of the 4 sides and the base. Thus $A = 4xh + x^2$ $= 4x\left(\frac{12}{x^2}\right) + x^2 = \frac{48}{x} + x^2, x > 0$.

17. Let h be the height in feet of the straight portion of the window. The circumference of the semicircle is $C = \frac{1}{2}\pi x$. Since the perimeter of the window is 30 feet we have $x + 2h + \frac{1}{2}\pi x = 30$. Solving for h we get $2h = 30 - x - \frac{1}{2}\pi x \quad \Leftrightarrow \quad h = 15 - \frac{1}{2}x - \frac{1}{4}\pi x$. The area of the window is $A = xh + \frac{1}{2}\pi\left(\frac{1}{2}x\right)^2 = x\left(15 - \frac{1}{2}x - \frac{1}{4}\pi x\right) + \frac{1}{8}\pi x^2 = 15x - \frac{1}{2}x^2 - \frac{1}{4}\pi x^2 + \frac{1}{8}\pi x^2$ $= 15x - \frac{1}{2}x^2 - \frac{1}{8}\pi x^2 = 15x - \frac{1}{8}(\pi + 4)x^2$, $x > 0$.

19. Let l be the length of the field in feet. Since the length of the fencing is 2400 feet, we have $2x + l = 2400 \quad \Leftrightarrow \quad l = 2400 - 2x$. So the area of the fenced region is $A = xl = x(2400 - 2x)$ $= 2x(1200 - x)$ where $0 < x < 1200$.

21. Let d_1 be the distance traveled south by the first ship and d_2 be the distance traveled east by the second ship. Since the first ship travels south for t hours at 15 mi/h, $d_1 = 15t$ and, similarly, $d_2 = 20t$. Since the ships are traveling at right angles to each other, we can apply the Pythagorean Theorem to get $d^2 = d_1^2 + d_2^2 = (15t)^2 + (20t)^2 = 225t^2 + 400t^2 = 625t^2$. Thus $d = 25t$, where $t \geq 0$.

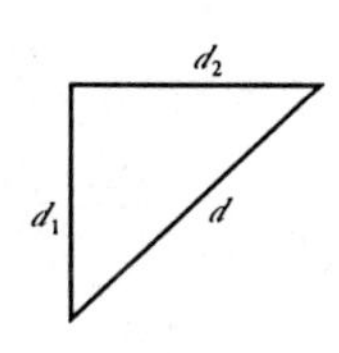

23. Viewing rectangle $[0, 10000]$ by $[0, 100]$.

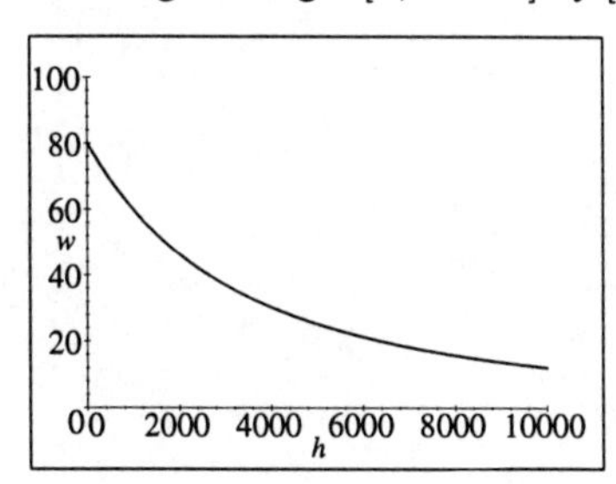

25. Viewing rectangle $[0, 20]$ by $[0, 40000]$.

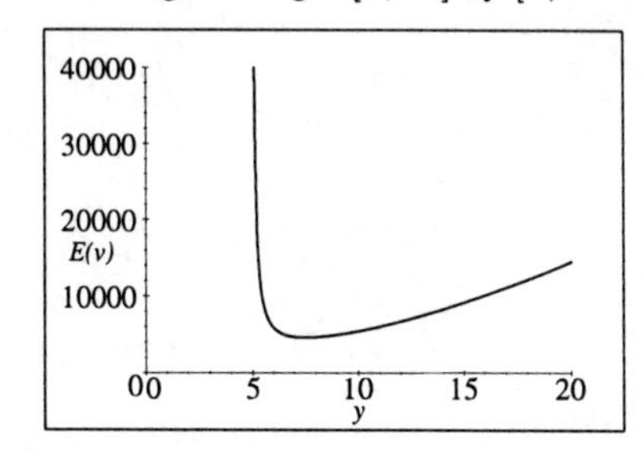

27. $R = kt$

29. $v = \dfrac{k}{z}$

31. $y = \dfrac{ks}{t}$

33. $z = k\sqrt{y}$

35. Since y is directly proportional to x, $y = kx$. Since $y = 72$ when $x = 4 \quad \Rightarrow \quad 72 = k(4) \quad \Leftrightarrow$ $k = 18$. So $y = 18x$.

37. Since M varies directly as x and inversely as y, $M = \dfrac{kx}{y}$. Since $M = 5$ when $x = 2$ and $y = 6$ $\Rightarrow \quad 5 = \dfrac{k(2)}{6} \quad \Leftrightarrow \quad k = 15$. Therefore $M = \dfrac{15x}{y}$.

39. Since W is inversely proportional to the square of r, $W = \dfrac{k}{r^2}$. Since $W = 10$ when $r = 6 \quad \Rightarrow$ $10 = \dfrac{k}{(6)^2} \quad \Leftrightarrow \quad k = 360$. So $W = \dfrac{360}{r^2}$.

41. (a) The force F needed is $F = kx$.

(b) Since $F = 40$ when $x = 5 \quad \Rightarrow \quad 40 = k(5) \quad \Leftrightarrow \quad k = 8$.

(c) From part (b), we have $F = 8x$. Substituting $x = 4$ into $F = 8x$ gives $F = 8(4) = 32$ N.

43. (a) Let p be the number of pages in the magazine, let n be the number of magazines to be printed and let C be the cost of printing, then $C = kpn$. Since $C = 60,000$ when $p = 120$ and $n = 4000$ we get $60000 = k(120)(4000) \quad \Leftrightarrow \quad k = \frac{1}{8}$. So $C = \frac{1}{8}\,pn$.

(b) Substituting $p = 92$ and $n = 5000$ we get $C = \frac{1}{8}(92)(5000) = \$57,500$.

45. (a) Let R be the resistance of the wire, let L be its length and let d be its diameter. Then $R = \dfrac{kL}{d^2}$.

Since $R = 140$ when $L = 1.2$ and $d = 0.005$ we get $140 = \dfrac{k(1.2)}{(0.005)^2} \quad \Leftrightarrow$

$k = \frac{7}{2400} = 0.00291\overline{6}$.

(b) Substituting $L = 3$ and $d = 0.008$ we have $R = \dfrac{7}{2400} \cdot \dfrac{3}{(0.008)^2} = \dfrac{4375}{32} \approx 137\,\Omega$.

Exercises 3.8

1. (a) Shift the graph of $y = f(x)$ 4 units downwards.
   (b) Shift the graph of $y = f(x)$ 4 units to the right.

3. (a) Stretch the graph of $y = f(x)$ vertically by a factor of 3.
   (b) Shrink the graph of $y = f(x)$ vertically by a factor of 3.

5. (a) Reflect the graph of $y = f(x)$ about the x-axis and then shift 5 units upward.
   (b) Reflect the graph of $y = f(x)$ about the y-axis and then shift 5 units upward.

7. (a) Shift the graph of $y = f(x)$ 2 units to the right and 3 units downward.
   (b) Shift the graph of $y = f(x)$ 3 units to the right, then stretch it vertically by a factor of 2.

9. (a) $y = f(x - 2)$

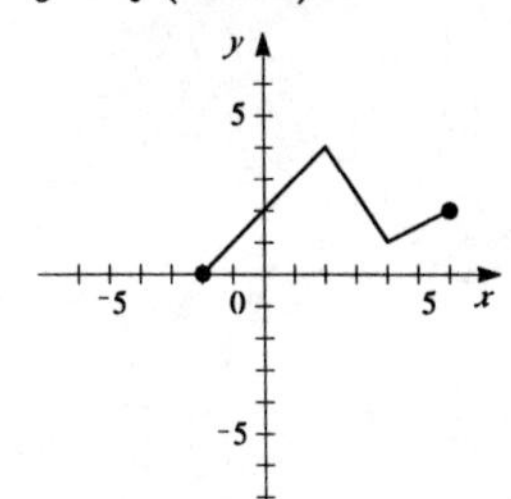

(b) $y = f(x) - 2$

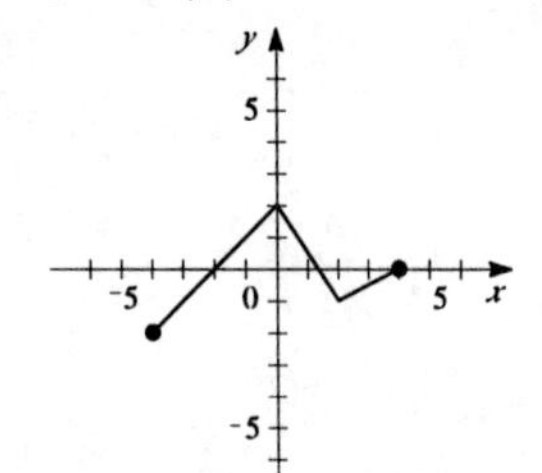

(c) $y = 2f(x)$

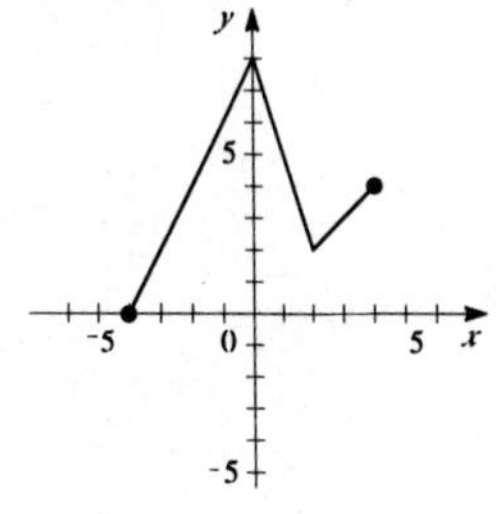

(d) $y = -f(x) + 3$

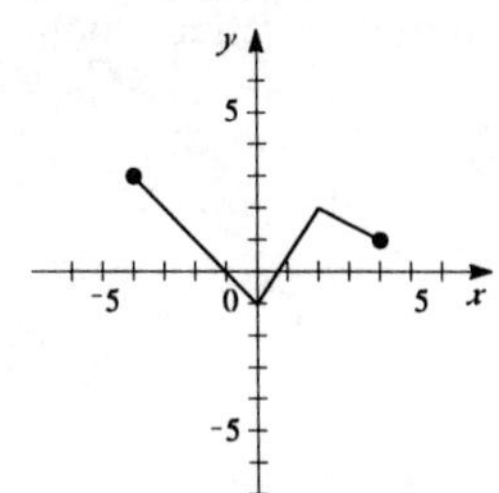

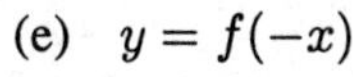
(e) $y = f(-x)$

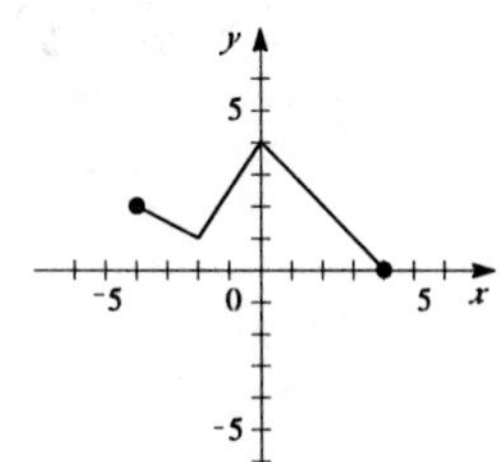

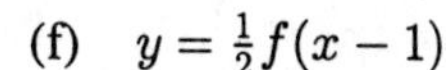
(f) $y = \frac{1}{2}f(x - 1)$

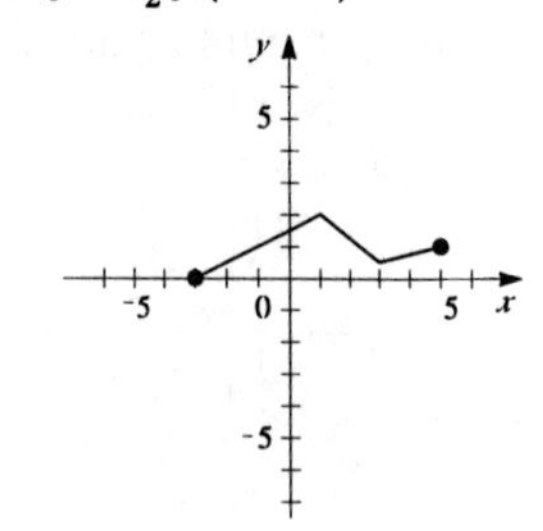

11. (a) $f(x) = \frac{1}{x}$

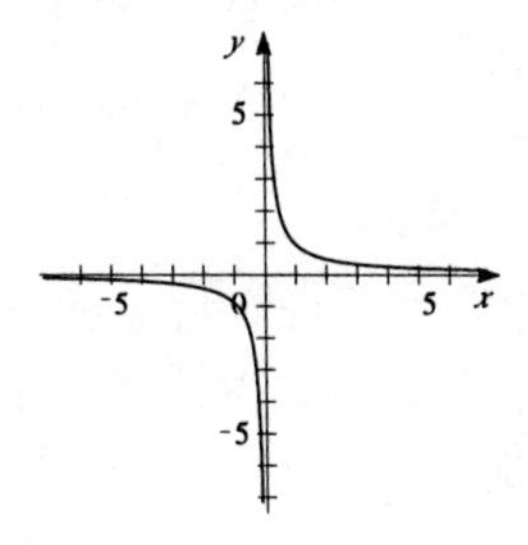

(b) (i) $y = -\frac{1}{x}$. Reflect graph of f about the x-axis.

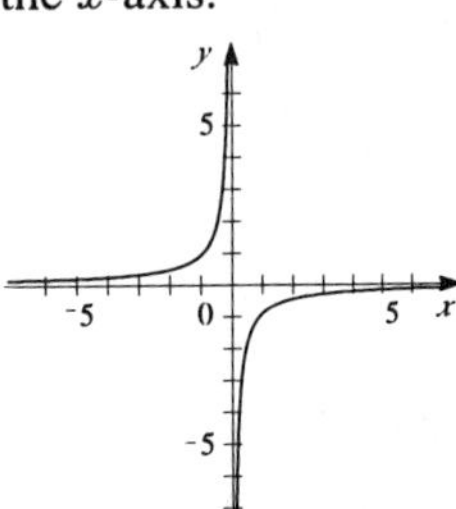

(ii) $y = \frac{1}{x-1}$. Shift graph of f 1 unit to the right.

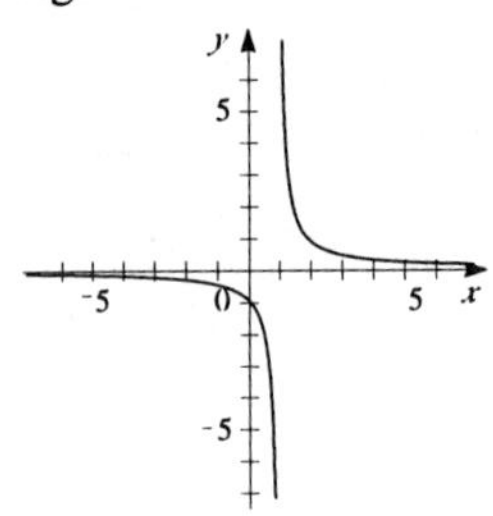

(iii) $y = \frac{2}{x+2}$. Shift graph of f left 2 and stretch vertically 2 units.

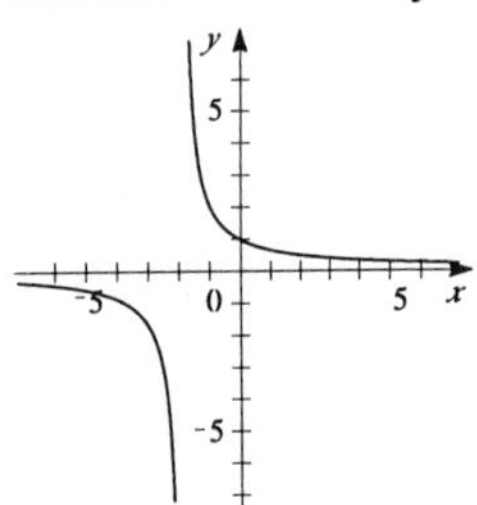

(iv) $y = 1 + \frac{1}{x-3}$. Shift the graph of f right 3 units and upward 1 unit.

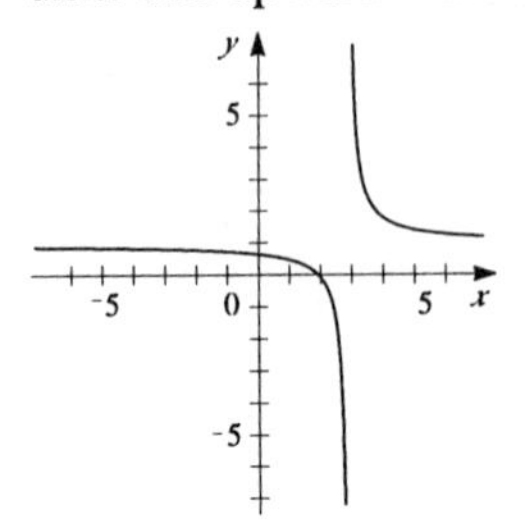

13. $f(x) = (x-2)^2$. Shift the graph of $y = x^2$ right 2 units.

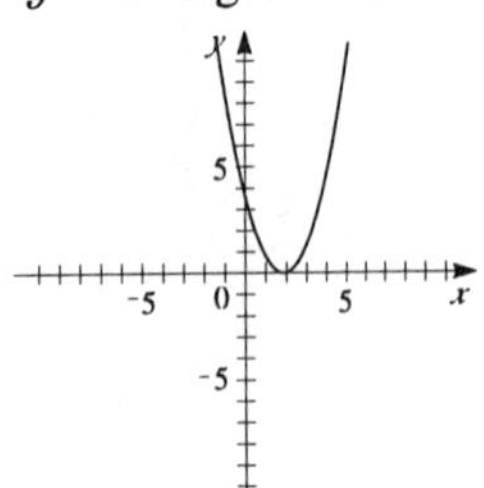

15. $f(x) = -(x+1)^2$. Shift the graph of $y = x^2$ left 1 unit then reflect about the x-axis.

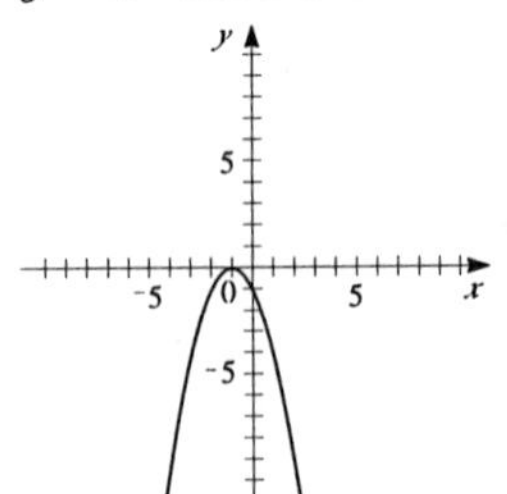

17. $f(x) = x^3 + 2$. Shift the graph of $y = x^3$ upward 2 units.

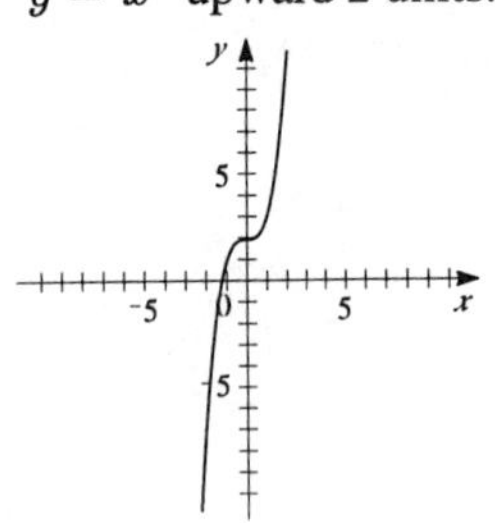

19. $y = 1 + \sqrt{x}$. Shift the graph of $y = \sqrt{x}$ upward 1 unit.

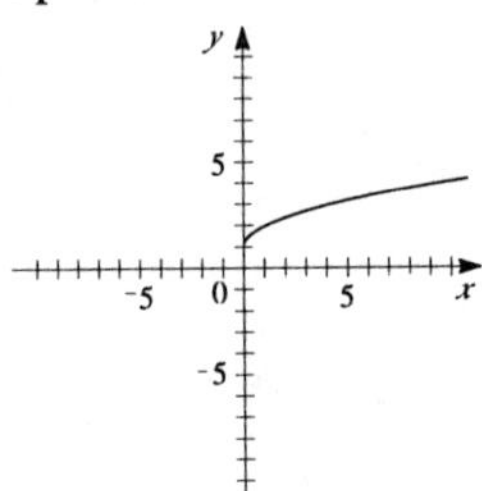

21. $y = \frac{1}{2}\sqrt{x+4} - 3$. Shift the graph of $y = \sqrt{x}$ left 4 units, shrink vertically by a factor of $\frac{1}{2}$, then shift downward 3 units.

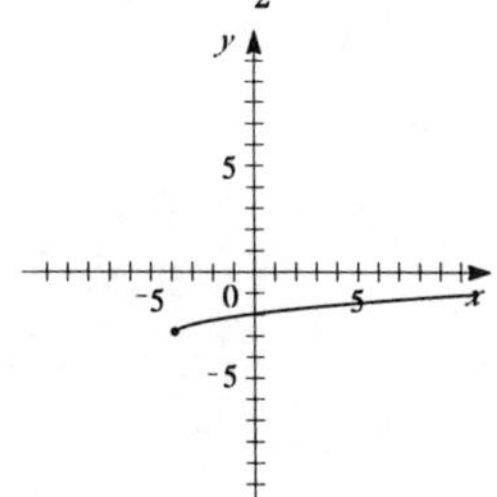

23. $y = 5 + (x+3)^2$. Shift the graph of $y = x^2$ left 3 units then upward 5 units.

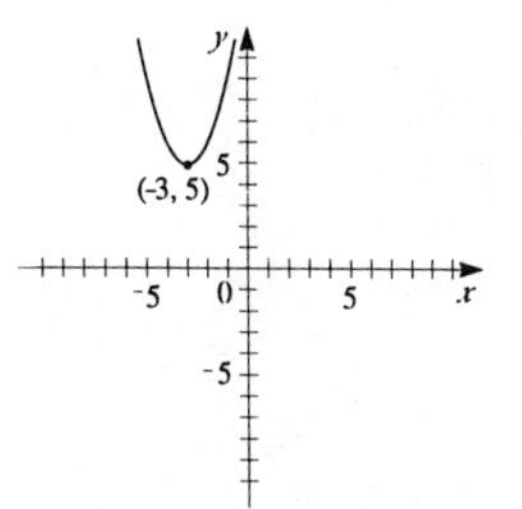

25. $y = |x| - 1$. Shift the graph of $y = |x|$ downward 1 unit.

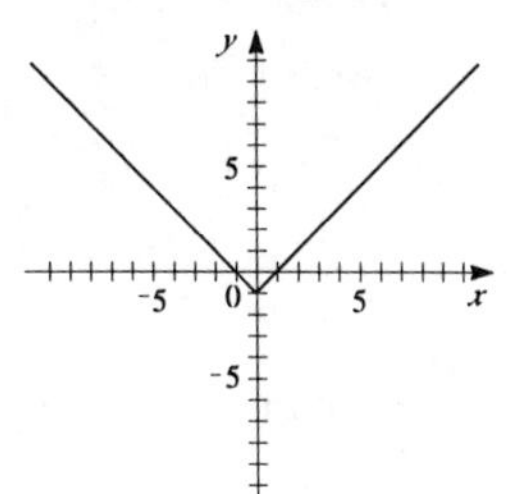

27. $y = |x+2| + 2$. Shift the graph of $y = |x|$ left 2 units and upward 2 units.

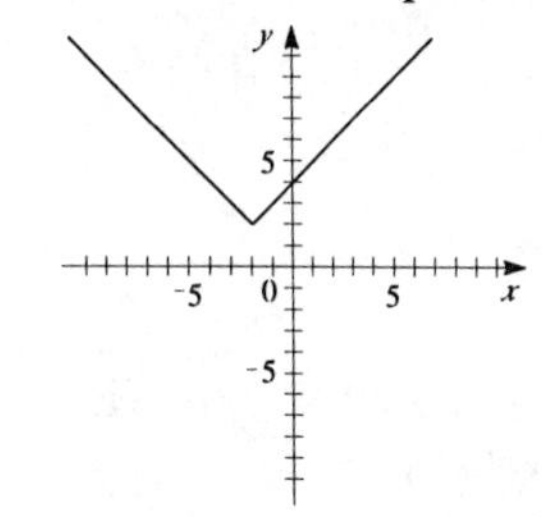

29.

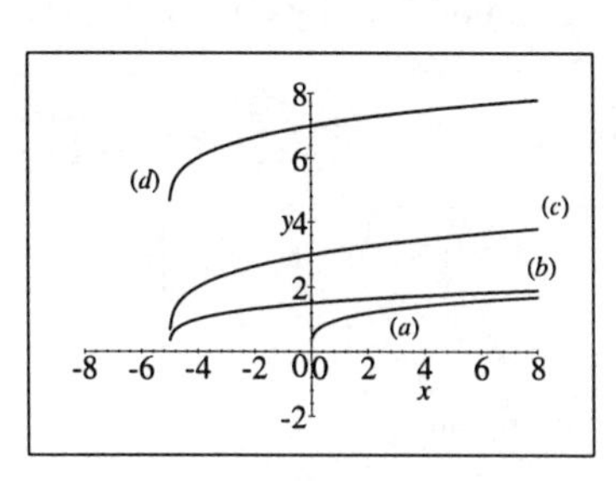

For part (b) shift the graph in (a) left 5 units; for part (c) shift the graph in (a) left 5 units and stretch vertically by a factor of 2; for part (d) shift the graph in (a) left 5 units, stretch vertically by a factor of 2, then shift upward 4 units.

31.

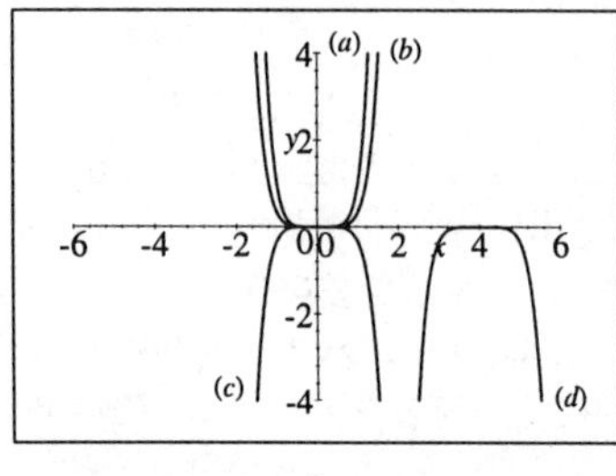

For part (b) shrink the graph in (a) vertically by a factor of 3; for part (c) shrink the graph in (a) vertically by a factor of 3 and reflect about the x-axis; for part (d) shift the graph in (a) left 4 units, shrink vertically by a factor of 3, and then reflect about the x-axis.

33. Since $f(x) = x^2 - 4 < 0$ for $-2 < x < 2$, the graph of $y = g(x)$ is found by sketching the graph of $y = f(x)$ for $x \leq -2$ and $x \geq 2$ then reflecting about the x-axis the part of the graph of $y = f(x)$ for $-2 < x < 2$.

35. (a) $f(x) = 4x - x^2$

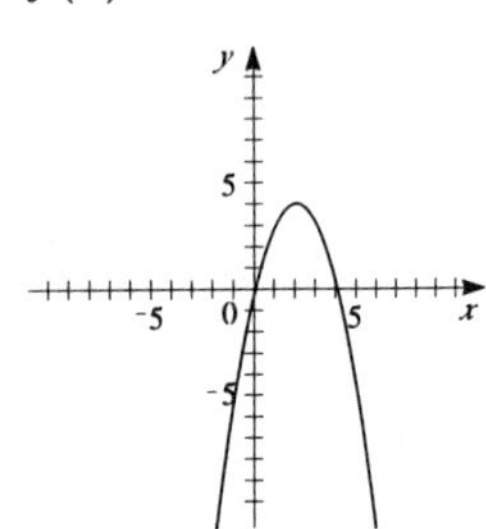

(b) $f(x) = |4x - x^2|$

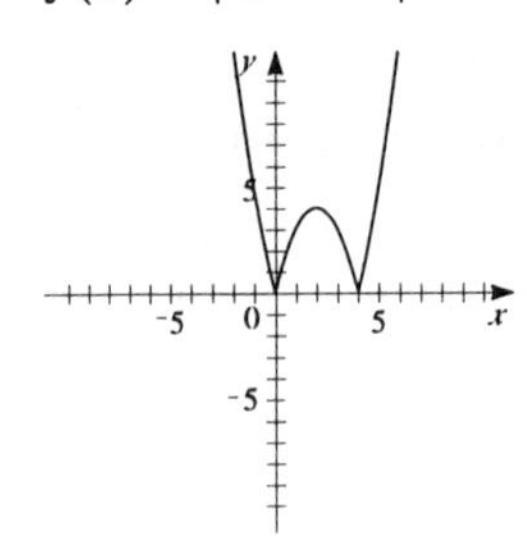

37. (a) $y = f(2x)$

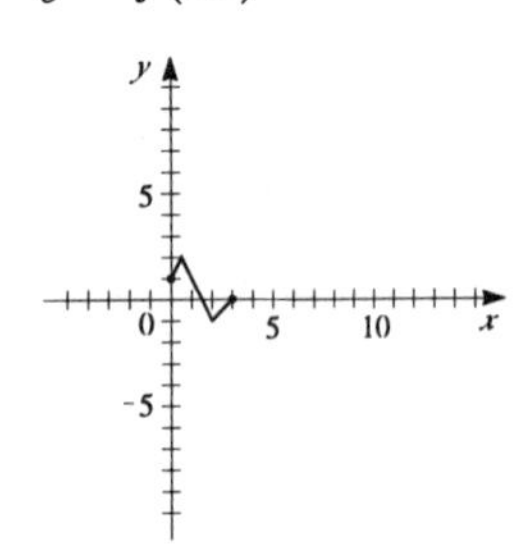

$y = f(\frac{1}{2}x)$

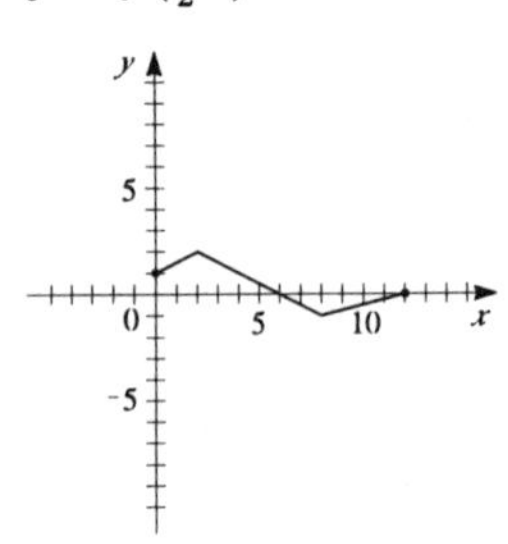

(b) When $a > 1$ the graph of $f(ax)$ is obtain from the graph of $f(x)$ by horizontally shrinking the graph by a factor of a.

(c) When $0 < a < 1$ the graph of $f(ax)$ is obtain from the graph of $f(x)$ by horizontally stretching the graph by a factor of a.

39. (a) $y = f(x) = \sqrt{2x - x^2}$

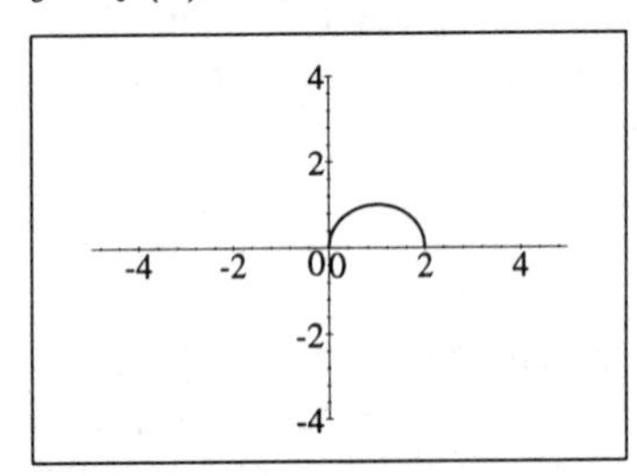

(b) $y = f(2x) = \sqrt{2(2x) - (2x)^2} = \sqrt{4x - 4x^2}$

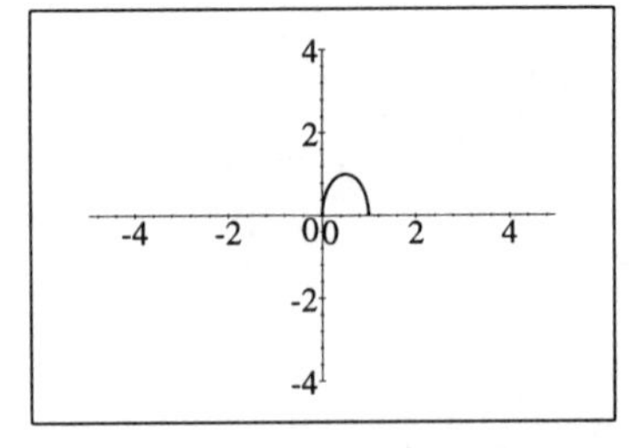

(c) $y = f(\frac{1}{2}x) = \sqrt{2(\frac{1}{2}x) - (\frac{1}{2}x)^2}$
$= \sqrt{x - \frac{1}{4}x^2}$

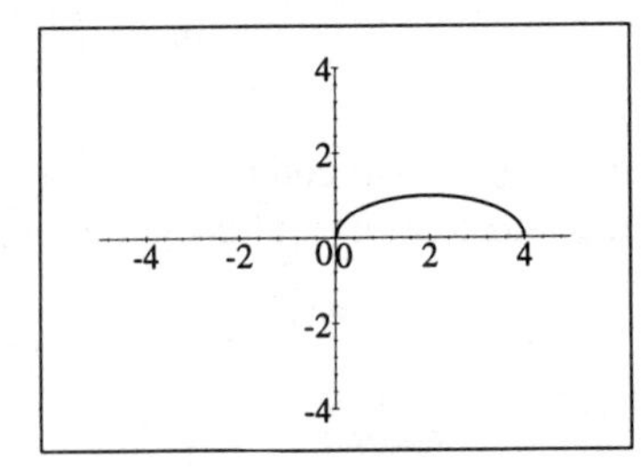

The graph in part (b) is obtained by horizontally shrinking the graph in part (a) by a factor of 2 (so the graph is $\frac{1}{2}$ as wide). The graph in part (c) is obtain by horizontally stretching the graph in part (a) by a factor of 2 (so the graph is twice as wide).

41. $f(-x) = (-x)^{-2} = x^{-2} = f(x)$. Thus $f(x)$ is even.

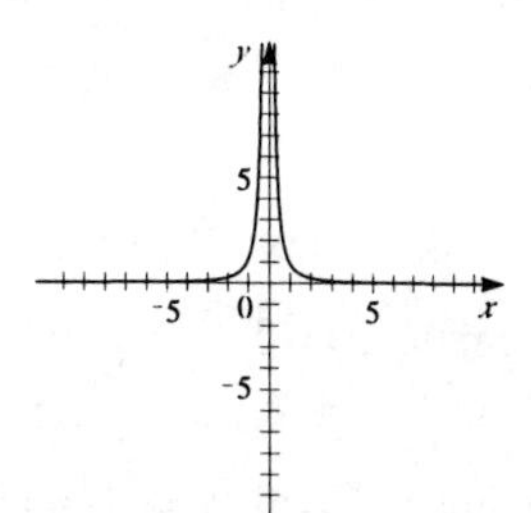

43. $f(-x) = (-x)^2 + (-x) = x^2 - x$. Thus $f(-x) \neq f(x)$. Also $f(-x) \neq -f(x)$. So $f(x)$ is neither odd nor even.

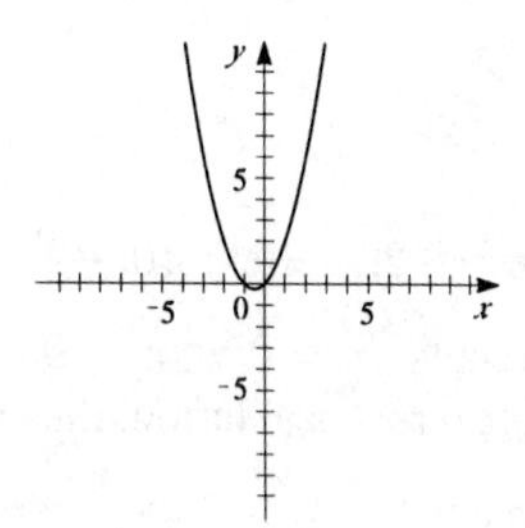

45. $f(-x) = (-x)^3 - (-x) = -x^3 + x$ $= -(x^3 - x) = -f(x)$. Thus $f(x)$ is odd.

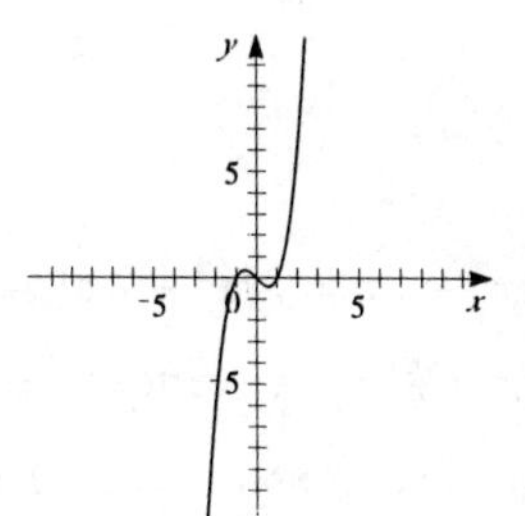

47. $f(-x) = 1 - \sqrt[3]{(-x)} = 1 + \sqrt[3]{x}$. Thus $f(-x) \neq f(x)$. Also $f(-x) \neq -f(x)$. So $f(x)$ is neither odd nor even.

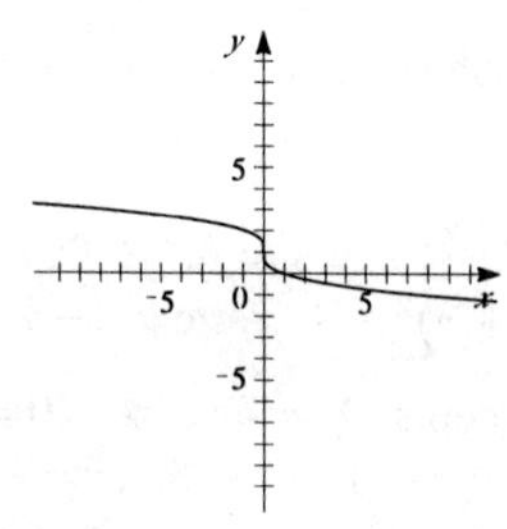

49. (a) f even implies $f(-x) = f(x)$; g even implies $g(-x) = g(x)$. So $(f+g)(-x) = f(-x) + g(-x) = f(x) + g(x) = (f+g)(x)$ and thus $f+g$ is even.

(b) f odd implies $f(-x) = -f(x)$; g odd implies $g(-x) = -g(x)$. So $(f+g)(-x) = f(-x) + g(-x) = -f(x) - g(x) = -(f+g)(x)$ and thus $f+g$ is odd.

Exercises 3.9

1. $y = x^2 - 8$. vertex: $(0, -8)$.

   x-intercepts: $y = 0 \quad \Rightarrow \quad 0 = x^2 - 8 \quad \Leftrightarrow \quad x^2 = 8 \quad \Leftrightarrow$
   $x = \pm\sqrt{8} = \pm 2\sqrt{2}$.

   y-intercepts: $x = 0 \quad \Rightarrow \quad y = -8$

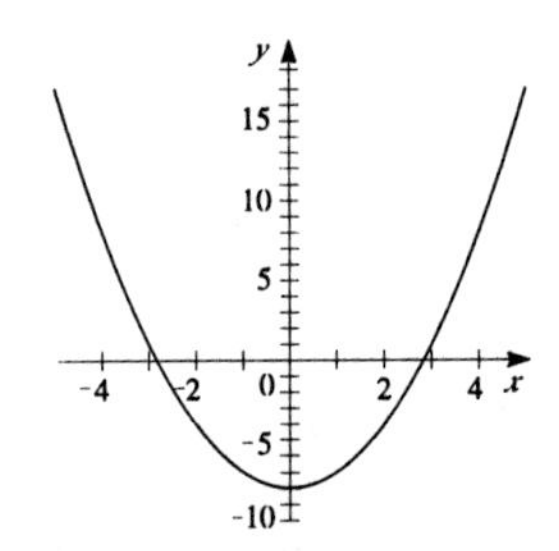

3. $y = -x^2 - 2$. vertex: $(0, -2)$.

   x-intercepts: $y = 0 \quad \Rightarrow \quad 0 = -x^2 - 2 \quad \Leftrightarrow \quad x^2 = -2 \quad \Rightarrow$
   there is no real x solution, thus no x-intercept.

   y-intercepts: $x = 0 \quad \Rightarrow \quad y = -2$.

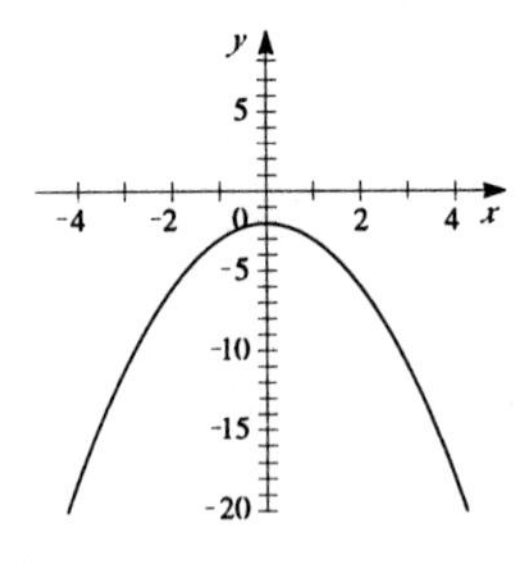

5. $y = 2x^2 - 6x = 2(x^2 - 3x) = 2\left[x^2 - 3x + \left(\frac{3}{2}\right)^2\right] - \frac{9}{2}$
   $= 2\left(x - \frac{3}{2}\right)^2 - \frac{9}{2}$. vertex: $\left(\frac{3}{2}, -\frac{9}{2}\right)$.

   x-intercepts: $y = 0 \quad \Rightarrow \quad 0 = 2x^2 - 6x \quad \Leftrightarrow \quad 0 = 2x(x - 3)$. So $x = 0$ or $x = 3$. The x-intercepts are at $x = 0$ and $x = 3$.

   y-intercepts: $x = 0 \quad \Rightarrow \quad y = 0$.

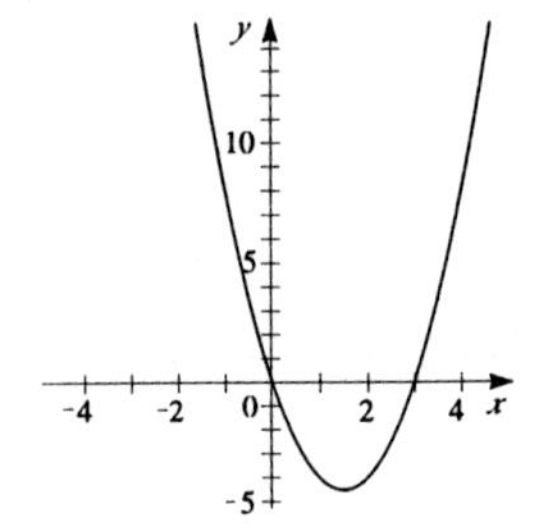

7. $y = x^2 + 6x + 8 = (x^2 + 6x) + 8 = (x^2 + 6x + 9) + 8 - 9$
   $= (x + 3)^2 - 1$. vertex: $(-3, -1)$.

   x-intercepts: $y = 0 \quad \Rightarrow \quad 0 = x^2 + 6x + 8 \quad \Leftrightarrow$
   $(x + 2)(x + 4) = 0$. So $x = -2$ or $x = -4$. The x-intercepts are at $x = -2$ and $x = -4$.

   y-intercepts: $x = 0 \quad \Rightarrow \quad y = 8$.

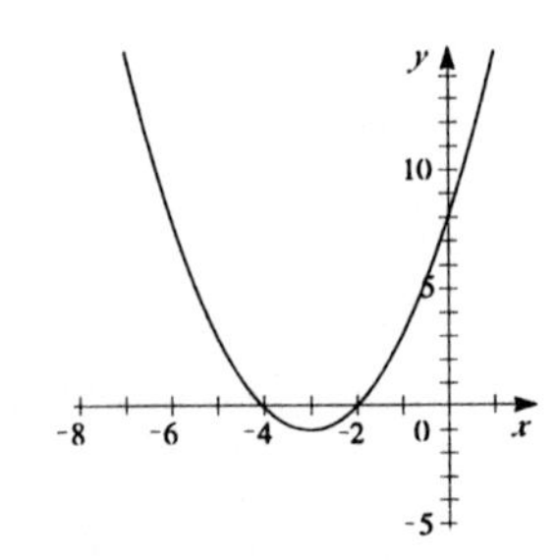

9. $y = 2x^2 + 4x + 3 = 2(x^2 + 2x) + 3 = 2(x^2 + 2x + 1) + 3 - 2$
$= 2(x+1)^2 + 1$. Vertex: $(-1, 1)$.

x-intercepts: $y = 0 \quad \Rightarrow \quad 0 = 2x^2 + 4x + 3 = 2(x+1)^2 + 1 \quad \Leftrightarrow$
$2(x+1)^2 = -1 \quad \Rightarrow$ there is no real x solution, thus no x-intercept.

y-intercepts: $x = 0 \quad \Rightarrow \quad y = 3$.

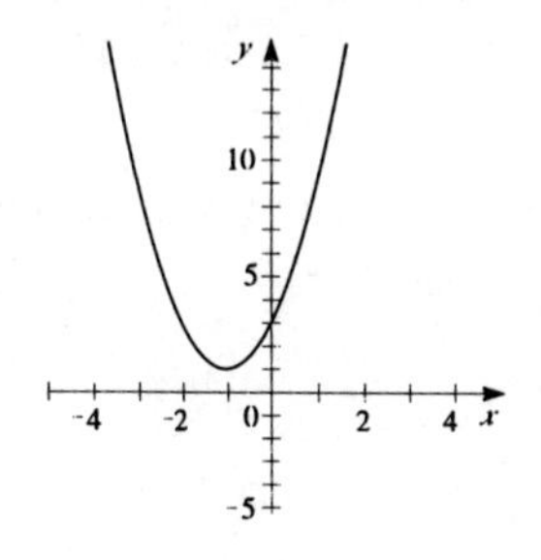

11. $y = 2x^2 - 20x + 57 = 2(x^2 - 10x) + 57$
$= 2(x^2 - 10x + 25) + 57 - 50 = 2(x-5)^2 + 7$

vertex: (5,7).

x-intercepts: $y = 0 \quad \Rightarrow \quad 0 = 2x^2 - 20x + 57 = 2(x-5)^2 + 7 \quad \Leftrightarrow$
$2(x-5)^2 = -7 \quad \Rightarrow$ there is no real x solution, thus no x-intercept.

y-intercepts: $x = 0 \quad \Rightarrow \quad y = 57$.

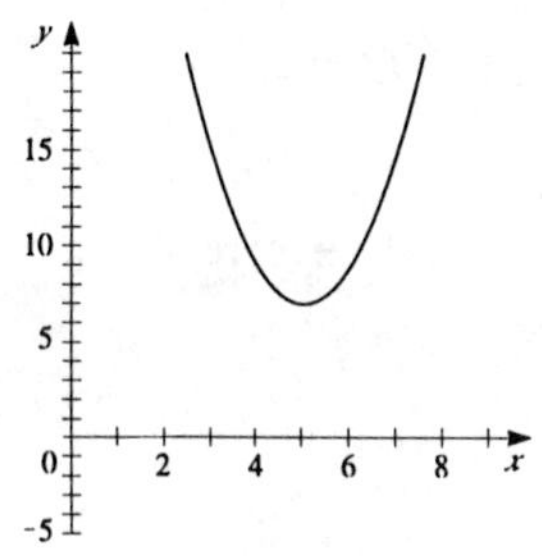

13. $f(x) = 2x - x^2 = -(x^2 - 2x) = -(x^2 - 2x + 1) + 1$
$= -(x-1)^2 + 1$.

Therefore, the maximum value is $f(1) = 1$.

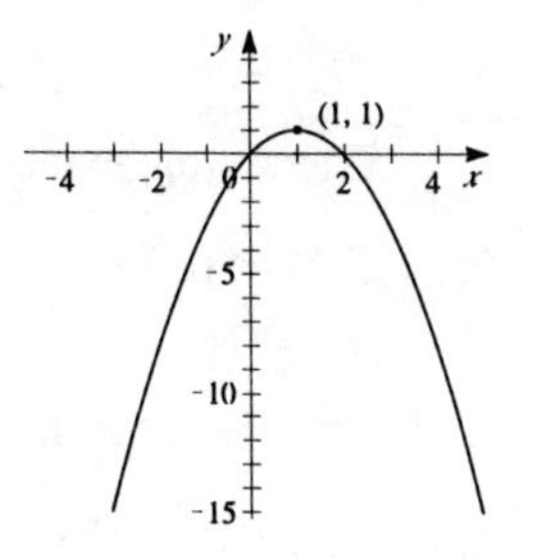

15. $f(x) = x^2 + 2x - 1 = (x^2 + 2x) - 1 = (x^2 + 2x + 1) - 1 - 1$
$= (x+1)^2 - 2$.

Therefore, the minimum value is $f(-1) = -2$.

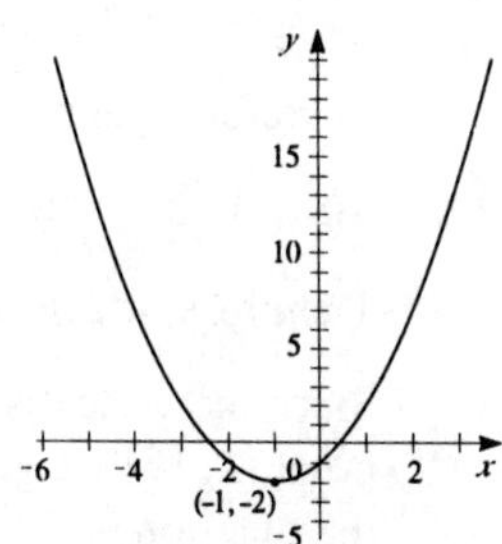

17. $f(x) = -x^2 - 3x + 3 = -(x^2 + 3x) + 3 = -\left(x^2 + 3x + \frac{9}{4}\right) + 3 + \frac{9}{4}$
$= -\left(x + \frac{3}{2}\right)^2 + \frac{21}{4}$.

Therefore, the maximum value is $f\left(-\frac{3}{2}\right) = \frac{21}{4}$.

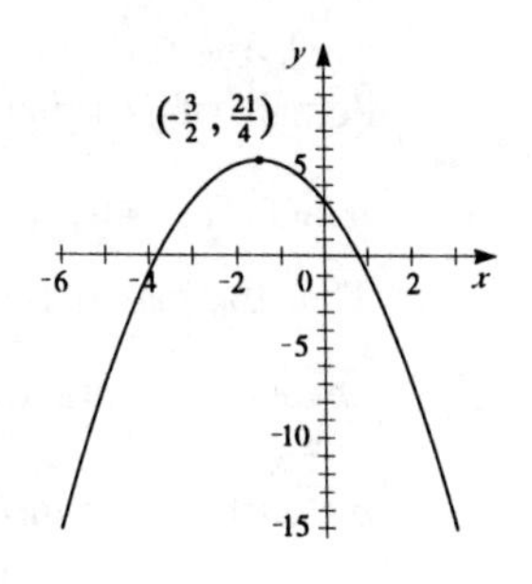

19. 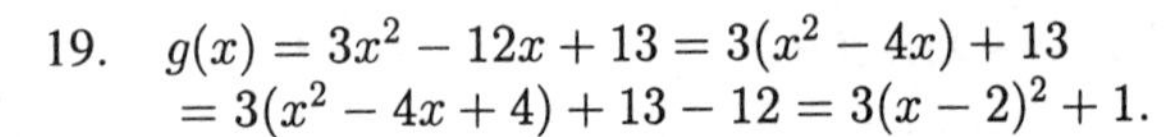
$g(x) = 3x^2 - 12x + 13 = 3(x^2 - 4x) + 13$
$= 3(x^2 - 4x + 4) + 13 - 12 = 3(x - 2)^2 + 1.$

Therefore, the minimum value is $g(2) = 1$.

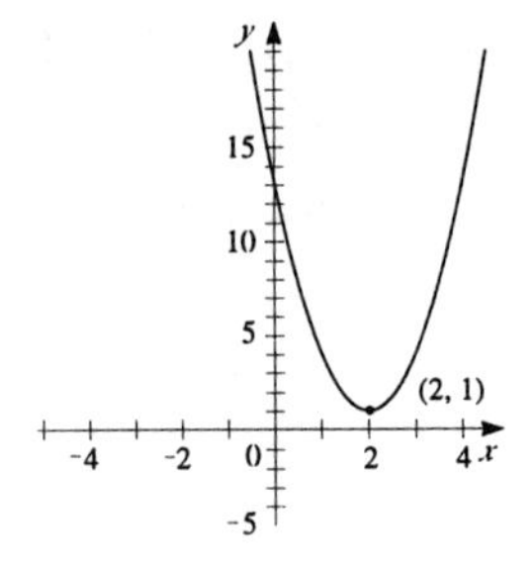

21. $h(x) = 1 - x - x^2 = -(x^2 + x) + 1 = -\left(x^2 + x + \frac{1}{4}\right) + 1 + \frac{1}{4}$
$= -\left(x + \frac{1}{2}\right)^2 + \frac{5}{4}.$

Therefore, the maximum value is $h\left(-\frac{1}{2}\right) = \frac{5}{4}$.

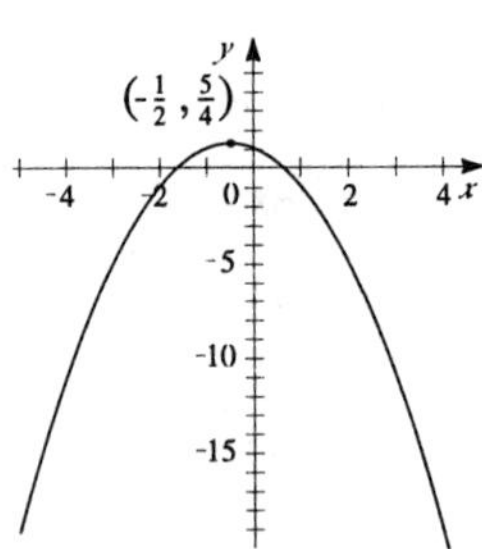

23. $f(x) = x^2 + x + 1 = (x^2 + x) + 1 = \left(x^2 + x + \frac{1}{4}\right) + 1 + \frac{1}{4} = \left(x + \frac{1}{2}\right)^2 + \frac{3}{4}.$
Therefore, the minimum value is $f\left(-\frac{1}{2}\right) = \frac{3}{4}$.

25. $f(t) = 100 - 49t - 7t^2 = -7(t^2 + 7t) + 100 = -7\left(t^2 + 7t + \frac{49}{4}\right) + 100 + \frac{343}{4}$
$= -7\left(t + \frac{7}{2}\right)^2 + \frac{743}{4}.$
Therefore, the maximum value is $f\left(-\frac{7}{2}\right) = \frac{743}{4}$.

27. $f(s) = s^2 - 1.2s + 16 = (s^2 - 1.2s) + 16 = (s^2 - 1.2s + 0.36) + 16 - 0.36$
$= (s - 0.6)^2 + 15.64.$
Therefore, the maximum value is $f(0.6) = 15.64$.

29. $h(x) = \frac{1}{2}x^2 + 2x - 6 = \frac{1}{2}(x^2 + 4x) - 6 = \frac{1}{2}(x^2 + 4x + 4) - 6 - 2 = \frac{1}{2}(x + 2)^2 - 8.$

Therefore, the maximum value is $h(-2) = -8$.

31. Since the vertex is at $(1, -2)$, the function is of the form $f(x) = a(x - 1)^2 - 2$. Substituting the point $(4, 16)$ we get $16 = a(4 - 1)^2 - 2 \quad \Leftrightarrow \quad 16 = 9a - 2 \quad \Leftrightarrow \quad 9a = 18 \quad \Leftrightarrow \quad a = 2$. So the function is $f(x) = 2(x - 1)^2 - 2 = 2x^2 - 4x$.

33. $f(x) = -x^2 + 4x - 3 = -(x^2 - 4x) - 3 = -(x^2 - 4x + 4) - 3 + 4 = -(x - 2)^2 + 1$. So the domain of $f(x)$ is $(-\infty, \infty)$. Since $f(x)$ has a maximum value of 1, the range is $(-\infty, 1]$.

35. $y = f(t) = 40t - 16t^2 = -16\left(t^2 - \frac{5}{2}\right) = -16\left[t^2 - \frac{5}{2}t + \left(\frac{5}{4}\right)^2\right] + 16\left(\frac{5}{4}\right)^2 = -16\left(t - \frac{5}{4}\right)^2 + 25.$
Thus the maximum height attained by the ball is $f\left(\frac{5}{4}\right) = 25$ feet.

37. Let x be the larger number and let y be the smaller number. Since the difference is 100 we have $x - y = 100 \quad \Leftrightarrow \quad y = x - 100$. The product of the two numbers is $P = xy = x(x - 100)$ which we wish to minimize. $P = x(x - 100) = x^2 - 100x = (x^2 - 100x + 2500) - 2500$
$= (x - 50)^2 - 2500$. Thus the minimum product is -2500 and it occurs when $x = 50$ so $y = 50 - 100 = -50$. Thus the two numbers are 50 and -50.

39. Let x and y be the two numbers. Since their sum is -24 we have $x + y = -24 \quad \Leftrightarrow$ $y = -x - 24$. The product of the two numbers is $P = xy = x(-x - 24) = -x^2 - 24x$ which we wish to maximize. $P = -x^2 - 24x = -(x^2 + 24x) = -(x^2 + 24x + 144) + 144$ $= -(x + 12)^2 + 144$. Thus the maximum product is 144 and it occurs when $x = -12$ and $y = -(-12) - 24 = -12$. Thus the two numbers are 12 and -12.

41. Let w be the width of the field (in feet) and l be the length of the field (in feet). Since the farmer has 2400 ft of fencing we must have $2w + l = 2400 \quad \Leftrightarrow \quad l = 2400 - 2w$. The area of the fenced-in field is given by $A = l \cdot 2w = (2400 - 2w)w = -2w^2 + 2400w = -2(w^2 - 1200w)$ $= -2(w^2 - 1200w + 600^2) + 2(600^2) = -2(w - 600)^2 + 720000$. So the maximum area occurs when $w = 600$ feet and $l = 2(1200 - 600) = 1200$ feet.

43. Let w be the width of the rectangular area (in feet) and l be the length of the field (in feet). Since the farmer has 750 feet of fencing we must have $5w + 2l = 750 \quad \Leftrightarrow \quad 2l = 750 - 5w \quad \Leftrightarrow$ $l = \frac{5}{2}(150 - w)$. The area of the four pens is $A = l \cdot w = \frac{5}{2}w(150 - w) = -\frac{5}{2}(w^2 - 150w)$ $= -\frac{5}{2}(w^2 - 150w + 75^2) + \left(\frac{5}{2}\right) \cdot 75^2 = -\frac{5}{2}(w - 75)^2 + 14062.5$. Therefore the largest possible area of the four pens is 14,062.5 square feet.

45. (a) $f(x) = x^2 + 1.79x - 3.21$ is shown in the viewing rectangle on the right. The minimum value of $f(x) \approx -4.01$.

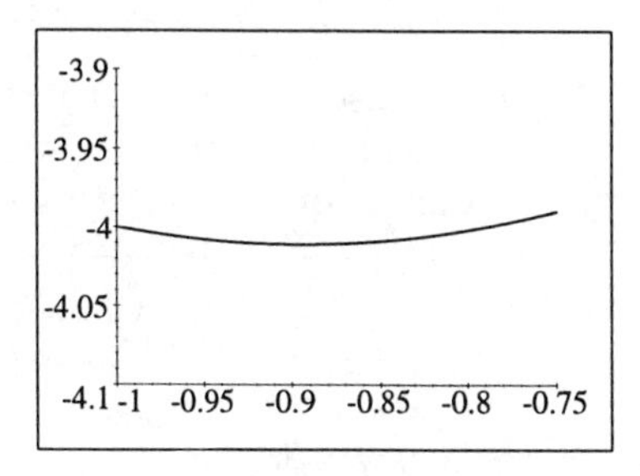

(b) $f(x) = x^2 + 1.79x - 3.21$
$= \left[x^2 + 1.79x + \left(\frac{1.79}{2}\right)^2\right] - 3.21 - \left(\frac{1.79}{2}\right)^2$
$= (x + 0.895)^2 - 4.011025$

Therefore the exact minimum of $f(x)$ is -4.011025.

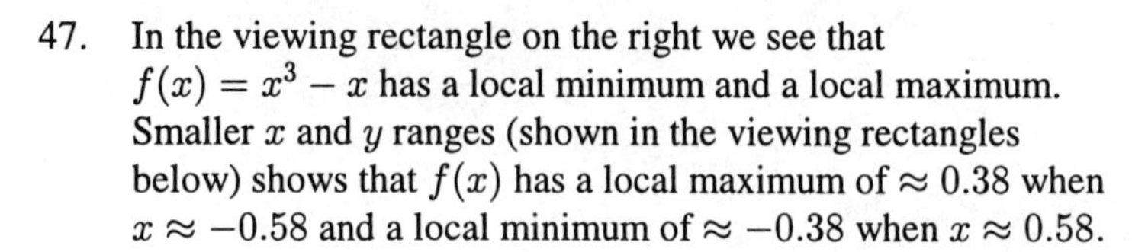

47. In the viewing rectangle on the right we see that $f(x) = x^3 - x$ has a local minimum and a local maximum. Smaller x and y ranges (shown in the viewing rectangles below) shows that $f(x)$ has a local maximum of ≈ 0.38 when $x \approx -0.58$ and a local minimum of ≈ -0.38 when $x \approx 0.58$.

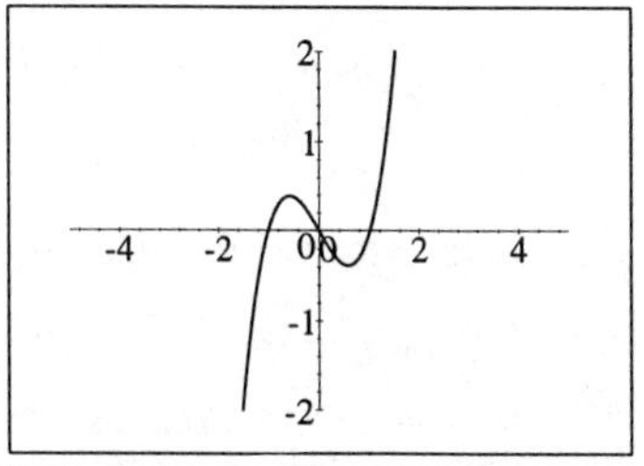

0.5 0.48 0.46 0.44 0.42 0.4 0.38 0.36 0.34 0.32 0.3 -0.6 -0.58 -0.56 -0.54 -0.52 -0.5

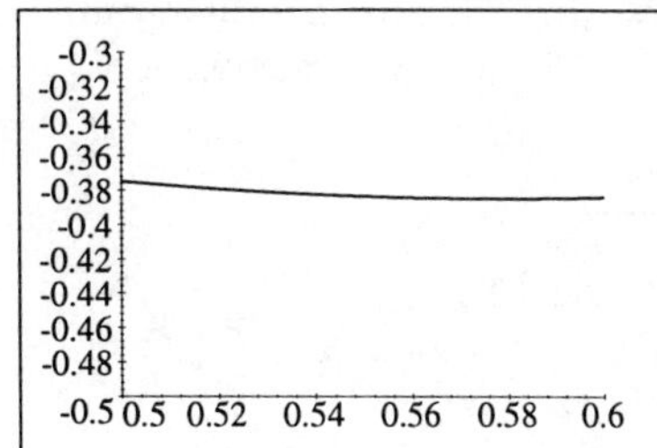

49. In the viewing rectangle on the right we see that $g(x) = x^4 - 2x^3 - 11x^2$ has two local minimums and a local maximum. The local maximum is $g(x) = 0$ when $x = 0$. Smaller x and y ranges (shown in the viewing rectangles below) shows that local minima are $g(x) \approx -13.61$ when $x \approx -1.71$ and $g(x) \approx -73.32$ when $x \approx 3.21$.

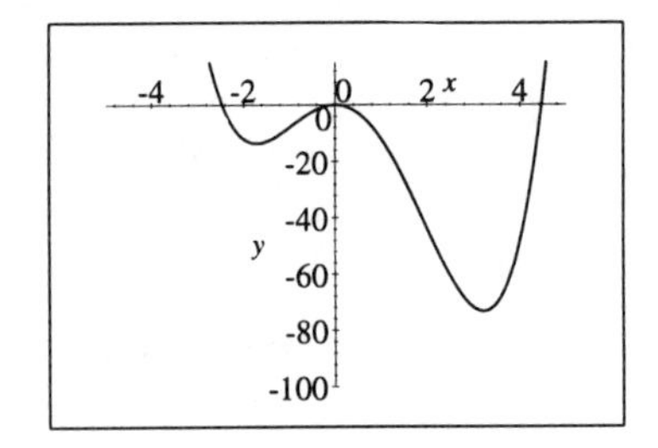

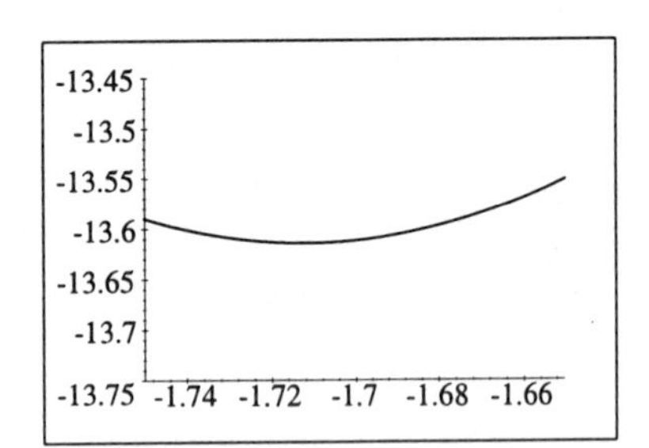

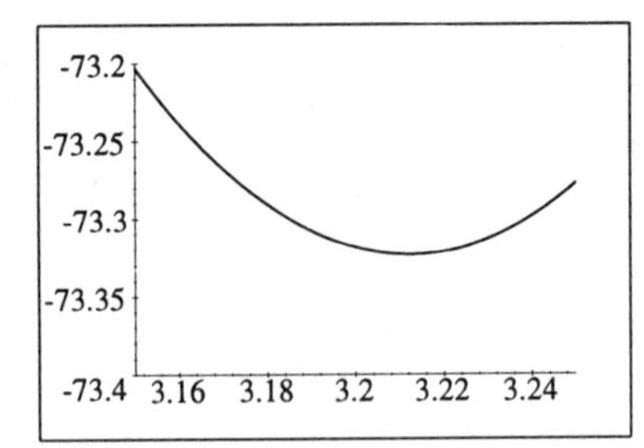

51. In the first viewing rectangle below we see that $U(x) = x\sqrt{6-x}$ only has a local maximum. Smaller x and y ranges in the second viewing rectangle below, shows that $U(x)$ has a local maximum of ≈ 5.66 when $x \approx 4.00$.

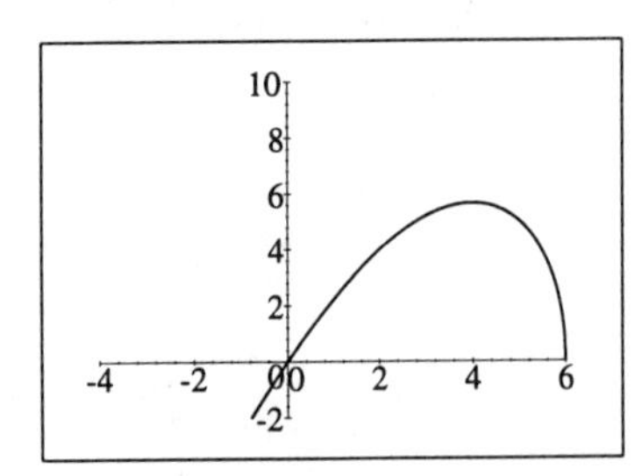

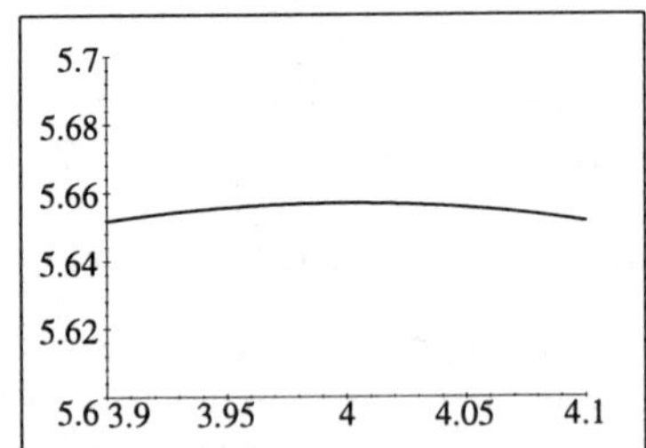

53. In the viewing rectangle on the right we see that $V(x) = \dfrac{1-x^2}{x^3}$ has a local minimum and a local maximum. Smaller x and y ranges (shown in the viewing rectangles below) shows that $V(x)$ has a local maximum of ≈ 0.38 when $x \approx -1.73$ and a local minimum of ≈ -0.38 when $x \approx 1.73$.

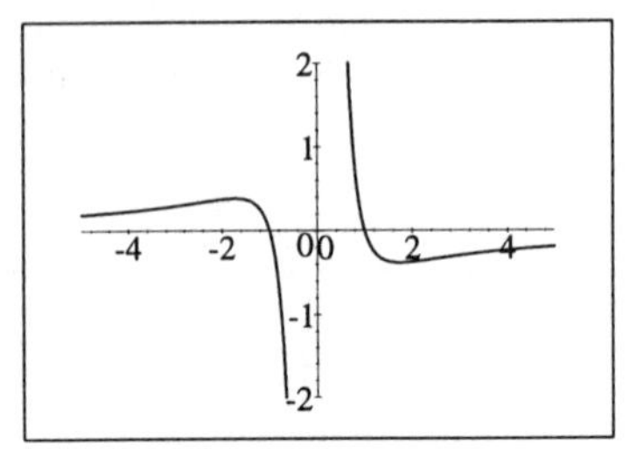

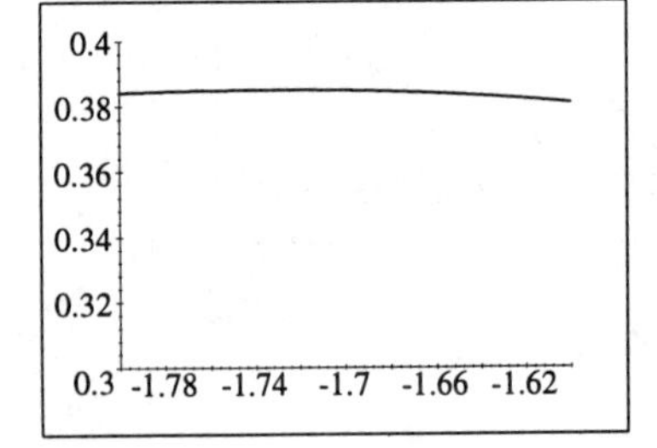

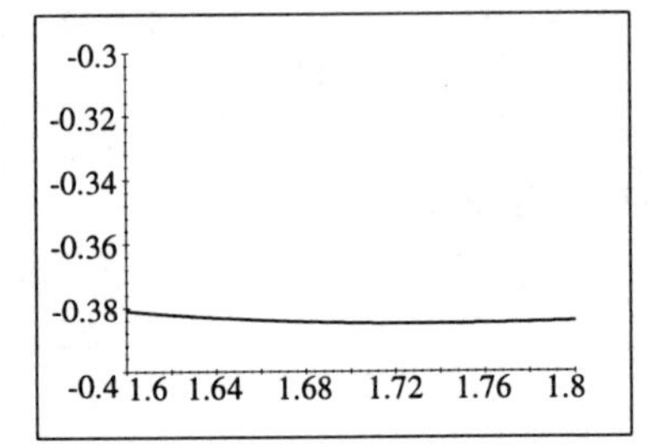

55. In the first viewing rectangle below we see the general location of the minimum of $E(v) = 2.73v^3 \frac{10}{v-5}$. In the second viewing rectangle we isolate the minimum and from this graph we see that energy is minimized when $v \approx 7.5$ mi/h.

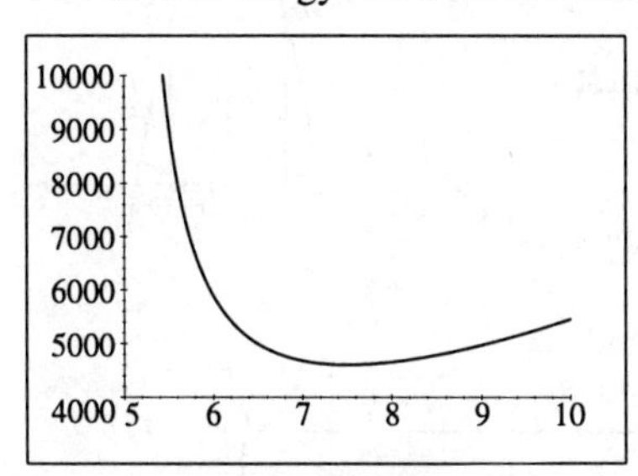

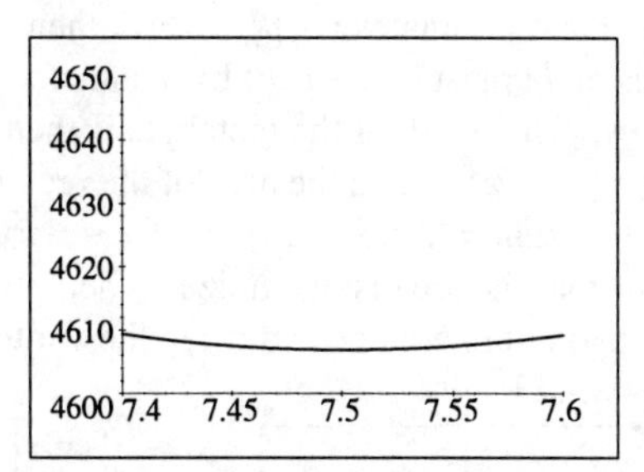

57. We saw in Exercise 15 in Section 3.7 that the surface area of the box is given by the formula $A(x) = \frac{48}{x} + x^2$ where x is the length of a side of the base and the height is $h = \frac{12}{x^2}$. The function $y = A(x)$ is shown in the first viewing rectangle below. In the second viewing rectangle we isolate the minimum and we see that the amount of material is minimized when x (the length and width) is 2.88 ft. Then the height is $h = \frac{12}{x^2} \approx 1.44$ ft.

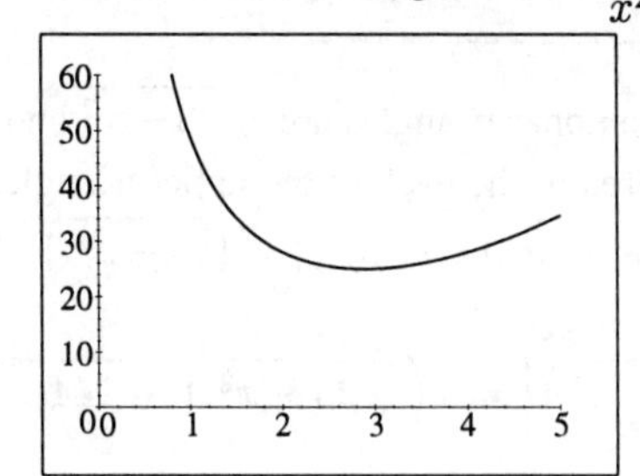

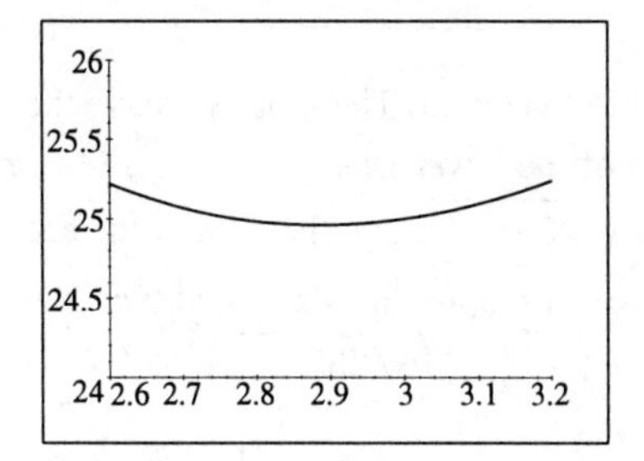

59. We saw in Exercise 17 in Section 3.7 that the area of the window is given by the formula $A(x) = 15x - \frac{x^2}{8}(\pi + 4)$ where x is the width of the window and the height of the straight portion of the window is $h = 15 - \frac{x}{2} - \frac{\pi x}{4}$. The function $y = A(x)$ is shown in the first viewing rectangle below. In the second viewing rectangle we isolate the maximum and we see that the area of the window is maximized (and hence the greatest amount of light admitted) when the width ≈ 8.40 ft. Then the straight height is $h \approx 15 - \frac{8.40}{2} - \frac{8.40\pi}{4} \approx 4.20$ ft.

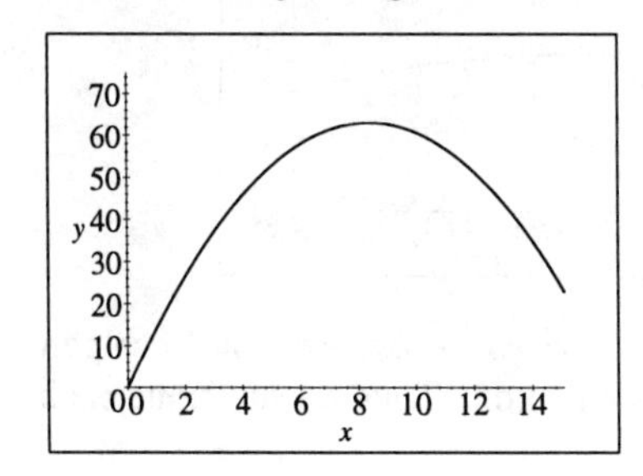

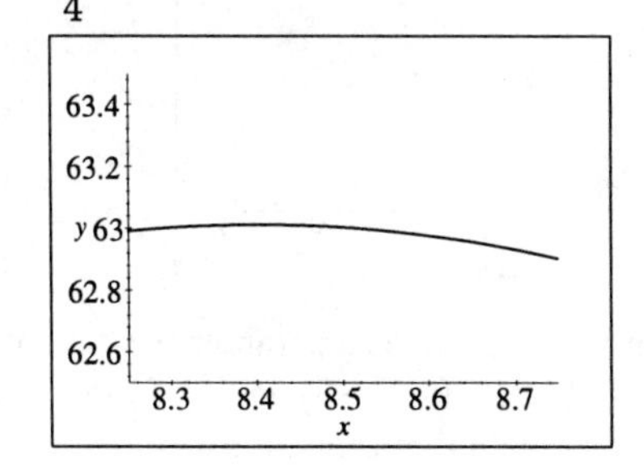

61. Let A, B, C, D be the vertices of the rectangle such that its base AB is on the x-axis and its other two vertices, C and D, are above the x-axis and lying on the parabola $y = 8 - x^2$. Let C have the coordinates (x, y), $x > 0$, then the coordinates of D must be $(-x, y)$ by symmetry. See the first graph below. The width of the rectangle is then $2x$ and the length is $y = 8 - x^2$. Thus the area of the rectangle is $A = \textit{length} \cdot \textit{width} = 2x(8 - x^2) = 16x - 2x^3$. The graphs below show that the area is maximized when $x \approx 1.63$. Hence the maximum area occurs when the width ≈ 3.27 and the length ≈ 5.33.

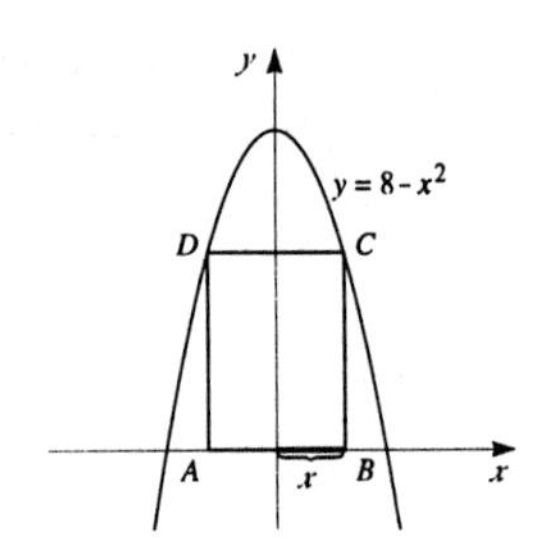

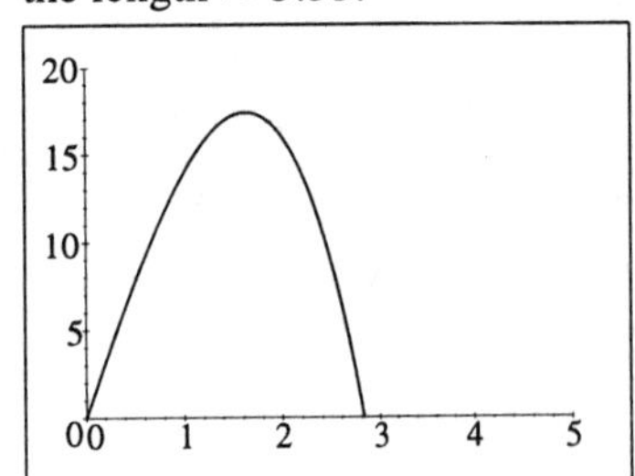

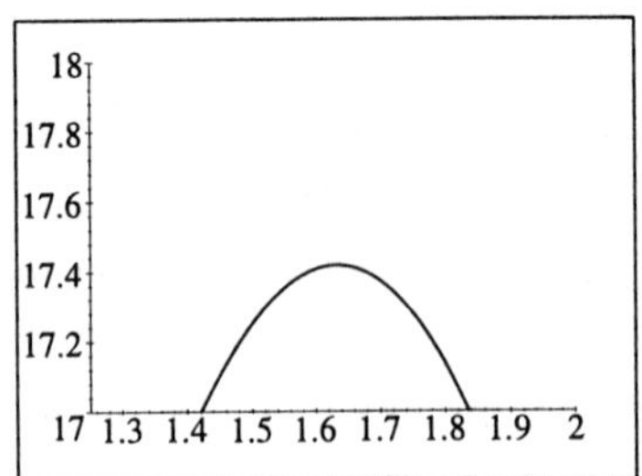

63. (a) Using Pythagorean Theorem, we have the height of the upper triangles are $\sqrt{25 - x^2}$ and the height of the lower triangles are $\sqrt{144 - x^2}$. So the area of the each of the upper triangles is $\frac{1}{2}(x)\left(\sqrt{25 - x^2}\right)$ and the area of the each of the lower triangles is $\frac{1}{2}(x)\left(\sqrt{144 - x^2}\right)$. Since there are two upper triangles and two lower triangles, we get $\textit{Area} = 2 \cdot \left[\frac{1}{2}(x)\left(\sqrt{25 - x^2}\right)\right] + 2 \cdot \left[\frac{1}{2}(x)\left(\sqrt{144 - x^2}\right)\right] = x\left(\sqrt{25 - x^2} + \sqrt{144 - x^2}\right)$.

(b) The function $y = A(x) = x\left(\sqrt{25 - x^2} + \sqrt{144 - x^2}\right)$ is shown in the first viewing rectangle below. In the second viewing rectangle we isolate the maximum and we see that the area of the kite is maximized when $x \approx 4.615$. So the length of the horizontal crosspiece must be $2 \cdot 4.615 = 9.23$. The length of the vertical crosspiece is $\sqrt{5^2 - (4.615)^2} + \sqrt{12^2 - (4.615)^2} \approx 13.00$.

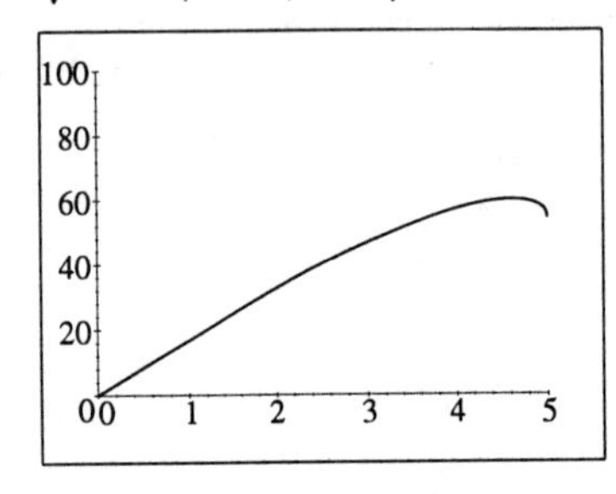

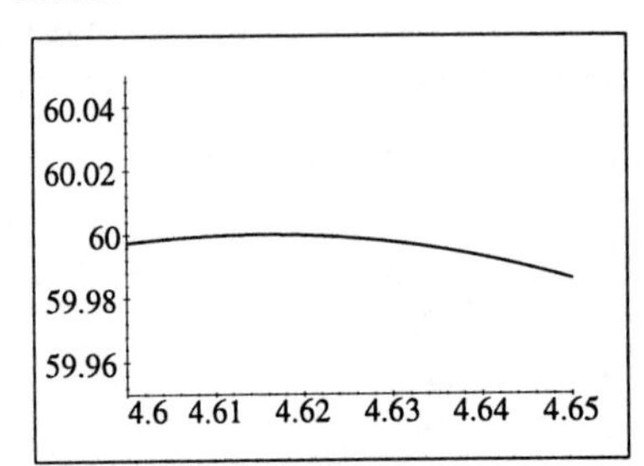

65. A function whose graph is a parabola has the form $y = f(x) = ax^2 + bx + c$. And each pair of coordinates $(1, -1)$, $(-1, -3)$ and $(3, 9)$ must satisfy this equation. Thus we substitute each coordinate to obtain three equations.

For $(1, -1)$: $-1 = a \cdot (1)^2 + b \cdot (1) + c \quad \Leftrightarrow \quad a + b + c = -1$.

For $(-1, -3)$: $-3 = a \cdot (-1)^2 + b \cdot (-1) + c \quad \Leftrightarrow \quad a - b + c = -3$.

For $(3, 9)$: $9 = a \cdot (3)^2 + b \cdot (3) + c \quad \Leftrightarrow \quad 9a + 3b + c = 9.$

Subtracting the second equation from the first gives

$$\begin{array}{r} a + b + c = -1 \\ -(a - b + c = -3) \\ \hline 2b = 2 \end{array} \qquad \Leftrightarrow \qquad b = 1.$$

Subtracting the second equation from the third gives

$$\begin{array}{r} 9a + 3b + c = 9 \\ -(a - b + c = -3) \\ \hline 8a + 4b = 12 \end{array}$$

Substituting $b = 1$ into this new equation we obtain: $8a + 4(1) = 12 \quad \Leftrightarrow \quad 8a + 4 = 12 \quad \Leftrightarrow$ $8a = 8 \quad \Leftrightarrow \quad a = 1$. Finally, substituting $a = b = 1$ into the second equation gives $(1) - (1) + c = -3 \quad \Leftrightarrow \quad c = -3$. So the equation of the parabola is $y = f(x) = x^2 + x - 3$.

Exercises 3.10

1. $f(x) = x^2 - x$ has domain $(-\infty, \infty)$. $g(x) = x + 5$ has domain $(-\infty, \infty)$. The intersection of the domains of f and g is $(-\infty, \infty)$. Then:
$(f+g)(x) = (x^2 - x) + (x + 5) = x^2 + 5$, domain is $(-\infty, \infty)$;
$(f-g)(x) = (x^2 - x) - (x + 5) = x^2 - 2x - 5$, domain is $(-\infty, \infty)$;
$(fg)(x) = (x^2 - x)(x + 5) = x^3 + 4x^2 - 5x$, domain is $(-\infty, \infty)$;
$\left(\frac{f}{g}\right)(x) = \dfrac{x^2 - x}{x + 5}$, domain is $\{ x \mid x \neq -5 \}$.

3. $f(x) = \sqrt{1 + x}$, domain is $[-1, \infty)$. $g(x) = \sqrt{1 - x}$, domain is $(-\infty, 1]$. The intersection of the domains of f and g is $[-1, 1]$. Then:
$(f+g)(x) = \sqrt{1+x} + \sqrt{1-x}$, domain is $[-1, 1]$;
$(f-g)(x) = \sqrt{1+x} - \sqrt{1-x}$, domain is $[-1, 1]$;
$(fg)(x) = \sqrt{1+x} \cdot \sqrt{1-x} = \sqrt{1-x^2}$, domain is $[-1, 1]$;
$\left(\frac{f}{g}\right)(x) = \dfrac{\sqrt{1+x}}{\sqrt{1-x}} = \sqrt{\dfrac{1+x}{1-x}}$, domain is $[-1, 1)$.

5. $f(x) = \frac{2}{x}$, domain is $x \neq 0$. $g(x) = -\frac{2}{x+4}$, domain is $x \neq -4$. The intersection of the domains of f and g is $x \neq 0, -4$, in interval notation, this is $(-\infty, -4) \cup (-4, 0) \cup (0, \infty)$. Then:
$(f+g)(x) = \frac{2}{x} + \left(-\frac{2}{x+4}\right) = \frac{2}{x} - \frac{2}{x+4} = \frac{8}{x(x+4)}$, domain is $(-\infty, -4) \cup (-4, 0) \cup (0, \infty)$;
$(f-g)(x) = \frac{2}{x} - \left(-\frac{2}{x+4}\right) = \frac{2}{x} + \frac{2}{x+4} = \frac{4x+8}{x(x+4)}$, domain is $(-\infty, -4) \cup (-4, 0) \cup (0, \infty)$;
$(fg)(x) = \frac{2}{x} \cdot \left(-\frac{2}{x+4}\right) = -\frac{4}{x(x+4)}$, domain is $(-\infty, -4) \cup (-4, 0) \cup (0, \infty)$;
$\left(\frac{f}{g}\right)(x) = \dfrac{\frac{2}{x}}{-\frac{2}{x+4}} = -\dfrac{x+4}{x}$, domain is $(-\infty, -4) \cup (-4, 0) \cup (0, \infty)$

7. $F(x) = \dfrac{\sqrt{4-x} + \sqrt{3+x}}{x^2 - 1}$. The domain of $\sqrt{4-x}$ is $(-\infty, 4]$ and the domain of $\sqrt{3+x}$ is $[-3, \infty)$. The denominator cannot equal 0, so $x^2 - 1 \neq 0 \Rightarrow x^2 \neq 1 \Rightarrow x \neq \pm 1$. In interval notation this is $(-\infty, -1) \cup (-1, 1) \cup (1, \infty)$. Thus the domain is $(-\infty, 4] \cap [-3, \infty) \cap \{(-\infty, -1) \cup (-1, 1) \cup (1, \infty)\} = [-3, -1) \cup (-1, 1) \cup (1, 4]$.

9.

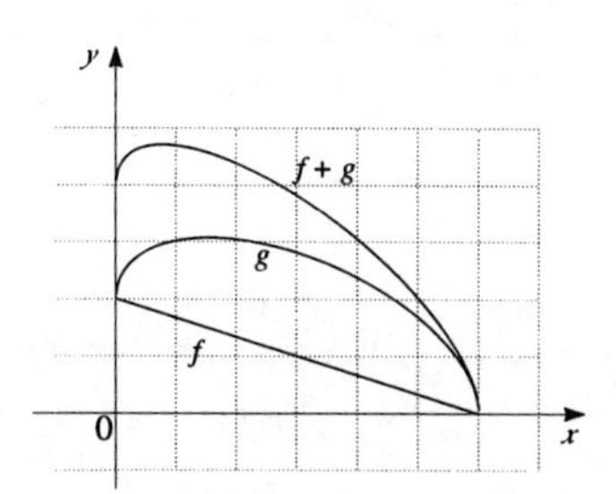

11.

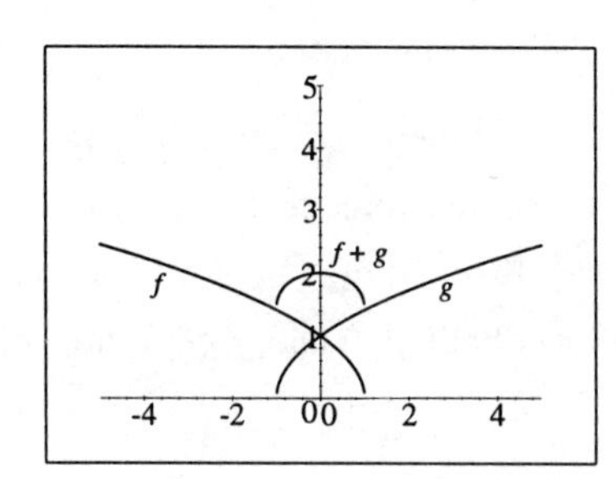

13.

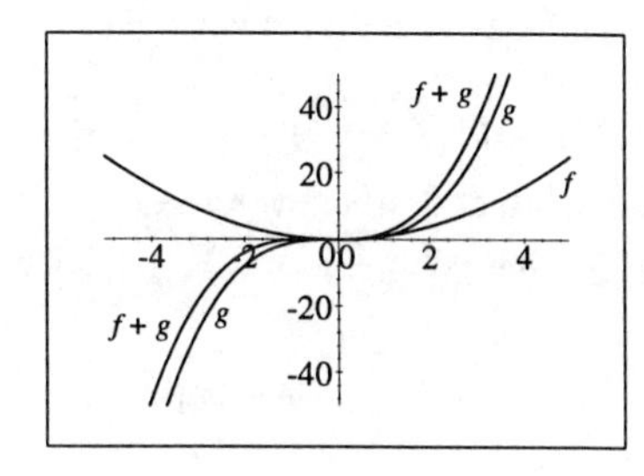

15. (a) $f(g(0)) = f(2-(0)^2) = f(2) = 3(2) - 5 = 1$

(b) $g(f(0)) = g(3(0) - 5) = g(-5) = 2 - (-5)^2 = -23$

17. (a) $(f \circ g)(-2) = f(g(-2)) = f(2 - (-2)^2) = f(-2) = 3(-2) - 5 = -11$

(b) $(g \circ f)(-2) = g(f(-2)) = g(3(-2) - 5) = g(-11) = 2 - (-11)^2 = -119$

19. (a) $(f \circ g)(x) = f(g(x)) = f(2 - x^2) = 3(2 - x^2) - 5 = 6 - 3x^2 - 5 = 1 - 3x^2$

(b) $(g \circ f)(x) = g(f(x)) = g(3x - 5) = 2 - (3x - 5)^2 = 2 - (9x^2 - 30x + 25)$
$= -9x^2 + 30x - 23$

21. $f(g(2)) = f(5) = 4$

23. $(g \circ f)(4) = g(f(4)) = g(2) = 5$

25. $(g \circ g)(-2) = g(g(-2)) = g(1) = 4$

27. $f(x) = 2x + 3,\ g(x) = 4x - 1.$

$(f \circ g)(x) = f(4x - 1) = 2(4x - 1) + 3 = 8x + 1$, domain is $(-\infty, \infty)$;

$(g \circ f)(x) = g(2x + 3) = 4(2x + 3) - 1 = 8x + 11$, domain is $(-\infty, \infty)$;

$(f \circ f)(x) = f(2x + 3) = 2(2x + 3) + 3 = 4x + 9$, domain is $(-\infty, \infty)$;

$(g \circ g)(x) = g(4x - 1) = 4(4x - 1) - 1 = 16x - 5$, domain is $(-\infty, \infty)$.

29. $f(x) = 2x^2 - x,\ g(x) = 3x + 2.$

$(f \circ g)(x) = f(3x + 2) = 2(3x + 2)^2 - (3x + 2) = 2(9x^2 + 12x + 4) - 3x - 2$
$= 18x^2 + 21x + 6$, domain is $(-\infty, \infty)$;

$(g \circ f)(x) = g(2x^2 - x) = 3(2x^2 - x) + 2 = 6x^2 - 3x + 2$, domain is $(-\infty, \infty)$;

$(f \circ f)(x) = f(2x^2 - x) = 2(2x^2 - x)^2 - (2x^2 - x) = 2(4x^4 - 4x^3 + x^2) - 2x^2 + x$
$= 8x^4 - 8x^3 + x$, domain is $(-\infty, \infty)$;

$(g \circ g)(x) = g(3x + 2) = 3(3x + 2) + 2 = 9x + 8$, domain is $(-\infty, \infty)$.

31. $f(x) = \sqrt{x - 1}$, $g(x) = x^2$.

$(f \circ g)(x) = f(x^2) = \sqrt{x^2 - 1}$. $(f \circ g)(x)$ is defined whenever both $g(x)$ and $f(g(x))$ are defined; that is whenever $x^2 - 1 \geq 0$. So, the domain is $\{\, x \mid x^2 - 1 \geq 0 \,\} = (-\infty, -1] \cup [1, \infty)$.

$(g \circ f)(x) = g(\sqrt{x - 1}) = (\sqrt{x - 1})^2 = x - 1$. $(g \circ f)(x)$ is defined whenever both $f(x)$ and $g(f(x))$ are defined; that is whenever $x - 1 \geq 0$. So, the domain is $[1, \infty)$.

$(f \circ f)(x) = f(\sqrt{x - 1}) = \sqrt{\sqrt{x - 1} - 1}$. $(f \circ f)(x)$ is defined whenever both $f(x)$ and $f(f(x))$ are defined; that is whenever $x - 1 \geq 0$ and $\sqrt{x - 1} - 1 \geq 0$. So, the domain is $\{\, x \mid x \geq 1 \text{ and } \sqrt{x - 1} \geq 1 \,\} = \{\, x \mid x \geq 1 \text{ and } x \geq 2 \,\} = [2, \infty)$.

$(g \circ g)(x) = g(x^2) = (x^2)^2 = x^4$. $(g \circ g)(x)$ is defined whenever both $g(x)$ and $g(g(x))$ are defined. Since $g(x)$ is defined everywhere, the domain is $(-\infty, \infty)$.

33. $f(x) = \dfrac{1}{x - 1}$, domain is $\{x \mid x \neq 1\}$. $g(x) = \dfrac{x - 1}{x + 1}$, domain is $\{x \mid x \neq -1\}$

$(f \circ g)(x) = f\left(\dfrac{x - 1}{x + 1}\right) = \dfrac{1}{\dfrac{x - 1}{x + 1} - 1} = -\dfrac{x + 1}{2}$. $(f \circ g)(x)$ is defined whenever both $g(x)$ and $f(g(x))$ are defined; that is whenever $x \neq -1$. So, the domain is $\{x \mid x \neq -1\}$.

$(g \circ f)(x) = g\left(\dfrac{1}{x - 1}\right) = \dfrac{\dfrac{1}{x - 1} - 1}{\dfrac{1}{x - 1} + 1} = \dfrac{\dfrac{1 - (x - 1)}{x - 1}}{\dfrac{1 + x - 1}{x - 1}} = \dfrac{2 - x}{x}$. $(g \circ f)(x)$ is defined whenever both $f(x)$ and $g(f(x))$ are defined; that is whenever $x \neq 1$ and $x \neq 0$. So, the domain is $\{x \mid x \neq 1, 0\}$.

$(f \circ f)(x) = f\left(\dfrac{1}{x - 1}\right) = \dfrac{1}{\dfrac{1}{x - 1} - 1} = \dfrac{1}{\dfrac{1 - (x - 1)}{x - 1}} = \dfrac{x - 1}{2 - x}$. $(f \circ f)(x)$ is defined whenever both $f(x)$ and $f(f(x))$ are defined; that is whenever $x \neq 1$ and $x \neq 2$. So, the domain is $\{x \mid x \neq 1, 2\}$.

$(g \circ g)(x) = g\left(\dfrac{x - 1}{x + 1}\right) = \dfrac{\dfrac{x - 1}{x + 1} - 1}{\dfrac{x - 1}{x + 1} + 1} = \dfrac{x - 1 - (x + 1)}{x - 1 + x + 1} = \dfrac{-2}{2x} = -\dfrac{1}{x}$. $(g \circ g)(x)$ is defined whenever both $g(x)$ and $g(g(x))$ are defined; that is whenever $x \neq -1$ and $x \neq 0$. So, the domain is $\{x \mid x \neq -1, 0\}$.

35. $f(x) = \sqrt[3]{x}$, domain is $(-\infty, \infty)$; $g(x) = 1 - \sqrt{x}$, domain is $[0, \infty)$.

$(f \circ g)(x) = f(1 - \sqrt{x}) = \sqrt[3]{1 - \sqrt{x}}$. $(f \circ g)(x)$ is defined whenever both $g(x)$ and $f(g(x))$ are defined. Since $f(x)$ has no restriction, the domain is $[0, \infty)$.

$(g \circ f)(x) = g(\sqrt[3]{x}) = 1 - \sqrt{\sqrt[3]{x}} = 1 - \sqrt[6]{x}$. $(g \circ f)(x)$ is defined whenever both $f(x)$ and $g(f(x))$ are defined; that is whenever $x \geq 0$. So, the domain is $[0, \infty)$.

$(f \circ f)(x) = f(\sqrt[3]{x}) = \sqrt[3]{\sqrt[3]{x}} = \sqrt[9]{x}$. $(f \circ f)(x)$ is defined whenever both $f(x)$ and $f(f(x))$ are defined. Since $f(x)$ is defined everywhere the domain is $(-\infty, \infty)$.

$(g \circ g)(x) = g(1 - \sqrt{x}) = 1 - \sqrt{1 - \sqrt{x}}$. $(g \circ g)(x)$ is defined whenever both $g(x)$ and $g(g(x))$ are defined; that is whenever $x \geq 0$ and $1 - \sqrt{x} \geq 0$. So the domain is $\{x \mid x \geq 0 \text{ and } 1 - \sqrt{x} \geq 0\} = \{ x \mid x \geq 0 \text{ and } x \leq 1 \} = [0, 1]$.

37. $f(x) = \dfrac{x+2}{2x+1}$, domain is $\{x \mid x \neq -\dfrac{1}{2} \}$. $g(x) = \dfrac{x}{x-2}$, domain is $\{ x \mid x \neq 2 \}$.

$(f \circ g)(x) = f\left(\dfrac{x}{x-2}\right) = \dfrac{\frac{x}{x-2} + 2}{2\left(\frac{x}{x-2}\right) + 1} = \dfrac{x + 2(x-2)}{2x + x - 2} = \dfrac{3x - 4}{3x - 2}$. $(f \circ g)(x)$ is defined whenever both $g(x)$ and $f(g(x))$ are defined; that is whenever $x \neq 2$ and $3x - 2 \neq 0$. So, the domain is $\{x \mid x \neq 2 \text{ and } 3x - 2 \neq 0\} = \{x \mid x \neq 2 \text{ and } x \neq \frac{2}{3}\}$.

$(g \circ f)(x) = g\left(\dfrac{x+2}{2x+1}\right) = \dfrac{\dfrac{x+2}{2x+1}}{\dfrac{x+2}{2x+1} - 2} = \dfrac{x+2}{x+2-2(2x+1)} = -\dfrac{x+2}{3x}$. $(g \circ f)(x)$ is defined whenever both $f(x)$ and $g(f(x))$ are defined; that is whenever $x \neq -\frac{1}{2}$ and $x \neq 0$. So, the domain is $\{x \mid x \neq -\frac{1}{2} \text{ and } x \neq 0\}$.

$(f \circ f)(x) = f\left(\dfrac{x+2}{2x+1}\right) = \dfrac{\dfrac{x+2}{2x+1} + 2}{2\left(\dfrac{x+2}{2x+1}\right) + 1} = \dfrac{x + 2 + 2\,(2x+1)}{2\,(x+2) + 2x + 1} = \dfrac{5x+4}{4x+5}$. $(f \circ f)(x)$ is defined whenever both $f(x)$ and $f(f(x))$ are defined; that is whenever $x \neq -\frac{1}{2}$ and $x \neq -\frac{5}{4}$. So the domain is $\{x \mid x \neq -\frac{1}{2} \text{ and } x \neq -\frac{5}{4}\}$.

$(g \circ g)(x) = g\left(\frac{x}{x-2}\right) = \dfrac{\frac{x}{x-2}}{\frac{x}{x-2} - 2} = \dfrac{x}{x - 2\,(x-2)} = \dfrac{x}{4-x}$. $(g \circ g)(x)$ is defined whenever both $g(x)$ and $g(g(x))$ are defined; that is whenever $x \neq 2$ and $x \neq 4$. So, the domain is $\{x \mid x \neq 2 \text{ and } x \neq 4\}$.

39. $(f \circ g \circ h)(x) = f(g(h(x))) = f(g(x-1) = f(\sqrt{x-1}) = \sqrt{x-1} - 1$

41. $(f \circ g \circ h)(x) = f(g(h(x))) = f(g(\sqrt{x})) = f(\sqrt{x} - 5) = (\sqrt{x} - 5)^4 + 1$

43. $F(x) = (x-9)^5$. Let $f(x) = x^5$ and $g(x) = x - 9$, then $F(x) = (f \circ g)(x)$.

45. $G(x) = \frac{x^2}{x^2+4}$. Let $f(x) = \frac{x}{x+4}$ and $g(x) = x^2$, then $G(x) = (f \circ g)(x)$.

47. $H(x) = |1 - x^3|$. Let $f(x) = |x|$ and $g(x) = 1 - x^3$, then $H(x) = (f \circ g)(x)$.

For Exercises 49 and 51 there are several possible solutions, only one of which is shown.

49. $F(x) = \dfrac{1}{x^2+1}$. Let $f(x) = \dfrac{1}{x}$, $g(x) = x + 1$, and $h(x) = x^2$, then $F(x) = (f \circ g \circ h)(x)$.

51. $G(x) = (4 + \sqrt[3]{x})^9$. Let $f(x) = x^9$, $g(x) = 4 + x$ and $h(x) = \sqrt[3]{x}$, then $G(x) = (f \circ g \circ h)(x)$.

53. Let r be the radius of the circular ripple in cm. Since the ripple travels at a speed of 60 cm/s, the distance traveled in t seconds is the radius, so $r = 60t$. Therefore the area of the circle can be written as $A(t) = \pi r^2 = \pi(60t)^2 = 3600\pi t^2$ cm^2.

55. Let t be the time since the plane flew over the radar station.

(a) Since *distance* = *rate* × *time* we have $d = g(t) = 350t$

(b) Using Pythagorean theorem, $s = f(d) = \sqrt{1 + d^2}$.

(c) $s(t) = (f \circ g)(t) = f(350t) = \sqrt{1 + (350t)^2} = \sqrt{1 + 122500t^2}$.

57. $f(x) = 3x + 5$ and $h(x) = 3x^2 + 3x + 2$. We wish to find g so that $(f \circ g)(x) = h(x)$. Thus $f(g(x)) = 3x^2 + 3x + 2 \iff 3(g(x)) + 5 = 3x^2 + 3x + 2 \iff 3(g(x)) = 3x^2 + 3x - 3$ $\iff g(x) = x^2 + x - 1$.

59. $g(x)$ is even $\Rightarrow g(-x) = g(x)$. Then $h(-x) = f(g(-x)) = f(g(x)) = h(x)$. So yes, h is always an even function.

Exercises 3.11

1. By the Horizontal Line Test, f is not one-to-one.

3. f is one-to-one.

5. f is not one-to-one.

7. $f(x) = 7x - 3$. If $x_1 \neq x_2$, then $7x_1 \neq 7x_2$ and $7x_1 - 3 \neq 7x_2 - 3$. So f is one-to-one.

9. $g(x) = \sqrt{x}$. If $x_1 \neq x_2$, then $\sqrt{x_1} \neq \sqrt{x_2}$ because two different numbers cannot have the same square root. Therefore g is one-to-one.

11. $h(x) = x^4 + 5$. Since every number and its negative have the same fourth power, for example, $(-1)^4 = 1 = (1)^4$. So $h(-1) = h(1)$ and h is not a one-to-one function.

13. (a) $f(2) = 7$. Since f is one-to-one, $f^{-1}(7) = 2$.

    (b) $f^{-1}(3) = -1$. Since f is one-to-one, $f(-1) = 3$.

15. $f(x) = 5 - 2x$. Since f is one-to-one and $f(1) = 5 - 2(1) = 3$, then $f^{-1}(3) = 1$. (Find 1 by solving the equation $5 - 2x = 3$.)

17. $f(g(x)) = f\left(\dfrac{x+5}{2}\right) = 2\left(\dfrac{x+5}{2}\right) - 5 = x + 5 - 5 = x$, for all x.

    $g(f(x)) = g(2x - 5) = \dfrac{(2x-5)+5}{2} = x$, for all x. Thus f and g are inverses of each other.

19. $f(g(x)) = f(\sqrt{x+4}) = (\sqrt{x+4})^2 - 4 = x + 4 - 4 = x$, for all $x \geq -4$.

    $g(f(x)) = g(x^2 - 4) = \sqrt{(x^2-4)+4} = \sqrt{x^2} = x$, for all $x \geq 0$. Thus f and g are inverses of each other.

21. $f(g(x)) = f\left(\dfrac{1}{x} + 1\right) = \dfrac{1}{\left(\dfrac{1}{x}+1\right) - 1} = x$, for all $x \neq 0$.

    $g(f(x)) = g\left(\dfrac{1}{x-1}\right) = \dfrac{1}{\left(\dfrac{1}{x-1}\right)} + 1 = (x-1) + 1 = x$, for all $x \neq 1$. Thus f and g are inverses of each other.

23. $f(x) = 2x + 1$. $y = 2x + 1 \;\Leftrightarrow\; 2x = y - 1 \;\Leftrightarrow\; x = \frac{1}{2}(y-1)$. So $f^{-1}(x) = \frac{1}{2}(x-1)$.

25. $f(x) = 4x + 7$. $y = 4x + 7 \;\Leftrightarrow\; 4x = y - 7 \;\Leftrightarrow\; x = \frac{1}{4}(y-7)$. Therefore $f^{-1}(x) = \frac{1}{4}(x-7)$.

27. $f(x) = \dfrac{1}{x+2}$, $x > -2$. $y = \dfrac{1}{x+2} \;\Leftrightarrow\; x + 2 = \dfrac{1}{y} \;\Leftrightarrow\; x = \dfrac{1}{y} - 2$. So $f^{-1}(x) = \dfrac{1}{x} - 2$, $x > 0$.

29. $f(x) = \dfrac{1+3x}{5-2x}$. $y = \dfrac{1+3x}{5-2x} \quad \Leftrightarrow \quad y\,(5-2x) = 1+3x \quad \Leftrightarrow \quad 5y - 2xy = 1+3x \quad \Leftrightarrow$
$3x + 2xy = 5y - 1 \quad \Leftrightarrow \quad x\,(3+2y) = 5y-1 \quad \Leftrightarrow \quad x = \dfrac{5y-1}{2y+3}$. So $f^{-1}(x) = \dfrac{5x-1}{2x+3}$.

31. $f(x) = \sqrt{2+5x}, x \geq -\frac{2}{5}$. $y = \sqrt{2+5x}, y \geq 0 \quad \Leftrightarrow \quad y^2 = 2+5x \quad \Leftrightarrow \quad 5x = y^2 - 2 \quad \Leftrightarrow$
$x = \frac{1}{5}(y^2-2)$ and $y \geq 0$. So $f^{-1}(x) = \frac{1}{5}(x^2-2)$, $x \geq 0$.

33. $f(x) = 4 - x^2$, $x \geq 0$. $y = 4 - x^2 \quad \Leftrightarrow \quad x^2 = 4 - y \quad \Leftrightarrow \quad x = \sqrt{4-y}$. So
$f^{-1}(x) = \sqrt{4-x}$. Note: $x \geq 0 \quad \Rightarrow \quad f(x) \leq 4$.

35. $f(x) = 4 + \sqrt[3]{x}$. $y = 4 + \sqrt[3]{x} \quad \Leftrightarrow \quad \sqrt[3]{x} = y - 4 \quad \Leftrightarrow \quad x = (y-4)^3$. So $f^{-1}(x) = (x-4)^3$.

37. $f(x) = 1 + \sqrt{1+x}$. $y = 1 + \sqrt{1+x}, y \geq 1 \quad \Leftrightarrow \quad \sqrt{1+x} = y - 1 \quad \Leftrightarrow \quad 1 + x = (y-1)^2$
$\Leftrightarrow \quad x = (y-1)^2 - 1 = y^2 - 2y$. So $f^{-1}(x) = x^2 - 2x$, $x \geq 1$.

39. $f(x) = x^4, x \geq 0$. $y = x^4, y \geq 0 \quad \Leftrightarrow \quad x = \sqrt[4]{y}$. So $f^{-1}(x) = \sqrt[4]{x}$, $x \geq 0$.

41. (a)

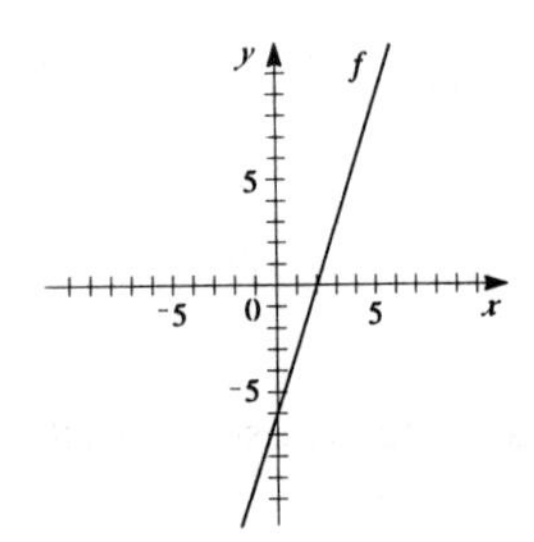

(b)

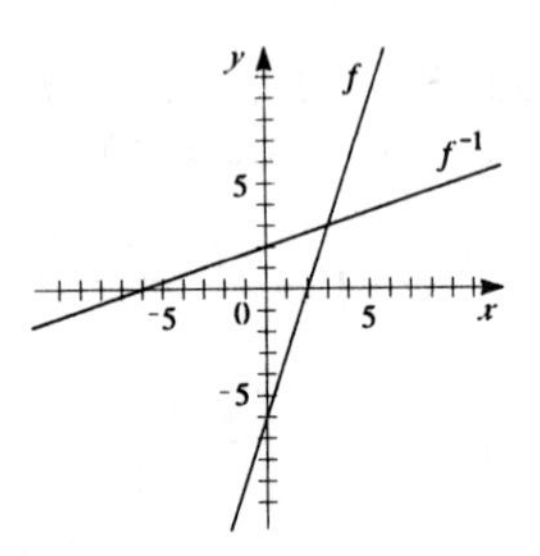

(c) $f(x) = 3x - 6$. $y = 3x - 6 \quad \Leftrightarrow \quad 3x = y + 6 \quad \Leftrightarrow \quad x = \frac{1}{3}(y+6)$. So
$f^{-1}(x) = \frac{1}{3}(x+6)$.

43. (a)

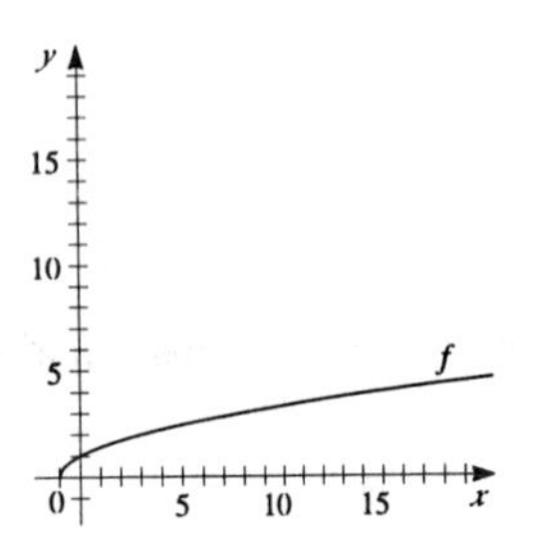

(b)

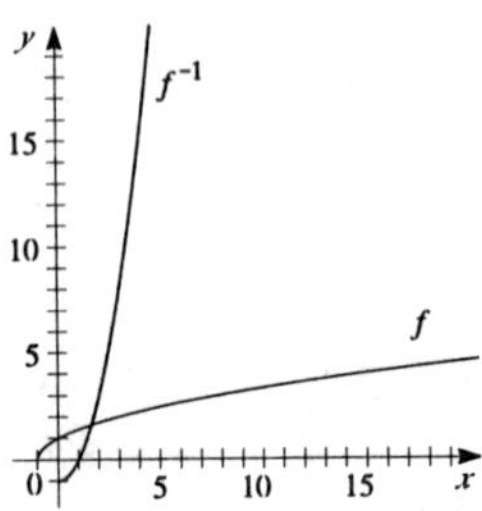

(c) $f(x) = \sqrt{x+1}, x \geq -1$. $y = \sqrt{x+1}, y \geq 0 \quad \Leftrightarrow \quad y^2 = x + 1 \quad \Leftrightarrow \quad x = y^2 - 1$ and
$y \geq 0$. So $f^{-1}(x) = x^2 - 1$, $x \geq 0$.

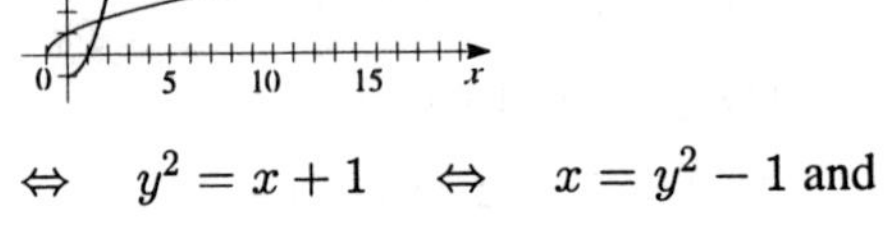

45. $f(x) = x^3 - x$. Using a graphing device and the Horizontal Line Test, we see that f is not a one-to-one function. For example, $f(0) = 0 = f(-1)$.

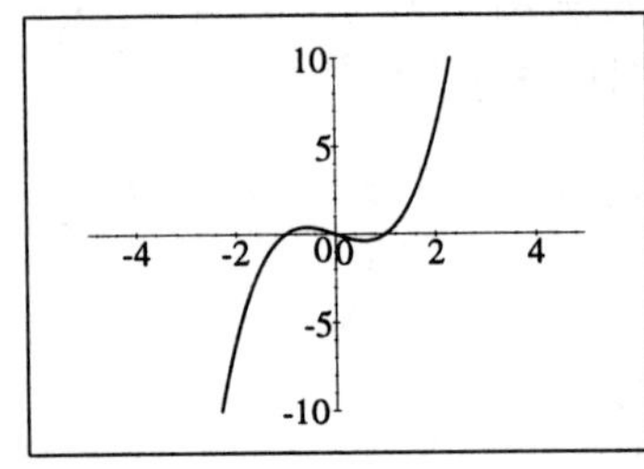

47. $f(x) = \dfrac{x+12}{x-6}$. Using a graphing device and the Horizontal Line Test, we see that f is a one-to-one function.

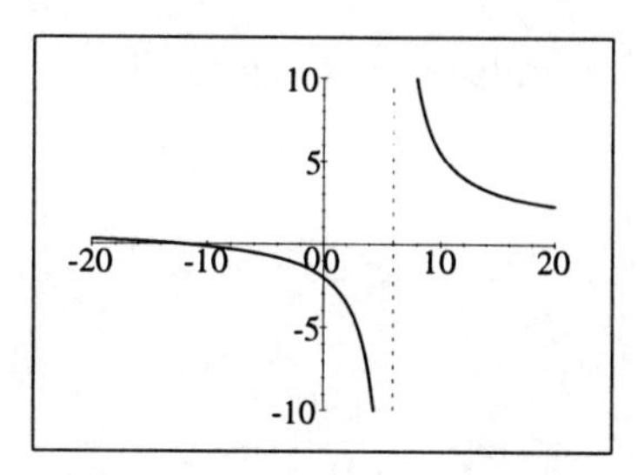

49. Consider two numbers x_1 and x_2. When $x_1 < x_2$, since f is increasing, $f(x_1) < f(x_2)$. Also, when $x_1 > x_2$, since f is increasing, $f(x_1) > f(x_2)$. Therefore, whenever $x_1 \neq x_2$ it follows that $f(x_1) \neq f(x_2)$. Thus, by definition, f is a one-to-one function.

51. $f(x) = mx + b$. $f(x_1) = f(x_2) \quad \Leftrightarrow \quad mx_1 + b = mx_2 + b \quad \Leftrightarrow \quad mx_1 = mx_2$. We can conclude that $x_1 = x_2$ only if $m \neq 0$. Therefore f is one-to-one if $m \neq 0$. If $m \neq 0$, $f(x) = mx + b \quad \Leftrightarrow \quad y = mx + b \quad \Leftrightarrow \quad mx = y - b \quad \Leftrightarrow \quad x = \dfrac{y-b}{m}$. So, $f^{-1}(x) = \frac{x-b}{m}$.

Review Exercises for Chapter 3

1. (a)

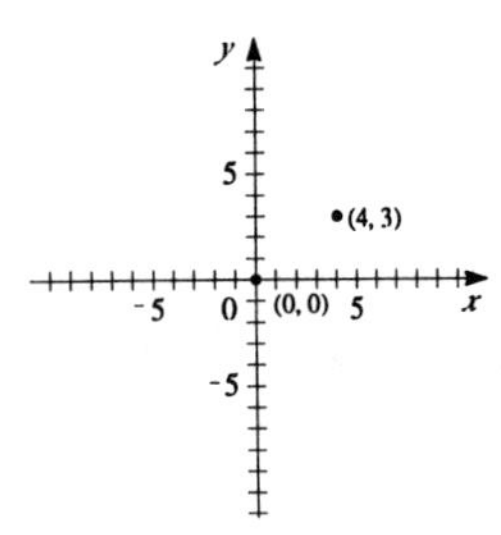

(b) The distance from P to Q is $d(PQ) = \sqrt{(4-0)^2 + (3-0)^2} = \sqrt{25} = 5$.

(c) Midpoint $= \left(\frac{4+0}{2}, \frac{3+0}{2}\right) = \left(2, \frac{3}{2}\right)$.

(d) The line has slope $m = \frac{3-0}{4-0} = \frac{3}{4}$ and its equation is $y - 3 = \frac{3}{4}(x-4) \quad \Leftrightarrow$ $y = \frac{3}{4}x$.

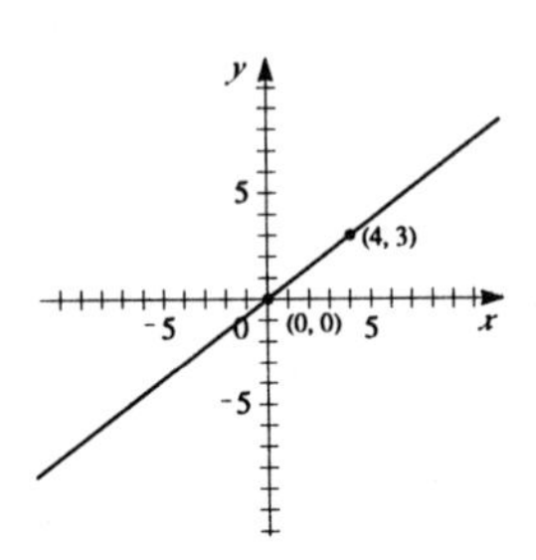

(e) The radius of this circle was found in part (a), it is $r = d(PQ) = 5$. So the equation is $(x-0)^2 + (y-0)^2 = 5^2$ $\Leftrightarrow \quad x^2 + y^2 = 25$.

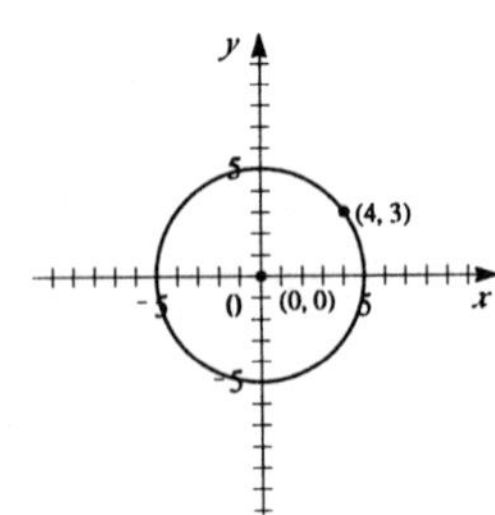

3. (a)

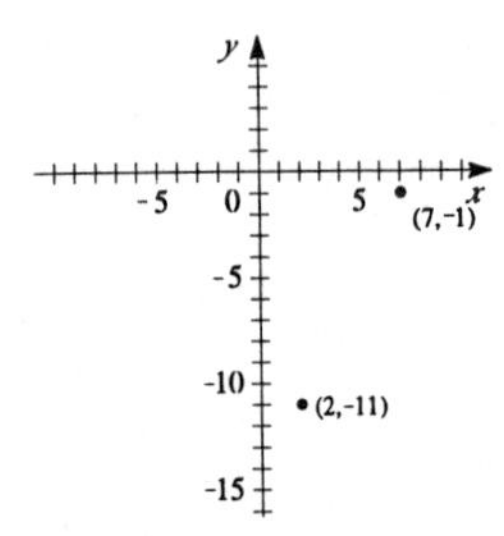

(b) The distance from P to Q is $d(PQ) = \sqrt{(2-7)^2 + (-11+1)^2} = \sqrt{25+100} = \sqrt{125}$ $= 5\sqrt{5}$.

(c) Midpoint $= \left(\frac{2+7}{2}, \frac{-11-1}{2}\right) = \left(\frac{9}{2}, -6\right)$.

(d) The line has slope $m = \frac{-11+1}{2-7} = \frac{-10}{-5} = 2$ and its equation is $y + 11 = 2(x - 2) \Leftrightarrow$ $y + 11 = 2x - 4 \Leftrightarrow y = 2x - 15$.

(e) The radius of this circle was found in part a, it is $r = |PQ| = 5\sqrt{5}$. So the equation is

$(x-7)^2 + (y+1)^2 = \left(5\sqrt{5}\right)^2 \Leftrightarrow$
$(x-7)^2 + (y+1)^2 = 125$.

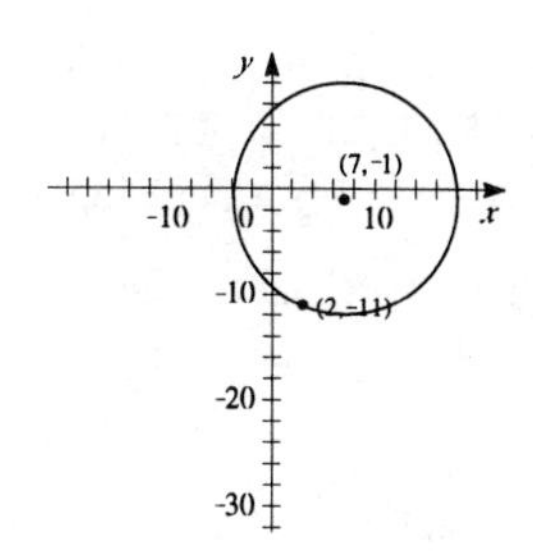

5.

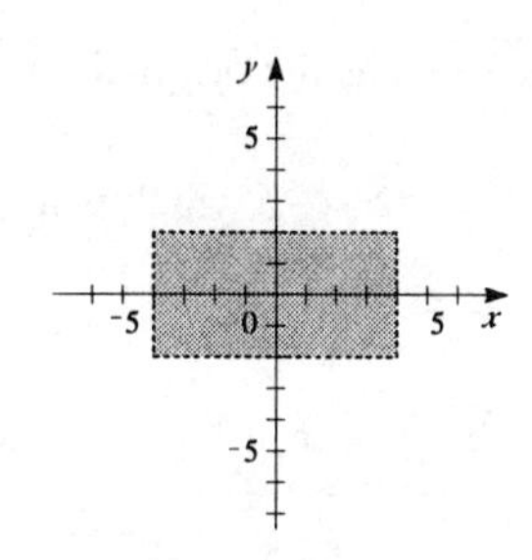

7. $d(AC) = \sqrt{(4+1)^2 + (4+3)^2} = \sqrt{74}$ and $d(BC) = \sqrt{(5+1)^2 + (3+3)^2} = \sqrt{72}$. Therefore B is closer to C.

9. The circle with center at $(2, -5)$ and radius $\sqrt{2}$ has equation $(x-2)^2 + (y+5)^2 = \left(\sqrt{2}\right)^2 \Leftrightarrow$ $(x-2)^2 + (y+5)^2 = 2$.

11. The midpoint of segment PQ is $\left(\frac{2-1}{2}, \frac{3+8}{2}\right) = \left(\frac{1}{2}, \frac{11}{2}\right)$ and the radius is $\frac{1}{2}$ of the distance from P to Q, or $r = \frac{1}{2}\sqrt{(2+1)^2 + (3-8)^2} \Leftrightarrow r = \frac{1}{2}\sqrt{34}$. Thus the equation is $(x - \frac{1}{2})^2 + (y - \frac{11}{2})^2 = \frac{17}{2}$.

13. $x^2 + y^2 + 2x - 6y + 9 = 0 \Leftrightarrow (x^2 + 2x) + (y^2 - 6y) = -9 \Leftrightarrow$ $(x^2 + 2x + 1) + (y^2 - 6y + 9) = -9 + 1 + 9 \Leftrightarrow (x+1)^2 + (y-3)^2 = 1$. This equation represents a circle with center at $(-1, 3)$ and radius 1.

15. $x^2 + y^2 + 72 = 12x \Leftrightarrow (x^2 - 12x) + y^2 = -72 \Leftrightarrow (x^2 - 12x + 36) + y^2 = -72 + 36$ $\Leftrightarrow (x-6)^2 + y^2 = -36$. Since the left side of this equation must be greater than or equal to zero, this equation has no graph.

17. The line has slope $m = \frac{-4+6}{2+1} = \frac{2}{3}$ and so, by the point-slope formula, the equation is $y + 4 = \frac{2}{3}(x - 2) \Leftrightarrow y = \frac{2}{3}x - \frac{16}{3} \Leftrightarrow 2x - 3y - 16 = 0$.

19. x-intercept is 4 and y-intercept is 12 $\Rightarrow m = \frac{12-0}{0-4} = -3$. Therefore, by the slope-intercept formula, the equation of the line is $y = -3x + 12$.

21. We first find the slope of the line $3x + 15y = 22$: $3x + 15y = 22 \Leftrightarrow 15y = -3x + 22 \Leftrightarrow$ $y = -\frac{1}{5}x + \frac{22}{15}$. So this line has slope $m = -\frac{1}{5}$ as does any line parallel to it. Then the parallel line passing through the origin has equation $y - 0 = -\frac{1}{5}(x - 0) \Leftrightarrow x + 5y = 0$.

23. Here the center is at $(0,0)$ and the circle passes through the point $(-5,12)$, so the radius is $r=\sqrt{(-5-0)^2+(12-0)^2}=\sqrt{25+144}=\sqrt{169}=13$. The equation of the circle is $x^2+y^2=13^2 \quad \Leftrightarrow \quad x^2+y^2=169$. The line shown is the tangent that passes through the point $(-5,12)$ so it is perpendicular to the line through the points $(0,0)$ and $(-5,12)$. This line has slope $m_1=\frac{12-0}{-5-0}=-\frac{12}{5}$. The slope of the line we seek is $m_2=-\dfrac{1}{m_1}=-\dfrac{1}{-12/5}=\dfrac{5}{12}$. Thus the equation of the tangent line is $y-12=\frac{5}{12}(x+5) \quad \Leftrightarrow \quad y-12=\frac{5}{12}x+\frac{25}{12} \quad \Leftrightarrow$ $y=\frac{5}{12}x+\frac{169}{12} \quad \Leftrightarrow \quad 5x-12y+169=0$.

25. $y=2x-3$
x-axis symmetry: $(-y)=2-3x \quad \Leftrightarrow \quad y=-2+3x$ which is not the same as the original equation. Not symmetric with respect to the x-axis.
y-axis symmetry: $y=2-3(-x) \quad \Leftrightarrow \quad y=2+3x$ which is not the same as the original equation. Not symmetric with respect to the y-axis.
Origin symmetry: $(-y)=2-3(-x) \quad \Leftrightarrow \quad -y=2+3x \quad \Leftrightarrow \quad y=-2-3x$ which is not the same as the original equation. Not symmetric with respect to the origin.
Hence the curve has no symmetry.

x	y
-2	8
0	2
$\frac{2}{3}$	0

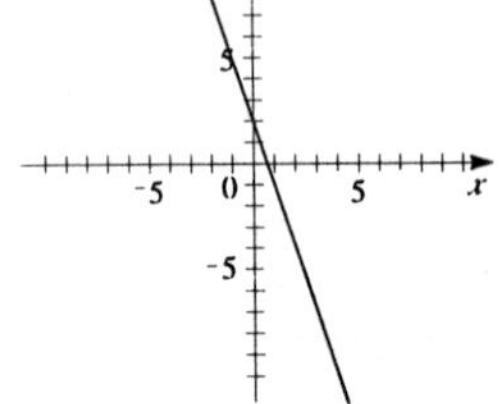

27. $x+3y=21 \quad \Leftrightarrow \quad y=-\frac{1}{3}x+7$
x-axis symmetry: $x+3(-y)=21 \quad \Leftrightarrow \quad x-3y=21$ which is not the same as the original equation. Not symmetric with respect to the x-axis.
y-axis symmetry: $(-x)+3y=21 \quad \Leftrightarrow \quad x-3y=-21$ which is not the same as the original equation. Not symmetric with respect to the y-axis.
Origin symmetry: $(-x)+3(-y)=21 \quad \Leftrightarrow \quad x+3y=-21$ which is not the same as the original equation. Not symmetric with respect to the origin.
Hence the graph has no symmetry.

x	y
-3	8
0	7
21	0

29. $\frac{x}{2}-\frac{y}{7}=1 \quad \Leftrightarrow \quad y=\frac{7}{2}x-7$
x-axis symmetry: $\frac{x}{2}-\dfrac{(-y)}{7}=1 \quad \Leftrightarrow \quad \frac{x}{2}+\frac{y}{7}=1$ which is not the same as the original equation. Not symmetric with respect to the x-axis.
y-axis symmetry: $\dfrac{(-x)}{2}-\frac{y}{7}=1 \quad \Leftrightarrow \quad \frac{x}{2}+\frac{y}{7}=-1$ which is not the same as the original equation. Not symmetric with respect to the y-axis.
Origin symmetry: $\dfrac{(-x)}{2}-\dfrac{(-y)}{7}=1 \quad \Leftrightarrow \quad \frac{x}{2}-\frac{y}{7}=-1$ which is not the same as the original

equation. Not symmetric with respect to the origin.
Hence the graph has no symmetry.

x	y
-2	-14
0	-7
2	0

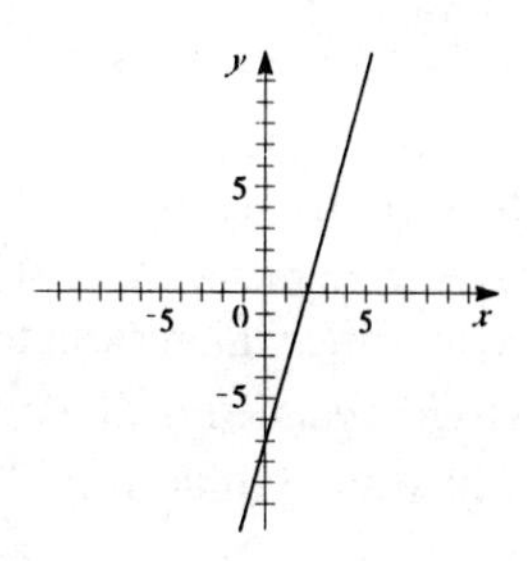

31. $y = 16 - x^2$
x-axis symmetry: $(-y) = 16 - x^2 \quad \Leftrightarrow \quad y = -16 + x^2$ which is not the same as the original equation. Not symmetric with respect to the x-axis.
y-axis symmetry: $y = 16 - (-x)^2 \quad \Leftrightarrow \quad y = 16 - x^2$ which is the same as the original equation. Symmetric with respect to the y-axis.
Origin symmetry: $(-y) = 16 - (-x)^2 \quad \Leftrightarrow \quad y = -16 + x^2$ which is not the same as the original equation. Not symmetric with respect to the origin.
Hence the graph has no symmetry.

x	y
-3	7
-1	15
0	16
1	15
3	7

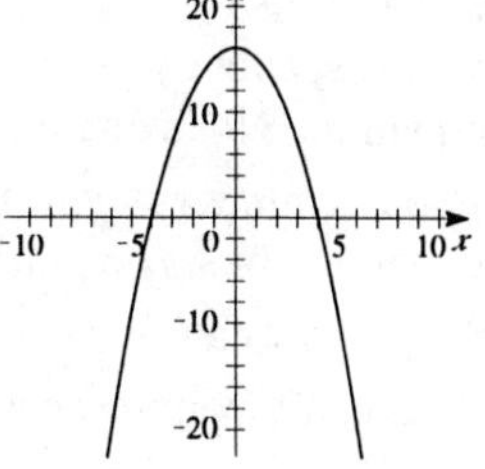

33. $x = \sqrt{y}$
x-axis symmetry: $x = \sqrt{-y}$ which is not the same as the original equation. Not symmetric with respect to the x-axis.

y-axis symmetry: $(-x) = \sqrt{y} \quad \Leftrightarrow \quad x = -\sqrt{y}$ which is not the same as the original equation. Not symmetric with respect to the y-axis.

Origin symmetry: $(-x) = \sqrt{-y}$ which is not the same as the original equation. Not symmetric with respect to the origin.
Hence the graph has no symmetry.

x	y
0	0
1	1
2	4
3	9

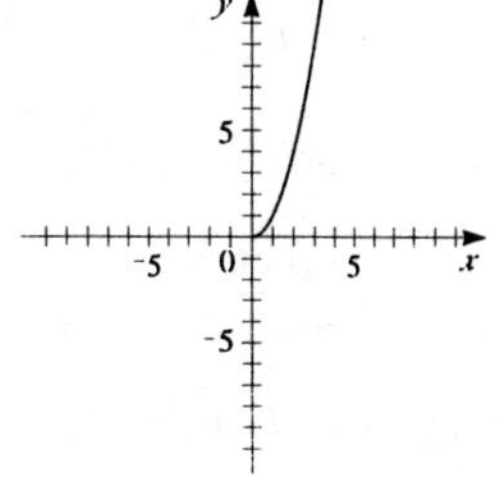

35. $2x^2 + 2y^2 = 5 \quad \Leftrightarrow \quad x^2 + y^2 = \frac{5}{2}$
x-axis symmetry: $2x^2 + 2(-y)^2 = 5 \quad \Leftrightarrow \quad 2x^2 + 2y^2 = 5$ which is the original equation. Symmetric with respect to the x-axis.
y-axis symmetry: $2(-x)^2 + 2y^2 = 5 \quad \Leftrightarrow \quad 2x^2 + 2y^2 = 5$ which is the original equation. Symmetric with respect to the y-axis.
Origin symmetry: $2(-x)^2 + 2(-y)^2 = 5 \quad \Leftrightarrow \quad 2x^2 + 2y^2 = 5$ which is not changed. Symmetric with respect to the origin
So the curve is symmetric with respect to the x-axis, the y-axis, and the origin.

x	y
$\pm\frac{5}{2}$	0
0	$\pm\frac{5}{2}$

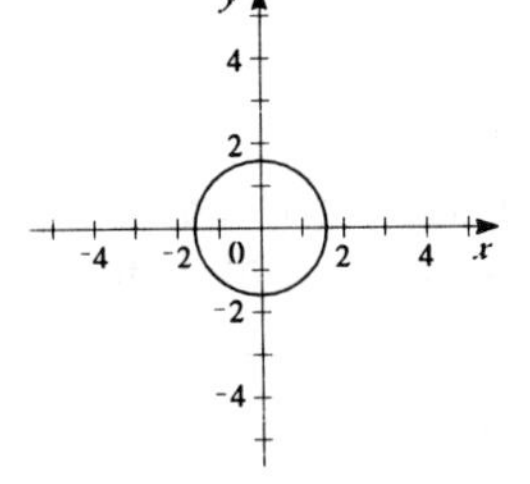

37. $y = x^3 - 4x$.
x-axis symmetry: $(-y) = x^3 - 4x \quad \Leftrightarrow \quad y = -x^3 + 4x$ which is not the same as the original equation. Not symmetric with respect to the x-axis.
y-axis symmetry: $y = (-x)^3 - 4(-x) \quad \Leftrightarrow \quad y = -x^3 + 4x$ which is not the same as the original equation. Not symmetric with respect to the y-axis.
Origin symmetry: $(-y) = (-x)^3 - 4(-x) \quad \Leftrightarrow \quad y = x^3 - 4x$ which is the original equation. Symmetric with respect to the origin.

x	y
-2	0
-1	3
0	0
1	-3
2	0

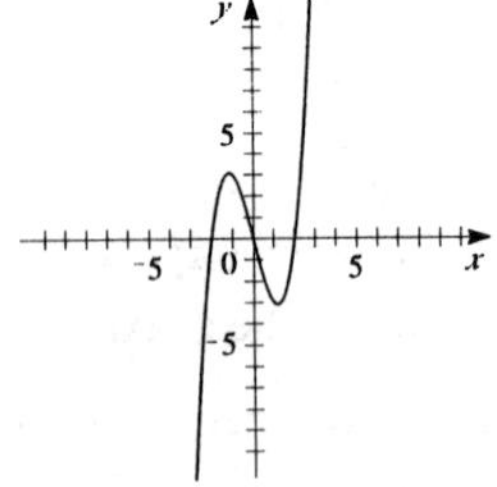

39. $f(x) = x^2 - x + 1$; $f(0) = (0)^2 - (0) + 1 = 1$; $f(2) = (2)^2 - (2) + 1 = 3$;
$f(-2) = (-2)^2 - (-2) + 1 = 7$; $f(a) = (a)^2 - (a) + 1 = a^2 - a + 1$;
$f(-a) = (-a)^2 - (-a) + 1 = a^2 + a + 1$;
$f(x+1) = (x+1)^2 - (x+1) + 1 = x^2 + 2x + 1 - x - 1 + 1 = x^2 + x + 1$;
$f(2x) = (2x)^2 - (2x) + 1 = 4x^2 - 2x + 1$;
$2f(x) - 2 = 2(x^2 - x + 1) - 2 = 2x^2 - 2x + 2 - 2 = 2x^2 - 2x$.

41. (a) $f(-2) = -1$. $f(2) = 2$.
(b) The domain of f is $[-4, 5]$.
(c) The range of f is $[-4, 4]$.
(d) f is increasing on $[-4, -2]$, $[-1, 4]$, and f is decreasing on $[-2, -1]$, $[4, 5]$.
(e) f is not one-to-on, for example, $f(-2) = -1 = f(0)$. There are many more examples.

43. Domain: $x + 3 \geq 0 \quad \Leftrightarrow \quad x \geq -3$. In interval notation the domain is $[-3, \infty)$.
Range: For x in the domain of f we have $x \geq -3 \quad \Leftrightarrow \quad x + 3 \geq 0 \quad \Leftrightarrow \quad \sqrt{x+3} \geq 0 \quad \Leftrightarrow$ $f(x) \geq 0$. So range is $[0, \infty)$.

45. $f(x) = 7x + 15$. Domain is all real numbers, $(-\infty, \infty)$

47. $f(x) = \sqrt{x+4}$. We require $x + 4 \geq 0 \quad \Leftrightarrow \quad x \geq -4$. Thus the domain is $[-4, \infty)$.

49. $f(x) = \frac{1}{x} + \frac{1}{x+1} + \frac{1}{x+2}$. The denominators cannot equal 0, therefore the domain is $\{x | x \neq 0, -1, -2\}$.

51. $h(x) = \sqrt{4-x} + \sqrt{x^2-1}$. We require the expression inside the radicals be nonnegative. So $4 - x \geq 0 \quad \Leftrightarrow \quad 4 \geq x$ and $x^2 - 1 \geq 0 \quad \Leftrightarrow \quad (x-1)(x+1) \geq 0$, using the methods from Chapter 2, we have:

Interval	$(-\infty, -1)$	$(-1, 1)$	$(1, \infty)$
Sign of $x - 1$	$-$	$-$	$+$
Sign of $x + 1$	$-$	$+$	$+$
Sign of $(x-1)(x+1)$	$+$	$-$	$+$

Thus the domain is $(-\infty, 4] \cap \{(-\infty, -1] \cup [1, \infty)\} = (-\infty, -1] \cup [1, 4]$

53. $f(x) = 1 - 2x$

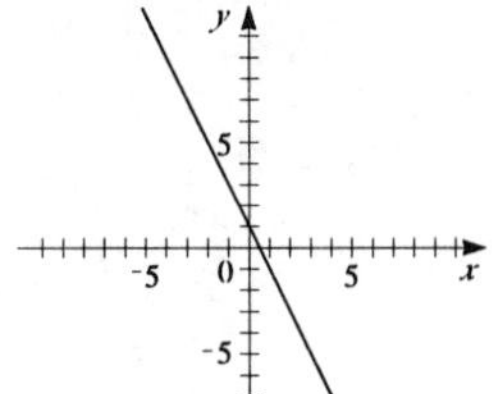
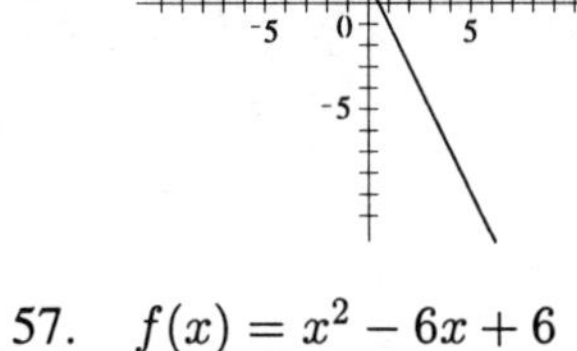

55. $f(t) = 1 - \frac{1}{2}t^2$

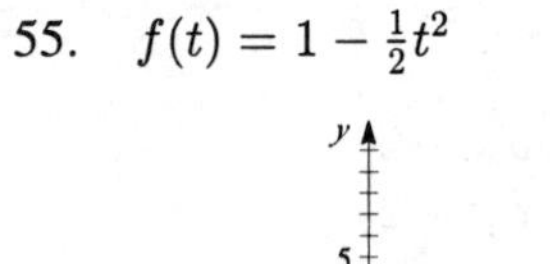
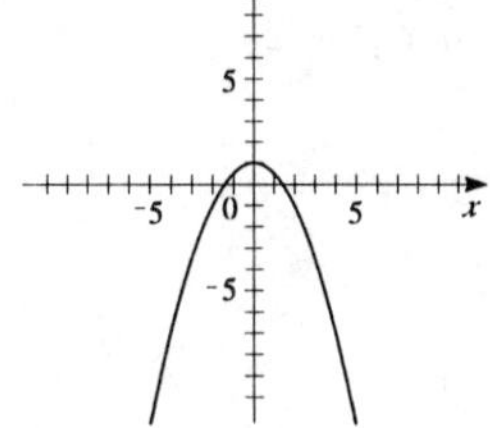

57. $f(x) = x^2 - 6x + 6$

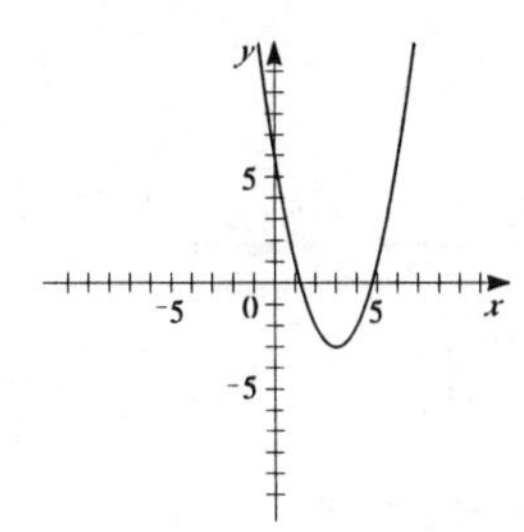

59. $y = 1 - \sqrt{x}$

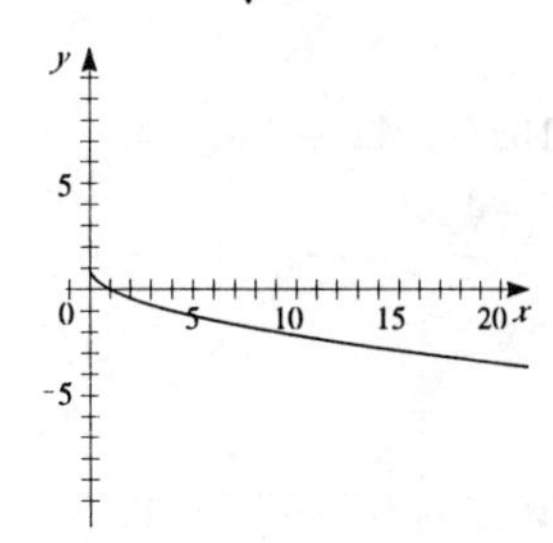

61. $y = \frac{1}{2}x^3$

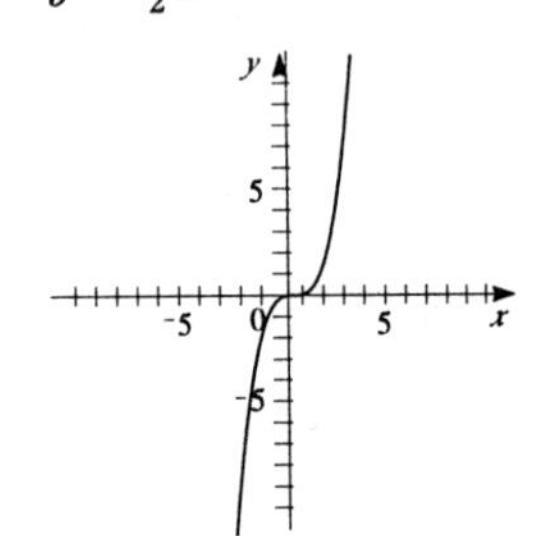

63. $h(x) = \sqrt[3]{x}$

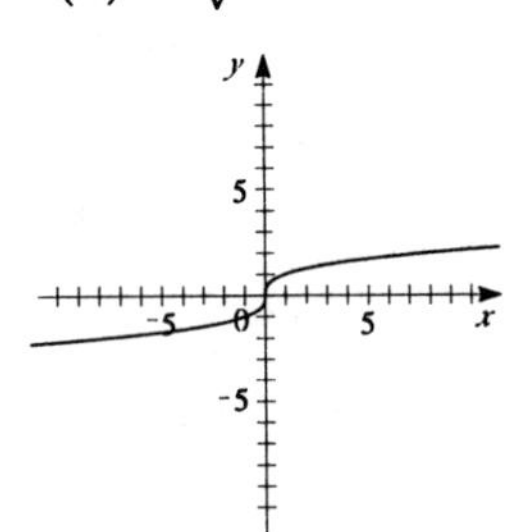

65. $g(x) = \dfrac{1}{x^2}$

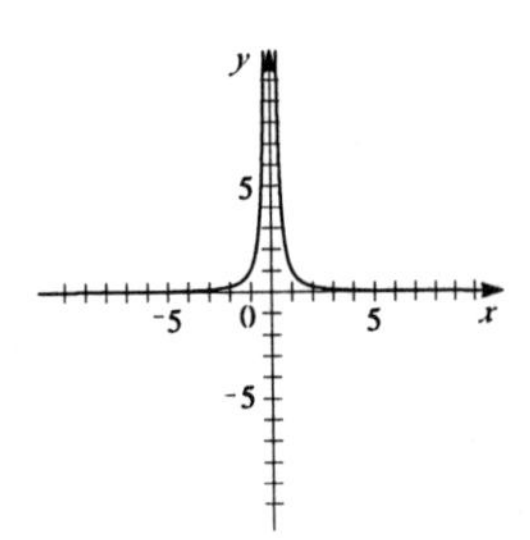

67. $f(x) = \begin{cases} 1 - x & \text{if } x < 0 \\ 1 & \text{if } x \geq 0 \end{cases}$

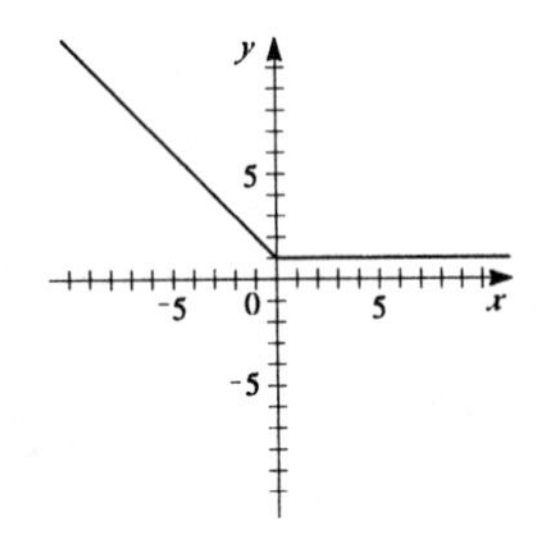

69. $f(x) = \begin{cases} x + 6 & \text{if } x < -2 \\ x^2 & \text{if } x \geq -2 \end{cases}$

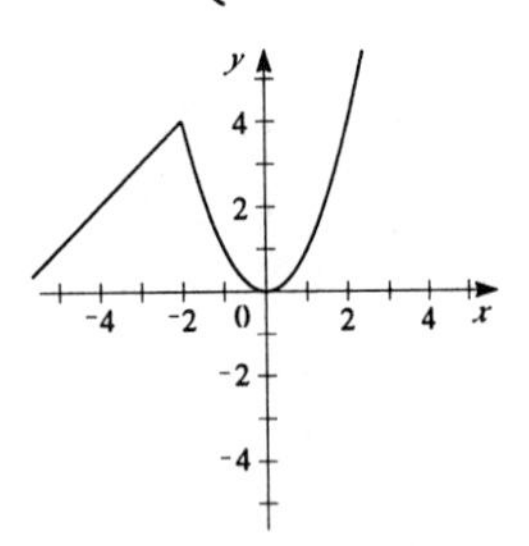

71. $f(x) = 6x^3 - 15x^2 + 4x - 1$

(i) $[-2, 2]$ by $[-4, 4]$

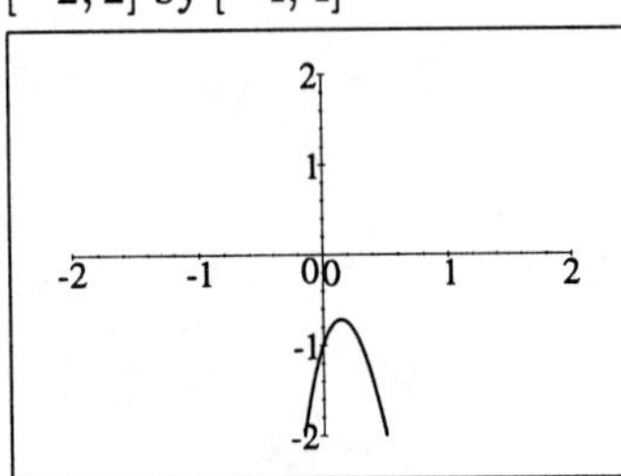

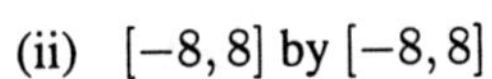

(ii) $[-8, 8]$ by $[-8, 8]$

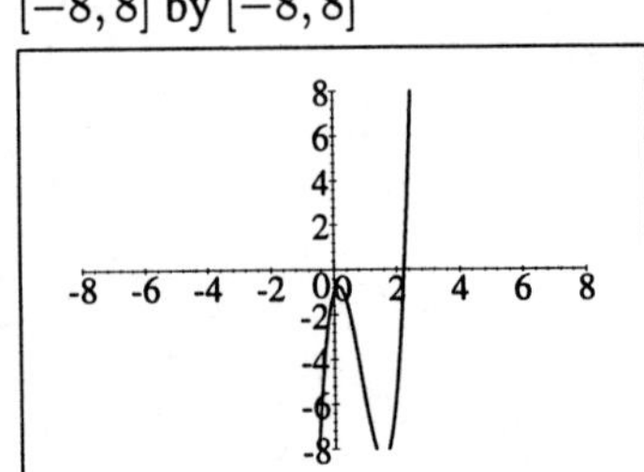

(iii) $[-4, 4]$ by $[-12, 12]$

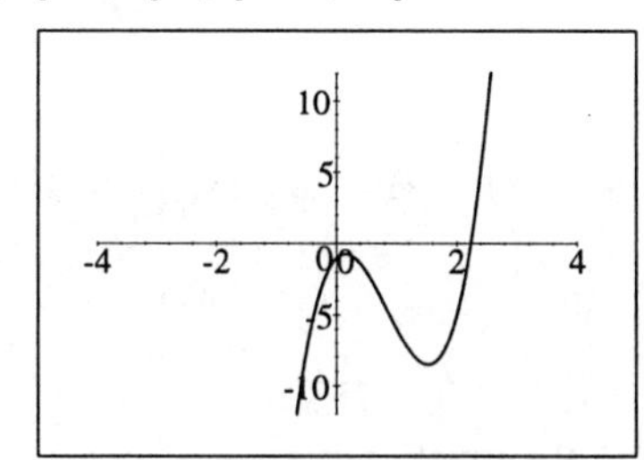

(iv) $[-100, 100]$ by $[-100, 100]$

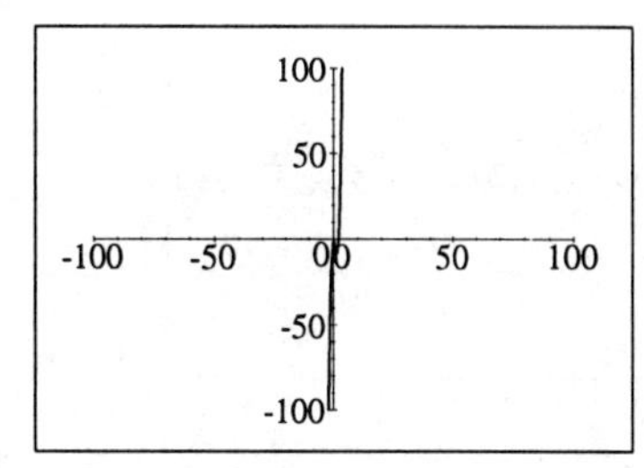

From the graphs we see that the viewing rectangle in (iii) produces the most appropriate graph.

73. $f(x) = x^2 + 25x + 173$
$= (x^2 + 25x + \frac{625}{4}) + 173 - \frac{625}{4} = (x + \frac{25}{2})^2 + \frac{67}{4}$.
Viewing rectangle $[-25, 0]$ by $[15, 40]$.

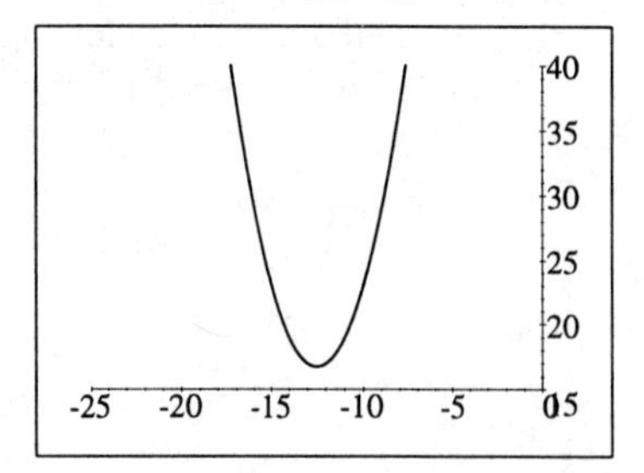

75. $y = \dfrac{x}{\sqrt{x^2 + 16}}$. Since $\sqrt{x^2 + 16} \geq \sqrt{x^2} = |x|$, so y should behave like $\frac{x}{|x|}$. Thus we just the viewing rectangle: $[-10, 10]$ by $[-2, 2]$.

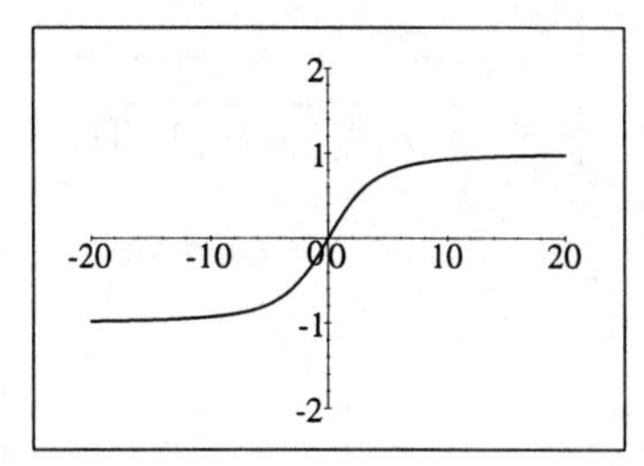

77. $x^3 - 3x - 1 = 0$. We graph the function $f(x) = x^3 - 3x - 1$ in the first viewing rectangle $[-2, 2]$ by $[-4, 4]$and observe that there are three solutions to the equation $f(x) = 0$. In the second viewing rectangle $[-1.55, -1.50]$ by $[-0.5, 0.5]$ we find that one solution is $x \approx -1.53$.

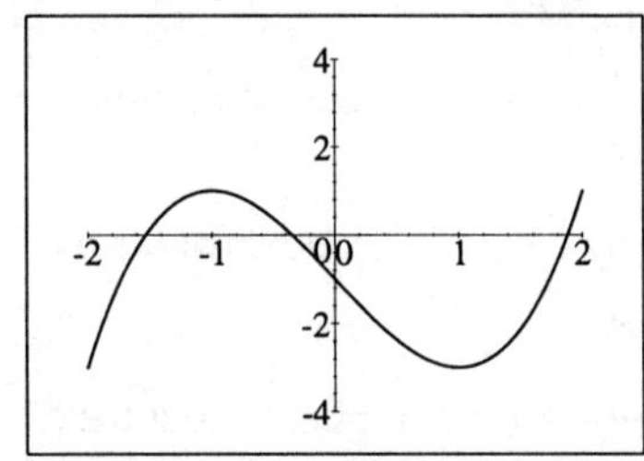

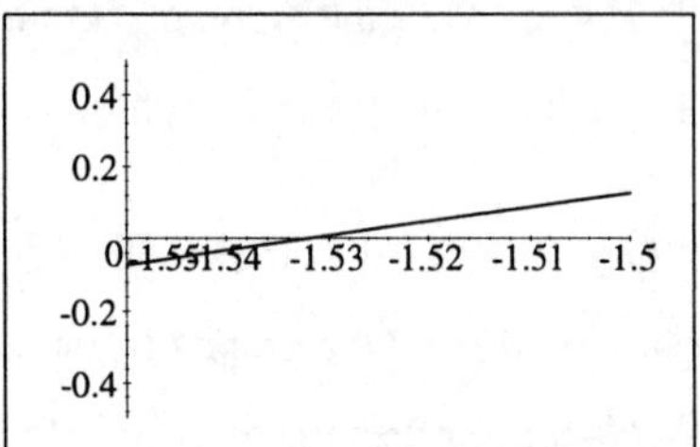

In the next viewing rectangle $[-0.40, -0.34]$ by $[-0.5, 0.5]$ we find the solution $x \approx -0.35$. And in the final viewing rectangle $[1.80, 1.90]$ by $[-0.5, 0.5]$ we find the solution $x \approx 1.88$. Thus the solutions are $x \approx -1.53,\ -0.35,\ 1.88$.

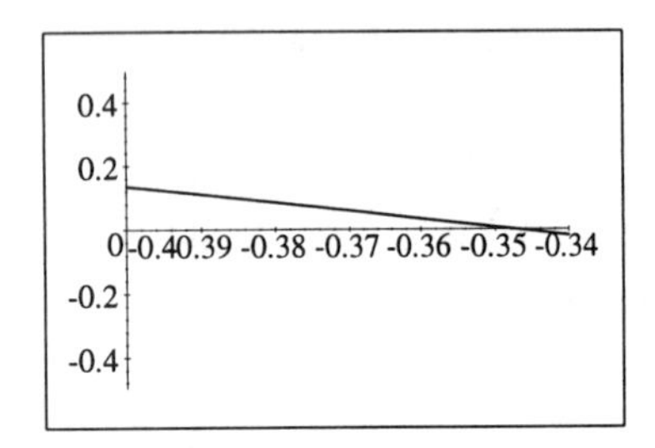

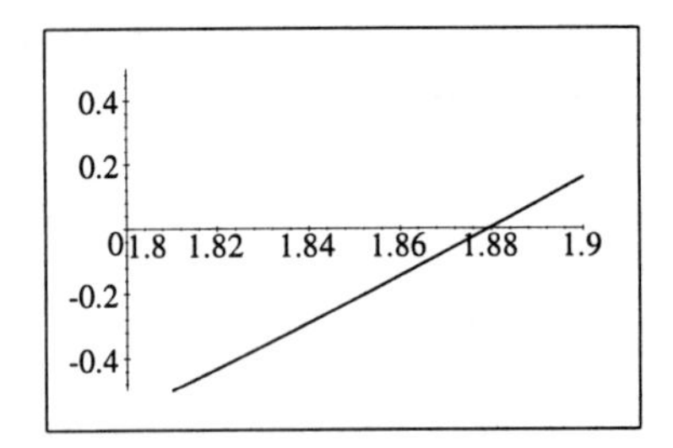

79. $0.1x + 1 \geq x^3 \quad \Leftrightarrow \quad -x^3 + 0.1x + 1 \geq 0$. We graph the function $f(x) = -x^3 + 0.1x + 1$ in the viewing rectangle $[-10, 10]$ by $[-10, 10]$and observe that there is one solution to the equation $f(x) = 0$. In the second viewing rectangle $[1.00, 1.05]$ by $[-0.1, 0.1]$ we find the solution $x \approx 1.03$. Furthermore, $f(x) \geq 0$ for $x \leq 1.03$. Thus the solution to the inequality is $(-\infty, 1.03]$.

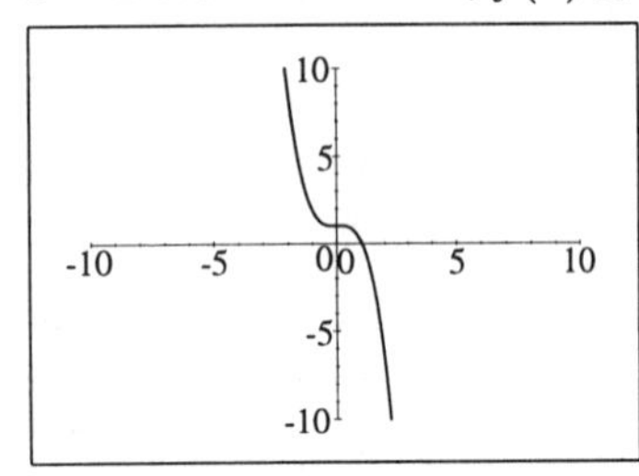

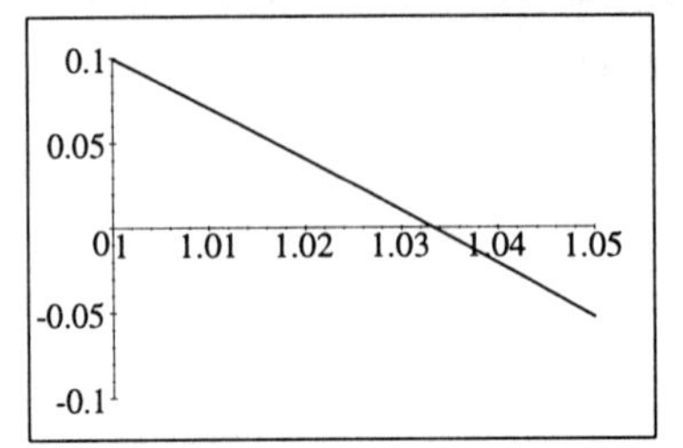

81. $f(x) = \sqrt{x^3 - 4x + 1}$. The domain consists of all x where $x^3 - 4x + 1 \geq 0$. Using a graphing device, the domain is approximately $[-2.1, 0.2] \cup [1.9, \infty)$.

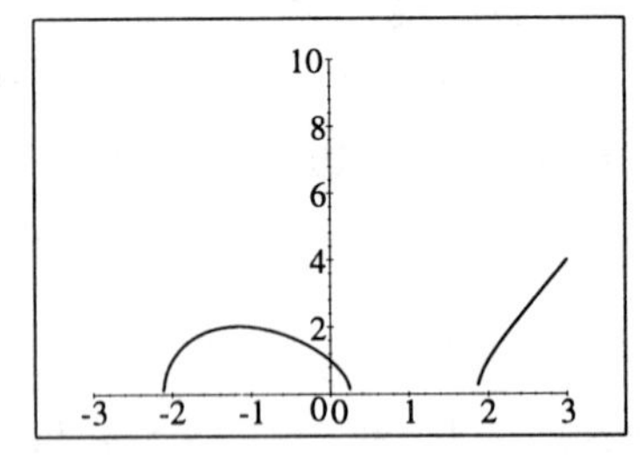

83. (a) $y = f(x) + 8$. Shift the graph of $f(x)$ upwards by 8 units.

(b) $y = f(x + 8)$. Shift the graph of $f(x)$ to the left by 8 units.

(c) $y = 1 + 2f(x)$. Stretch the graph of $f(x)$ vertically by a factor of 2, then shift upwards by 1 unit.

(d) $y = f(x - 2) - 2$. Shift the graph of $f(x)$ to the right by 2 units, then downwards by 2 units.

(e) $y = f(-x)$. Reflect the graph of $f(x)$ about the y-axis.

(f) $y = -f(-x)$. Reflect the graph of $f(x)$ first about the y-axis, then reflect about the x-axis.

(g) $y = -f(x)$. Reflect the graph of $f(x)$ about the x-axis.

(h) $y = f^{-1}(x)$. Reflect the graph of $f(x)$ about the line $y = x$.

85. (a) $f(x) = 2x^5 - 3x^2 + 2$.

$f(-x) = 2(-x)^5 - 3(-x)^2 + 2 = -2x^5 - 3x^2 + 2$. Since $f(x) \neq f(-x)$, f is not even.

$-f(x) = -2x^5 + 3x^2 - 2$. Since $-f(x) \neq f(-x)$, f is not odd.

(b) $f(x) = x^3 - x^7$.

$f(-x) = (-x)^3 - (-x)^7 = -(x^3 - x^7) = -f(x)$, hence f is odd.

(c) $f(x) = \dfrac{1 - x^2}{1 + x^2}$. $f(-x) = \dfrac{1 - (-x)^2}{1 + (-x)^2} = \dfrac{1 - x^2}{1 + x^2} = f(x)$. Since $f(x) = f(-x)$, f is even.

(d) $f(x) = \dfrac{1}{x + 2}$. $f(-x) = \dfrac{1}{(-x) + 2} = \dfrac{1}{2 - x}$. $-f(x) = -\dfrac{1}{x + 2}$. Since $f(x) \neq f(-x)$, f is not even, and since $f(-x) \neq -f(x)$, f is not odd.

87. $f(x) = x^2 + 4x + 1 = (x^2 + 4x + 4) + 1 - 4 = (x + 2)^2 - 3$

89. $g(x) = 2x^2 + 4x - 5 = 2(x^2 + 2x) - 5 = 2(x^2 + 2x + 1) - 5 - 2 = 2(x + 1)^2 - 7$. So the minimum value is $g(-1) = -7$.

91. $f(x) = 3.3 + 1.6x - 2.5x^3$. In the first viewing rectangle, $[-2, 2]$ by $[-4, 8]$ we see that $f(x)$ has a local maximum and a local minimum.

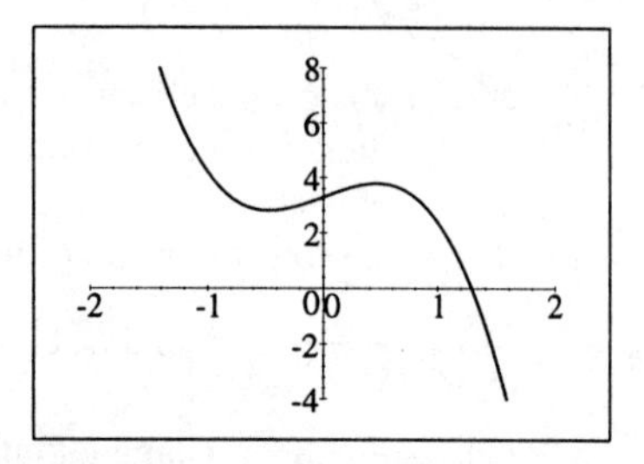

In the next viewing rectangle, $[0.4, 0.5]$ by $[3.78, 3.80]$ we isolate the local maximum value as approximately 3.79 when $x \approx 0.46$. In the last viewing rectangle, $[-0.5, -0.4]$ by $[2.80, 2.82]$, we isolate the local minimum value as 2.81 when $x \approx -0.46$.

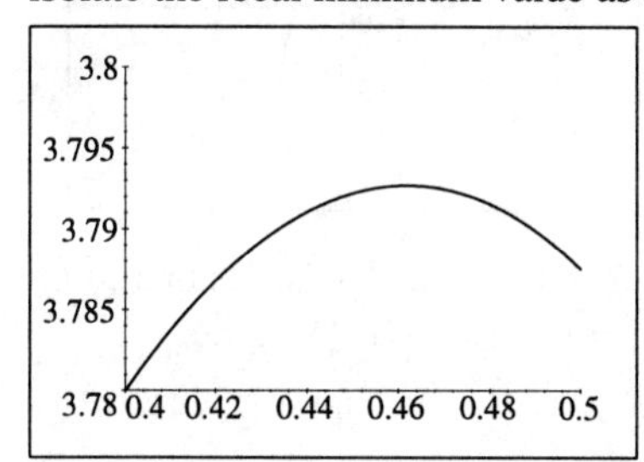

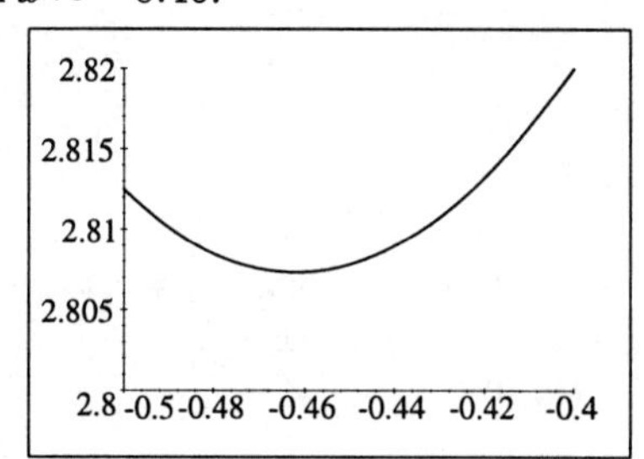

93. $f(x) = x^2 - 3x + 2$ and $g(x) = 4 - 3x$.

(a) $(f + g)(x) = (x^2 - 3x + 2) + (4 - 3x) = x^2 - 6x + 6$

(b) $(f - g)(x) = (x^2 - 3x + 2) - (4 - 3x) = x^2 - 2$

(c) $(fg)(x) = (x^2 - 3x + 2)(4 - 3x) = 4x^2 - 12x + 8 - 3x^3 + 9x^2 - 6x$
$= -3x^3 + 13x^2 - 18x + 8$

(d) $\left(\dfrac{f}{g}\right)(x) = \dfrac{x^2 - 3x + 2}{4 - 3x}$, $x \neq \dfrac{4}{3}$

(e) $(f \circ g)(x) = f(4 - 3x) = (4 - 3x)^2 - 3(4 - 3x) + 2 = 16 - 24x + 9x^2 - 12 + 9x + 2$
$= 9x^2 - 15x + 6$

(f) $(g \circ f)(x) = g(x^2 - 3x + 2) = 4 - 3(x^2 - 3x + 2) = -3x^2 + 9x - 2$

95. $f(x) = 3x - 1$ and $g(x) = 2x - x^2$.

$(f \circ g)(x) = f(2x - x^2) = 3(2x - x^2) - 1 = -3x^2 + 6x - 1$, domain is $(-\infty, \infty)$;

$(g \circ f)(x) = g(3x - 1) = 2(3x - 1) - (3x - 1)^2 = 6x - 2 - 9x^2 + 6x - 1 = -9x^2 + 12x - 3$, domain is $(-\infty, \infty)$;

$(f \circ f)(x) = f(3x - 1) = 3(3x - 1) - 1 = 9x - 4$, domain is $(-\infty, \infty)$;

$(g \circ g)(x) = g(2x - x^2) = 2(2x - x^2) - (2x - x^2)^2 = 4x - 2x^2 - 4x^2 + 4x^3 - x^4$

$= -x^4 + 4x^3 - 6x^2 + 4x$, domain is $(-\infty, \infty)$.

97. $f(x) = \sqrt{1 - x}$, $g(x) = 1 - x^2$ and $h(x) = 1 + \sqrt{x}$.

$(f \circ g \circ h)(x) = f(g(h(x))) = f\left(g\left(1 + \sqrt{x}\right)\right) = f\left(1 - \left(1 + \sqrt{x}\right)^2\right) = f(-x - 2\sqrt{x})$

$= \sqrt{1 - (-x - 2\sqrt{x})} = \sqrt{1 + 2\sqrt{x} + x} = \sqrt{(1 + \sqrt{x})^2} = 1 + \sqrt{x}$

99. $f(x) = 3 + x^3$. If $x_1 \neq x_2$ then $x_1^3 \neq x_2^3$ (unequal numbers have unequal cubes) and therefore $3 + x_1^3 \neq 3 + x_2^3$. Thus f is a one-to-one function.

101. $h(x) = \dfrac{1}{x^4}$. Since the fourth powers of a number and its negative are equal, h is not one-to-one. For example $h(-1) = \dfrac{1}{(-1)^4} = 1$ and $h(1) = \dfrac{1}{(1)^4} = 1$ so $h(-1) = h(1)$.

103. $p(x) = 3.3 + 1.6x - 2.5x^3$. Using a graphing device and the Horizontal Line Test, we see that p is not a one-to-one function.

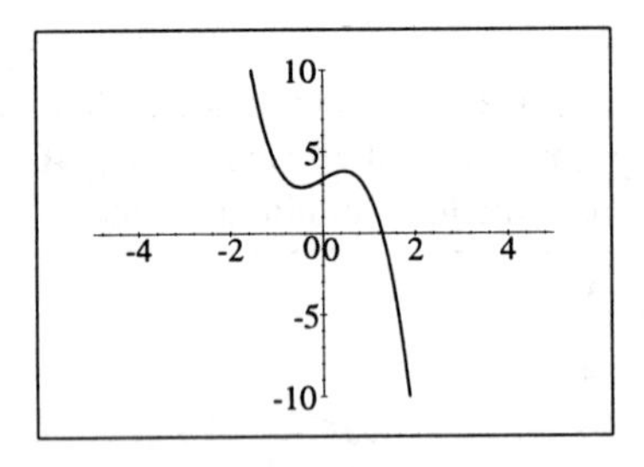

105. $f(x) = 3x - 2$. If $x_1 \neq x_2$ then $3x_1 \neq 3x_2$ and so $3x_1 - 2 \neq 3x_2 - 2$. Thus f is one-to-one. $f(x) = 3x - 2 \quad \Leftrightarrow \quad y = 3x - 2 \quad \Leftrightarrow \quad 3x = y + 2 \quad \Leftrightarrow \quad x = \frac{1}{3}(y + 2)$. So $f^{-1}(x) = \frac{1}{3}(x + 2)$.

107. $f(x) = x^2 - 4$, $x \geq 0$.

(a)

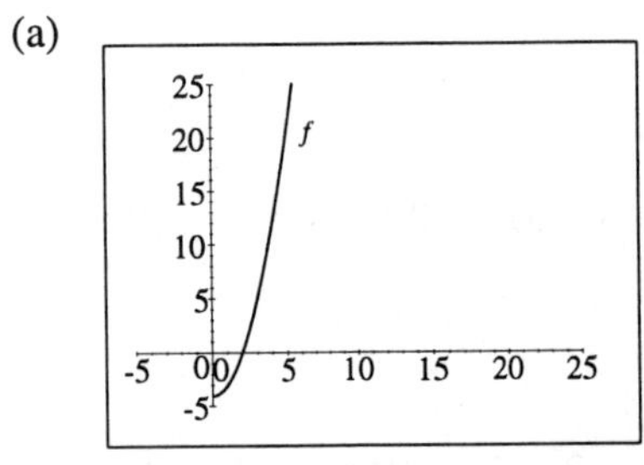

(b)

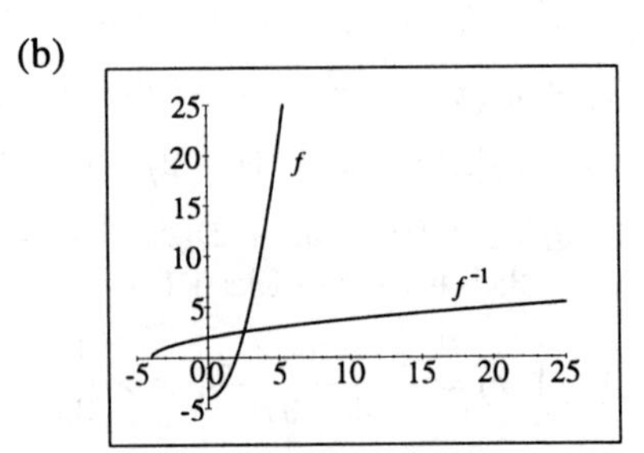

(c) $f(x) = x^2 - 4 \quad \Leftrightarrow \quad y = x^2 - 4,\ y \geq -4 \quad \Leftrightarrow \quad x^2 = y + 4 \quad \Leftrightarrow \quad x = \sqrt{y + 4}$. So $f^{-1}(x) = \sqrt{x + 4}$, $x \geq -4$.

109. M varies directly y as $z \quad \Rightarrow \quad M = kz$. Substituting $M = 120$ when $z = 15$ we find: $120 = k(15) \quad \Leftrightarrow \quad k = 8$. Therefore $M = 8z$.

111. (a) The intensity I varies inversely as the square of the distance d $\Rightarrow$ $I = \dfrac{k}{d^2}$.

(b) Substituting $I = 1000$ when $d = 8$we get: $1000 = \frac{k}{(8)^2}$ $\Leftrightarrow$ $k = 64,000$.

(c) From parts (a) and (b) we have $I = \dfrac{64000}{d^2}$. Substituting $d = 20$ we get: $I = \dfrac{64000}{(20)^2} = 160$ candles.

113. Let b be the length of the base, l be the length of the equal sides, and h be the height in cm. Since the perimeter is 8, $2l + b = 8$ $\Leftrightarrow$ $2l = 8 - b$ $\Leftrightarrow$ $l = \frac{1}{2}(8 - b)$. By the Pythagorean Theorem, $h^2 + (\frac{1}{2}b)^2 = l^2$ $\Leftrightarrow$ $h^2 = l^2 - (\frac{1}{2}b)^2$ $\Leftrightarrow$ $h = \sqrt{l^2 - \frac{1}{4}b^2}$. Therefore the area of the triangle is $A = \frac{1}{2} \cdot b \cdot h = \frac{1}{2} \cdot b\sqrt{l^2 - \frac{1}{4}b^2} = \frac{b}{2}\sqrt{\frac{1}{4}(8 - b)^2 - \frac{1}{4}b^2}$
$= \frac{b}{4}\sqrt{64 - 16b + b^2 - b^2} = \frac{b}{4} \cdot 4\sqrt{4 - b} = b\sqrt{4 - b}$.

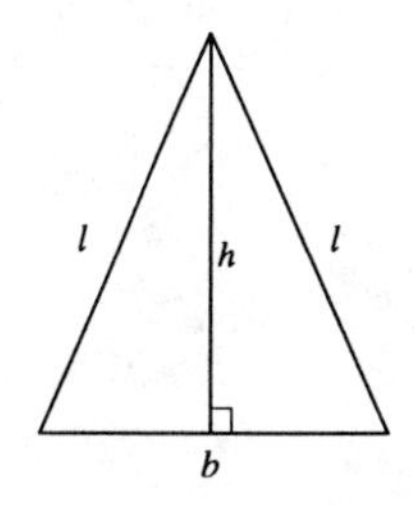

115. Let x be the length of wire in cm that is bent into a square. So $10 - x$ is the length of wire in cm that is bent into an equilateral triangle. The area of the square is $A_1 = \left(\dfrac{x}{4}\right)^2 = \dfrac{x^2}{16}$ and the area of the triangle is $A_2 = \frac{1}{2} \cdot b \cdot h = \frac{1}{2} \cdot \dfrac{10 - x}{3} \cdot \dfrac{\sqrt{3}\,(10 - x)}{6}$, where the height is found by using the Pythagorean Theorem. So $A_2 = \dfrac{\sqrt{3}\,(10 - x)^2}{36}$.

(a) The total area is $A = A_1 + A_2 = \dfrac{x^2}{16} + \dfrac{\sqrt{3}\,(10 - x)^2}{36}$, $0 \le x \le 10$.

(b) $A = \dfrac{x^2}{16} + \dfrac{\sqrt{3}\,(10 - x)^2}{36} = \dfrac{9x^2}{144} + \dfrac{4\sqrt{3}\,(10 - x)^2}{144}$
$= \frac{1}{144}\left(9x^2 + 400\sqrt{3} - 80\sqrt{3}\,x + 4\sqrt{3}\,x^2\right)$

$= \frac{1}{144}\left(9x^2 + 4\sqrt{3}\,x^2 - 80\sqrt{3}\,x\right) + \frac{25\sqrt{3}}{9} = \dfrac{9 + 4\sqrt{3}}{144}\left(x^2 - \dfrac{80\sqrt{3}\,x}{9 + 4\sqrt{3}}\right) + \dfrac{25\sqrt{3}}{9}$

$= \dfrac{9 + 4\sqrt{3}}{144}\left(x - \dfrac{40\sqrt{3}}{9 + 4\sqrt{3}}\right)^2 + \dfrac{25\sqrt{3}}{9} - \dfrac{4800}{144\left(9 + 4\sqrt{3}\right)}$. Therefore the area is a minimum when $x = \frac{40\sqrt{3}}{9+4\sqrt{3}} \approx 4.35$ m.

117. Let w be the width of the pen and l be the length in meters. We first use the area to establish a relationship between w and l, then use it to replace. Since the area is 100 m^2 we have: $l \cdot w = 100$ $\Leftrightarrow$ $l = \dfrac{100}{w}$. So the amount of fencing used is $F = 2l + 2w = 2\left(\frac{100}{w}\right) + 2w = \dfrac{200 + 2w^2}{w}$.
Using a graphing device we first graph F in the viewing rectangle, $[0, 40]$ by $[0, 100]$ and locate the approximate location of the minimum value. In the second viewing rectangle, $[8, 12]$ by $[39, 41]$, we see that the minimum value of F occurs when $w = 10$. Therefore the pen should be a square with side 10 m.

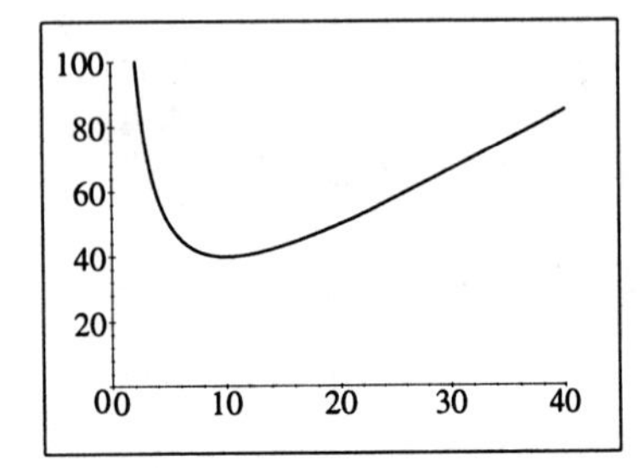
100
80
60
40
20
00
10
20
30
40

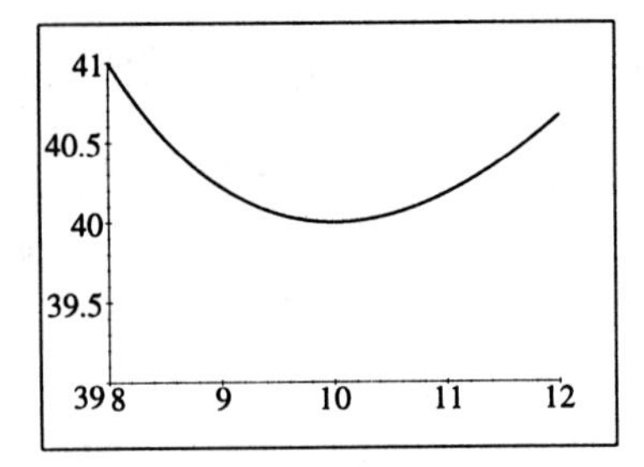
41
40.5
40
39.5
39
8
9
10
11
12

Chapter 3 Test

1. (a) The distance from A to B is $d(AB) = \sqrt{(-7-5)^2 + (4+12)^2} = \sqrt{144+256} = \sqrt{400}$ $= 20$.

   (b) Midpoint $= \left(\frac{-7+5}{2}, \frac{4-12}{2}\right) = (-1, -4)$.

   (c) The line has slope $m = \frac{4+12}{-7-5} = \frac{16}{-12} = -\frac{4}{3}$ and its equation is $y - 4 = -\frac{4}{3}(x+7) \quad \Leftrightarrow$ $y = -\frac{4}{3}x - \frac{28}{3} + 4 = -\frac{4}{3}x - \frac{16}{3}$. Thus the equation is $y = -\frac{4}{3}x - \frac{16}{3} \quad \Leftrightarrow$ $4x + 3y + 16 = 0$.

   (d) The perpendicular bisector has slope $m = -\dfrac{1}{-(4/3)} = \dfrac{3}{4}$ and contains the midpoint $(-1, -4)$. Thus from the point-slope formula the equation is $y + 4 = \frac{3}{4}(x+1) \quad \Leftrightarrow \quad y + 4 = \frac{3}{4}x + \frac{3}{4}$ $\Leftrightarrow \quad y = \frac{3}{4}x + \frac{3}{4} - 4 = \frac{3}{4}x - \frac{13}{4}$. Thus the equation is $y = \frac{3}{4}x - \frac{13}{4} \quad \Leftrightarrow$ $3x - 4y - 13 = 0$.

   (e) The length of a diameter of this circle was found in part (a), so the radius is $r = \frac{20}{2} = 10$. The center is the midpoint found in part (b), so the equation is $(x+1)^2 + (y+4)^2 = 10^2 \quad \Leftrightarrow$ $(x+1)^2 + (y+4)^2 = 100$.

2. (a) $x^2 + y^2 - 6x + 10y + 9 = 0 \quad \Leftrightarrow$
   $(x^2 - 6x) + (y^2 + 10y) = -9 \quad \Leftrightarrow$
   $(x^2 - 6x + 9) + (y^2 + 10y + 25) = -9 + 9 + 25$
   $\Leftrightarrow \quad (x-3)^2 + (y+5)^2 = 5^2$

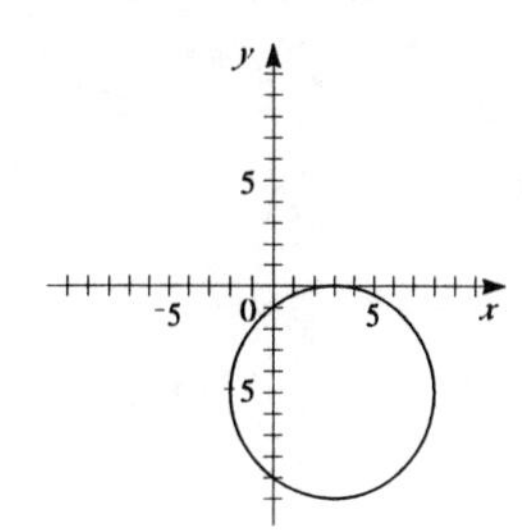

   (b) $2x^2 + 2y^2 + 6x + 10y + 17 = 0 \quad \Leftrightarrow$
   $2(x^2 + 3x) + 2(y^2 + 5y) = -17 \quad \Leftrightarrow$
   $(x^2 + 3x + \frac{9}{4}) + (y^2 + 5y + \frac{25}{4}) = -\frac{17}{2} + \frac{9}{4} + \frac{25}{4}$
   $\Leftrightarrow \quad \left(x + \frac{3}{2}\right)^2 + \left(y + \frac{5}{2}\right)^2 = 0$. This graph consists of a single point, $\left(-\frac{3}{2}, -\frac{5}{2}\right)$

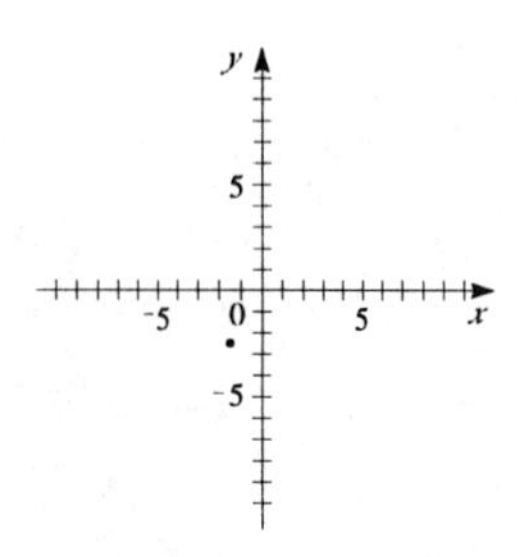

3. (a) We first find the slope of the line $2x + 6y = 17$: $2x + 6y = 17 \quad \Leftrightarrow \quad 6y = -2x + 17 \quad \Leftrightarrow$ $y = -\frac{1}{3}x + \frac{17}{6}$. Thus this line has slope $m = -\frac{1}{3}$ as does any line parallel to it. Then the parallel line passing through the point $(-2, 3)$ has the equation $y - 3 = -\frac{1}{3}(x+2) \quad \Leftrightarrow$ $3y - 9 = -x - 2 \quad \Leftrightarrow \quad x + 3y = 7$. (You can also find this equation by substituting the point $(-2, 3)$ into the equation $2x + 6y = c$ and solve for c.)

   (b) This line contains the points $(-3, 0)$ and $(0, 12)$. Thus the slope of the line containing these points is $m = \frac{12-0}{0+3} = 4$. Thus the equation is $y = 4x + 12$. (you can also use the intercept form of a line: $\dfrac{x}{a} + \dfrac{y}{b} = 1 \quad \Rightarrow \quad \dfrac{x}{-3} + \dfrac{y}{12} = 1$.)

4. (a) $-3x + 4y = 24$

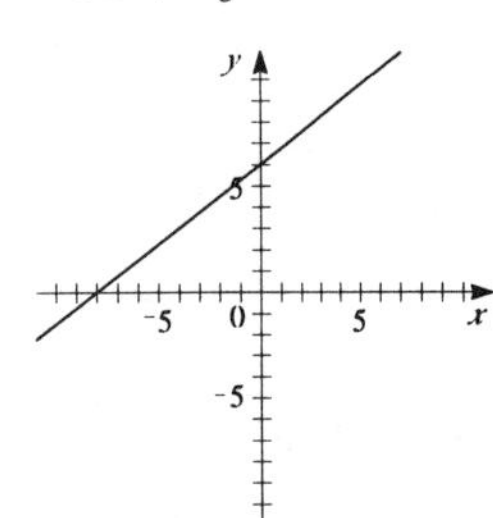

(b) $x = y^2 - 1$

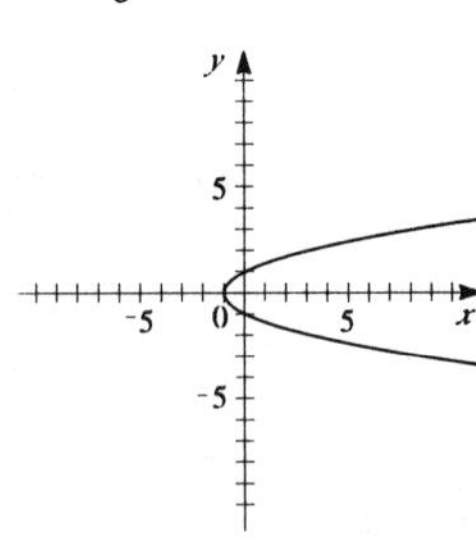

5. By the Vertical Line Test, figures (a) and (b) are graphs of functions. By the Horizontal Line Test, only figure (a) is the graph of a one-to-one function.

6. $f(x) = \dfrac{\sqrt{x}}{x-1}$. Our restrictions are that the input to the radical is nonnegative and the denominator must not be equal to zero. Thus $x \geq 0$ and $x - 1 \neq 0 \quad \Leftrightarrow \quad x \neq 1$. In interval notation, the domain is $[0, 1) \cup (1, \infty)$.

7. (a) $f(x) = x^3$

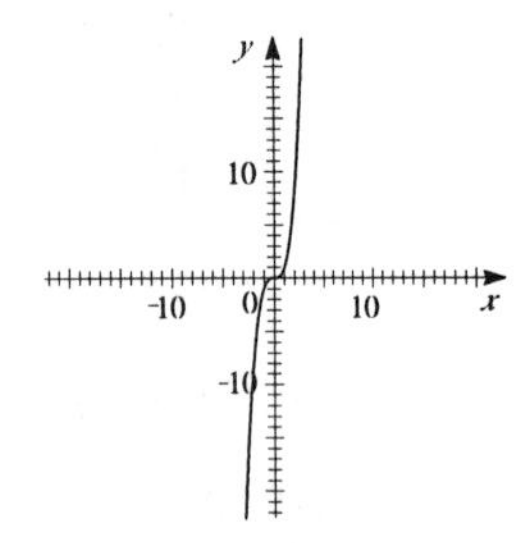

(b) $g(x) = (x-1)^3 - 2$. To obtain the graph of g shift the graph of f to the right 1 unit and downward 2 units.

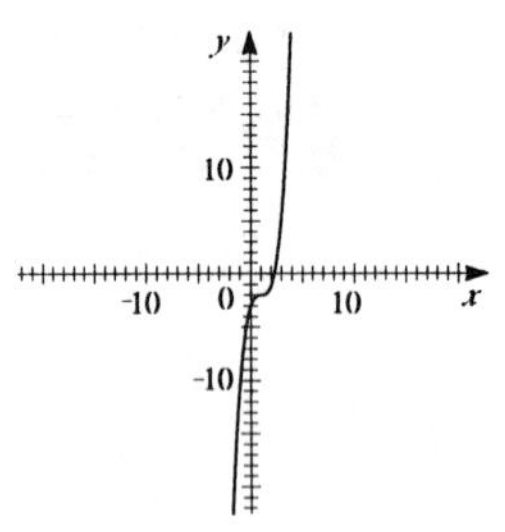

8. (a) $y = 2 - f(x+3)$. Shift the graph of $f(x)$ left by 3 units, then reflect about the y-axis and finally shift the graph upwards by 2 units..

(b) $y = -f(-x)$. Reflect the graph of $f(x)$ first about the y-axis, then reflect about the x-axis. (This is equivalent to reflecting the graph about the origin.)

9. (a) $f(x) = 2x^2 - 8x + 13$

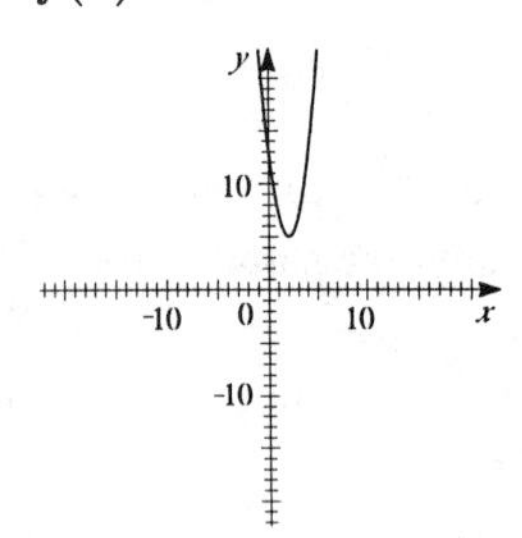

(b) $$\begin{aligned} f(x) &= 2x^2 - 8x + 13 \\ &= 2(x^2 - 4x) + 13 \\ &= 2(x^2 - 4x + 4) + 13 - 8 \\ &= 2(x-2)^2 + 5. \end{aligned}$$

Thus the minimum value of f is $f(2) = 5$.

10. (a) $f(-2) = 1 - (-2)^2 = 1 - 4 = -3$ (since $-2 \le 0$)

$f(1) = 2(1) + 1 = 2 + 1 = 3$ (since $1 > 0$)

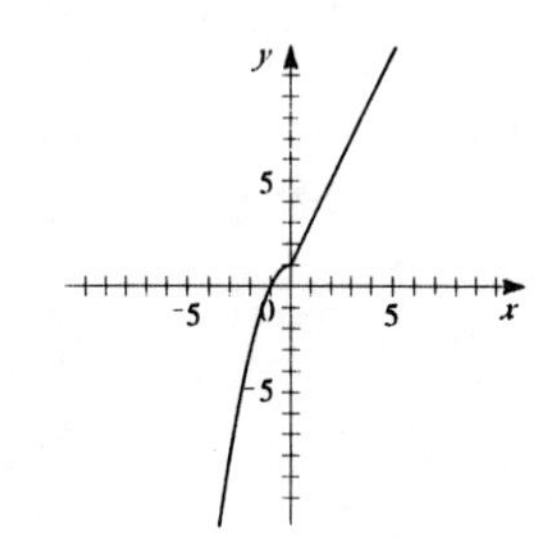

11. $f(x) = x^2 + 2x - 1$; $g(x) = 2x - 3$.

(a) $(f \circ g)(x) = f(g(x)) = f(2x - 3) = (2x - 3)^2 + 2(2x - 3) - 1$
$= 4x^2 - 12x + 9 + 4x - 6 - 1 = 4x^2 - 8x + 2$

(b) $(g \circ f)(x) = g(f(x)) = g(x^2 + 2x - 1) = 2(x^2 + 2x - 1) - 3 = 2x^2 + 4x - 2 - 3$
$= 2x^2 + 4x - 5$

(c) $f(g(2)) = f(1) = (1)^2 + 2(1) - 1 = 2$. (We have used the fact that $g(2) = 2(2) - 3 = 1$.)

(d) $g(f(2)) = g(7) = 2(7) - 3 = 11$. (We have used the fact that $f(2) = 2^2 + 2(2) - 1 = 7$.)

(e) $(g \circ g \circ g)(x) = g(g(g(x))) = g(g(2x - 3)) = g(4x - 9) = 2(4x - 9) - 3 = 8x - 18 - 3$
$= 8x - 21$. We have used $g(2x - 3) = 2(2x - 3) - 3 = 4x - 6 - 3 = 4x - 9$.

12. (a) $f(x) = \sqrt{3 - x}, x \le 3 \quad \Leftrightarrow$
$y = \sqrt{3 - x} \quad \Leftrightarrow \quad y^2 = 3 - x$
$\Leftrightarrow \quad x = 3 - y^2$. Thus
$f^{-1}(x) = 3 - x^2,\ x \ge 0$

(b) $f(x) = \sqrt{3 - x}, x \le 3$ and $f^{-1}(x) = 3 - x^2,\ x \ge 0$

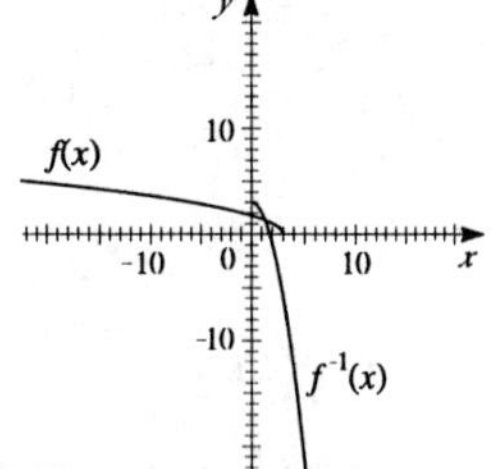

13. (a) Since 800 ft of fencing are available, $4x + 2y = 800 \quad \Leftrightarrow \quad 2y = 800 - 4x \quad \Leftrightarrow$
$y = 400 - 2x$. Thus the area of the pens is $A = xy = x(400 - 2x) = 400x - 2x^2$.

(b) $A = 400x - 2x^2 = -2(x^2 - 200x) = -2(x^2 - 200x + 100^2) + 20000$
$= -2(x - 100)^2 + 20000$. So, when $x = 100$ ft the total area will be maximized.

14. (a) $f(x) = 3x^4 - 14x^2 + 5x - 3$. The graph is shown in the viewing rectangle $[-10, 10]$ by $[-30, 10]$.

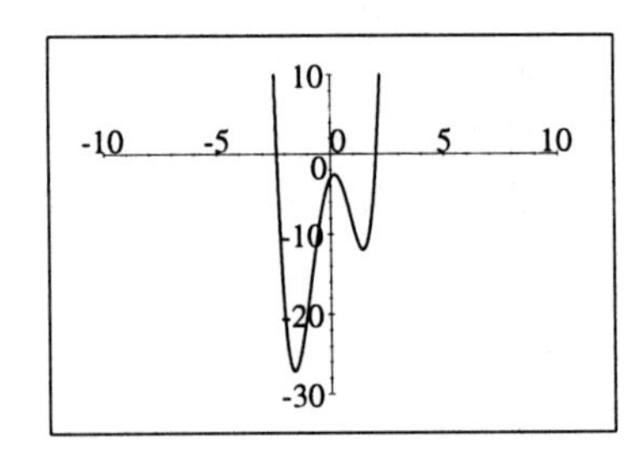

(b) No.

(c) Local minimum ≈ -27.18 when $x \approx -1.61$. Shown is the viewing rectangle $[-1.65, -1.55]$ by $[-27.5, -27]$.

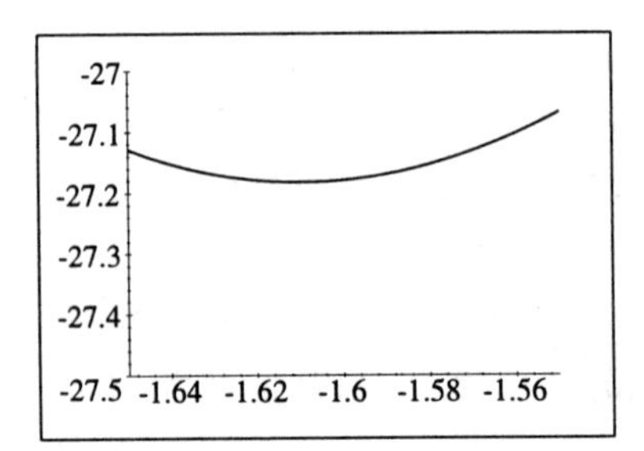

Local maximum ≈ -2.55 when $x \approx 0.18$. Shown is the viewing rectangle $[0.15, 0.25]$ by $[-2.6, -2.5]$.

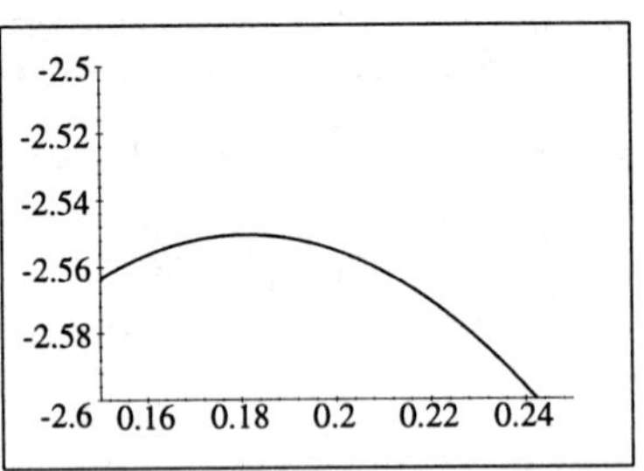

Local minimum ≈ -11.93 when $x \approx 1.43$. Shown is the viewing rectangle $[1.4, 1.5]$ by $[-12, -11.9]$.

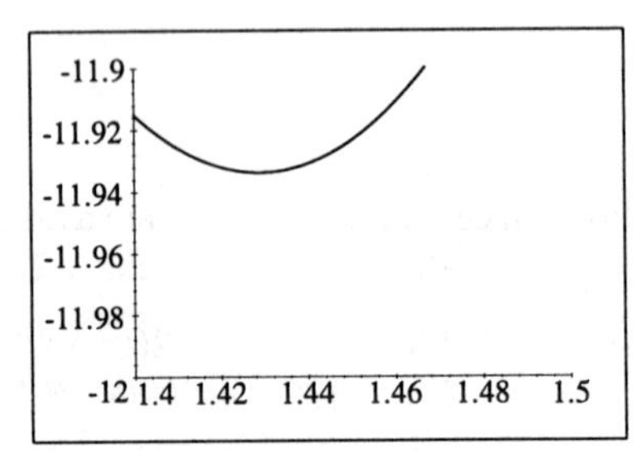

(d) Because the local minimum is approximately -27.18 the range is $[-27.18, \infty)$

(e) Using the information from part (c) and the graph in part (a), $f(x)$ is increasing on the intervals $[-1.61, 0.18]$ and $[1.43, \infty)$ and decreasing on the intervals $(-\infty, -1.61]$ and $[0.18, 1.43]$.

Focus on Problem Solving

1. The final digit in 947^{362} is determined by the final digit in 7^{362}. Looking at the first few powers of 7 we see: $7^1 = 7$, $7^2 = 49$, $7^3 = 343$, $7^4 = 2401$, $7^5 = 16807$, $7^6 = 117649$ and it appears that the final digit cycles in a pattern, namely $7 \to 9 \to 3 \to 1 \to 7 \to 9 \to 3 \to 1$, of length 4. Since $362 = 4 \times 90 + 2$, the final digit is the second in the cycle, namely 9.

3. $f_0(x) = x^2$ and $f_{n+1}(x) = f_0(f_n(x))$ for $n = 0, 1, 2, \ldots$.
$f_1(x) = f_0(f_0(x)) = f_0(x^2) = (x^2)^2 = x^4$, $\quad f_2(x) = f_0(f_1(x)) = f_0(x^4) = (x^4)^2 = x^8$,
$f_3(x) = f_0(f_2(x)) = f_0(x^8) = (x^8)^2 = x^{16}$, ... Thus a general formula is $f_n(x) = x^{2^{n+1}}$.

5. Let us see what happens when we square similar numbers with fewer 9's:

$$39^2 = 1521,\ 399^2 = 159201,\ 3999^2 = 15992001,\ 39999^2 = 1599920001.$$

The pattern is that the square always seems to start with 15 and end with 1, and if $39\cdots9$ has n 9's, then the 2 in the middle of its square is preceded by $(n-1)$ 9's and followed by $(n-1)$ 0's. From this pattern, we make the guess that

$$3{,}999{,}999{,}999{,}999^2 = 15{,}999{,}999{,}999{,}992{,}000{,}000{,}000{,}001$$

This can be verified by writing the number as follows:

$3{,}999{,}999{,}999{,}999^2 = (4{,}000{,}000{,}000{,}000 - 1)^2$
$= 4{,}000{,}000{,}000{,}000^2 - 2 \cdot 4{,}000{,}000{,}000{,}000 + 1$
$= 16{,}000{,}000{,}000{,}000{,}000{,}000{,}000{,}000 - 8{,}000{,}000{,}000{,}000 + 1$
$= 15{,}999{,}999{,}999{,}992{,}000{,}000{,}000{,}001$

7. $f(x) = |x^2 - 4|x| + 3|$. If $x \geq 0$, then $f(x) = |x^2 - 4x + 3| = |(x-1)(x-3)|$
Case (i): If $0 < x \leq 1$, then $f(x) = x^2 - 4x + 3$.
Case (ii): If $1 < x \leq 3$, then
$f(x) = -(x^2 - 4x + 3) = -x^2 + 4x - 3$.
Case (iii): If $x > 3$, then $f(x) = x^2 - 4x + 3$.
This enables us to sketch the graph for $x \geq 0$.
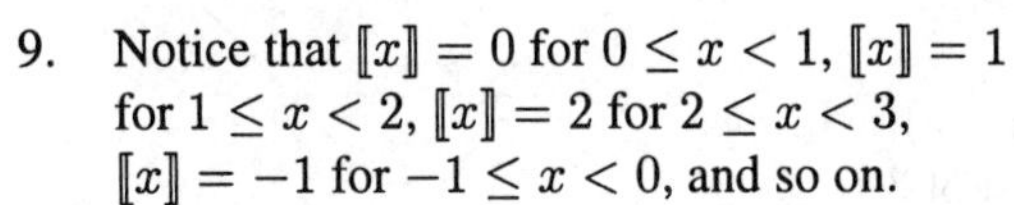
Then we use the fact that f is an even function to reflect this part of the graph about the y-axis to obtain the entire graph. (Or: consider also the cases $x < -3$, $-3 \leq x < -1$, and $-1 \leq x < 0$.)

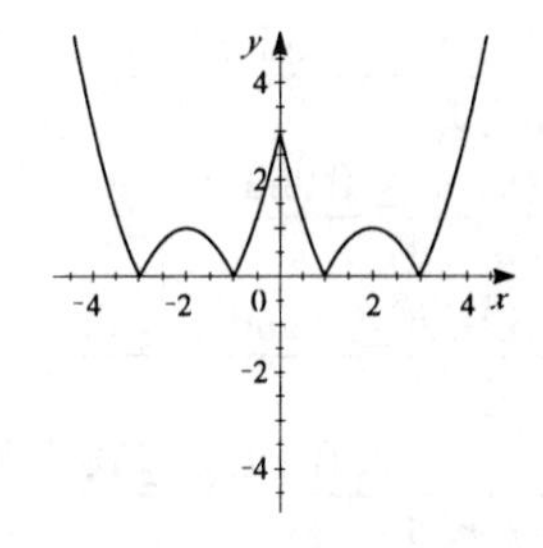

9. Notice that $[\![x]\!] = 0$ for $0 \leq x < 1$, $[\![x]\!] = 1$ for $1 \leq x < 2$, $[\![x]\!] = 2$ for $2 \leq x < 3$, $[\![x]\!] = -1$ for $-1 \leq x < 0$, and so on.

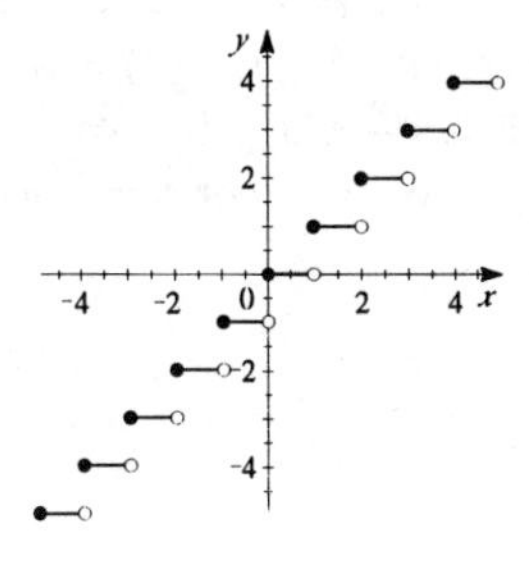

11. $[\![x]\!]^2 + [\![y]\!]^2 = 1$.
Since $[\![x]\!]^2$ and $[\![y]\!]^2$ are positive integers or 0, there are only 4 cases:
Case (i): $[\![x]\!] = 1$, $[\![y]\!] = 0 \quad \Rightarrow \quad 1 \le x < 2$ and $0 \le y < 1$
Case (ii): $[\![x]\!] = -1$, $[\![y]\!] = 0 \quad \Rightarrow \quad -1 \le x < 0$ and $0 \le y < 1$
Case (iii): $[\![x]\!] = 0$, $[\![y]\!] = 1 \quad \Rightarrow \quad 0 \le x < 1$ and $1 \le y < 2$
Case (iv): $[\![x]\!] = 0$, $[\![y]\!] = -1 \quad \Rightarrow \quad 0 \le x < 1$ and $-1 \le y < 0$

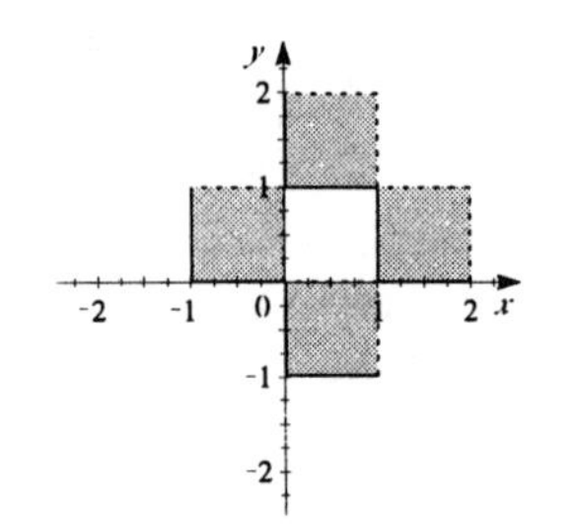

13. Since $\sqrt[3]{1729} \approx 12.0023$ we start with $n = 12$ and find the other perfect cube.

n	$1729 - n^3$
12	$1729 - (12)^3 = 1$
11	$1729 - (11)^3 = 398$
10	$1729 - (10)^3 = 729$

Since 1 and 729 are perfect cubes the two representations we seek are $(1)^3 + (12)^3 = 1729$ and $(9)^3 + (10)^3 = 1729$.

15. You can see infinitely far in this forest, because if your vision is blocked by a tree (at the rational point (a, b), say) then the slope of the line of sight from you to this tree is $\frac{b-0}{a-0} = \frac{b}{a}$, which is a rational number. Thus if you look along a line of irrational slope, you will see infinitely far.

17. We continue the pattern. Three parallel cuts produce 10 pieces. Thus each new cut produces an additional 3 pieces. Since the first cut produces 4 pieces we get the formula $f(n) = 4 + 3(n - 1)$, $n \ge 1$. Since $f(142) = 4 + 3(141) = 427$, we see that 142 parallel cuts produce 427 pieces.

19. We consider four cases corresponding to each quadrant.
Case (i), $x \ge 0$ and $y \ge 0$: $|x| + |y| \le 1$ becomes $x + y \le 1$.
Case (ii), $x \le 0$ and $y \ge 0$: $|x| + |y| \le 1$ becomes $-x + y \le 1$.
Case (iii), $x \ge 0$ and $y \le 0$: $|x| + |y| \le 1$ becomes $x - y \le 1$.
Case (iv), $x \le 0$ and $y \le 0$: $|x| + |y| \le 1$ becomes $-x - y \le 1 \quad \Leftrightarrow \quad x + y \ge -1$.

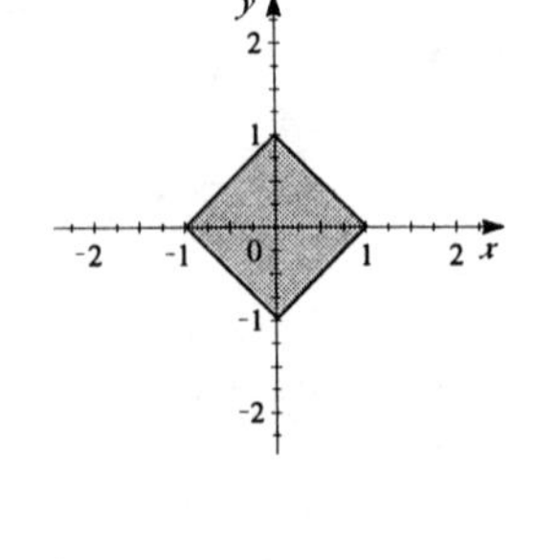

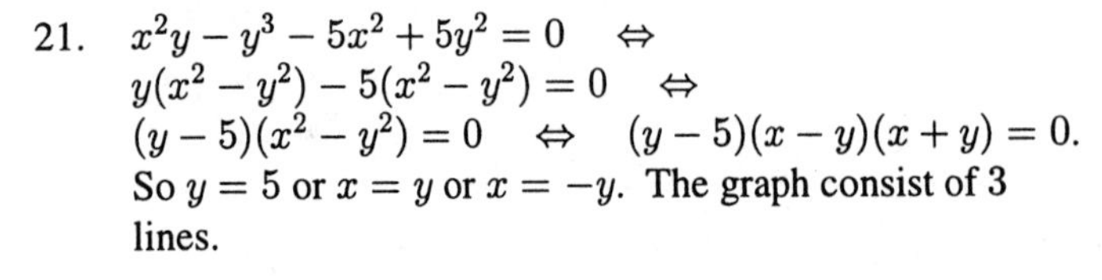

21. $x^2y - y^3 - 5x^2 + 5y^2 = 0 \quad \Leftrightarrow$
$y(x^2 - y^2) - 5(x^2 - y^2) = 0 \quad \Leftrightarrow$
$(y - 5)(x^2 - y^2) = 0 \quad \Leftrightarrow \quad (y - 5)(x - y)(x + y) = 0$.
So $y = 5$ or $x = y$ or $x = -y$. The graph consist of 3 lines.

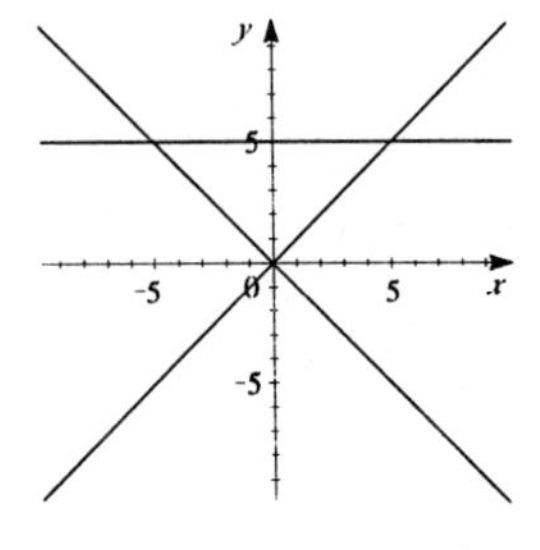

Chapter Four

Exercises 4.1

1. $y = x^3 - 8$

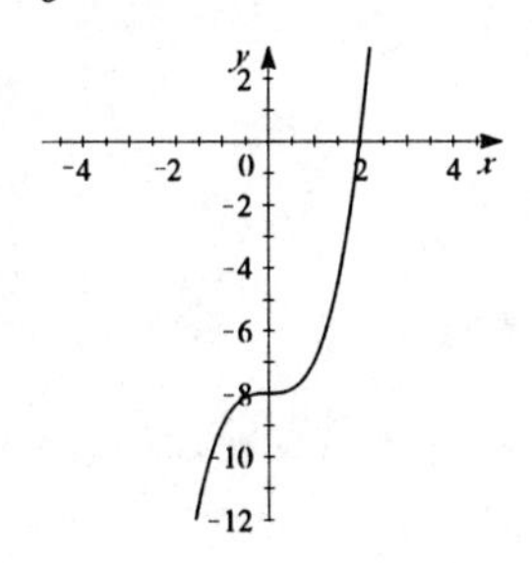

3. $y = -x^4 + 16$

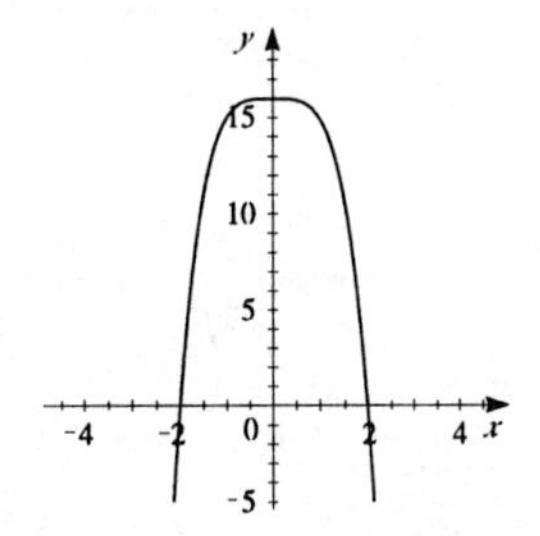

5. $y = -(x-1)^4 + 1$

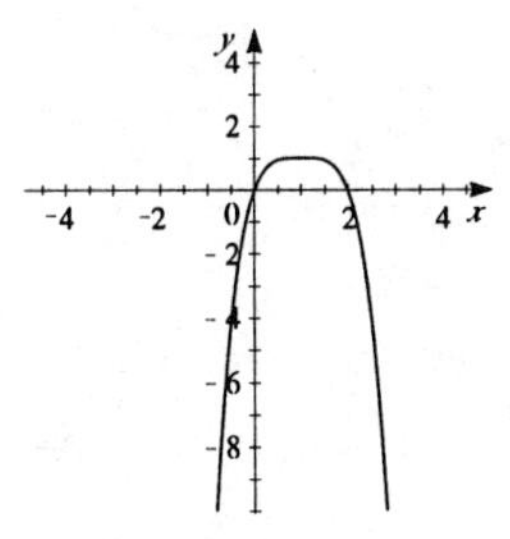

7. $y = 4(x-2)^5 - 4$

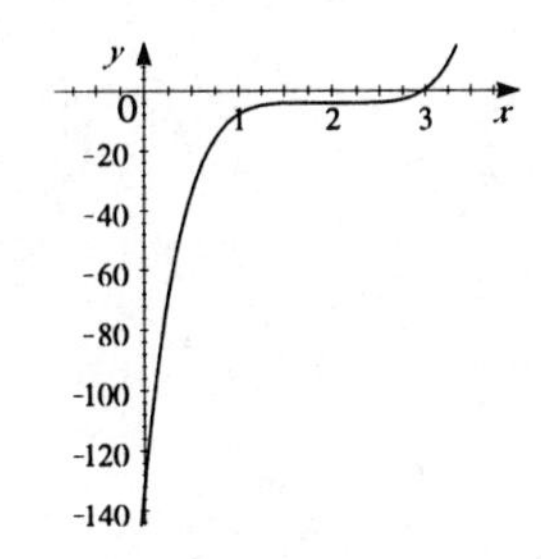

9. $y = (x-3)(x+1)$

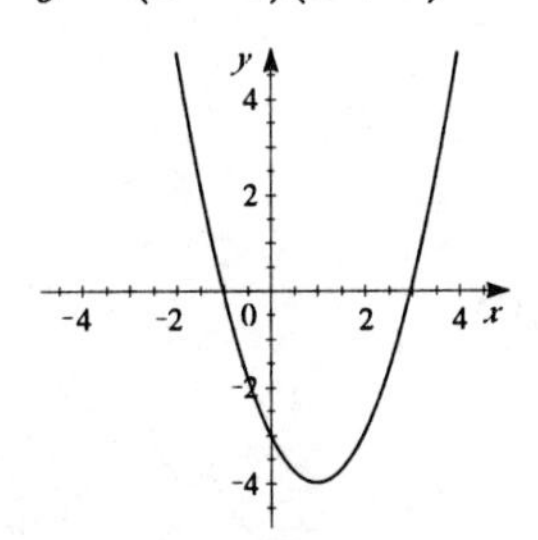

11. $y = x(x-2)(x+1)$

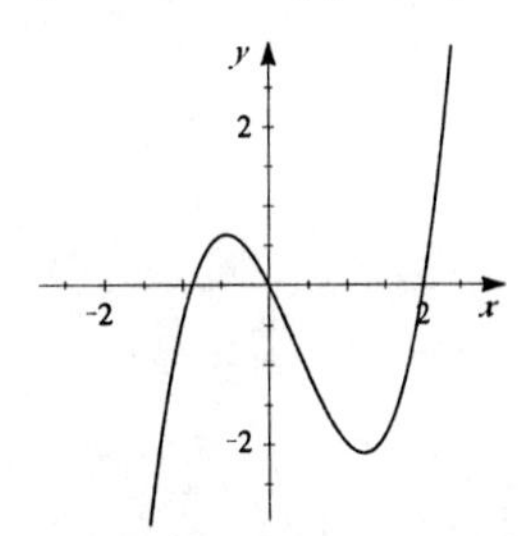

13. $y = (x-1)^2(x-3)$

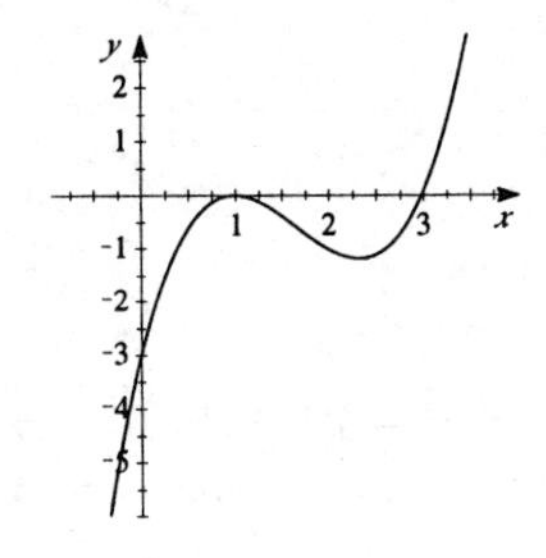

15. $y = \frac{1}{12}(x+2)^2(x-3)$

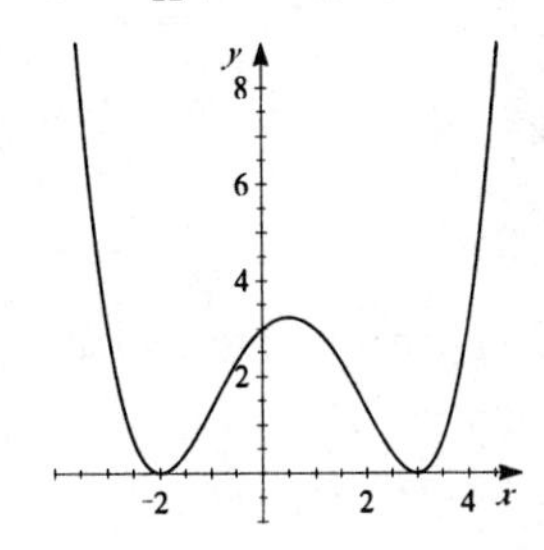

17. $y = x^3 + 3x^2 - 4x - 12$

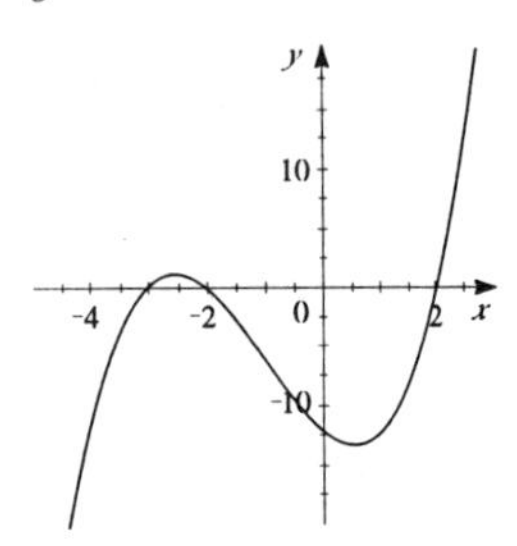

19. $y = x^3 - x^2 - 6x$

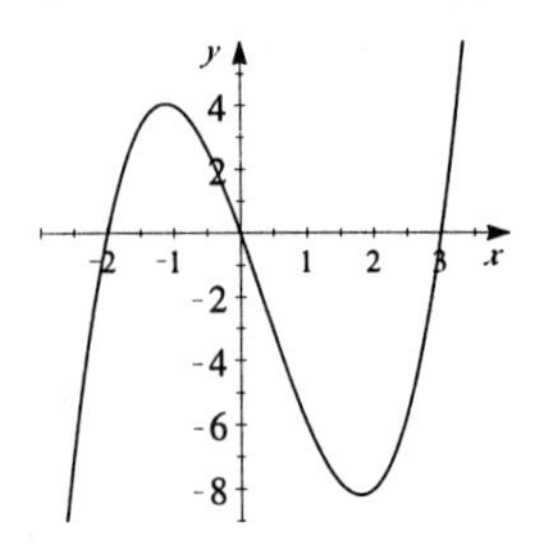

21. $y = \frac{1}{8}(2x^4 + 3x^3 - 16x - 24)$

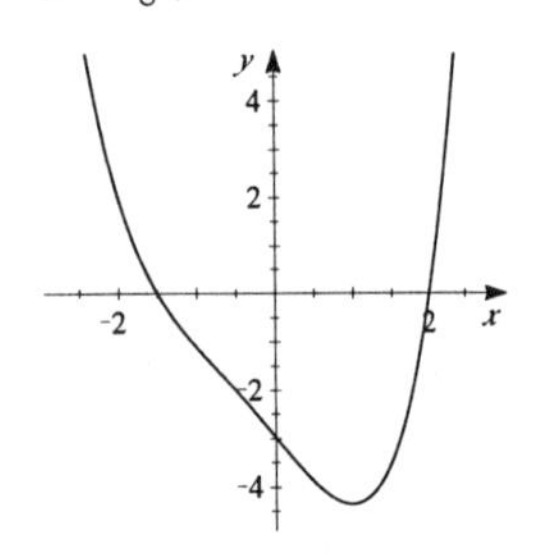

23. $y = x^4 - 2x^3 + 8x - 16$

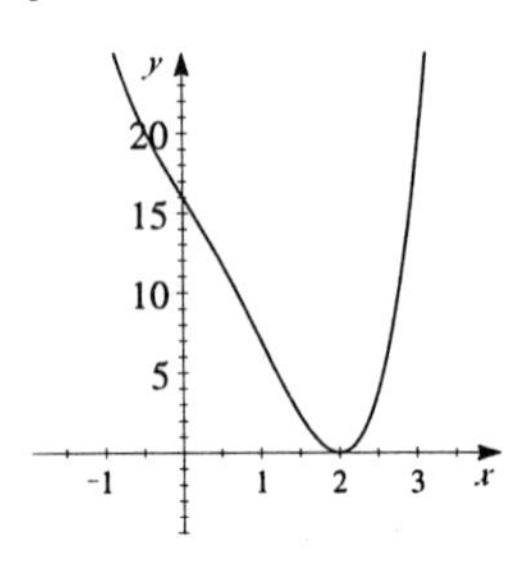

25. $y = x^5 - 9x^3$

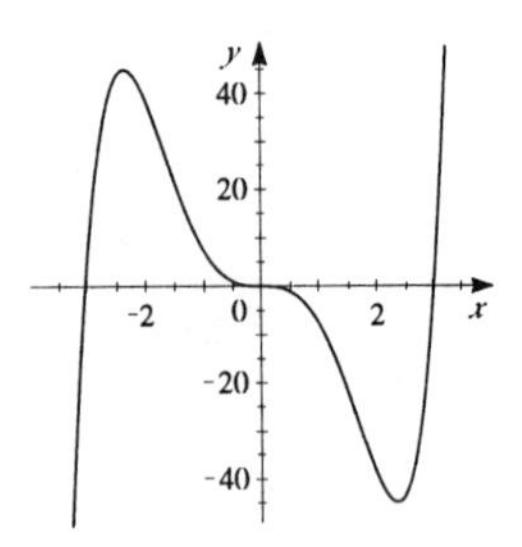

27. $(-2, 0) \cup (2, \infty)$

29. $(-\infty, -3] \cup \{0\}$

31. (a)

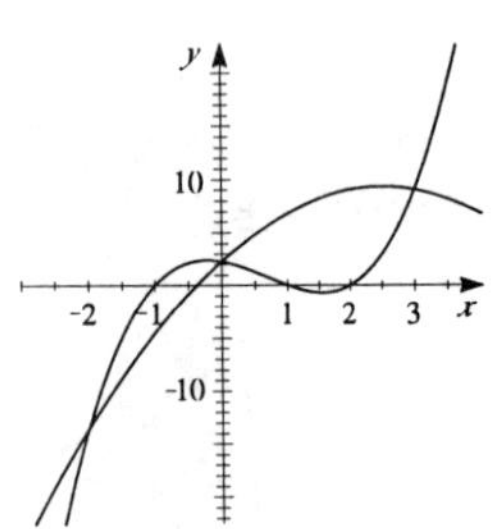

(b) The two graphs appear to intersect at 3 points.

(c) $x^3 - 2x^2 - x + 2 = -x^2 + 5x + 2 \Leftrightarrow x^3 - x^2 - 6x = 0 \Leftrightarrow x(x^2 - x - 6) = 0 \Leftrightarrow x(x-3)(x+2) = 0$. Hence, intersection points occur when $x = 0 \Rightarrow y = 2$, when $x = 3 \Rightarrow y = 8$, and when $x = -2 \Rightarrow y = -12$. Therefore, the points of intersection are $P_1(0, 2)$, $P_2(3, 8)$, and $P_3(-2, -12)$.

33. (a) Since f and g are odd, $f(-x) = -f(x)$ and $g(-x) = -g(x)$. Therefore, $(f+g)(-x) = f(-x) + g(-x) = -f(x) - g(x) = -[f(x) + g(x)] = -(f+g)(x)$, and so the function $f+g$ is also odd.

(b) Since f and g are even $f(-x) = f(x)$ and $g(-x) = g(x)$. Therefore, $(f+g)(-x) = f(-x) + g(-x) = f(x) + g(x) = (f+g)(x)$, and so the function $f+g$ is also even.

(c) Since f is odd and g is even, $f(-x) = -f(x)$ and $g(-x) = g(x)$. Therefore, $(f+g)(-x) = f(-x) + g(-x) = -f(x) + g(x) \neq -[f(x) + g(x)]$ (since $g(x) \neq -g(x)$ if x is a point for which $g(x) \neq 0$) and so the function $f+g$ is not odd. Also, $-f(x) + g(x) \neq f(x) + g(x)$ (for a similar reason) and so the function $f+g$ is not even. Thus, $f+g$ is neither even nor odd.

(d) Let $P(x)$ be a polynomial containing only odd powers of x. Then, each term of $P(x)$ can be written as Cx^{2n+1} for some constant C and integer n. Since $C(-x)^{2n+1} = -Cx^{2n+1}$, each term of $P(x)$ is an odd function. Thus by part (a), $P(x)$ is an odd function.

(e) Let $P(x)$ be a polynomial containing only even powers of x. Then, each term of $P(x)$ can be written as Cx^{2n} for some constant C and integer n. Since $C(-x)^{2n} = Cx^{2n}$, each term of $P(x)$ is an even function. Thus by part (b), $P(x)$ is an even function.

(f) Since $P(x)$ contains both even and odd powers of x, we can write it in the form $P(x) = R(x) + Q(x)$, where $R(x)$ contains all the even-powered terms in $P(x)$ and $Q(x)$ contains all the odd. By part (d), $Q(x)$ is an odd function, and by part (e), $R(x)$ is an even function. Thus, since neither $Q(x)$ nor $R(x)$ are constantly 0 (by assumption), $P(x) = R(x) + Q(x)$ is neither even nor odd, by part (c).

(g) $P(x) = x^5 + 6x^3 - x^2 - 2x + 5 = (x^5 + 6x^3 - 2x) + (-x^2 + 5) = Q(x) + R(x)$ where $Q(x) = x^5 + 6x^3 - 2x$ and $R(x) = -x^2 + 5$. Since $Q(x)$ contains only odd powers of x, it is an odd function, and since $R(x)$ contains only even powers of x, it is an even function.

35. (a) Let h = height of the box. Then the total length of all 12 edges is $(8x + 4h) = 144$ in. Thus, $8x + 4h = 144 \iff 2x + h = 36 \iff h = 36 - 2x$. Now, the volume of the box is equal to (area of base) $\times$ (height) $= (x^2) \times (36 - 2x) = -2x^3 + 36x^2$. Therefore, the volume of the box is $V(x) = -2x^3 + 36x^2$.

(b) To find x-intercepts, set $V(x) = -2x^3 + 36x^2 = 0 \iff x^2(36 - 2x) = 0$, so $x = 0$ or $2x = 36 \Rightarrow x = 18$. Therefore, graph crosses x-axis when $x = 0, 18$. For y-intercepts, set $x = 0$, thus $y = V(0) = 0$. Therefore, graph passes through origin. Note that $V(x) > 0$ when $x \in (-\infty, 0) \cup (0, 18)$. Plotting some more points allows us to graph the function:

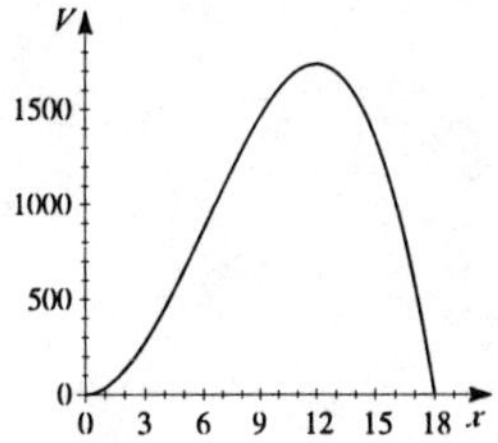

V
1500
1000
500
0
0 3 6 9 12 15 18 x

(c) Since x and V both represent positive quantities, we must restrict $V(x) > 0$ and $x > 0 \Rightarrow x \in (0, 18)$.

Exercises 4.2

1. $y = -x^2 + 8x$, $[-4, 12]$ by $[-50, 30]$
   x-intercepts: $0, 8$
   y-intercept: 0
   Local maximum: $(4, 16)$

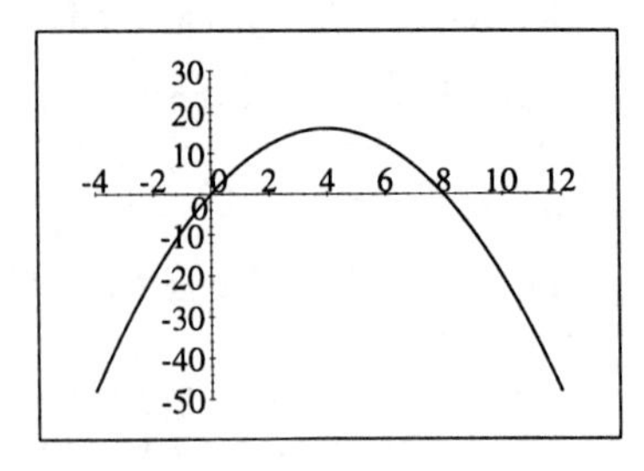

3. $y = x^3 - 12x + 9$, $[-5, 5]$ by $[-30, 30]$.
   x-intercepts: $3, 0.79, -3.79$
   y-intercept: 9
   Local maximum: $(-2, 25)$
   Local minimum: $(2, -7)$

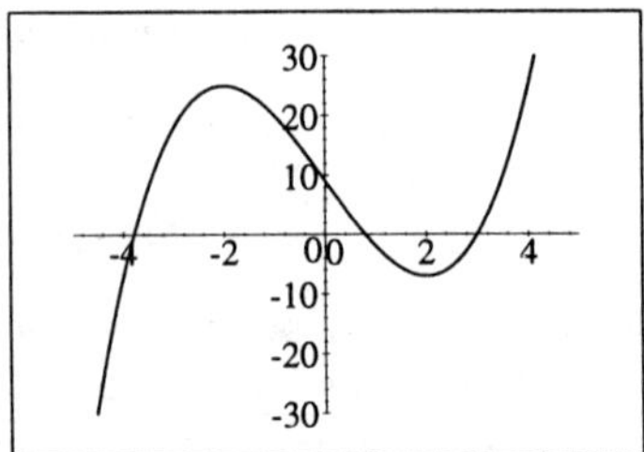

5. $y = x^4 + 4x^3$, $[-5, 5]$ by $[-30, 30]$.
   x-intercepts: $-4, 0$
   y-intercept: 0
   Local minimum: $(-3, -27)$

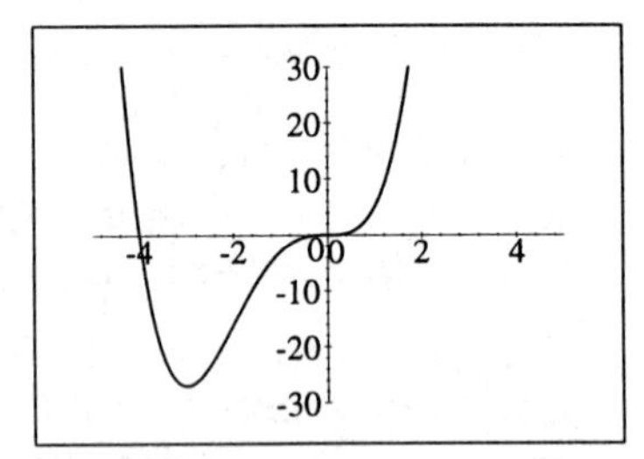

7. $y = 3x^5 - 5x^3 + 3$, $[-3, 3]$ by $[-5, 10]$.
   x-intercept: -1.42
   y-intercept: 3
   Local minimum: $(1, 1)$

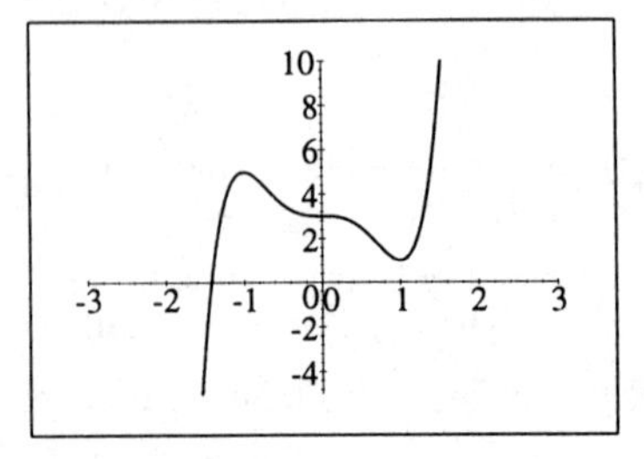

9. $y = -2x^2 + 3x + 5.$

Local maximum: $(0.75, 6.13)$

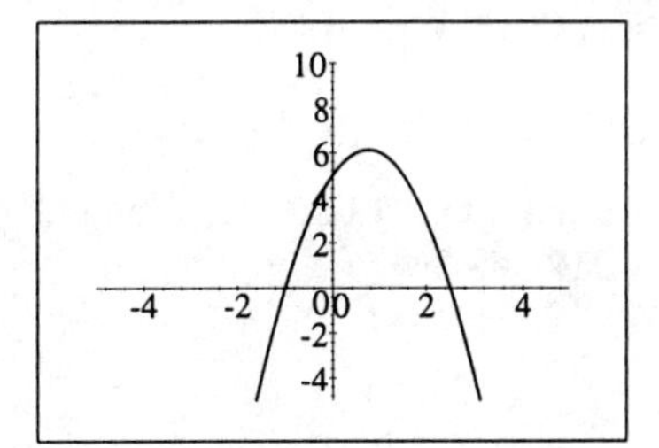

11. $y = x^3 - x^2 - x.$

Local maximum: $(-0.33, 0.19)$;

local minimum: $(1.00, -1.00)$.

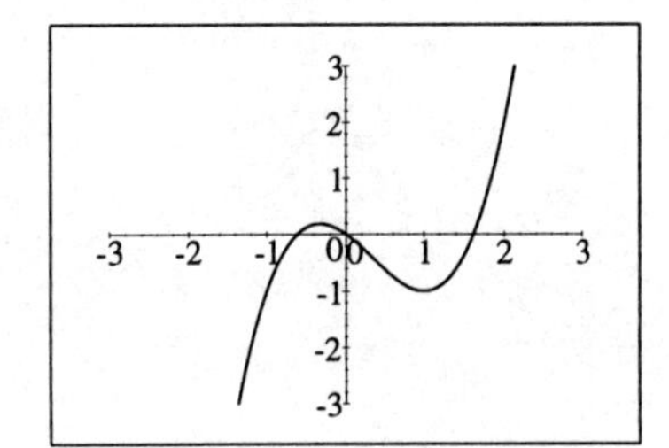

13. $y = x^4 - 5x^2 + 4.$

Local maximum: (0,4);

Local minima:
$(-1.58, -2.25), (1.58, -2.25)$.

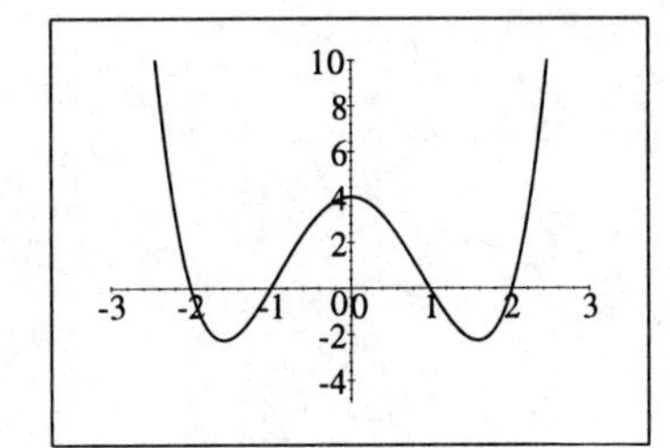

15. $y = (x - 2)^5 + 32.$

No extrema.

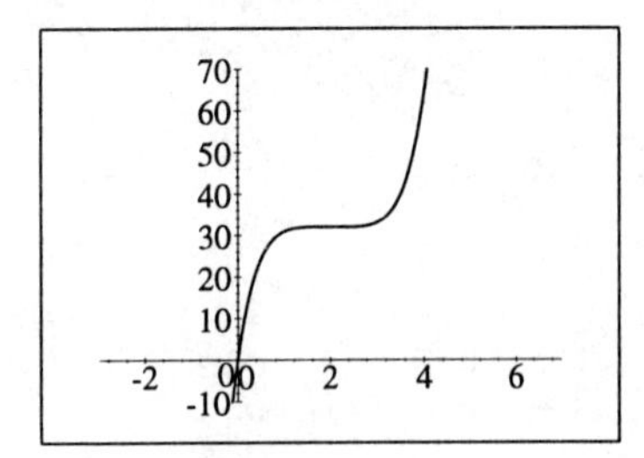

17. $y = (x - 4)(2x + 1)^2.$

Local maximum: $(-0.50, 0)$;

local minimum: $(2.50, -54)$.

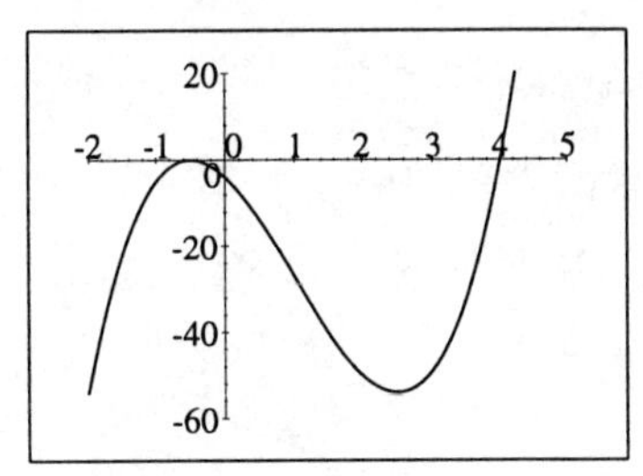

19. $y = x^8 - 3x^4 + x$.

Local maximum: $(0.44, 0.33)$;

local minima: $(1.09, -1.15)$ and $(-1.12, -3.36)$.

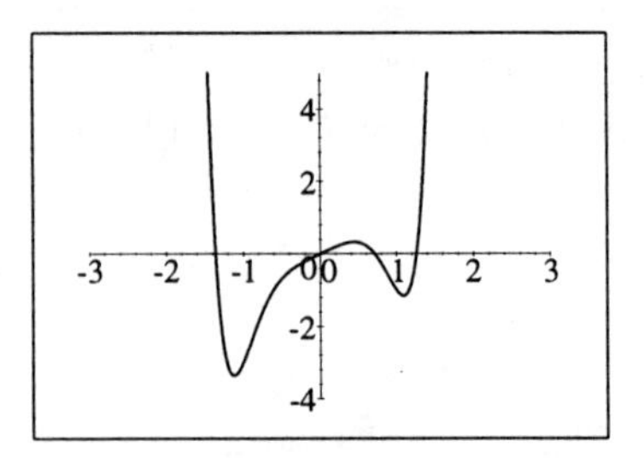

21. $y = 3x^3 - x^2 + 5x + 1$.

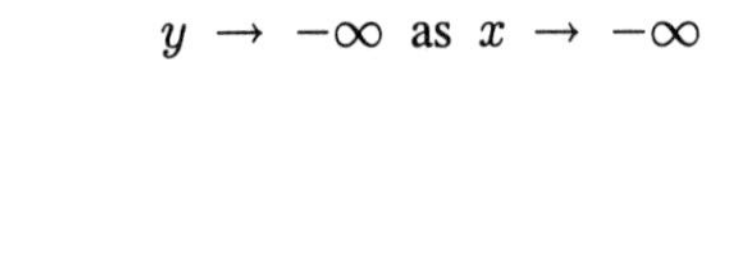

$y \to \infty$ as $x \to \infty$

$y \to -\infty$ as $x \to -\infty$

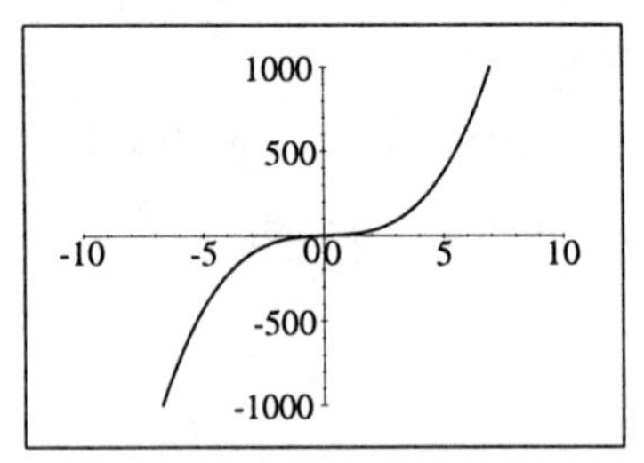

23. $y = x^4 - 7x^2 + 5x + 5$.

$y \to \infty$ as $x \to \pm\infty$

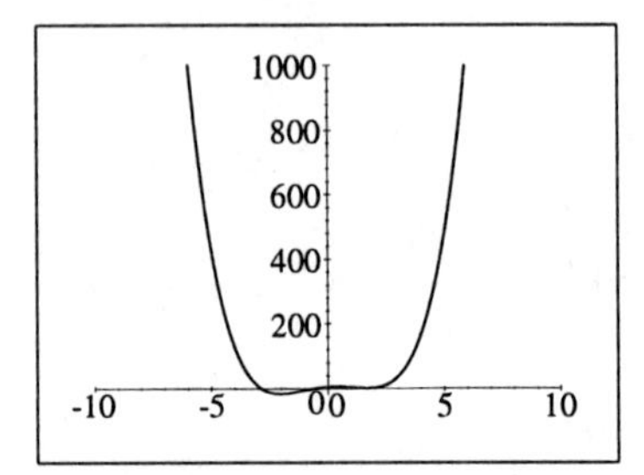

25. $y = x^{11} - 9x^9$.

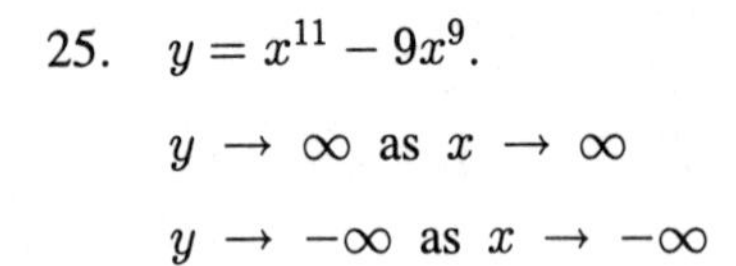

$y \to \infty$ as $x \to \infty$

$y \to -\infty$ as $x \to -\infty$

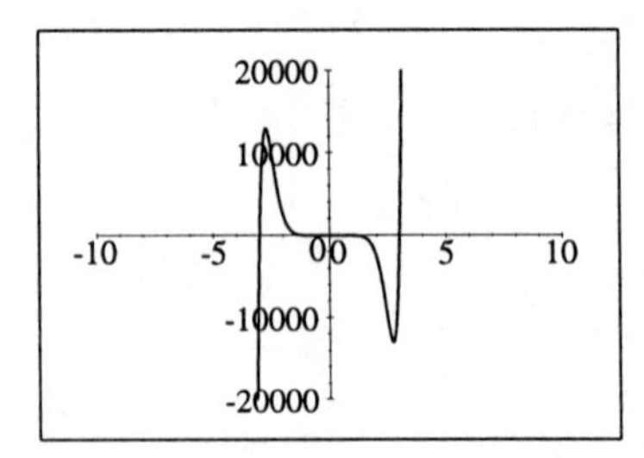

27. $y = 200x^2 - 0.001x^5$

$y \to -\infty$ as $x \to \infty$

$y \to \infty$ as $x \to -\infty$

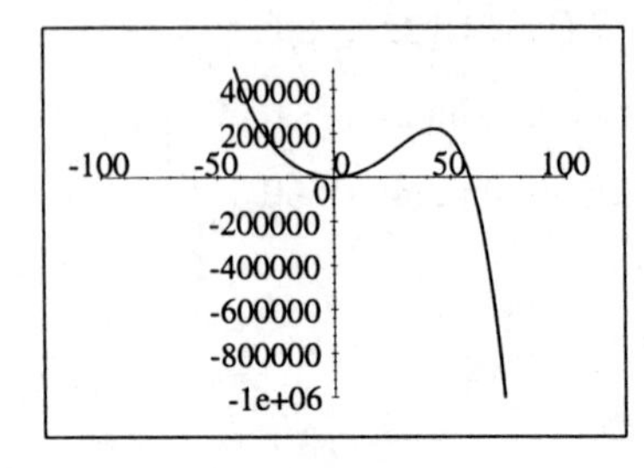

29. $3x^2 + 6x - 14 < 0$. The solution is $(-3.38, 1.38)$.

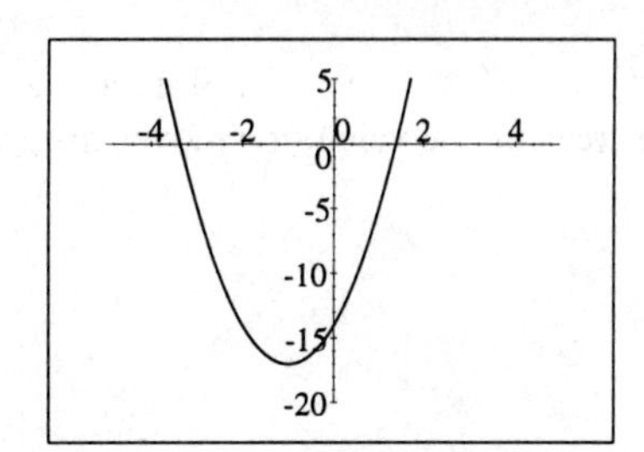

31. $x^3 - 2x^2 - 5x + 6 \geq 0$. The solution is $[-2, 1] \cup [3, \infty)$.

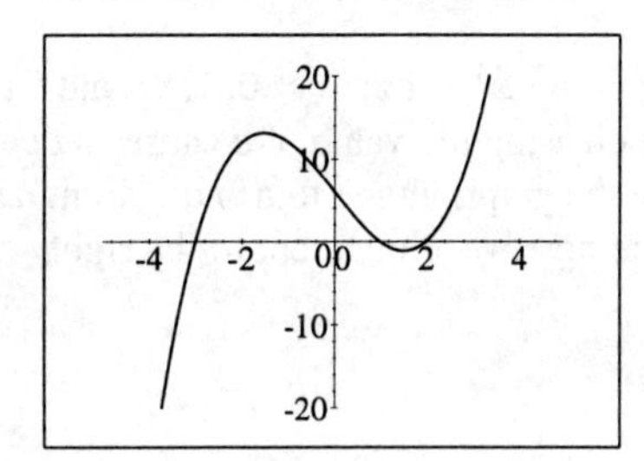

33. $2x^3 - 3x + 1 < 0$. The solution is $(-\infty, -1.37) \cup (0.37, 1.00)$.

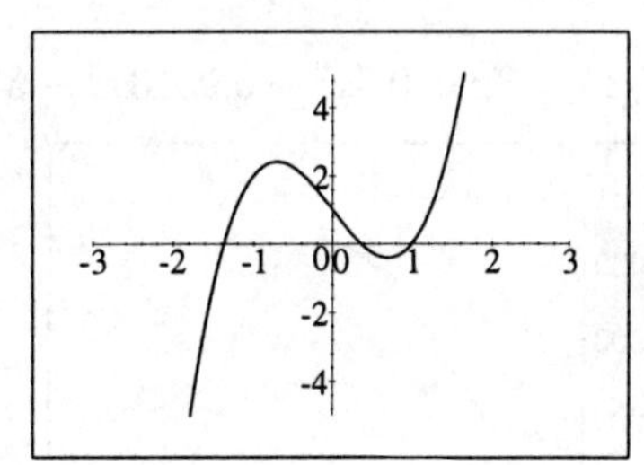

35. $5x^4 < 8x^3 \quad \Leftrightarrow \quad 5x^4 - 8x^3 < 0$. The solution is $(0, 1.6)$.

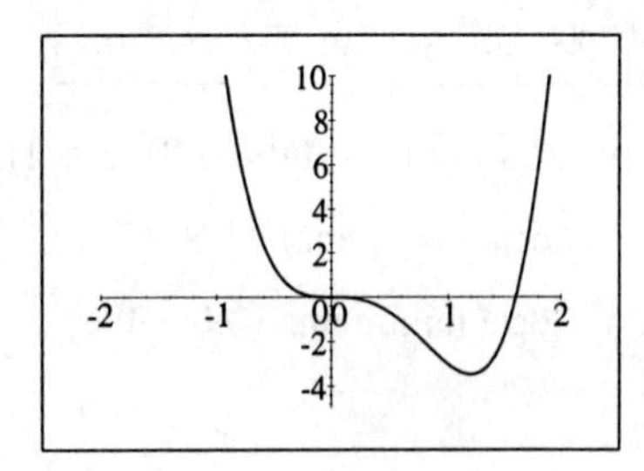

37. $P(x) = cx^3$; $c = 0, 2, 5$, and $\frac{1}{2}$. Increasing the value of c stretches the graph vertically.

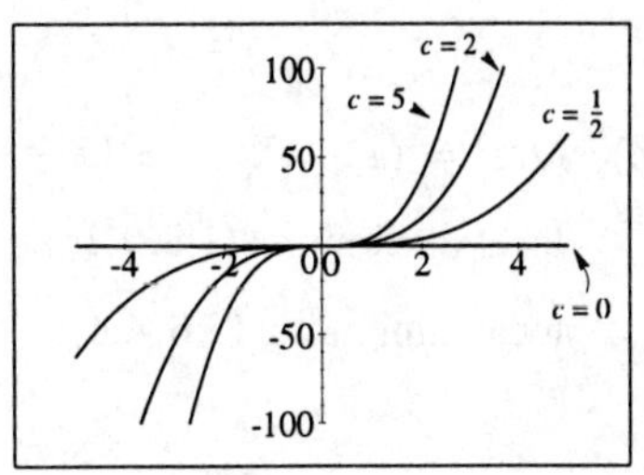

39. $P(x) = x^4 + c$; $c = -1, 0, 1$, and 2.
Increasing the value of c moves the graph up.

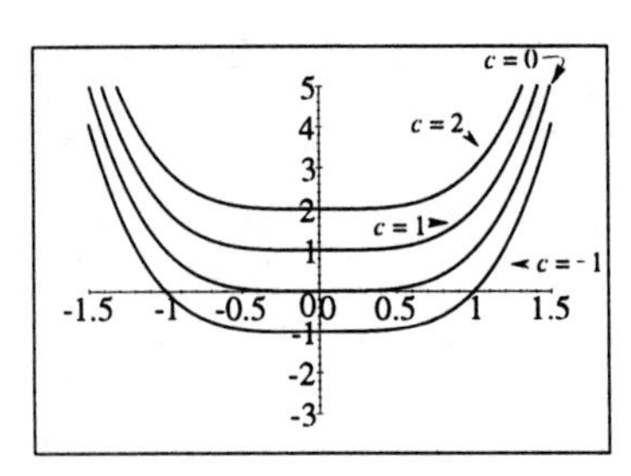

41. $P(x) = x^4 - cx$; $c = 0, 1, 8$, and 27.
Increasing the value of c causes a deeper dip in the graph, in the fourth quadrant, and moves the positive x-intercept to the right.

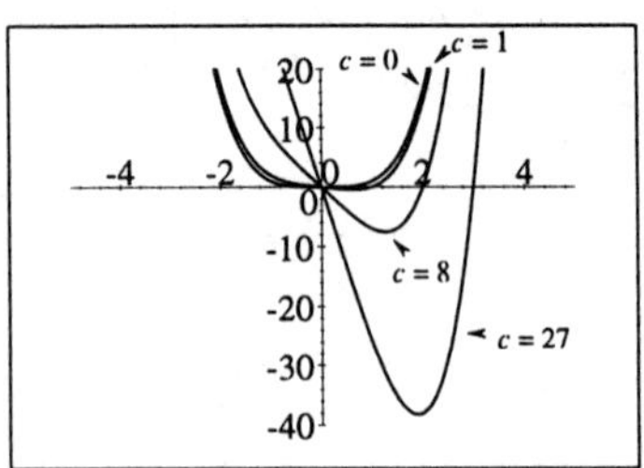

43. $P(x) = 8x + 0.3x^2 - 0.0013x^3 - 372$

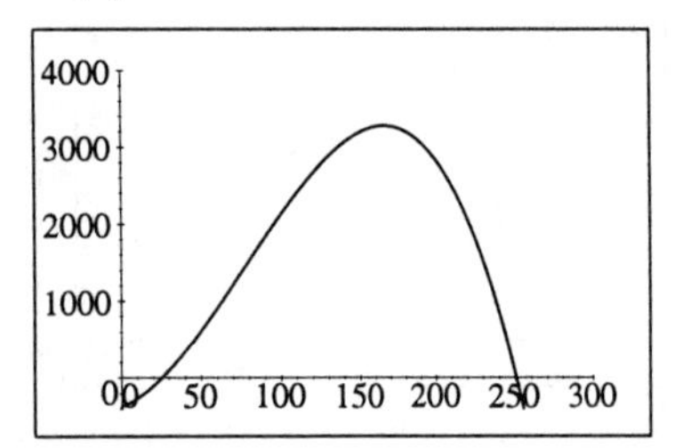

(a) For the firm to break even, $P(x) = 0 \Rightarrow x \approx 25.2$. Of course, the firm cannot produce fractions of a blender, so the manufacturer must produce at least 26 blenders a year.

(b) No, the profit does not increase indefinitely; the largest profit is approximately \$3276, which occurs when the firm produces 166 blenders per year.

45. (a) $P(x) = (x - 1)(x - 3)(x - 4)$.

local maximum: $(1.8, 2.1)$

local minimum: $(3.6, -0.6)$

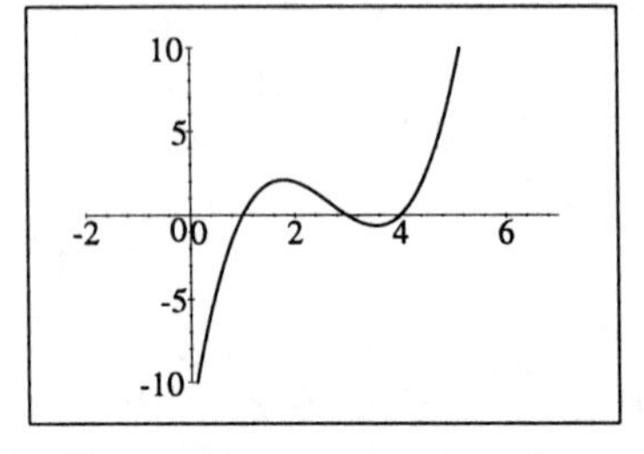

(b) $Q(x) = (x - 1)(x - 3)(x - 4) + 5$.

local maximum: $(1.8, 7.1)$

local minimum: $(3.6, 4.4)$

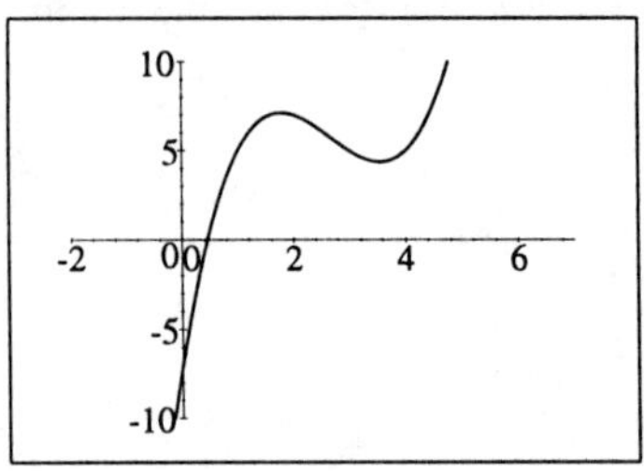

(c) Since $P(a) = P(b) = 0$, and $P(x) > 0$ for $a < x < b$ (see the table below), the graph of P must first rise and then fall on the interval (a, b) and so P must have at least one local maximum between a and b. Using similar reasoning, the fact that $P(b) = P(c) = 0$ and $P(x) < 0$ for $b < x < c$ shows that P must have at least one local minimum between b and c. Thus P has at least two local extrema.

Interval	$(-\infty, a)$	(a, b)	(b, c)	(c, ∞)
Sign of $x - a$	$-$	$+$	$+$	$+$
Sign of $x - b$	$-$	$-$	$+$	$+$
Sign of $x - c$	$-$	$-$	$-$	$+$
Sign of $(x - a)(x - b)(x - c)$	$-$	$+$	$-$	$+$

(d) $Q(x) = (x - a)(x - b)(x - c) + d$, where $a < b < c$. Since the graph of Q is obtained by shifting the graph of $P(x) = (x - a)(x - b)(x - c)$ vertically by d units, then Q has the same number of local extrema as P, which was found to be 2 in part (c).

47. (a) $P(x) = x^3 - 4x = x(x - 2)(x + 2)$ has 3 x-intercepts and 2 local extrema.

(b) $Q(x) = x^3 + 4x = x(x^2 + 4)$ has 1 x-intercept and no local extrema.

(c) For the x-intercepts of $P(x) = x^3 - ax$, we solve $x^3 - ax = 0 \quad \Leftrightarrow \quad x(x^2 - a) = 0 \quad \Leftrightarrow$ $x = 0$ or $x^2 = a \quad \Rightarrow \quad x = \pm\sqrt{a}$. So P has 3 x-intercepts. Since $P(x) = x(x^2 - a) = (x + \sqrt{a})(x - \sqrt{a})$, then by part (c) of problem 47, P has 2 local extrema. For the x-intercepts of $Q(x) = x^3 + ax$, we solve $x^3 + ax = 0 \quad \Leftrightarrow \quad x(x^2 + a) = 0 \quad \Leftrightarrow \quad x = 0$ or $x^2 = -a$ which has no real solutions because $a > 0$. So Q has 1 x-intercept. We now show that Q is always increasing and hence has no extrema. If $x_1 < x_2$, then $ax_1 < ax_2$ (because $a > 0$) and $x_1^3 < x_2^3$, and so $x_1^3 + ax_1 < x_2^3 + ax_2 \quad \Leftrightarrow \quad Q(x_1) < Q(x_2)$. Thus Q is increasing, that is, its graph always rises, and so it has no local extrema.

Exercises 4.3

1. The synthetic division table for this problem takes the following form.

$$\begin{array}{r|rrr} 3 & 1 & -5 & 4 \\ \hline & & 3 & -6 \\ \hline & 1 & -2 & -2 \end{array}$$

Thus the quotient is $x - 2$, and the remainder is -2.

3. Since $x + 2 = x - (-2)$, the synthetic division table for this problem takes the following form.

$$\begin{array}{r|rrrr} -2 & 1 & 2 & 2 & 1 \\ \hline & & -2 & 0 & -4 \\ \hline & 1 & 0 & 2 & -3 \end{array}$$

Thus the quotient is $x^2 + 2$, and the remainder is -3.

5. Since $x + 3 = x - (-3)$ and $x^3 - 8x + 2 = x^3 + 0x^2 - 8x + 2$, the synthetic division table for this problem takes the following form.

$$\begin{array}{r|rrrr} -3 & 1 & 0 & -8 & 2 \\ \hline & & -3 & 9 & -3 \\ \hline & 1 & -3 & 1 & -1 \end{array}$$

Thus the quotient is $x^2 - 3x + 1$, and the remainder is -1.

7. Since $x^5 + 3x^3 - 6 = x^5 + 0x^4 + 3x^3 + 0x^2 + 0x - 6$, the synthetic division table for this problem takes the following form.

$$\begin{array}{r|rrrrrr} 1 & 1 & 0 & 3 & 0 & 0 & -6 \\ \hline & & 1 & 1 & 4 & 4 & 4 \\ \hline & 1 & 1 & 4 & 4 & 4 & -2 \end{array}$$

Thus the quotient is $x^4 + x^3 + 4x^2 + 4x + 4$, and the remainder is -2.

9.

$$\begin{array}{r l} & x + 2 \\ \cline{2-2} x^2 - 2x + 2 \,\big) & x^3 + 0x^2 + 6x + 2 \\ & x^3 - 2x^2 + 2x \\ \cline{2-2} & \quad 2x^2 + 4x + 3 \\ & \quad 2x^2 - 4x + 4 \\ \cline{2-2} & \qquad\quad 8x - 1 \end{array}$$

Thus the quotient is $x + 2$, and the remainder is $8x - 1$.

11.

$$\begin{array}{r l} & 3x + 1 \\ \cline{2-2} x^2 + 0x + 5 \,\big) & 6x^3 + 2x^2 + 22x + 0 \\ & 6x^3 \qquad\quad + 15x \\ \cline{2-2} & \quad 2x^2 + 7x + 0 \\ & \quad 2x^2 \qquad + 5 \\ \cline{2-2} & \qquad\quad 7x - 5 \end{array}$$

Thus the quotient is $3x + 1$, and the remainder is $7x - 5$.

13.
$$\begin{array}{r}
x^3 \qquad\qquad +1 \\
x^2+x+1\overline{\big)\, x^5+x^4+x^3+x^2+x+1} \\
\underline{x^5+x^4+x^3 \qquad\qquad\qquad} \\
x^2+x+1 \\
\underline{x^2+x+1} \\
0
\end{array}$$

Thus the quotient is x^3+1, and the remainder is 0.

15. Since $x = x-(0)$, the synthetic division table for this problem takes the following form.

$$\begin{array}{r|rrrr}
0 & 1 & 0 & -4 & 7 \\
\hline
 & & 0 & 0 & 0 \\
\hline
 & 1 & 0 & -4 & 7
\end{array}$$

Thus the quotient is x^2-4, and the remainder is 7.

17.
$$\begin{array}{r}
x^3 \qquad\qquad +x \qquad\qquad \\
x^2-1\overline{\big)\, x^5+0x^4+0x^3+0x^2+0x-1} \\
\underline{x^5 \qquad\quad -x^3 \qquad\qquad\qquad} \\
x^3 \qquad\qquad\qquad \\
\underline{x^3 \qquad\quad -x \quad} \\
x-1
\end{array}$$

Thus the quotient is x^3+x, and the remainder is $x-1$.

19. The synthetic division table for this problem takes the following form.

$$\begin{array}{r|rrrr}
\frac{1}{2} & 2 & 3 & -2 & 1 \\
\hline
 & & 1 & 2 & 0 \\
\hline
 & 2 & 4 & 0 & 1
\end{array}$$

Thus the quotient is $2x^2+4x$, and the remainder is 1.

21.
$$\begin{array}{l}
\qquad\quad x^{100} \quad +x^{99} \quad +x^{98}+\cdots+1 \\
x-1\overline{\big)\, x^{101}+0x^{100}+0x^{99}+\cdots+0x-1} \\
\qquad\quad \underline{x^{101} \quad -x^{100}} \\
\qquad\qquad\qquad x^{100}+0x^{99} \\
\qquad\qquad\qquad \underline{x^{100} \quad -x^{99}} \\
\qquad\qquad\qquad\qquad x^{99}\ldots
\end{array}$$

Thus the quotient is $x^{100}+x^{99}+x^{98}+\cdots+x^2+x+1$, and the remainder is 0.

23. $\dfrac{x^2-x-3}{2x-4} = \dfrac{1}{2}\left(\dfrac{x^2-x-3}{x-2}\right)$. Thus the synthetic division table for this problem is:

$$\begin{array}{r|rrr}
2 & 1 & -1 & -3 \\
\hline
 & & 2 & 2 \\
\hline
 & 1 & 1 & -1
\end{array}$$

So $\dfrac{x^2-x-3}{2x-4} = \dfrac{1}{2}\left(\dfrac{x^2-x-3}{x-2}\right) = \dfrac{1}{2}\left(x+1-\dfrac{1}{x-2}\right) = \dfrac{1}{2}x+\dfrac{1}{2}-\dfrac{1}{2x-4}$. Therefore the quotient is $\frac{1}{2}x+\frac{1}{2}$ and the remainder is -1.

25. Since $2x - 1 = 2\left(x - \frac{1}{2}\right)$ we have $\dfrac{x^4 - 2x^3 + x + 2}{2x - 1} = \dfrac{1}{2}\left(\dfrac{x^4 - 2x^3 + x + 2}{x - \frac{1}{2}}\right)$

$$\begin{array}{r|rrrrr} \frac{1}{2} & 1 & -2 & 0 & 1 & 2 \\ & & \frac{1}{2} & -\frac{3}{4} & -\frac{3}{8} & \frac{5}{16} \\ \hline & 1 & -\frac{3}{2} & -\frac{3}{4} & \frac{5}{8} & \frac{37}{16} \end{array}$$

Therefore, $\dfrac{x^4 - 2x^3 + x + 2}{2x - 1} = \dfrac{1}{2}\left(\dfrac{x^4 - 2x^3 + x + 2}{x - \frac{1}{2}}\right) = \dfrac{1}{2}\left(x^3 - \dfrac{3}{2}x^2 - \dfrac{3}{4}x + \dfrac{5}{8} + \dfrac{\frac{37}{16}}{x - \frac{1}{2}}\right)$

$= \frac{1}{2}x^3 - \frac{3}{4}x^2 - \frac{3}{8}x + \frac{5}{16} + \dfrac{\frac{37}{16}}{2x - 1}$. Thus the quotient is $\frac{1}{2}x^3 - \frac{3}{4}x^2 - \frac{3}{8}x + \frac{5}{16}$ and the remainder is $\frac{37}{16}$.

27. $P(x) = 4x^2 + 12x + 5$, $c = -1$

$$\begin{array}{r|rrr} -1 & 4 & 12 & 5 \\ & & -4 & -8 \\ \hline & 4 & 8 & -3 \end{array}$$

Therefore $P(-1) = -3$.

29. $P(x) = 2x^2 + 9x + 1$, $c = 0.1$

$$\begin{array}{r|rrr} 0.1 & 2 & 9 & 1 \\ & & 0.2 & 0.92 \\ \hline & 2 & 9.2 & 1.92 \end{array}$$

Therefore $P(0.1) = 1.92$.

31. $P(x) = 2x^3 - 21x^2 + 9x - 200$, $c = 11$

$$\begin{array}{r|rrrr} 11 & 2 & -21 & 9 & -200 \\ & & 22 & 11 & 220 \\ \hline & 2 & 1 & 20 & 20 \end{array}$$

Therefore $P(11) = 20$.

33. $P(x) = 6x^5 + 10x^3 + x + 1$, $c = -2$

$$\begin{array}{r|rrrrrr} -2 & 6 & 0 & 10 & 0 & 1 & 1 \\ & & -12 & 24 & -68 & 136 & -274 \\ \hline & 6 & -12 & 34 & -68 & 137 & -273 \end{array}$$

Therefore $P(-2) = -273$.

35. $P(x) = -2x^6 + 7x^5 + 40x^4 - 7x^2 + 10x + 112$, $c = -3$

$$\begin{array}{r|rrrrrrr} -3 & -2 & 7 & 40 & 0 & -7 & 10 & 112 \\ & & 6 & -39 & -3 & 9 & -6 & -12 \\ \hline & -2 & 13 & 1 & -3 & 2 & 4 & 100 \end{array}$$

Therefore $P(-3) = 100$.

37. $P(x) = x^3 - x + 1$, $c = \frac{1}{4}$

$$\begin{array}{r|rrrr} \frac{1}{4} & 1 & 0 & -1 & 1 \\ & & \frac{1}{4} & \frac{1}{16} & -\frac{15}{64} \\ \hline & 1 & \frac{1}{4} & -\frac{15}{16} & \frac{49}{64} \end{array}$$

Therefore $P\left(\frac{1}{4}\right) = \frac{49}{64}$.

39. $P(x) = -2x^3 + 3x^2 + 4x + 6$, $c = 1 + \sqrt{2}$

$$\begin{array}{r|rrrr} 1+\sqrt{2} & -2 & 3 & 4 & 6 \\ & & -2-\sqrt{2} & -3-\sqrt{2} & -1 \\ \hline & -2 & 1-\sqrt{2} & 1-\sqrt{2} & 5 \end{array}$$

Therefore $P\left(1+\sqrt{2}\right) = 5$.

41. $P(x) = x^3 - 3x^2 + 3x - 1$, $c = 1$

$$\begin{array}{r|rrrr} 1 & 1 & -3 & 3 & -1 \\ & & 1 & -2 & 1 \\ \hline & 1 & -2 & 1 & 0 \end{array}$$

Since the remainder is 0, $x - 1$ is a factor.

43. $P(x) = 2x^3 + 7x^2 + 6x - 5$, $c = \frac{1}{2}$

$$\begin{array}{r|rrrr} \frac{1}{2} & 2 & 7 & 6 & -5 \\ & & 1 & 4 & 5 \\ \hline & 2 & 8 & 10 & 0 \end{array}$$

Since the remainder is 0, $x - \frac{1}{2}$is a factor.

45. $P(x) = x^3 - x^2 - 11x + 15$, $c = 3$

$$\begin{array}{r|rrrr} 3 & 1 & -1 & -11 & 15 \\ & & 3 & 6 & -15 \\ \hline & 1 & 2 & -5 & 0 \end{array}$$

Since the remainder is 0, $x - 3$ is a factor and $x^3 - x^2 - 11x + 15 = (x-3)(x^2+2x-5)$. Now $x^2 + 2x - 5 = 0$ when $x = \dfrac{-2 \pm \sqrt{2^2 + 4(1)(5)}}{2} = -1 \pm \sqrt{6}$.

Hence, the zeros are $-1-\sqrt{6}$, $-1+\sqrt{6}$, and 3.

47. Since the zeros of the polynomial are 1, -2, and 3, it follows that $P(x) = C(x-1)(x+2)(x-3)$ $= C(x^3 - 2x^2 - 5x + 6) = Cx^3 - 2Cx^2 - 5Cx + 6C$. Since the coefficient of x^2 to be 3, $-2C = 3$ so $C = -\frac{3}{2}$. Therefore, $P(x) = -\frac{3}{2}(x^3 - 2x^2 - 5x + 6) = -\frac{3}{2}x^3 + 3x^2 + \frac{15}{2}x - 9$ is the polynomial.

49. To ensure integer coefficients, let $P(x) = (x+k)(x-3)(x-4)(x+2)$, where k is an integer. The constant coefficient of $P(x)$ is then $k(-3)(-4)(2) = 24k$. Since we want the constant coefficient to be 24, we want $24 = 24k \quad \Leftrightarrow \quad k = 1$. Thus $P(x) = (x+1)(x-3)(x-4)(x+2)$ $= (x+1)(x-3)(x^2-2x-8) = (x+1)(x^3-5x^2-2x+24) = x^4 - 4x^3 - 7x^2 + 22x + 24$. That is $P(x) = x^4 - 4x^3 - 7x^2 + 22x + 24$.

51. If $x - 1$ is a factor of $P(x) = x^{567} - 3x^{400} + x^9 + 2$, then $P(1) = 0$. However, $P(1) = (1)^{567} - 3(1)^{400} + (1)^9 + 2 = 1 - 3 + 1 + 2 = 1 \neq 0$. Therefore, $x - 1$ is not a factor.

53. $x - 3$ will be a factor of $P(x) = k^2x^2 + 2kx - 12$ when $P(3) = 0$. Now, $P(3) = 9k^2 + 6k - 12 = 0 \quad \Leftrightarrow \quad 3k^2 + 2k - 4 = 0 \quad \Leftrightarrow \quad k = \dfrac{-2 \pm \sqrt{2^2 + 4(3)(4)}}{2(3)}$ $= \dfrac{-2 \pm \sqrt{52}}{6} = \dfrac{-2 \pm 2\sqrt{13}}{6} = \dfrac{-1 \pm \sqrt{13}}{3}$.

Exercises 4.4

1. $P(x) = x^3 - 2x^2 - 5x + 3 = 0$ has possible rational roots ± 1 and ± 3.

3. $R(x) = 2x^4 - 3x^3 - x + 6 = 0$ has possible rational roots $\pm 1, \pm 2, \pm 3, \pm 6, \pm \frac{1}{2}, \pm \frac{3}{2}$.

5. Since we can factor this quadratic, we have $2x^2 + x - 6 = (2x - 3)(x + 2) = 0$. Therefore, the roots are $x = -2, \frac{3}{2}$.

7. We factor by grouping, $x^3 - x^2 + x - 1 = x^2(x - 1) + (x - 1) = (x - 1)(x^2 + 1) = 0$. Therefore, $x = 1$ or $x^2 = -1$. The only real root is $x = 1$.

9. $x^3 - 3x^2 - 4x + 12 = 0$. Using synthetic division, we see that $(x - 2)$ is a factor:

$$\begin{array}{r|rrrr} & 1 & -3 & -4 & 12 \\ 2 & 1 & -1 & -6 & 0 \end{array} \quad \Rightarrow \quad x - 2 \text{ is a factor}$$

So $x^3 - x^2 + x - 1 = (x - 2)(x^2 - x - 6) = (x - 2)(x + 2)(x - 3) = 0$. Therefore, the roots are $x = 3, \pm 2$.

11. $x^3 - 4x^2 + x + 6 = 0$. Using synthetic division, we see that $(x + 1)$ is a factor:

$$\begin{array}{r|rrrr} & 1 & -4 & 1 & 6 \\ -1 & 1 & -5 & 6 & 0 \end{array} \quad \Rightarrow \quad x + 1 \text{ is a factor}$$

So $x^3 - 4x^2 + x + 6 = (x + 1)(x^2 - 5x + 6) = (x + 1)(x - 3)(x - 2) = 0$. Therefore, the roots are $x = -1, 2, 3$.

13. $x^3 - 7x^2 + 14x - 8 = 0$. Using synthetic division, we see that $(x - 1)$ is a factor:

$$\begin{array}{r|rrrr} & 1 & -7 & 14 & -8 \\ 1 & 1 & -6 & 8 & 0 \end{array} \quad \Rightarrow \quad x - 1 \text{ is a factor}$$

So $x^3 - 7x^2 + 14x - 8 = (x - 1)(x^2 - 6x + 8) = (x - 1)(x - 2)(x - 4) = 0$. Therefore, the roots are $x = 1, 2, 4$.

15. $x^3 + 4x^2 - 11x + 6 = 0$. Using synthetic division, we see that $(x - 1)$ is a factor:

$$\begin{array}{r|rrrr} & 1 & 4 & -11 & 6 \\ 1 & 1 & 5 & -6 & 0 \end{array} \quad \Rightarrow \quad x - 1 \text{ is a factor}$$

So $x^3 + 4x^2 - 11x + 6 = (x - 1)(x^2 + 5x - 6) = (x - 1)(x - 1)(x + 6) = (x - 1)^2 (x + 6) = 0$. Therefore, the roots are $x = -6, 1$.

17. $x^3 + 3x^2 + 6x + 4 = 0$. Using synthetic division, we see that $(x + 1)$ is a factor:

$$\begin{array}{r|rrrr} & 1 & 3 & 6 & 4 \\ -1 & 1 & 2 & 4 & 0 \end{array} \quad \Rightarrow \quad x + 1 \text{ is a factor}$$

So $x^3 + 3x^2 + 6x + 4 = (x + 1)(x^2 + 2x + 4) = 0$. Now, $x^2 + 2x + 4 = 0$ has no real roots, since the discriminant of this quadratic is $b^2 - 4ac = (2)^2 - 4(1)(4) = -12 < 0$. Thus, the only real root is $x = -1$.

19. Notice that this polynomial has only terms of even degree, so we can use the substitution $x^2 = u$ to factor: $x^4 + 3x^2 - 4 = (x^2 - 1)(x^2 + 4) = (x - 1)(x + 1)(x^2 + 4) = 0$. Since $x^2 + 4 = 0$ has no real roots, the only real roots are $x = \pm 1$.

21. $P(x) = x^4 - 2x^3 - 3x^2 + 8x - 4$. Using synthetic division, we see that $(x - 1)$ is a factor of $P(x)$:

$$\begin{array}{r|rrrrr} & 1 & -2 & -3 & 8 & -4 \\ \hline 1 & 1 & -1 & -4 & 4 & 0 \end{array} \Rightarrow \quad x - 1 \text{ is a factor}$$

Continue by factoring the quotient, we see that $(x - 1)$ is again a factor:

$$\begin{array}{r|rrrr} & 1 & -1 & -4 & 4 \\ \hline 1 & 1 & 0 & -4 & 0 \end{array} \Rightarrow \quad x - 1 \text{ is a factor}$$

$x^4 - 2x^3 - 3x^2 + 8x - 4 = (x - 1)(x - 1)(x^2 - 4) = (x - 1)^2 (x - 2)(x + 2) = 0$. Therefore, the roots are $x = 1, \pm 2$.

23. $P(x) = x^4 + 6x^3 + 7x^2 - 6x - 8$. The possible rational roots are $\pm 1, \pm 2, \pm 4, \pm 8$. $P(x)$ has 1 variation in sign and hence 1 positive real root. $P(-x) = x^4 - 6x^3 + 7x^2 + 6x - 8$ has 3 variations in sign and hence 1 or 3 negative real roots.

$$\begin{array}{r|rrrrr} & 1 & 6 & 7 & -6 & -8 \\ \hline 1 & 1 & 7 & 14 & 8 & 0 \end{array} \Rightarrow \quad x = 1 \text{ is a root and there are no other positive roots}$$

$x^4 + 6x^3 + 7x^2 - 6x - 8 = (x - 1)(x^3 + 7x^2 + 14x + 8)$. Continue by factoring the quotient, we have:

$$\begin{array}{r|rrrr} & 1 & 7 & 14 & 8 \\ \hline -1 & 1 & 6 & 8 & 0 \end{array} \Rightarrow \quad x = -1 \text{ is a root}$$

$x^4 + 6x^3 + 7x^2 - 6x - 8 = (x - 1)(x + 1)(x^2 + 6x + 8) = (x - 1)(x + 1)(x + 2)(x + 4) = 0$. Therefore, the roots are $\pm 1, -2$ and -4.

25. $P(x) = 4x^4 - 25x^2 + 36$ has possible rational roots $\pm 1, \pm 2, \pm 3, \pm 4, \pm 6, \pm 9, \pm 12, \pm 18, \pm 36, \pm \frac{1}{2}, \pm \frac{1}{4}, \pm \frac{3}{2}, \pm \frac{3}{4}, \pm \frac{9}{2}, \pm \frac{9}{4}$. Since $P(x)$ has 2 variations in sign, there are 0 or 2 positive real roots. Since $P(-x) = 4x^4 - 25x^2 + 36$ has 2 variations in sign, there are 0 or 2 negative real roots.

$$\begin{array}{r|rrrrr} & 4 & 0 & -25 & 0 & 36 \\ \hline 1 & 4 & 4 & -21 & -21 & 15 \\ 2 & 4 & 8 & -8 & -18 & 0 \end{array} \Rightarrow \quad x = 2 \text{ is a root}$$

$P(x) = (x - 2)(4x^3 + 8x^2 - 9x - 18)$

$$\begin{array}{r|rrrr} & 4 & 8 & -9 & -18 \\ \hline 2 & 4 & 16 & 23 & 28 \\ \frac{1}{2} & 4 & 10 & -4 & -20 \\ \frac{1}{4} & 4 & 9 & -\frac{27}{4} & -\frac{315}{16} \\ \frac{3}{2} & 4 & 14 & 12 & 0 \end{array}$$

For 2: $\Rightarrow$ all positive, $x = 2$ is an upper bound. For $\frac{3}{2}$: $\Rightarrow$ $x = \frac{3}{2}$ is a root

$P(x) = (x - 2)(2x - 3)(2x^2 + 7x + 6) = (x - 2)(2x - 3)(2x + 3)(x + 2) = 0$ Therefore, the roots are $\pm 2, \pm \frac{3}{2}$.

Note: Since $P(x)$ has only even terms factoring by substation also works. Let $x^2 = u$ then $P(x) = 4u^2 - 25u + 36 = (u - 4)(4u - 9) = (x^2 - 4)(4x^2 - 9)$ which yields the same results.

27. $P(x) = 4x^3 - 7x + 3$. The possible rational roots are $\pm 1, \pm 3, \pm \frac{1}{2}, \pm \frac{3}{2}, \pm \frac{1}{4}, \pm \frac{3}{4}$. Since $P(x)$ has 2 variations in sign, there are 0 or 2 positive roots. Since $P(-x) = -4x^3 + 7x + 3$ has 1 variation in sign, there is 1 negative root.

$$\begin{array}{r|rrrr} & 4 & 0 & -7 & 3 \\ \frac{1}{2} & 4 & 2 & -6 & 0 \end{array} \quad \Rightarrow \quad x = \tfrac{1}{2} \text{ is a root}$$

$P(x) = (x - \frac{1}{2})(4x^2 + 2x - 6) = (2x - 1)(2x^2 + x - 3) = (2x - 1)(x - 1)(2x + 3) = 0$. Thus, the roots are $-\frac{3}{2}, \frac{1}{2}$, and 1.

29. Factoring by grouping can be applied to this exercise. $4x^3 + 4x^2 - x - 1 = 4x^2(x + 1) - (x + 1)$ $= (x + 1)(4x^2 - 1) = (x + 1)(2x + 1)(2x - 1)$. Therefore, the roots are $x = -1, \pm \frac{1}{2}$.

31. We use the difference of squares to factor this polynomial. $P(x) = x^8 - 1 = (x^4 - 1)(x^4 + 1)$ $= (x^2 - 1)(x^2 + 1)(x^4 + 1) = (x - 1)(x + 1)(x^2 + 1)(x^4 + 1)$. We can use Descartes' Rule of Signs to verify that the factors $x^2 + 1$ and $x^4 + 1$ both have no real zeroes. Therefore, the only roots are $x = \pm 1$.

33. $P(x) = x^3 - x^2 - x - 3$. Since $P(x)$ has 1 variation in sign, $P(x)$ has 1 positive real zero. Since $P(-x) = -x^3 - x^2 + x - 3$ has 2 variations in sign, $P(x)$ has 2 or 0 negative real zeros. Thus, $P(x)$ has 1 or 3 real zeros.

35. $P(x) = 2x^6 + 5x^4 - x^3 - 5x - 1$. Since $P(x)$ has 1 variation in sign, $P(x)$ has 1 positive real zero. Since $P(-x) = 2x^6 + 5x^4 + x^3 + 5x - 1$ has 1 variation in sign, $P(x)$ has 1 negative real zero. Therefore, $P(x)$ has 2 real zeros.

37. $P(x) = x^5 + 4x^3 - x^2 + 6x$. Since $P(x)$ has 2 variations in sign, $P(x)$ has 2 or 0 positive real zeros. Since $P(-x) = -x^5 - 4x^3 - x^2 - 6x$ has 0 variations in sign, $P(x)$ has 0 negative real zeros. Therefore, $P(x)$ has a total of 1 or 3 real zeros (since $x = 0$ is a zero, but is neither positive nor negative).

39. $P(x) = 4x^7 - 3x^5 + 5x^4 + x^3 - 3x^2 + 2x - 5$. Since $P(x)$ has 5 variations in sign, $P(x)$ has 5, 3, or 1 positive real zeros. Since $P(-x) = -4x^7 + 3x^5 + 5x^4 - x^3 - 3x^2 - 2x - 5$ has 2 variations in sign, $P(x)$ has 2 or 0 negative real zeros. Therefore, $P(x)$ has 1, 3, 5, or 7 real zeros.

41. $2x^3 + 5x^2 + x - 2 = 0$; $a = -3, b = 1$

$$\begin{array}{r|rrrr} & 2 & 5 & 1 & -2 \\ -3 & 2 & -1 & 4 & -14 \\ 1 & 2 & 7 & 8 & 6 \end{array} \quad \begin{array}{lcl} \text{alternating signs} & \Rightarrow & \text{lower bound} \\ \text{all non-negative} & \Rightarrow & \text{upper bound} \end{array}$$

Therefore $a = -3, b = 1$ are lower and upper bounds, respectively.

43. $8x^3 + 10x^2 - 39x + 9 = 0$; $a = -3, b = 2$

$$\begin{array}{r|rrrr} & 8 & 10 & -39 & 9 \\ -3 & 8 & -14 & 3 & 0 \\ 2 & 8 & 26 & 13 & 35 \end{array} \quad \begin{array}{lcl} \text{alternating signs} & \Rightarrow & \text{lower bound} \\ \text{all non-negative} & \Rightarrow & \text{upper bound} \end{array}$$

Therefore $a = -3, b = 2$ are lower and upper bounds, respectively. $x = -3$ is also a zero.

There are many possible solutions to Exercises 45 and 47 since we only asked to find 'an upper bound' and 'a lower bound'

45. Set $P(x) = x^3 - 3x^2 + 4$ and use the Upper and Lower Bounds Theorem:

$$\begin{array}{r|rrrr ll}
 & 1 & -3 & 0 & 4 & & \\ \hline
-1 & 1 & -4 & 4 & 0 & \text{alternating signs} \Rightarrow & \text{lower bound} \\
3 & 1 & 0 & 0 & 4 & \text{all non-negative} \Rightarrow & \text{upper bound}
\end{array}$$

Therefore $a = -1$ is a lower bound (and a zero) and $b = 3$ is an upper bound.

47. Set $P(x) = x^4 - 2x^3 + x^2 - 9x + 2$.

$$\begin{array}{r|rrrrr ll}
 & 1 & -2 & 1 & -9 & 2 & & \\ \hline
1 & 1 & -1 & 0 & -9 & -7 & & \\
2 & 1 & 0 & 1 & -7 & -12 & & \\
3 & 1 & 1 & 4 & 3 & 11 & \text{all positive} \Rightarrow & \text{upper bound} \\
-1 & 1 & -3 & 4 & -13 & 15 & \text{alternating signs} \Rightarrow & \text{lower bound}
\end{array}$$

Therefore -1 is a lower bound and 3 is an upper bound.

49. $P(x) = 2x^4 + 3x^3 - 4x^2 - 3x + 2$

$$\begin{array}{r|rrrrr l}
 & 2 & 3 & -4 & -3 & 2 & \\ \hline
1 & 2 & 5 & 1 & -2 & 0 & \Rightarrow \quad x = 1 \text{ is a root, } P(x) = (x-1)(2x^3 + 5x^2 + x - 2)
\end{array}$$

$$\begin{array}{r|rrrr l}
 & 2 & 5 & 1 & -2 & \\ \hline
-1 & 2 & 3 & -2 & 0 & \Rightarrow \quad x = -1 \text{ is a root}
\end{array}$$

$P(x) = (x-1)(x+1)(2x^2 + 3x - 2) = (x-1)(x+1)(2x-1)(x+2)$. Therefore, the roots are $x = -2, \frac{1}{2}, \pm 1$.

51. <u>Method #1</u>: $P(x) = 4x^4 - 21x^2 + 5$ has 2 variations in sign, so by Descartes' rule of signs there are either 2 or 0 positive roots. If we replace x with $(-x)$, the function does not change, so there are either 2 or 0 negative roots. Possible rational roots are $\pm 1, \pm \frac{1}{2}, \pm \frac{1}{4}, \pm 5, \pm \frac{5}{2}, \pm \frac{5}{4}$. By inspection, ± 1 and ± 5 are not roots, so we must look for non-integer solutions:

$$\begin{array}{r|rrrrr l}
 & 4 & 0 & -21 & 0 & 5 & \\ \hline
\frac{1}{2} & 4 & 2 & -20 & -10 & 0 & \Rightarrow \quad x = \frac{1}{2} \text{ is a root}
\end{array}$$

$P(x) = (x - \frac{1}{2})(4x^3 + 2x^2 - 20x - 10)$, continuing with the quotient, we have:

$$\begin{array}{r|rrrr l}
 & 4 & 2 & -20 & -10 & \\ \hline
-\frac{1}{2} & 4 & 0 & -20 & 0 & \Rightarrow \quad x = -\frac{1}{2} \text{ is a root}
\end{array}$$

$P(x) = (x - \frac{1}{2})(x + \frac{1}{2})(4x^2 - 20) = 0$. So $x = \pm \frac{1}{2}$ or $4x^2 - 20 = 0 \Rightarrow x = \pm\sqrt{5}$. Thus the roots are $x = \pm \frac{1}{2}, \pm\sqrt{5}$.

<u>Method #2</u>: Substituting $u = x^2$ the equation becomes $4u^2 - 21u + 5 = 0$, which factors: $4u^2 - 21u + 5 = (4u-1)(u-5) = (4x^2 - 1)(x^2 - 5) \Rightarrow x^2 = 5 \Rightarrow x = \pm\sqrt{5}$, or $x^2 = \frac{1}{4} \Rightarrow x = \pm\sqrt{\frac{1}{4}} = \pm\frac{1}{2}$.

53. $P(x) = x^5 - 7x^4 + 9x^3 + 23x^2 - 50x + 24$. The possible rational roots are $\pm 1, \pm 2, \pm 3, \pm 4, \pm 6, \pm 8, \pm 12, \pm 24$. $P(x)$ has 4 variations in sign and hence 0, 2, or 4 positive real roots. $P(-x) = -x^5 - 7x^4 - 9x^3 + 23x^2 + 50x + 24$ has 1 variation in sign, and hence 1 negative real root.

$$\begin{array}{r|rrrrrr} & 1 & -7 & 9 & 23 & -50 & 24 \\ \hline 1 & 1 & -6 & 3 & 26 & -24 & 0 \end{array} \quad \Rightarrow \quad x = 1 \text{ is a root}$$

$P(x) = (x-1)(x^4 - 6x^3 + 3x^2 + 26x - 24)$, continuing with the quotient, we try 1 again.

$$\begin{array}{r|rrrrr} & 1 & -6 & 3 & 26 & -24 \\ \hline 1 & 1 & -5 & -2 & 24 & 0 \end{array} \quad \Rightarrow \quad x = 1 \text{ is a root}$$

$P(x) = (x-1)^2(x^3 - 5x^2 - 2x + 24)$, continuing with the quotient, we start by trying 1 again.

$$\begin{array}{r|rrrr} & 1 & -5 & -2 & 24 \\ \hline 1 & 1 & -4 & -6 & 18 \\ 2 & 1 & -3 & -8 & 8 \\ 3 & 1 & -2 & -8 & 0 \end{array} \quad \Rightarrow \quad x = 3 \text{ is a root}$$

$P(x) = (x-1)^2(x-3)(x^2-2x-8) = (x-1)^2(x-3)(x-4)(x+2)$. Therefore, the roots are $-2, 1, 3$ and 4.

55. We start by modifying the polynomial to obtain integer coefficients.
$x^4 - \frac{11}{4}x^3 - \frac{11}{2}x^2 + \frac{5}{4}x + 3 = 0 \;\Leftrightarrow\; 4x^4 - 11x^3 - 22x^2 + 5x + 12 = 0$. Set $P(x) = 4x^4 - 11x^3 - 22x^2 + 5x + 12$. The possible rational roots are $\pm 1, \pm 2, \pm 3, \pm 4, \pm 6, \pm 12, \pm \frac{1}{2}, \pm \frac{3}{2}, \pm \frac{1}{4}, \pm \frac{3}{4}$. Since $P(x)$ has 2 variations in sign, there are 0 or 2 positive roots. Since $P(-x) = 4x^4 + 11x^3 - 22x^2 - 5x + 12$ has 2 variations in sign, there are 0 or 2 negative real roots.

$$\begin{array}{r|rrrrr} & 4 & -11 & -22 & 5 & 12 \\ \hline 1 & 4 & -7 & -29 & 24 & -12 \\ 2 & 4 & -3 & -28 & -51 & -90 \\ 3 & 4 & 1 & -19 & -52 & -144 \\ 4 & 4 & 5 & -2 & -3 & 0 \end{array} \quad \Rightarrow \quad x = 4 \text{ is a root}$$

$P(x) = (x-4)(4x^3 + 5x^2 - 2x - 3) = 0$, continuing with the quotient, we try 4 again

$$\begin{array}{r|rrrr} & 4 & 5 & -2 & -3 \\ \hline 4 & 4 & 21 & 82 & 325 \\ \frac{3}{4} & 4 & 8 & 4 & 0 \end{array} \quad \Rightarrow \quad x = \tfrac{3}{4} \text{ is a root}$$

$P(x) = (x-4)(4x-3)(x^2+2x+1) = (x-4)(4x-3)(x+1)^2 = 0$. Therefore, the roots are $-1, \frac{3}{4}$, and 4.

57. $P(x) = x^3 - x - 2$. The only possible rational roots of $P(x)$ are ± 1 and ± 2.

$$\begin{array}{r|rrrr} & 1 & 0 & -1 & -2 \\ \hline 1 & 1 & 1 & 0 & -2 \\ 2 & 1 & 2 & 3 & 4 \\ -1 & 1 & -1 & 0 & -2 \end{array}$$

Since the row that contains -1 alternates between non-negative and non-positive, -1 is a lower bound and there is no need to try -2. Therefore, $P(x)$ does not have a rational zero.

59. $P(x) = 3x^3 - x^2 - 6x + 12$ has possible rational zeros $\pm 1, \pm 2, \pm 3, \pm 4, \pm 6, \pm 12, \pm \frac{1}{3}, \pm \frac{2}{3}, \pm \frac{4}{3}$.

	3	-1	-6	12	
1	3	2	-4	8	
2	3	5	4	20	all positive $\Rightarrow$ $x = 2$ is an upper bound
$\frac{1}{3}$	3	0	-6	10	
$\frac{2}{3}$	3	1	$-\frac{16}{3}$	$\frac{76}{9}$	
$\frac{4}{3}$	3	3	-2	$\frac{28}{3}$	
-1	3	-4	-2	14	
-2	3	-7	8	-4	alternating signs $\Rightarrow$ $x = -2$ is a lower bound
$-\frac{1}{3}$	3	-2	$-\frac{16}{3}$	$\frac{124}{9}$	
$-\frac{2}{3}$	3	-3	-4	$\frac{44}{3}$	
$-\frac{4}{3}$	3	-5	$\frac{2}{3}$	$\frac{100}{9}$	

Therefore, there are no rational zeros.

61. (a)

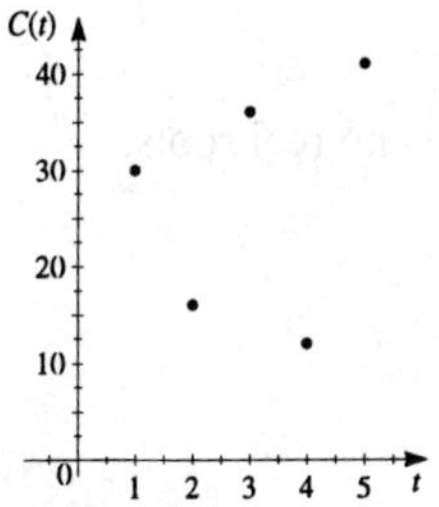

(b) It is impossible for a third degree polynomial to pass through all five points since any function whose graph contains these points will have at least three turning points, and a cubic has at most two. (A quadratic has one.)

(c) The lowest degree polynomial that will pass through the points is a degree four polynomial.

63. (a) Since $z > b$, $z - b > 0$. Since all the coefficients of $Q(x)$ are nonnegative, and since $z > 0$, then $Q(z) > 0$ (being a sum of positive terms). Thus, $P(z) = (z - b) \cdot Q(z) + r > 0$, since it is the sum of a positive number and a nonnegative number.

(b) In part (a) we showed that if b satisfies the conditions of the first part of the Upper and Lower Bounds Theorem, then $P(z) > 0$. This means that no real root of P can be larger than b, so b is an upper bound for the real roots.

65. $P(x) = x^5 - x^4 - x^3 - 5x^2 - 12x - 6$ has possible rational roots $\pm 1, \pm 2, \pm 3, \pm 6$. Since $P(x)$ has 1 variation in sign, there is 1 positive real root. Since $P(-x) = -x^5 - x^4 + x^3 - 5x^2 + 12x - 6$ has 4 variations in sign, there are 0, 2, or 4 negative real roots.

	1	-1	-1	-5	-12	-6	
1	1	0	-1	-6	-18	-24	
2	1	1	1	-3	-18	-42	
3	1	2	5	10	18	48	$\Rightarrow$ 3 is an upper bound
-1	1	-2	1	-6	-6	0	$\Rightarrow$ $x = -1$ is a root

$P(x) = (x + 1)(x^4 - 2x^3 + x^2 - 6x - 6)$, continuing with the quotient we have

	1	-2	1	-6	-6	
-1	1	-3	4	-10	4	$\Rightarrow$ -1 is a lower bound

Therefore, there is 1 rational root, namely -1. Since there are 1, 3 or 5 real roots, and we found 1 rational root, there must be 0, 2 or 4 irrational roots. However, since 1 root must be positive, there cannot be 0 irrational roots. Therefore, there is exactly 1 rational root and 2 or 4 irrational roots.

67. $P(x) = x^3 - kx^2 + k^2x - 3$ has possible rational roots $\pm 1, \pm 3$.

$$\begin{array}{c|cccc} & 1 & -k & k^2 & -3 \\ 1 & 1 & -k+1 & k^2-k+1 & k^2-k-2 \end{array}$$

Hence, 1 is a zero if $k^2 - k - 2 = 0 \quad \Leftrightarrow \quad (k-2)(k+1) = 0 \quad \Leftrightarrow \quad k = 2$ or $k = -1$.

$$\begin{array}{c|cccc} & 1 & -k & k^2 & -3 \\ 3 & 1 & -k+3 & k^2-3k+9 & 3k^2-9k+24 \end{array}$$

Thus, 3 is a zero if $3k^2 - 9k + 24 = 0 \quad \Leftrightarrow \quad k = \dfrac{9 \pm \sqrt{81 - (4)(3)(24)}}{6} = \dfrac{9 \pm \sqrt{-207}}{6}$ which is not a real number.

$$\begin{array}{c|cccc} & 1 & -k & k^2 & -3 \\ -1 & 1 & -k-1 & k^2+k+1 & -k^2-k-4 \end{array}$$

So -1 is a zero if $k^2 + k + 4 = 0$. But this equation has no real roots.

$$\begin{array}{c|cccc} & 1 & -k & k^2 & -3 \\ -3 & 1 & -k-3 & k^2+3k+9 & -k^2-9k-30 \end{array}$$

Thus, -3 is a zero if $3k^2 + 9k + 30 = 0$. But this equation also has no real roots.

Thus, the equation has a rational root for $k = -1$ or $k = 2$.

Exercises 4.5

1. $P(x) = x^3 + x - 1$; between 0 and 1; to the nearest tenth.

x	$x^3 + x - 1$
0	-1
1	1
0.5	-0.375
0.6	-0.184
0.7	0.043
0.65	-0.754

Thus the root is 0.7.

3. $P(x) = x^3 + x^2 + x - 2$; between 0 and 1; to one decimal

x	$x^3 + x^2 + x - 2$
0	-2
1	1
0.9	0.439
0.8	-0.046
0.85	0.1866

Thus the root is 0.8

5. $P(x) = x^3 - 4x^2 + 2$; between 0 and 1; to two decimal places.

x	$x^3 - 4x^2 + 2$
0	2
1	-1
0.5	1.125
0.7	0.383
0.8	-0.048
0.75	0.171875
0.79	-0.003361
0.78	0.040952
0.785	0.018837

Thus the root is 0.79.

7. $P(x) = 2x^4 - 4x^2 + 1$; between -1 and 0, to the nearest hundredth.

x	$2x^4 - 4x^2 + 1$
-1	-1
0	1
-0.5	0.125
-0.6	-0.1808
-0.55	-0.026988
-0.54	0.003661
-0.545	-0.011652

Thus the root is -0.54.

9. $2x^3 - x^2 - 6 = 0$. By Descartes' Rule of Signs, there is only one positive root.

x	$2x^3 - x^2 - 6$	
0	-6	
1	-5	
2	6	$\Rightarrow$ root between 1 and 2.
1.5	-1.5	
1.6	-0.368	
1.7	0.936	
1.65	0.26175	
1.61	-0.245538	
1.62	-0.121344	
1.63	0.004594	
1.625	-0.0585938	

Thus the one positive root to two decimal places is 1.63.

11. The smaller positive zero of $x^3 - 3x + 1$. By Descartes' Rule of Signs, there are 2 or 0 positive real roots.

x	$x^3 - 3x + 1$
0	1
1	-1
2	3
1.5	-1.5
1.6	-0.368
1.7	0.936
1.65	0.26175
1.61	-0.245538
1.62	-0.121344
1.63	0.004594
1.625	-0.0585938

Thus the root we seek is between 0 and 1.

Thus the smaller positive root to two decimal places is 1.63.

13. $x^4 + 2x^3 + x^2 - 1 = 0$. By Descartes' Rule, there is one positive root.

x	$x^4 + 2x^3 + x^2 - 1$
0	-1
1	3
0.7	0.4161
0.6	-0.0784
0.61	-0.03556
0.62	0.00882
0.615	-0.0135

Hence, the positive root is 0.62.

15. $P(x) = x^3 + 2x^2 - 6x - 4$. The possible rational roots are ± 1, ± 2, and ± 4.

	1	2	-6	-4
1	1	3	-3	-7
2	1	4	2	0

$\Rightarrow$ $x = 2$ is a root

$P(x) = (x - 2)(x^2 + 4x + 2)$. Using the quadratic formula to find the other roots we have $x = \dfrac{-4 \pm \sqrt{16 - 4(2)(1)}}{2} = \dfrac{-4 \pm 2\sqrt{2}}{2} = -2 \pm \sqrt{2}$. Hence, the roots of $P(x) = x^3 + 2x^2 - 6x - 4$ are 2 and $-2 \pm \sqrt{2}$.

17. $P(x) = 2x^5 + 3x^4 + x^2 + x - 1$ has possible rational roots ± 1, $\pm \frac{1}{2}$. By Descartes' Rule of Signs, there is 1 positive root, and 2 or 0 negative roots.

	2	3	0	1	1	-1	
1	2	5	5	6	7	6	all positive $\Rightarrow$ $x = 1$ is an upper bound
$\frac{1}{2}$	2	4	2	2	2	0	$\Rightarrow$ $x = \frac{1}{2}$ is a root

$P(x) = \left(x - \frac{1}{2}\right)(2x^4 + 4x^3 + 2x^2 + 2x + 2) = 2\left(x - \frac{1}{2}\right)(x^4 + 2x^3 + x^2 + x + 1)$

	1	2	1	1	1
-1	1	1	0	1	0

$\Rightarrow$ $x = -1$ is a root

$P(x) = 2\left(x - \frac{1}{2}\right)(x + 1)(x^3 + x^2 + 1)$. By Descartes' Rule of Signs, $x^3 + x^2 + 1$ has no positive and one negative real root. Since $P(-1) = 1$, the negative real root is irrational.

x	$x^3 + x^2 + 1$
-1	1
-2	-3
-1.5	-0.125
-1.4	0.216
-1.47	-0.0156
-1.46	0.01946
-1.465	0.0020

Thus the root we seek is between -2 and -1.

$\Rightarrow$ -1.47 is a root.

Hence, the roots of $P(x) = 2x^5 + 3x^4 + x^2 + x - 1$ are -1, $\frac{1}{2}$ and -1.47.

19. $P(x) = x^3 - 3x^2 + 3$ has possible rational roots ± 1, ± 3. By Descartes' Rule of Signs, there are 0 or 2 positive roots, and 1 negative root.

x	$x^3 - 3x^2 + 3$
-3	-51
-1	-1
0	3
1	1
2	-1
3	3

There are no rational roots. Since there is a sign change between -1 and 0, another sign change between 1 and 2, and between 2 and 3, there are 3 irrational roots.

x	$x^3 - 3x^2 + 3$
0	3
-0.9	-0.159
-0.8	0.568
-0.88	-0.004672
-0.87	0.070797
-0.875	0.033203125

Root is -0.88

x	$x^3 - 3x^2 + 3$
1	1
1.4	-0.136
1.3	0.127
1.35	-0.007125
1.34	0.019304
1.345	0.00606

Root is 1.35

x	$x^3 - 3x^2 + 3$
3	3
2.5	-0.125
2.6	0.296
2.53	-0.008423
2.54	0.032264
2.535	0.01181

Root is 2.53

Therefore, the roots are -0.88, 1.35 and 2.53 correct to two decimals.

21. $P(x) = x^5 + x^4 - 4x^3 - x^2 + 5x - 2$ has possible rational roots ± 1, ± 2.

$$\begin{array}{r|rrrrrr} & 1 & 1 & -4 & -1 & 5 & -2 \\ \hline 1 & 1 & 2 & -2 & -3 & 2 & 0 \end{array} \quad \Rightarrow \quad x = 1 \text{ is a root}$$

$P(x) = (x - 1)(x^4 + 2x^3 - 2x^2 - 3x + 2)$

$$\begin{array}{r|rrrrr} & 1 & 2 & -2 & -3 & 2 \\ \hline 1 & 1 & 3 & 1 & -2 & 0 \end{array} \quad \Rightarrow \quad x = 1 \text{ is a root}$$

$P(x) = (x-1)^2(x^3+3x^2+x-2)$

$$\begin{array}{r|rrrr} & 1 & 3 & 1 & -2 \\ \hline 1 & 1 & 4 & 5 & 3 \\ -1 & 1 & 2 & -1 & -1 \\ -2 & 1 & 1 & -1 & 0 \end{array}$$

$\Rightarrow$ no more positive rational roots (row 1); $\Rightarrow$ $x=-2$ is a root (row -2)

$P(x) = (x-1)^2(x+2)(x^2+x-1)$. Using the quadratic formula on the factor x^2+x-1 we find the irrational roots $x = \dfrac{-1 \pm \sqrt{1-4(1)(-1)}}{2} = \dfrac{-1 \pm \sqrt{5}}{2}$. Therefore, the roots are 1, -2 and $\frac{-1\pm\sqrt{5}}{2}$.

23. $P(x) = 2x^4 - 7x^3 + 9x^2 + 5x - 4$ has possible rational roots $\pm 1, \pm 2, \pm 4, \pm \frac{1}{2}$. By Descartes' Rule of Signs, there are 1 or 3 positive roots, and 1 negative root.

$$\begin{array}{r|rrrrr} & 2 & -7 & 9 & 5 & -4 \\ \hline \frac{1}{2} & 2 & -6 & 6 & 8 & 0 \end{array} \quad \Rightarrow \quad x = \tfrac{1}{2} \text{ is a root}$$

$P(x) = \left(x - \frac{1}{2}\right)(2x^3 - 6x^2 + 6x + 8) = 2\left(x - \frac{1}{2}\right)(x^3 - 3x^2 + 3x + 4)$. Continuing with the quotient, $Q(x) = x^3 - 3x^2 + 3x + 4$, and the possible roots $\pm 1, \pm 2, \pm 4$:

$$\begin{array}{r|rrrr} & 1 & -3 & 3 & 4 \\ \hline 1 & 1 & -2 & 1 & 5 \\ 2 & 1 & -1 & 1 & 6 \\ 4 & 1 & 1 & 7 & 32 \\ -1 & 1 & -4 & 7 & -3 \end{array}$$

$\Rightarrow$ 4 is an upper bound (row 4)

Since $Q(0) = 4$ and $Q(-1) = -3$ the sign change indicates root between -1 and 0.

x	$x^3 - 3x^2 + 3x + 4$
-0.5	1.625
-0.7	0.087
-0.8	-0.832
-0.71	-0.000211
-0.705	0.0435

$\Rightarrow$ Root is -0.71

Therefore, the roots are $\frac{1}{2}$ and -0.71 (correct to 2 decimals).

25. Let x and y be the lengths of the sides in feet. Since the diagonal is 2 feet longer than x, it follows from the Pythagorean Theorem that $y = \sqrt{(x+2)^2 - x^2} = \sqrt{x^2+4x+4-x^2} = 2\sqrt{x+1}$. Since the area is 10 ft^2, we get: $2x\sqrt{x+1} = 10 \Leftrightarrow 4x^2(x+1) = 100 \Leftrightarrow$ $x^3 + x^2 - 25 = 0$. Since we are determining the lengths of sides we are interested in positive roots only.

x	$x^3 + x^2 - 25$
2	-13
3	11
2.6	-0.664
2.62	-0.150872
2.63	0.108347
2.626	0.0044
2.625	-0.021484
2.6255	-0.009

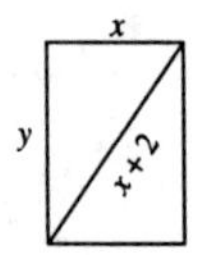

Root is $x \approx 2.626$ (correct to 3 decimals). Therefore, $y = 2\sqrt{x+1} \approx 3.808$, and so the dimensions of the rectangle are 2.626 ft by 3.808 ft (correct to 3 decimals).

27. Let $r =$ the radius of the cone and cylinder. Let $h =$ the height of the cone. Since the height and diameter are equal, we get $h = 2r$. So the volume of the cylinder is $V_1 = \pi r^2 \cdot (\textit{cylinder height}) = 20\pi r^2$, and the volume of the cone is $V_2 = \frac{1}{3}\pi r^2 h = \frac{1}{3}\pi r^2(2r) = \frac{2}{3}\pi r^3$. Since the total volume is $\frac{500\pi}{3}$, it follows that $\frac{2}{3}\pi r^3 + 20\pi r^2 = \frac{500\pi}{3} \quad \Leftrightarrow \quad r^3 + 30r^2 - 250 = 0$. By Descartes' Rule of Signs there is 1 positive root.

r	$r^3 + 30r^2 - 250$
1	-219
2	-122
3	47
2.7	-11.617
2.8	7.152
2.765	1.440932999

Therefore, the radius should be 2.76 m (correct to two decimals).

Exercises 4.6

1. $x^3 - 3x^2 - 4x + 12 = 0$; $[-4, 4]$ by $[-15, 15]$.
   The possible rational solutions are ± 1, ± 2, ± 3, ± 4, ± 6, ± 12.
   The solutions of the given equation are $-2, 2$ and 3.

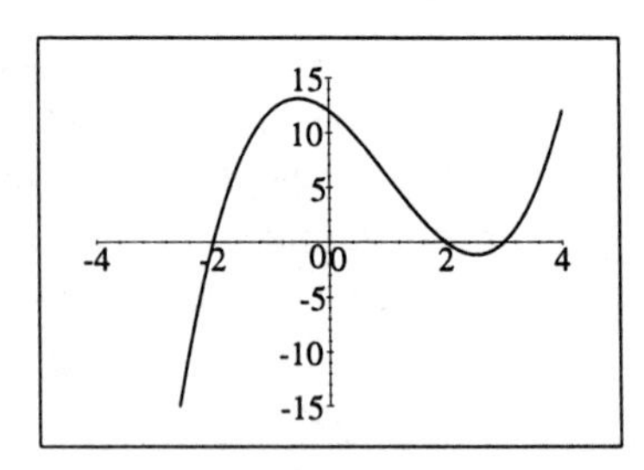

3. $2x^3 + x^2 + 6x + 3 = 0$; $[-2, 2]$ by $[-10, 10]$.
   The possible rational solutions are ± 1, ± 2, ± 3, $\pm \frac{1}{2}$, $\pm \frac{3}{2}$.
   The solution of the given equation is $-\frac{1}{2}$.

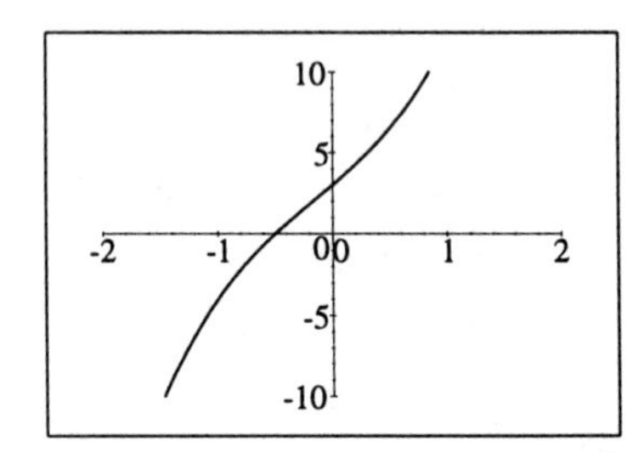

5. $2x^4 - 5x^3 - 14x^2 + 5x + 12 = 0$; $[-2, 5]$ by $[-40, 40]$.
   The possible rational solutions are ± 1, ± 2, ± 3, ± 4, ± 6, ± 12, $\pm \frac{1}{2}$, $\pm \frac{3}{2}$.
   The solutions of the given equation are $-\frac{3}{2}, -1, 1$, and 4.

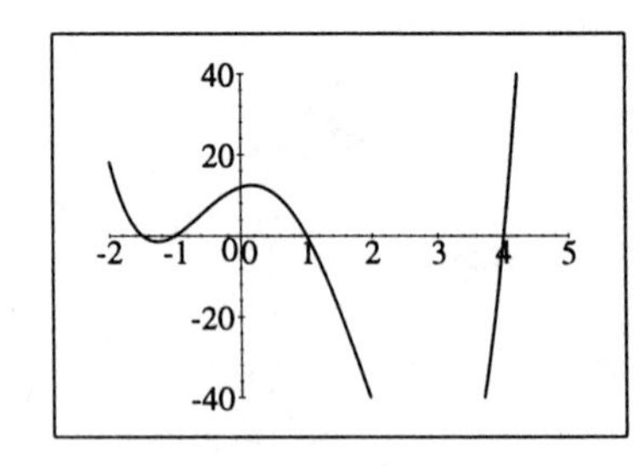

7. $2x^3 - x^2 = 8x + 5 \quad \Leftrightarrow \quad 2x^3 - x^2 - 8x - 5 = 0$;
   $[-2, 3]$ by $[-30, 30]$.
   The possible rational solutions are ± 1, ± 5, $\pm \frac{1}{2}$, $\pm \frac{5}{2}$.
   The solutions of the given equation are -1 and $\frac{5}{2}$.

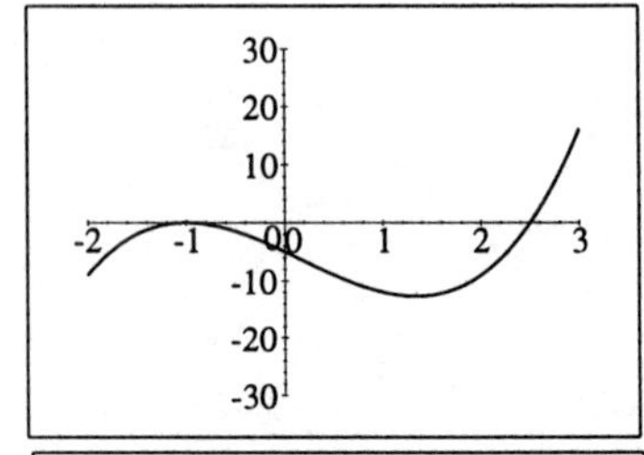

9. $3x^3 + 8x^2 + 5x + 2 = 0$; $[-3, 3]$ by $[-10, 10]$
   The possible rational solutions are ± 1, ± 2, $\pm \frac{1}{3}$, $\pm \frac{2}{3}$.
   The only solution of the given equation is -2.

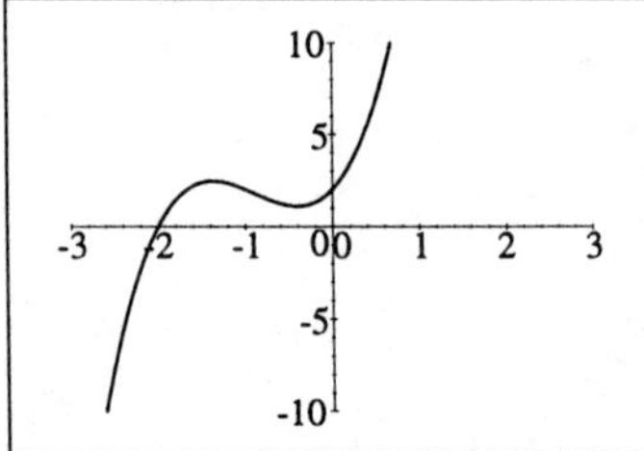

11. $x^3 - 5x^2 - 4 = 0$. By Descartes' Rule, there is 1 positive solution and 0 negative solutions Thus $x = 0$ is a lower bound.

$$\begin{array}{r|rrrr} & 1 & -5 & 0 & -4 \\ \hline 5 & 1 & 0 & 0 & -4 \\ 6 & 1 & 1 & 6 & 32 \end{array} \quad \Rightarrow \quad x = 6 \text{ is an upper bound}$$

Therefore, we graph the function in the viewing rectangle $[0, 6]$ by $[-25, 10]$ and see that there is only one solution. In the second viewing rectangle. $[5.1, 5.2]$ by $[-1, 1]$, we find the solution $x \approx 5.15$

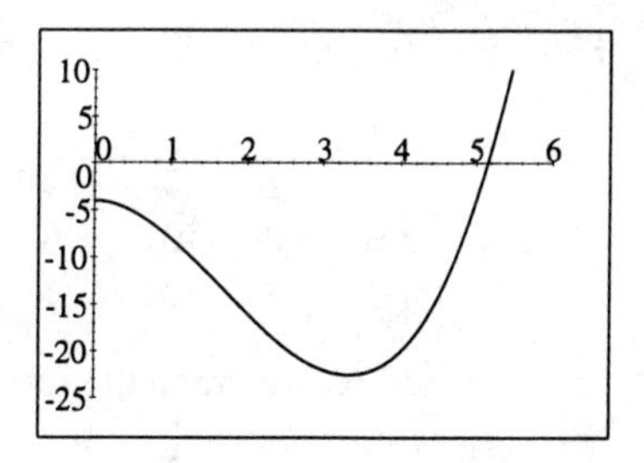

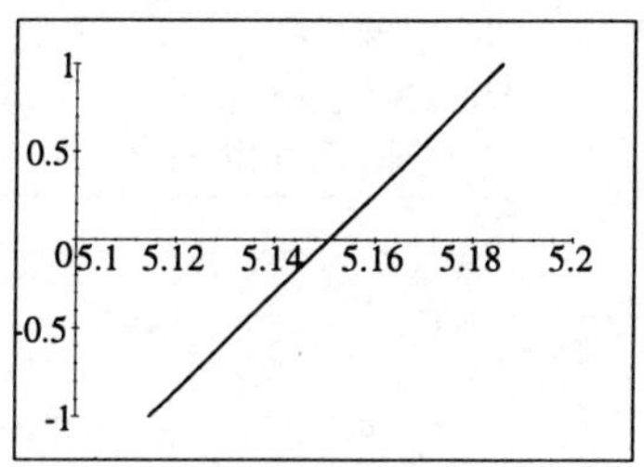

13. $3x^3 + x^2 + x - 2 = 0$.

$$\begin{array}{r|rrrr} & 3 & 1 & 1 & -2 \\ \hline 1 & 3 & 4 & 5 & 4 \\ -1 & 3 & -2 & 3 & -5 \end{array} \quad \begin{array}{l} \Rightarrow \quad x = 1 \text{ is an upper bound} \\ \Rightarrow \quad x = -1 \text{ is a lower bound} \end{array}$$

Therefore, we graph the function in the viewing rectangle $[-1, 1]$ by $[-5, 5]$. There is one solution between 0.5 and 1. Zooming in, we see that, correct to two decimals, the solution is $x \approx 0.67$.

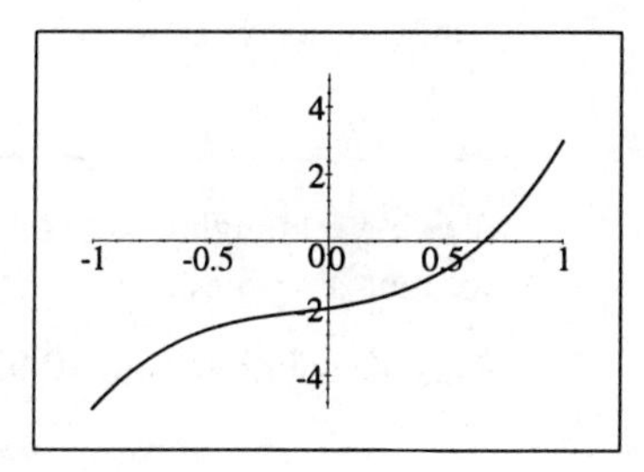

15. $10x^4 - 9x^3 - 11x^2 + 5x - 3 = 0$.

$$\begin{array}{r|rrrrr} & 10 & -9 & -11 & 5 & -3 \\ \hline 2 & 10 & 11 & 11 & 27 & 51 \\ -1 & 10 & -19 & 8 & -3 & 0 \end{array} \quad \begin{array}{l} \Rightarrow \quad x = 2 \text{ is an upper bound} \\ \Rightarrow \quad x = -1 \text{ is a root} \end{array}$$

Also, since the last row alternates between non-negative and non-positive, $x = -1$ is a lower bound. ($x = 2$ is still an upper bound.) We graph $y = 10x^4 - 9x^3 - 11x^2 + 5x - 3$ in the viewing rectangle $[-2, 2]$ by $[-10, 5]$. And see that there are two solutions, one of which we already got, $x = -1$.

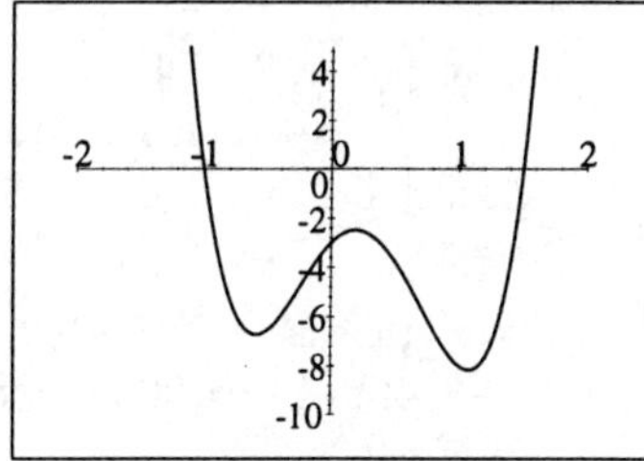

Viewing rectangle: $[1.45, 1.55]$ by $[-1, 1]$.
Solution $x \approx 1.50$.

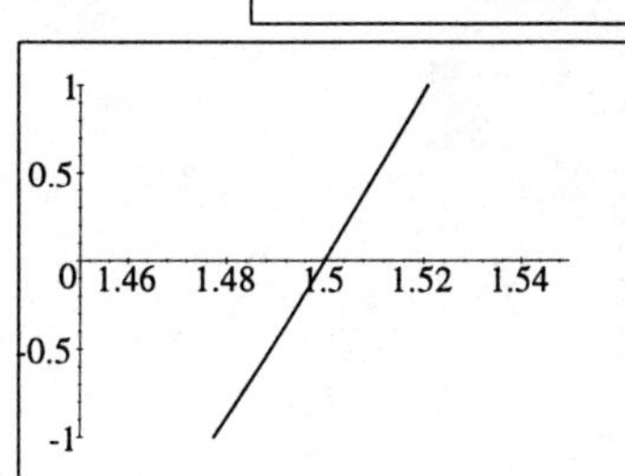

Thus the solutions are -1 and 1.50.

17. $x^3 + 6 = 6x^2 \quad \Leftrightarrow \quad x^3 - 6x^2 + 6 = 0$.

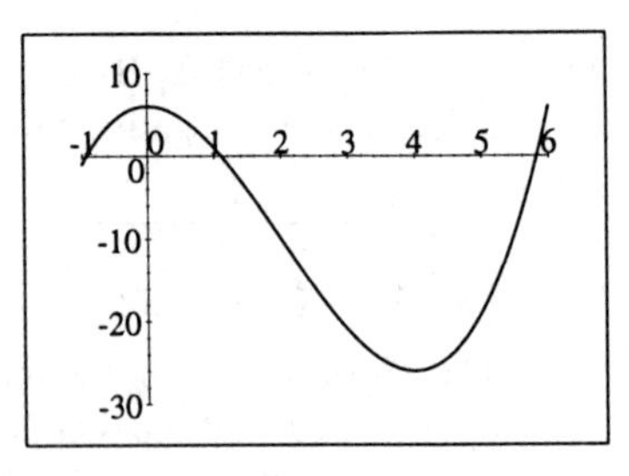

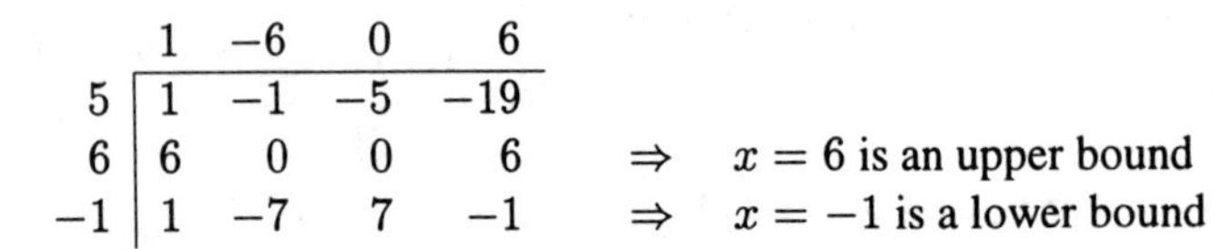

	1	−6	0	6		
5	1	−1	−5	−19		
6	6	0	0	6	$\Rightarrow$	$x = 6$ is an upper bound
−1	1	−7	7	−1	$\Rightarrow$	$x = -1$ is a lower bound

Therefore, we graph the function in the viewing rectangle $[-1, 6]$ by $[-30, 10]$. There are three solutions.

Viewing rectangle: $[-0.95, -0.85]$ by $[-1, 1]$. Solution $x \approx -0.93$.

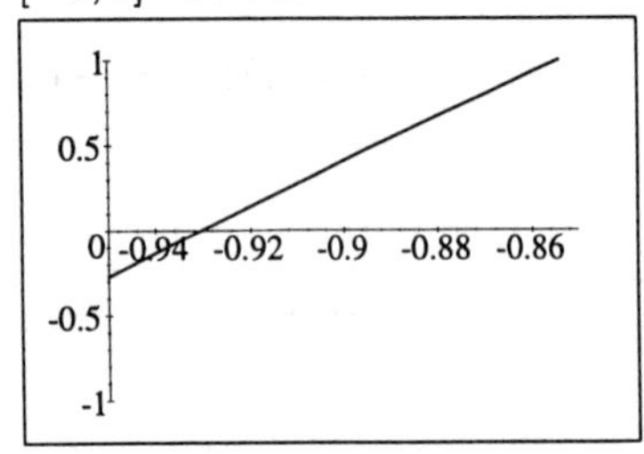

Viewing rectangle: $[1.1, 1.2]$ by $[-1, 1]$. Solution $x \approx 1.11$.

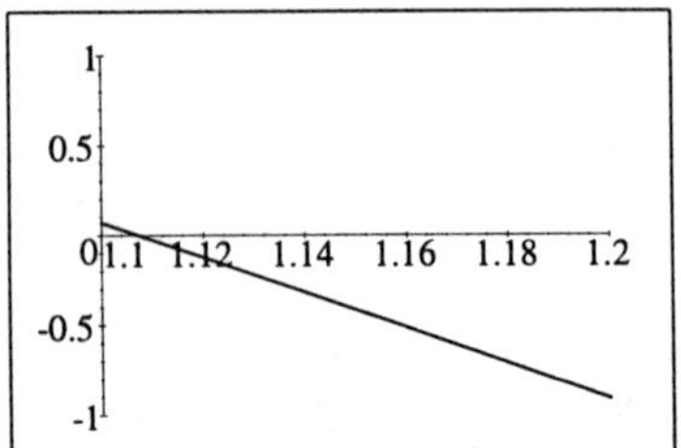

Viewing rectangle: $[5.8, 5.9]$ by $[-1, 1]$. Solution $x \approx 5.82$.

Thus the solutions are $-0.93, 1.11,$ and 5.82.

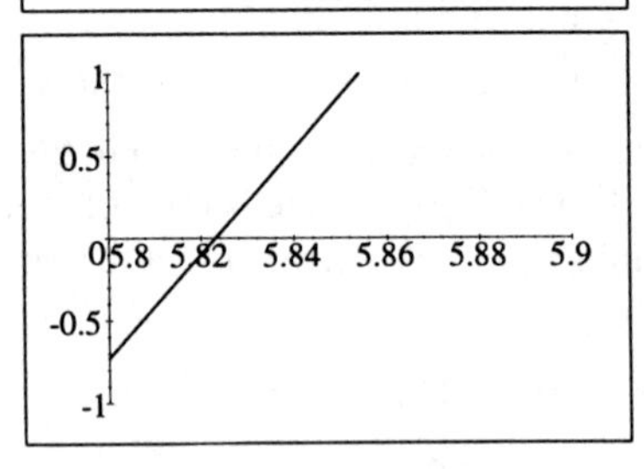

19. $x^4 + 8x + 16 = 2x^3 + 8x^2 \quad \Leftrightarrow \quad x^4 - 2x^3 - 8x^2 + 8x + 16 = 0$.

	1	−2	−8	8	16		
3	1	1	−5	−7	−5		
4	1	2	0	8	2	$\Rightarrow$	$x = 4$ is an upper bound
−1	1	−3	−5	13	3		
−2	1	−4	0	8	0	$\Rightarrow$	$x = -2$ is a root
−3	1	−5	7	−13	23	$\Rightarrow$	$x = -3$ is a lower bound

We graph $x^4 - 2x^3 - 8x^2 + 8x + 16)$ in the viewing rectangle $[-3, 4]$ by $[-10, 20]$. There are for solutions, one of which is $x = -2$.

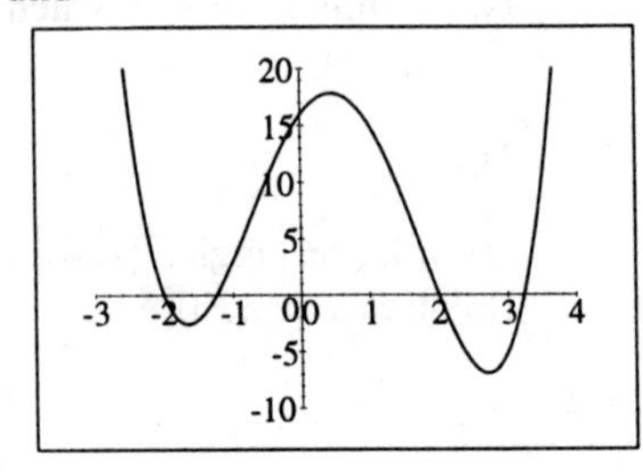

Viewing rectangle: $[-1.3, 1.2]$ by $[-1, 1]$. Solution $x \approx -1.24$.

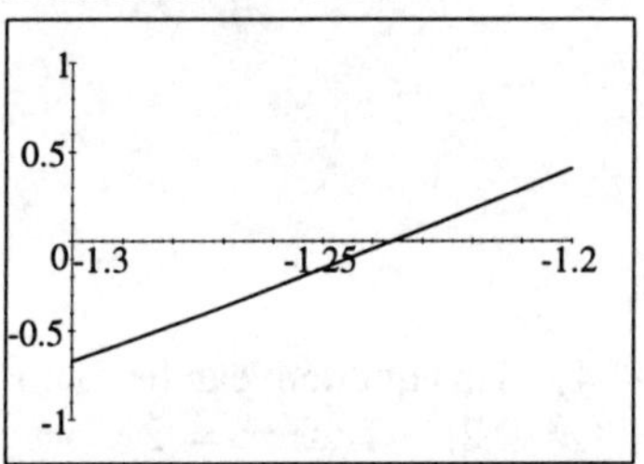

Viewing rectangle: $[1.95, 2.05]$ by $[-1, 1]$. Solution $x \approx 2.00$.

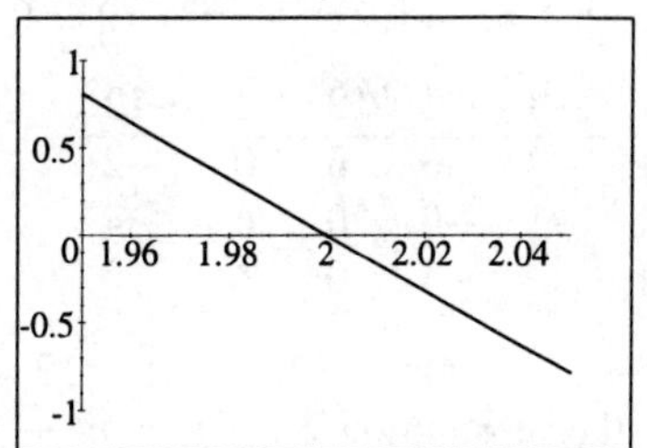

Viewing rectangle: $[3.2, 3.3]$ by $[-1, 1]$. Solution $x \approx 3.24$.

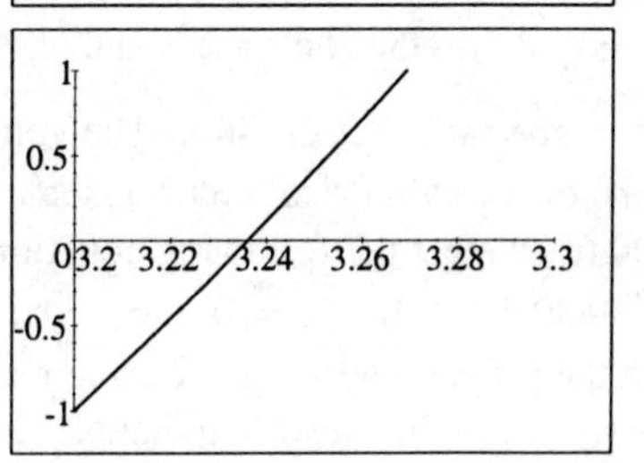

Therefore the solutions are -2, -1.24, 2.00 and 3.24.

21. $4.00x^4 + 4.00x^3 - 10.96x^2 - 5.88x + 9.09 = 0$.

	4	4	-10.96	-5.88	9.09	
1	4	8	-2.96	-8.84	0.25	
2	4	12	13.04	20.2	49.48	$\Rightarrow$ $x = 2$ is an upper bound
-2	4	-4	-2.96	0.04	9.01	
-3	4	-8	13.04	-45	144.09	$\Rightarrow$ $x = -3$ is a lower bound

Therefore, we graph the function in the viewing rectangle $[-3, 2]$ by $[-10, 40]$. There appear to be two solutions.

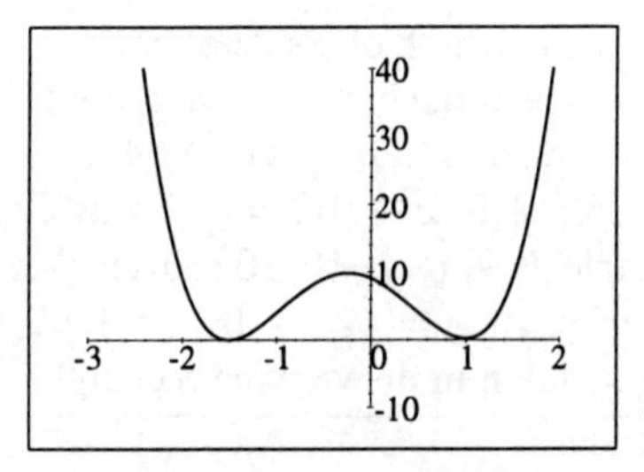

Viewing rectangle: $[-1.6, -1.4]$ by $[-0.1, 0.1]$. Solution $x \approx -1.50$.

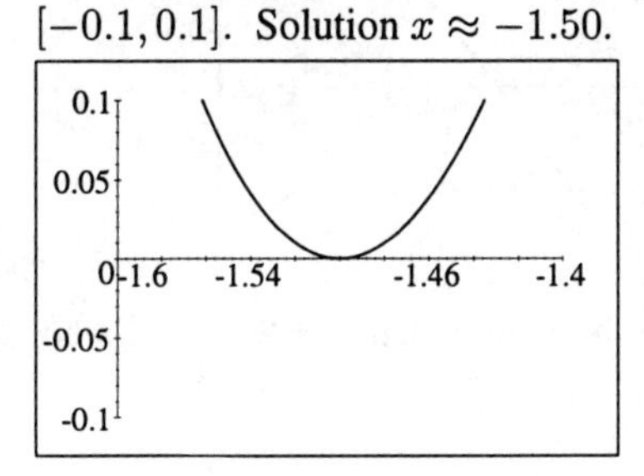

Viewing rectangle: $[0.8, 1.2]$ by $[0, 1]$. The graph comes close but does go through the x-axis, thus there is no solution here.

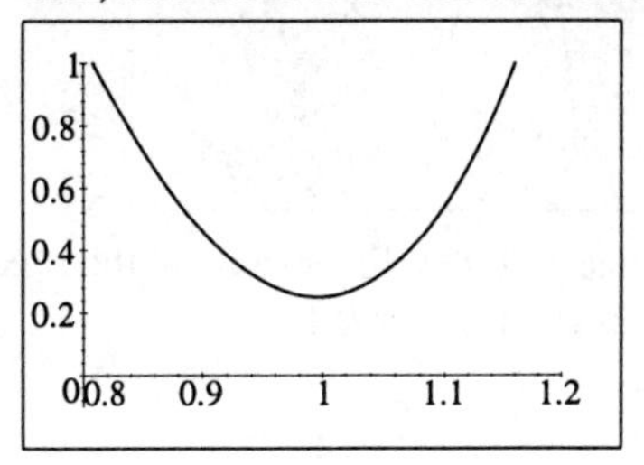

Thus the only solution is $x \approx 1.50$.

23. $x^5 - 3x^4 - 4x^3 + 12x^2 + 4x - 12 = 0$

	1	−3	−4	12	4	−12
1	1	−2	−6	6	10	−2
2	1	−1	−6	0	4	−4
3	1	0	−4	0	4	0

$\Rightarrow$ $x = 3$ is a root

So $x^5 - 3x^4 - 4x^3 + 12x^2 + 4x - 12 = (x-3)(x^4 - 4x^2 + 4)$. This quotient can be factored using the substitution $x^2 = u$, so $u^2 - 4u + 4 = (u-2)^2 = (x^2-2)^2$. Thus $x^2 = 2 \Leftrightarrow x = \pm\sqrt{2}$. So the roots are 3 and $\pm\sqrt{2}$.

25. Let $r =$ the radius of the silo. The volume of the hemispherical roof $\frac{1}{2}\left(\frac{4}{3}\pi r^3\right) = \frac{2}{3}\pi r^3$. The volume of the cylindrical section is $\pi(r^2)(30) = 30\pi r^2$. Because the total volume of the silo is 15000 ft^3, we get the following equation: $\frac{2}{3}\pi r^3 + 30\pi r^2 = 15000 \Leftrightarrow \frac{2}{3}\pi r^3 + 30\pi r^2 - 15000 = 0 \Leftrightarrow \pi r^3 + 45\pi r^2 - 22500 = 0$. Using a graphing device, we first graph the polynomial in the viewing rectangle $[0, 15]$ by $[-10000, 10000]$. The solution, $r \approx 11.28$ ft., is shown in the viewing rectangle $[11.2, 11.4]$ by $[-1, 1]$.

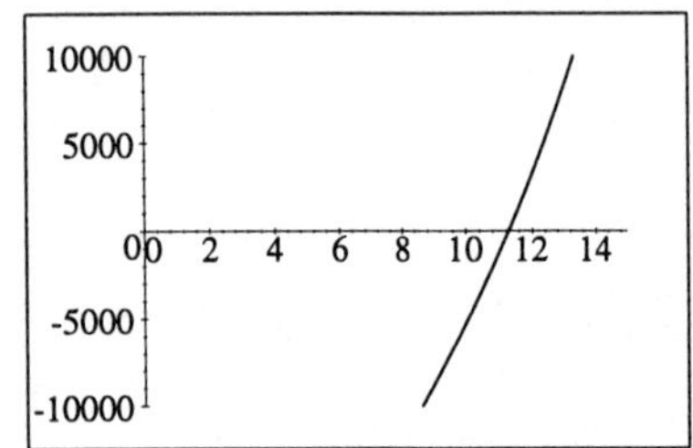

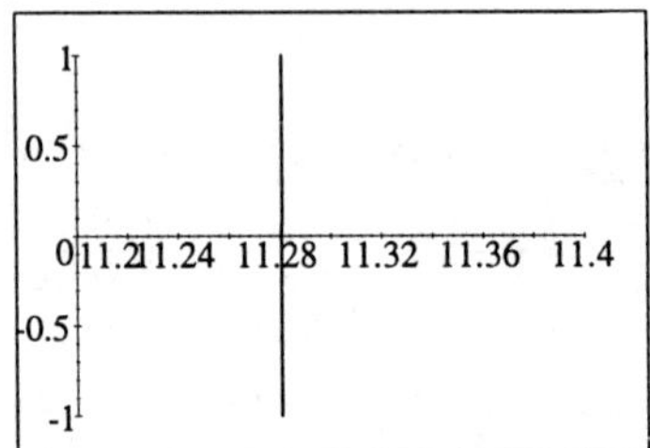

27. Let $x =$ the length of a side of each of the equal squares to be cut from the corners. Then the dimensions of the box, in inches, are $(18-2x)$ by $(18-2x)$ by x. Since the volume is 400 in^3, we get the equation $x(18-2x)^2 = 400 \Leftrightarrow 324x - 72x^2 + 4x^3 = 400 \Leftrightarrow x^3 - 18x^2 + 81x - 100 = 0$. Using a graphing device, we first graph the polynomial in the viewing rectangle $[0, 9]$ by $[-10, 10$ and see that there is 2 solution. One solution appears to be $x = 4$, which we verify, $P(4) = (4)^3 - 18(4)^2 + 81(4) - 100 = 64 - 288 + 324 - 100 = 0$. We isolate the other solution in the viewing rectangle $[2, 2.2]$ by $[-1, 1, x \approx 2.1$.

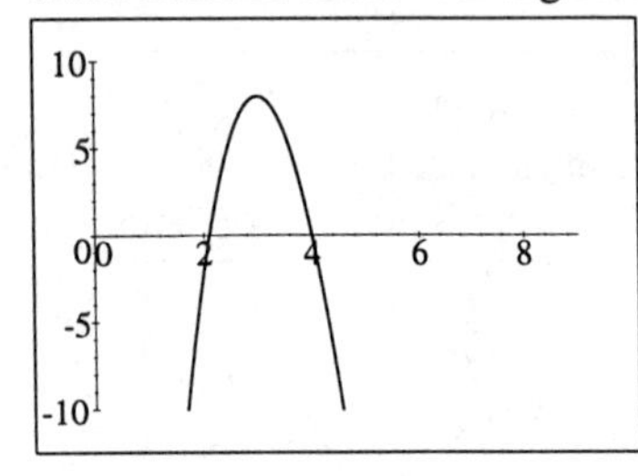

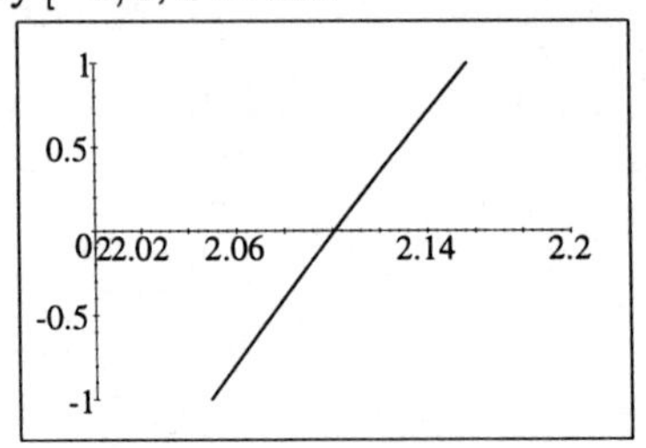

When $x = 4$, the dimensions of the box is 10 in $\times$ 10 in $\times$ 4 in and when $x = 2.1$ the dimensions are 13.8 in $\times$ 13.8 in $\times$ 2.1 in.

29. $h(t) = 11.60t - 12.41t^2 + 6.20t^3 - 1.58t^4 + 0.20t^5 - 0.01t^6$ is shown in the viewing rectangle $[0, 10]$ by $[0, 6]$.

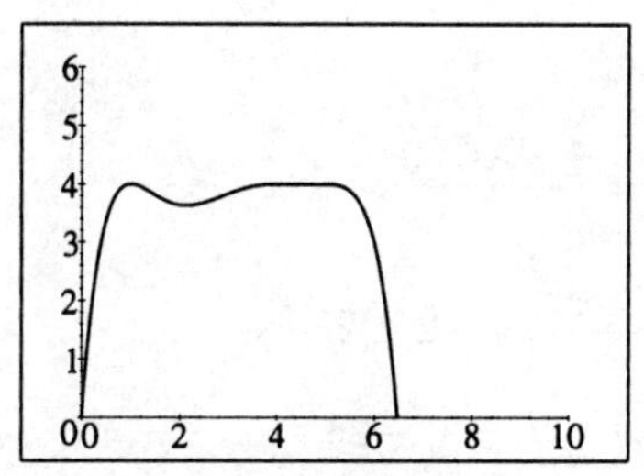

(a) It started to snow again.

(b) No, $h(t) \leq 4$.

(c) The function $h(t)$ is shown in the viewing rectangle $[6, 6.5]$ by $[0, 0.5]$. The x-intercept of the function is a little less than 6.5, which means that the snow melted just before midnight Sunday night.

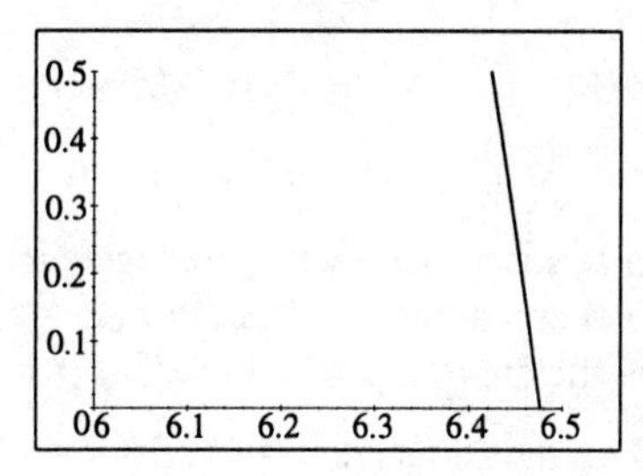

Exercises 4.7

1. $x^2+4=0 \quad \Leftrightarrow \quad x^2=-4 \quad \Leftrightarrow \quad x=\pm 2i$

3. $x^2-x+1=0 \quad \Leftrightarrow \quad x=\dfrac{-(-1)\pm\sqrt{(-1)^2-4(1)(1)}}{2(1)}=\dfrac{1\pm\sqrt{-3}}{2}=\frac{1}{2}\pm i\frac{\sqrt{3}}{2}$

5. $x^2+4x+8=0 \quad \Leftrightarrow \quad x=\dfrac{-4\pm\sqrt{16-4(1)(8)}}{2}=\dfrac{-4\pm 4i}{2}=-2\pm 2i$

7. $3x^2-5x+4=0 \quad \Leftrightarrow \quad x=\dfrac{-(-5)\pm\sqrt{(-5)^2-4(3)(4)}}{2(3)}=\dfrac{5\pm\sqrt{-23}}{6}=\frac{5}{6}\pm i\frac{\sqrt{23}}{6}$

9. $x^2-8x+17=0 \quad \Leftrightarrow \quad x=\dfrac{8\pm\sqrt{(-8)^2-4(17)}}{2}=\dfrac{8\pm\sqrt{-4}}{2}=4\pm i$

11. $t+3+\frac{3}{t}=0 \quad \Leftrightarrow \quad t^2+3t+3=0,\ t\neq 0 \quad \Leftrightarrow \quad t=\dfrac{-3\pm\sqrt{(3)^2-4(3)}}{2}=-\frac{3}{2}\pm i\frac{\sqrt{3}}{2}$

13. $z^2-iz=0 \quad \Leftrightarrow \quad z(z-i)=0 \quad \Leftrightarrow \quad z=0,\ i.$

15. Since i is a root, by the Conjugate Roots Theorem, $-i$ is also a root. So the factorization of the polynomial must be $P(x)=a(x-2)(x-i)(x+i)=a(x^3-2x^2+x-2)$. Since the leading coefficient is 1, $a=1$ and $P(x)=x^3-2x^2+x-2$.

17. Since the zeros are $1+i$ and $3-4i$, by the Conjugate Roots Theorem, the other zeros are $1-i$ and $3+4i$. So a factorization is
$S(x)=C(x-[1+i])(x-[1-i])(x-[3-4i])(x-[3+4i])$
$=C(x^2-2x+[1-i][1+i])(x^2-6x+[3-4i][3+4i])=C(x^2-2x+2)(x^2-6x+25)$
$=C(x^4-6x^3+25x^2-2x^3+12x^2-50x+2x^2-12x+50)$
$=C(x^4-8x^3+39x^2-62x+50)=Cx^4-8Cx^3+39Cx^2-62Cx+50C$
Since the coefficient of x^2 is 39, it follows that $C=\frac{39}{39}=1$, and so
$S(x)=x^4-8x^3+39x^2-62x+50$.

19. Since $1-i$ is a zero of multiplicity 2, the zeros of $T(x)$ are $1+i$, $1+i$, $1-i$, and $1-i$. Thus, a factorization is $T(x)=C(x-[1+i])^2(x-[1-i])^2=C([x-1]-i)^2([x-1]+i)^2$
$=C[([x-1]-i)([x-1]+i)]^2=C([x-1]^2-i^2)^2=C[x^2-2x+2]^2$
$=C(x^4-4x^3+8x^2-8x+4)$. Since the constant coefficient is 8, it follows that $4C=8 \quad \Leftrightarrow$
$C=2$, and so $T(x)=2(x^4-4x^3+8x^2-8x+4)=2x^4-8x^3+16x^2-16x+8$.

21. Let $P(x)=x^3-2x^2+4x-8$. Then, $P(2i)=(2i)^3-2(2i)^2+4(2i)-8=0$. Thus, $x=2i$ is a solution. By the Conjugate Roots Theorem, $x=-2i$ must also be a solution. Thus, $(x-2i)$ and $(x+2i)$ are factors of $P(x)$, and so $(x-2i)(x+2i)=x^2-4i^2=x^2+4$ must divide the polynomial.

$$
\begin{array}{r|l}
 & x \qquad\qquad -2 \\
x^2+4 & x^3-2x^2+4x-8 \\
 & \underline{x^3 \qquad\quad +4x} \\
 & -2x^2 \qquad\quad -8 \\
 & \underline{-2x^2 \qquad\quad -8} \\
 & 0
\end{array}
$$

So $P(x)=(x-2)(x^2+4)$ and $x=2$ is a solution. Since $P(x)$ is degree 3, it has at most three distinct roots. We have found 3 roots; thus, all the solutions of $P(x)=0$ are $x=2,\ \pm 2i$. (Note that we could have also used the Factor Theorem to show that $x=2$ is a solution.)

23. Let $P(x) = 2x^4 + 9x^2 + 4$. Then $P(2i) = 2(2i)^4 + 9(2i)^2 + 4 = 32i^4 + 36i^2 + 4 = 0$. Thus, $x = 2i$ is a solution. Therefore, $x = -2i$ is also a solution, and $(x - 2i)(x + 2i) = x^2 + 4$ is a factor of $P(x)$:

$$\begin{array}{r|l} & 2x^2 \qquad +1 \\ \hline x^2+4 & 2x^4 + 9x^2 + 4 \\ & 2x^4 + 8x^2 \\ \hline & \qquad x^2 + 4 \\ & \qquad x^2 + 4 \\ \hline & \qquad\qquad 0 \end{array}$$

So $P(x) = (x^2 + 4)(2x^2 + 1)$. Setting the other factor equal to zero we have $2x^2 + 1 = 0 \quad \Rightarrow \quad x = \pm \frac{1}{\sqrt{2}} i$. So the solutions to $P(x) = 0$ are $x = \pm 2i, \pm \frac{1}{\sqrt{2}} i$.

25. $x^4 - 1 = 0 \quad \Leftrightarrow \quad 0 = (x^2 - 1)(x^2 + 1) = (x - 1)(x + 1)(x^2 + 1)$. So $x = \pm 1$ or $x^2 = -1$ $\Leftrightarrow \quad x = \pm i$. Therefore the four solutions are $x = \pm 1, \pm i$.

27. $x^3 + 8 = 0 \quad \Leftrightarrow \quad (x + 2)(x^2 - 2x + 4) = 0$. So $x = -2$, or $x^2 - 2x + 4 = 0 \quad \Rightarrow$ $x = \dfrac{2 \pm \sqrt{4 - 16}}{2} = \dfrac{2 \pm \sqrt{-12}}{2} = \dfrac{2 \pm 2i\sqrt{3}}{2} = 1 \pm i\sqrt{3}$. Thus, $x = -2, 1 \pm i\sqrt{3}$.

29. $x^4 - 16 = 0 \quad \Leftrightarrow \quad 0 = (x^2 - 4)(x^2 + 4) = (x - 2)(x + 2)(x^2 + 4)$. So $x = \pm 2$ or $x^2 + 4 = 0$ $\Rightarrow \quad x^2 = -4 \quad \Rightarrow \quad x = \pm 2i$. Therefore the four solutions are $x = \pm 2, \pm 2i$.

31. $x^6 - 729 = 0 \quad \Leftrightarrow \quad 0 = (x^3 - 27)(x^3 + 27) = (x - 3)(x^2 + 3x + 9)(x + 3)(x^2 - 3x + 9)$. Clearly, $x = \pm 3$ are solutions. If $x^2 + 3x + 9 = 0$, then $x = \dfrac{-3 \pm \sqrt{9 - 4(1)(9)}}{2}$ $= \dfrac{-3 \pm \sqrt{-27}}{2} = -\frac{3}{2} \pm \frac{\sqrt{-27}}{2}$ so $x = -\frac{3}{2} \pm \frac{3\sqrt{3}}{2} i$. If $x^2 - 3x + 9 = 0$, then

$x = \dfrac{3 \pm \sqrt{9 - 4(1)(9)}}{2} = \dfrac{3 \pm \sqrt{-27}}{2} = \frac{3}{2} \pm \frac{\sqrt{-27}}{2} = \frac{3}{2} \pm \frac{3\sqrt{3}}{2} i$. Therefore, the solutions are

$x = \pm 3, -\frac{3}{2} \pm \frac{3\sqrt{3}}{2} i, \frac{3}{2} \pm \frac{3\sqrt{3}}{2} i$.

33. $x^4 + 10x^2 + 25 = 0 \quad \Leftrightarrow \quad (x^2 + 5)^2 = 0 \quad \Leftrightarrow \quad x^2 = -5 \Rightarrow x = \pm i\sqrt{5}$.

35. $x^3 + 2x^2 + 4x + 8 = 0 \quad \Leftrightarrow \quad x^2(x + 2) + 4(x + 2) = 0 \quad \Leftrightarrow \quad (x + 2)(x^2 + 4) = 0$. So $x = -2$ or $x^2 + 4 = 0 \quad \Rightarrow \quad x^2 = -4 \quad \Rightarrow \quad x = \pm 2i$. Solutions are $x = -2, \pm 2i$.

37. $x^3 - 2x^2 + 2x - 1 = 0$. By inspection, $P(1) = 1 - 2 + 2 - 1 = 0$ and hence $x = 1$ is a solution.

$$\begin{array}{r|rrrr} & 1 & -2 & 2 & -1 \\ 1 & 1 & -1 & 1 & 0 \end{array}$$

Thus $P(x) = (x - 1)(x^2 - x + 1)$. So $x = 1$ or $x^2 - x + 1 = 0$

Using the quadratic formula we have $x = \dfrac{1 \pm \sqrt{1 - 4(1)(1)}}{2} = \dfrac{1 \pm i\sqrt{3}}{2}$. Hence, the solutions are $1, \frac{1 \pm i\sqrt{3}}{2}$.

39. $x^3 - 3x^2 + 3x - 2 = 0$

$$\begin{array}{r|rrrr} & 1 & -3 & 3 & -2 \\ 1 & 1 & -1 & 1 & 0 \end{array}$$

Thus $P(x) = (x - 2)(x^2 - x + 1)$. So $x = 2$ or $x^2 - x + 1 = 0$

Using the quadratic formula we have $x = \dfrac{1 \pm \sqrt{1 - 4(1)(1)}}{2} = \dfrac{1 \pm i\sqrt{3}}{2}$. Hence, the solutions are $2, \frac{1 \pm i\sqrt{3}}{2}$

41. $P(x) = x^4 + x^3 + 7x^2 + 9x - 18 = 0$. Since $P(x)$ has one change in sign, we are guaranteed a positive root and since $P(-x) = x^4 - x^3 + 7x^2 - 9x - 18$, there are 1 or 3 negative roots.

$$\begin{array}{r|rrrrr} & 1 & 1 & 7 & 9 & -18 \\ \hline 1 & 1 & 2 & 9 & 18 & 0 \end{array} \Rightarrow \quad P(x) = (x-1)(x^3 + 2x^2 + 9x + 18)$$

Continuing with the quotient, we try negative roots.

$$\begin{array}{r|rrrr} & 1 & 2 & 9 & 18 \\ \hline -1 & 1 & 1 & 8 & 10 \\ -2 & 1 & 0 & 9 & 0 \end{array} \Rightarrow \quad P(x) = (x-1)(x+2)(x^2+9)$$

So, $x = -2$ or 1, or $x^2 + 9 = 0 \Rightarrow x^2 = -9 \Rightarrow x = \pm 3i$. Therefore, $x = -2, 1, \pm 3i$.

43. $P(x) = x^3 + 27 = x^3 + 3^3 = (x+3)(x^2 - 3x + 9)$ and $x^2 - 3x + 9$ has roots $x = \dfrac{3 \pm \sqrt{9 - 4(1)(9)}}{2} = \dfrac{3 \pm 3i\sqrt{3}}{2}$. So $P(x) = (x+3)\left(x - \frac{3+3i\sqrt{3}}{2}\right)\left(x - \frac{3-3i\sqrt{3}}{2}\right)$.

45. $P(x) = x^6 - 64 = (x^3 - 8)(x^3 + 8) = (x-2)(x^2 + 2x + 4)(x+2)(x^2 - 2x + 4)$. Now $x^2 + 2x + 4$ has roots $x = \dfrac{-2 \pm \sqrt{4 - 4(1)(4)}}{2} = \dfrac{-2 \pm 2i\sqrt{3}}{2} = -1 \pm i\sqrt{3}$ and $x^2 - 2x + 4$ has roots $x = \dfrac{2 \pm \sqrt{4 - 4(1)(4)}}{2} = \dfrac{2 \pm 2i\sqrt{3}}{2} = 1 \pm i\sqrt{3}$. Therefore,
$P(x) = (x-2)(x+2)(x - 1 - i\sqrt{3})(x - 1 + i\sqrt{3})(x + 1 - i\sqrt{3})(x + 1 + i\sqrt{3})$.

47. $P(x) = 2x^3 + 7x^2 + 12x + 9$ has possible zeros $\pm 1, \pm 3, \pm 9, \pm \frac{1}{2}, \pm \frac{3}{2}, \pm \frac{9}{2}$. Since all coefficients are positive, there are no positive real roots.

$$\begin{array}{r|rrrr} & 2 & 7 & 12 & 9 \\ \hline -1 & 2 & 5 & 7 & 2 \\ -2 & 2 & 3 & 6 & -3 \\ -\frac{3}{2} & 2 & 4 & 6 & 0 \end{array} \quad \begin{array}{l} \\ \\ \Rightarrow \text{ There is a root between } -1 \text{ and } -2. \\ \Rightarrow x = -\tfrac{3}{2} \text{ is a root} \end{array}$$

$P(x) = \left(x + \frac{3}{2}\right)(2x^2 + 4x + 6) = 2\left(x + \frac{3}{2}\right)(x^2 + 2x + 3)$. Now $x^2 + 2x + 3$ has roots $x = \dfrac{-2 \pm \sqrt{4 - 4(3)(1)}}{2} = \dfrac{-2 \pm 2\sqrt{-2}}{2} = -1 \pm i\sqrt{2}$. Therefore,
$P(x) = 2\left(x + \frac{3}{2}\right)\left(x - \left[-1 + i\sqrt{2}\right]\right)\left(x - \left[-1 - i\sqrt{2}\right]\right)$
$= 2\left(x + \frac{3}{2}\right)\left(x + 1 - i\sqrt{2}\right)\left(x + 1 + i\sqrt{2}\right)$.

49. Suppose $P(x)$ has only imaginary roots. Since P has real coefficients, the imaginary roots come in pairs: $a \pm b\,i$ (by the Conjugate Roots Theorem), where $b \neq 0$. Thus there must be an even number of roots, and hence $P(x)$ has even degree. This proves that if the degree of P is odd, it can't have only imaginary roots. Thus at least one of the roots must be real.

51. $P(2i) = 3\,(2i)^3 - (2i)^2 + 2i - 4 = 24i^3 - 4i^2 + 2i - 4 = -22i$.

53. (a) Since i and $1 + i$ are zeros, $-i$ and $1 - i$ are also zeros. So
$P(x) = C(x - i)(x + i)(x - [1 + i])(x - [1 - i]) = C(x^2 + 1)(x^2 - 2x + 2)$
$= C(x^4 - 2x^3 + 2x^2 + x^2 - 2x + 2) = C(x^4 - 2x^3 + 3x^2 - 2x + 2)$. Since $C = 1$, the polynomial is $P(x) = x^4 - 2x^3 + 3x^2 - 2x + 2$.

(b) Since i and $1 + i$ are zeros, $P(x) = C(x - i)(x - [i + 1]) = C(x^2 - xi - x - xi - 1 + i)$
$= C[x^2 - (1 + 2i)x - 1 + i]$. Since $C = 1$, the polynomial is
$P(x) = x^2 - (1 + 2i)x - 1 + i$.

55. (a) $x^4 - 2x^3 - 11x^2 + 12x = x(x^3 - 2x^2 - 11x + 12) = 0$. We first find the bounds for our viewing rectangle.

$$\begin{array}{r|rrrr} & 1 & -2 & -11 & 12 \\ \hline 5 & 1 & 3 & 4 & 32 \\ -4 & 1 & -6 & 13 & -50 \end{array} \quad \begin{array}{l} \\ \Rightarrow \quad x = 5 \text{ is an upper bound} \\ \Rightarrow \quad x = -4 \text{ is a lower bound} \end{array}$$

The viewing rectangle $[-4, 5]$ by $[-50, 10]$ shows that $P(x) = x^4 - 2x^3 - 11x^2 + 12x$ has 4 real solutions. Since this matches the degree of $P(x)$, $P(x)$ has no imaginary solutions.

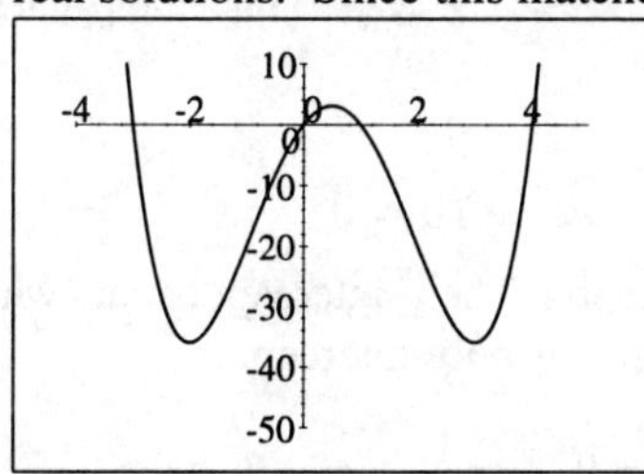

(b) $x^4 - 2x^3 - 11x^2 + 12x - 5 = 0$. We use the same bounds for our viewing rectangle, $[-4, 5]$ by $[-50, 10]$ and see that $R(x) = x^4 - 2x^3 - 11x^2 + 12x - 5$ has 2 real solutions. Since the degree of $R(x)$ is 4, $R(x)$ must have 2 imaginary solutions.

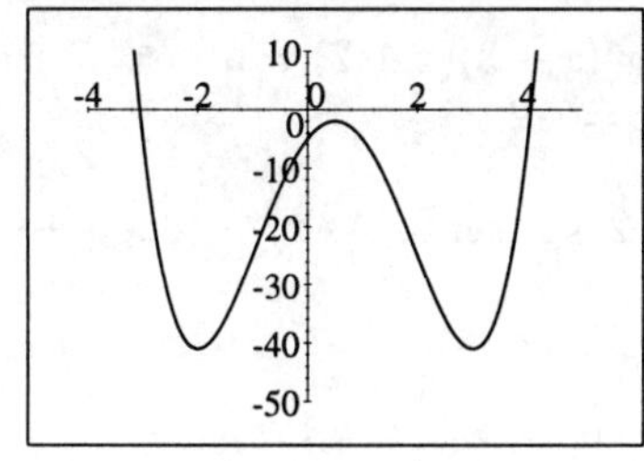

(c) $x^4 - 2x^3 - 11x^2 + 12x + 40 = 0$. We use the same bounds for our viewing rectangle, $[-4, 5]$ by $[-10, 50]$ and see that $T(x) = x^4 - 2x^3 - 11x^2 + 12x + 40$ has no real solutions. Since the degree of $T(x)$ is 4, $T(x)$ must have 4 imaginary solutions.

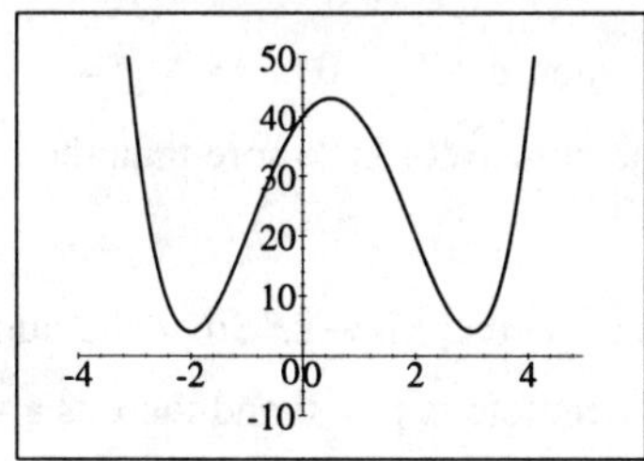

Exercises 4.8

1. $y = \dfrac{x-6}{x+1}$. The x-intercept occurs when $y = 0 \Leftrightarrow x - 6 = 0 \Leftrightarrow x = 6$. The y-intercept occurs when $x = 0 \Leftrightarrow y = -6$.

3. $y = \dfrac{x}{x^2 - 2x - 15}$. The x-intercept occurs when $y = 0 \Leftrightarrow x = 0$. Thus, the y-intercept is also zero, since the function passes through (0, 0).

5. $y = \dfrac{x^2 + 10}{2x}$. The x-intercept occurs when $y = 0 \Leftrightarrow x^2 + 10 = 0 \Leftrightarrow x^2 = -10$, which is impossible unless x is imaginary; there is no real x-intercept. The y-intercept occurs when $x = 0$, but then y is undefined (division by zero). Therefore, there is no y-intercept.

7. $y = \dfrac{5}{x+3}$. There is a vertical asymptote where $x + 3 = 0 \Leftrightarrow x = -3$. As $x \to \pm\infty$, $y \to 0$, so the horizontal asymptote is $y = 0$.

9. $y = \dfrac{x^2}{x^2 - x - 6} = \dfrac{x^2}{(x-3)(x+2)} = \dfrac{1}{1 - \frac{1}{x} - \frac{6}{x^2}} \to 1$ as $x \to \pm\infty$. Hence, the horizontal asymptote is $y = 1$. The vertical asymptotes occur when $(x-3)(x+2) = 0 \Leftrightarrow x = 3$ or $x = -2$, and so the vertical asymptotes are $x = 3$ and $x = -2$.

11. $y = \dfrac{6}{x^2 + 2}$. There is no vertical asymptote since $x^2 + 2$ is never 0. As $x \to \infty$, $y \to 0$, so the horizontal asymptote is $y = 0$.

13. $y = \dfrac{x^2 + 2}{x - 1}$. A vertical asymptote occurs when $x - 1 = 0 \Leftrightarrow x = 1$. There are no horizontal asymptotes. Using long division, $y = x + 1 + \dfrac{4}{x-1}$, so the slant asymptote is $y = x + 1$.

$$\begin{array}{r|rrr} & 1 & 0 & 2 \\ \hline 1 & 1 & 1 & 4 \end{array}$$

15. $y = \dfrac{2x^3 - x^2 - 8x + 4}{x + 3}$. There is a vertical asymptote when $x + 3 = 0 \Leftrightarrow x = -3$. There are no horizontal or slant asymptotes, since the degree of the numerator is 2 more than the degree of the denominator.

17. $y = \dfrac{4}{x - 2}$. When $x = 0$, $y = \dfrac{4}{0 - 2} = -2$, and so the y-intercept is -2. Since the numerator can never be zero, there is no x-intercept. The horizontal asymptote is $y = 0$ and there is a vertical asymptote when $x - 2 = 0 \Leftrightarrow x = 2$.

as $x \to$	-2^+	-2^-
sign of $y = \dfrac{4}{x+2}$	$\dfrac{(+)}{(+)}$	$\dfrac{(+)}{(-)}$
$y \to$	∞	$-\infty$

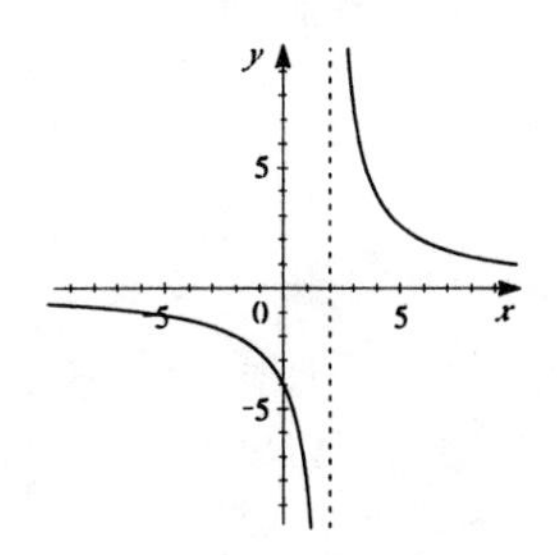

19. $y = \dfrac{x-1}{x-2}$. When $x = 0$, $y = \dfrac{1}{2}$, so the y-intercept is $\dfrac{1}{2}$. When $y = 0$, $x - 1 = 0 \quad \Leftrightarrow \quad x = 1$, so the x-intercept is 1. The horizontal asymptote is $y = 1$. A vertical asymptote occurs when $x - 2 = 0 \quad \Leftrightarrow \quad x = 2$.

as $x \to$	2^+	2^-
sign of $y = \dfrac{x-1}{x-2}$	$\dfrac{(+)}{(+)}$	$\dfrac{(+)}{(-)}$
$y \to$	∞	$-\infty$

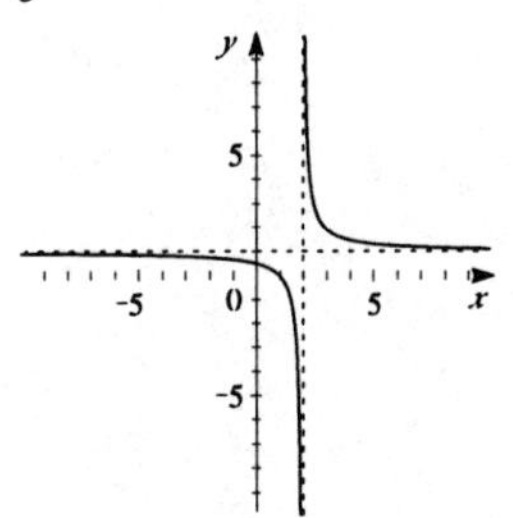

21. $y = \dfrac{4x-4}{x+2}$. When $x = 0$, $y = -2$, so the y-intercept is -2. When $y = 0$, $4x - 4 = 0 \quad \Leftrightarrow$ $x = 1$, so the x-intercept is 1. The horizontal asymptote is $y = 4$. A vertical asymptote occurs when $x = -2$.

as $x \to$	-2^+	-2^-
sign of $y = \dfrac{4x-4}{x+2}$	$\dfrac{(+)}{(+)}$	$\dfrac{(+)}{(-)}$
$y \to$	∞	$-\infty$

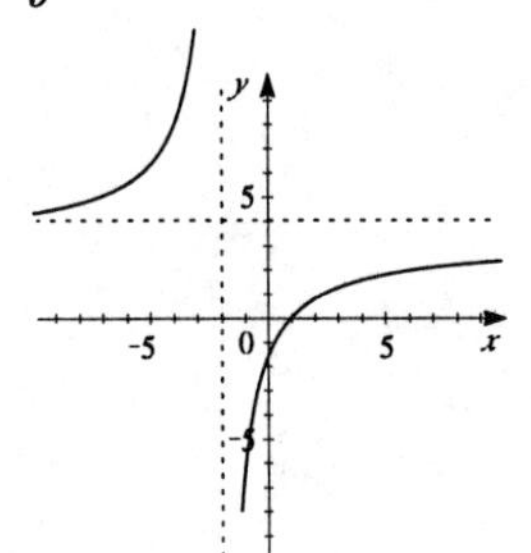

23. $y = \dfrac{2x-4}{x}$. A vertical asymptote occurs when $x = 0$. Since there is an asymptote at $x = 0$, there is no y-intercept. The x-intercept occurs when $y = 0 \quad \Leftrightarrow \quad 2x - 4 = 0 \quad \Leftrightarrow \quad x = 2$. The horizontal asymptote is $y = 2$.

as $x \to$	0^-	0^+
sign of $y = \dfrac{2x-4}{x}$	$\dfrac{(-)}{(-)}$	$\dfrac{(-)}{(+)}$
$y \to$	∞	$-\infty$

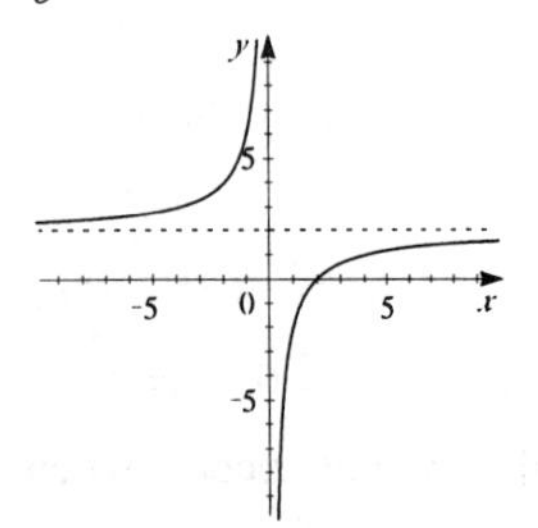

25. $y = \dfrac{18}{(x-3)^2}$. When $x = 0$, $y = \dfrac{18}{9} = 2$, and so the y-intercept is 2. Since the numerator can never be zero, there is no x-intercept. There is a vertical asymptote when $x - 3 = 0 \quad \Leftrightarrow \quad x = 3$, and the horizontal asymptote is $y = 0$.

as $x \to$	3^-	3^+
sign of $y = \dfrac{18}{(x-3)^2}$	$\dfrac{(+)}{(+)}$	$\dfrac{(+)}{(+)}$
$y \to$	∞	∞

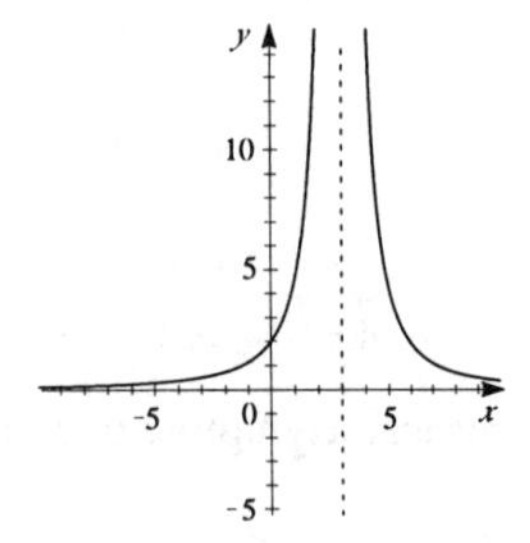

27. $y = \dfrac{4x+8}{(x-4)(x+1)}$. When $x = 0$, $y = \dfrac{8}{(-4)(1)} = -2$, so the y-intercept is -2. When $y = 0$, $4x + 8 = 0 \quad \Leftrightarrow \quad x = -2$, so the x-intercept is -2. The vertical asymptotes are $x = -1$ and $x = 4$, and the horizontal asymptote is $y = 0$.

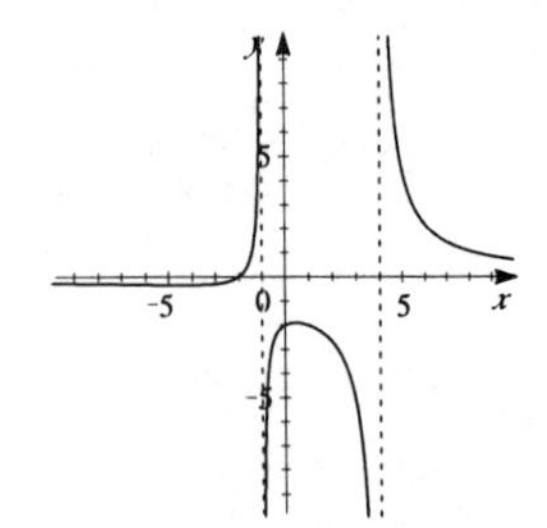

29. $y = \dfrac{(x-1)(x+2)}{(x+1)(x-3)}$. When $x = 0$, $y = \dfrac{2}{3}$, so the y-intercept is $\dfrac{2}{3}$. When $y = 0$, $(x-1)(x+2) = 0 \quad \Rightarrow \quad x = -2, 1$. Thus, there are two x-intercepts. The vertical asymptotes are $x = -1$ and $x = 3$, while the horizontal asymptote is $y = 1$.

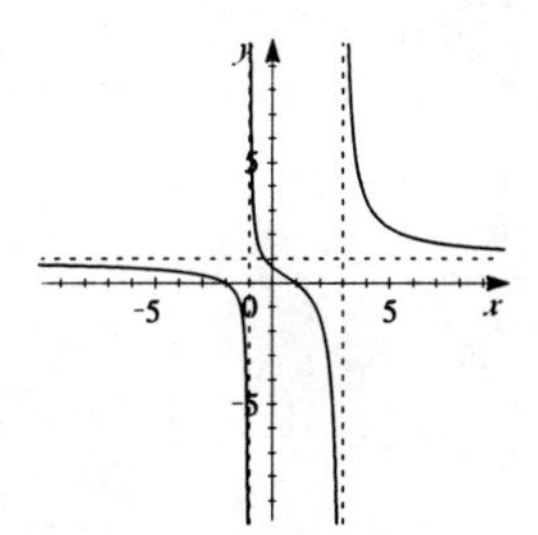

31. $y = \dfrac{x^2 - 2x + 1}{x^2 + 2x + 1} = \dfrac{(x-1)^2}{(x+1)^2} = \left(\dfrac{x-1}{x+1}\right)^2$. When $x = 0$, $y = 1$, so the y-intercept is 1. When $y = 0$, $x = 1$, so the x-intercept is 1. A vertical asymptote occurs at $x + 1 = 0 \quad \Leftrightarrow \quad x = -1$, with $y \to \infty$ as $x \to -1$ from either side. The horizontal asymptote is $y = 1$.

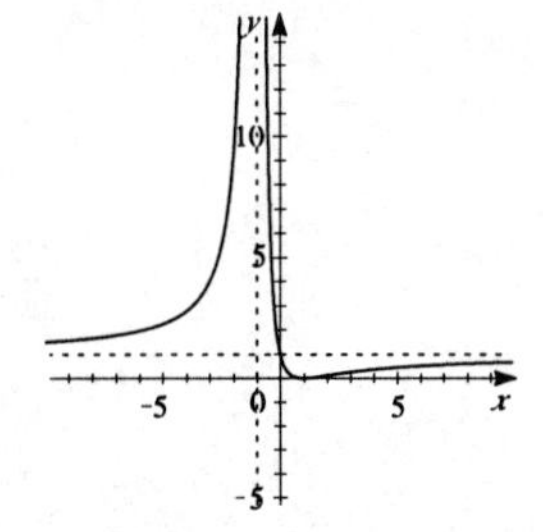

33. $y = \dfrac{2x^2 + 10x - 12}{x^2 + x - 6} = \dfrac{2(x-1)(x+6)}{(x-2)(x+3)}$. When $x = 0$, $y = \dfrac{2(-1)(6)}{(-2)(3)} = 2$, so the y-intercept is 2. When $y = 0$, $2(x-1)(x+6) = 0 \quad \Rightarrow \quad x = -6, 1$, so there are two x-intercepts. Vertical asymptotes occur when $(x-2)(x+3) = 0 \quad \Leftrightarrow \quad x = -3$ or $x = 2$. As $x \to \pm\infty$, $y \to 2$, so the horizontal asymptote is $y = 2$.

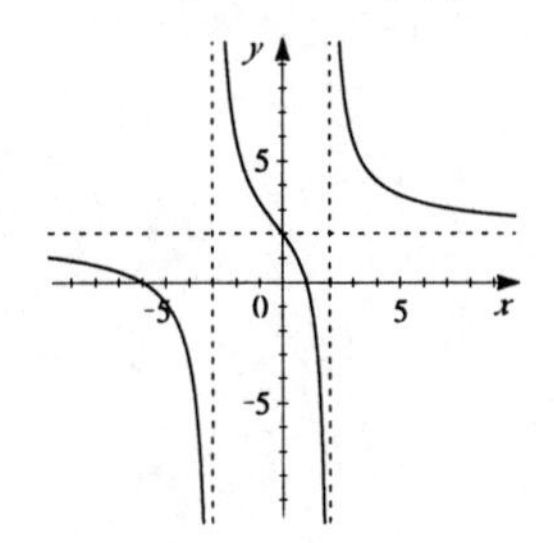

35. $y = \dfrac{x^2 - x - 6}{x^2 + 3x} = \dfrac{(x-3)(x+2)}{x(x+3)}$. The vertical asymptotes are $x = 0$ and $x = -3$. Since an asymptote occurs when $x = 0$, there is no y-intercept. When $y = 0$, $(x-3)(x+2) = 0 \quad \Rightarrow$ $x = -2, 3$, so there are two x-intercepts. As $x \to \pm\infty$, $y \to 1$, so the horizontal asymptote is $y = 1$.

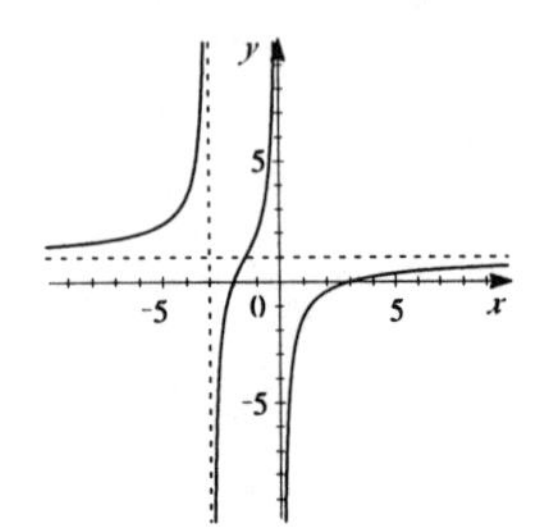

37. $y = \frac{3x^2+6}{x^2-2x-3} = \frac{3(x^2+2)}{(x-3)(x+1)}$. When $x = 0$, $y = -2$, so the y-intercept is -2. Since the numerator can never equal zero, there is no x-intercept. Vertical asymptotes occur when $x = -1, 3$. As $x \to \pm\infty$, $y \to 3$, so $y = 3$ is the horizontal asymptote.

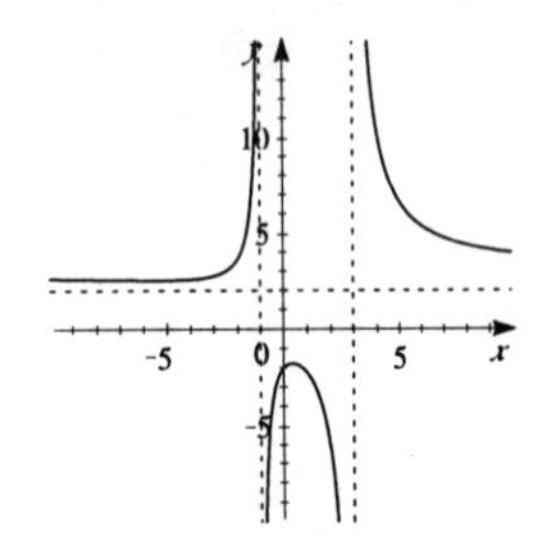

39. $y = \frac{x^2}{x-2}$. When $x = 0$, $y = 0$, so the graph passes through the origin. There is a vertical asymptote when $x - 2 = 0 \quad \Leftrightarrow \quad x = 2$, with $y \to \infty$ as $x \to 2^+$, and $y \to -\infty$ as $x \to 2^-$. There is no horizontal asymptote. By using long division, we see that $y = x + 2 + \frac{4}{x-2}$, so $y = x + 2$ is a slant asymptote.

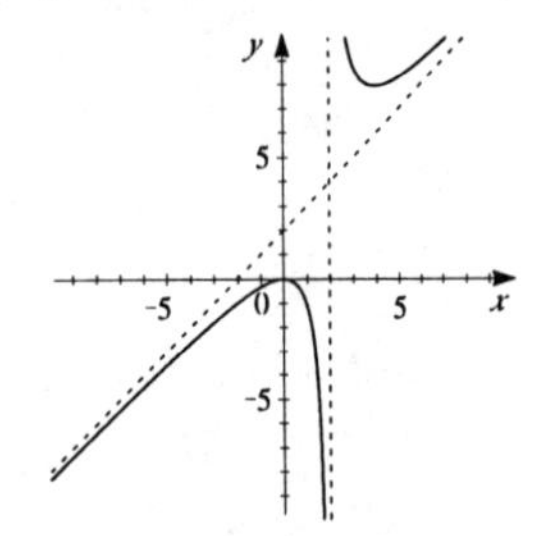

41. $y = \frac{x^2-2x-8}{x} = \frac{(x-4)(x+2)}{x}$. The vertical asymptote is $x = 0$; thus, there is no y-intercept. If $y = 0$ then $(x-4)(x+2) = 0 \quad \Rightarrow \quad x = -2, 4$. Thus, the x-intercepts are -2 and 4. There are no horizontal asymptotes, but there is a slant asymptote $y = x - 2$, since $y = x - 2 - \frac{8}{x}$.

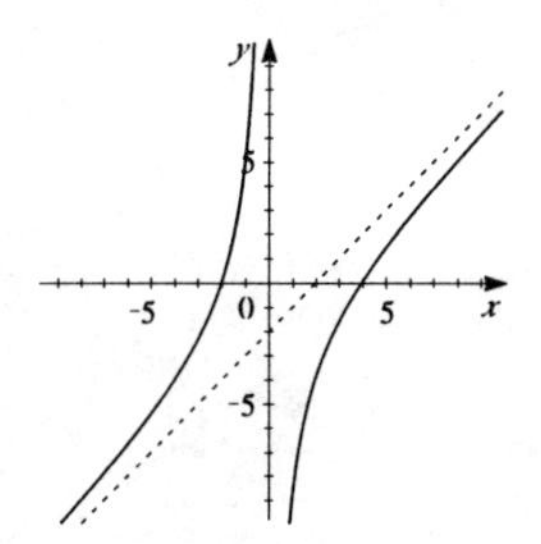

43. $y = \dfrac{x^2 + 5x + 4}{x - 3} = \dfrac{(x+4)(x+1)}{x-3}$. When $x = 0$, $y = -\dfrac{4}{3}$, so the y-intercept is $-\dfrac{4}{3}$. When $y = 0$, $(x+4)(x+1) = 0 \quad \Leftrightarrow \quad x = -4, -1$, so there are two x-intercepts. A vertical asymptote occurs when $x = 3$, with $y \to \infty$ as $x \to 3^+$, and $y \to -\infty$ as $x \to 3^-$. Using long division, we see that the line $y = x + 8$ is a slant asymptote.

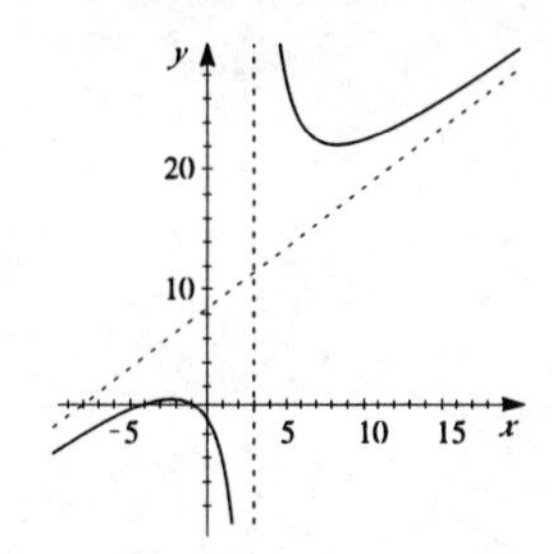

45. $y = \dfrac{x^3 + x^2}{x^2 - 4} = \dfrac{x^2(x+1)}{(x-2)(x+2)}$. When $x = 0$, $y = 0$, so the graph passes through the origin. Moreover, when $y = 0$, $x^2(x+1) = 0 \quad \Rightarrow \quad x = 0, -1$, so there are actually two x-intercepts. Vertical asymptotes occur when $x = \pm 2$, and there are no horizontal asymptotes. Using long division, we see that $y = x + 1$ is a slant asymptote.

as $x \to$	-2^-	-2^+	2^-	2^+
sign of $y = \dfrac{x^2(x+1)}{(x-2)(x+2)}$	$\dfrac{(+)(-)}{(-)(-)}$	$\dfrac{(+)(-)}{(-)(+)}$	$\dfrac{(+)(+)}{(-)+)}$	$\dfrac{(+)(+)}{(+)(+)}$
$y \to$	$-\infty$	∞	$-\infty$	∞

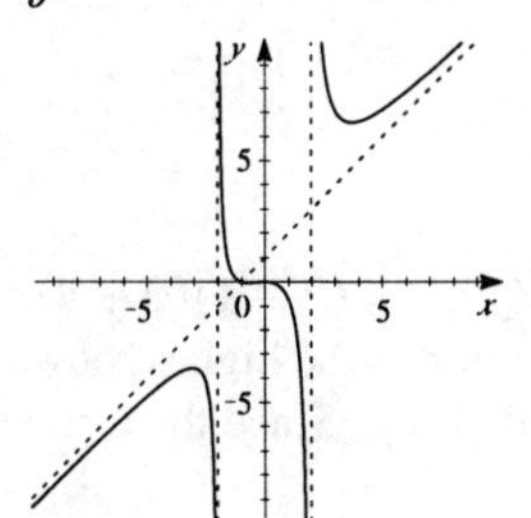

47. $y = \dfrac{x^3}{x - 2}$. When $x = 0$, $y = 0$, so the graph passes through the origin. There is a vertical asymptote when $x - 2 = 0 \quad \Leftrightarrow \quad x = 2$, with $y \to \infty$ as $x \to 2^+$, and $y \to -\infty$ as $x \to 2^-$. The graph has no horizontal or slant asymptotes.

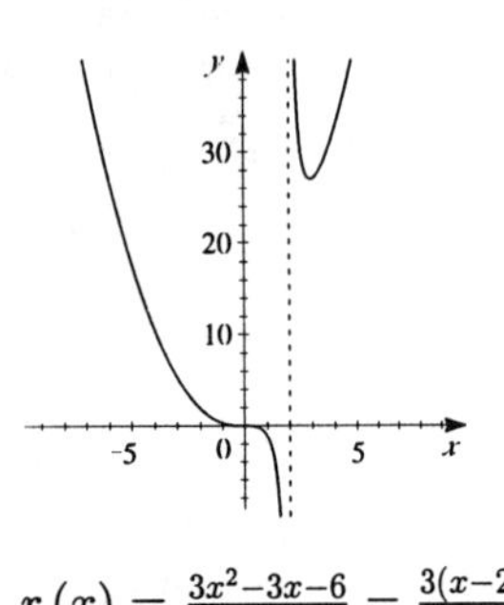

49. $r(x) = \frac{3x^2-3x-6}{x-2} = \frac{3(x-2)(x+1)}{x-2} = 3(x+1)$, for $x \neq 2$. Therefore, $r(x) = 3x+3$, $x \neq 2$. Since $3(2)+3 = 9$, the graph is the line $y = 3x+3$ with the point $(2, 9)$ removed.

51. $y = \dfrac{2x^2 - x - 1}{x-1} = \dfrac{(2x+1)(x-1)}{x-1} = 2x+1$, for $x \neq 1$. Therefore, the graph is the line $y = 2x+1$ with the point $(1, 3)$ removed.

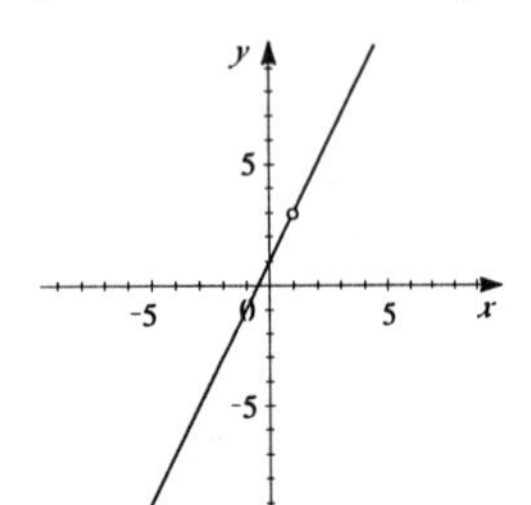

53. $y = \dfrac{2x^2 - 5x - 3}{x^2 - 2x - 3} = \dfrac{(2x+1)(x-3)}{(x+1)(x-3)} = \dfrac{2x+1}{x+1}$, for $x - 3 \neq 0 \quad \Leftrightarrow \quad x \neq 3$. Therefore, the graph is $y = \dfrac{2x+1}{x+1}$ with the point $(3, \frac{7}{4})$ removed.

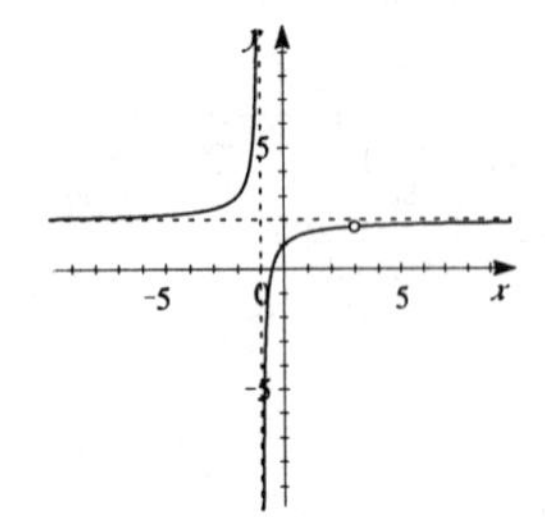

55. Since $x = 1$ and $x = -4$ are the vertical asymptotes, it follows that $Q(x) = (x-1)(x+4)$. The horizontal asymptote is $y = 1$, and so the highest power of the numerator = the highest power of the denominator = 2, and the coefficients of x^2 are identical in $P(x)$ and $Q(x)$. Since the x-intercepts are 2 and 3, then $P(x) = (x-2)(x-3)$. Thus, the function is
$$y = \frac{(x-2)(x-3)}{(x-1)(x+4)} = \frac{x^2 - 5x + 6}{x^2 + 3x - 4}.$$

57. Since $x = \frac{1}{2}$ is the vertical asymptote, it follows that $Q(x) = 2x - 1$. The slant asymptote is $y = 3x - 6$, and so $P(x) = r(x) \cdot Q(x) + R(x) = (3x-6)(2x-1) + R(x) \quad \Leftrightarrow$ $P(x) = 6x^2 - 15x + 6 + R(x)$. However, the graph passes through the origin and so

$P(0) = 6(0)^2 - 15(0) + 6 + R(0) = 0 \quad \Leftrightarrow \quad R(0) = -6$. Since $R(x)$ must be degree 0, $R(x) = -6$. Therefore, the function is $y = \dfrac{6x^2 - 15x}{2x - 1}$.

59. (a)

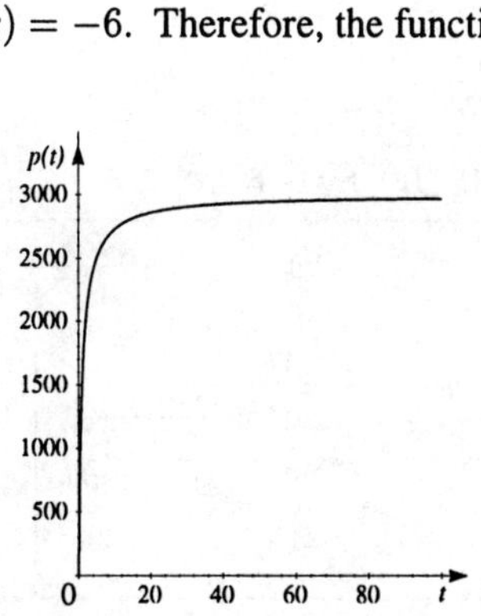

(b) $p(t) = \dfrac{3000t}{t+1} = 3000 - \dfrac{3000}{t+1}$. So as $t \to \infty$ we have $p(t) \to 3000$.

Exercises 4.9

1. $y = \dfrac{2x}{x-3}$. Vertical asymptote: $x = 3$; Horizontal asymptote: $y = 2$.

$[-4, 8]$ by $[-5, 10]$

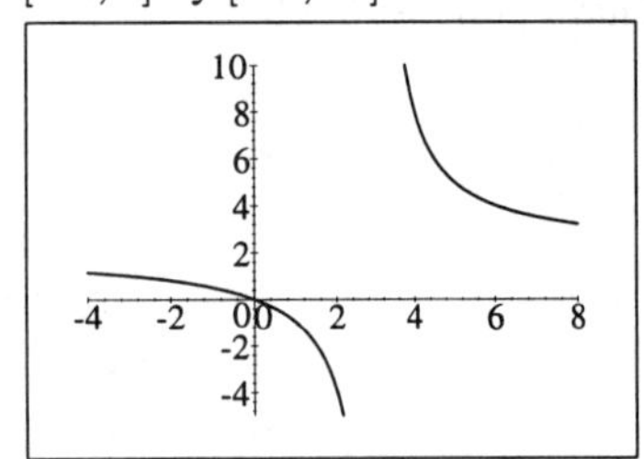

$[-100, 100]$ by $[-5, 10]$

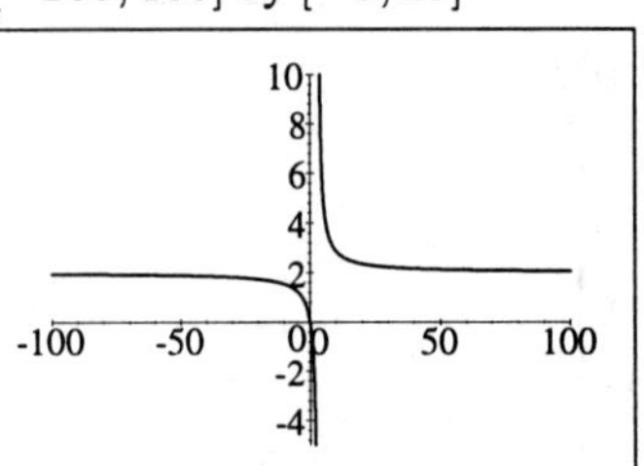

3. $y = \dfrac{4-2x}{x}$. Vertical asymptote: $x = 0$; Horizontal asymptote: $y = -2$.

$[-8, 8]$ by $[-20, 10]$

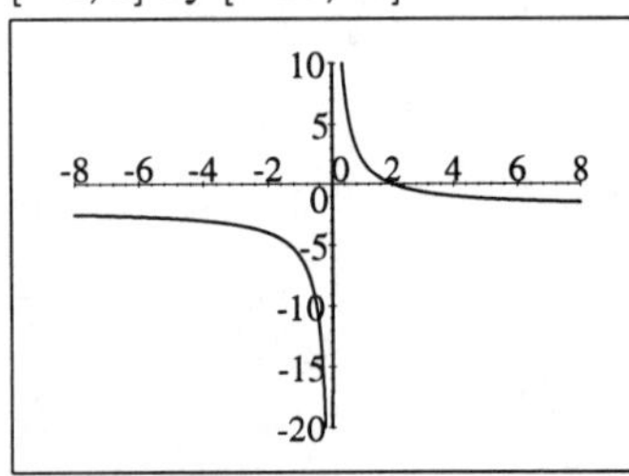

$[-100, 100]$ by $[-20, 10]$

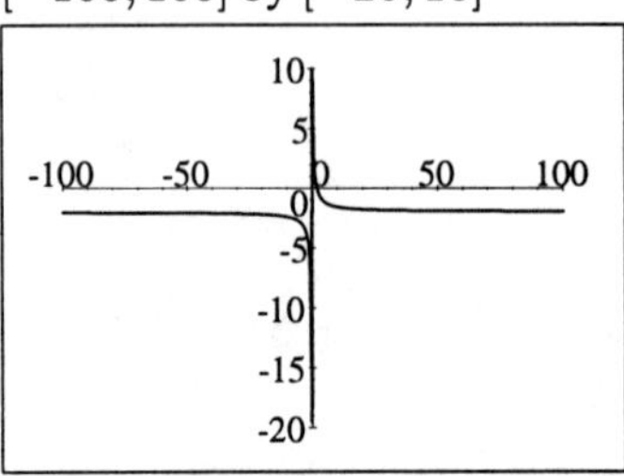

5. $y = \dfrac{3x^2+1}{x^2-9}$. Vertical asymptotes: $x = -3$, $x = 3$; Horizontal asymptote: $y = 3$.

$[-10, 10]$ by $[-10, 10]$

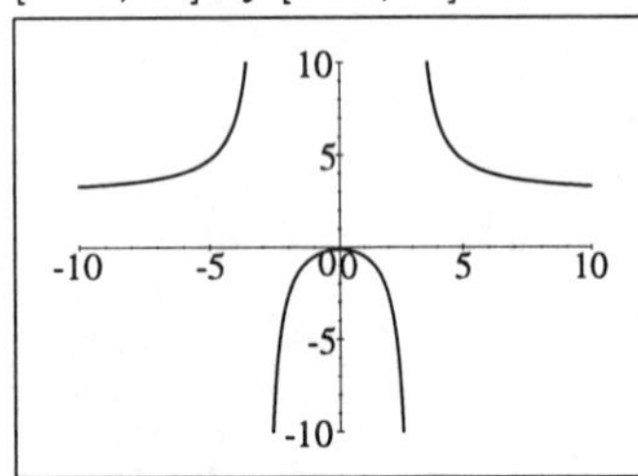

$[-100, 100]$ by $[-10, 10]$

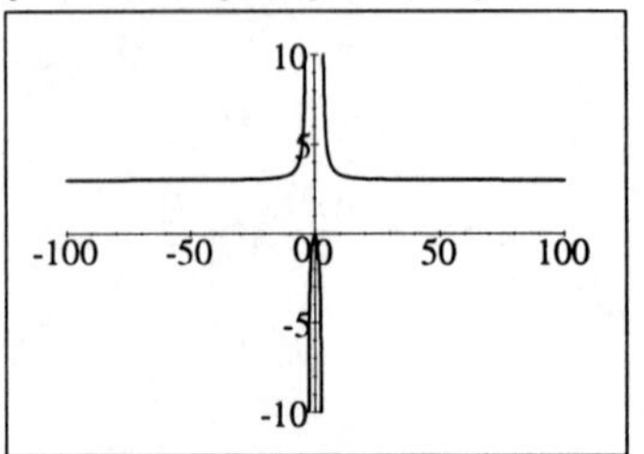

7. $y = \dfrac{x+1}{x-1}$.

Vertical asymptote: $x = 1$;

Horizontal asymptote: $y = 1$.

x-intercept: -1

y-intercept: -1

No extrema.

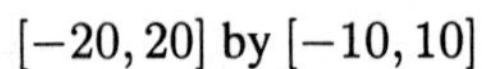

$[-20, 20]$ by $[-10, 10]$

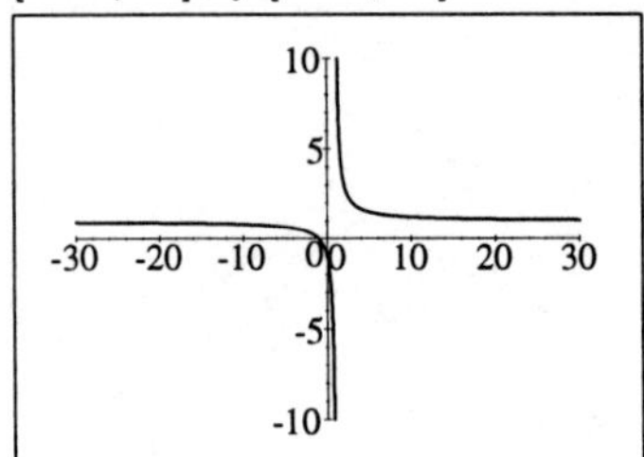

9. $y = \dfrac{7x - 14}{x}$.

Vertical asymptote: $x = 0$;

Horizontal asymptote: $y = 7$.

x-intercept: 2

y-intercept: none

No extrema.

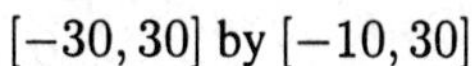

$[-30, 30]$ by $[-10, 30]$

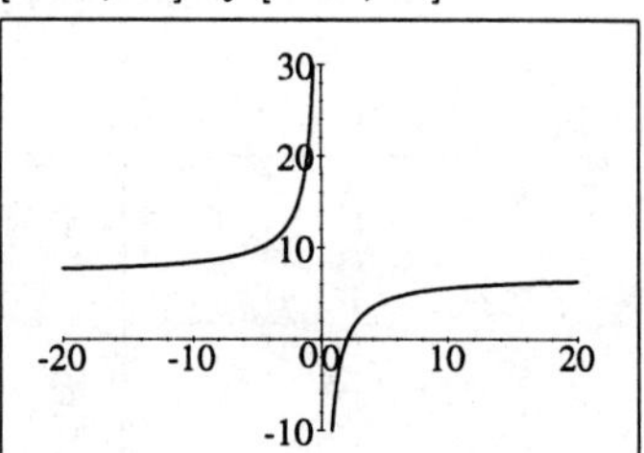

11. $y = \dfrac{4x}{x^2 - 4}$.

Vertical asymptotes: $x = -2$, $x = 2$;

Horizontal asymptote: $y = 0$.

x-intercept: 0

y-intercept: 0

No extrema.

$[-5, 5]$ by $[-10, 10]$

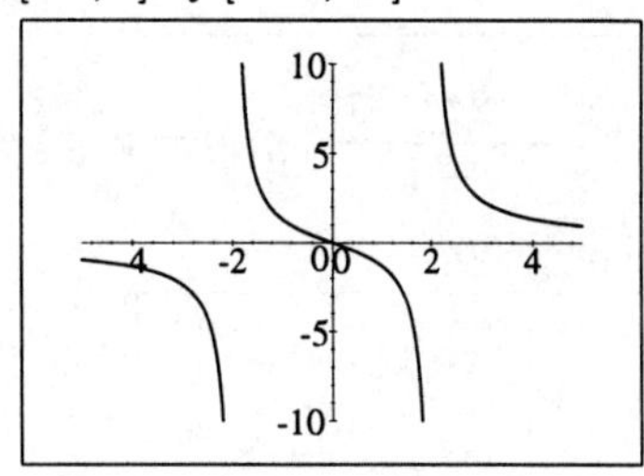

13. $y = \dfrac{6x^2 - 6}{x^2 + 2}$.

Vertical asymptotes: none;

Horizontal asymptote: $y = 6$.

x-intercept: ± 1

y-intercept: -3

Minimum: $(0, -3)$.

$[-20, 20]$ by $[-10, 10]$

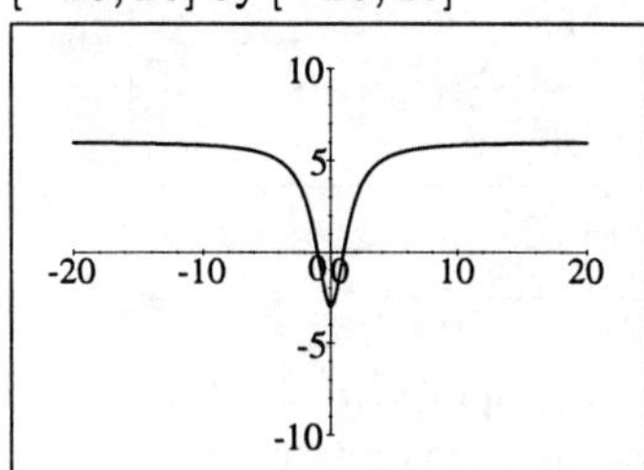

15. $y = \dfrac{4}{(x - 1)^2}$.

Vertical asymptote: $x = 1$

Horizontal asymptote: $y = 0$.

x-intercept: none

y-intercept: 4

No extrema.

$[-20, 20]$ by $[-20, 20]$

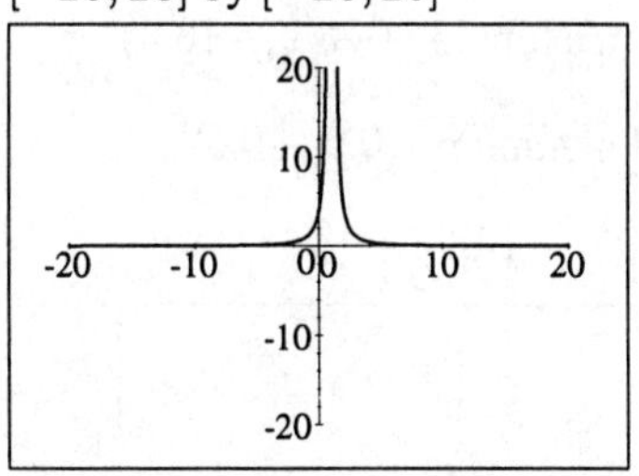

17. $y = \dfrac{2x^2 + 3x - 2}{x^2}$.

Vertical asymptote: $x = 0$

Horizontal asymptote: $y = 2$.

x-intercept: $-2, \frac{1}{2}$

y-intercept: none

Maximum: $(1.33, 3.13)$

$[-10, 10]$ by $[-5, 5]$

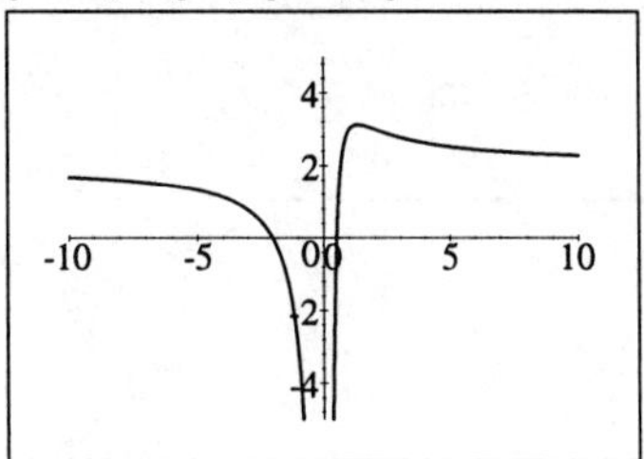

19. $f(x) = \dfrac{2x^2 + 6x + 6}{x + 3}$, $g(x) = 2x$. Vertical asymptote: $x = -3$.

$[-10, 5]$ by $[-20, 10]$

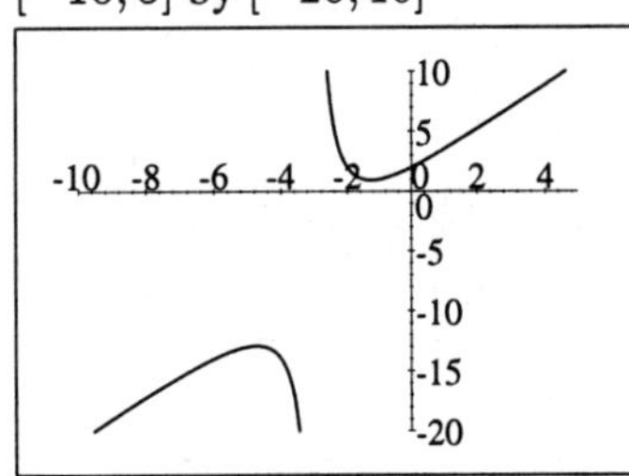

$[-20, 20]$ by $[-50, 50]$

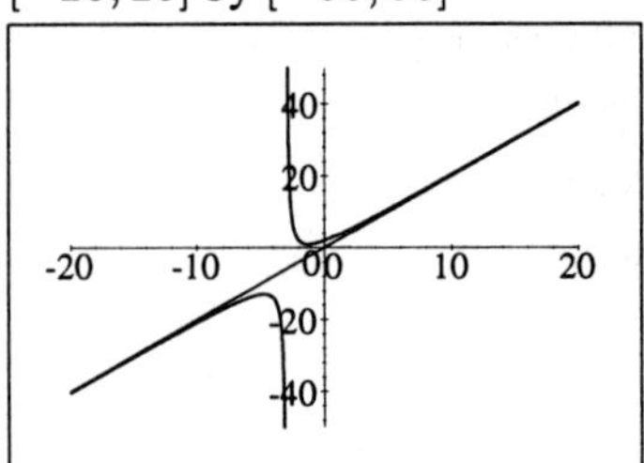

21. $f(x) = \dfrac{x^3 - 2x^2 + 16}{x - 2}$, $g(x) = x^2$. Vertical asymptote: $x = 2$.

$[-4, 8]$ by $[-30, 30]$

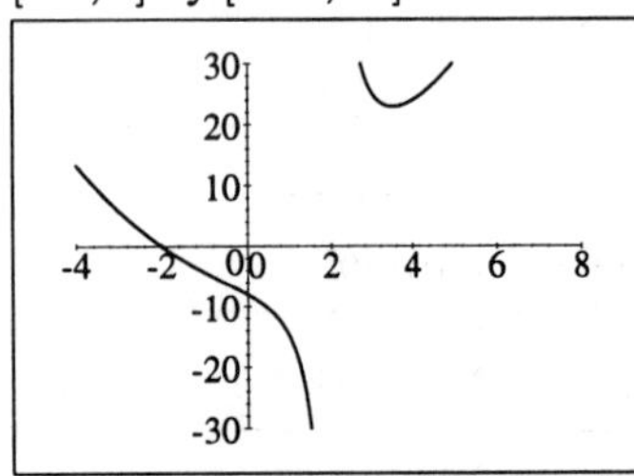

$[-20, 20]$ by $[-50, 50]$

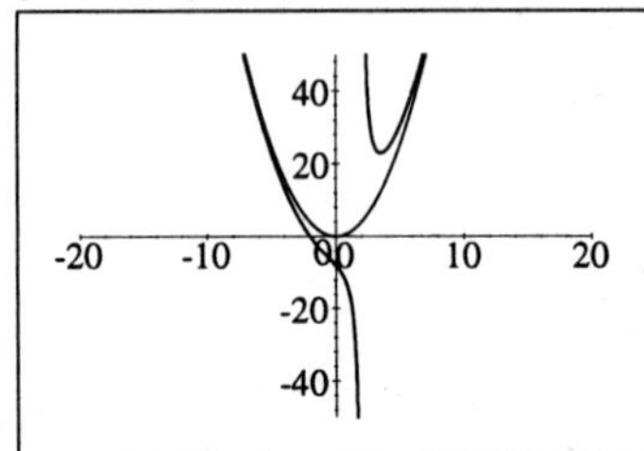

23. $y = \dfrac{2x^2 - 5x}{2x + 3}$.

Vertical asymptote: $x = -1.5$

x-intercepts: $0, 2.5$

y-intercept: 0

Local maximum: $(-3.9, -10.4)$

Local minimum: $(0.9, -0.6)$

$$\begin{array}{r|l} & x \quad -4 \\ \hline 2x+3 & 2x^2 - 5x \\ & \underline{2x^2 + 3x} \\ & \quad -8x \\ & \quad \underline{-8x - 12} \\ & \qquad +12 \end{array}$$

Using long division, we get $y = x - 4 + \dfrac{12}{2x + 3}$ and the end behavior is like $y = x - 4$.

$[-10, 10]$ by $[-30, 30]$

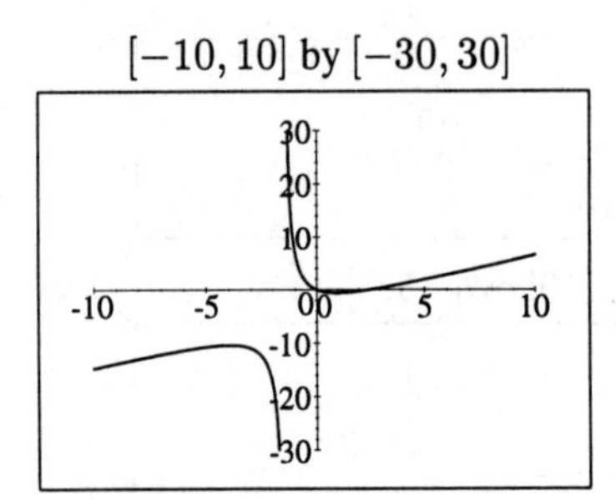

$[-20, 20]$ by $[-50, 50]$

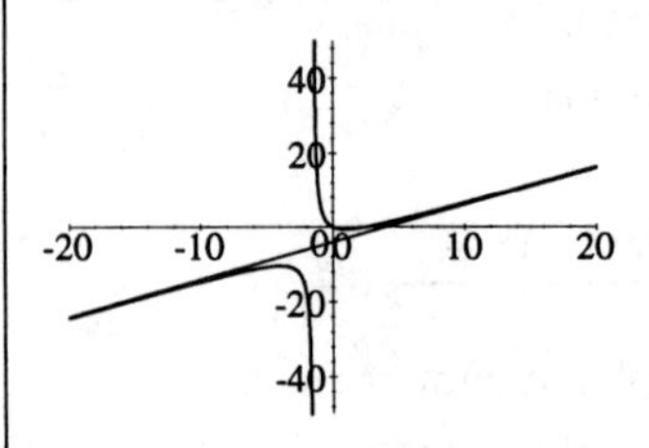

25. $y = \dfrac{x^5}{x^3 - 1}$.
Vertical asymptote: $x = 1$
x-intercept: 0
y-intercept: 0
Local minimum: $(1.4, 3.1)$

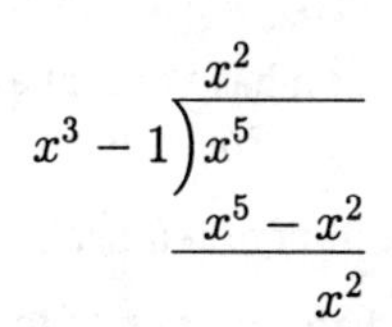

Thus $y = x^2 + \dfrac{x^2}{x^3 - 1}$ and the end behavior is like $y = x^2$.

$[-5, 5]$ by $[-10, 10]$

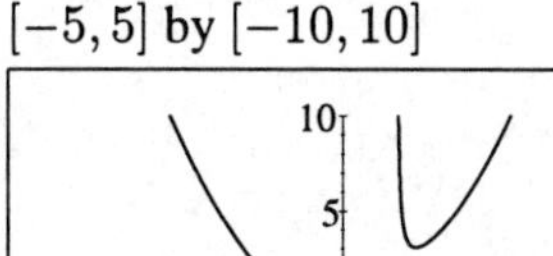
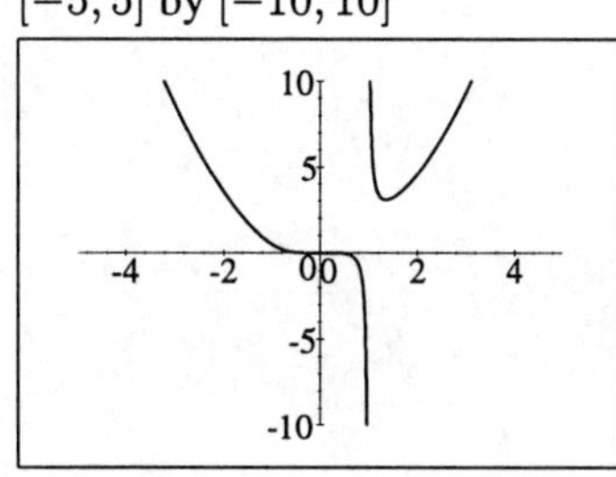

$[-10, 10]$ by $[-10, 10]$

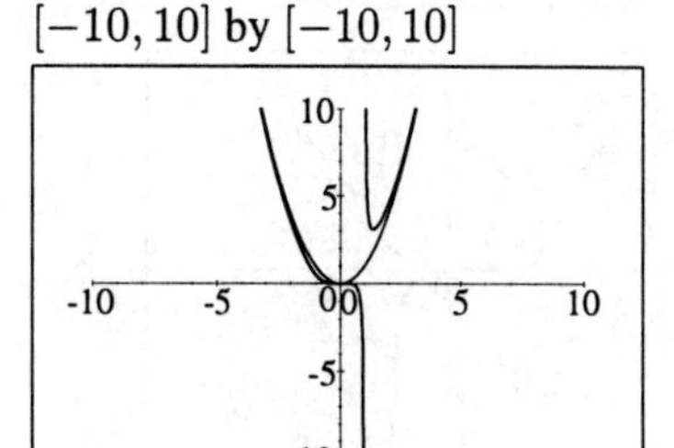

27. $[0, 20]$ by $[0, 3]$

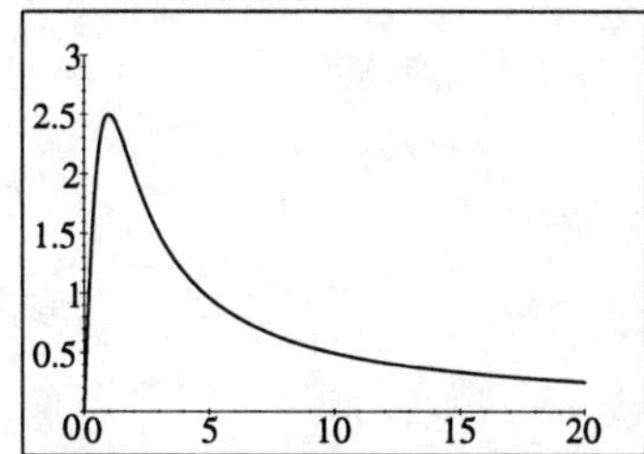

(a) The highest concentration of drug is 2.50 mg/L and it is reached 1 hour after the drug is administered.

(b) The concentration of the drug in the bloodstream goes to 0.

(c) From the first viewing rectangle we see that an approximate solution is near $t = 15$. Thus we graph $y = \dfrac{5t}{t^2 + 1}$ and $y = 0.3$ in the viewing rectangle $[14, 18]$ by $[0, 0.5]$. So it takes about 16.61 hours for the concentration to drop below 0.3 mg/L.

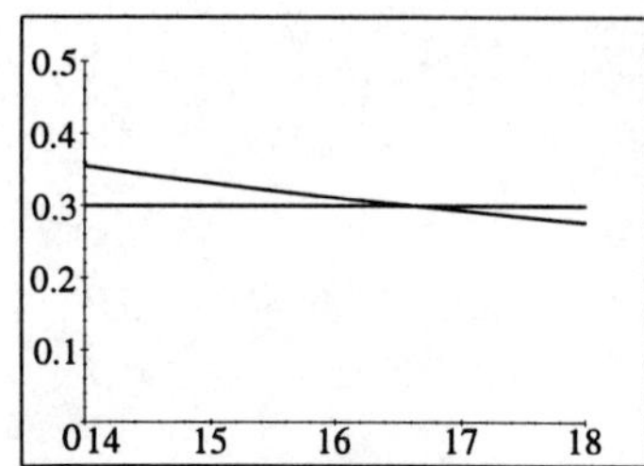

29. $P(v) = P_0\left(\dfrac{s_0}{s_0 - v}\right) \Rightarrow$
$P(v) = 440\left(\dfrac{332}{332 - v}\right)$

If the speed of the train approaches the speed of sound, the pitch of the whistle becomes very large. This would be experienced as a "sonic boom"— an effect seldom heard with trains.

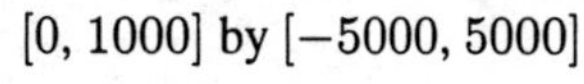
$[0, 1000]$ by $[-5000, 5000]$

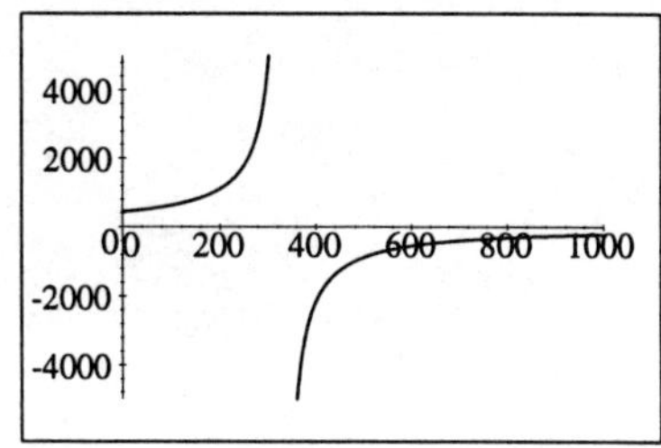

31. Since the rational function has vertical asymptotes $x = \pm 1$, its denominator contains $(x-1)(x+1) = x^2 - 1$. Since its end behavior is the same as the end behavior of $y = x^2$, this means that the rational function must be of the form of $y = x^2 + \dfrac{R(x)}{x^2 - 1}$. Since the x-intercept and the y-intercept are 0, therefore one of the rational functions that satisfies the above conditions is $y = x^2 + \dfrac{x}{x^2 - 1} = \dfrac{x^4 - x^2 + x}{x^2 - 1}$.

$[-4, 4]$ by $[-8, 8]$

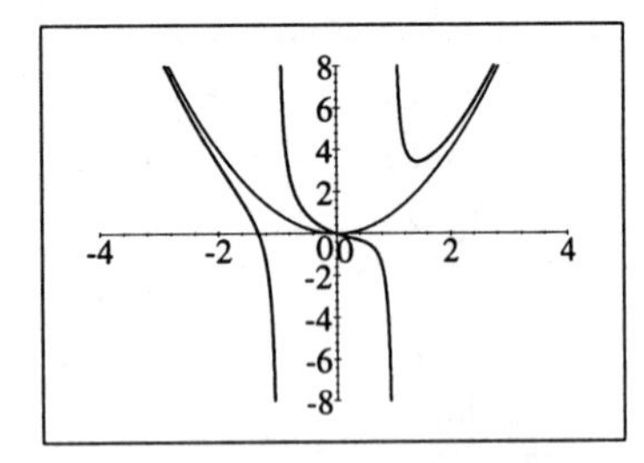

Review Exercises for Chapter 4

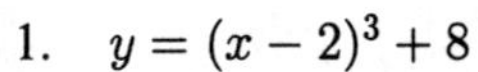

1. $y = (x-2)^3 + 8$

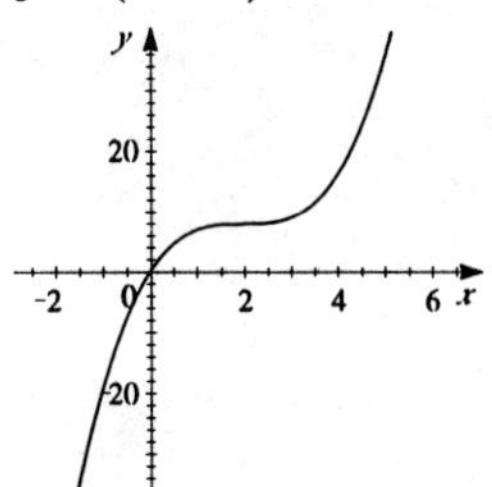

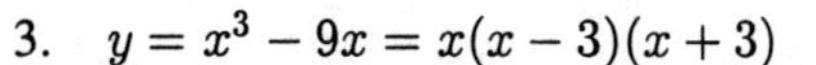

3. $y = x^3 - 9x = x(x-3)(x+3)$

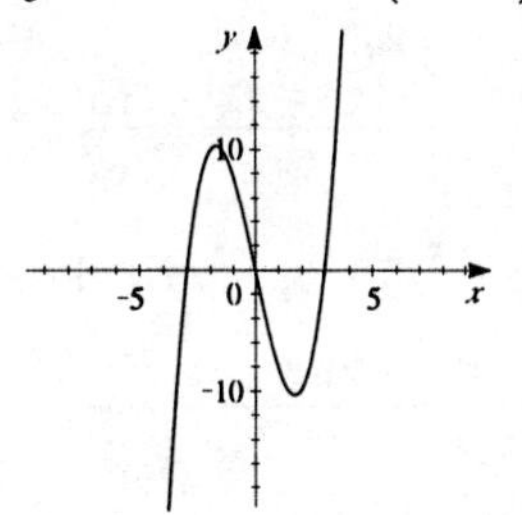

5. $y = x^3 - 5x^2 - 4x + 20$
$= (x-5)(x-2)(x+2)$

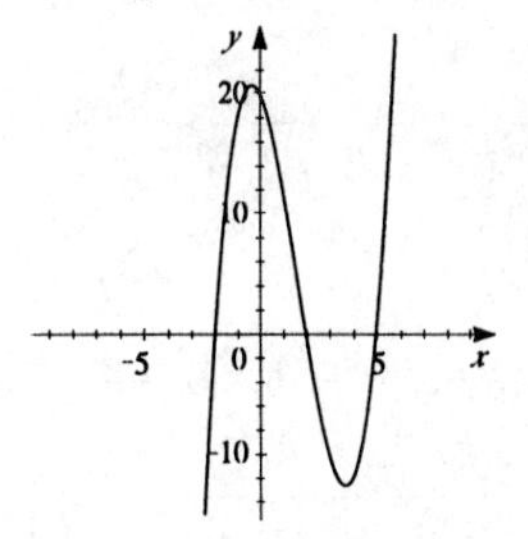

7. $y = 2x^3 + x^2 - 18x - 9$.
x-intercepts: $-3, -0.5, 3$
y-intercept: -9
local maximum: $(-1.9, 15.1)$
local minimum: $(1.6, -27.1)$
$y \to \infty$ as $x \to \infty$; $y \to -\infty$ as $x \to -\infty$

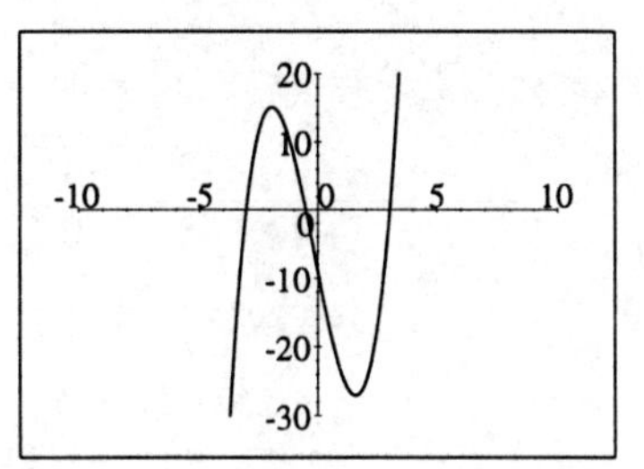

9. $y = x^5 + x^2 - 5$.
x-intercept: 1.3
y-intercept: -5
local maximum: $(-0.7, -4.7)$
local minimum: $(0, -5)$
$y \to \infty$ as $x \to \infty$; $y \to -\infty$ as $x \to -\infty$

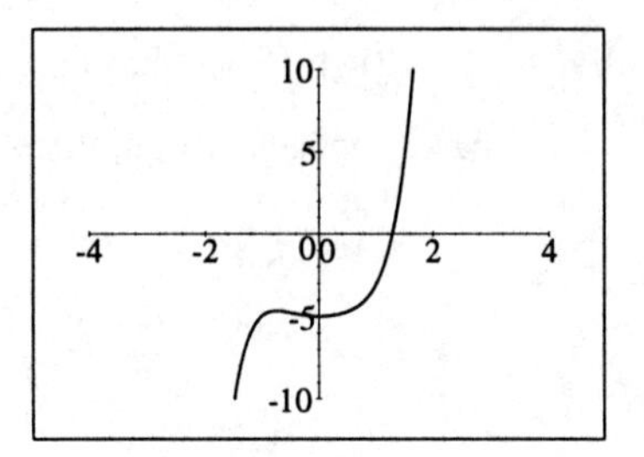

11.
$$\begin{array}{r|rrrr} 3 & 1 & -1 & 1 & 11 \\ & & 3 & 6 & 21 \\ \hline & 1 & 2 & 7 & 10 \end{array}$$

Therefore, $Q(x) = x^2 + 2x + 7$, and $R(x) = 10$.

13.
$$\begin{array}{rrrrr} & & x & -3 & \\ \cline{2-5} x^2+2x-5 \big) & x^3 & -x^2 & -11x & +6 \\ & x^3 & +2x^2 & -5x & \\ \cline{2-4} & & -3x^2 & -6x & +6 \\ & & -3x^2 & -6x & +15 \\ \cline{3-5} & & & & -9 \end{array}$$

Therefore, $Q(x) = x - 3$, and $R(x) = -9$.

15.
$$\begin{array}{r|rrrrr} & 1 & 0 & -25 & 4 & 15 \\ \hline -5 & 1 & -5 & 0 & 4 & -5 \end{array}$$

Therefore, $Q(x) = x^3 - 5x^2 + 4$, and $R(x) = -5$.

17.
$$\begin{array}{r|rrrrr} & 1 & 1 & -2 & -3 & -1 \\ \hline \sqrt{3} & 1 & 1+\sqrt{3} & 1+\sqrt{3} & \sqrt{3} & 2 \end{array}$$

Therefore, $Q(x) = x^3 + (1+\sqrt{3})\,x^2 + (1+\sqrt{3})\,x + \sqrt{3}$, and $R(x) = 2$.

19. $P(x) = 2x^3 - 9x^2 - 7x + 13$; find $P(5)$.

$$\begin{array}{r|rrrr} & 2 & -9 & -7 & 13 \\ \hline 5 & 2 & 1 & -2 & 3 \end{array}$$

Therefore, $P(5) = 3$.

21. $\frac{1}{2}$ is a zero of $P(x) = 2x^4 + x^3 - 5x^2 + 10x - 4$ if $P\left(\frac{1}{2}\right) = 0$.

$$\begin{array}{r|rrrrr} & 2 & 1 & -5 & 10 & -4 \\ \hline \frac{1}{2} & 2 & 2 & -4 & 8 & 0 \end{array}$$

Since $P\left(\frac{1}{2}\right) = 0$, $\frac{1}{2}$ is a zero of the polynomial.

23. $P(x) = x^{500} + 6x^{201} - x^2 - 2x + 4$. The remainder on division by $x - 1$ is $P(1) = (1)^{500} + 6(1)^{201} - (1)^2 - 2(1) + 4 = 8$.

25. $x^5 - 6x^3 - x^2 + 2x + 18 = 0$ has possible rational roots $\pm 1, \pm 2, \pm 3, \pm 6, \pm 9, \pm 18$.

Since $P(x)$ has 2 changes in sign, there are either 0 or 2 positive real roots. Since $P(-x) = -x^5 + 6x^3 - x^2 - 2x + 18$ has 3 variations in sign, there are 1 or 3 negative real roots.

27. $x^3 - 2x - 1 = 0$

x	$x^3 - 2x - 1$
1	-2
2	3
1.5	-0.625
1.6	-0.104
1.7	0.513
1.65	0.192

Hence the root is 1.6.

29. $x^6 - 5x^4 + 10 = 0$

x	$x^6 - 5x^4 + 10$
1	6
2	-6
1.3	0.546309
1.4	-1.678464
1.32	0.110064
1.33	-0.1101351962
1.325	0.00012527

Hence the root is 1.33.

31. Since the zeros are $-\frac{1}{2}$, 2, and 3, a factorization is $P(x) = C\left(x + \frac{1}{2}\right)(x-2)(x-3)$ $= \frac{C}{2}(2x+1)(x^2-5x+6) = \frac{C}{2}(2x^3 - 10x^2 + 12x + x^2 - 5x + 6) = \frac{C}{2}(2x^3 - 9x^2 + 7x + 6)$.

Since the constant coefficient is 12, $\frac{C}{2}(6) = 12 \quad \Leftrightarrow \quad C = 4$, and so the polynomial is $P(x) = 4x^3 - 18x^2 + 14x + 12$.

33. No, there is no polynomial of degree 4 with integer coefficients that has zeros i, $2i$,$3i$ and $4i$. Since the imaginary roots of polynomial equations with real coefficients come in complex conjugate pairs, there would have to be 8 zeros, which is impossible for a polynomial of degree 4.

35. $x^3 - 3x^2 - 13x + 15 = 0$ has possible rational roots $\pm 1, \pm 3, \pm 5, \pm 15$.

$$\begin{array}{r|rrrr} & 1 & -3 & -13 & 15 \\ \hline 1 & 1 & -2 & -15 & 0 \end{array} \quad \Rightarrow \quad x = 1 \text{ is a root.}$$

$x^3 - 3x^2 - 13x + 15 = (x-1)(x^2 - 2x - 15) = 0 \quad \Leftrightarrow \quad (x-1)(x-5)(x+3) = 0$. Therefore, the roots are -3, 1, and 5.

37. $x^4 + 6x^3 + 17x^2 + 28x + 20 = 0$ has possible rational roots $\pm 1, \pm 2, \pm 4, \pm 5, \pm 10, \pm 20$. Since all of the coefficients are positive, there are no positive real roots.

$$\begin{array}{r|rrrrr} & 1 & 6 & 17 & 28 & 20 \\ \hline -1 & 1 & 5 & 12 & 16 & 4 \\ -2 & 1 & 4 & 9 & 10 & 0 \end{array} \quad \Rightarrow \quad x = -2 \text{ is a root.}$$

$x^4 + 6x^3 + 17x^2 + 28x + 20 = (x+2)(x^3 + 4x^2 + 9x + 10) = 0$. Continuing with the quotient we have

$$\begin{array}{r|rrrr} & 1 & 4 & 9 & 10 \\ \hline -2 & 1 & 2 & 5 & 0 \end{array} \quad \Rightarrow \quad x = -2 \text{ is a root.}$$

Thus $x^4 + 6x^3 + 17x^2 + 28x + 20 = (x+2)^2(x^2 + 2x + 5) = 0$. Now $x^2 + 2x + 5 = 0$ when $x = \frac{-2 \pm \sqrt{4 - 4(5)(1)}}{2} = \frac{-2 \pm 4i}{2} = -1 \pm 2i$. Thus, the roots are -2 (multiplicity 2) and $-1 \pm 2i$.

39. $x^5 - 3x^4 - x^3 + 11x^2 - 12x + 4 = 0$ has possible rational roots $\pm 1, \pm 2, \pm 4$.

$$\begin{array}{r|rrrrrr} & 1 & -3 & -1 & 11 & -12 & 4 \\ \hline 1 & 1 & -2 & -3 & 8 & -4 & 0 \end{array} \quad \Rightarrow \quad x = 1 \text{ is a root.}$$

$x^5 - 3x^4 - x^3 + 11x^2 - 12x + 4 = (x-1)(x^4 - 2x^3 - 3x^2 + 8x - 4) = 0$. Continuing with the quotient we have

$$\begin{array}{r|rrrrr} & 1 & -2 & -3 & 8 & -4 \\ \hline 1 & 1 & -1 & -4 & 4 & 0 \end{array} \quad \Rightarrow \quad x = 1 \text{ is a root.}$$

So $x^5 - 3x^4 - x^3 + 11x^2 - 12x + 4 = (x-1)^2(x^3 - x^2 - 4x + 4) = (x-1)^3(x^2 - 4)$
$= (x-1)^3(x-2)(x+2) = 0$. Therefore, the roots are 1 (multiplicity 3), -2, and 2.

41. $x^6 = 64 \quad \Leftrightarrow \quad x^6 - 64 = 0 \quad \Leftrightarrow \quad (x^3 - 8)(x^3 + 8) =$
$(x-2)(x^2 + 2x + 4)(x+2)(x^2 - 2x + 4) = 0$. Now $x^2 + 2x + 4 = 0$ when

$x = \frac{-2 \pm \sqrt{4 - 4(4)(1)}}{2} = \frac{-2 \pm 2\sqrt{3}\,i}{2} = -1 \pm \sqrt{3}\,i$, and $x^2 - 2x + 4 = 0$ when

$x = \frac{2 \pm \sqrt{4 - 4(4)(1)}}{2} = \frac{2 \pm 2\sqrt{3}\,i}{2} = 1 \pm \sqrt{3}\,i$. Therefore, the roots are 2, -2, $1 \pm \sqrt{3}\,i$, and

$-1 \pm \sqrt{3}\,i$

43. $6x^4 - 18x^3 + 6x^2 - 30x + 36 = 6(x^4 - 3x^3 + x^2 - 5x + 6) = 0$ has possible rational roots ± 1, ± 2, ± 3, ± 6.

$$\begin{array}{r|rrrrr} & 6 & -18 & 6 & -30 & 36 \\ \hline 1 & 6 & -12 & -6 & -36 & 0 \end{array} \Rightarrow \quad x = 1 \text{ is a root.}$$

So $6x^4 - 18x^3 + 6x^2 - 30x + 36 = (x-1)(6x^3 - 12x^2 - 6x - 36)$
$= 6(x-1)(x^3 - 2x^2 - x - 6) = 0$. Continuing with the quotient we have

$$\begin{array}{r|rrrr} & 1 & -2 & -1 & -6 \\ \hline 1 & 1 & -1 & -2 & -8 \\ 2 & 1 & 0 & -1 & -8 \\ 3 & 1 & 1 & 2 & 0 \end{array} \Rightarrow \quad x = 3 \text{ is a root.}$$

So, $6x^4 - 18x^3 + 6x^2 - 30x + 36 = 6(x-1)(x-3)(x^2 + x + 2) = 0$. Now, $x^2 + x + 2 = 0$ when $x = \dfrac{-1 \pm \sqrt{1 - 4(1)(2)}}{2} = \dfrac{-1 \pm \sqrt{7}\,i}{2}$, and so the roots are 1, 3, and $\dfrac{-1 \pm \sqrt{7}\,i}{2}$.

45. $2 - \sqrt{2}\,i$ will be a root if and only if $2 + \sqrt{2}\,i$ is a root. Thus $(x - 2 + \sqrt{2}\,i)(x - 2 - \sqrt{2}\,i)$ $= x^2 - 4x + 6$ will divide the polynomial evenly if these numbers are roots.

$$\begin{array}{r} x^2 \quad - x \quad - 2 \\ x^2 - 4x + 6 \overline{\big)\, x^4 - 5x^3 + 8x^2 + 2x - 12} \\ \underline{x^4 - 4x^3 + 6x^2} \qquad\qquad \\ -x^3 + 2x^2 + 2x \qquad \\ \underline{-x^3 + 4x^2 - 6x} \qquad \\ -2x^2 + 8x - 12 \\ \underline{-2x^2 + 8x - 12} \\ 0 \end{array}$$

So $x^4 - 5x^3 + 8x^2 + 2x - 12 = (x^2 - 4x + 6)(x^2 - x - 2) = (x^2 - 4x + 6)(x - 2)(x + 1)$.
Thus the other roots are 2, -1, and $2 + \sqrt{2}\,i$.

47. $x^4 - x^2 - x - 2 = 0$ has possible rational roots ± 1, ± 2 and one positive real root.

$$\begin{array}{r|rrrrr} & 1 & 0 & -1 & -1 & -2 \\ \hline 1 & 1 & 1 & 0 & -1 & -3 \\ 2 & 1 & 2 & 3 & 5 & 8 \end{array}$$

Therefore, the positive real root is irrational and lies between 1 and 2.

x	$x^4 - x^2 - x - 2$
1.5	-0.6875
1.6	0.3936
1.56	-0.07119104
1.57	0.04083201
1.565	-0.015521899

Therefore, the positive irrational root is 1.57 (correct to two decimals).

49. $2x^2 = 5x + 3 \quad \Leftrightarrow \quad 2x^2 - 5x - 3 = 0$.
The solutions are $x = -0.5, 3$.

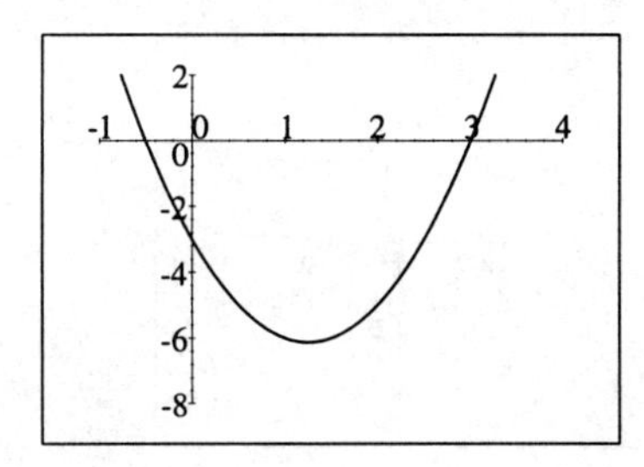

51. $x^4 - 3x^3 - 3x^2 - 9x - 2 = 0$ has solutions $x \approx -0.24, 4.24$.

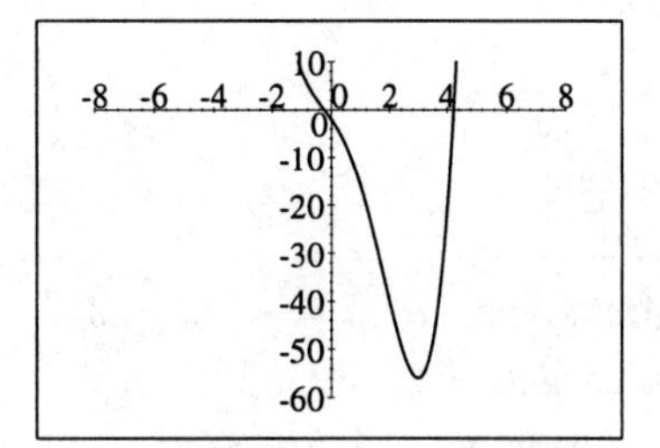

53. $y = \dfrac{3x - 12}{x + 1}$

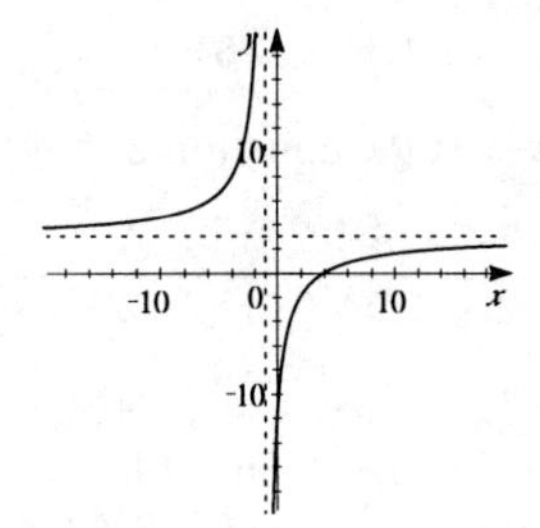

55. $y = \dfrac{x - 2}{x^2 - 2x - 8} = \dfrac{x - 2}{(x + 2)(x - 4)}$

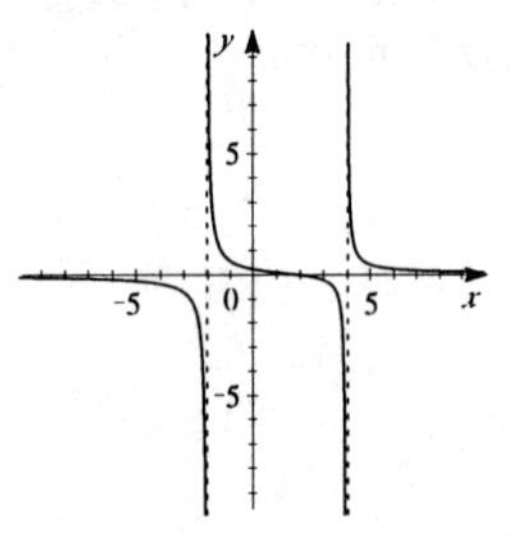

57. $y = \dfrac{x^2 - 9}{2x^2 + 1} = \dfrac{(x + 3)(x - 3)}{2x^2 + 1}$

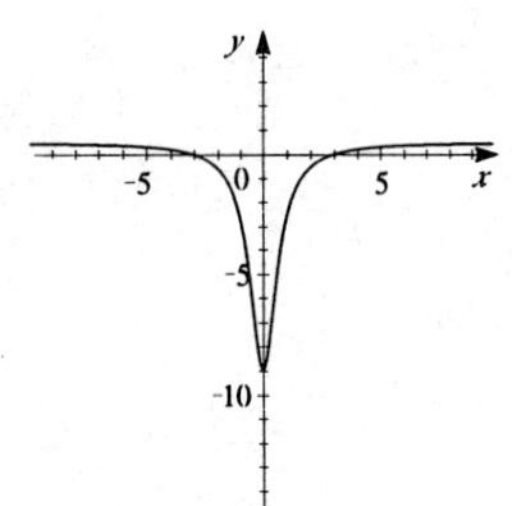

59. $y = \dfrac{x-3}{2x+6}$.

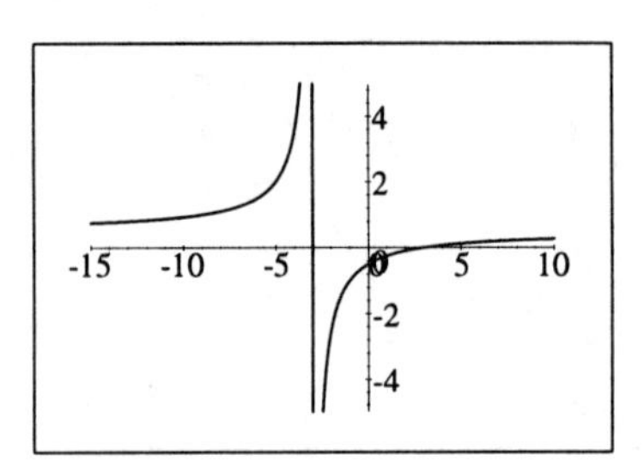

x-intercept: 3

y-intercept: -0.5

Vertical asymptote: $x = -3$;

Horizontal asymptote: $y = 0.5$

No local extrema.

61. 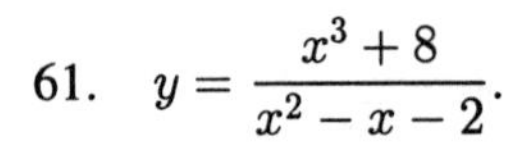
$y = \dfrac{x^3+8}{x^2-x-2}$.

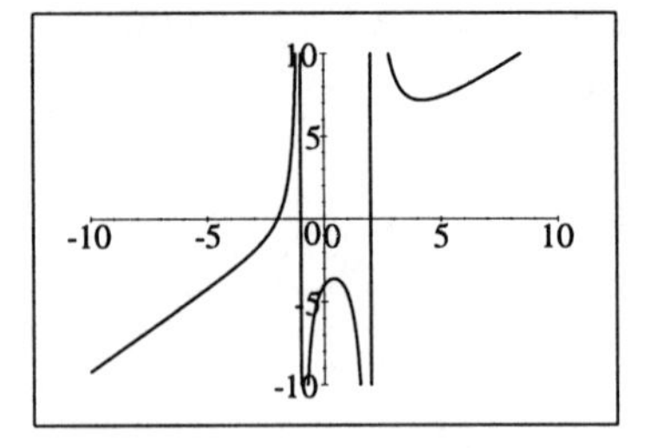

x-intercept: -2

y-intercept: -4

Vertical asymptote: $x = -1, x = 2$

Slant asymptote: $y = x + 1$

local maximum: $(0.425, -3.599)$

local minimum: $(4.216, 7.175)$

63. (a) -1 is a root of the equation $2x^4 + 5x^3 + x + 4 = 0$ if $P(-1) = 0$.

	2	5	0	1	4	
-1	2	3	-3	4	0	$\Rightarrow$ $x = -1$ is a root.

(b) So $2x^4 + 5x^3 + x + 4 = (x+1)(2x^3 + 3x^2 - 3x + 4) = 0$. Since $P(x) = 2x^4 + 5x^3 + x + 4$ has no changes in sign, there are no positive real roots. Further, since $Q(x) = 2x^3 + 3x^2 - 3x + 4$ is a factor of $P(x)$, it follows that $Q(x)$ also cannot have any positive real roots.

Chapter 4 Test

1. $f(x) = x^3 - x^2 - 9x + 9$
   $= (x-1)(x^2-9) = (x-1)(x-3)(x+3)$.

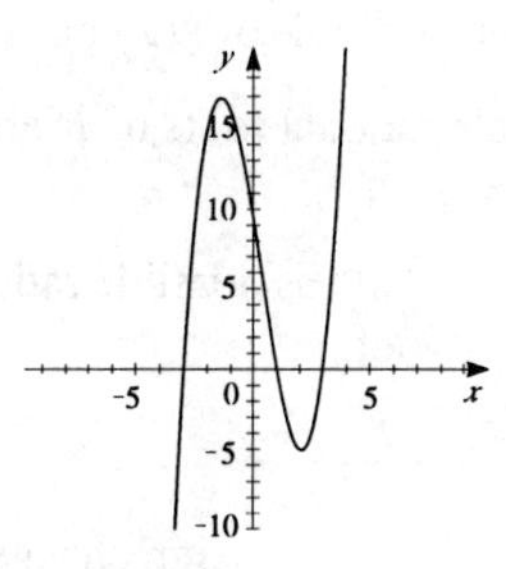

2. $P(x) = 2x^4 - 17x^3 + 53x^2 - 72x + 36$.

   (a) The possible rational zeros of P are: $\pm 1, \pm 2, \pm 3, \pm 4, \pm 6, \pm 9, \pm 12, \pm 18, \pm 36, \pm\frac{1}{2}, \pm\frac{3}{2}$, and $\pm\frac{9}{2}$.

   (b) Since $P(x)$ has 4 changes in sign, $P(x) = 0$ can have 4 or 2 or 0 positive real roots. Since $P(-x) = 2x^4 + 17x^3 + 53x^2 + 72x + 36$ has no changes in sign, there are no negative real roots.

   (c)

	2	−17	53	−72	36
9	2	1	62	486	4374

   Therefore, since the last row contains no negative entries, then 9 is an upper bound for the real roots of $P(x) = 0$.

   (d) Using the information from the answers to parts (a), (b) and (c), a new shorter list of possible rational zeros is $1, 2, 3, 4, 6, \frac{1}{2}, \frac{3}{2}$, and $\frac{9}{2}$.

   (e)

	2	−17	53	−72	36	
1	2	−15	38	−34	2	
2	2	−13	27	−18	0	$\Rightarrow$ $x = 2$ is a root.

   $P(x) = (x-2)(2x^3 - 13x^2 + 27x - 18)$

	2	−13	27	−18	
2	2	−9	9	0	$\Rightarrow$ $x = 2$ is a root.

   $P(x) = (x-2)^2(2x^2 - 9x + 9) = (x-2)^2(x-3)(2x-3)$. Therefore, the rational zeros are 2, 3 and $\frac{3}{2}$.

   (f) $P(x) = (x-2)^2(x-3)(2x-3) = 2(x-\frac{3}{2})(x-2)^2(x-3)$

3. Since $1+2i$ is a zero of $P(x)$, then $1-2i$ is also a zero of $P(x)$. And since -1 is a zero of multiplicity 3, then $P(x) = (x+1)^3(x-[1+2i])(x-[1-2i])$
   $= (x+1)^3([x-1]-2i)([x-1]+2i) = (x+1)^3([x-1])^2+4)$
   $= (x^3+3x^2+3x+1)(x^2-2x+5) = x^5+x^4+2x^3+10x^2+13x+5$.

4. $P(x) = x^{23} - 5x^{12} + 8x - 1$, $Q(x) = 3x^4 + x^2 - x - 15$, and $R(x) = 4x^6 + x^4 + 2x^2 + 16$.

   (a) By the Rational Roots Theorem, only an odd integer could be a root of $P(x)$ and $Q(x)$. Since neither $R(x)$ nor $R(-x)$ have any changes in sign, by Descartes' Rule of signs, $R(x)$ has no real roots at all.

(b) No, by Descartes' Rule of Signs.

(c) $Q(x)$ has 1 change in sign and so has 1 positive real root. $Q(-x) = 3x^4 + x^2 + x - 15$ has 1 variation in sign and so $Q(x)$ has 1 negative real root. Therefore $Q(x)$ has 2 real roots.

(d) The possible rational roots of P are ± 1. Since $P(1) \neq 0$ and $P(-1) \neq 0$, then P has no rational zeros.

5. $x^4 + 2x^2 - x - 4 = 0$ has possible rational roots ± 1, ± 2, and ± 4. By Descartes' Rule of Signs, there is 1 positive root.

x	$x^4 + 2x^2 - x - 4$	
1	-2	
2	18	sign change $\Rightarrow$ root between 1 and 2
1.2	-0.2464	
1.3	0.9361	
1.25	0.316406	

Thus the root is 1.2.

6. $2x^4 - 17x^3 + 42x^2 - 25x + 4 = 0$ has possible rational roots ± 1, ± 2, ± 4 and $\pm \frac{1}{2}$. By Descartes' Rule of Signs, there 4 or 2 or 0 positive roots, and there are no negative roots.

	2	-17	42	-25	4	
1	2	-15	27	2	6	
2	2	-13	16	7	18	
4	2	-9	6	-1	0	$\Rightarrow$ $x = 4$ is a root.

So $2x^4 - 17x^3 + 42x^2 - 25x + 4 = (x-4)(2x^3 - 8x^2 + 6x - 1) = 0$

	2	-9	6	-1	
4	2	-1	2	7	
$\frac{1}{2}$	2	-8	2	0	$\Rightarrow$ $x = \frac{1}{2}$ is a root.

So $2x^4 - 17x^3 + 42x^2 - 25x + 4 = (x-4)(x-\frac{1}{2})(2x^2 - 9x + 2) = 0$. Using the quadratic formula on the last factor we have $x = \dfrac{8 \pm \sqrt{48}}{4} = 2 \pm \sqrt{3}$. Therefore, the roots are 4, $\frac{1}{2}$, $2 + \sqrt{3}$ and $2 - \sqrt{3}$.

7. $x^4 + x^3 + 5x^2 + 4x + 4 = 0$. Since $-2i$ is a root, then $2i$ is also a root. So $(x+2i)(x-2i) = x^2 + 4$ divides the polynomial evenly.

$$
\begin{array}{r|l}
 & x^2 + x + 1 \\
\hline
x^2 + 4 & x^4 + x^3 + 5x^2 + 4x + 4 \\
 & \underline{x^4 \qquad\ + 4x^2} \\
 & \quad x^3 + x^2 + 4x \\
 & \quad \underline{x^3 \qquad\ + 4x} \\
 & \qquad\quad x^2 \qquad + 4 \\
 & \qquad\quad \underline{x^2 \qquad + 4} \\
 & \qquad\qquad\qquad 0
\end{array}
$$

So $x^4 + x^3 + 5x^2 + 4x + 4 = (x^2 + 4)(x^2 + x + 1) = 0$. Solving the last factor we have $x = \dfrac{-1 \pm \sqrt{1 - 4(1)(1)}}{2} = \dfrac{-1 \pm \sqrt{3}\,i}{2}$. Thus the roots are $\pm 2i$ and $\dfrac{-1 \pm \sqrt{3}\,i}{2}$.

8. $r(x) = \dfrac{2x-1}{x^2-x-2}$, $s(x) = \dfrac{x^3+27}{x^2+4}$, $t(x) = \dfrac{x^3-9x}{x+2}$ and $u(x) = \dfrac{x^2+x-6}{x^2-25}$.

(a) As $x \to \pm\infty$, $r(x) \to 0$, so r has the horizontal asymptote $y = 0$. As $x \to \pm\infty$, $u(x) \to 1$, so u has the horizontal asymptote $y = 1$.

(b) Since the degree of the numerator of $s(x)$ is one more than the degree of the denominator of $s(x)$, then $s(x)$ has a slant asymptote.

(c) Since the denominator of s(x) is never 0, it has no vertical asymptotes.

(d) $y = u(x) = \dfrac{(x+3)(x-2)}{(x-5)(x+5)}$.

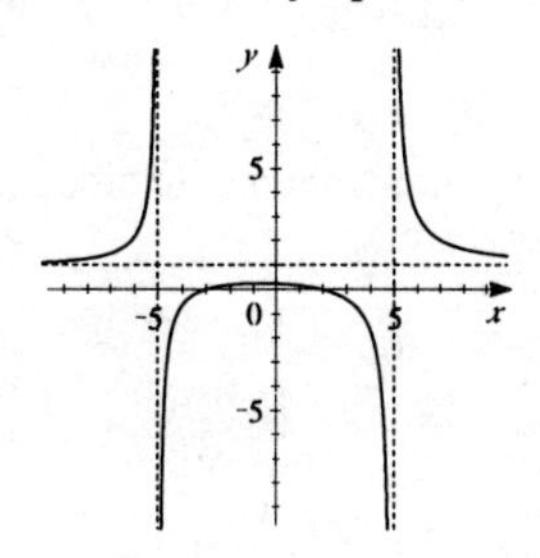

9. $P(x) = x^4 - 4x^3 + 8x$.
x-intercepts: $-1.24, 0, 2$ and 3.24.
Local maximum: $(1, 5)$
Local minima: $(-0.73, -4.00)$, $(2.73, -4.00)$.

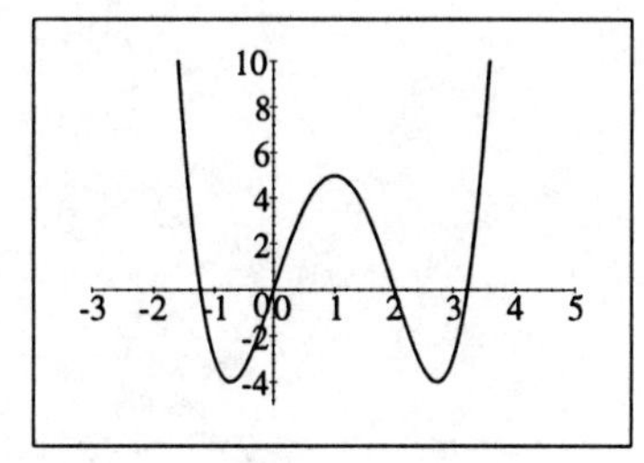

Focus on Problem Solving

1. Since there are exactly 2 vertices of odd order, the network is traversable. For example, we can choose the following path:

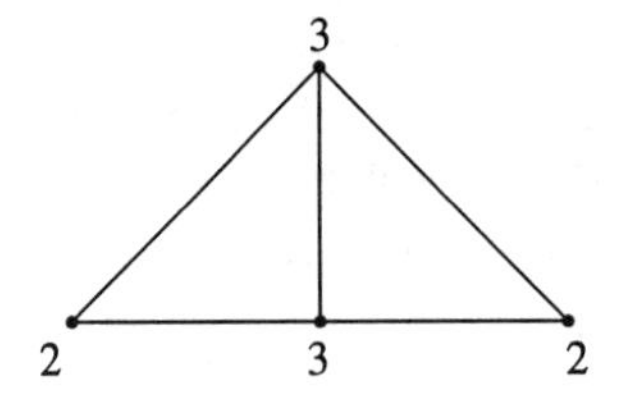

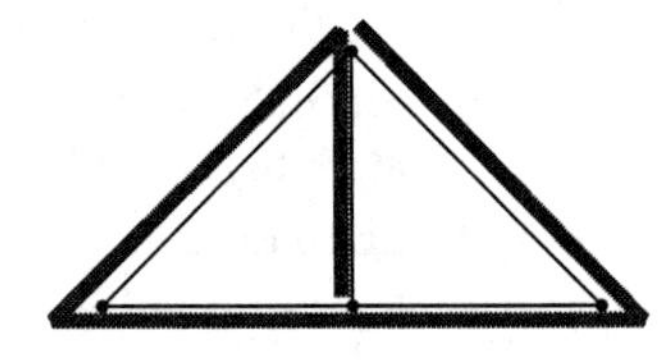

3. No, the network is not traversable since it has 4 vertices with odd order.

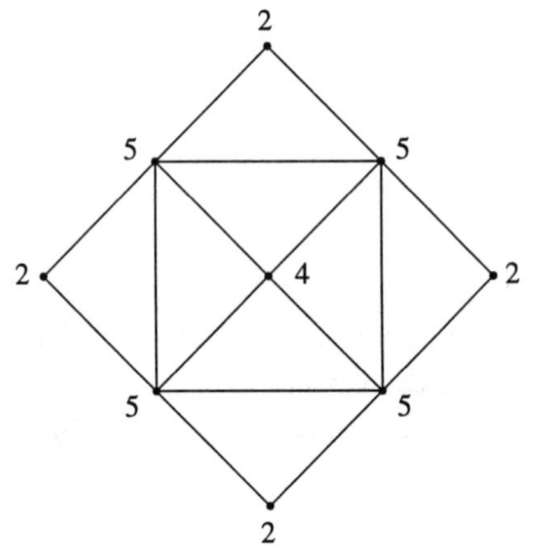

5. Since there are no vertices of odd order, the network is traversable For example, we can choose the following path:

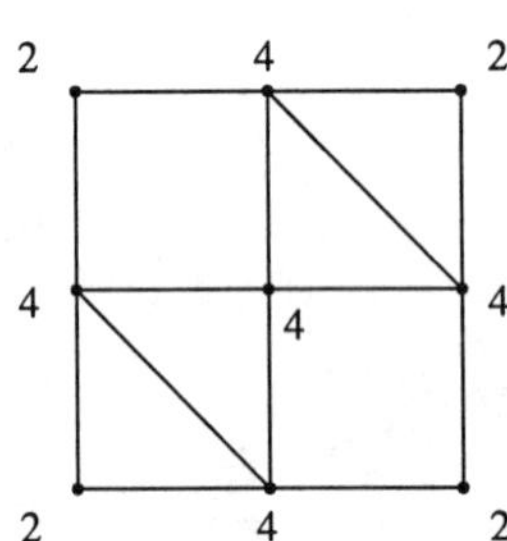

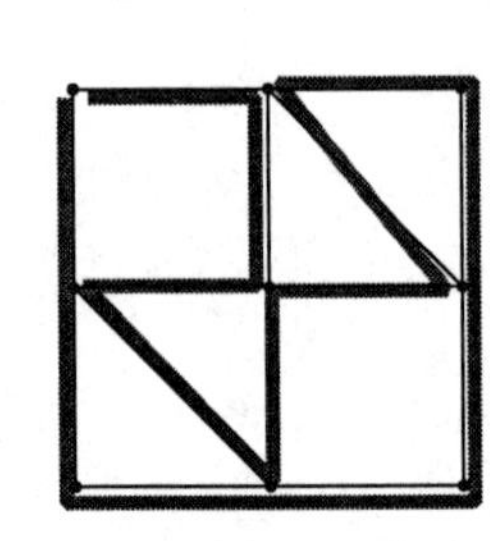

7. The diagram shows that a museum tour is possible if there are 5 rooms to a side. For an $n \times n$ museum, where n is an odd number, a similar right-left path works. If n is even, an argument similar to the one in the text (WBWB$\cdots$WB) shows that such a tour is impossible.

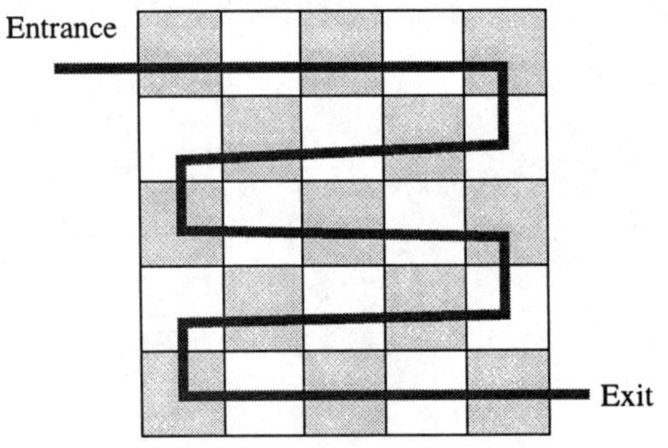

9. Consider the line joining B to A. Each time we cross the curve we move from outside to inside or from inside to outside. So we count the number of crossings in going from B to A along the line segment. If the number n of crossings is odd, A is inside; if n is even, A is outside. In the case illustrated, $n = 5$, so A is inside.

11. The equation will have four roots, so let the other root be k. Then $x^4 + ax^2 + bx + c = (x-1)(x-2)(x-3)(x-k)$. The coefficient of x^3 on the right will be $-1-2-3-k = -6-k$. But this must equal the coefficient of x^3 on the left which is 0. Therefore $k = -6$. Equating constant coefficients gives $c = (-1)(-2)(-3) \cdot 6 = -36$.

13. Color the grid as a checkerboard, that is, with squares alternating black and white. Now each domino covers two squares, one black and one white. For the dominoes to cover the grid without overlapping, an equal number of black and white squares must be on the grid. However, the grid has 32 of one color and 30 of the other color. Therefore, it is impossible to cover the grid without overlapping the dominoes.

15. (a) Let n, $n+1$, $n+2$, and $n+3$ be four consecutive integers. Thus $n(n+1)(n+2)(n+3)+1 = n^4 + 6n^3 + 11n^2 + 6n + 1$. To show that this is a perfect square we start by looking at some cases of n. For $n = 1$ we have $n(n+1)(n+2)(n+3)+1 = 1 \cdot 2 \cdot 3 \cdot 4 + 1 = 25 = 5^2$ and for $n = 2$ we have $n(n+1)(n+2)(n+3)+1 = 2 \cdot 3 \cdot 4 \cdot 5 + 1 = 121 = 11^2$. We next need to find a relation between n and the perfect square. After some experimentation we find that $5 = (1)(4) + 1$ and $11 = (2)(5) + 1$. Thus the perfect square is $[n(n+3)+1]^2 = (n^2+3n+1)^2$ $= n^4 + 6n^3 + 11n^2 + 6n + 1$. Putting these together we have $n(n+1)(n+2)(n+3)+1 = (n^2+3n+1)^2$.

    (b) Let n, $n+1$, and $n+2$ be three consecutive integers. Then the product of three consecutive integers plus the middle is $n(n+1)(n+2) + (n+1) = n^3 + 3n^2 + 2n + n + 1$ $= n^3 + 3n^2 + 3n + 1 = (n+1)^3$.

17. Label the figures as shown. The unshaded area is c^2 in the first figure and $a^2 + b^2$ in the second figure. These unshaded areas are equal since in each case they equal the area of the large square minus the areas of the four shaded triangles. Thus $a^2 + b^2 = c^2$. Finally note that a, b, and c are the legs and hypotenuse of the shaded triangle.

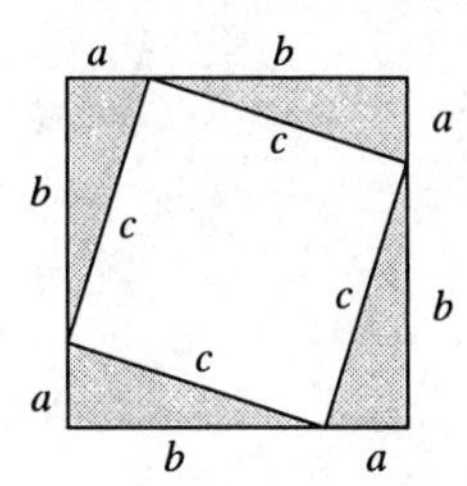

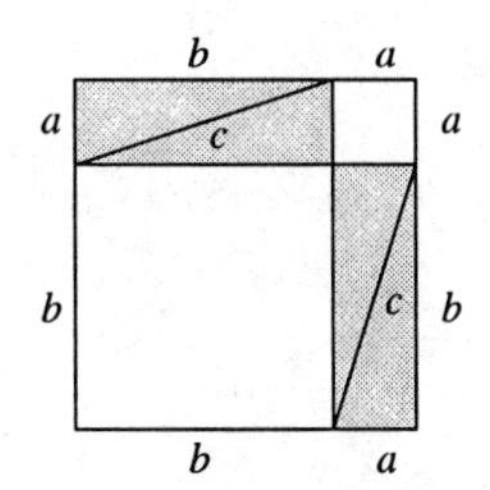

Chapter Five

Exercises 5.1

1. $f(x) = 2^x$

x	y
-4	$\frac{1}{16}$
-2	$\frac{1}{4}$
0	1
2	4
4	16

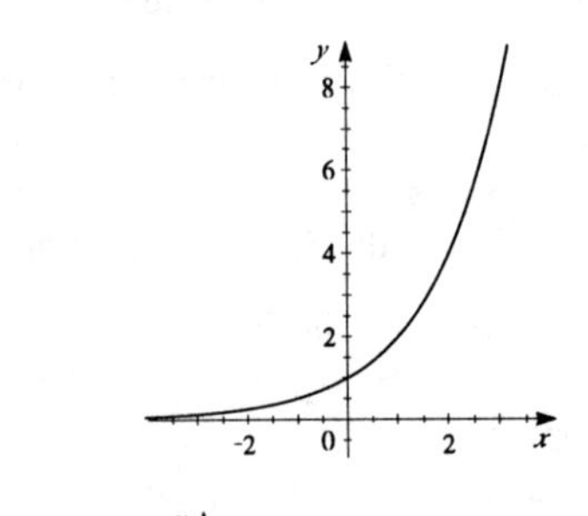

3. $f(x) = 6^x$

x	y
-2	$\frac{1}{36}$
-1	$\frac{1}{6}$
0	1
1	6
2	36

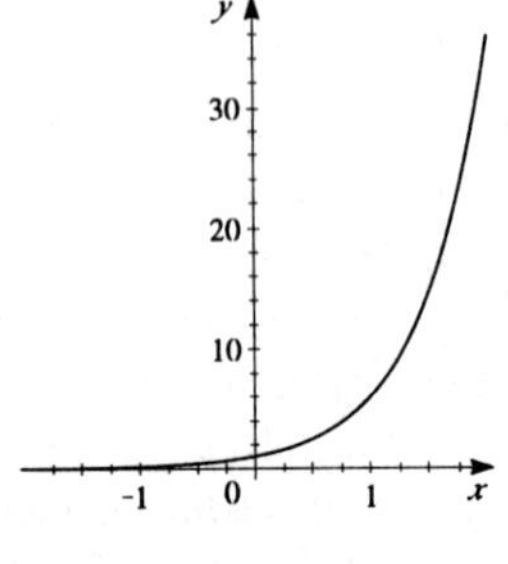

5. $f(x) = \left(\frac{1}{3}\right)^x$

x	y
-2	9
-1	3
0	1
1	$\frac{1}{3}$
2	$\frac{1}{9}$

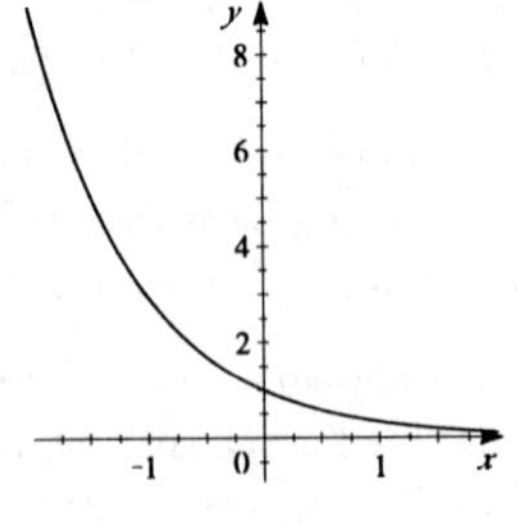

7. $g(x) = \left(\frac{1}{4}\right)^x$

x	y
-2	16
-1	4
0	1
1	$\frac{1}{4}$
2	$\frac{1}{16}$

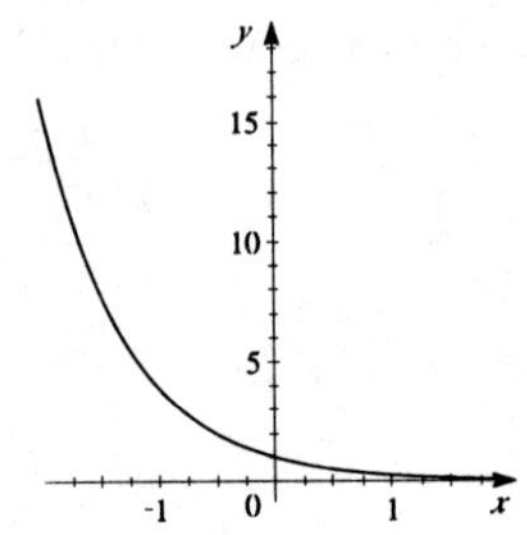

9. $y = 4^x$ and $y = 7^x$.

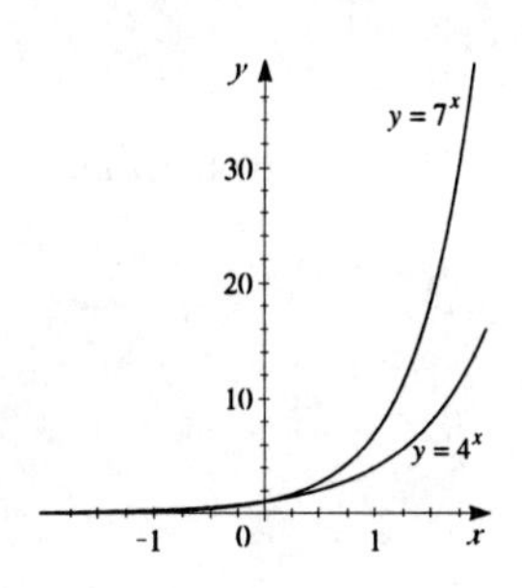

11. From the graph $f(2) = a^2 = 9$. So $a = 3$. Thus $f(x) = 3^x$

13. From the graph $f(2) = a^2 = \frac{1}{16}$. So $a = \frac{1}{4}$. Thus $f(x) = \left(\frac{1}{4}\right)^x$

15. III

17. I

19. II

21. $f(x) = -3^x$
The graph of f is obtained by reflecting the graph of $y = 3^x$ about the x-axis.
Domain: $(-\infty, \infty)$
Range: $(-\infty, 0)$
Asymptote: $y = 0$

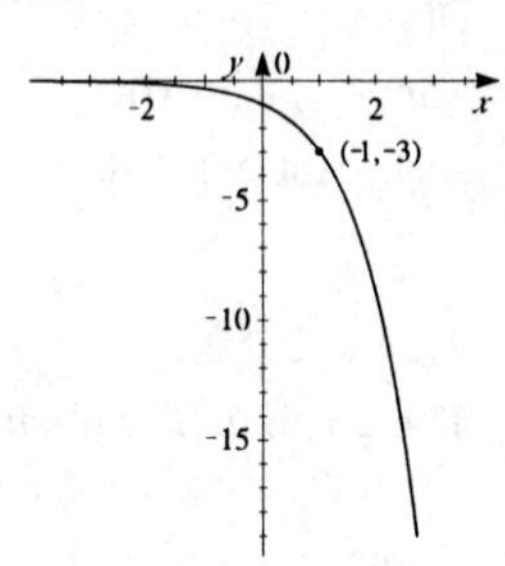

23. $g(x) = 2^x - 3$
The graph of g is obtained by shifting the graph of $y = 2^x$ downward 3 units.
Domain: $(-\infty, \infty)$
Range: $(-3, \infty)$
Asymptote: $y = -3$

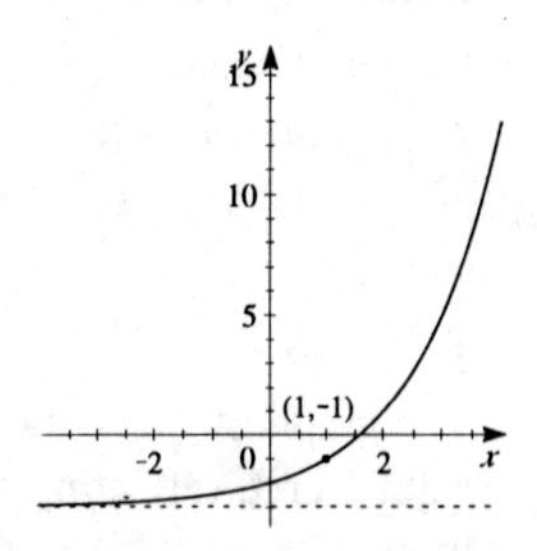

25. $h(x) = 4 + \left(\frac{1}{2}\right)^x$
The graph of h is obtained by shifting the graph of $y = \left(\frac{1}{2}\right)^x$ upward 4 units.
Domain: $(-\infty, \infty)$
Range: $(4, \infty)$
Asymptote: $y = 4$

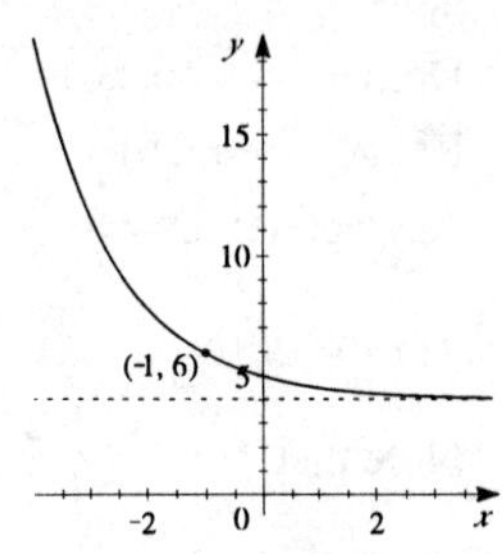

27. $f(x) = 10^{x+3}$

The graph of f is obtained by shifting the graph of $y = 10^x$ to the left 3 units.

Domain: $(-\infty, \infty)$

Range: $(0, \infty)$

Asymptote: $y = 0$

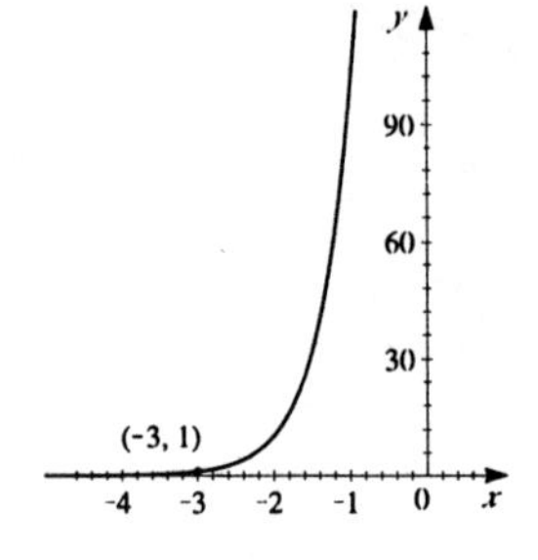

29. $f(x) = -3^{-x}$

The graph of f is obtained by reflecting the graph of $y = 3^x$ about the y-axis and then reflecting about the x-axis.

Domain: $(-\infty, \infty)$

Range: $(-\infty, 0)$

Asymptote: $y = 0$

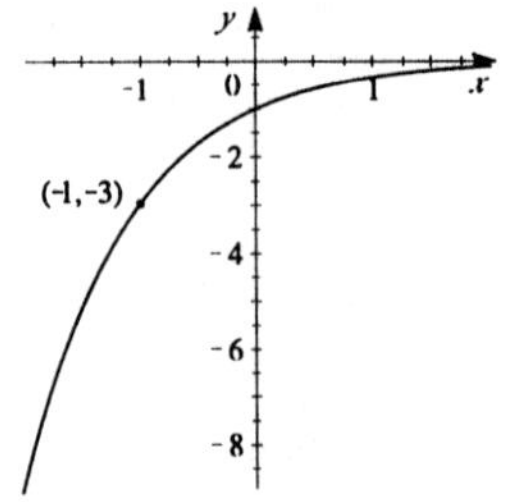

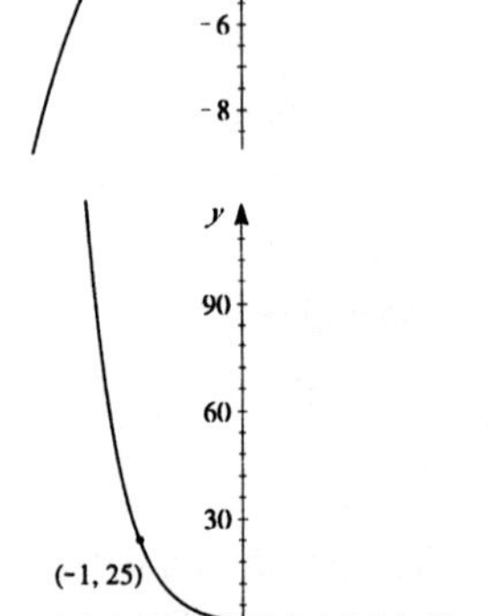

31. $f(x) = 5^{-2x}$

The graph of f is obtained by reflecting the graph of $y = 5^x$ about the x-axis and by shrinking it horizontally by a factor of 2.

Domain: $(-\infty, \infty)$

Range: $(0, \infty)$

Asymptote: $y = 0$

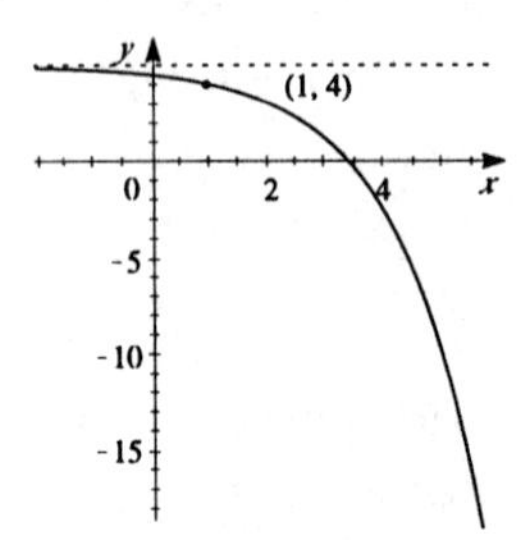

33. $f(x) = 5 - 2^{x-1}$

The graph of f is obtained by shifting the graph of $y = 2^x$ to the right 1 unit, reflecting about the x-axis, then shifting upward 5 units.

Domain: $(-\infty, \infty)$

Range: $(-\infty, 5)$

Asymptote: $y = 5$

35. $f(x) = 2^{|x|}$

Note that $f(x) = \begin{cases} 2^x & x \geq 0 \\ 2^{-x} & x < 0. \end{cases}$

So, for $x \geq 0$ the graph of f is just the graph of $y = 2^x$; for $x < 0$ the graph of f is obtained by reflecting the graph of $y = 2^x$ about the y-axis.

Domain: $(-\infty, \infty)$

Range: $[1, \infty)$

Asymptote: None

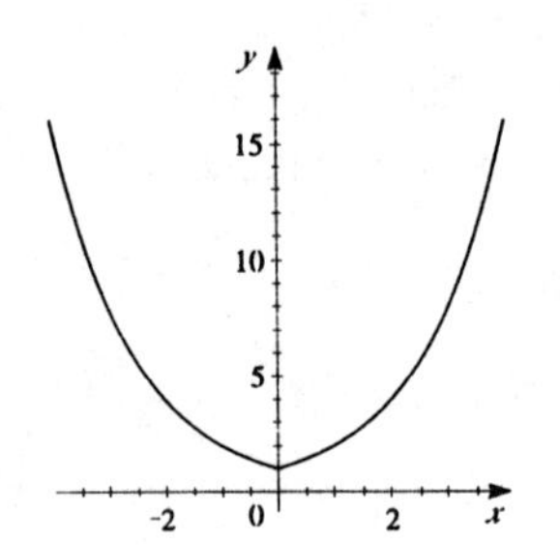

37. Using the points $(0, 3)$ and $(2, 12)$ we have $f(0) = Ca^0 = 3 \quad \Leftrightarrow \quad C = 3$. $f(2) = 3a^2 = 12$ $\Leftrightarrow \quad a^2 = 4 \quad \Leftrightarrow \quad a = 2$ (recall that for an exponential function $f(x) = a^x$ we require $a > 0$). Thus $f(x) = 3 \cdot 2^x$

39. (a)

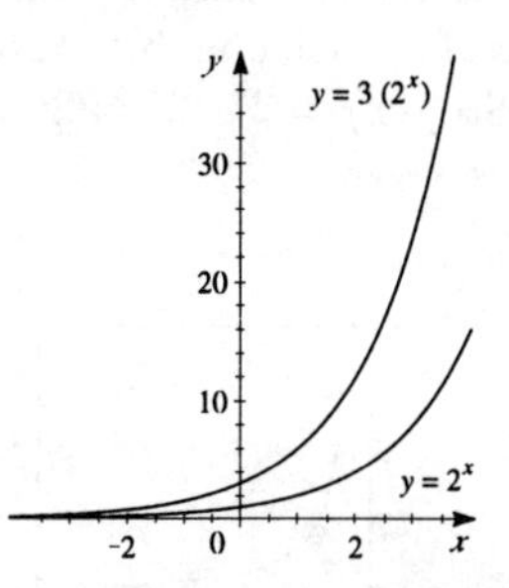

(b) Since $g(x) = 3(2^x) = 3f(x)$ and $f(x) > 0$ the height of the graph of $g(x)$ is always three times the height of the graph of $f(x) = 2^x$ above the x-axis.

41. $f(x) = 10^x$. $\dfrac{f(x+h) - f(x)}{h} = \dfrac{10^{x+h} - 10^x}{h} = \dfrac{10^x \cdot 10^h - 10^x}{h} = 10^x \dfrac{10^h - 1}{h}$

43. Calculating the pay for method (ii), we have $pay = 2 + 2^2 + 2^3 + \cdots + 2^{30} > 2^{30}$ cents $= \$10,737,418.24$. Since this is much more than method (i), method (ii) is more profitable.

45. (a) From the graphs in (i) - (iii) below we see that the graph of f ultimately increases much more quickly than the graph of g.

(i) $[0, 5]$ by $[0, 20]$

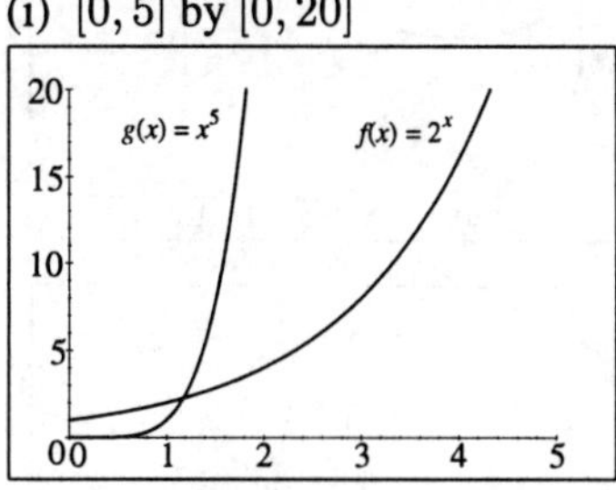

(ii) $[0, 20]$ by $[0, 10^7]$

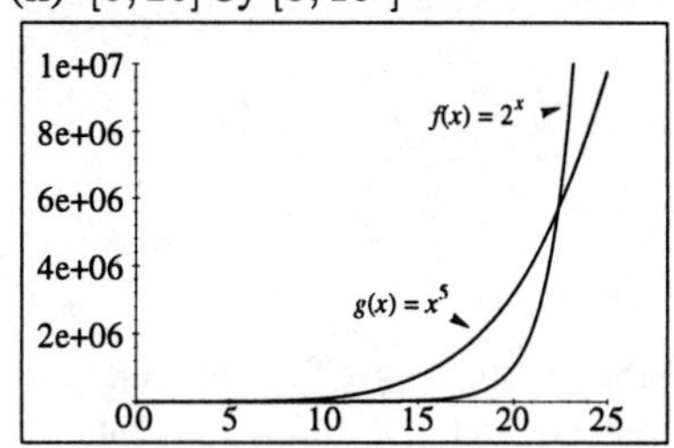

(iii) $[0, 5]$ by $[0, 10^8]$

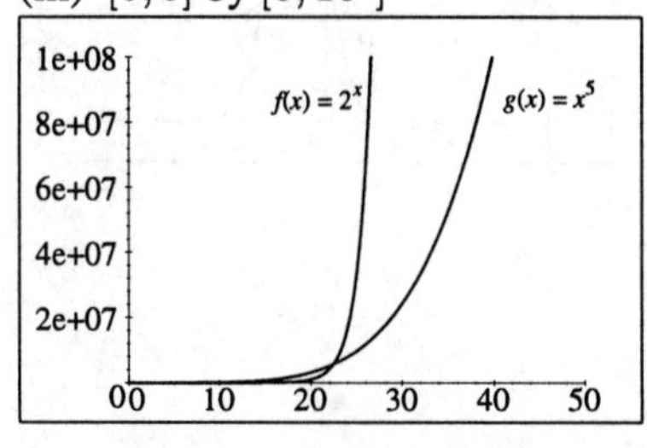

(b) From the graphs in parts (i) and (ii), we see that the approximate solutions are $x \approx 1.2$ and $x \approx 22.4$.

47.

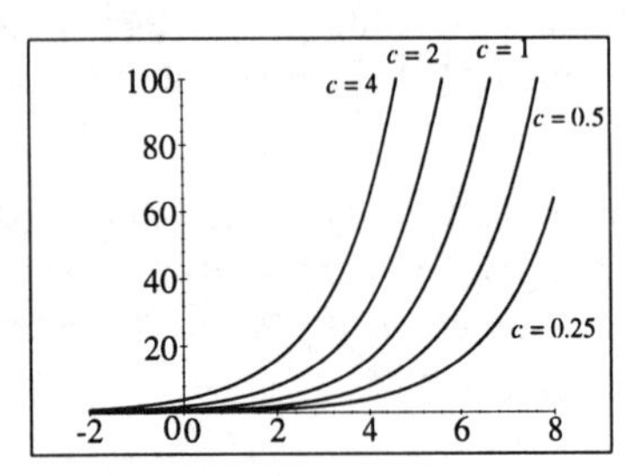

The larger the value of c the more rapidly the graph of $f(x) = c\,2^x$ increases. Also, some students might notice that the graphs are just shifted horizontally 1 unit. This is because of our choice of c; each c in this exercise is of the form 2^k. So $f(x) = 2^k \cdot 2^x = 2^{x+k}$.

49. $y = 2^{1/x}$

Vertical Asymptote: $x = 0$

Horizontal Asymptote: $y = 1$

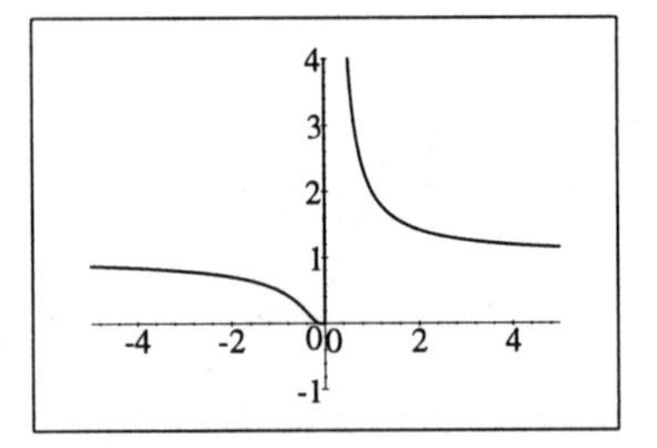

51. $g(x) = x^x$. $g(x)$ is only defined for $x \geq 0$. Shown is the viewing rectangle $[0, 1.5]$ by $[0, 1.5]$. From the graphs, we see that there is a local minimum ≈ 0.69 when $x \approx 0.37$.

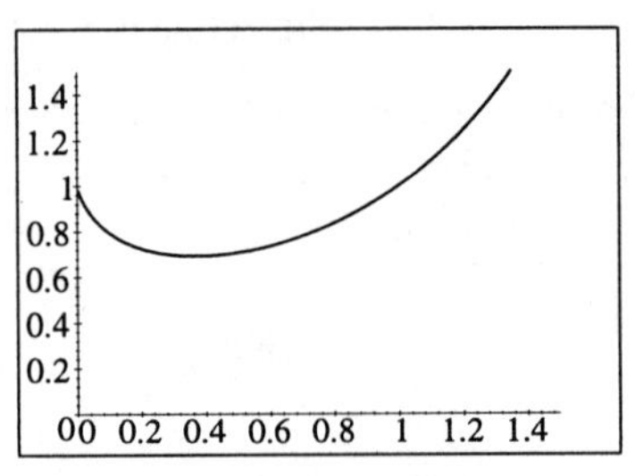

53. $y = 10^{x-x^2}$

(a) From the graph, we see that the function is increasing on $(-\infty, 0.50]$ and that it is decreasing on $[0.50, \infty)$.

(b) From the graph, we see that the range is approximately $(0, 1.78]$.

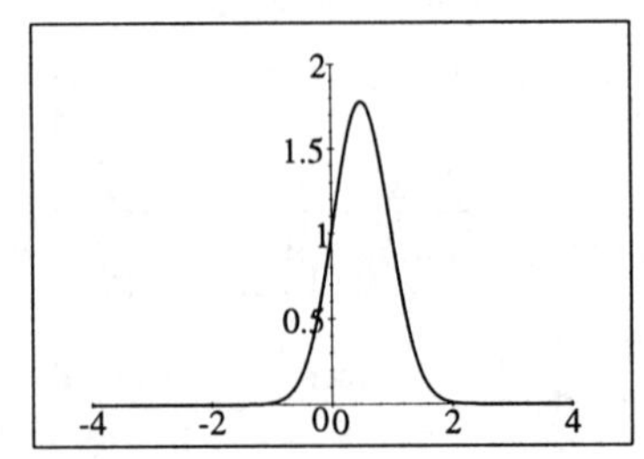

Exercises 5.2

1. $f(x) = 3e^x$

x	$f(x) = 3e^x$
-2	0.406006
-1.5	0.66939
-1	1.10364
-0.5	1.81959
0	3
0.5	4.94616
1	8.15485
1.5	13.4451
2	22.1672

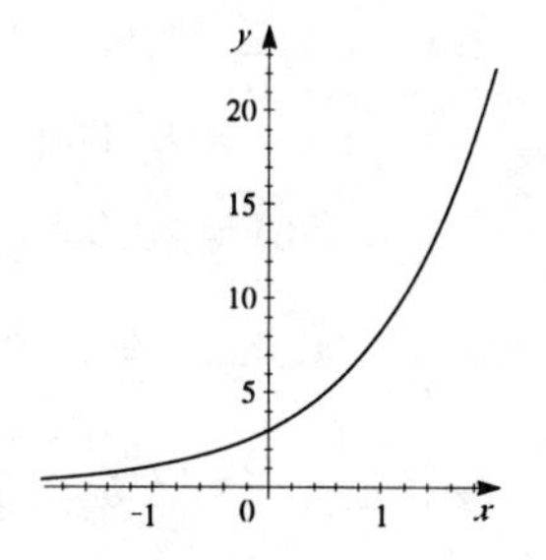

3. $y = -e^x$

The graph of $y = -e^x$ is obtained from the graph of $y = e^x$ by reflecting it about the x-axis.

Domain: $(-\infty, \infty)$

Range: $(-\infty, 0)$

Asymptote: $y = 0$

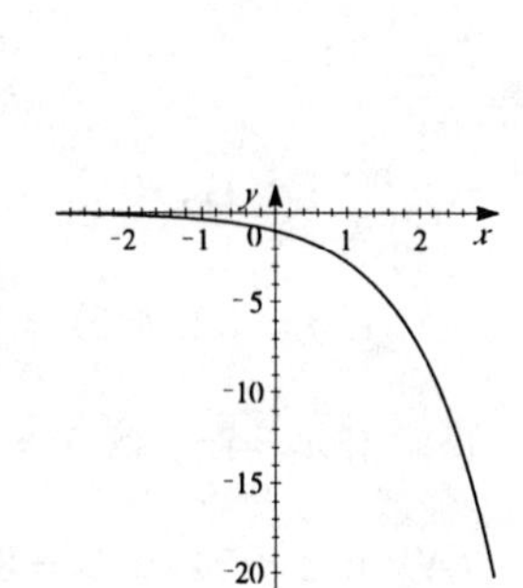

5. $y = e^{-x} - 1$

The graph of $y = e^{-x} - 1$ is obtained from the graph of $y = e^x$ by reflecting it about the y-axis then shifting downward 1 unit.

Domain: $(-\infty, \infty)$

Range: $(-1, \infty)$

Asymptote: $y = -1$

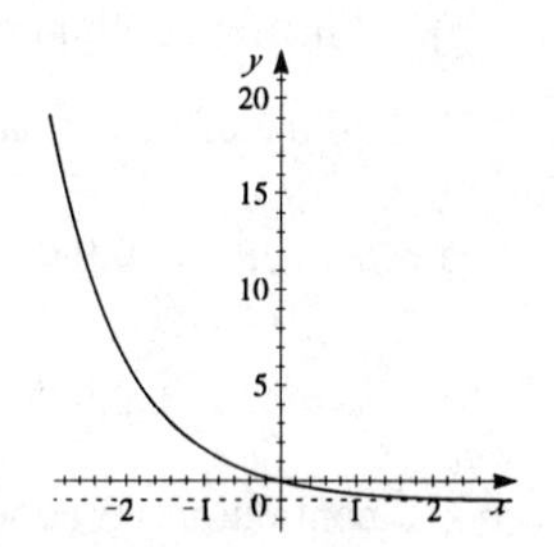

7. $y = e^{x-2}$

The graph of $y = e^{x-2}$ is obtained from the graph of $y = e^x$ by shifting it to the right 2 units.

Domain: $(-\infty, \infty)$

Range: $(0, \infty)$

Asymptote: $y = 0$

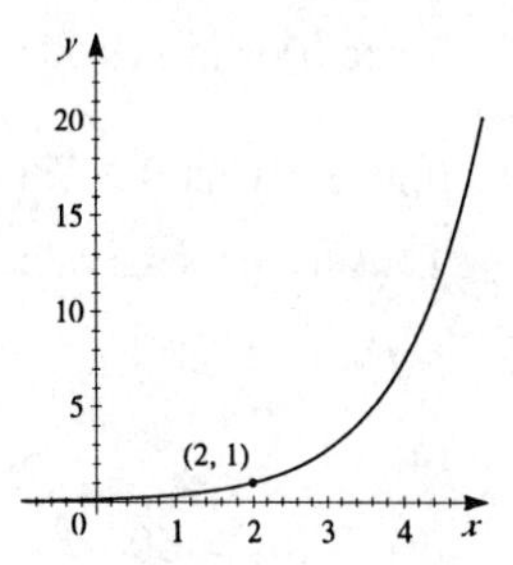

9. $P = 10,000$, $r = 0.10$, and $n = 2$. Then $A(t) = 10{,}000\left(1 + \frac{0.10}{2}\right)^{2t} = 10{,}000 \cdot 1.05^{2t}$.

(a) $A(5) = 10,000 \cdot 1.05^{10} \approx 16{,}288.95$ and so the value of the investment is \$16,288.95.

(b) $A(10) = 10,000 \cdot 1.05^{20} \approx 26{,}532.98$. So the value of the investment is \$26,532.98.

(c) $A(15) = 10,000 \cdot 1.05^{30} \approx 43{,}219.42$. So the value of the investment is \$43,219.42.

11. $P = 3000$ and $r = 0.09$. Then $A(t) = 3000\left(1 + \frac{0.09}{n}\right)^{nt}$, and so $A(5) = 3000\left(1 + \frac{0.09}{n}\right)^{5n}$.

(a) If $n = 1$, $A(5) = 3000\left(1 + \frac{0.09}{1}\right)^{5} = 3000 \cdot 1.09^5 \approx \$4{,}615.87$.

(b) If $n = 2$, $A(5) = 3000\left(1 + \frac{0.09}{2}\right)^{10} = 3000 \cdot 1.045^{10} \approx \$4{,}658.91$.

(c) If $n = 12$, $A(5) = 3000\left(1 + \frac{0.09}{12}\right)^{60} = 3000 \cdot 1.0075^{60} \approx \$4{,}697.04$.

(d) If $n = 52$, $A(5) = 3000\left(1 + \frac{0.09}{52}\right)^{260} \approx \$4{,}703.11$.

(e) If $n = 365$, $A(5) = 3000(1 + \frac{0.09}{365})^{1825} \approx \$4{,}704.68$.

(f) If $n = 24 \cdot 365 = 8760$, $A(5) = 3000\left(1 + \frac{0.09}{8760}\right)^{43800} \approx \$4{,}704.93$.

(g) If interest is compounded continuously, $A(5) = 3000 \cdot e^{0.45} \approx \$4{,}704.94$.

13. We find the effective rate, with $P = 1$, and $t = 1$. So $A = (1 + \frac{r}{n})^n$

(i) $n = 2$, $r = 0.085$; $A = \left(1 + \frac{0.085}{2}\right)^2 = (1.0425)^2 \approx 1.0868$.

(ii) $n = 4$, $r = 0.0825$; $A = \left(1 + \frac{0.0825}{4}\right)^4 = (1.020625)^4 \approx 1.0851$

(iii) continuous compounding, $r = 0.08$; $A = e^{0.08} = 1.0833$.

Since (i) is larger than the others, the best investment is the 8.5% compounded semiannually.

15. Find P. $10000 = P\left(1 + \frac{0.09}{2}\right)^{2(3)} = P(1.045)^6 \quad \Leftrightarrow \quad 10000 = 1.3023P \quad \Leftrightarrow \quad P = 7678.96$.
Thus the present value is \$7,678.96.

17. $n(t) = 500\,e^{0.45t}$

(a) 45%

(b) $n(0) = 500\,e^0 = 500$. Thus the initial population was 500 bacteria.

(c) $n(5) = 500\,e^{0.45(5)} = 500\,e^{2.25} \approx 4744$. Thus the culture will contain 4744 bacteria at $t = 5$.

19. (a) $r = 0.08$ and $n(0) = 18000$. Thus the population is given by the formula $n(t) = 18000\,e^{0.08t}$.

(b) $t = 2000 - 1992 = 8$. Thus $n(8) = 18000\,e^{0.08(8)} = 18000\,e^{0.64} \approx 34,137$. Thus there should be 34, 137 foxes in the region by the year 2000.

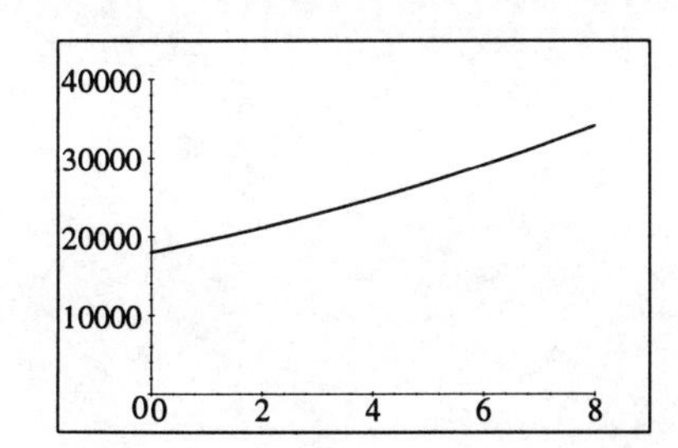

21. $n(t) = n_0\, e^{rt}$; $n_0 = 110$ million, $t = 2020 - 1995 = 25$

(a) $r = 0.03$; $n(25) = 110,000,000\, e^{0.03(25)} = 110,000,000\, e^{0.75} \approx 232,870,000$. Thus at a 3% growth rate, the projected population will be approximately 223 million people by 2020.

(b) $r = 0.02$; $n(25) = 110,000,000\, e^{0.02(25)} = 110,000,000\, e^{0.50} \approx 181,359,340$. Thus at a 2% growth rate, the projected population will be approximately 181 million people by 2020.

23. $n(t) = n_0\, e^{rt}$; $n_0 = 5$ billion, $t = 1995 - 1987 = 8$, $r = 0.02$. Thus

$n(8) = (5 \text{ billion})e^{0.02(8)} \approx 5.87$ billion .

25. $r = 2.20$ so $n(t) = n_0\, e^{2.2t}$

(a) $n(2) = n_0\, e^{2.2(2)} \approx 40000 \quad \Leftrightarrow \quad n_0\,(81.45) \approx 40000 \quad \Leftrightarrow \quad n_0 \approx 491$. Thus about 500 bacteria were initially introduced into the food.

(b) $n(3) = 491\, e^{2.2(3)} = 491\, e^{6.6} \approx 360,932$. Then there will be approximately $361,000$ bacteria in 3 hours.

27. (a) $m(0) = 6\, e^{-0.087(0)} \approx 6$ grams

(b) $m(20) = 6\, e^{-0.087(20)} \approx 6(0.1755) = 1.053$. Thus approximately 1 gram of radioactive iodine remains after 20 days.

29. (a) $Q(5) = 15\left(1 - e^{-0.04(5)}\right) \approx 15(0.1813)$ $= 2.7345$. Thus approximately 2.7 lb of salt are in the barrel after 5 minutes.

(b) $Q(10) = 15\left(1 - e^{-0.04(10)}\right) \approx 15(0.3297)$ $= 4.946$. Thus approximately 4.9 lb of salt are in the barrel after 10 minutes.

(c)

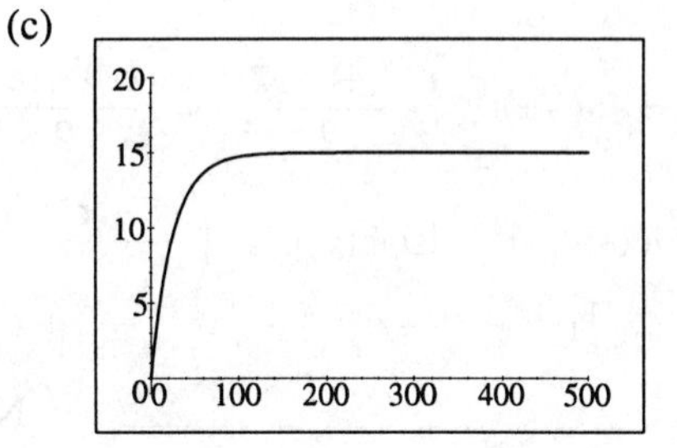

(d) 15. Yes, since 50 gal $\times$ 0.3 lb/gal $=$ 15 lb

31. (a) Substituting $n_0 = 50$ and $t = 12$, we have $n(12) = \dfrac{300}{0.05 + \left(\frac{300}{50} - 0.05\right)e^{-0.55(12)}} \approx 5164$

(b)

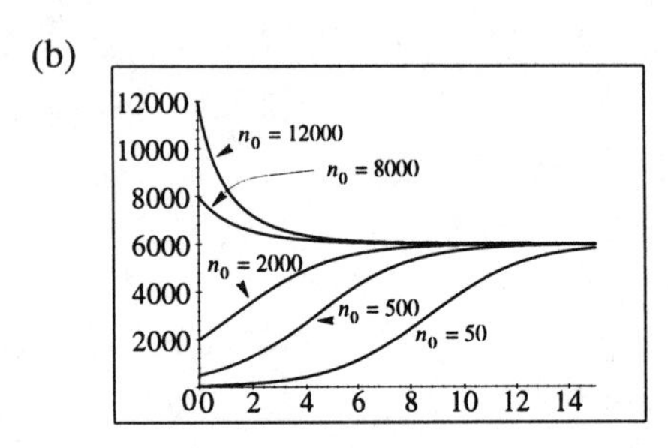

(c) 6000 rabbits.

33.

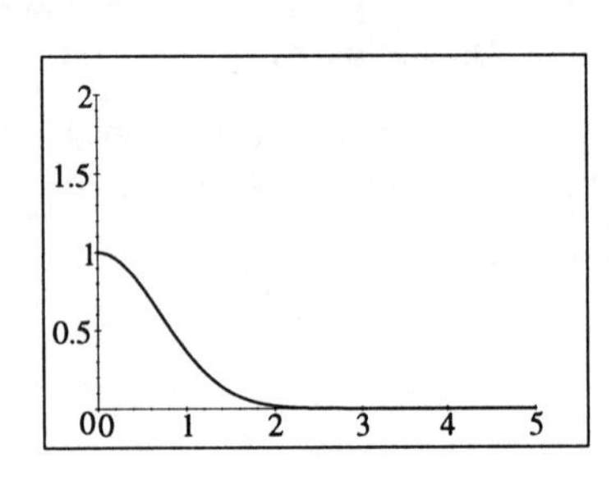

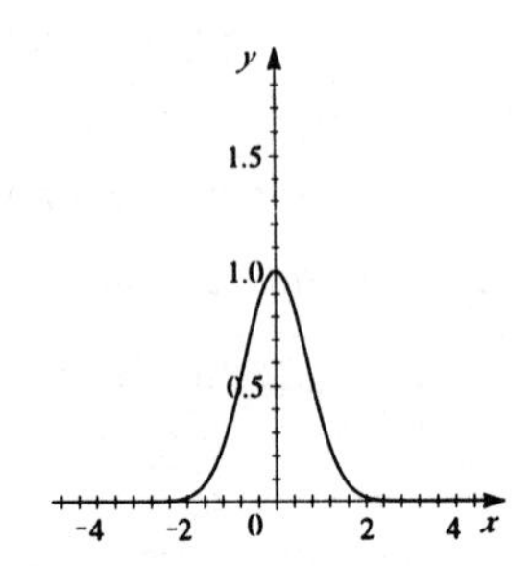

35.

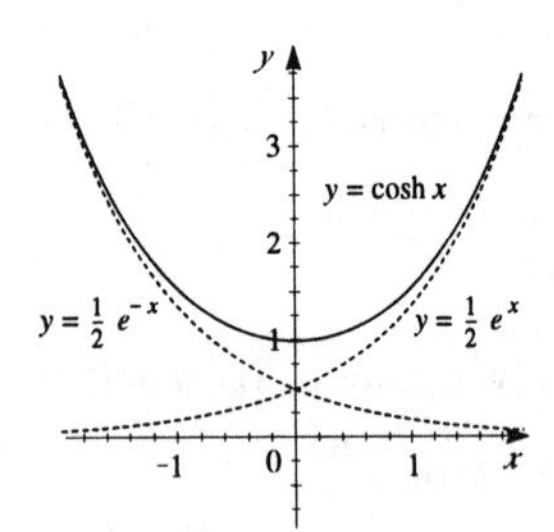

37. $\cosh(-x) = \dfrac{e^{-x} + e^{-(-x)}}{2} = \dfrac{e^{-x} + e^{x}}{2} = \dfrac{e^{x} + e^{-x}}{2} = \cosh(x)$

39. $[\cosh(x)]^2 - [\sinh(x)]^2 = \left(\dfrac{e^x + e^{-x}}{2}\right)^2 - \left(\dfrac{e^x - e^{-x}}{2}\right)^2$
$= \frac{1}{4}(e^{2x} + 2 + e^{-2x}) - \frac{1}{4}(e^{2x} - 2 + e^{-2x}) = \frac{2}{4} + \frac{2}{4} = 1.$

41.

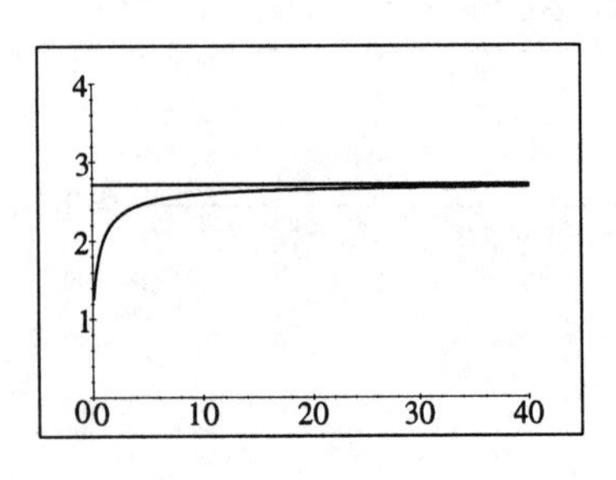

Note from the graph that $y = \left(1 + \frac{1}{x}\right)^x$ approaches e as x get large.

43. (a)

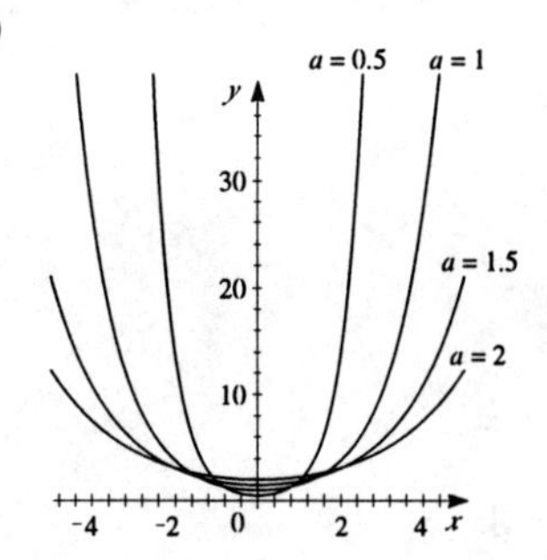

(b) As a increases the curve $y = \frac{a}{2}\left(e^{x/a} + e^{-x/a}\right)$ flattens out and the y intercept increases.

45. $f(x) = x\,e^{-x}$

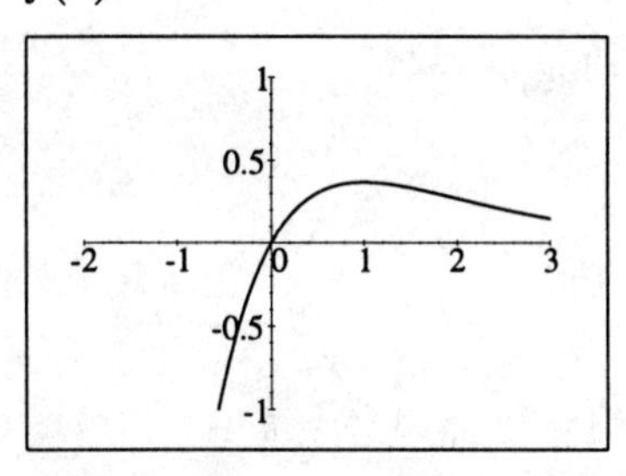

Shown is the viewing rectangle $[-2, 3]$ by $[-1, 1]$. From the graph, we see that there is a local maximum ≈ 0.37 when $x \approx 1.00$.

Exercises 5.3

1. (a) $2^5 = 32$ (b) $5^0 = 1$

3. (a) $4^{1/2} = 2$ (b) $2^{-4} = \frac{1}{16}$

5. (a) $e^x = 5$ (b) $e^5 = y$

7. (a) $\log_2 8 = 3$ (b) $\log_{10} 0.001 = -3$

9. (a) $\log_4 0.125 = -\frac{3}{2}$ (b) $\log_7 343 = 3$

11. (a) $\ln 2 = x$ (b) $\ln y = 3$

13. (a) $\log_5 5^4 = 4$ (b) $\log_4 64 = \log_4 4^3 = 3$
    (c) $\log_9 9 = 1$

15. (a) $\log_8 64 = \log_8 8^2 = 2$ (b) $\log_7 49 = \log_7 7^2 = 2$
    (c) $\log_7 7^{10} = 10$

17. (a) $\log_3 \left(\frac{1}{27}\right) = \log_3 3^{-3} = -3$ (b) $\log_{10} \sqrt{10} = \log_{10} 10^{1/2} = \frac{1}{2}$
    (c) $\log_5 0.2 = \log_5 \left(\frac{1}{5}\right) = \log_5 5^{-1} = -1$

19. (a) $2^{\log_2 37} = 37$ (b) $3^{\log_3 8} = 8$
    (c) $e^{\ln \sqrt{5}} = \sqrt{5}$

21. (a) $\log_8 0.25 = \log_8 8^{-2/3} = -\frac{2}{3}$ (b) $\ln e^4 = 4$
    (c) $\ln \left(\frac{1}{e}\right) = \ln e^{-1} = -1$

23. (a) $\log_2 x = 5 \quad \Leftrightarrow \quad x = 2^5 = 32$ (b) $x = \log_2 16 = \log_2 2^4 = 4$

25. (a) $\log_{10} x = 2 \quad \Leftrightarrow \quad x = 10^2 = 100$ (b) $\log_5 x = 2 \quad \Leftrightarrow \quad x = 5^2 = 25$

27. (a) $\log_x 16 = 4 \quad \Leftrightarrow \quad x^4 = 16 \quad \Leftrightarrow \quad x = 2$
    (b) $\log_x 8 = \frac{3}{2} \quad \Leftrightarrow \quad x^{3/2} = 8 \quad \Leftrightarrow \quad x = 8^{2/3} = 4$

29. (a) $\log 2 \approx 0.3010$ (b) $\log 35.2 \approx 1.5465$
    (c) $\log\left(\frac{2}{3}\right) \approx -0.1761$

31. (a) $\ln 5 \approx 1.6094$ (b) $\ln 25.3 \approx 3.2308$
    (c) $\ln\left(1 + \sqrt{3}\right) \approx 1.0051$

33. Since the point $(5, 1)$ is on the graph, we have $1 = \log_a 5 \quad \Leftrightarrow \quad a^1 = 5$. Thus the function is $y = \log_5 x$.

35. Since the point $\left(3, \frac{1}{2}\right)$ is on the graph, we have $\frac{1}{2} = \log_a 3 \quad \Leftrightarrow \quad a^{1/2} = 3 \quad \Leftrightarrow \quad a = 9$. Thus the function is $y = \log_9 x$.

37. II

39. III

41. VI

43. The graph of $y = \log_4 x$ is obtained from the graph of $y = 4^x$ by reflecting it about the line $y = x$.

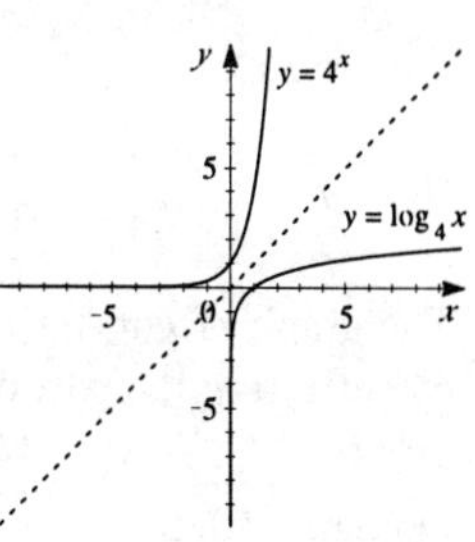

45. $f(x) = \log_2(x - 4)$

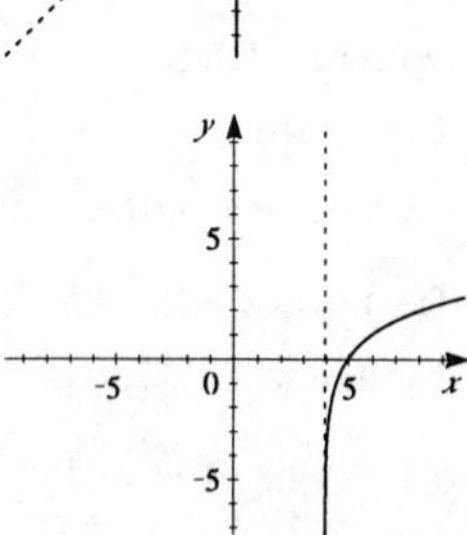

The graph of f is obtained from the graph of $y = \log_2 x$ by shifting it right 4 units.
Domain: $(4, \infty)$
Range: $(-\infty, \infty)$
Vertical asymptote: $x = 4$

47. 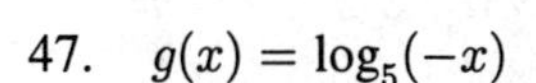
$g(x) = \log_5(-x)$

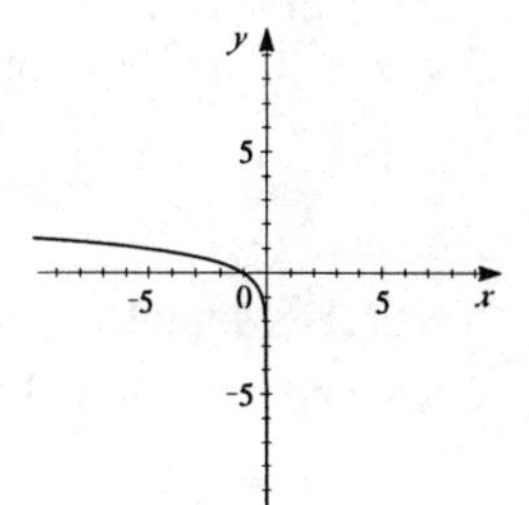

The graph of g is obtained from the graph of $y = \log_5 x$ by reflecting it about the y-axis.
Domain: $(-\infty, 0)$
Range: $(-\infty, \infty)$
Vertical asymptote: $x = 0$

49. 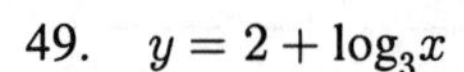
$y = 2 + \log_3 x$

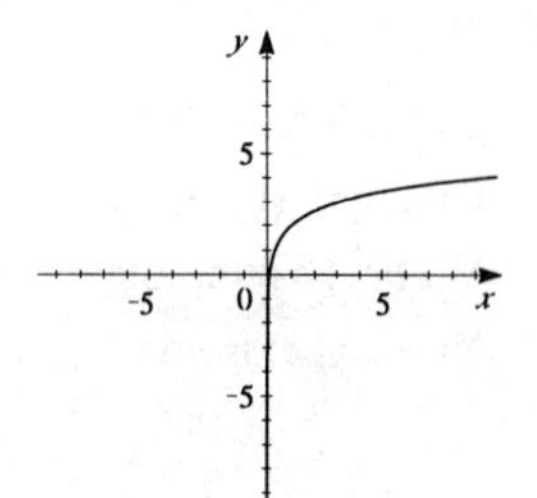

The graph of $y = 2 + \log_3 x$ is obtained from the graph of $y = \log_3 x$ by shifting it upward 2 units.
Domain: $(0, \infty)$
Range: $(-\infty, \infty)$
Vertical asymptote: $x = 0$

51. 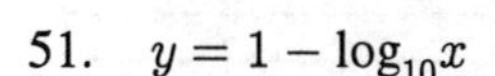
$y = 1 - \log_{10} x$

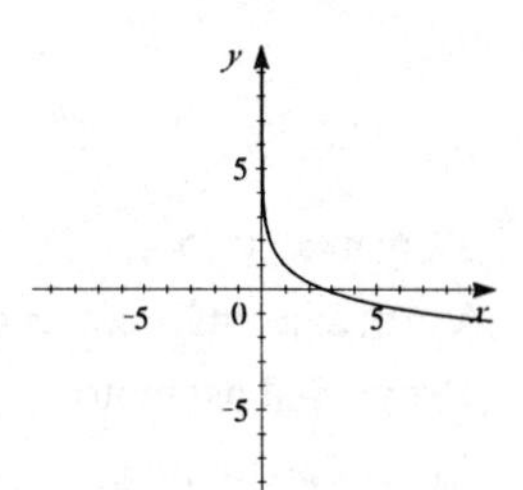

The graph of $y = 1 - \log_{10} x$ is obtained from the graph of $y = \log_{10} x$ by reflecting it about the x-axis, then shifting it upward 1 unit.
Domain: $(0, \infty)$
Range: $(-\infty, \infty)$
Vertical asymptote: $x = 0$

53. $y = |\ln x|$

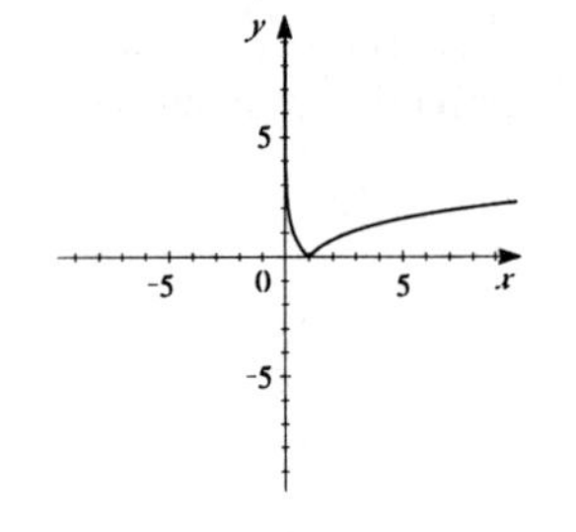

Note $y = \begin{cases} \ln x & x \geq 1 \\ -\ln x & 0 < x < 1. \end{cases}$

The graph of $y = |\ln x|$ is obtained from the graph of $y = \ln x$ by reflecting the part of the graph for $0 < x < 1$ about the x-axis.

Domain: $(0, \infty)$

Range: $[0, \infty)$

Vertical asymptote: $x = 0$

55. $f(x) = \log_{10}(2 + 5x)$. We require that $2 + 5x > 0 \quad \Leftrightarrow \quad 5x > -2 \quad \Leftrightarrow \quad x > -\frac{2}{5}$. So the domain is $\left(-\frac{2}{5}, \infty\right)$.

57. $g(x) = \log_3(x^2 - 1)$. We require that $x^2 - 1 > 0 \quad \Leftrightarrow \quad x^2 > 1 \quad \Rightarrow \quad x < -1$ or $x > 1$. So the domain is $(-\infty, -1) \cup (1, \infty)$.

59. $h(x) = \ln x + \ln(2 - x)$. We require that $x > 0$ and $2 - x > 0 \quad \Leftrightarrow \quad x > 0$ and $x < 2 \quad \Leftrightarrow$ $0 < x < 2$. So the domain is $(0, 2)$.

61. $y = \log_{10}(1 - x^2)$

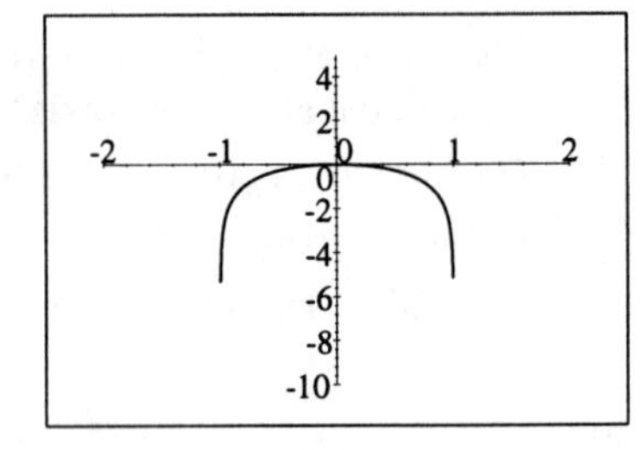

Domain: $(-1, 1)$

Vertical asymptote: $x = -1$ and $x = 1$

Local maximum $y = 0$ at $x = 0$

63. $y = x + \ln x$

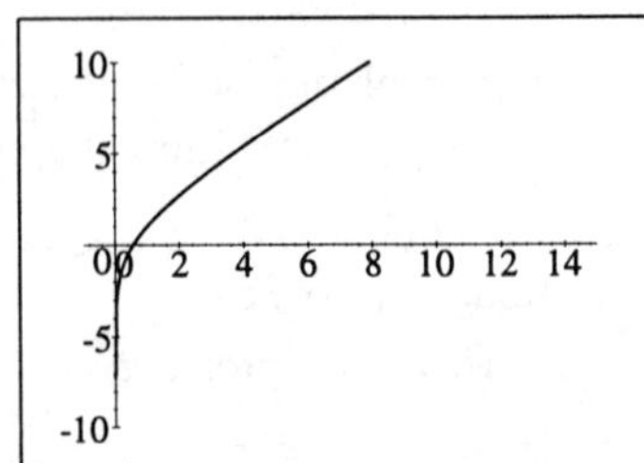

Domain: $(0, \infty)$

Vertical asymptote: $x = 0$

No local extrema.

65. $y = \dfrac{\ln x}{x}$

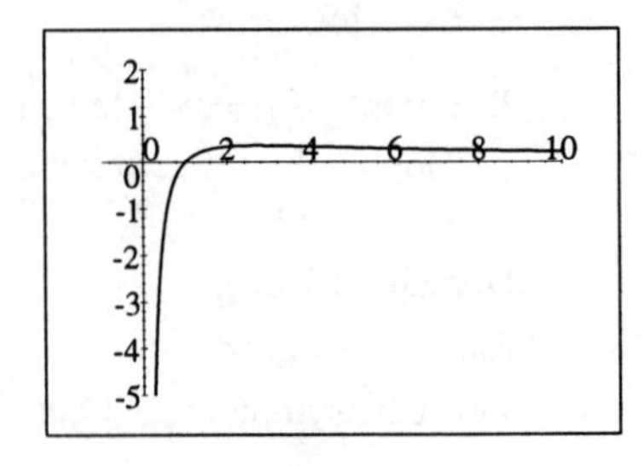

Domain: $(0, \infty)$

Vertical asymptote: $x = 0$

Horizontal asymptote: $y = 0$

Local maximum $y \approx 0.37$ at $x \approx 2.72$

67. $g(x) = \sqrt{x}$ grows faster than $f(x) = \ln x$ for $x > 1$.

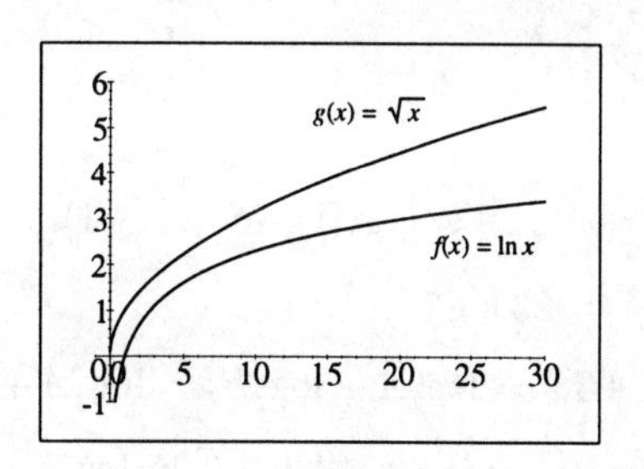

69. (a)

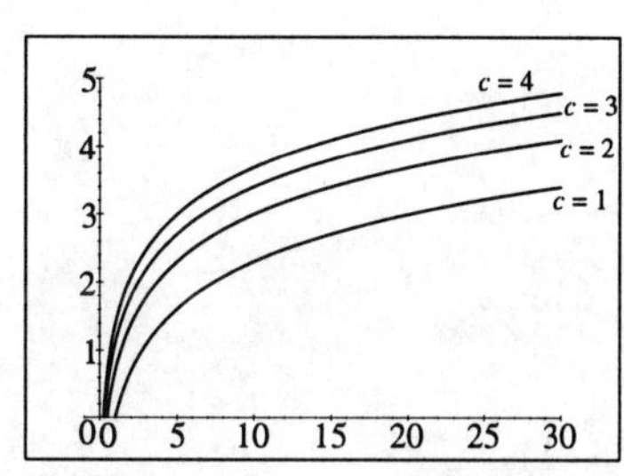

(b) Since $f(x) = \log(cx) = \log c + \log x$, as c increases the graph of $f(x) = \log(cx)$ is shifted upward $\log c$ units.

71. (a) Since 2 feet $= 24$ inches, the height of the graph is $2^{24} = 1677216$ inches. Now, since there are 12 inches per foot and 5280 feet per mile, there are $12(5280) = 63360$ inches per mile. So the height of the graph is $\frac{1677216}{63360} \approx 264.8$, or about 265 miles.

(b) Since $\log_2(264.8 \text{ miles}) \approx \log_2(2^{24} \text{ inches}) = 24$inches, we must be about 265 miles to the right of the origin before the height of $y = \log_2 x$ reaches 2 feet.

73. (a) $f(x) = \log_2(\log_{10} x)$. Since the input to $\log_2()$ must be positive, we have: $\log_{10} x > 0 \quad \Leftrightarrow \quad x > 10^0 = 1$. Thus the domain is $(1, \infty)$

(b) $y = \log_2(\log_{10} x) \quad \Leftrightarrow \quad 2^y = \log_{10} x \quad \Leftrightarrow \quad 10^{2^y} = x$. Thus $f^{-1}(x) = 10^{2^x}$.

75. (a) $f(x) = \dfrac{2^x}{1+2^x}$. $y = \dfrac{2^x}{1+2^x} \quad \Leftrightarrow \quad y + y\,2^x = 2^x \quad \Leftrightarrow \quad y = 2^x - y\,2^x = 2^x(1-y) \quad \Leftrightarrow$

$2^x = \dfrac{y}{1-y} \quad \Leftrightarrow \quad x = \log_2\left(\dfrac{y}{1-y}\right)$. Thus $f^{-1}(x) = \log_2\left(\dfrac{x}{1-x}\right)$.

(b) $\dfrac{x}{1-x} > 0$. Solving this using the methods from Chapter 2 we start with the endpoints of the potential intervals, 0 and 1.

Interval	$(-\infty, 0)$	$(0, 1)$	$(1, \infty)$
Sign of x	$-$	$+$	$+$
Sign of $1-x$	$+$	$+$	$-$
Sign of $\dfrac{x}{1-x}$	$-$	$+$	$-$

Thus the domain of $f^{-1}(x)$ is $(0, 1)$.

Exercises 5.4

1. $\log_2[x(x-1)] = \log_2 x + \log_2(x-1)$

3. $\log 7^{23} = 23\log 7$

5. $\log_2(AB^2) = \log_2 A + \log_2 B^2 = \log_2 A + 2\log_2 B$

7. $\log_3(x\sqrt{y}) = \log_3 x + \log_3\sqrt{y} = \log_3 x + \frac{1}{2}\log_3 y$

9. $\log_5\sqrt[3]{x^2+1} = \frac{1}{3}\log_5(x^2+1)$

11. $\ln\sqrt{ab} = \frac{1}{2}\ln ab = \frac{1}{2}(\ln a + \ln b)$

13. $\log\left(\frac{x^3y^4}{z^6}\right) = \log(x^3y^4) - \log z^6 = 3\log x + 4\log y - 6\log z$

15. $\log_2\left(\frac{x(x^2+1)}{\sqrt{x^2-1}}\right) = \log_2 x + \log_2(x^2+1) - \frac{1}{2}\log_2(x^2-1)$

17. $\ln\left(x\sqrt{\frac{y}{z}}\right) = \ln x + \frac{1}{2}\ln\left(\frac{y}{z}\right) = \ln x + \frac{1}{2}(\ln y - \ln z)$

19. $\log\sqrt[4]{x^2+y^2} = \frac{1}{4}\log(x^2+y^2)$

21. $\log\sqrt[3]{\frac{x^2+4}{(x^2+1)(x^3-7)^2}} = \frac{1}{3}\log\frac{x^2+4}{(x^2+1)(x^3-7)^2} = \frac{1}{3}[\log(x^2+4) - \log(x^2+1)(x^3-7)^2]$

$= \frac{1}{3}[\log(x^2+4) - \log(x^2+1) - 2\log(x^3-7)]$

23. $\ln\frac{z^4\sqrt{x}}{\sqrt[3]{y^2+6y+17}} = \ln(z^4\sqrt{x}) - \ln\sqrt[3]{y^2+6y+17} = 4\ln z + \frac{1}{2}\ln x - \frac{1}{3}\ln(y^2+6y+17)$

25. $\log_5\sqrt{125} = \log_5 5^{3/2} = \frac{3}{2}$

27. $\log 2 + \log 5 = \log 10 = 1$

29. $\log_4 192 - \log_4 3 = \log_4\frac{192}{3} = \log_4 64 = \log_4 4^3 = 3$

31. $\ln 6 - \ln 15 + \ln 20 = \ln\frac{6}{15} + \ln 20 = \ln(\frac{2}{5}\cdot 20) = \ln 8$

33. $10^{2\log 4} = (10^{\log 4})^2 = 4^2 = 16$

35. $\log_3 5 + 5\log_3 2 = \log_3 5 + \log_3 2^5 = \log_3(5\cdot 2^5) = \log_3 160$

37. $\log_2 A + \log_2 B - 2\log_2 C = \log_2(AB) - \log_2(C^2) = \log_2\left(\frac{AB}{C^2}\right)$

39. $4\log x - \frac{1}{3}\log(x^2+1) + 2\log(x-1) = \log x^4 - \log\sqrt[3]{x^2+1} + \log(x-1)^2$

$= \log\left(\frac{x^4}{\sqrt[3]{x^2+1}}\right) + \log(x-1)^2 = \log\left(\frac{x^4(x-1)^2}{\sqrt[3]{x^2+1}}\right)$

41. $\ln 5 + 2\ln x + 3\ln(x^2+5) = \ln(5x^2) + \ln(x^2+5)^3 = \ln[5x^2(x^2+5)^3]$

43. $\frac{1}{3}\log(2x+1) + \frac{1}{2}\left[\log(x-4) - \log(x^4 - x^2 - 1)\right] =$

$$\log\sqrt[3]{2x+1} + \frac{1}{2}\log\frac{x-4}{x^4-x^2-1} = \log\left(\sqrt[3]{2x+1}\cdot\sqrt{\frac{x-4}{x^4-x^2-1}}\right)$$

45. $\log_2 x - \log_2 y = \log_2\left(\frac{x}{y}\right) \neq \log_2(x-y)$ and so the equation is NOT an identity.

47. $\log 2^z = z\log 2$ and so the equation is an identity.

49. $\log a - \log b = \log\left(\frac{a}{b}\right) \neq \dfrac{\log a}{\log b}$ and so the equation is NOT an identity.

51. $\log_a a^a = a\log_a a = a\cdot 1 = a$ and so the equation is an identity.

53. $-\ln\left(\frac{1}{A}\right) = -\ln A^{-1} = -1(-\ln A) = \ln A$ and so the equation is an identity.

55. $\log_2 7 = \frac{\log 7}{\log 2} \approx 2.807355$

57. $\log_3 11 = \frac{\log 11}{\log 3} \approx 2.182658$

59. $\log_7 3.58 = \frac{\log 3.58}{\log 7} \approx 0.655407$

61. $\log_4 322 = \frac{\log 322}{\log 4} \approx 4.165458$

63. $\log_3 x = \frac{\log_e x}{\log_e 3} = \frac{\ln x}{\ln 3} = \frac{1}{\ln 3}\ln x$

65. $\log e = \frac{\ln e}{\ln 10} = \frac{1}{\ln 10}$

67. $-\ln(x - \sqrt{x^2-1}) = \ln\left(\dfrac{1}{x-\sqrt{x^2-1}}\right) = \ln\left(\dfrac{1}{x-\sqrt{x^2-1}}\cdot\dfrac{x+\sqrt{x^2-1}}{x+\sqrt{x^2-1}}\right)$

$$= \ln\left(\frac{x+\sqrt{x^2-1}}{x^2-(x^2-1)}\right) = \ln\left(x+\sqrt{x^2-1}\right)$$

Exercises 5.5

1. $5^x = 16 \quad \Leftrightarrow \quad \log 5^x = \log 16 \quad \Leftrightarrow \quad x \log 5 = \log 16 \quad \Leftrightarrow \quad x = \frac{\log 16}{\log 5} = 1.7227$

3. $2^{1-x} = 3 \quad \Leftrightarrow \quad \log 2^{1-x} = \log 3 \quad \Leftrightarrow \quad (1-x)\log 2 = \log 3 \quad \Leftrightarrow \quad 1 - x = \frac{\log 3}{\log 2} \quad \Leftrightarrow$ $x = 1 - \frac{\log 3}{\log 2} \approx -0.5850$

5. $3\,e^x = 10 \quad \Leftrightarrow \quad e^x = \frac{10}{3} \quad \Leftrightarrow \quad x = \ln\left(\frac{10}{3}\right) \approx 1.2040$

7. $e^{1-4x} = 2 \quad \Leftrightarrow \quad 1 - 4x = \ln 2 \quad \Leftrightarrow \quad -4x = -1 + \ln 2 \quad \Leftrightarrow \quad x = \frac{1-\ln 2}{4} = 0.0767$

9. $4 + 3^{5x} = 8 \quad \Leftrightarrow \quad 3^{5x} = 4 \quad \Leftrightarrow \quad \log 3^{5x} = \log 4 \quad \Leftrightarrow \quad 5x \log 3 = \log 4 \quad \Leftrightarrow \quad 5x = \frac{\log 4}{\log 3}$ $\Leftrightarrow \quad x = \frac{\log 4}{5 \log 3} \approx 0.2524$

11. $8^{0.4x} = 5 \quad \Leftrightarrow \quad \log 8^{0.4x} = 0.4x \log 8 = \log 5 \quad \Leftrightarrow \quad 0.4x = \frac{\log 5}{\log 8} \quad \Leftrightarrow \quad x = \frac{\log 5}{0.4 \log 8} \approx 1.9349$

13. $5^{-x/100} = 2 \quad \Leftrightarrow \quad \log 5^{-x/100} = -\frac{x}{100} \log 5 = \log 2 \quad \Leftrightarrow \quad x = -\frac{100 \log 2}{\log 5} \approx -43.0677$

15. $e^{2x+1} = 200 \quad \Leftrightarrow \quad 2x + 1 = \ln 200 \quad \Leftrightarrow \quad 2x = -1 + \ln 200 \quad \Leftrightarrow \quad x = \frac{-1+\ln 200}{2} \approx 2.1492$

17. $5^x = 4^{x+1} \quad \Leftrightarrow \quad \log 5^x = \log 4^{x+1} \quad \Leftrightarrow \quad x \log 5 = (x+1)\log 4 = x \log 4 + \log 4 \quad \Leftrightarrow$ $x \log 5 - x \log 4 = \log 4 \quad \Leftrightarrow \quad x(\log 5 - \log 4) = \log 4 \quad \Leftrightarrow \quad x = \frac{\log 4}{\log 5 - \log 4} \approx 6.2126$

19. $2^{3x+1} = 3^{x-2} \quad \Leftrightarrow \quad \log 2^{3x+1} = \log 3^{x-2} \quad \Leftrightarrow \quad (3x+1)\log 2 = (x-2)\log 3 \quad \Leftrightarrow \quad 3x \log 2 + \log 2 = x \log 3 - 2 \log 3 \quad \Leftrightarrow \quad 3x\log 2 - x\log 3 = -\log 2 - 2\log 3 \quad \Leftrightarrow$ $x(3 \log 2 - \log 3) = -(\log 2 + 2 \log 3) \quad \Leftrightarrow \quad s = -\frac{\log 2 + 2\log 3}{3 \log 2 - \log 3} \approx -2.9469$

21. $\dfrac{50}{1 + e^{-x}} = 4 \quad \Leftrightarrow \quad 50 = 4 + 4e^{-x} \quad \Leftrightarrow \quad 46 = 4e^{-x} \quad \Leftrightarrow \quad 11.5 = e^{-x} \quad \Leftrightarrow \quad \ln 11.5 = -x$ $\Leftrightarrow \quad x = -\ln 11.5 \approx -2.4423$

23. $100(1.04)^{2t} = 300 \quad \Leftrightarrow \quad 1.04^{2t} = 3 \quad \Leftrightarrow \quad \log 1.04^{2t} = 2t \log 1.04 = \log 3 \quad \Leftrightarrow$ $t = \frac{\log 3}{2 \log 1.04} \approx 14.0055$

25. $x^2 2^x - 2^x = 0 \quad \Leftrightarrow \quad 2^x(x^2 - 1) = 0 \quad \Rightarrow \quad 2^x = 0$ (never) or $x^2 - 1 = 0 \quad \Leftrightarrow \quad x^2 = 1 \quad \Rightarrow$ $x = \pm 1$. So the only solutions are $x = \pm 1$.

27. $4x^3 e^{-3x} - 3x^4 e^{-3x} = 0 \quad \Leftrightarrow \quad x^3 e^{-3x}(4x - 3) = 0 \quad \Rightarrow \quad x^3 = 0 \quad \Rightarrow \quad x = 0$ or $e^{-3x} = 0$ (never) or $4x - 3 = 0 \quad \Leftrightarrow \quad 4x = 3 \quad \Leftrightarrow \quad x = \frac{3}{4}$. So the solutions are $x = 0$ and $x = \frac{3}{4}$.

29. $e^{2x} - 3e^x + 2 = 0 \quad \Leftrightarrow \quad (e^x - 1)(e^x - 2) = 0 \quad \Rightarrow \quad e^x - 1 = 0 \quad \Leftrightarrow \quad e^x = 1 \quad \Leftrightarrow$ $x = \ln 1 = 0$ or $e^x - 2 = 0 \quad \Leftrightarrow \quad e^x = 2 \quad \Leftrightarrow \quad x = \ln 2 \approx 0.6931$. So the solutions are $x = 0$ and $x \approx 0.6931$.

31. $e^{4x} + 4e^{2x} - 21 = 0 \quad \Leftrightarrow \quad (e^{2x} + 7)(e^{2x} - 3) = 0 \quad \Rightarrow \quad e^{2x} = -7$ or $e^{2x} = 3$. Now $e^{2x} = -7$ has no solution, but $e^{2x} = 3 \quad \Leftrightarrow \quad 2x = \ln 3 \quad \Leftrightarrow \quad x = \frac{1}{2}\ln 3 \approx 0.5493$.

33. $\ln x = 10 \quad \Leftrightarrow \quad x = e^{10} \approx 22026$

35. $\log x = -2 \quad \Leftrightarrow \quad x = 10^{-2} = 0.01$

37. $\log(3x+5)=2 \quad\Leftrightarrow\quad 3x+5=10^2=100 \quad\Leftrightarrow\quad 3x=95 \quad\Leftrightarrow\quad x=\frac{95}{3}\approx 31.6667$

39. $2-\ln(3-x)=0 \quad\Leftrightarrow\quad 2=\ln(3-x) \quad\Leftrightarrow\quad e^2=3-x \quad\Leftrightarrow\quad x=3-e^2\approx -4.3891$

41. $\log_2 3+\log_2 x=\log_2 5+\log_2(x-2) \quad\Leftrightarrow\quad \log_2(3x)=\log_2(5x-10) \quad\Leftrightarrow\quad 3x=5x-10$ $\Leftrightarrow\quad 2x=10 \quad\Leftrightarrow\quad x=5$

43. $\log x+\log(x-1)=\log(4x) \quad\Leftrightarrow\quad \log[x(x-1)]=\log(4x) \quad\Leftrightarrow\quad x^2-x=4x \quad\Leftrightarrow$ $x^2-5x=0 \quad\Leftrightarrow\quad x(x-5)=0 \quad\Rightarrow\quad x=0$ or $x=5$. So the *possible* solutions are $x=0$ and $x=5$. However, when $x=0$, $\log x$ is undefined. Thus the only solution is $x=5$.

45. $\log_5(x+1)-\log_5(x-1)=2 \quad\Leftrightarrow\quad \log_5\left[\dfrac{x+1}{x-1}\right]=2 \quad\Leftrightarrow\quad \dfrac{x+1}{x-1}=5^2 \quad\Leftrightarrow$ $x+1=25x-25 \quad\Leftrightarrow\quad 24x=26 \quad\Leftrightarrow\quad x=\frac{13}{12}$

47. $\log_9(x-5)+\log_9(x+3)=1 \quad\Leftrightarrow\quad \log_9[(x-5)(x+3)]=1 \quad\Leftrightarrow\quad (x-5)(x+3)=9^1$ $\Leftrightarrow\quad x^2-2x-24=0 \quad\Leftrightarrow\quad (x-6)(x+4)=0 \quad\Rightarrow\quad x=6$ or -4. However, $x=-4$ is inadmissible, so $x=6$ is the only solution.

49. $\log(x+3)=\log x+\log 3 \quad\Leftrightarrow\quad \log(x+3)=\log(3x) \quad\Leftrightarrow\quad x+3=3x \quad\Leftrightarrow\quad 2x=3 \quad\Leftrightarrow$ $x=\frac{3}{2}$

51. $2^{2/\log_5 x}=\frac{1}{16} \quad\Leftrightarrow\quad \log_2 2^{2/\log_5 x}=\log_2\left(\frac{1}{16}\right) \quad\Leftrightarrow\quad \frac{2}{\log_5 x}=-4 \quad\Leftrightarrow\quad \log_5 x=-\frac{1}{2} \quad\Leftrightarrow$ $x=5^{-1/2}=\frac{1}{\sqrt{5}}\approx 0.4472$

53. $15e^{-0.087t}=5 \quad\Leftrightarrow\quad e^{-0.087t}=\frac{1}{3} \quad\Leftrightarrow\quad -0.087t=\ln\left(\frac{1}{3}\right)=-\ln 3 \quad\Leftrightarrow\quad t=\frac{\ln 3}{0.087}\approx 12.6277.$ So only 5 gram remain after approximately 13 days.

55. (a) $I=\frac{60}{13}\left(1-e^{-13t/5}\right) \quad\Leftrightarrow\quad \frac{13}{60}I=1-e^{-13t/5} \quad\Leftrightarrow\quad e^{-13t/5}=1-\frac{13}{60}I \quad\Leftrightarrow$ $-\frac{13}{5}t=\ln\left(1-\frac{13}{60}I\right) \quad\Leftrightarrow\quad t=-\frac{5}{13}\ln\left(1-\frac{13}{60}I\right)$

(b) Substituting $I=2$ we have $t=-\frac{5}{13}\ln\left[1-\frac{13}{60}(2)\right]\approx 0.218$ seconds.

57. $\ln x=3-x \quad\Leftrightarrow\quad \ln x+x-3=0$. Let $f(x)=\ln x+x-3$. We need to solve the equation $f(x)=0$. From the graph of f we get $x\approx 2.21$.

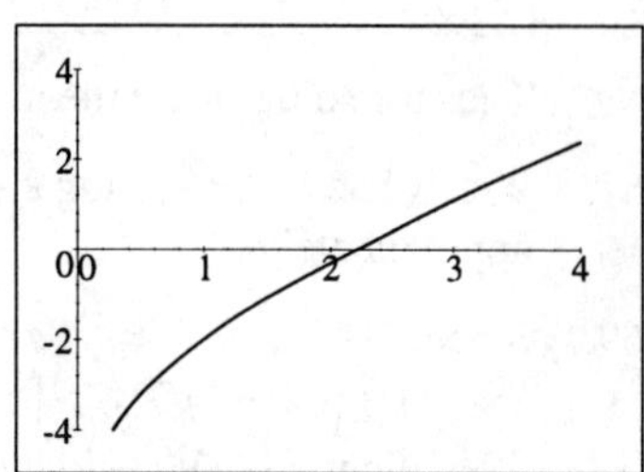

59. $x^3-x=\log_{10}(x+1) \quad\Leftrightarrow\quad x^3-x-\log_{10}(x+1)=0$. Let $f(x)=x^3-x-\log_{10}(x+1)$. We need to solve the equation $f(x)=0$. From the graph of f we get $x=0$ or $x\approx 1.14$

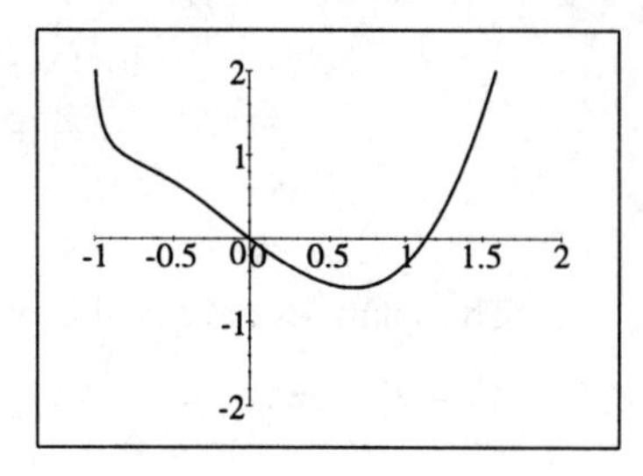

61. $e^x = -x \quad \Leftrightarrow \quad e^x + x = 0$. Let $f(x) = e^x + x$. We need to solve the equation $f(x) = 0$. From the graph of f we get $x \approx -0.57$

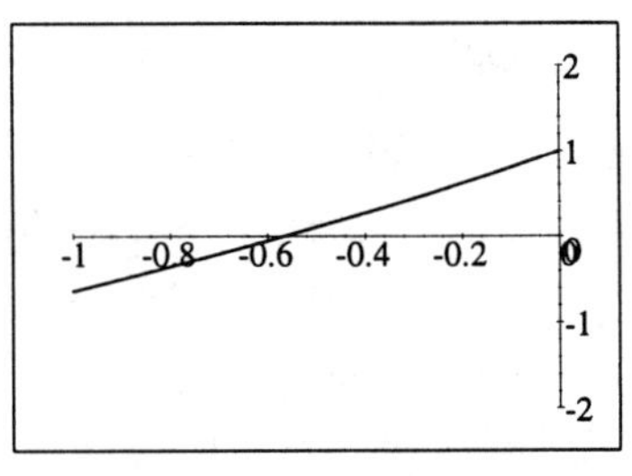

63. $4^{-x} = \sqrt{x} \quad \Leftrightarrow \quad 4^{-x} - \sqrt{x} = 0$. Let $f(x) = 4^{-x} - \sqrt{x}$. We need to solve the equation $f(x) = 0$. From the graph of f we get $x \approx 0.36$.

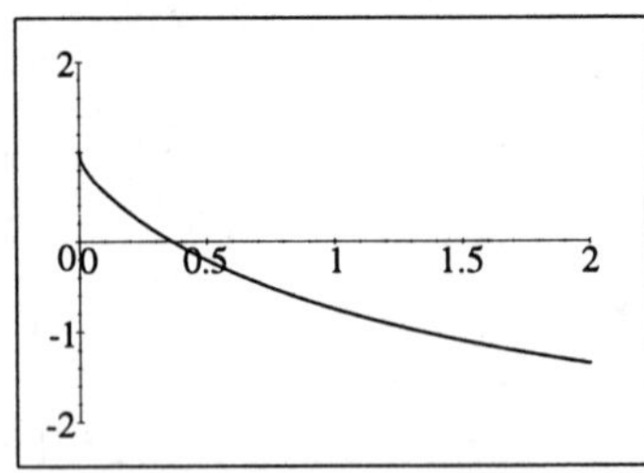

65. $\log(x-2) + \log(9-x) < 1 \quad \Leftrightarrow \quad \log[(x-2)(9-x)] < 1 \quad \Leftrightarrow \quad \log(-x^2 + 11x - 18) < 1 \quad \Rightarrow \quad -x^2 + 11x - 18 < 10^1 \quad \Leftrightarrow \quad 0 < x^2 - 11x + 28 \quad \Leftrightarrow \quad 0 < (x-7)(x-4)$, Also, since the domain of a logarithm is positive we must have $0 < -x^2 + 11x - 18 \quad \Leftrightarrow \quad 0 < (x-2)(9-x)$. Using the method from Chapter 2 with the endpoints $2, 4, 7, 9$ for the intervals we have

Interval	$(-\infty, 2)$	$(2,4)$	$(4,7)$	$(7,9)$	$(9,\infty)$
Sign of $x-7$	$-$	$-$	$-$	$+$	$+$
Sign of $x-4$	$-$	$-$	$+$	$+$	$+$
Sign of $x-2$	$-$	$+$	$+$	$+$	$+$
Sign of $9-x$	$+$	$+$	$+$	$+$	$-$
Sign of $(x-7)(x-4)$	$+$	$+$	$-$	$+$	$+$
Sign of $(x-2)(9-x)$	$-$	$+$	$+$	$+$	$-$

Thus the solution are the intervals $(2,4) \cup (7,9)$.

67. $2 < 10^x < 5 \quad \Leftrightarrow \quad \log 2 < x < \log 5 \quad \Leftrightarrow \quad 0.3010 < x < 0.6990$ or $(0.3010, 0.6990)$, approximately..

69. Since $x - 1 > 0 \quad \Leftrightarrow \quad x > 1$, we can take the log of both sides, thus $(x-1)^{\log(x-1)} = 100(x-1)$ $\Leftrightarrow \quad \log\left[(x-1)^{\log(x-1)}\right] = \log[100(x-1)] \quad \Leftrightarrow \quad [\log(x-1)][\log(x-1)] = 2 + \log(x-1)$ $\Leftrightarrow \quad [\log(x-1)]^2 - \log(x-1) - 2 = 0 \quad \Leftrightarrow \quad [\log(x-1)+1][\log(x-1)-2] = 0 \quad \Rightarrow$

$$\begin{array}{lrcl r}
& \log(x-1)+1=0 & \text{or} & & \log(x-1)-2=0 \\
\Leftrightarrow & \log(x-1)=-1 & & \Leftrightarrow & \log(x-1)=2 \\
\Leftrightarrow & x-1=\frac{1}{10} & & \Leftrightarrow & x-1=10^2 \\
\Leftrightarrow & x=\frac{11}{10} & & \Leftrightarrow & x=101
\end{array}$$

The solutions are $x = 1.1$ or $x = 101$.

71. $4^x - 2^{x+1} = 3 \quad \Leftrightarrow \quad 2^{2x} - 2 \cdot 2^x = 3 \quad \Leftrightarrow \quad (2^x)^2 - 2 \cdot 2^x - 3 = 0 \quad \Leftrightarrow \quad (2^x - 3)(2^x + 1) = 0 \quad \Leftrightarrow \quad 2^x = 3$ or $2^x = -1$. Now $2^x = -1$ has no solution since $2^x > 0$ for all x. But $2^x = 3 \quad \Leftrightarrow \quad x \log 2 = \log 3 \quad \Leftrightarrow \quad x = \frac{\log 3}{\log 2} \approx 1.58$.

Exercises 5.6

1. (a) $A(3) = 10000\left(1 + \frac{0.085}{4}\right)^{4(3)} = 10000(1.02125^{12}) = 12870.19$. Thus the amount after 3 years is \$12,870.19.

   (b) $20000 = 10000\left(1 + \frac{0.085}{4}\right)^{4t} = 10000(1.02125^{4t}) \quad \Leftrightarrow \quad 2 = 1.02125^{4t} \quad \Leftrightarrow \quad \log 2 = 4t\log 1.02125 \quad \Leftrightarrow \quad t = \frac{\log 2}{4\log 1.02125} \approx 8.24$ years. Thus the investment will double in about 8.24 years.

3. $8000 = 5000\left(1 + \frac{0.095}{4}\right)^{4t} = 5000(1.02375^{4t}) \quad \Leftrightarrow \quad 1.6 = 1.02375^{4t} \quad \Leftrightarrow$
   $\log 1.6 = 4t\log 1.02375 \quad \Leftrightarrow \quad t = \frac{\log 1.6}{4\log 1.02375} \approx 5.0059$ years. Approximately 5 years.

5. $2 = e^{0.085t} \quad \Leftrightarrow \quad \ln 2 = 0.085t \quad \Leftrightarrow \quad t = \frac{\ln 2}{0.085} \approx 8.15$ years. Thus the investment will double in about 8.15 years.

7. (a) $n(0) = 500$

   (b) $0.45 = 45\%$. The projected population is about $142,000$. The projected population is about 142,000.

   (c) $n(3) = 500e^{0.45(3)} \approx 1929$

   (d) $10000 = 500\,e^{0.45t} \quad \Leftrightarrow \quad 20 = e^{0.45t} \quad \Leftrightarrow \quad 0.45t = \ln 20 \quad \Leftrightarrow \quad t = \frac{\ln 20}{0.45} \approx 6.66$ hours, or 6 hours 40 minutes.

9. (a) $n(t) = 112000e^{0.04t}$

   (b) $t = 2000 - 1994 = 6$ and $n(6) = 112000\,e^{0.04(6)} \approx 142380$

   (c) $200000 = 112000\,e^{0.04t} \quad \Leftrightarrow \quad \frac{25}{14} = e^{0.04t} \quad \Leftrightarrow \quad 0.04t = \ln\left(\frac{25}{14}\right) \quad \Leftrightarrow$

   $t = 25\ln\left(\frac{25}{14}\right) \approx 14.5$. Since $1994 + 14.5 = 2008.5$, so the population will reach $200,000$ during the year 2008.

11. (a) $20,000$

   (b) Using the model $n(t) = 20000\,e^{rt}$ and the point $(4, 31000)$, we have $31000 = 20000\,e^{4r} \quad \Leftrightarrow \quad 1.55 = e^{4r} \quad \Leftrightarrow \quad 4r = \ln 1.55 \quad \Leftrightarrow \quad r = \frac{1}{4}\ln 1.55 \approx 0.1096$. Thus $n(t) = 20000\,e^{0.1096t}$

   (c) $n(8) = 20000\,e^{0.1096(8)} \approx 48218$. The projected dear population in 1998 is about $48,000$.

   (d) $100000 = 20000\,e^{0.1096t} \quad \Leftrightarrow \quad 5 = e^{0.1096t} \quad \Leftrightarrow \quad 0.1096t = \ln 5 \quad \Leftrightarrow \quad t = \frac{\ln 5}{0.1096} \approx 14.63$. Since $1990 + 14.63 = 2004.63$, the deer population will reach $100,000$ during the year 2004.

13. (a) From the formula for population growth we have $n(t) = 1500e^{rt}$, where t is given in minutes. Since the bacteria doubles in number when $t = 30$, we have $3000 = 1500e^{r \cdot 30} \quad \Leftrightarrow \quad 2 = e^{30r} \quad \Leftrightarrow \quad 30r = \ln 2 \quad \Leftrightarrow \quad r = \frac{\ln 2}{30} \approx 0.0231$. Thus the number of bacteria after t minutes is given by $n(t) = 1500e^{0.0231t}$.

   (b) Since 2 hours is 120 minutes we have $n(120) = 1500e^{(0.0231)(120)} \approx 23986$. Thus there are about 24000 bacteria after 2 hours.

(c) $4000 = 1500e^{0.0231t} \quad\Leftrightarrow\quad \frac{8}{3} = e^{0.0231t} \quad\Leftrightarrow\quad 0.0231t = \ln\left(\frac{8}{3}\right) \quad\Leftrightarrow\quad t = \frac{\ln(8/3)}{0.031} \approx 42.46$. Thus the bacteria count will reach 4000 in about 42.5 minutes.

15. (a) Using the formula $n(t) = n_0\, e^{rt}$ with $n_0 = 8600$ and $n(1) = 10000$ we solve for r. $10000 = 8600\, e^{r} \quad\Leftrightarrow\quad \frac{50}{43} = e^{r} \quad\Leftrightarrow\quad r = \ln\left(\frac{50}{43}\right) \approx 0.1508$. Thus $n(t) = 8600\, e^{0.1508\, t}$.

(b) $n(2) = 8600\, e^{0.1508(2)} \approx 11627$. Thus the number of bacteria after two hours is about 11,600.

(c) $17200 = 8600e^{0.1508\, t} \quad\Leftrightarrow\quad 2 = e^{0.1508\, t} \quad\Leftrightarrow\quad 0.1508t = \ln 2 \quad\Leftrightarrow\quad t = \frac{\ln 2}{0.1508} \approx 4.596$. Thus the number of bacteria will double in about 4.6 hours

17. (a) $2 = e^{0.02t} \quad\Leftrightarrow\quad 0.02t = \ln 2 \quad\Leftrightarrow\quad t = 50 \ln 2 \approx 34.65$. $t = 1995 + 34.65 = 2029.65$, so at the current growth rate the population will double by 2029.

(b) $3 = e^{0.02t} \quad\Leftrightarrow\quad 0.02t = \ln 3 \quad\Leftrightarrow\quad t = 50 \ln 3 \approx 54.93$. $t = 1995 + 54.93 = 2049.93$, so at the current growth rate the population will triple by 2049.

19. $n(t) = n_0 e^{2t}$. When $n_0 = 1$, the critical level is $n(24) = e^{2(24)} = e^{48}$. Using this critical level with $n_0 = 10$, we solve for t. $e^{48} = 10\, e^{2t} \quad\Leftrightarrow\quad 48 = \ln 10 + 2t \quad\Leftrightarrow\quad 2t = 48 - \ln 10 \quad\Leftrightarrow$ $t = \frac{1}{2}(48 - \ln 10) \approx 22.85$ hours.

21. (a) Using $m(t) = m_0\, 2^{-t/h}$ with $m_0 = 10$ and $h = 30$, we have $m(t) = 10(2^{-t/30})$

(b) $m(80) = 10\left(2^{-80/30}\right) \approx 1.5749$ grams

(c) $2 = 10\left(2^{-t/30}\right) \quad\Leftrightarrow\quad \frac{1}{5} = 2^{-t/30} \quad\Leftrightarrow\quad \log\left(\frac{1}{5}\right) = \left(-\frac{t}{30}\right)\log 2 \quad\Leftrightarrow\quad t = \frac{30 \log 5}{\log 2} \approx 69.66$ years.

23. By the formula in the text $m(t) = m_0\, e^{-rt}$ where $r = \frac{\ln 2}{h}$. So $m(t) = 50e^{-\frac{\ln 2}{25}\cdot t}$. We need to solve for t in the equation $32 = 50e^{-\frac{\ln 2}{25}t}$: $e^{-\frac{\ln 2}{25}t} = \frac{32}{50} \quad\Leftrightarrow\quad -\frac{\ln 2}{25}t = \ln\left(\frac{32}{50}\right) \quad\Leftrightarrow\quad t = -\frac{25}{\ln 2} \cdot \ln\left(\frac{32}{50}\right)$ ≈ 16.09. So it takes about 16 years.

25. By the formula for radioactive decay we have $m(t) = m_0\, e^{-rt}$ where $r = \frac{\ln 2}{h}$, in other words $m(t) = m_0 e^{-\frac{\ln 2}{h}\cdot t}$. In this exercise we have to solve for h in the equation $200 = 250e^{-\frac{\ln 2}{h}\cdot 48} \quad\Leftrightarrow$ $0.8 = e^{-\frac{\ln 2}{h}\cdot 48} \quad\Leftrightarrow\quad \ln(0.8) = -\frac{\ln 2}{h} \cdot 48 \quad\Leftrightarrow\quad h = -\frac{\ln 2}{\ln 0.8} \cdot 48 \approx 149.1$ hours. So the half-life is approximately 149 hours.

27. By the formula $m(t) = m_0\, e^{-\frac{\ln 2}{h}\cdot t}$ in the text we have $0.65 = 1 \cdot e^{-\frac{\ln 2}{5730}\cdot t} \quad\Leftrightarrow\quad \ln(0.65) = -\frac{\ln 2}{5730}t$ $\Leftrightarrow\quad t = -\frac{5730 \ln 0.65}{\ln 2} \approx 3561$. Thus the artifact is about 3560 years old.

29. (a) $T(0) = 65 + 145\, e^{-0.05(0)} = 65 + 145 = 210°\text{F}$

(b) $T(10) = 65 + 145\, e^{-0.05(10)} \approx 152.9$. Thus the temperature after 10 minutes about 153°F.

(c) $100 = 65 + 145\, e^{-0.05t} \quad\Leftrightarrow\quad 35 = 145\, e^{-0.05t} \quad\Leftrightarrow\quad 0.2414 = e^{-0.05t} \quad\Leftrightarrow\quad \ln 0.2414 = -0.05t \quad\Leftrightarrow\quad t = -\frac{\ln 0.2414}{0.05} \approx 28.4$. Thus the temperature will be 100°F in about 28 minutes.

31. Using Newton's Law of Cooling, $T(t) = T_s + D_0\, e^{-kt}$ with $T_s = 75$ and $D_0 = 185 - 75 = 110$. So $T(t) = 75 + 110\, e^{-kt}$. Since $T(30) = 150$, therefore $75 + 110\, e^{-30k} = 150 \quad\Leftrightarrow\quad 110\, e^{-30k} = 75$ $\Leftrightarrow\quad e^{-30k} = \frac{15}{22} \quad\Leftrightarrow\quad -30k = \ln\left(\frac{15}{22}\right) \quad\Leftrightarrow\quad k = -\frac{1}{30}\ln\left(\frac{15}{22}\right)$.

(a) $T(45) = 75 + 110\,e^{(45/30)\ln(15/22)} \approx 136.9$ and so the temperature of the turkey after 45 minutes is about 137° F.

(b) The temperature will be 100° F when $75 + 110\,e^{(t/30)\ln(15/22)} = 100 \quad \Leftrightarrow$ $e^{(t/30)\ln(15/22)} = \frac{25}{110} = \frac{5}{22} \quad \Leftrightarrow \quad \left(\frac{t}{30}\right)\ln\left(\frac{15}{22}\right) = \ln\left(\frac{5}{22}\right) \quad \Leftrightarrow \quad t = 30\frac{\ln\left(\frac{5}{22}\right)}{\ln\left(\frac{15}{22}\right)} \approx 116.1$ and so the temperature will be 100° F after 116 minutes.

33. (a) $\text{pH} = -\log[\text{H}^+] = -\log(5.0 \times 10^{-3}) \approx 2.3$

(b) $\text{pH} = -\log[\text{H}^+] = -\log(3.2 \times 10^{-4}) \approx 3.5$

(c) $\text{pH} = -\log[\text{H}^+] = -\log(5.0 \times 10^{-9}) \approx 8.3$

35. (a) $\text{pH} = -\log[\text{H}^+] = 3.0 \quad \Leftrightarrow \quad [\text{H}^+] = 10^{-3}\text{ M}$

(b) $\text{pH} = -\log[\text{H}^+] = 6.5 \quad \Leftrightarrow \quad [\text{H}^+] = 10^{-6.5} \approx 3.2 \times 10^{-7}\text{ M}$

37. $4.0 \times 10^{-7} \le [\text{H}^+] \le 1.6 \times 10^{-5} \quad \Leftrightarrow \quad \log(4.0 \times 10^{-7}) \le \log[\text{H}^+] \le \log(1.6 \times 10^{-5}) \quad \Leftrightarrow$ $-\log(4.0 \times 10^{-7}) \ge \text{pH} \ge -\log(1.6 \times 10^{-5}) \quad \Leftrightarrow \quad 6.4 \ge \text{pH} \ge 4.8$. Therefore the range of pH readings for cheese is approximately 4.8 to 6.4.

39. Let I_0 be the intensity of the smaller earthquake and I_1 the intensity of the larger earthquake. Then $I_1 = 20\,I_0$. Since $M_0 = \log\left(\frac{I_0}{S}\right) = \log I_0 - \log S$ and $M_1 = \log\left(\frac{I_1}{S}\right) = \log\left(\frac{20\,I_0}{S}\right) = \log 20 + \log I_0 - \log S$ then $M_1 - M_0 = \log 20 + \log I_0 - \log S - \log I_0 + \log S = \log 20 \approx 1.3$. Therefore the magnitude is 1.3 times larger.

41. Let the subscript A represent the Alaska earthquake, SF represent the San Francisco earthquake. Then $M_A = \log\left(\frac{I_A}{S}\right) = 8.6 \quad \Leftrightarrow \quad I_A = S \cdot 10^{8.6}$. So $\frac{I_A}{I_{SF}} = \frac{S \cdot 10^{8.6}}{S \cdot 10^{8.3}} = 10^{0.3} \approx 1.995$ and so the Alaskan earthquake was roughly twice as intense as the San Francisco earthquake.

43. Let the subscript MC represent the Mexico City earthquake, and T represent the Tangshan earthquake. $\dfrac{I_T}{I_{MC}} = 1.26 \quad \Leftrightarrow \quad \log 1.26 = \log\dfrac{I_T}{I_{MC}} = \log\dfrac{I_T/S}{I_{MC}/S} = \log\dfrac{I_T}{S} - \log\dfrac{I_{MC}}{S}$ $= M_T - M_{MC}$. Therefore $M_T = M_{MC} + \log 1.26 \approx 8.1 + 0.1 = 8.2$. Thus the magnitude of the Tangshan earthquake was roughly 8.2.

45. $98 = 10\log\left(\frac{I}{10^{-12}}\right) \quad \Leftrightarrow \quad \log(I \cdot 10^{12}) = 9.8 \quad \Leftrightarrow \quad \log I = 9.8 - \log 10^{12} = -2.2 \quad \Leftrightarrow$ $I = 10^{-2.2} \approx 6.3 \times 10^{-3}$ and so the intensity was 6.3×10^{-3} watts/m^2.

47. (a) $\beta_1 = 10\log\left(\dfrac{I_1}{I_0}\right)$ and $I_1 = \dfrac{k}{d_1^2} \quad \Leftrightarrow \quad \beta_1 = 10\log\left(\dfrac{k}{d_1^2\,I_0}\right) = 10\left[\log\left(\dfrac{k}{I_0}\right) - 2\log d_1\right]$ $= 10\log\left(\dfrac{k}{I_0}\right) - 20\log d_1$. Similarly $\beta_2 = 10\log\left(\dfrac{I_2}{I_0}\right)$ and $I_2 = \dfrac{k}{d_2^2}$ which implies $\beta_2 = 10 \cdot \log\left(\dfrac{k}{d_2^2\,I_0}\right) = 10\left[\log\left(\dfrac{k}{I_0}\right) - 2\log d_2\right] = 10\log\left(\dfrac{k}{I_0}\right) - 20\log d_2$. Substituting the expression for β_1 gives $\beta_2 = 10\log\left(\dfrac{k}{I_0}\right) - 20\log d_1 + 20\log d_1 - 20\log d_2$

$$= \beta_1 + 20\log d_1 - 20\log d_2 = \beta_1 + 20\log\left(\frac{d_1}{d_2}\right).$$

(b) $\beta_1 = 120$, $d_1 = 2$, and $d_2 = 10$. Then $\beta_2 = \beta_1 + 20\log\left(\frac{d_1}{d_2}\right) = 120 + 20\log\left(\frac{2}{10}\right)$ $= 120 + 20\log 0.2 \approx 106$ and so the intensity level at 10 m is approximately 106 dB.

Review Exercises for Chapter 5

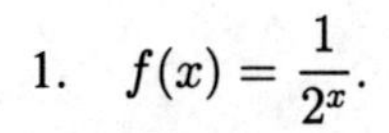

1. $f(x) = \dfrac{1}{2^x}$.

   Domain: $(-\infty, \infty)$

   Range: $(0, \infty)$

   Asymptote: $y = 0$.

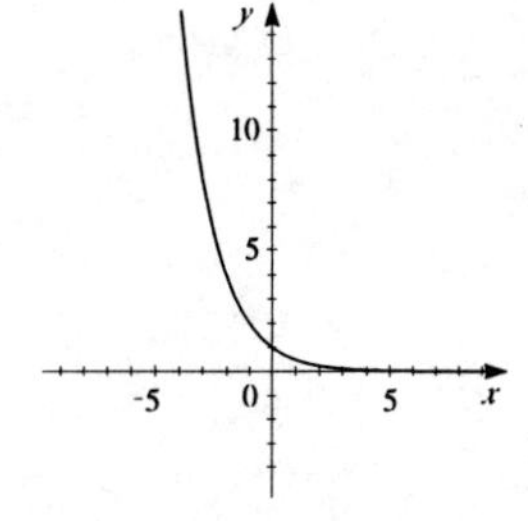

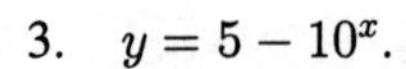

3. $y = 5 - 10^x$.

   Domain: $(-\infty, \infty)$

   Range: $(-\infty, 5)$

   Asymptote: $y = 5$.

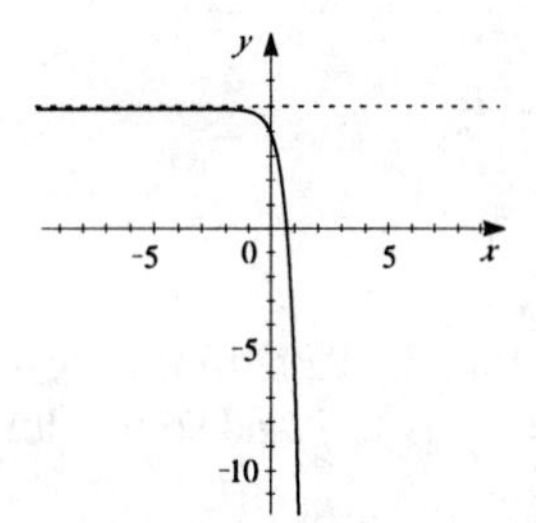

5. $f(x) = \log_3(x - 1)$.

   Domain: $(1, \infty)$

   Range: $(-\infty, \infty)$

   Asymptote: $x = -1$.

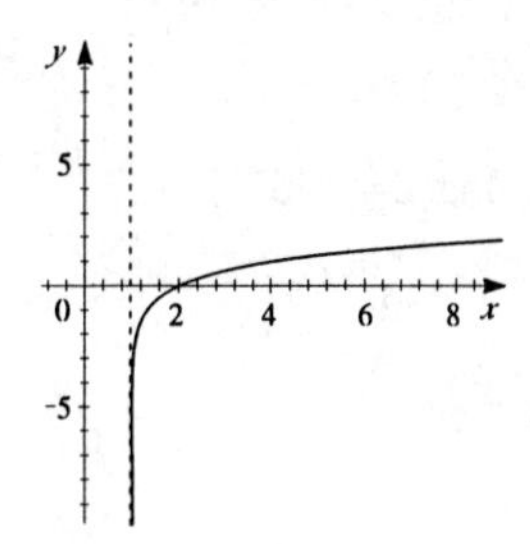

7. $y = 2 - \log_2 x$.

   Domain: $(0, \infty)$

   Range: $(-\infty, \infty)$

   Asymptote: $x = 0$.

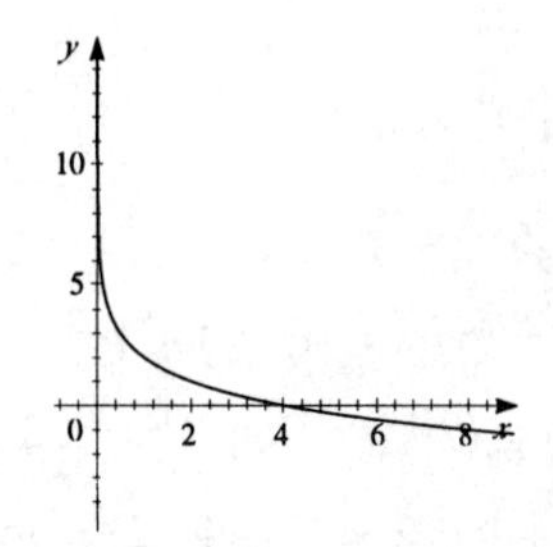

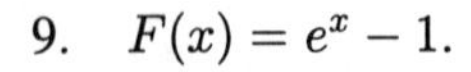

9. $F(x) = e^x - 1$.

Domain: $(-\infty, \infty)$

Range: $(-1, \infty)$

Asymptote: $y = -1$.

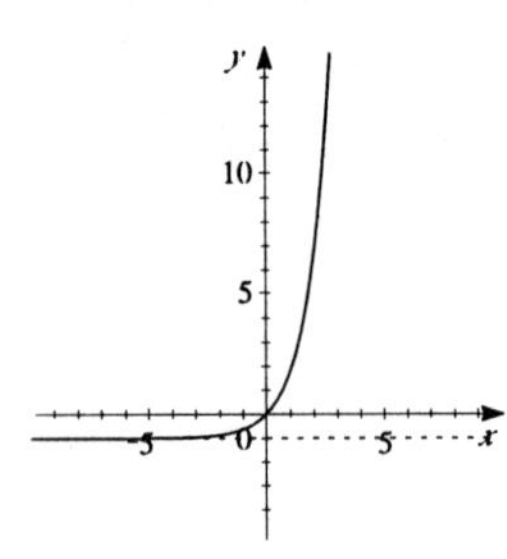

11. $y = 2\ln x$.

Domain: $(0, \infty)$

Range: $(-\infty, \infty)$

Asymptote: $x = 0$.

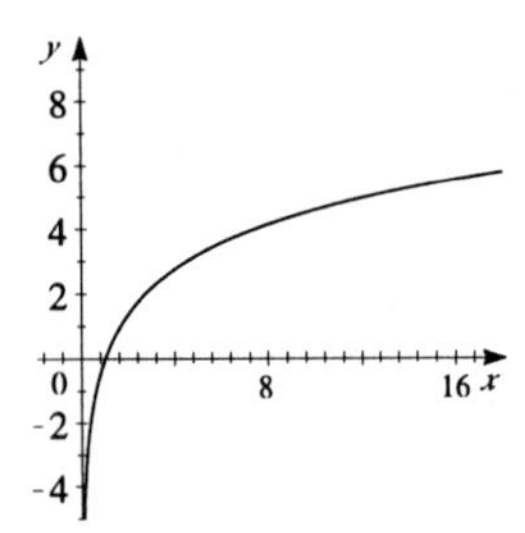

13. $f(x) = 10^{x^2} + \log(1 - 2x)$. Since $\log u$ is defined only for $u > 0$, we require $1 - 2x > 0 \Leftrightarrow -2x > -1 \Leftrightarrow x < \frac{1}{2}$ and so the domain is $\left(-\infty, \frac{1}{2}\right)$.

15. $\log_2 1024 = 10 \Leftrightarrow 2^{10} = 1024$

17. $\log x = y \Leftrightarrow 10^y = x$

19. $2^6 = 64 \Leftrightarrow \log_2 64 = 6$

21. $10^x = 74 \Leftrightarrow \log_{10} 74 = x$

23. $\log_2 128 = \log_2(2^7) = 7$

25. $10^{\log 45} = 45$

27. $\ln(e^6) = 6$

29. $\log_3 \frac{1}{27} = \log_3 3^{-3} = -3$

31. $\log_5 \sqrt{5} = \log_5 5^{1/2} = \frac{1}{2}$

33. $\log 25 + \log 4 = \log(25 \cdot 4) = \log 10^2 = 2$

35. $\log_2(16^{23}) = \log_2(2^4)^{23} = \log_2 2^{92} = 92$

37. $\log_8 6 - \log_8 3 + \log_8 2 = \log_8\left(\frac{6}{3} \cdot 2\right) = \log_8 4 = \log_8 8^{2/3} = \frac{2}{3}$

39. $\log(AB^2C^3) = \log A + 2\log B + 3\log C$

41. $\ln\sqrt{\dfrac{x^2 - 1}{x^2 + 1}} = \frac{1}{2}\ln\left(\dfrac{x^2 - 1}{x^2 + 1}\right) = \frac{1}{2}[\ln(x^2 - 1) - \ln(x^2 + 1)]$

43. $\log_5\left(\dfrac{x^2(1 - 5x)^{3/2}}{\sqrt{x^3 - x}}\right) = \log_5 x^2(1 - 5x)^{3/2} - \log_5\sqrt{x(x^2 - 1)}$

$= 2\log_5 x + \frac{3}{2}\log_5(1 - 5x) - \frac{1}{2}\log_5(x^3 - x)$

45. $\log 6 + 4\log 2 = \log 6 + \log 2^4 = \log(6 \cdot 2^4) = \log 96$

47. $\frac{3}{2}\log_2(x-y) - 2\log_2(x^2+y^2) = \log_2(x-y)^{3/2} - \log_2(x^2+y^2)^2 = \log_2\left(\frac{(x-y)^{3/2}}{(x^2+y^2)^2}\right)$

49. $\log(x-2) + \log(x+2) - \frac{1}{2}\log(x^2+4) = \log[(x-2)(x+2)] - \log\sqrt{x^2+4} =$
$\log\left(\frac{x^2-4}{\sqrt{x^2+4}}\right)$

51. $\log_2(1-x) = 4 \quad\Leftrightarrow\quad 1-x = 2^4 \quad\Leftrightarrow\quad x = 1-2^4 = -15$

53. $5^{5-3x} = 26 \quad\Leftrightarrow\quad \log_5 26 = 5-3x \quad\Leftrightarrow\quad 3x = 5 - \log_5 26 \quad\Leftrightarrow\quad x = \frac{1}{3}(5-\log_5 26) \approx 0.99$

55. $e^{3x/4} = 10 \quad\Leftrightarrow\quad \ln e^{3x/4} = \ln 10 \quad\Leftrightarrow\quad \frac{3x}{4} = \ln 10 \quad\Leftrightarrow\quad x = \frac{4}{3}\ln 10 \approx 3.07$

57. $\log x + \log(x+1) = \log 12 \quad\Leftrightarrow\quad \log[x(x+1)] = \log 12 \quad\Leftrightarrow\quad x(x+1) = 12 \quad\Leftrightarrow$
$x^2 + x - 12 = 0 \quad\Leftrightarrow\quad (x+4)(x-3) = 0 \quad\Rightarrow\quad x = 3$ or -4. Because $\log x$ and $\log(x+1)$ are undefined at $x = -4$, it follows that $x = 3$ is the only solution.

59. $x^2e^{2x} + 2x\,e^{2x} = 8\,e^{2x} \quad\Leftrightarrow\quad e^{2x}(x^2+2x-8) = 0 \quad\Leftrightarrow\quad x^2+2x-8 = 0$ (since $e^{2x} > 0$ for all x) $\quad\Leftrightarrow\quad (x+4)(x-2) = 0 \quad\Leftrightarrow\quad x = 2, -4$

61. $5^{-2x/3} = 0.63 \quad\Leftrightarrow\quad \frac{-2x}{3}\log 5 = \log 0.63 \quad\Leftrightarrow\quad x = -\frac{3\log 0.63}{2\log 5} \approx 0.430618$

63. $5^{2x+1} = 3^{4x-1} \quad\Leftrightarrow\quad (2x+1)\log 5 = (4x-1)\log 3 \quad\Leftrightarrow\quad 2x\log 5 + \log 5 = 4x\log 3 - \log 3$
$\Leftrightarrow\quad x(2\log 5 - 4\log 3) = -\log 3 - \log 5 \quad\Leftrightarrow\quad x = \dfrac{\log 3 + \log 5}{4\log 3 - 2\log 5} \approx 2.303600$

65. $y = e^{x/(x+2)}$

Vertical Asymptote: $x = -2$

Horizontal Asymptote: $y = 2.72$

No maximum or minimum.

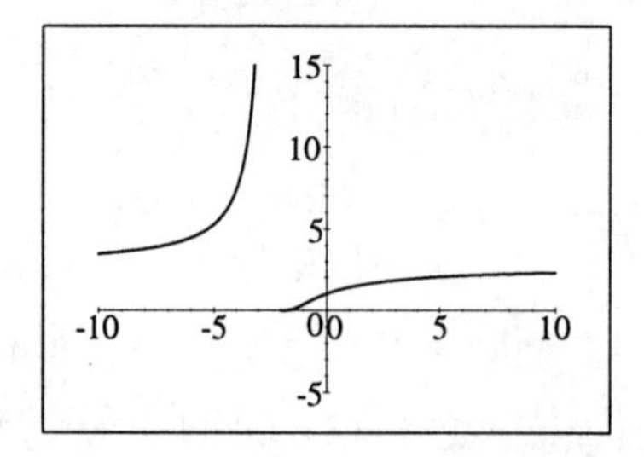

67. $y = \log(x^3 - x)$

Vertical Asymptotes: $x = -1,\ x = 0,\ x = 1$

Horizontal Asymptote: none

Local maximum ≈ -0.41 when $x \approx -0.58$

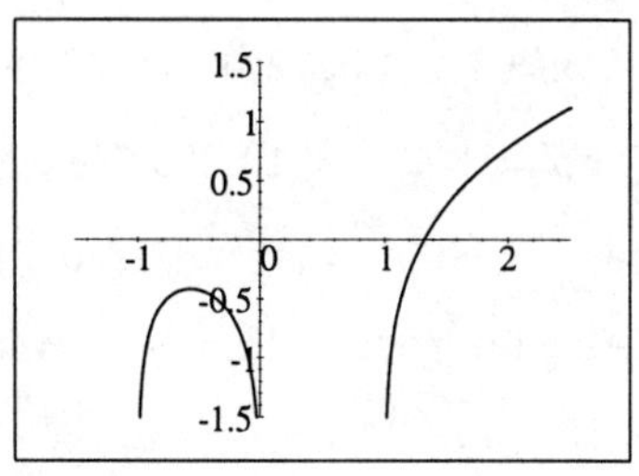

69. $3\log x = 6 - 2x$

We graph $y = 3\log x$ and $y = 6 - 2x$ in the same viewing rectangle. The solution occurs where the two graphs intersect. From the graphs, we see that the solution is $x \approx 2.42$.

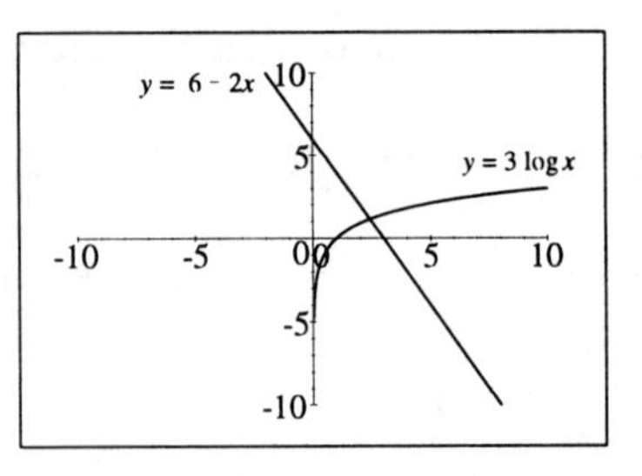

71. $\ln x > x - 2$. We graph the function $f(x) = \ln x - x + 2$ and we see that the graph lies above the x-axis for $0.16 < x < 3.15$. So the approximate solution of the given inequality is $0.16 < x < 3.15$.

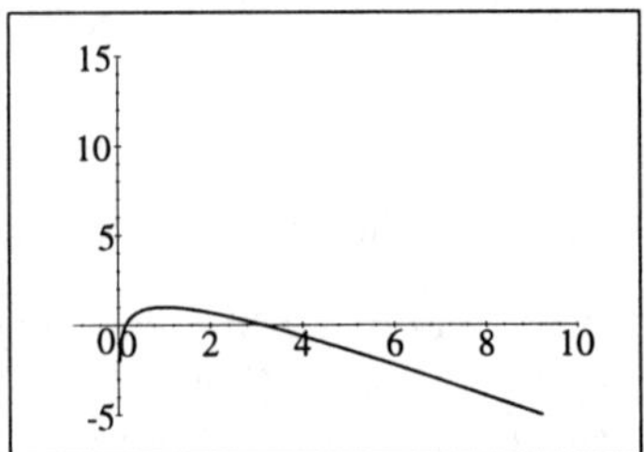

73. $f(x) = e^x - 3\,e^{-x} - 4x$. We see that the function is increasing on $(-\infty, 0]$ and $[1.10, \infty)$, and that it is decreasing on $[0, 1.10]$.

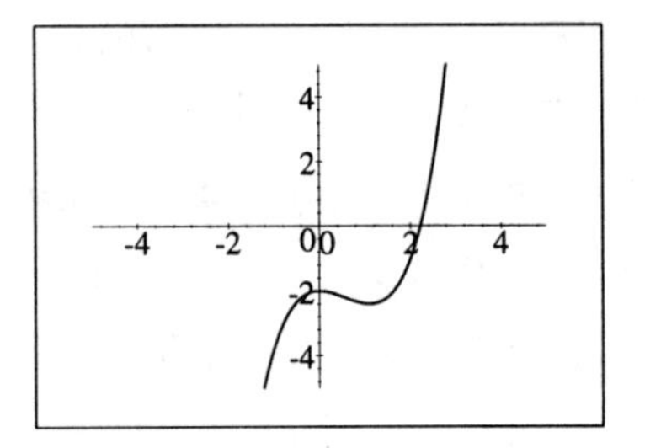

75. $\log_4 15 = \frac{\log 15}{\log 4} = 1.953445$

77. $\log_4 258 > \log_4 256 = \log_4 4^4 = 4$ and so $\log_4 258 > 4$. $\log_5 620 < \log_5 625 = \log_5 5^4 = 4$ and so $\log_5 620 < 4$. Since $\log_4 258 > 4 > \log_5 620$, then $\log_4 258$ is larger.

79. $P = 12,000$, $r = 0.10$, and $t = 3$. Then $A = P(1 + \frac{r}{n})^{nt}$.

(a) For $n = 2$, $A = 12,000(1 + \frac{0.10}{2})^{2(3)} = 12{,}000(1.05^6) \approx \$16,081.15$.

(b) For $n = 12$, $A = 12,000(1 + \frac{0.10}{12})^{12(3)} \approx \$16,178.18$.

(c) For $n = 365$, $A = 12,000(1 + \frac{0.10}{365})^{365(3)} \approx \$16,197.64$.

(d) For $n = \infty$, $A = P\,e^{rt} = 12,000\,e^{0.10(3)} \approx \$16,198.31$

81. (a) Using the model $n(t) = n_0\,e^{rt}$, with $n_0 = 30$ and $r = 0.15$ we have the formula

$$n(t) = 30\,e^{0.15t}$$

(b) $n(4) = 30\,e^{0.15(4)} \approx 55$

(c) $500 = 30\,e^{0.15t} \Leftrightarrow \frac{50}{3} = e^{0.15t} \Leftrightarrow 0.15t = \ln\left(\frac{50}{3}\right) \Leftrightarrow t = \frac{1}{0.15}\ln\left(\frac{50}{3}\right) \approx 18.76$.
So the stray cat population will reach 500 in about 19 years.

83. (a) From the formula for radioactive decay we have $m(t) = 10e^{-rt}$ where $r = -\frac{\ln 2}{2.7\times 10^5}$. So after 1000 years the amount remaining is $m(1000) = 10 \cdot e^{-\frac{\ln 2}{2.7\times 10^5}\cdot 1000} = 10e^{-\frac{\ln 2}{2.7\times 10^2}}$ $= 10e^{-\frac{\ln 2}{270}} \approx 9.97$. So the amount remaining is about 9.97 mg.

(b) We solve for t in the equation $7 = 10e^{-\frac{\ln 2}{2.7\times 10^5}\cdot t} \Leftrightarrow 0.7 = e^{-\frac{\ln 2}{2.7\times 10^5}\cdot t} \Leftrightarrow$ $\ln 0.7 = -\frac{\ln 2}{2.7\times 10^5}\cdot t \Leftrightarrow t = -\frac{\ln 0.7}{\ln 2}\cdot 2.7\times 10^5 \approx 138,934.75$. Thus it takes about 139,000 years.

85. (a) From the formula for radioactive decay $r = \frac{\ln 2}{1590} \approx 0.0004359$ and $m(t) = 150 \cdot e^{-0.0004359\,t}$

(b) $m(1000) = 150 \cdot e^{-0.0004359\cdot 1000} \approx 97.00$ and so the amount remaining is about 97.00 mg.

(c) Find t so that $50 = 150 \cdot e^{-0.0004359\,t} \Leftrightarrow t = -\frac{1}{0.0004359}\ln\left(\frac{1}{3}\right) \approx 2520$. Thus only 50 mg remain after about 2520 years.

87. (a) Using $n_0 = 1500$ and $n(5) = 3200$ in the formula $n(t) = n_0\, e^{rt}$ we have $3200 = 1500\, e^{5r}$ $\Leftrightarrow e^{5r} = \frac{32}{15} \Leftrightarrow 5r = \ln\left(\frac{32}{15}\right) \Leftrightarrow r = \frac{1}{5}\ln\left(\frac{32}{15}\right) \approx 0.1515$. Thus $n(t) = 1500 \cdot e^{0.1515t}$

(b) $t = 1999 - 1988 = 11$ so $n(11) = 1500e^{0.1515\cdot 11} \approx 7940$. Thus in 1999 the bird population should be about 7940 birds.

89. $[H^+] = 1.3 \times 10^{-8}$ M. Then $\text{pH} = -\log[H^+] = -\log(1.3 \times 10^{-8}) \approx 7.9$ and so fresh egg whites are basic.

91. Let I_0 be the intensity of the smaller earthquake and I_1 be the intensity of the larger earthquake. Then $I_1 = 35\, I_0$. Since $M = \log\left(\frac{I}{S}\right)$ then $M_0 = \log\left(\frac{I_0}{S}\right) = 6.5$ and $M_1 = \log\left(\frac{I_1}{S}\right)$ $= \log\left(\frac{35\, I_0}{S}\right) = \log 35 + \log\left(\frac{I_0}{S}\right) = \log 35 + M_0 = \log 35 + 6.5 \approx 8.04$. So the magnitude on the Richter scale is approximately 8.0.

Chapter 5 Test

1. $y = 4^x$ and $y = \log_4 x$.

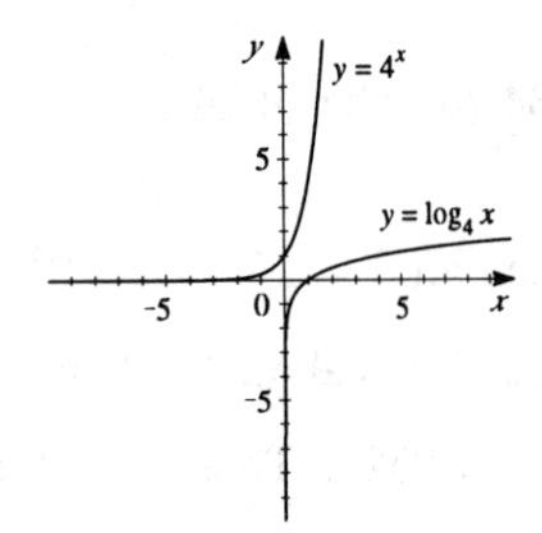

2. $f(x) = \log(x + 2)$.

   Domain: $(-2, \infty)$

   Range: $(-\infty, \infty)$

   Vertical asymptote: $x = -2$.

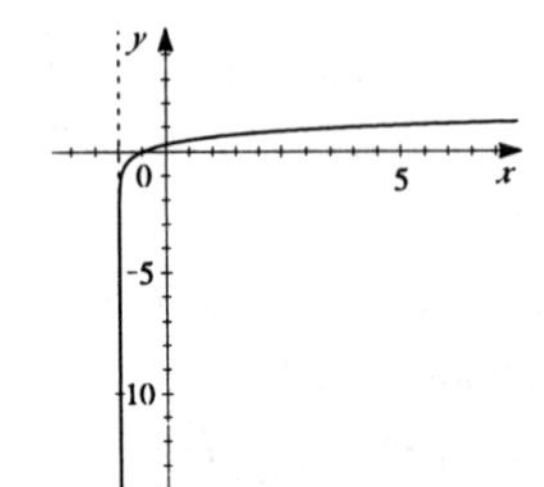

3. (a) $\log_3 \sqrt{27} = \log_3(3^3)^{\frac{1}{2}} \log_3 3^{3/2} = \frac{3}{2}$

   (b) $\log_2 56 - \log_2 7 = \log_2\left(\frac{56}{7}\right) = \log_2 8 = \log_2 2^3 = 3$

   (c) $\log_8 4 = \log_8 8^{2/3} = \frac{2}{3}$

   (d) $\log_6 4 + \log_6 9 = \log_6(4 \cdot 9) = \log_6 6^2 = 2$

4. $\log\sqrt{\dfrac{x^2 - 1}{x^3(y^2 + 1)^5}} = \frac{1}{2}\log\left(\dfrac{x^2 - 1}{x^3(y^2 + 1)^5}\right) = \frac{1}{2}\left[\log(x^2 - 1) - (3\log x + 5\log(y^2 + 1))\right]$

   $= \frac{1}{2}[\log(x^2 - 1) - 3\log x - 5\log(y^2 + 1)]$

5. $\ln x - 2\ln(x^2 + 1) + \frac{1}{2}\ln(3 - x^4) = \ln\left(x\sqrt{3 - x^4}\right) - \ln(x^2 + 1)^2 = \ln\left(\dfrac{x\sqrt{3 - x^4}}{(x^2 + 1)^2}\right)$

6. (a) $2^{x-1} = 10 \Leftrightarrow \log 2^{x-1} = \log 10 = 1 \Leftrightarrow (x - 1)\log 2 = 1 \Leftrightarrow x - 1 = \frac{1}{\log 2}$ $\Leftrightarrow x = 1 + \frac{1}{\log 2} \approx 4.32$

   (b) $5\ln(3 - x) = 4 \Leftrightarrow \ln(3 - x) = \frac{4}{5} \Leftrightarrow e^{\ln(3-x)} = e^{4/5} \Leftrightarrow 3 - x = e^{4/5} \Leftrightarrow$ $x = 3 - e^{4/5} \approx 0.77$

   (c) $10^{x+3} = 6^{2x} \Leftrightarrow \log 10^{x+3} = \log 6^{2x} \Leftrightarrow x + 3 = 2x\log 6 \Leftrightarrow 2x\log 6 - x = 3$ $\Leftrightarrow x(2\log 6 - 1) = 3 \Leftrightarrow x = \dfrac{3}{2\log 6 - 1} \approx 5.39$

(d) $\log_2(x+2) + \log_2(x-1) = 2 \quad \Leftrightarrow \quad \log_2((x+2)(x-1)) = 2 \quad \Leftrightarrow \quad x^2 + x - 2 = 2^2$ $\Leftrightarrow \quad x^2 + x - 6 = 0 \quad \Leftrightarrow \quad (x+3)(x-2) = 0 \quad \Rightarrow \quad x = -3$ or $x = 2$. However both logarithms are undefined at $x = -3$, so the only solution is $x = 2$.

7. (a) From the formula for population growth we have $8000 = 1000e^{r \cdot 1} \quad \Leftrightarrow \quad 8 = e^r \quad \Leftrightarrow$ $r = \ln 8 \approx 2.07944$. Thus $n(t) = 1000e^{2.07944t}$.

(b) $n(1.5) = 1000e^{2.07944(1.5)} \approx 22{,}627$

(c) $15000 = 1000e^{2.07944t} \quad \Leftrightarrow \quad 15 = e^{2.07944t} \quad \Leftrightarrow \quad \ln 15 = 2.07944t \quad \Leftrightarrow$ $t = \frac{\ln 15}{2.07944} \approx 1.3$. Thus the population will reach 15,000 after approximately 1.3 hours.

(d)

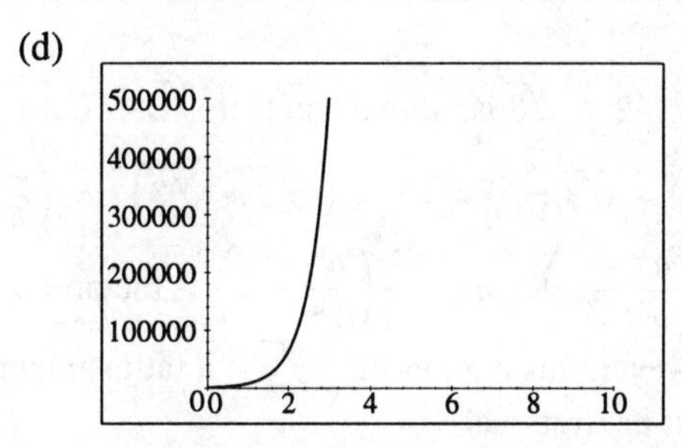

8. $2P = P\left(1 + \frac{0.085}{2}\right)^{2t} \quad \Leftrightarrow \quad 2 = 1.0425^{2t} \quad \Leftrightarrow \quad \log 2 = 2t \log 1.0425 \quad \Leftrightarrow$ $t = \dfrac{\log 2}{2 \log 1.0425} \approx 8.33$. Since it is compounded only twice a year, it will take 8 and $\frac{1}{2}$ years for the investment to double.

9. $f(x) = \log(x+4) + \log(8-5x)$. Since $\log u$ is defined only when $u > 0$, we require $x + 4 > 0$ and $8 - 5x > 0 \quad \Leftrightarrow \quad x > -4$ and $x < \frac{8}{5} \quad \Leftrightarrow \quad -4 < x < \frac{8}{5}$. So the domain is $\left(-4, \frac{8}{5}\right)$.

10. $f(x) = \dfrac{e^x}{x^3}$

(a)

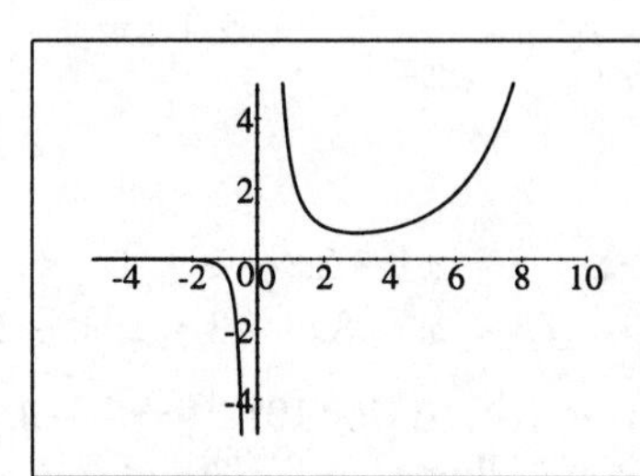

(b) Vertical asymptote: $x = 0$; Horizontal asymptote: $y = 0$.

(c) Local minimum ≈ 0.74 when $x \approx 3.00$

(d) Range $\approx (-\infty, 0) \cup [0.74, \infty)$

(e) $\dfrac{e^x}{x^3} = 2x + 1$.

We see that the graphs intersect at $x \approx -0.85, 0.96$, and 9.92.

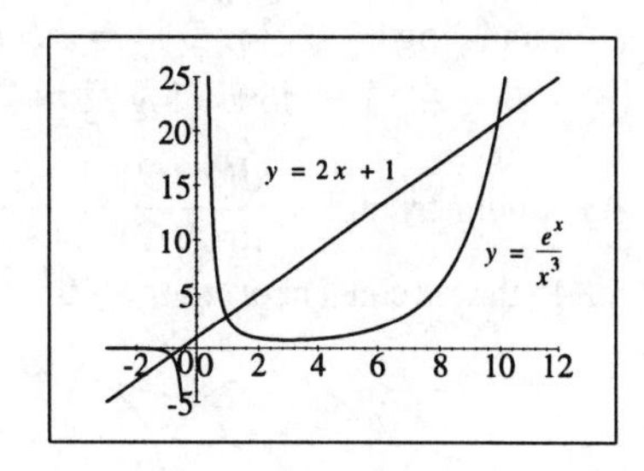

Focus on Problem Solving

1. (a) We mimic the proof in the text that $\sqrt{2}$ is irrational. Suppose that $\sqrt{6}$ is rational, so that $\sqrt{6}=\frac{a}{b}$, where a and b are natural numbers with no common factor. Then $a=\sqrt{6}\,b \Rightarrow a^2=6b^2 \Rightarrow a^2$ is a multiple of 6 (since a and b have no common factors). Thus a is a multiple of 6 $\Rightarrow a=6m$, for some integer m. Then $a^2=6b^2 \Rightarrow (6m)^2=6b^2 \Rightarrow 36m^2=6b^2 \Rightarrow b^2=6m^2 \Rightarrow b^2$ is a multiple of 6 $\Rightarrow b$ is also a multiple of 6. Thus a and b have 6 as a common factor. This contradicts our assumption that a and b have no common factor. So $\sqrt{6}$ must be irrational.

   (b) We prove this by contradiction. Suppose that $\sqrt{2}+\sqrt{3}$ is rational, so that $\sqrt{2}+\sqrt{3}=\frac{a}{b}$, where a and b are natural numbers. Then $\sqrt{2}+\sqrt{3}=\frac{a}{b} \Rightarrow \left(\sqrt{2}+\sqrt{3}\right)^2=\left(\frac{a}{b}\right)^2 \Leftrightarrow 2+2\sqrt{6}+3=\frac{a^2}{b^2} \Leftrightarrow \sqrt{6}=\frac{1}{2}\left(\frac{a^2}{b^2}-5\right)$. Since $\frac{1}{2}\left(\frac{a^2}{b^2}-5\right)$ is the product of rational numbers, it is a rational number. However, this implies that $\sqrt{6}$ is a rational number, which contradicts part (a). So $\sqrt{2}+\sqrt{3}$ must be irrational.

3. (a) Note that $12=2^2\cdot 3$. To get the smallest perfect cube, we have to multiply by $2\cdot 3^2=18$. Then $12\cdot 18=2^2\cdot 3\cdot 2\cdot 3^2=2^3\cdot 3^3=6^3$.

   (b) Since $15=3\cdot 5$, we have to multiply by $3^2\cdot 5^2=225$: $15\cdot 225=3^3\cdot 5^3=15^3$.

   (c) Write the number as a product of powers of prime factors: $p_1^m\,p_2^n\cdots$. For each prime factor p_i, we multiply by as many p_i's as are necessary to make the power of p_i divisible by 3. For instance, for the number $2\cdot 3^2\cdot 5^7\cdot 11^3$, we would have to multiply by $2^2\cdot 3\cdot 5^2$ and get $2^3\cdot 3^3\cdot 5^9\cdot 11^3=(2\cdot 3\cdot 125\cdot 11)^3$.

5. Using the Change of Base formula, we have $\frac{1}{\log_2 x}+\frac{1}{\log_3 x}+\frac{1}{\log_5 x}=\log_x 2+\log_x 3+\log_x 5=\log_x(2\cdot 3\cdot 5)=\log_x 30=\frac{1}{\log_{30} x}$

7. $\log(x^2-2x-2)\le 0 \Leftrightarrow 0<x^2-2x-2\le 1 \Leftrightarrow 3<x^2-2x+1\le 4 \Leftrightarrow 3<(x-1)^2\le 4 \Leftrightarrow \sqrt{3}<|x-1|\le 2 \Leftrightarrow 1+\sqrt{3}<x\le 3$ or $-3\le x<-1-\sqrt{3}$.

9. (a) Let n be a k-digit positive integer. Then $10^{k-1}\le n<10^k$, so $[\![\log 10^{k-1}]\!]=k-1\le[\![\log n]\!]$ and $[\![\log n]\!]<[\![\log 10^k]\!]=k$. Thus $[\![\log n]\!]=k-1$ and $[\![\log n]\!]+1=k$.

   (b) $[\![\log 2^{500}]\!]=[\![500\cdot\log 2]\!]\approx[\![150.5]\!]=151$.

11. By similar triangles, $\frac{d(A,Q)}{d(Q,C)}=\frac{d(A,P)}{x}$. So $\frac{50}{10}=\frac{30}{x} \Leftrightarrow 50x=300 \Leftrightarrow x=6$. Using the Pythagorean Theorem $y^2+6^2=10^2 \Leftrightarrow y=8$ and so $|DE|=16$ cm.

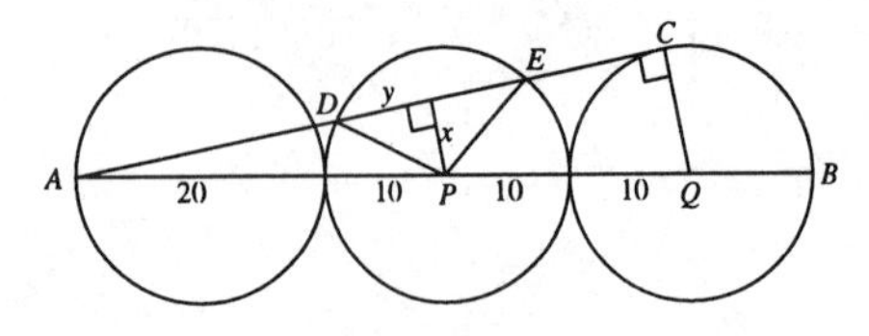

Chapter Six

Exercises 6.1

1. $\begin{cases} x + y = 3 \\ 2x - y = 0 \end{cases}$

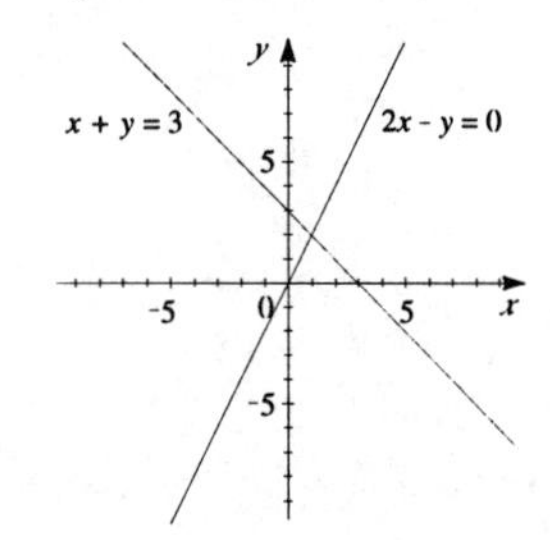

Solution: $x = 1, y = 2$

3. $\begin{cases} 2x + 3y = 12 \\ x - y = 1 \end{cases}$

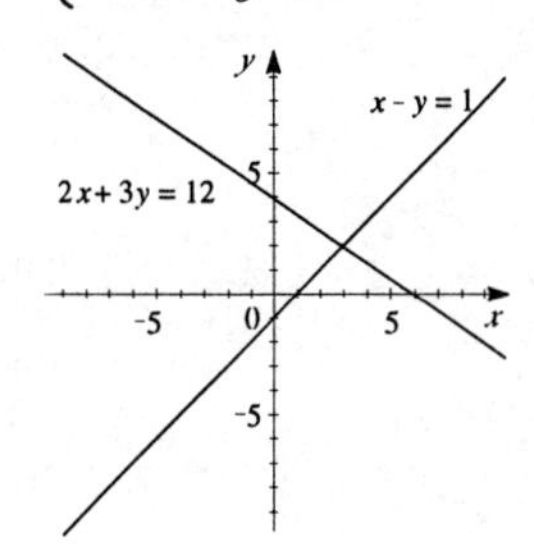

Solution: $x = 3, y = 2$

5. $\begin{cases} 2x + 5y = 15 \\ 4x + 8y = 22 \end{cases}$

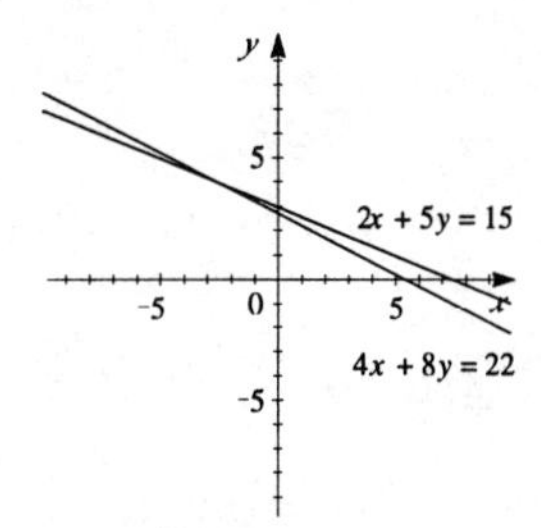

Solution: $x = -\frac{5}{2}$, $y = 4$

7. $-x + y = 2 \quad \Leftrightarrow \quad y = x + 2$. Substituting for y into $4x - 3y = -3$ gives $4x - 3(x + 2) = -3$ $\Leftrightarrow \quad 4x - 3x - 6 = -3 \quad \Leftrightarrow \quad x = 3$, and so $y = (3) + 2 = 5$. Hence, the solution is $(x, y) = (3, 5)$.

9. $x + 2y = 7 \quad \Leftrightarrow \quad x = 7 - 2y$. Substituting for x into $5x - y = 2$ gives $5(7 - 2y) - y = 2 \quad \Leftrightarrow$ $35 - 10y - y = 2 \quad \Leftrightarrow \quad -11y = -33 \quad \Leftrightarrow \quad y = 3$, and so $x = 7 - 2(3) = 1$. Hence, the solution is $(1, 3)$.

11. $\frac{1}{2}x + \frac{1}{3}y = 2 \quad \Leftrightarrow \quad x + \frac{2}{3}y = 4 \quad \Leftrightarrow \quad x = 4 - \frac{2}{3}y$. Substituting for x into $\frac{1}{5}x - \frac{2}{3}y = 8$ gives $\frac{1}{5}(4 - \frac{2}{3}y) - \frac{2}{3}y = 8 \quad \Leftrightarrow \quad \frac{4}{5} - \frac{2}{15}y - \frac{10}{15}y = 8 \quad \Leftrightarrow \quad 12 - 2y - 10y = 120 \quad \Leftrightarrow \quad y = -9$, and so $x = 4 - \frac{2}{3}(-9) = 10$. Thus, the solution is $(10, -9)$.

13. Adding gives
$$\begin{array}{r} 3x + 2y = 8 \\ x - 2y = 0 \\ \hline 4x \qquad = 8 \end{array} \quad \Leftrightarrow \quad x = 2$$

So $x - 2y = (2) - 2y = 0 \quad \Leftrightarrow \quad 2y = 2 \quad \Leftrightarrow \quad y = 1$. Thus, the solution is $(2, 1)$.

15. $\begin{cases} 2x - 6y = 10 \\ -3x + 9y = -15 \end{cases}$ Adding 3 times equation 1 to 2 times equation 2 gives $\begin{array}{r} 6x - 18y = 30 \\ -6x + 18y = -30 \\ \hline 0 = 0 \end{array}$

Write the equation in slope-intercept form, we have $y = \frac{1}{3}x - \frac{5}{3}$, so a solution is any pair of the form $\left(x, \frac{1}{3}x - \frac{5}{3}\right)$, where x is any real number.

17. $$\begin{array}{lll} 3x + 5y = 17 & \times 7 & \\ 7x + 9y = 29 & \times -3 & \Rightarrow \end{array} \quad \begin{array}{r} 21x + 35y = 119 \\ -21x - 27y = -87 \\ \hline 8y = 32 \end{array} \quad \Leftrightarrow \quad y = 4$$

So $3x + 5(4) = 17 \quad \Leftrightarrow \quad x = -1$. Thus, the solution is $(-1, 4)$.

19. $$\begin{array}{lll} 8s - 3t = -3 & \times 2 & \\ 5s - 2t = -1 & \times 3 & \Rightarrow \end{array} \quad \begin{array}{r} 16s - 6t = -6 \\ 15s - 6t = -3 \\ \hline s = -3 \end{array}$$

So $8(-3) - 3t = -3 \quad \Leftrightarrow \quad -24 - 3t = -3 \quad \Leftrightarrow \quad t = -7$. Thus, the solution is $(s, t) = (-3, -7)$.

21. $$\begin{array}{lll} \frac{1}{2}x + \frac{3}{5}y = 3 & \times 10 & \\ \frac{1}{3}x + 2y = -6 & \times -3 & \Rightarrow \end{array} \quad \begin{array}{r} 5x + 6y = 30 \\ -x - 6y = 18 \\ \hline 4x \qquad = 48 \end{array} \quad \Leftrightarrow \quad x = 12$$

So $5(12) + 6y = 30 \quad \Leftrightarrow \quad y = -5$. Thus, the solution is $(12, -5)$.

23. $$\begin{array}{lll} 0.2r + 0.3s = 0.16 & \times 30 & \\ -1.2r + 4s = 1.36 & \times 5 & \Rightarrow \end{array} \quad \begin{array}{r} 6r + 9s = 4.8 \\ -6r + 20s = 6.8 \\ \hline 29s = 11.6 \end{array} \quad \Leftrightarrow \quad s = \frac{2}{5}$$

So $-6r + 20\left(\frac{2}{5}\right) = 6.8 \quad \Leftrightarrow \quad 6r = 1.2 \quad \Leftrightarrow \quad r = \frac{1}{5}$. The solution is $(r, s) = \left(\frac{1}{5}, \frac{2}{5}\right)$.

25. $$\begin{array}{lll} \sqrt{3}x + \sqrt{2}y = 5 & \times 2\sqrt{2} & \\ 2\sqrt{6}x + 4y = \sqrt{5} & \times -1 & \Rightarrow \end{array} \quad \begin{array}{r} 2\sqrt{6}x + 4y = 10\sqrt{2} \\ -2\sqrt{6}x - 4y = -\sqrt{5} \\ \hline 0 = 10\sqrt{2} - \sqrt{5} \end{array} \quad \text{which is false.}$$

Hence there is no solution to this system.

27. Let $x = \dfrac{1}{u}$, $y = \dfrac{1}{v}$. Then

$$\begin{array}{l} \dfrac{2}{u} + \dfrac{1}{v} = 1 \\ \dfrac{3}{u} - \dfrac{2}{v} = 1 \end{array} \quad \Rightarrow \quad \begin{array}{ll} 2x + y = 1 & \times 2 \\ 3x - 2y = 1 & \times 1 \end{array} \quad \Rightarrow \quad \begin{array}{r} 4x + 2y = 2 \\ 3x - 2y = 1 \\ \hline 7x \qquad = 3 \end{array} \quad \Leftrightarrow \quad x = \frac{3}{7}$$

So $2\left(\frac{3}{7}\right) + y = 1 \quad \Leftrightarrow \quad y = \frac{1}{7}$. Therefore, the solution is $(u, v) = \left(\frac{1}{x}, \frac{1}{y}\right) = \left(\frac{7}{3}, 7\right)$.

29. Let $x = z^3$, $y = w^3$. Then

$$\begin{array}{l} 2z^3 + \frac{1}{2}w^3 = 2 \\ -3z^3 + \dfrac{3}{2}w^3 = 15 \end{array} \quad \Rightarrow \quad \begin{array}{ll} 2x + \frac{1}{2}y = 2 & \times 3 \\ -3x + \frac{3}{2}y = 15 & \times -1 \end{array} \quad \Rightarrow \quad \begin{array}{r} 6x + \frac{3}{2}y = 6 \\ 3x + \frac{3}{2}y = -15 \\ \hline 9x \qquad = -9 \end{array} \quad \Leftrightarrow \quad x = -1$$

So $2(-1) + \frac{1}{2}y = 2 \quad \Leftrightarrow \quad \frac{1}{2}y = 4 \quad \Leftrightarrow \quad y = 8$. Thus, the solution is $(z, w) = (\sqrt[3]{x}, \sqrt[3]{y}) = (-1, 2)$.

31. $$\begin{array}{rl} \frac{2x-5}{3} + \frac{y-1}{6} = \frac{1}{2} & \times 6 \\ \frac{x}{5} + \frac{3y-6}{12} = 1 & \times 60 \end{array} \quad \Rightarrow \quad \begin{array}{c} 4x - 10 + y - 1 = 3 \\ 12x + 15y - 30 = 60 \end{array}$$

$$\begin{array}{rl} 4x + y = 14 & \times 3 \\ 12x + 15y = 90 & \times -1 \end{array} \quad \Rightarrow \quad \begin{array}{r} 2x + 3y = 42 \\ -12x - 15y = -90 \\ \hline -12y = -48 \end{array} \quad \Leftrightarrow \quad y = 4$$

So $4x + (4) = 14 \quad \Leftrightarrow \quad 4x = 10 \quad \Leftrightarrow \quad x = \frac{5}{2}$. Thus, the solution is $\left(\frac{5}{2}, 4\right)$.

33. $$\begin{array}{rl} x - 2y = 1 & \\ 2x + 2y = 1 & \\ \hline 3x \quad = 2 & \end{array} \quad \Leftrightarrow \quad x = \frac{2}{3}$$

So $\left(\frac{2}{3}\right) - 2y = 1 \quad \Leftrightarrow \quad 2y = -\frac{1}{3} \quad \Leftrightarrow \quad y = -\frac{1}{6}$. $2x + 2y = 1$. Therefore, the solution is $\left(\frac{2}{3}, -\frac{1}{6}\right)$.

35. $$\begin{array}{rl} x + y = 0 & \times -1 \\ x + ay = 1 & \times 1 \end{array} \quad \Rightarrow \quad \begin{array}{r} -x - y = 0 \\ x + ay = 1 \\ \hline ay - y = 1 \end{array} \quad \Leftrightarrow \quad y(a-1) = 1 \quad \Leftrightarrow \quad y = \frac{1}{a-1}, a \neq 1$$

So $x + \left(\frac{1}{a-1}\right) = 0 \quad \Leftrightarrow \quad x = \frac{1}{1-a}$. Thus, the solution is $\left(\frac{1}{1-a}, \frac{1}{a-1}\right)$.

37. $$\begin{array}{rl} ax + by = 1 & \times -b \\ bx + ay = 1 & \times a \end{array} \quad \Rightarrow \quad \begin{array}{r} -abx - b^2y = -b \\ abx + a^2y = a \\ \hline (a^2 - b^2)y = a - b \end{array} \quad \Leftrightarrow \quad y = \frac{a-b}{a^2-b^2} = \frac{1}{a+b},\ a^2 - b^2 \neq 0$$

So $ax + \frac{b}{a+b} = 1 \quad \Leftrightarrow \quad ax = \frac{a}{a+b} \quad \Leftrightarrow \quad x = \frac{1}{a+b}$. Thus, the solution is $\left(\frac{1}{a+b}, \frac{1}{a+b}\right)$.

39. Let the two numbers be x and y. $$\begin{array}{r} x + y = 34 \\ x - y = 10 \\ \hline 2x \quad = 44 \end{array} \quad \Leftrightarrow \quad x = 22$$

So $22 + y = 34 \quad \Leftrightarrow \quad y = 12$. Therefore, the two numbers are 22 and 12.

41. Let d be the number of dimes and q be the number of quarters.

$$\begin{array}{rl} d + q = 14 & \times -1 \\ 0.10d + 0.25q = 2.75 & \times 10 \end{array} \quad \Rightarrow \quad \begin{array}{r} -d - q = -14 \\ d + 2.5q = 27.5 \\ \hline 1.5q = 13.5 \end{array} \quad \Leftrightarrow \quad q = 9$$

So $d + (9) = 14 \quad \Leftrightarrow \quad d = 5$. So, the number of dimes is 5 and the number of quarters is 9.

43. Let x be the speed of the plane in still air and y be the air speed.

$$\begin{array}{rl} 2x - 2y = 180 & \times -6 \\ 1.2x + 1.2y = 180 & \times 10 \end{array} \quad \Rightarrow \quad \begin{array}{r} 12x - 12y = 1080 \\ 12x + 12y = 1800 \\ \hline 24x \quad = 2880 \end{array} \quad \Leftrightarrow \quad x = 120$$

So $2(120) - 2y = 180 \Leftrightarrow -2y = -60 \Leftrightarrow y = 30$. Therefore, the speed of the plane is 120 mi/h and the wind speed is 30 mi/h.

45. Let x be the cycling speed and y be the running speed. (Remember to divide by 60 to convert minutes to decimal hours.)

$$\begin{array}{rl} 0.5x + 0.5y = 12.5 & \times -2 \\ 0.75x + 0.2y = 16 & \times 5 \end{array} \Rightarrow \begin{array}{r} -x - y = -25 \\ 3.75x + y = 80 \\ \hline 2.75x \quad = 55 \end{array} \Leftrightarrow x = 20$$

So $20 + y = 25 \Leftrightarrow y = 5$. Thus, the cycling speed is 20 mi/h and the running speed is 5 mi/h.

47. Let a be the grams of food A and b be the grams of food B.

$$\begin{array}{rl} 0.12a + 0.20b = 32 & \times -250 \\ 100a + 50b = 22000 & \times 1 \end{array} \Rightarrow \begin{array}{r} -30a - 50b = -8000 \\ 100a + 50b = 22000 \\ \hline 70a \quad = 14000 \end{array} \Leftrightarrow a = 200$$

So $0.12(200) + 0.20b = 32 \Leftrightarrow 0.20b = 8 \Leftrightarrow b = 40$. Thus, she should use 200 grams of food A and 40 grams of food B.

49. Let x and y be the sulfuric acid concentrations in the first and second containers.

$$\begin{array}{rl} 300x + 600y = 900(0.15) & \times -1 \\ 100x + 500y = 600(0.125) & \times 3 \end{array} \Rightarrow \begin{array}{r} -300x - 600y = -135 \\ 300x + 1500y = 225 \\ \hline 900y = 90 \end{array} \Leftrightarrow y = 0.10$$

So $100x + 500(0.10) = 75 \Leftrightarrow x = 0.25$. The concentrations of sulfuric acid are 25% in the first container and 10% in the second.

51. Let x be the usual speed, t be the usual time (hours), and d be the distance home, so $d = xt$. On Monday the time traveled is $t + \frac{15}{60} - \frac{19}{60} = t - \frac{1}{15}$ hours and on Tuesday the time traveled is $t - \frac{20}{60} + \frac{36}{60} = t + \frac{4}{15}$. Thus the *distance home* $= d = 30(t - \frac{1}{15})$ and $d = 15(t + \frac{4}{15})$. Setting these expressions equal, we have $30(t - \frac{1}{15}) = 15(t + \frac{4}{15}) \Leftrightarrow 30t - 2 = 15t + 4 \Leftrightarrow 15t = 6 \Leftrightarrow t = \frac{6}{15}$. Thus, $d = 15(\frac{6}{15} + \frac{4}{15}) = 10$, and $x = \frac{d}{t} = 10 \cdot \frac{15}{6} = 25$. Therefore, her usual speed is 25 mi/h, and she usually arrives home at 5:00 PM + 24 min = 5:24 PM (since $\frac{6}{15}$h = 24 min).

53. Let x be the tens digit and y be the ones digit of the number.

$$\begin{array}{r} x + y = 9 \\ 10y + x = \frac{3}{8}(10x + y) \end{array} \Rightarrow \begin{array}{rl} x + y = 9 & \times 22 \\ 22x - 77y = 0 & \times -1 \end{array} \Rightarrow \begin{array}{r} 22x + 22y = 198 \\ -22x + 77y = 0 \\ \hline 99y = 198 \end{array} \Leftrightarrow y = 2$$

So $x + 2 = 9 \Leftrightarrow x = 7$. Therefore, the number is 72.

55. $y = ax^2 + bx + c$. Since the parabola passes through the origin : $0 = a(0)^2 + b(0) + c \Leftrightarrow c = 0$. Since the parabola passes through $(1, 12)$: $12 = a(1)^2 + b(1) = a + b$. And since the parabola passes through $(3, 6)$, $6 = a(3)^2 + b(3) = 9a + 3b$

$$\begin{array}{rl} 12 = a + b & \times -3 \\ 6 = 9a + 3b & \times 1 \end{array} \Rightarrow \begin{array}{r} -36 = -3a - 3b \\ 6 = 9a + 3b \\ \hline -30 = 6a \end{array} \Leftrightarrow a = -5$$

So $12 = -5 + b \quad \Leftrightarrow \quad b = 17$. Thus, the equation of the parabola is $y = -5x^2 + 17x$.

57. 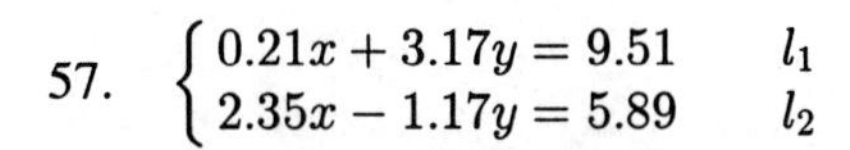

$$\begin{cases} 0.21x + 3.17y = 9.51 & l_1 \\ 2.35x - 1.17y = 5.89 & l_2 \end{cases}$$

The solution is approximately $(3.87, 2.74)$.

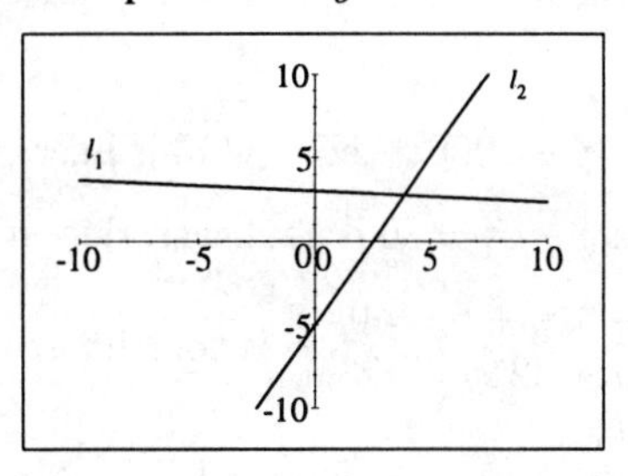

59. $$\begin{cases} 2371x - 6552y = 13,591 & l_1 \\ 9815x + 992y = 618,555 & l_2 \end{cases}$$

The solution is approximately $(61.00, 20.00)$.

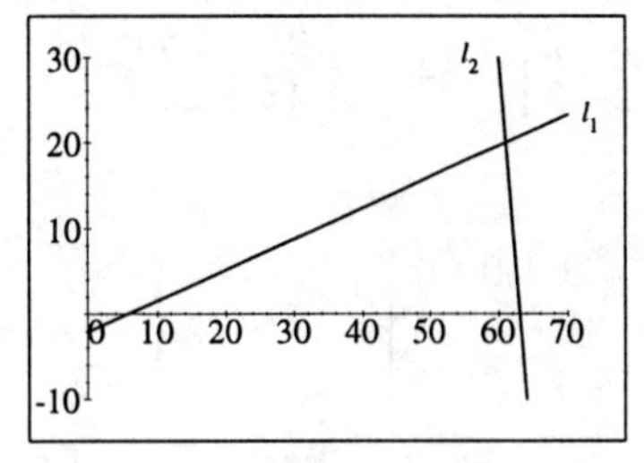

61. $$\begin{cases} \sqrt{3}\,x + \sqrt{5}\,y = \sqrt{7} & l_1 \\ -\sqrt{2}\,x + \sqrt{11}\,y = -\sqrt{13} & l_2 \end{cases}$$

The solution is approximately $(1.89, -0.28)$.

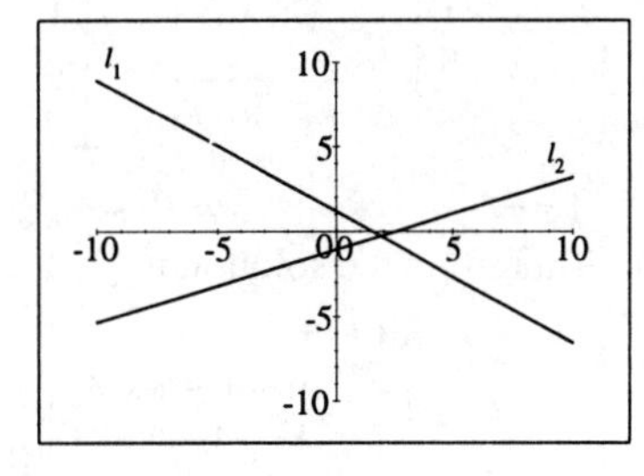

Exercises 6.2

1. $6x - 3y + 1000z - w = \sqrt{13}$ is linear.

3. $x_1^2 + x_2^2 + x_3^2 = 36$ is not linear, since it contains squares of variables.

5. $\begin{cases} x - 3xy + 5y = 0 \\ 12x + 321y = 123 \end{cases}$ is not a linear system, since the first equation contains a product of variables.

7. $\begin{bmatrix} 2 & 3 & 1 \\ 4 & 2 & 3 \end{bmatrix} \Leftrightarrow \begin{cases} 2x + 3y = 1 \\ 4x + 2y = 3 \end{cases}$

9. $\begin{bmatrix} 0 & 1 & 0 & 0 \\ 1 & 0 & 1 & 0 \\ 0 & -2 & 2 & 7 \end{bmatrix} \Leftrightarrow \begin{cases} y = 0 \\ x + z = 0 \\ -2y + 2z = 7 \end{cases}$

11. $\begin{bmatrix} 1 & -2 & 1 & 1 \\ 0 & 1 & 2 & 5 \\ 1 & 1 & 3 & 8 \end{bmatrix} \xrightarrow[R_3 \to R_3 - R_1]{} \begin{bmatrix} 1 & -2 & 1 & 1 \\ 0 & 1 & 2 & 5 \\ 0 & 3 & 2 & 7 \end{bmatrix} \xrightarrow{R_3 \to R_3 - 3R_2} \begin{bmatrix} 1 & -2 & 1 & 1 \\ 0 & 1 & 2 & 5 \\ 0 & 0 & -4 & -8 \end{bmatrix}$

Thus, $-4z = -8 \Leftrightarrow z = 2$, so $y + 2(2) = 5 \Leftrightarrow y = 1$, so $x - 2(1) + (2) = 1 \Leftrightarrow x = 1$. Therefore, the solution is $(1, 1, 2)$.

13. $\begin{bmatrix} 1 & 1 & 1 & 2 \\ 2 & -3 & 2 & 4 \\ 4 & 1 & -3 & 1 \end{bmatrix} \xrightarrow[R_3 \to R_3 - 4R_1]{R_2 \to R_2 - 2R_1} \begin{bmatrix} 1 & 1 & 1 & 2 \\ 0 & -5 & 0 & 0 \\ 0 & -3 & -7 & -7 \end{bmatrix} \xrightarrow{R_3 \to R_3 - \frac{3}{5} R_2} \begin{bmatrix} 1 & 1 & 1 & 2 \\ 0 & -5 & 0 & 0 \\ 0 & 0 & -7 & -7 \end{bmatrix}$

Thus, $-7z = -7 \Leftrightarrow z = 1$. Also, $-5y = 0 \Leftrightarrow y = 0$. Thus, $x + 0 + 1 = 2 \Leftrightarrow x = 1$. Therefore, the solution is $(x, y, z) = (1, 0, 1)$.

15. $\begin{bmatrix} 1 & 2 & -1 & -2 \\ 1 & 0 & 1 & 0 \\ 2 & -1 & -1 & -3 \end{bmatrix} \xrightarrow[R_3 \to R_3 - 2R_1]{R_2 \to R_2 - R_1} \begin{bmatrix} 1 & 2 & -1 & -2 \\ 0 & -2 & 2 & 2 \\ 0 & -5 & 1 & 1 \end{bmatrix} \xrightarrow{R_2 \to -\frac{1}{2} R_2}$

$\begin{bmatrix} 1 & 2 & -1 & -2 \\ 0 & 1 & 1 & 1 \\ 0 & -5 & 1 & 1 \end{bmatrix} \xrightarrow[R_3 \to R_3 + 5R_1]{} \begin{bmatrix} 1 & 2 & -1 & -2 \\ 0 & 1 & 1 & 1 \\ 0 & 0 & 6 & 6 \end{bmatrix}$

Thus, $6z = 6 \Leftrightarrow z = 1$. Also, $y + (1) = 1 \Leftrightarrow y = 0$. Thus, $x + 2(0) - (1) = -2 \Leftrightarrow x = -1$. Therefore, the solution is $(x, y, z) = (-1, 0, 1)$.

17. $\begin{bmatrix} 1 & 2 & -1 & 9 \\ 2 & 0 & -1 & -2 \\ 3 & 5 & 2 & 22 \end{bmatrix} \xrightarrow[R_3 \to R_3 - 3R_1]{R_2 \to R_2 - 2R_1} \begin{bmatrix} 1 & 2 & -1 & 9 \\ 0 & -4 & 1 & -20 \\ 0 & -1 & 5 & -5 \end{bmatrix} \xrightarrow{R_3 \to 4R_3 - R_2}$

$\begin{bmatrix} 1 & 2 & -1 & 9 \\ 0 & -4 & 1 & -20 \\ 0 & 0 & 19 & 0 \end{bmatrix}$

Then, $19x_3 = 0 \quad \Leftrightarrow \quad x_3 = 0$; $-4x_2 = -20 \quad \Leftrightarrow \quad x_2 = 5$; $x_1 + 2(5) = 9 \quad \Leftrightarrow \quad x_1 = -1$. Therefore, the solution is $(x_1, x_2, x_3) = (-1, 5, 0)$.

19. $$\begin{bmatrix} 2 & -3 & -1 & 13 \\ -1 & 2 & -5 & 6 \\ 5 & -1 & -1 & 49 \end{bmatrix} \xrightarrow[R_3 \to 2R_3 - 5R_1]{R_2 \to 2R_2 + R_1} \begin{bmatrix} 2 & -3 & -1 & 13 \\ 0 & 1 & -11 & 25 \\ 0 & 13 & 3 & 33 \end{bmatrix} \xrightarrow{R_3 \to R_3 - 13R_2} \begin{bmatrix} 2 & -3 & -1 & 13 \\ 0 & 1 & -11 & 25 \\ 0 & 0 & 146 & -292 \end{bmatrix}$$

Then, $146z = -292 \quad \Leftrightarrow \quad z = -2$; $y - 11(-2) = 25 \quad \Leftrightarrow \quad y = 3$; $2x - 3 \cdot 3 + 2 = 13 \quad \Leftrightarrow \quad x = 10$. Thus the solution is $(10, 3, -2)$.

21. $$\begin{bmatrix} 0.1 & 1 & -0.2 & 0.6 \\ -0.2 & 1.1 & 0.6 & -1.6 \\ 0.3 & 0.2 & 1 & -1.4 \end{bmatrix} \xrightarrow[R_3 \to R_3 - 3R_1]{R_2 \to R_2 + 2R_1} \begin{bmatrix} 0.1 & 1 & -0.2 & 0.6 \\ 0 & 3.1 & 0.2 & -0.4 \\ 0 & -2.8 & 1.6 & -3.2 \end{bmatrix}$$

$$\xrightarrow{R_3 \to 3.1R_3 + 2.8R_2} \begin{bmatrix} 0.1 & 1 & -0.2 & 0.6 \\ 0 & 3.1 & 0.2 & -0.4 \\ 0 & 0 & 5.52 & -11.04 \end{bmatrix}$$

Thus, $5.52z = -11.04 \quad \Leftrightarrow \quad z = -2$; $3.1y - 0.4 = -0.4 \quad \Leftrightarrow \quad y = 0$; $0.1x + 0.4 = 0.6 \quad \Leftrightarrow \quad x = 2$. So the solution is $(2, 0, -2)$.

23. $$\begin{bmatrix} 3 & 1 & 1 & \frac{3}{2} \\ 3 & 0 & 12 & -5 \\ 0 & 2 & -4 & 4 \end{bmatrix} \xrightarrow{R_2 \to R_2 - R_1} \begin{bmatrix} 3 & 1 & 1 & \frac{3}{2} \\ 0 & -1 & 11 & -\frac{13}{2} \\ 0 & 2 & -4 & 4 \end{bmatrix} \xrightarrow{R_3 \to R_3 + 2R_2} \begin{bmatrix} 3 & 1 & 1 & \frac{3}{2} \\ 0 & -1 & 11 & -\frac{13}{2} \\ 0 & 0 & 18 & -9 \end{bmatrix}$$

So, $18z = -9 \quad \Leftrightarrow \quad z = -\frac{1}{2}$; $-y - \frac{11}{2} = -\frac{13}{2} \quad \Leftrightarrow \quad y = 1$; $3x + 1 - \frac{1}{2} = \frac{3}{2} \quad \Leftrightarrow \quad x = \frac{1}{3}$. Thus, the solution is $\left(\frac{1}{3}, 1, -\frac{1}{2}\right)$.

25. $$\begin{bmatrix} 1 & 1 & 0 & -2 \\ 0 & 2 & 1 & 1 \\ 1 & 0 & -3 & -20 \end{bmatrix} \xrightarrow{R_3 \to R_3 - R_1} \begin{bmatrix} 1 & 1 & 0 & -2 \\ 0 & 2 & 1 & 1 \\ 0 & -1 & -3 & -18 \end{bmatrix} \xrightarrow{R_3 \to 2R_3 + R_2} \begin{bmatrix} 1 & 1 & 0 & -2 \\ 0 & 2 & 1 & 1 \\ 0 & 0 & -5 & -35 \end{bmatrix}$$

Thus, $-5z = -35 \quad \Leftrightarrow \quad z = 7$, $2y + 7 = 1 \quad \Leftrightarrow \quad y = -3$, and $x - 3 = -2 \quad \Leftrightarrow \quad x = 1$, and so the solution is $(1, -3, 7)$.

27. $$\begin{bmatrix} 1 & 0 & 7 & -20 \\ 2 & -5 & 0 & 7 \\ 0 & -3 & 1 & 0 \end{bmatrix} \xrightarrow{R_2 \to R_2 - 2R_1} \begin{bmatrix} 1 & 0 & 7 & -20 \\ 0 & -5 & -14 & 47 \\ 0 & -3 & 1 & 0 \end{bmatrix} \xrightarrow{R_3 \to 5R_3 - 3R_2}$$

$$\begin{bmatrix} 1 & 0 & 7 & -20 \\ 0 & -5 & -14 & 47 \\ 0 & 0 & 47 & -141 \end{bmatrix}$$

Thus, $47z = -141 \quad \Leftrightarrow \quad z = -3$, $-5y + 42 = 47 \quad \Leftrightarrow \quad y = -1$, and $x - 21 = -20 \quad \Leftrightarrow$ $x = 1$, and so the solution is $(1, -1, -3)$.

29. $$\begin{bmatrix} -1 & 2 & 1 & -3 & 3 \\ 3 & -4 & 1 & 1 & 9 \\ -1 & -1 & 1 & 1 & 0 \\ 2 & 1 & 4 & -2 & 3 \end{bmatrix} \xrightarrow{-R_1} \begin{bmatrix} 1 & -2 & -1 & 3 & -3 \\ 3 & -4 & 1 & 1 & 9 \\ -1 & -1 & 1 & 1 & 0 \\ 2 & 1 & 4 & -2 & 3 \end{bmatrix}$$

$$\xrightarrow[R_4 \to R_4 - 2R_1]{R_2 \to R_2 - 3R_1,\ R_3 \to R_3 + R_1} \begin{bmatrix} 1 & -2 & -1 & 3 & -3 \\ 0 & 2 & 4 & -8 & 18 \\ 0 & -3 & 0 & 4 & -3 \\ 0 & 5 & 6 & -8 & 9 \end{bmatrix} \xrightarrow{\frac{1}{2}R_2}$$

$$\begin{bmatrix} 1 & -2 & -1 & 3 & -3 \\ 0 & 1 & 2 & -4 & 9 \\ 0 & -3 & 0 & 4 & -3 \\ 0 & 5 & 6 & -8 & 9 \end{bmatrix} \xrightarrow[R_4 \to R_4 - 5R_2]{R_3 \to R_3 + 3R_2} \begin{bmatrix} 1 & -2 & -1 & 3 & -3 \\ 0 & 1 & 2 & -4 & 9 \\ 0 & 0 & 6 & -8 & 24 \\ 0 & 0 & -4 & 12 & -36 \end{bmatrix}$$

$$\xrightarrow{R_4 \to 3R_4 + 2R_3} \begin{bmatrix} 1 & -2 & -1 & 3 & -3 \\ 0 & 1 & 2 & -4 & 9 \\ 0 & 0 & 6 & -8 & 24 \\ 0 & 0 & 0 & 20 & -60 \end{bmatrix}$$

Therefore, $20w = -60 \quad \Leftrightarrow \quad w = -3$; $6z + 24 = 24 \quad \Leftrightarrow \quad z = 0$; $y + 12 = 9 \quad \Leftrightarrow$ $y = -3$; and $x + 6 - 9 = -3 \quad \Leftrightarrow \quad x = 0$. Hence, the solution is $(0, -3, 0, -3)$.

31. $$\begin{bmatrix} 1 & -1 & 1 & 2 & 3 & 0 \\ -1 & -2 & 1 & -2 & 1 & 7 \\ -1 & 1 & 0 & 1 & -1 & -4 \\ 2 & -2 & 3 & -1 & 0 & 12 \\ 1 & 0 & 1 & -1 & -5 & 5 \end{bmatrix} \xrightarrow[\substack{R_4 \to R_4 - 2R_1 \\ R_5 \to R_5 - R_1}]{\substack{R_2 \to R_2 + R_1 \\ R_3 \to R_3 + R_1}} \begin{bmatrix} 1 & -1 & 1 & 2 & 3 & 0 \\ 0 & -3 & 2 & 0 & 4 & 7 \\ 0 & 0 & 1 & 3 & 2 & -4 \\ 0 & 0 & 1 & -5 & -6 & 12 \\ 0 & 1 & 0 & -3 & -8 & 5 \end{bmatrix}$$

$$\xrightarrow{R_2 \leftrightarrow R_5} \begin{bmatrix} 1 & -1 & 1 & 2 & 3 & 0 \\ 0 & 1 & 0 & -3 & -8 & 5 \\ 0 & 0 & 1 & 3 & 2 & -4 \\ 0 & 0 & 1 & -5 & -6 & 12 \\ 0 & -3 & 2 & 0 & 4 & 7 \end{bmatrix} \xrightarrow{R_5 \to R_5 + 3R_2} \begin{bmatrix} 1 & -1 & 1 & 2 & 3 & 0 \\ 0 & 1 & 0 & -3 & -8 & 5 \\ 0 & 0 & 1 & 3 & 2 & -4 \\ 0 & 0 & 1 & -5 & -6 & 12 \\ 0 & 0 & 2 & -9 & -20 & 22 \end{bmatrix}$$

$$\xrightarrow[R_5 \to R_5 - 2R_3]{R_4 \to R_4 - R_3} \begin{bmatrix} 1 & -1 & 1 & 2 & 3 & 0 \\ 0 & 1 & 0 & -3 & -8 & 5 \\ 0 & 0 & 1 & 3 & 2 & -4 \\ 0 & 0 & 0 & -8 & -8 & 16 \\ 0 & 0 & 0 & -15 & -24 & 30 \end{bmatrix} \xrightarrow[R_5 \to R_5 + 15R_4]{-\frac{1}{8}R_4}$$

$$\begin{bmatrix} 1 & -1 & 1 & 2 & 3 & 0 \\ 0 & 1 & 0 & -3 & -8 & 5 \\ 0 & 0 & 1 & 3 & 2 & -4 \\ 0 & 0 & 0 & 1 & 1 & -2 \\ 0 & 0 & 0 & 0 & -9 & 0 \end{bmatrix}$$

Therefore, $x_5 = 0$, $x_4 + 0 = -2 \quad \Leftrightarrow \quad x_4 = -2$, $x_3 - 6 = -4 \quad \Leftrightarrow \quad x_3 = 2$, $x_2 + 6 = 5$ $\Leftrightarrow \quad x_2 = -1$, and $x_1 + 1 + 2 - 4 = 0 \quad \Leftrightarrow \quad x_1 = 1$. So, the solution is $(1, -1, 2, -2, 0)$.

33. Let x, y, z represent the number of VitaMax, Vitron, and VitaPlus pills taken daily. The matrix representation for the system of equations is:

$$\begin{bmatrix} 5 & 10 & 15 & 50 \\ 15 & 20 & 0 & 50 \\ 10 & 10 & 10 & 50 \end{bmatrix} \xrightarrow[\frac{1}{5}R_3]{\frac{1}{5}R_1,\ \frac{1}{5}R_2} \begin{bmatrix} 1 & 2 & 3 & 10 \\ 3 & 4 & 0 & 10 \\ 2 & 2 & 2 & 10 \end{bmatrix} \xrightarrow[R_3 \to R_3 - 2R_1]{R_2 \to R_2 - 3R_1} \begin{bmatrix} 1 & 2 & 3 & 10 \\ 0 & -2 & -9 & -20 \\ 0 & -2 & -4 & -10 \end{bmatrix}$$

$$\xrightarrow{R_3 \to R_3 - R_2} \begin{bmatrix} 1 & 2 & 3 & 10 \\ 0 & -2 & -9 & -20 \\ 0 & 0 & 5 & 10 \end{bmatrix}$$

Thus, $5z = 10 \quad \Leftrightarrow \quad z = 2$, $-2y - 18 = -20 \quad \Leftrightarrow \quad y = 1$, and $x + 2 + 6 = 10 \quad \Leftrightarrow$ $x = 2$. Hence, he should take 2 VitaMax, 1 Vitron, and 2 VitaPlus pills daily.

35. Let p, n, d, and q represent the number of pennies, nickels, dimes, and quarters respectively. Then, the system of equations is:

$$\begin{cases} p + n + d + q = 30 \\ .01p + .05n + .1d + .25q = 3.31 \\ p + n = d + q \\ .25q = 5\,(.1d) \end{cases}$$

which has the following matrix representation:

$$\begin{bmatrix} 1 & 1 & 1 & 1 & 30 \\ .01 & .05 & .1 & .25 & 3.31 \\ 1 & 1 & -1 & -1 & 0 \\ 0 & 0 & .5 & -.25 & 0 \end{bmatrix} \xrightarrow[4R_4]{100R_2} \begin{bmatrix} 1 & 1 & 1 & 1 & 30 \\ 1 & 5 & 10 & 25 & 331 \\ 1 & 1 & -1 & -1 & 0 \\ 0 & 0 & 2 & -1 & 0 \end{bmatrix}$$

$$\xrightarrow[R_3 \to R_3 - R_1]{R_2 \to R_2 - R_1} \begin{bmatrix} 1 & 1 & 1 & 1 & 30 \\ 0 & 4 & 9 & 24 & 301 \\ 0 & 0 & -2 & -2 & -30 \\ 0 & 0 & 2 & -1 & 0 \end{bmatrix} \xrightarrow{R_4 \to R_4 + R_3} \begin{bmatrix} 1 & 1 & 1 & 1 & 30 \\ 0 & 4 & 9 & 24 & 301 \\ 0 & 0 & -2 & -2 & -30 \\ 0 & 0 & 0 & -3 & -30 \end{bmatrix}$$

Therefore, $-3q = -30 \quad \Leftrightarrow \quad q = 10$, $-2d - 20 = -30 \quad \Leftrightarrow \quad d = 5$, $4n + 45 + 240 = 301$ $\Leftrightarrow \quad n = 4$, and $p + 4 + 5 + 10 = 30 \quad \Leftrightarrow \quad p = 11$. Hence, there are 11 pennies, 4 nickels, 5 dimes, and 10 quarters.

37. Let s, d, and f represent the daily rate of standard, deluxe, and first-class rooms respectively, where $s, d, f \geq 0$. Then, the system of equations is

$$\begin{cases} 6s + 2d + 2f = 530 \Rightarrow 3s + d + f = 265 \\ 5s + 4d + f = 510 \\ 3s + 3d + 4f = 645 \end{cases}$$ This has the following matrix representation:

$$\begin{bmatrix} 3 & 1 & 1 & 265 \\ 5 & 4 & 1 & 510 \\ 3 & 3 & 4 & 645 \end{bmatrix} \xrightarrow[\text{R}_3 \to \text{R}_3 - \text{R}_1]{\text{R}_2 \to 3\text{R}_2 - 5\text{R}_1} \begin{bmatrix} 3 & 1 & 1 & 265 \\ 0 & 7 & -2 & 205 \\ 0 & 2 & 3 & 380 \end{bmatrix} \xrightarrow{\text{R}_3 \to 7\text{R}_3 - 2\text{R}_2}$$

$$\begin{bmatrix} 3 & 1 & 1 & 265 \\ 0 & 7 & -2 & 205 \\ 0 & 0 & 25 & 2250 \end{bmatrix}$$

Therefore, $25f = 2250 \Leftrightarrow f = 90$, $7d - 180 = 205 \Leftrightarrow d = 55$, and $3s + 55 + 90 = 265 \Leftrightarrow s = 40$. Thus, the standard room rate is \$40, the deluxe room rate is \$55, and the first-class room rate is \$90.

39. For $(-2, 24)$: $4a - 2b + c = 24$. For $(1, 3)$: $a + b + c = 3$. For $(3, 9)$: $9a + 3b + c = 9$. Therefore, the system is

$$\begin{bmatrix} 1 & 1 & 1 & 3 \\ 4 & -2 & 1 & 24 \\ 9 & 3 & 1 & 9 \end{bmatrix} \xrightarrow[\text{R}_3 \to \text{R}_3 - 9\text{R}_1]{\text{R}_2 \to \text{R}_2 - 4\text{R}_1} \begin{bmatrix} 1 & 1 & 1 & 3 \\ 0 & -6 & -3 & 12 \\ 0 & -6 & -8 & -18 \end{bmatrix} \xrightarrow{\text{R}_3 \to \text{R}_3 - \text{R}_2}$$

$$\begin{bmatrix} 1 & 1 & 1 & 3 \\ 0 & -6 & -3 & 12 \\ 0 & 0 & -5 & -30 \end{bmatrix}$$

Therefore, $-5c = -30 \Leftrightarrow c = 6$, $-6b - 18 = 12 \Leftrightarrow b = -5$, and $a - 5 + 6 = 3 \Leftrightarrow a = 2$. So, the equation of the parabola is $y = 2x^2 - 5x + 6$.

41. $$\begin{cases} 5x + 2y = 1 \\ 4x - z = 1 \\ 4y + 3z = 1 \end{cases}$$ which has the following matrix representation:

$$\begin{bmatrix} 5 & 2 & 0 & 1 \\ 4 & 0 & -1 & 1 \\ 0 & 4 & 3 & 1 \end{bmatrix} \xrightarrow{\text{R}_2 \to 5\text{R}_2 - 4\text{R}_1} \begin{bmatrix} 5 & 2 & 0 & 1 \\ 0 & -8 & -5 & 1 \\ 0 & 4 & 3 & 1 \end{bmatrix} \xrightarrow{\text{R}_3 \to 2\text{R}_3 + \text{R}_2}$$

$$\begin{bmatrix} 5 & 2 & 0 & 1 \\ 0 & -8 & -5 & 1 \\ 0 & 0 & 1 & 3 \end{bmatrix}$$

Therefore, $z = 3$, $-8y - 15 = 1 \Leftrightarrow y = -2$, and $5x - 4 = 1 \Leftrightarrow x = 1$. Hence, the solution is $(1, -2, 3)$.

Exercises 6.3

1. This matrix is in reduced echelon form.

3. This matrix is not in echelon form, since first row does not have a leading 1.

5. This matrix is not in echelon form, since the row of zeros is not at the bottom.

7. $$\begin{bmatrix} 1 & 1 & 1 & 2 \\ 0 & 1 & -3 & 1 \\ 2 & 1 & 5 & 0 \end{bmatrix} \xrightarrow{R_3 \to R_3 - 2R_1} \begin{bmatrix} 1 & 1 & 1 & 2 \\ 0 & 1 & -3 & 1 \\ 0 & -1 & 3 & -4 \end{bmatrix} \xrightarrow{R_3 \to R_3 + R_2} \begin{bmatrix} 1 & 1 & 1 & 3 \\ 0 & 1 & -3 & 1 \\ 0 & 0 & 0 & -3 \end{bmatrix}$$

The third row of the matrix states $0 = -3$, which is impossible. Hence, the system is inconsistent and there is no solution.

9. $$\begin{bmatrix} 1 & -1 & 3 & 3 \\ 4 & -8 & 32 & 24 \\ 2 & -3 & 11 & 4 \end{bmatrix} \xrightarrow[R_3 \to R_3 - 2R_1]{R_2 \to R_2 - 4R_1} \begin{bmatrix} 1 & -1 & 3 & 3 \\ 0 & -4 & 20 & 12 \\ 0 & -1 & 5 & -2 \end{bmatrix} \xrightarrow[R_3 \to R_3 + R_2]{-\frac{1}{4}R_2}$$

$$\begin{bmatrix} 1 & -1 & 3 & 3 \\ 0 & 1 & -5 & -3 \\ 0 & 0 & 0 & -5 \end{bmatrix}$$

The third row of the matrix states $0 = -5$, which is impossible. Hence, the system is inconsistent and there is no solution.

11. $$\begin{bmatrix} 1 & 5 & 12 \\ 3 & -7 & 14 \\ 2 & -4 & 10 \end{bmatrix} \xrightarrow[R_3 \to R_3 - 2R_1]{R_2 \to R_2 - 3R_1} \begin{bmatrix} 1 & 5 & 12 \\ 0 & -22 & -22 \\ 0 & -14 & -14 \end{bmatrix} \xrightarrow[R_3 \to R_3 + 14R_2]{-\frac{1}{22}R_2} \begin{bmatrix} 1 & 5 & 12 \\ 0 & 1 & 1 \\ 0 & 0 & 0 \end{bmatrix}$$

Therefore, $y = 1$ and $x + 5 \cdot 1 = 12 \quad \Leftrightarrow \quad x = 7$. Hence, the solution is $(7, 1)$.

13. $$\begin{bmatrix} 1 & -3 & 1 \\ 3 & -1 & 5 \\ 4 & -8 & 3 \end{bmatrix} \xrightarrow[R_3 \to R_3 - 4R_1]{R_2 \to R_2 - 3R_1} \begin{bmatrix} 1 & -3 & 1 \\ 0 & 8 & 2 \\ 0 & 4 & -1 \end{bmatrix} \xrightarrow[R_3 \to R_3 - R_2]{\frac{1}{2}R_2} \begin{bmatrix} 1 & -3 & 1 \\ 0 & 4 & 1 \\ 0 & 0 & -2 \end{bmatrix}$$

The third row of the matrix states $0 = -2$, which is impossible. Therefore, the system is inconsistent and has no solution.

15. $$\begin{bmatrix} 1 & -1 & -1 & 0 \\ 4 & -3 & 8 & 12 \end{bmatrix} \xrightarrow{R_2 \to R_2 - 4R_1} \begin{bmatrix} 1 & -1 & -1 & 0 \\ 0 & 1 & 12 & 12 \end{bmatrix}$$ Therefore, $y + 12z = 12 \quad \Leftrightarrow$ $y = 12 - 12z$ and $x - (12 - 12z) - z = 0 \quad \Leftrightarrow \quad x = 12 - 11z$, and so the solutions are $(12 - 11z, 12 - 12z, z)$, where z is any real number.

17. $$\begin{bmatrix} 2 & -1 & 5 & 12 \\ 1 & 4 & -2 & -3 \\ 8 & 5 & 11 & 30 \end{bmatrix} \xrightarrow{R_1 \leftrightarrow R_2} \begin{bmatrix} 1 & 4 & -2 & -3 \\ 2 & -1 & 5 & 12 \\ 8 & 5 & 11 & 30 \end{bmatrix} \xrightarrow[R_3 \to R_3 - 8R_1]{R_2 \to R_2 - 2R_1}$$

$$\begin{bmatrix} 1 & 4 & -2 & -3 \\ 0 & -9 & 9 & 18 \\ 0 & -27 & 27 & 54 \end{bmatrix} \xrightarrow{R_3 \to R_3 - 3R_2} \begin{bmatrix} 1 & 4 & -2 & -3 \\ 0 & -9 & 9 & 18 \\ 0 & 0 & 0 & 0 \end{bmatrix}$$

Therefore, this system has infinitely many solutions: $-9y + 9z = 18 \quad \Leftrightarrow \quad y = -2 + z$, and $x + 4(-2 + z) - 2z = -3 \quad \Leftrightarrow \quad x = 5 - 2z$. Hence, the solutions are $(5 - 2z, -2 + z, z)$, where z is any real number.

19. $$\begin{bmatrix} 2 & 1 & -2 & 12 \\ -1 & -\frac{1}{2} & 1 & -6 \\ 3 & \frac{3}{2} & -3 & 18 \end{bmatrix} \xrightarrow[-R_1]{R_1 \leftrightarrow R_2} \begin{bmatrix} 1 & \frac{1}{2} & -1 & 6 \\ 2 & 1 & -2 & 12 \\ 3 & \frac{3}{2} & -3 & 18 \end{bmatrix} \xrightarrow[R_3 \to R_3 - 3R_1]{R_2 \to R_2 - 2R_1} \begin{bmatrix} 1 & \frac{1}{2} & -1 & 6 \\ 0 & 0 & 0 & 0 \\ 0 & 0 & 0 & 0 \end{bmatrix}$$

Therefore, this system has infinitely many solutions: $x + \frac{1}{2}y - z = 6 \quad \Leftrightarrow \quad x = 6 - \frac{1}{2}y + z$. Hence, the solutions are $(6 - \frac{1}{2}y + z, y, z)$, where y and z are any real numbers.

21. $$\begin{bmatrix} 1 & 1 & 1 & 1 & 8 \\ 0 & 1 & 0 & -1 & 0 \\ 3 & 2 & 1 & 0 & 12 \\ -3 & -2 & 1 & 4 & 0 \end{bmatrix} \xrightarrow[R_4 \to R_4 + 3R_1]{R_3 \to R_3 - 3R_1} \begin{bmatrix} 1 & 1 & 1 & 1 & 8 \\ 0 & 1 & 0 & -1 & 0 \\ 0 & -1 & -2 & -3 & -12 \\ 0 & 1 & 4 & 7 & 24 \end{bmatrix} \xrightarrow[R_4 \to R_4 - R_2]{R_3 \to R_3 + R_2}$$

$$\begin{bmatrix} 1 & 1 & 1 & 1 & 8 \\ 0 & 1 & 0 & -1 & 0 \\ 0 & 0 & -2 & -4 & -12 \\ 0 & 0 & 4 & 8 & 24 \end{bmatrix} \xrightarrow[-\frac{1}{2}R_3]{R_4 \to R_4 + 2R_3} \begin{bmatrix} 1 & 1 & 1 & 1 & 8 \\ 0 & 1 & 0 & -1 & 0 \\ 0 & 0 & 1 & 2 & 6 \\ 0 & 0 & 0 & 0 & 0 \end{bmatrix}$$

Thus, the system has infinitely many solutions: $z + 2w\mathrm{w} = 6 \quad \Leftrightarrow \quad z = 6 - 2w$, $y - w = 0 \quad \Leftrightarrow \quad y = w$, $x + w + 6 - 2w + w = 8 \quad \Leftrightarrow \quad x = 2$, and so the solutions are $(2, w, 6 - 2w, w)$, where w is any real number.

23. $$\begin{bmatrix} 2 & -1 & 2 & 1 & 5 \\ -1 & 1 & 4 & -1 & 3 \\ 3 & -2 & -1 & 0 & 0 \end{bmatrix} \xrightarrow[-R_1]{R_1 \leftrightarrow R_2} \begin{bmatrix} 1 & -1 & -4 & 1 & -3 \\ 2 & -1 & 2 & 1 & 5 \\ 3 & -2 & -1 & 0 & 0 \end{bmatrix} \xrightarrow[R_3 \to R_3 - 3R_1]{R_2 \to R_2 - 2R_1}$$

$$\begin{bmatrix} 1 & -1 & -4 & 1 & -3 \\ 0 & 1 & 10 & -1 & 11 \\ 0 & 1 & 11 & -3 & 9 \end{bmatrix} \xrightarrow{R_3 \to R_3 - R_2} \begin{bmatrix} 1 & -1 & -4 & 1 & -3 \\ 0 & 1 & 10 & -1 & 11 \\ 0 & 0 & 1 & -2 & -2 \end{bmatrix}$$

Therefore, the system has infinitely many solutions: $z - 2w = -2 \quad \Leftrightarrow \quad z = -2 + 2w$, $y + 10(-2 + 2w) - w = 11 \quad \Leftrightarrow \quad y = 31 - 19w$, and $x - (31 - 19w) - 4(-2 + 2w) + w = -3 \quad \Leftrightarrow \quad x = 20 - 12w$. Hence, the solutions are $(20 - 12w, 31 - 19w, -2 + 2w, w)$, where w is any real number.

25. $$\begin{bmatrix} 1 & -1 & 0 & 1 & 0 \\ 3 & 0 & -1 & 2 & 0 \\ 1 & -4 & 1 & 2 & 0 \end{bmatrix} \xrightarrow[R_3 \to R_3 - R_1]{R_2 \to R_2 - 3R_1} \begin{bmatrix} 1 & -1 & 0 & 1 & 0 \\ 0 & 3 & -1 & -1 & 0 \\ 0 & -3 & 1 & 1 & 0 \end{bmatrix} \xrightarrow{R_3 \to R_3 + R_2}$$

$$\begin{bmatrix} 1 & -1 & 0 & 1 & 0 \\ 0 & 3 & -1 & -1 & 0 \\ 0 & 0 & 0 & 0 & 0 \end{bmatrix}$$

Therefore, the system has infinitely many solutions: $3y - z - w = 0 \quad \Leftrightarrow \quad y = \frac{1}{3}(z + w)$, $x - \frac{1}{3}(z + w) + w = 0 \quad \Leftrightarrow \quad x = \frac{1}{3}(z - 2w)$. So, the solutions are $(\frac{1}{3}[z - 2w], \frac{1}{3}[z + w], z, w)$, where z and w are arbitrary.

27. $\begin{bmatrix} 2 & -1 & 1 & 5 \\ 3 & -4 & -2 & 1 \\ 1 & -2 & 4 & 9 \\ 2 & -3 & 5 & 0 \end{bmatrix} \xrightarrow{R_1 \leftrightarrow R_3} \begin{bmatrix} 1 & -2 & 4 & 9 \\ 3 & -4 & -2 & 1 \\ 2 & -1 & 1 & 5 \\ 2 & -3 & 5 & 0 \end{bmatrix} \xrightarrow[R_4 \to R_4 - 2R_1]{R_2 \to R_2 - 3R_1,\ R_3 \to R_3 - 2R_1}$

$\begin{bmatrix} 1 & -2 & 4 & 9 \\ 0 & 2 & -14 & -26 \\ 0 & 3 & -7 & -13 \\ 0 & 1 & -3 & -18 \end{bmatrix} \xrightarrow{R_4 \leftrightarrow R_2} \begin{bmatrix} 1 & -2 & 4 & 9 \\ 0 & 1 & -3 & -18 \\ 0 & 3 & -7 & -13 \\ 0 & 2 & -14 & -26 \end{bmatrix} \xrightarrow[R_4 \to R_4 - 2R_2]{R_3 \to R_3 - 3R_2}$

$\begin{bmatrix} 1 & -2 & 4 & 9 \\ 0 & 1 & -3 & -18 \\ 0 & 0 & 2 & 41 \\ 0 & 0 & -8 & 10 \end{bmatrix} \xrightarrow{R_4 \to R_4 + 4R_3} \begin{bmatrix} 1 & -2 & 4 & 9 \\ 0 & 1 & -3 & -18 \\ 0 & 0 & 2 & 41 \\ 0 & 0 & 0 & 174 \end{bmatrix}$

The fourth row states $0 = 174$, which is impossible. Therefore, the system is inconsistent and has no solution.

29. Let x represent the amount of soya powder, y the amount of ground millet, and z the amount of dried milk. Then, the system of equations is

$$\begin{cases} 0.2x + 0.5y + 0.4z = 1.1 \\ x + y + z = 3.5 \\ .2x + 2.0y + 1.4z = 3.1 \end{cases} \Leftrightarrow \begin{cases} 2x + 5y + 4z = 11 \\ x + y + z = 3.5 \\ 2x + 20y + 14z = 31 \end{cases}$$

and the matrix representation is:

$\begin{bmatrix} 1 & 1 & 1 & 3.5 \\ 2 & 5 & 4 & 11 \\ 2 & 20 & 14 & 31 \end{bmatrix} \xrightarrow[R_3 \to R_3 - 2R_1]{R_2 \to R_2 - 2R_1} \begin{bmatrix} 1 & 1 & 1 & 3.5 \\ 0 & 3 & 2 & 4 \\ 0 & 18 & 12 & 24 \end{bmatrix} \xrightarrow{R_3 \to R_3 - 6R_2} \begin{bmatrix} 1 & 1 & 1 & 3.5 \\ 0 & 3 & 2 & 4 \\ 0 & 0 & 0 & 0 \end{bmatrix}$

Therefore, there are infinitely many solutions: $3y + 2z = 4 \Leftrightarrow y = \frac{2}{3}(2 - z)$, and $x + \frac{2}{3}(2 - z) + z = \frac{7}{2} \Leftrightarrow x = \frac{1}{3}(\frac{13}{2} - z) = -\frac{1}{3}z + \frac{13}{6}$. Since $x \geq 0$, $y \geq 0$, and $z \geq 0$, we must have $-\frac{1}{3}z + \frac{13}{6} \geq 0$ and $-\frac{2}{3}z + \frac{4}{3} \geq 0$ and $z \geq 0 \Leftrightarrow z \leq \frac{13}{2}$ and $z \leq 2$ and $z \geq 0$ $\Leftrightarrow 0 \leq z \leq 2$. Thus she may use $-\frac{1}{3}z + \frac{16}{3}$ ounces of soya, $-\frac{2}{3}z + \frac{4}{3}$ ounces of ground millet, and z ounces of dried milk, where $0 \leq z \leq 2$.

31. Let t be the number of tables produced, c the number of chairs, and a the number of armoires. Then, the system of equations is

$$\begin{cases} \frac{1}{2}t + c + a = 300 \\ \frac{1}{2}t + \frac{3}{2}c + a = 400 \\ t + \frac{3}{2}c + 2a = 590 \end{cases} \Leftrightarrow \begin{cases} t + 2c + 2a = 600 \\ t + 3c + 2a = 800 \\ 2t + 3c + 4a = 1180 \end{cases}$$ and the matrix representation is:

$\begin{bmatrix} 1 & 2 & 2 & 600 \\ 1 & 3 & 2 & 800 \\ 2 & 3 & 4 & 1180 \end{bmatrix} \xrightarrow[R_3 \to R_3 - 2R_1]{R_2 \to R_2 - R_1} \begin{bmatrix} 1 & 2 & 2 & 600 \\ 0 & 1 & 0 & 200 \\ 0 & -1 & 0 & -20 \end{bmatrix} \xrightarrow{R_3 \to R_3 + R_2} \begin{bmatrix} 1 & 2 & 2 & 600 \\ 0 & 1 & 0 & 200 \\ 0 & 0 & 0 & 180 \end{bmatrix}$

The third row states $0 = 180$, which is impossible and so the system is inconsistent. Therefore, it is impossible to use all of the available labor-hours.

33. Let p be the number of pennies that I have, n the number of nickels, and d the number of dimes, such that $p, d, n \geq 0$. Then, the system of equations is: $0.01p + 0.05n + 0.1d = 0.72$

$$\begin{cases} p + 5n + 10d = 72 \\ \qquad\qquad d = \frac{1}{3}(n+p) \\ p + n - 3d = 0 \end{cases}$$

Therefore, subtracting equation 2 from equation 1, we get $4n + 13d = 72 \quad \Leftrightarrow$ $n = \frac{1}{4}(72 - 13d)$, and $p = 72 - 10d - \frac{5}{4}(72 - 13d) = -18 + \frac{25}{4}d$ which must be a non-negative integer, and so d must be divisible by 4. Thus d could be 4, 8, 12, $\ldots$. But since $n \geq 0$, we must have $\frac{1}{4}(72 - 13d) \geq 0 \;\Rightarrow\; d \leq \frac{72}{13} \approx 5.54$. Thus $d = 4$ is the only possibility, and so I have 7 pennies, 5 nickels, and 4 dimes.

35. (a) We begin by substituting $\left(\frac{x_0 + x_1}{2}, \frac{y_0 + y_1}{2}, \frac{z_0 + z_1}{2}\right)$ into the left side of the first equation which gives: $a_1\left(\frac{x_0 + x_1}{2}\right) + b_1\left(\frac{y_0 + y_1}{2}\right) + c_1\left(\frac{z_0 + z_1}{2}\right)$ $= \frac{1}{2}[(a_1x_0 + b_1y_0 + c_1z_0) + (a_1x_1 + b_1y_1 + c_1z_1)] = \frac{1}{2}[d_1 + d_1] = d_1$. Thus the given ordered triple satisfies the first equation. We can show that it satisfies the second and the third in exactly the same way. Thus it is a solution of the system.

(b) We have shown in part (a) that if the system has two different solutions, we can find a third one by averaging the two solutions. But then we can find a fourth and a fifth solution by averaging the new one with each of the previous two. Then we can find four more by repeating this process with these new solutions, and so on. Clearly this process can continue indefinitely, so there will be infinitely many solutions.

Exercises 6.4

In Exercises 1 – 21, the matrices A, B, C, D, E, F, and G are defined as follows:

$$A = \begin{bmatrix} 2 & -5 \\ 0 & 7 \end{bmatrix} \quad B = \begin{bmatrix} 3 & \frac{1}{2} & 5 \\ 1 & -1 & 3 \end{bmatrix} \quad C = \begin{bmatrix} 2 & -\frac{5}{2} & 0 \\ 0 & 2 & -3 \end{bmatrix} \quad D = \begin{bmatrix} 7 & 3 \end{bmatrix}$$

$$E = \begin{bmatrix} 1 \\ 2 \\ 0 \end{bmatrix} \quad F = \begin{bmatrix} 1 & 0 & 0 \\ 0 & 1 & 0 \\ 0 & 0 & 1 \end{bmatrix} \quad G = \begin{bmatrix} 5 & -3 & 10 \\ 6 & 1 & 0 \\ -5 & 2 & 2 \end{bmatrix}$$

1. $B + C = \begin{bmatrix} 3 & \frac{1}{2} & 5 \\ 1 & -1 & 3 \end{bmatrix} + \begin{bmatrix} 2 & -\frac{5}{2} & 0 \\ 0 & 2 & -3 \end{bmatrix} = \begin{bmatrix} 5 & -2 & 5 \\ 1 & 1 & 0 \end{bmatrix}$

3. $C - B = \begin{bmatrix} 2 & -\frac{5}{2} & 0 \\ 0 & 2 & -3 \end{bmatrix} - \begin{bmatrix} 3 & \frac{1}{2} & 5 \\ 1 & -1 & 3 \end{bmatrix} = \begin{bmatrix} -1 & -3 & -5 \\ -1 & 3 & -6 \end{bmatrix}$

5. $3B + 2C = 3\begin{bmatrix} 3 & \frac{1}{2} & 5 \\ 1 & -1 & 3 \end{bmatrix} + 2\begin{bmatrix} 2 & -\frac{5}{2} & 0 \\ 0 & 2 & -3 \end{bmatrix} = \begin{bmatrix} 13 & -\frac{7}{2} & 15 \\ 3 & 1 & 3 \end{bmatrix}$

7. $2C - 6B = 2\begin{bmatrix} 2 & -\frac{5}{2} & 0 \\ 0 & 2 & -3 \end{bmatrix} - 6\begin{bmatrix} 3 & \frac{1}{2} & 5 \\ 1 & -1 & 3 \end{bmatrix} = \begin{bmatrix} -14 & -8 & -30 \\ -6 & 10 & -24 \end{bmatrix}$

9. AD is undefined because A (2×2) and D (1×2) have incompatible dimensions.

11. $BF = \begin{bmatrix} 3 & \frac{1}{2} & 5 \\ 1 & -1 & 3 \end{bmatrix} \begin{bmatrix} 1 & 0 & 0 \\ 0 & 1 & 0 \\ 0 & 0 & 1 \end{bmatrix} = \begin{bmatrix} 3 & \frac{1}{2} & 5 \\ 1 & -1 & 3 \end{bmatrix}$

13. $(DA)B = \begin{bmatrix} 7 & 3 \end{bmatrix} \begin{bmatrix} 2 & -5 \\ 0 & 7 \end{bmatrix} \begin{bmatrix} 3 & \frac{1}{2} & 5 \\ 1 & -1 & 3 \end{bmatrix} = \begin{bmatrix} 14 & -14 \end{bmatrix} \begin{bmatrix} 3 & \frac{1}{2} & 5 \\ 1 & -1 & 3 \end{bmatrix} = \begin{bmatrix} 28 & 21 & 28 \end{bmatrix}$

15. $GE = \begin{bmatrix} 5 & -3 & 10 \\ 6 & 1 & 0 \\ -5 & 2 & 2 \end{bmatrix} \begin{bmatrix} 1 \\ 2 \\ 0 \end{bmatrix} = \begin{bmatrix} -1 \\ 8 \\ -1 \end{bmatrix}$

17. $A^3 = \begin{bmatrix} 2 & -5 \\ 0 & 7 \end{bmatrix} \begin{bmatrix} 2 & -5 \\ 0 & 7 \end{bmatrix} \begin{bmatrix} 2 & -5 \\ 0 & 7 \end{bmatrix} = \begin{bmatrix} 4 & -45 \\ 0 & 49 \end{bmatrix} \begin{bmatrix} 2 & -5 \\ 0 & 7 \end{bmatrix} = \begin{bmatrix} 8 & -335 \\ 0 & 343 \end{bmatrix}$

19. B^2 is undefined because the dimensions $(2 \times 3$ and $2 \times 3)$ are incompatible.

21. $BF + FE$ is undefined because the dimensions, $(2 \times 3) \cdot (3 \times 3) = (2 \times 3)$ and $(3 \times 3) \cdot (3 \times 1) = (3 \times 1)$, are incompatible.

23. $\begin{cases} 2x - 5y = 7 \\ 3x + 2y = 4 \end{cases}$ as a matrix equation is $\begin{bmatrix} 2 & -5 \\ 3 & 2 \end{bmatrix} \begin{bmatrix} x \\ y \end{bmatrix} = \begin{bmatrix} 7 \\ 4 \end{bmatrix}$

25. $\begin{cases} 3x_1 + 2x_2 - x_3 + x_4 = 0 \\ x_1 - x_3 = 5 \\ 3x_2 + x_3 - x_4 = 4 \end{cases}$ as a matrix equation is

$$\begin{bmatrix} 3 & 2 & -1 & 1 \\ 1 & 0 & -1 & 0 \\ 0 & 3 & 1 & -1 \end{bmatrix} \begin{bmatrix} x_1 \\ x_2 \\ x_3 \\ x_4 \end{bmatrix} = \begin{bmatrix} 0 \\ 5 \\ 4 \end{bmatrix}$$

27. $\begin{bmatrix} 6 & x \\ 1 & 0 \end{bmatrix} \begin{bmatrix} y & 2 \\ -1 & 2 \end{bmatrix} = \begin{bmatrix} 6y - x & 12 + 2x \\ y & 2 \end{bmatrix} = \begin{bmatrix} 4 & 16 \\ 1 & 2 \end{bmatrix}$. Thus we must solve the system

$\begin{cases} 6y - x = 4 \\ 12 + 2x = 16 \\ y = 1 \\ 2 = 2 \end{cases} \Rightarrow \quad y = 1;\ 12 + 2x = 16 \quad \Leftrightarrow \quad x = 2$. Thus the solution is $x = 2, y = 1$.

29. $2X - A = B \quad \Leftrightarrow \quad X = \frac{1}{2}(A + B) = \frac{1}{2}\left(\begin{bmatrix} 4 & 6 \\ 1 & 3 \end{bmatrix} + \begin{bmatrix} 2 & 5 \\ 3 & 7 \end{bmatrix}\right) = \frac{1}{2}\begin{bmatrix} 6 & 11 \\ 4 & 10 \end{bmatrix} = \begin{bmatrix} 3 & \frac{11}{2} \\ 2 & 5 \end{bmatrix}$

31. $3X + B = C \quad \Leftrightarrow \quad 3X = C - B \quad \Leftrightarrow \quad X = \frac{1}{3}(C - B)$ which is undefined because the dimensions, (3×2) and (2×2), are incompatible.

33. No, consider the following counterexample: $A = \begin{bmatrix} 0 & 1 \\ 0 & 0 \end{bmatrix}$ and $B = \begin{bmatrix} 0 & 2 \\ 0 & 0 \end{bmatrix}$. Then, $AB = O$, but neither $A = O$ nor $B = O$.

35. Let $A = \begin{bmatrix} a & b \\ c & d \end{bmatrix}$ and $B = \begin{bmatrix} e & f \\ g & h \end{bmatrix}$. Then, $A + B = \begin{bmatrix} a+e & b+f \\ c+g & d+h \end{bmatrix}$, and

$$(A + B)^2 = \begin{bmatrix} a+e & b+f \\ c+g & d+h \end{bmatrix} \begin{bmatrix} a+e & b+f \\ c+g & d+h \end{bmatrix}$$
$$= \begin{bmatrix} (a+e)^2 + (b+f)(c+g) & (a+e)(b+f) + (b+f)(d+h) \\ (c+g)(a+e) + (d+h)(c+g) & (c+g)(b+f) + (d+h)^2 \end{bmatrix}$$

$$A^2 = \begin{bmatrix} a & b \\ c & d \end{bmatrix} \begin{bmatrix} a & b \\ c & d \end{bmatrix} = \begin{bmatrix} a^2 + bc & ab + bd \\ ac + cd & bc + d^2 \end{bmatrix}$$

$$B^2 = \begin{bmatrix} e & f \\ g & h \end{bmatrix} \begin{bmatrix} e & f \\ g & h \end{bmatrix} = \begin{bmatrix} e^2 + fg & ef + fh \\ eg + gh & fg + h^2 \end{bmatrix}$$

$$AB = \begin{bmatrix} a & b \\ c & d \end{bmatrix} \begin{bmatrix} e & f \\ g & h \end{bmatrix} = \begin{bmatrix} ae + bg & af + bh \\ ce + dg & cf + dh \end{bmatrix}$$

$$BA = \begin{bmatrix} e & f \\ g & h \end{bmatrix} \begin{bmatrix} a & b \\ c & d \end{bmatrix} = \begin{bmatrix} ae + cf & be + df \\ ag + ch & bg + dh \end{bmatrix}$$

Then, $A^2 + AB + BA + B^2$

$$= \begin{bmatrix} a^2 + bc + ae + bg + ae + cf + e^2 + fg & ab + bd + ef + fh + af + bh + be + df \\ ac + cd + eg + gh + ce + dg + ag + ch & bc + d^2 + fg + h^2 + cf + dh + bg + dh \end{bmatrix}$$

$$= \begin{bmatrix} a^2 + 2ae + e^2 + b(c+g) + f(c+g) & a(b+f) + e(b+f) + b(d+h) + f(d+h) \\ c(a+e) + g(a+e) + d(c+g) + h(c+g) & c(b+f) + g(b+f) + d^2 + 2dh + h^2 \end{bmatrix}$$

$$= \begin{bmatrix} (a+e)^2 + (b+f)(c+g) & (a+e)(b+f) + (b+f)(d+h) \\ (c+g)(a+e) + (d+h)(c+g) & (c+g)(b+f) + (d+h)^2 \end{bmatrix}$$

$= (A+B)^2.$

37. $A = \begin{bmatrix} 1 & 1 \\ 0 & 1 \end{bmatrix}$

(a) $A^2 = \begin{bmatrix} 1 & 1 \\ 0 & 1 \end{bmatrix} \begin{bmatrix} 1 & 1 \\ 0 & 1 \end{bmatrix} = \begin{bmatrix} 1 & 2 \\ 0 & 1 \end{bmatrix}$

$A^3 = A \cdot A^2 = \begin{bmatrix} 1 & 1 \\ 0 & 1 \end{bmatrix} \begin{bmatrix} 1 & 2 \\ 0 & 1 \end{bmatrix} = \begin{bmatrix} 1 & 3 \\ 0 & 1 \end{bmatrix}$

$A^4 = A \cdot A^3 = \begin{bmatrix} 1 & 1 \\ 0 & 1 \end{bmatrix} \begin{bmatrix} 1 & 3 \\ 0 & 1 \end{bmatrix} = \begin{bmatrix} 1 & 4 \\ 0 & 1 \end{bmatrix}$

(b) $A^n = \begin{bmatrix} 1 & n \\ 0 & 1 \end{bmatrix}$

39. (a) $BA = [\$0.90 \quad \$0.80 \quad \$1.10] \begin{bmatrix} 4000 & 1000 & 3500 \\ 400 & 300 & 200 \\ 700 & 500 & 9000 \end{bmatrix} = [\$4690 \quad \$1690 \quad \$13{,}310]$

(b) The entries in the product matrix represent the total food sales in Santa Monica, Long Beach, and Anaheim respectively.

41. $A = \begin{bmatrix} 1 & 0 & 6 & -1 \\ 2 & \frac{1}{2} & 4 & 0 \end{bmatrix}, \quad B = [1 \quad 7 \quad -9 \quad 2], \; C = \begin{bmatrix} 1 \\ 0 \\ -1 \\ -2 \end{bmatrix}.$

ABC is undefined because the dimensions of A (2×4) and B (1×4) are not compatible.

$ACB = \begin{bmatrix} -3 \\ -2 \end{bmatrix} [1 \quad 7 \quad -9 \quad 2] = \begin{bmatrix} -3 & -21 & 27 & -6 \\ -2 & -14 & 18 & -4 \end{bmatrix}$

BAC is undefined because the dimensions of B (1×4) and A (2×4) are not compatible.

BCA is undefined because the dimensions of C (4×1) and A (2×4) are not compatible.

CAB is undefined because the dimensions of C (4×1) and A (2×4) are not compatible.

CBA is undefined because the dimensions of B (1×4) and A (2×4) are not compatible.

Exercises 6.5

1. $A = \begin{bmatrix} 7 & 4 \\ 3 & 2 \end{bmatrix} \Leftrightarrow A^{-1} = \frac{1}{14-12}\begin{bmatrix} 2 & -4 \\ -3 & 7 \end{bmatrix} = \begin{bmatrix} 1 & -2 \\ -\frac{3}{2} & \frac{7}{2} \end{bmatrix}$

   Then, $AA^{-1} = \begin{bmatrix} 7 & 4 \\ 3 & 2 \end{bmatrix}\begin{bmatrix} 1 & -2 \\ -\frac{3}{2} & \frac{7}{2} \end{bmatrix} = \begin{bmatrix} 1 & 0 \\ 0 & 1 \end{bmatrix}$

   and $A^{-1}A = \begin{bmatrix} 1 & -2 \\ -\frac{3}{2} & \frac{7}{2} \end{bmatrix}\begin{bmatrix} 7 & 4 \\ 3 & 2 \end{bmatrix} = \begin{bmatrix} 1 & 0 \\ 0 & 1 \end{bmatrix}$

3. $\begin{bmatrix} 5 & 3 \\ 3 & 2 \end{bmatrix}^{-1} = \frac{1}{10-9}\begin{bmatrix} 2 & -3 \\ -3 & 5 \end{bmatrix} = \begin{bmatrix} 2 & -3 \\ -3 & 5 \end{bmatrix}$

5. $\begin{bmatrix} 2 & 5 \\ -5 & -13 \end{bmatrix}^{-1} = \frac{1}{-26+25}\begin{bmatrix} -13 & -5 \\ 5 & 2 \end{bmatrix} = \begin{bmatrix} 13 & 5 \\ -5 & -2 \end{bmatrix}$

7. $\begin{bmatrix} 6 & -3 \\ -8 & 4 \end{bmatrix}^{-1} = \frac{1}{24-24}\begin{bmatrix} 4 & 3 \\ 8 & 6 \end{bmatrix}$ which is not defined, and so there is no inverse.

9. $\begin{bmatrix} 0.4 & -1.2 \\ 0.3 & 0.6 \end{bmatrix}^{-1} = \frac{1}{0.24+0.36}\begin{bmatrix} 0.6 & 1.2 \\ -0.3 & 0.4 \end{bmatrix} = \begin{bmatrix} 1 & 2 \\ -\frac{1}{2} & \frac{2}{3} \end{bmatrix}$

11. $\begin{bmatrix} 2 & 4 & 1 & 1 & 0 & 0 \\ -1 & 1 & -1 & 0 & 1 & 0 \\ 1 & 4 & 0 & 0 & 0 & 1 \end{bmatrix} \xrightarrow[R_3 \to 2R_3 - R_1]{R_2 \to 2R_2 + R_1} \begin{bmatrix} 2 & 4 & 1 & 1 & 0 & 0 \\ 0 & 6 & -1 & 1 & 2 & 0 \\ 0 & 4 & -1 & -1 & 0 & 2 \end{bmatrix}$

    $\xrightarrow[R_1 \to 3R_1 - 2R_2]{R_3 \to 3R_3 - 2R_2} \begin{bmatrix} 6 & 0 & 5 & 1 & -4 & 0 \\ 0 & 6 & -1 & 1 & 2 & 0 \\ 0 & 0 & -1 & -5 & -4 & 6 \end{bmatrix} \xrightarrow[R_2 \to R_2 - R_3]{R_1 \to R_1 + 5R_3}$

    $\begin{bmatrix} 6 & 0 & 0 & -24 & -24 & 30 \\ 0 & 6 & 0 & 6 & 6 & -6 \\ 0 & 0 & -1 & -5 & -4 & 6 \end{bmatrix} \xrightarrow[R_2 \to \frac{1}{6}R_2,\ R_3 \to -R_3]{R_1 \to \frac{1}{6}R_1} \begin{bmatrix} 1 & 0 & 0 & -4 & -4 & 5 \\ 0 & 1 & 0 & 1 & 1 & -1 \\ 0 & 0 & 1 & 5 & 4 & -6 \end{bmatrix}$

    Therefore, the inverse matrix is $\begin{bmatrix} -4 & -4 & 5 \\ 1 & 1 & -1 \\ 5 & 4 & -6 \end{bmatrix}$.

13. $\begin{bmatrix} 1 & 2 & 3 & 1 & 0 & 0 \\ 4 & 5 & -1 & 0 & 1 & 0 \\ 1 & -1 & -10 & 0 & 0 & 1 \end{bmatrix} \xrightarrow[R_3 \to R_3 - R_1]{R_2 \to R_2 - 4R_1} \begin{bmatrix} 1 & 2 & 3 & 1 & 0 & 0 \\ 0 & -3 & -13 & -4 & 1 & 0 \\ 0 & -3 & -13 & -1 & 0 & 1 \end{bmatrix}$

    $\xrightarrow{R_3 \to R_3 - R_2} \begin{bmatrix} 1 & 2 & 3 & 1 & 0 & 0 \\ 0 & -3 & -13 & -4 & 1 & 0 \\ 0 & 0 & 0 & 3 & -1 & 1 \end{bmatrix}$

    Since the left half of the last row consists entirely of zeros, there is no inverse matrix.

15. $\begin{bmatrix} 0 & -2 & 2 & 1 & 0 & 0 \\ 3 & 1 & 3 & 0 & 1 & 0 \\ 1 & -2 & 3 & 0 & 0 & 1 \end{bmatrix} \xrightarrow{R_1 \leftrightarrow R_3} \begin{bmatrix} 1 & -2 & 3 & 0 & 0 & 1 \\ 3 & 1 & 3 & 0 & 1 & 0 \\ 0 & -2 & 2 & 1 & 0 & 0 \end{bmatrix} \xrightarrow{R_2 \to R_2 - 3R_1}$

$\begin{bmatrix} 1 & -2 & 3 & 0 & 0 & 1 \\ 0 & 7 & -6 & 0 & 1 & -3 \\ 0 & -2 & 2 & 1 & 0 & 0 \end{bmatrix} \xrightarrow[R_2 \to R_2 + 3R_3]{R_1 \to R_1 - R_3} \begin{bmatrix} 1 & 0 & 1 & -1 & 0 & 1 \\ 0 & 1 & 0 & 3 & 1 & -3 \\ 0 & -2 & 2 & 1 & 0 & 0 \end{bmatrix}$

$\xrightarrow{R_3 \to R_3 + 2R_2} \begin{bmatrix} 1 & 0 & 1 & -1 & 0 & 1 \\ 0 & 1 & 0 & 3 & 1 & -3 \\ 0 & 0 & 2 & 7 & 2 & -6 \end{bmatrix} \xrightarrow{R_3 \to \frac{1}{2} R_3} \begin{bmatrix} 1 & 0 & 1 & -1 & 0 & 1 \\ 0 & 1 & 0 & 3 & 1 & -3 \\ 0 & 0 & 1 & \frac{7}{2} & 1 & -3 \end{bmatrix}$

$\xrightarrow{R_1 \to R_1 - R_3} \begin{bmatrix} 1 & 0 & 0 & -\frac{9}{2} & -1 & 4 \\ 0 & 1 & 0 & 3 & 1 & -3 \\ 0 & 0 & 1 & \frac{7}{2} & 1 & -3 \end{bmatrix}$

Therefore, the inverse matrix is $\begin{bmatrix} -\frac{9}{2} & -1 & 4 \\ 3 & 1 & -3 \\ \frac{7}{2} & 1 & -3 \end{bmatrix}$.

17. $\begin{bmatrix} 1 & 2 & 0 & 3 & 1 & 0 & 0 & 0 \\ 0 & 1 & 1 & 1 & 0 & 1 & 0 & 0 \\ 0 & 1 & 0 & 1 & 0 & 0 & 1 & 0 \\ 1 & 2 & 0 & 2 & 0 & 0 & 0 & 1 \end{bmatrix} \xrightarrow[R_4 \to R_4 - R_1]{R_3 \to R_3 - R_2} \begin{bmatrix} 1 & 2 & 0 & 3 & 1 & 0 & 0 & 0 \\ 0 & 1 & 1 & 1 & 0 & 1 & 0 & 0 \\ 0 & 0 & -1 & 0 & 0 & -1 & 1 & 0 \\ 0 & 0 & 0 & -1 & -1 & 0 & 0 & 1 \end{bmatrix}$

$\xrightarrow[R_4 \to -R_4]{R_3 \to -R_3} \begin{bmatrix} 1 & 2 & 0 & 3 & 1 & 0 & 0 & 0 \\ 0 & 1 & 1 & 1 & 0 & 1 & 0 & 0 \\ 0 & 0 & 1 & 0 & 0 & 1 & -1 & 0 \\ 0 & 0 & 0 & 1 & 1 & 0 & 0 & -1 \end{bmatrix} \xrightarrow[R_2 \to R_2 - R_3]{R_1 \to R_1 - 2R_2}$

$\begin{bmatrix} 1 & 0 & -2 & 1 & 1 & -2 & 0 & 0 \\ 0 & 1 & 0 & 1 & 0 & 0 & 1 & 0 \\ 0 & 0 & 1 & 0 & 0 & 1 & -1 & 0 \\ 0 & 0 & 0 & 1 & 1 & 0 & 0 & -1 \end{bmatrix} \xrightarrow[R_2 \to R_2 - R_4]{R_1 \to R_1 + 2R_3}$

$\begin{bmatrix} 1 & 0 & 0 & 1 & 1 & 0 & -2 & 0 \\ 0 & 1 & 0 & 0 & -1 & 0 & 1 & 1 \\ 0 & 0 & 1 & 0 & 0 & 1 & -1 & 0 \\ 0 & 0 & 0 & 1 & 1 & 0 & 0 & -1 \end{bmatrix} \xrightarrow{R_1 \to R_1 - R_4} \begin{bmatrix} 1 & 0 & 0 & 0 & 0 & 0 & -2 & 1 \\ 0 & 1 & 0 & 0 & -1 & 0 & 1 & 1 \\ 0 & 0 & 1 & 0 & 0 & 1 & -1 & 0 \\ 0 & 0 & 0 & 1 & 1 & 0 & 0 & -1 \end{bmatrix}$

Therefore, the inverse matrix is $\begin{bmatrix} 0 & 0 & -2 & 1 \\ -1 & 0 & 1 & 1 \\ 0 & 1 & -1 & 0 \\ 1 & 0 & 0 & -1 \end{bmatrix}$

19. $\begin{cases} 5x + 3y = 4 \\ 3x + 2y = 0 \end{cases}$ is equivalent to the matrix equation $\begin{bmatrix} 5 & 3 \\ 3 & 2 \end{bmatrix} \begin{bmatrix} x \\ y \end{bmatrix} = \begin{bmatrix} 4 \\ 0 \end{bmatrix}$

Using the inverse from Exercise 3, $\begin{bmatrix} x \\ y \end{bmatrix} = \begin{bmatrix} 2 & -3 \\ -3 & 5 \end{bmatrix} \begin{bmatrix} 4 \\ 0 \end{bmatrix} = \begin{bmatrix} 8 \\ -12 \end{bmatrix}$

Therefore, $x = 8$ and $y = -12$.

21. $\begin{cases} 2x + 5y = 2 \\ -5x - 13y = 20 \end{cases}$ is equivalent to the matrix equation $\begin{bmatrix} 2 & 5 \\ -5 & -13 \end{bmatrix} \begin{bmatrix} x \\ y \end{bmatrix} = \begin{bmatrix} 2 \\ 20 \end{bmatrix}$

Using the inverse from Exercise 5, $\begin{bmatrix} x \\ y \end{bmatrix} = \begin{bmatrix} 13 & 5 \\ -5 & -2 \end{bmatrix} \begin{bmatrix} 2 \\ 20 \end{bmatrix} = \begin{bmatrix} 126 \\ -50 \end{bmatrix}$

Therefore, $x = 126$ and $y = -50$.

23. $\begin{cases} 2x + 4y + z = 7 \\ -x + y - z = 0 \\ x + 4y = -2 \end{cases}$ is equivalent to $\begin{bmatrix} 2 & 4 & 1 \\ -1 & 1 & -1 \\ 1 & 4 & 0 \end{bmatrix} \begin{bmatrix} x \\ y \\ z \end{bmatrix} = \begin{bmatrix} 7 \\ 0 \\ -2 \end{bmatrix}$

Using the inverse from Exercise 11, $\begin{bmatrix} x \\ y \\ z \end{bmatrix} = \begin{bmatrix} -4 & -4 & 5 \\ 1 & 1 & -1 \\ 5 & 4 & -6 \end{bmatrix} \begin{bmatrix} 7 \\ 0 \\ -2 \end{bmatrix} = \begin{bmatrix} -38 \\ 9 \\ 47 \end{bmatrix}$

Thus, $x = -38$, $y = 9$, and $z = 47$.

25. $\begin{cases} 3x - 2y = 6 \\ 5x + y + z = 12 \\ 2x - 2y = 18 \end{cases}$ is equivalent to $\begin{bmatrix} 3 & -2 & 0 \\ 5 & 1 & 1 \\ 2 & -2 & 0 \end{bmatrix} \begin{bmatrix} x \\ y \\ z \end{bmatrix} = \begin{bmatrix} 6 \\ 12 \\ 18 \end{bmatrix}$

Using the inverse from Exercise 16, $\begin{bmatrix} x \\ y \\ z \end{bmatrix} = \begin{bmatrix} 1 & 0 & -1 \\ 1 & 0 & -\frac{3}{2} \\ -6 & 1 & \frac{13}{2} \end{bmatrix} \begin{bmatrix} 6 \\ 12 \\ 18 \end{bmatrix} = \begin{bmatrix} -12 \\ -21 \\ 93 \end{bmatrix}$

Thus, $x = -12$ $y = -21$ and $z = 93$.

27. $\begin{bmatrix} 3 & -2 \\ -4 & 3 \end{bmatrix}^{-1} = \frac{1}{9-8} \begin{bmatrix} 3 & 2 \\ 4 & 3 \end{bmatrix} = \begin{bmatrix} 3 & 2 \\ 4 & 3 \end{bmatrix}$

Hence, $\begin{bmatrix} x & y & z \\ u & v & w \end{bmatrix} = \begin{bmatrix} 3 & 2 \\ 4 & 3 \end{bmatrix} \begin{bmatrix} 1 & 0 & -1 \\ 2 & 1 & 3 \end{bmatrix} = \begin{bmatrix} 7 & 2 & 3 \\ 10 & 3 & 5 \end{bmatrix}$

29. (a) $\begin{bmatrix} 3 & 1 & 3 & 1 & 0 & 0 \\ 4 & 2 & 4 & 0 & 1 & 0 \\ 3 & 2 & 4 & 0 & 0 & 1 \end{bmatrix} \xrightarrow[R_1 \leftrightarrow R_2]{R_3 \to R_3 - R_1} \begin{bmatrix} 4 & 2 & 4 & 0 & 1 & 0 \\ 3 & 1 & 3 & 1 & 0 & 0 \\ 0 & 1 & 1 & -1 & 0 & 1 \end{bmatrix} \xrightarrow{R_1 \to R_1 - R_2}$

$\begin{bmatrix} 1 & 1 & 1 & -1 & 1 & 0 \\ 3 & 1 & 3 & 1 & 0 & 0 \\ 0 & 1 & 1 & -1 & 0 & 1 \end{bmatrix} \xrightarrow{R_2 \to R_2 - 3R_1} \begin{bmatrix} 1 & 1 & 1 & -1 & 1 & 0 \\ 0 & -2 & 0 & 4 & -3 & 0 \\ 0 & 1 & 1 & -1 & 0 & 1 \end{bmatrix}$

$\xrightarrow[R_3 \to R_3 - R_2]{R_1 \to R_1 - R_2,\ R_2 \to -\frac{1}{2} R_2} \begin{bmatrix} 1 & 0 & 1 & 1 & -\frac{1}{2} & 0 \\ 0 & 1 & 0 & -2 & \frac{3}{2} & 0 \\ 0 & 0 & 1 & 1 & -\frac{3}{2} & 1 \end{bmatrix} \xrightarrow{R_1 \to R_1 - R_3}$

$\begin{bmatrix} 1 & 0 & 0 & 0 & 1 & -1 \\ 0 & 1 & 0 & -2 & \frac{3}{2} & 0 \\ 0 & 0 & 1 & 1 & -\frac{3}{2} & 1 \end{bmatrix}$

Therefore, the inverse of the matrix is $\begin{bmatrix} 0 & 1 & -1 \\ -2 & \frac{3}{2} & 0 \\ 1 & -\frac{3}{2} & 1 \end{bmatrix}$

(b) $\begin{bmatrix} A \\ B \\ C \end{bmatrix} = \begin{bmatrix} 0 & 1 & -1 \\ -2 & \frac{3}{2} & 0 \\ 1 & -\frac{3}{2} & 1 \end{bmatrix} \begin{bmatrix} 10 \\ 14 \\ 13 \end{bmatrix} = \begin{bmatrix} 1 \\ 1 \\ 2 \end{bmatrix}$

Therefore, he should feed the rats 1 oz of A, 1 oz of B, and 2 oz of C.

(c) $\begin{bmatrix} A \\ B \\ C \end{bmatrix} = \begin{bmatrix} 0 & 1 & -1 \\ -2 & \frac{3}{2} & 0 \\ 1 & -\frac{3}{2} & 1 \end{bmatrix} \begin{bmatrix} 9 \\ 12 \\ 10 \end{bmatrix} = \begin{bmatrix} 2 \\ 0 \\ 1 \end{bmatrix}$

Therefore, he should feed the rats 2 oz of A, 0 oz of B, and 1 oz of C.

(d) $\begin{bmatrix} A \\ B \\ C \end{bmatrix} = \begin{bmatrix} 0 & 1 & -1 \\ -2 & \frac{3}{2} & 0 \\ 1 & -\frac{3}{2} & 1 \end{bmatrix} \begin{bmatrix} 2 \\ 4 \\ 11 \end{bmatrix} = \begin{bmatrix} -7 \\ 2 \\ 7 \end{bmatrix}$

Since $A < 0$, there is no combination of foods giving the required supply.

31. (a) $\begin{cases} x + y + 2z = 675 \\ 2x + y + z = 600 \\ x + 2y + z = 625 \end{cases}$

(b) $\begin{bmatrix} 1 & 1 & 2 \\ 2 & 1 & 1 \\ 1 & 2 & 1 \end{bmatrix} \begin{bmatrix} x \\ y \\ z \end{bmatrix} = \begin{bmatrix} 675 \\ 600 \\ 625 \end{bmatrix}$

(c) $\begin{bmatrix} 1 & 1 & 2 & 1 & 0 & 0 \\ 2 & 1 & 1 & 0 & 1 & 0 \\ 1 & 2 & 1 & 0 & 0 & 1 \end{bmatrix} \xrightarrow[R_3 \to R_3 - R_1]{R_2 \to R_2 - 2R_1} \begin{bmatrix} 1 & 1 & 2 & 1 & 0 & 0 \\ 0 & -1 & -3 & -2 & 1 & 0 \\ 0 & 1 & -1 & -1 & 0 & 1 \end{bmatrix}$

$\xrightarrow[R_3 \to R_3 + R_2]{R_1 \to R_1 + R_2} \begin{bmatrix} 1 & 0 & -1 & -1 & 1 & 0 \\ 0 & -1 & -3 & -2 & 1 & 0 \\ 0 & 0 & -4 & -3 & 1 & 1 \end{bmatrix} \xrightarrow[R_3 \to -\frac{1}{4}R_3]{R_2 \to -R_2}$

$\begin{bmatrix} 1 & 0 & -1 & -1 & 1 & 0 \\ 0 & 1 & 3 & 2 & -1 & 0 \\ 0 & 0 & 1 & \frac{3}{4} & -\frac{1}{4} & -\frac{1}{4} \end{bmatrix} \xrightarrow[R_2 \to R_2 - 3R_2]{R_1 \to R_1 + R_3} \begin{bmatrix} 1 & 0 & 0 & -\frac{1}{4} & \frac{3}{4} & -\frac{1}{4} \\ 0 & 1 & 0 & -\frac{1}{4} & -\frac{1}{4} & \frac{3}{4} \\ 0 & 0 & 1 & \frac{3}{4} & -\frac{1}{4} & -\frac{1}{4} \end{bmatrix}$

Therefore, the inverse of the matrix is $\begin{bmatrix} -\frac{1}{4} & \frac{3}{4} & -\frac{1}{4} \\ -\frac{1}{4} & -\frac{1}{4} & \frac{3}{4} \\ \frac{3}{4} & -\frac{1}{4} & -\frac{1}{4} \end{bmatrix}$ and

$$\begin{bmatrix} x \\ y \\ z \end{bmatrix} = \begin{bmatrix} -\frac{1}{4} & \frac{3}{4} & -\frac{1}{4} \\ -\frac{1}{4} & -\frac{1}{4} & \frac{3}{4} \\ \frac{3}{4} & -\frac{1}{4} & -\frac{1}{4} \end{bmatrix} \begin{bmatrix} 675 \\ 600 \\ 625 \end{bmatrix} = \begin{bmatrix} 125 \\ 150 \\ 200 \end{bmatrix}.$$

Thus he earns \$125 on a standard set, \$150 on a deluxe set, and \$200 or a leather-bound set.

33. $\begin{bmatrix} x & 1 \\ -1 & \frac{1}{x} \end{bmatrix}^{-1} = \frac{1}{1+1}\begin{bmatrix} \frac{1}{x} & -1 \\ 1 & x \end{bmatrix} = \begin{bmatrix} \frac{1}{2x} & -\frac{1}{2} \\ \frac{1}{2} & \frac{x}{2} \end{bmatrix}$

35. $\begin{bmatrix} 1 & e^x & 0 & 1 & 0 & 0 \\ e^x & -e^{2x} & 0 & 0 & 1 & 0 \\ 0 & 0 & 2 & 0 & 0 & 1 \end{bmatrix} \xrightarrow{R_2 \to R_2 - e^x R_1} \begin{bmatrix} 1 & e^x & 0 & 1 & 0 & 0 \\ 0 & -2e^{2x} & 0 & -e^x & 1 & 0 \\ 0 & 0 & 2 & 0 & 0 & 1 \end{bmatrix} \xrightarrow[R_3 \to \frac{1}{2}R_3]{R_2 \to -\frac{1}{2}e^{-2x}R_2}$

$\begin{bmatrix} 1 & e^x & 0 & 1 & 0 & 0 \\ 0 & 1 & 0 & \frac{1}{2}e^{-x} & -\frac{1}{2}e^{-2x} & 0 \\ 0 & 0 & 1 & 0 & 0 & \frac{1}{2} \end{bmatrix} \xrightarrow{R_1 \to R_1 - e^x R_2} \begin{bmatrix} 1 & 0 & 0 & \frac{1}{2} & \frac{1}{2}e^{-x} & 0 \\ 0 & 1 & 0 & \frac{1}{2}e^{-x} & -\frac{1}{2}e^{-2x} & 0 \\ 0 & 0 & 1 & 0 & 0 & \frac{1}{2} \end{bmatrix}$

Therefore, the inverse matrix is $\begin{bmatrix} \frac{1}{2} & \frac{1}{2}e^{-x} & 0 \\ \frac{1}{2}e^{-x} & -\frac{1}{2}e^{-2x} & 0 \\ 0 & 0 & \frac{1}{2} \end{bmatrix}$

37. $\begin{bmatrix} a & 0 & 0 & 0 & 1 & 0 & 0 & 0 \\ 0 & b & 0 & 0 & 0 & 1 & 0 & 0 \\ 0 & 0 & c & 0 & 0 & 0 & 1 & 0 \\ 0 & 0 & 0 & d & 0 & 0 & 0 & 1 \end{bmatrix} \xrightarrow[R_3 \to \frac{1}{c}R_3,\, R_4 \to \frac{1}{d}R_4]{R_1 \to \frac{1}{a}R_1,\, R_2 \to \frac{1}{b}R_2} \begin{bmatrix} 1 & 0 & 0 & 0 & \frac{1}{a} & 0 & 0 & 0 \\ 0 & 1 & 0 & 0 & 0 & \frac{1}{b} & 0 & 0 \\ 0 & 0 & 1 & 0 & 0 & 0 & \frac{1}{c} & 0 \\ 0 & 0 & 0 & 1 & 0 & 0 & 0 & \frac{1}{d} \end{bmatrix}$

Thus the matrix $\begin{bmatrix} a & 0 & 0 & 0 \\ 0 & b & 0 & 0 \\ 0 & 0 & c & 0 \\ 0 & 0 & 0 & d \end{bmatrix}$ has inverse $\begin{bmatrix} \frac{1}{a} & 0 & 0 & 0 \\ 0 & \frac{1}{b} & 0 & 0 \\ 0 & 0 & \frac{1}{c} & 0 \\ 0 & 0 & 0 & \frac{1}{d} \end{bmatrix}$.

Exercises 6.6

1. $[\,3\,]$ has determinant $|D| = 3$.

3. $\begin{bmatrix} 4 & 5 \\ 0 & -1 \end{bmatrix}$ has determinant $|D| = (4)(-1) - (5)(0) = -4$.

5. $[2 \quad 5]$ does not have a determinant because the matrix is not square.

7. $\begin{bmatrix} \frac{1}{2} & \frac{1}{8} \\ 1 & \frac{1}{2} \end{bmatrix}$ has determinant $|D| = \frac{1}{2} \cdot \frac{1}{2} - 1 \cdot \frac{1}{8} = \frac{1}{4} - \frac{1}{8} = \frac{1}{8}$.

In Exercises 9 – 13, the matrix is $A = \begin{bmatrix} 1 & 0 & \frac{1}{2} \\ -3 & 5 & 2 \\ 0 & 0 & 4 \end{bmatrix}$

9. $M_{11} = 5 \cdot 4 - 0 \cdot 2 = 20$, $A_{11} = (-1)^2\, M_{11} = 20$

11. $M_{12} = -3 \cdot 4 - 0 \cdot 2 = -12$, $A_{12} = (-1)^3\, M_{12} = 12$

13. $M_{23} = 1 \cdot 0 - 0 \cdot 0 = 0$, $A_{23} = (-1)^5\, M_{23} = 0$

15. $M = \begin{bmatrix} 1 & 3 & 7 \\ 2 & 0 & -1 \\ 0 & 2 & 6 \end{bmatrix}$. Therefore, expanding by the third row, $|M| = -2\begin{vmatrix} 1 & 7 \\ 2 & -1 \end{vmatrix} + 6\begin{vmatrix} 1 & 3 \\ 2 & 0 \end{vmatrix}$
$= -2(-1-14) + 6(0-6) = 30 - 36 = -6$. Since $|M| \neq 0$, the matrix has an inverse.

17. $M = \begin{bmatrix} 30 & 0 & 20 \\ 0 & -10 & -20 \\ 40 & 0 & 10 \end{bmatrix}$. Therefore, expanding by the first row,

$$|M| = 30\begin{vmatrix} -10 & -20 \\ 0 & 10 \end{vmatrix} + 20\begin{vmatrix} 0 & -10 \\ 40 & 0 \end{vmatrix} = 30(-100+0) + 20(0+400)$$
$= -3000 + 8000 = 5000$, and so M^{-1} exists.

19. $M = \begin{bmatrix} 1 & 3 & 3 & 0 \\ 0 & 2 & 0 & 1 \\ -1 & 0 & 0 & 2 \\ 1 & 6 & 4 & 1 \end{bmatrix}$. Therefore, expanding by the third row,

$$|M| = -1\begin{vmatrix} 3 & 3 & 0 \\ 2 & 0 & 1 \\ 6 & 4 & 1 \end{vmatrix} - 2\begin{vmatrix} 1 & 3 & 3 \\ 0 & 2 & 0 \\ 1 & 6 & 4 \end{vmatrix} = 1\begin{vmatrix} 3 & 3 \\ 6 & 4 \end{vmatrix} - 1\begin{vmatrix} 3 & 3 \\ 2 & 0 \end{vmatrix} - 4\begin{vmatrix} 1 & 3 \\ 1 & 4 \end{vmatrix} = -6 + 6 - 4 = -4,$$
and so M^{-1} exists.

21. $|M| = \begin{vmatrix} 0 & 0 & 4 & 6 \\ 2 & 1 & 1 & 3 \\ 2 & 1 & 2 & 3 \\ 3 & 0 & 1 & 7 \end{vmatrix} = \begin{vmatrix} 0 & 0 & 4 & 6 \\ 2 & 1 & 1 & 3 \\ 0 & 0 & 1 & 0 \\ 3 & 0 & 1 & 7 \end{vmatrix}$ by replacing R_3 by $R_3 - R_2$. Then, expanding by the third row, $|M| = 1 \begin{vmatrix} 0 & 0 & 6 \\ 2 & 1 & 3 \\ 3 & 0 & 7 \end{vmatrix} = 6 \begin{vmatrix} 2 & 1 \\ 3 & 0 \end{vmatrix} = 6(2 \cdot 0 - 3 \cdot 1) = -18.$

23. $M = \begin{bmatrix} 1 & 2 & 3 & 4 & 5 \\ 0 & 2 & 4 & 6 & 8 \\ 0 & 0 & 3 & 6 & 9 \\ 0 & 0 & 0 & 4 & 8 \\ 0 & 0 & 0 & 0 & 5 \end{bmatrix}$ Then, expanding by the fifth row $|M| = 5 \begin{vmatrix} 1 & 2 & 3 & 4 \\ 0 & 2 & 4 & 6 \\ 0 & 0 & 3 & 6 \\ 0 & 0 & 0 & 4 \end{vmatrix}$

$= 5 \cdot 4 \begin{vmatrix} 1 & 2 & 3 \\ 0 & 2 & 4 \\ 0 & 0 & 3 \end{vmatrix} = 20 \cdot 3 \begin{vmatrix} 1 & 2 \\ 0 & 2 \end{vmatrix} = 60 \cdot 2 = 120.$

25. $B = \begin{bmatrix} 4 & 1 & 0 \\ -2 & -1 & 1 \\ 4 & 0 & 3 \end{bmatrix}$

(a) $|B| = 2 \begin{vmatrix} 1 & 0 \\ 0 & 3 \end{vmatrix} - 1 \begin{vmatrix} 4 & 0 \\ 4 & 3 \end{vmatrix} - 1 \begin{vmatrix} 4 & 1 \\ 4 & 0 \end{vmatrix} = 6 - 12 + 4 = -2$

(b) $|B| = -1 \begin{vmatrix} 4 & 1 \\ 4 & 0 \end{vmatrix} + 3 \begin{vmatrix} 4 & 1 \\ -2 & -1 \end{vmatrix} = 4 - 6 = -2$

(c) Yes, as expected, the results agree.

27. $\begin{cases} 2x - y = -9 \\ x + 2y = 8 \end{cases}$ Then, $|D| = \begin{vmatrix} 2 & -1 \\ 1 & 2 \end{vmatrix} = 5$, $|D_x| = \begin{vmatrix} -9 & -1 \\ 8 & 2 \end{vmatrix} = -10$, and $|D_y| = \begin{vmatrix} 2 & -9 \\ 1 & 8 \end{vmatrix} = 25$. Hence, $x = \dfrac{|D_x|}{|D|} = \dfrac{-10}{5} = -2$, $y = \dfrac{|D_y|}{|D|} = \dfrac{25}{5} = 5$, and so the solution is $(-2, 5)$.

29. $\begin{cases} x - 6y = 3 \\ 3x + 2y = 1 \end{cases}$ Then, $|D| = \begin{vmatrix} 1 & -6 \\ 3 & 2 \end{vmatrix} = 20$, $|D_x| = \begin{vmatrix} 3 & -6 \\ 1 & 2 \end{vmatrix} = 12$, and $|D_y| = \begin{vmatrix} 1 & 3 \\ 3 & 1 \end{vmatrix} = -8$. Hence, $x = \dfrac{|D_x|}{|D|} = \dfrac{12}{20} = 0.6$, $y = \dfrac{|D_y|}{|D|} = \dfrac{-8}{20} = -0.4$, and so the solution is $(0.6, -0.4)$.

31. $\begin{cases} 0.4x + 1.2y = 0.4 \\ 1.2x + 1.6y = 3.2 \end{cases}$ Then, $|D| = \begin{vmatrix} 0.4 & 1.2 \\ 1.2 & 1.6 \end{vmatrix} = -0.8$, $|D_x| = \begin{vmatrix} 0.4 & 1.2 \\ 3.2 & 1.6 \end{vmatrix} = -3.2$, and $|D_y| = \begin{vmatrix} 0.4 & 0.4 \\ 1.2 & 3.2 \end{vmatrix} = 0.8$. Hence, $x = \dfrac{|D_x|}{|D|} = \dfrac{-3.2}{-0.8} = 4$, $y = \dfrac{|D_y|}{|D|} = \dfrac{0.8}{-0.8} = -1$, and so the solution is $(4, -1)$.

33. $\begin{cases} x - y + 2z = 0 \\ 3x + z = 11 \\ -x + 2y = 0 \end{cases}$ Then, expanding by the third row

$$|D| = \begin{vmatrix} 1 & -1 & 2 \\ 3 & 0 & 1 \\ -1 & 2 & 0 \end{vmatrix} = -3\begin{vmatrix} -1 & 2 \\ 2 & 0 \end{vmatrix} - 1\begin{vmatrix} 1 & -1 \\ -1 & 2 \end{vmatrix} = 12 - 1 = 11,$$

$$|D_x| = \begin{vmatrix} 0 & -1 & 2 \\ 11 & 0 & 1 \\ 0 & 2 & 0 \end{vmatrix} = -11\begin{vmatrix} -1 & 2 \\ 2 & 0 \end{vmatrix} = 44, \ |D_y| = \begin{vmatrix} 1 & 0 & 2 \\ 3 & 11 & 1 \\ -1 & 0 & 0 \end{vmatrix} = 11\begin{vmatrix} 1 & 2 \\ -1 & 0 \end{vmatrix} = 22,$$

and $|D_z| = \begin{vmatrix} 1 & -1 & 0 \\ 3 & 0 & 11 \\ -1 & 2 & 0 \end{vmatrix} = -11\begin{vmatrix} 1 & -1 \\ -1 & 2 \end{vmatrix} = -11$. Therefore, $x = \frac{44}{11} = 4$, $y = \frac{22}{11} = 2$, $z = \frac{-11}{11} = -1$, and so the solution is $(4, 2, -1)$.

35. $\begin{cases} 2x_1 + 3x_2 - 5x_3 = 1 \\ x_1 + x_2 - x_3 = 2 \\ 2x_2 + x_3 = 8 \end{cases}$ Then, expanding by the third row

$$|D| = \begin{vmatrix} 2 & 3 & -5 \\ 1 & 1 & -1 \\ 0 & 2 & 1 \end{vmatrix} = -2\begin{vmatrix} 2 & -5 \\ 1 & -1 \end{vmatrix} + \begin{vmatrix} 2 & 3 \\ 1 & 1 \end{vmatrix} = -6 - 1 = -7,$$

$$|D_{x_1}| = \begin{vmatrix} 1 & 3 & -5 \\ 2 & 1 & -1 \\ 8 & 2 & 1 \end{vmatrix} = \begin{vmatrix} 1 & -1 \\ 2 & 1 \end{vmatrix} - 3\begin{vmatrix} 2 & -1 \\ 8 & 1 \end{vmatrix} - 5\begin{vmatrix} 2 & 1 \\ 8 & 2 \end{vmatrix} = 3 - 30 + 20 = -7,$$

$$|D_{x_2}| = \begin{vmatrix} 2 & 1 & -5 \\ 1 & 2 & -1 \\ 0 & 8 & 1 \end{vmatrix} = -8\begin{vmatrix} 2 & -5 \\ 1 & -1 \end{vmatrix} + \begin{vmatrix} 2 & 1 \\ 1 & 2 \end{vmatrix} = -24 + 3 = -21, \text{ and}$$

$|D_{x_3}| = \begin{vmatrix} 2 & 3 & 1 \\ 1 & 1 & 2 \\ 0 & 2 & 8 \end{vmatrix} = -2\begin{vmatrix} 2 & 1 \\ 1 & 2 \end{vmatrix} + 8\begin{vmatrix} 2 & 3 \\ 1 & 1 \end{vmatrix} = -6 - 8 = -14$. Thus, $x_1 = \frac{-7}{-7} = 1$, $x_2 = \frac{-21}{-7} = 3$, $x_3 = \frac{-14}{-7} = 2$ and so the solution is $(1, 3, 2)$.

37. $\begin{cases} \frac{1}{3}x - \frac{1}{5}y + \frac{1}{2}z = \frac{7}{10} \\ -\frac{2}{3}x + \frac{2}{5}y + \frac{3}{2}z = \frac{11}{10} \\ x - \frac{4}{5}y + z = \frac{9}{5} \end{cases} \Leftrightarrow \begin{cases} 10x - 6y + 15z = 21 \\ -20x + 12y + 45z = 33 \\ 5x - 4y + 5z = 9 \end{cases}$

Then, $|D| = \begin{vmatrix} 10 & -6 & 15 \\ -20 & 12 & 45 \\ 5 & -4 & 5 \end{vmatrix} = 10\begin{vmatrix} 12 & 45 \\ -4 & 5 \end{vmatrix} + 6\begin{vmatrix} -20 & 45 \\ 5 & 5 \end{vmatrix} + 15\begin{vmatrix} -20 & 12 \\ 5 & -4 \end{vmatrix}$
$= 2400 - 1950 + 300 = 750$,

$$|D_x| = \begin{vmatrix} 21 & -6 & 15 \\ 33 & 12 & 45 \\ 9 & -4 & 5 \end{vmatrix} = 21\begin{vmatrix} 12 & 45 \\ -4 & 5 \end{vmatrix} + 6\begin{vmatrix} 33 & 45 \\ 9 & 5 \end{vmatrix} + 15\begin{vmatrix} 33 & 12 \\ 9 & -4 \end{vmatrix}$$
$= 5040 - 1440 - 3600 = 0$,

$$|D_y| = \begin{vmatrix} 10 & 21 & 15 \\ -20 & 33 & 45 \\ 5 & 9 & 5 \end{vmatrix} = 10\begin{vmatrix} 33 & 45 \\ 9 & 5 \end{vmatrix} - 21\begin{vmatrix} -20 & 45 \\ 5 & 5 \end{vmatrix} + 15\begin{vmatrix} -20 & 33 \\ 5 & 9 \end{vmatrix}$$

$= -2400 + 6825 - 5175 = -750$, and

$$|D_z| = \begin{vmatrix} 10 & -6 & 21 \\ -20 & 12 & 33 \\ 5 & -4 & 9 \end{vmatrix} = 10\begin{vmatrix} 12 & 33 \\ -4 & 9 \end{vmatrix} + 6\begin{vmatrix} -20 & 33 \\ 5 & 9 \end{vmatrix} + 21\begin{vmatrix} -20 & 12 \\ 5 & -4 \end{vmatrix}$$

$= 2400 - 2070 + 420 = 750$.

Therefore, $x = 0$, $y = -1$, $z = 1$, and so the solution is $(0, -1, 1)$.

39. $\begin{cases} 3y + 5z = 4 \\ 2x - z = 10 \\ 4x + 7y = 0 \end{cases}$ Then, $|D| = \begin{vmatrix} 0 & 3 & 5 \\ 2 & 0 & -1 \\ 4 & 7 & 0 \end{vmatrix} = -3\begin{vmatrix} 2 & -1 \\ 4 & 0 \end{vmatrix} + 5\begin{vmatrix} 2 & 0 \\ 4 & 7 \end{vmatrix} = -12 + 70 = 58$,

$$|D_x| = \begin{vmatrix} 4 & 3 & 5 \\ 10 & 0 & -1 \\ 0 & 7 & 0 \end{vmatrix} = -7\begin{vmatrix} 4 & 5 \\ 10 & -1 \end{vmatrix} = 378,$$

$$|D_y| = \begin{vmatrix} 0 & 4 & 5 \\ 2 & 10 & -1 \\ 4 & 0 & 0 \end{vmatrix} = 4\begin{vmatrix} 4 & 5 \\ 10 & -1 \end{vmatrix} = -216, \text{ and}$$

$$|D_z| = \begin{vmatrix} 0 & 3 & 4 \\ 2 & 0 & 10 \\ 4 & 7 & 0 \end{vmatrix} = 4\begin{vmatrix} 3 & 4 \\ 0 & 10 \end{vmatrix} - 7\begin{vmatrix} 0 & 4 \\ 2 & 10 \end{vmatrix} = 120 + 56 = 176.$$

Thus, $x = \frac{189}{29}$, $y = -\frac{108}{29}$, and $z = \frac{88}{29}$, and so the solution is $\left(\frac{189}{29}, -\frac{108}{29}, \frac{88}{29}\right)$.

41. $\begin{cases} 3r - s + 3t = 7 \\ 4r + 5s - 2t = 0 \\ 9r + s + t = 0 \end{cases}$ Then,

$$|D| = \begin{vmatrix} 3 & -1 & 3 \\ 4 & 5 & -2 \\ 9 & 1 & 1 \end{vmatrix} = 9\begin{vmatrix} -1 & 3 \\ 5 & -2 \end{vmatrix} - 1\begin{vmatrix} 3 & 3 \\ 4 & -2 \end{vmatrix} + \begin{vmatrix} 3 & -1 \\ 4 & 5 \end{vmatrix} = -117 + 18 + 19 = -80,$$

$$|D_r| = \begin{vmatrix} 7 & -1 & 3 \\ 0 & 5 & -2 \\ 0 & 1 & 1 \end{vmatrix} = 7\begin{vmatrix} 5 & -2 \\ 1 & 1 \end{vmatrix} = 49, |D_s| = \begin{vmatrix} 3 & 7 & 3 \\ 4 & 0 & -2 \\ 9 & 0 & 1 \end{vmatrix} = -7\begin{vmatrix} 4 & -2 \\ 9 & 1 \end{vmatrix} = -154,$$

and $|D_t| = \begin{vmatrix} 3 & -1 & 7 \\ 4 & 5 & 0 \\ 9 & 1 & 0 \end{vmatrix} = 7\begin{vmatrix} 4 & 5 \\ 9 & 1 \end{vmatrix} = -287$. Thus, $r = -\frac{49}{80}$, $s = \frac{77}{40}$, and $t = \frac{287}{80}$, and so the solution is $\left(-\frac{49}{80}, \frac{77}{40}, \frac{287}{80}\right)$.

43. $\begin{cases} x + y + z + w = 0 \\ 2z + w = 0 \\ y - z = 0 \\ x + 2z = 1 \end{cases}$. Then $|D| = \begin{vmatrix} 1 & 1 & 1 & 1 \\ 2 & 0 & 0 & 1 \\ 0 & 1 & -1 & 0 \\ 1 & 0 & 2 & 0 \end{vmatrix} = -1\begin{vmatrix} 2 & 0 & 1 \\ 0 & -1 & 0 \\ 1 & 2 & 0 \end{vmatrix} - 1\begin{vmatrix} 1 & 1 & 1 \\ 2 & 0 & 1 \\ 1 & 2 & 0 \end{vmatrix}$

$$= -\left(2\begin{vmatrix} -1 & 0 \\ 2 & 0 \end{vmatrix} + 1\begin{vmatrix} 0 & 1 \\ -1 & 0 \end{vmatrix}\right) - \left(-1\begin{vmatrix} 2 & 1 \\ 1 & 0 \end{vmatrix} - 2\begin{vmatrix} 1 & 1 \\ 2 & 1 \end{vmatrix}\right) = -2(0) - 1(1) + 1(-1) + 2(-1)$$
$= -4.$

$$|D_x| = \begin{vmatrix} 0 & 1 & 1 & 1 \\ 0 & 0 & 0 & 1 \\ 0 & 1 & -1 & 0 \\ 1 & 0 & 2 & 0 \end{vmatrix} = -1\begin{vmatrix} 1 & 1 & 1 \\ 0 & 0 & 1 \\ 1 & -1 & 0 \end{vmatrix} = -1(-1)\begin{vmatrix} 1 & 1 \\ 1 & -1 \end{vmatrix} = -2;$$

$$|D_y| = \begin{vmatrix} 1 & 0 & 1 & 1 \\ 2 & 0 & 0 & 1 \\ 0 & 0 & -1 & 0 \\ 1 & 1 & 2 & 0 \end{vmatrix} = 1\begin{vmatrix} 1 & 1 & 1 \\ 2 & 0 & 1 \\ 0 & -1 & 0 \end{vmatrix} = 1\begin{vmatrix} 0 & 1 \\ -1 & 0 \end{vmatrix} - 2\begin{vmatrix} 1 & 1 \\ -1 & 0 \end{vmatrix} = 1 - 2(1) = -1;$$

$$|D_z| = \begin{vmatrix} 1 & 1 & 0 & 1 \\ 2 & 0 & 0 & 1 \\ 0 & 1 & 0 & 0 \\ 1 & 0 & 1 & 0 \end{vmatrix} = -1\begin{vmatrix} 1 & 1 & 1 \\ 2 & 0 & 1 \\ 0 & 1 & 0 \end{vmatrix} = -1\begin{vmatrix} 0 & 1 \\ 1 & 0 \end{vmatrix} + 2\begin{vmatrix} 1 & 1 \\ 1 & 0 \end{vmatrix} = -1(-1) + 2(-1) = -1;$$

$$|D_w| = \begin{vmatrix} 1 & 1 & 1 & 0 \\ 2 & 0 & 0 & 0 \\ 0 & 1 & -1 & 0 \\ 1 & 0 & 2 & 1 \end{vmatrix} = 1\begin{vmatrix} 1 & 1 & 1 \\ 2 & 0 & 0 \\ 0 & 1 & -1 \end{vmatrix} = -2\begin{vmatrix} 1 & 1 \\ 1 & -1 \end{vmatrix} = -2(-2) = 4.$$ Hence the solution is

$$x = \frac{|D_x|}{|D|} = \frac{-2}{-4} = \frac{1}{2};\ y = \frac{|D_y|}{|D|} = \frac{-1}{-4} = \frac{1}{4};\ z = \frac{|D_z|}{|D|} = \frac{-1}{-4} = \frac{1}{4};\ w = \frac{|D_w|}{|D|} = \frac{4}{-4} = -1.$$

45. (a) Let $|M| = \begin{vmatrix} x_1 & y_1 & 1 \\ x_2 & y_2 & 1 \\ x & y & 1 \end{vmatrix}$. Then, expanding by the third column,

$$|M| = \begin{vmatrix} x_2 & y_2 \\ x & y \end{vmatrix} - \begin{vmatrix} x_1 & y_1 \\ x & y \end{vmatrix} + \begin{vmatrix} x_1 & y_1 \\ x_2 & y_2 \end{vmatrix} = x_2y - xy_2 - x_1y + xy_1 + x_1y_2 - x_2y_1$$
$= (x_2 - x_1)y + (-y_2 + y_1)x + x_1y_2 - x_2y_1$. So $|M| = 0 \Leftrightarrow$
$(x_2 - x_1)y + (-y_2 + y_1)x + x_1y_2 - x_2y_1 = 0 \Leftrightarrow$
$(x_2 - x_1)y = (y_2 - y_1)x - x_1y_2 + x_2y_1 \Leftrightarrow$
$(x_2 - x_1)y = (y_2 - y_1)x - x_1y_2 + x_1y_1 - x_1y_1 + x_2y_1 \Leftrightarrow$
$y = \dfrac{y_2 - y_1}{x_2 - x_1}x - \dfrac{x_1(y_2 - y_1)}{x_2 - x_1} + \dfrac{y_1(x_2 - x_1)}{x_2 - x_1} \Leftrightarrow y = \dfrac{y_2 - y_1}{x_2 - x_1}(x - x_1) + y_1 \Leftrightarrow$
$y - y_1 = \dfrac{y_2 - y_1}{x_2 - x_1}(x - x_1)$, which is the "two-point" form of the equation for the line passing through the points (x_1, y_1) and (x_2, y_2).

(b) Using the result of part (a), the line has the equation

$$\begin{vmatrix} 20 & 50 & 1 \\ -10 & 25 & 1 \\ x & y & 1 \end{vmatrix} = 0 \Rightarrow 20(25 - y) - 50(-10 - x) + (-10y - 25x) = 0 \Leftrightarrow$$

$$500 - 20y + 500 + 50x - 10y - 25x = 0 \quad \Leftrightarrow \quad 25x - 30y + 1000 = 0 \quad \Leftrightarrow$$

$$5x - 6y + 200 = 0.$$

47. $\begin{vmatrix} x & 12 & 13 \\ 0 & x-1 & 23 \\ 0 & 0 & x-2 \end{vmatrix} = 0 \quad \Leftrightarrow \quad (x-2)\begin{vmatrix} x & 12 \\ 0 & x-1 \end{vmatrix} = 0 \quad \Leftrightarrow \quad (x-2) \cdot x(x-1) = 0 \quad \Leftrightarrow$

$x = 0, 1$ or 2.

49. $\begin{vmatrix} 1 & 0 & x \\ x^2 & 1 & 0 \\ x & 0 & 1 \end{vmatrix} = 0 \quad \Leftrightarrow \quad 1\begin{vmatrix} 1 & 0 \\ 0 & 1 \end{vmatrix} + x\begin{vmatrix} x^2 & 1 \\ x & 0 \end{vmatrix} = 0 \quad \Leftrightarrow \quad 1 - x^2 = 0 \quad \Leftrightarrow \quad x^2 = 1 \quad \Leftrightarrow$

$x = \pm 1$.

Exercises 6.7

1. $\begin{cases} y = x^2 \\ y = x + 6 \end{cases}$ Subtracting the second equation from the first equation gives $0 = x^2 - x - 6 = (x-3)(x+2) \quad \Rightarrow \quad x = 3$ or $x = -2$. So since $y = x^2$, the solutions are $(-2, 4)$ and $(3, 9)$.

3. $\begin{cases} x^2 + y^2 = 8 \\ x + y = 0 \end{cases} \quad \Rightarrow \quad y = -x$, and substituting into the first equation gives $x^2 + (-x)^2 = 8$ $\Leftrightarrow \quad 2x^2 = 8 \quad \Leftrightarrow \quad x = \pm 2$. So, the solutions are $(2, -2)$ and $(-2, 2)$.

5. $\begin{cases} y + x^2 = 4x \\ y + 4x = 16 \end{cases}$ Subtracting the second equation from the first gives $x^2 - 4x = 4x - 16 \quad \Leftrightarrow$ $x^2 - 8x + 16 = 0 \quad \Leftrightarrow \quad (x-4)^2 = 0 \quad \Leftrightarrow \quad x = 4$. Substituting this value for x into either of the original equations gives $y = 0$. Therefore, the solution is $(4, 0)$.

7. $\begin{cases} x - 2y = 2 \\ y^2 - x^2 = 2x + 4 \end{cases}$ Now $x - 2y = 2 \quad \Leftrightarrow \quad x = 2y + 2$. Substituting for x gives
$y^2 - x^2 = 2x + 4 \quad \Leftrightarrow \quad y^2 - (2y+2)^2 = 2(2y+2) + 4 \quad \Leftrightarrow$
$y^2 - 4y^2 - 8y - 4 = 4y + 4 + 4 \quad \Leftrightarrow \quad y^2 + 4y + 4 = 0 \quad \Leftrightarrow \quad (y+2)^2 = 0 \quad \Leftrightarrow \quad y = -2.$
$x = 2(-2) + 2 = -2$. So, the solution is $(-2, -2)$.

9. $\begin{cases} x - y = 4 \\ xy = 12 \end{cases}$ Now $x - y = 4 \quad \Leftrightarrow \quad x = 4 + y$. Substituting for x we have $xy = 12 \quad \Leftrightarrow$ $(4+y)y = 12 \quad \Leftrightarrow \quad y^2 + 4y - 12 = 0 \quad \Leftrightarrow \quad (y+6)(y-2) = 0 \quad \Leftrightarrow \quad y = -6, y = 2$. Thus, the solutions are $(-2, -6)$ and $(6, 2)$.

11. $\begin{cases} 3x^2 - y^2 = 11 \\ x^2 + 4y^2 = 8 \end{cases} \quad \Leftrightarrow \quad \begin{cases} 12x^2 - 4y^2 = 44 \\ x^2 + 4y^2 = 8 \end{cases}$ Adding the equations gives $13x^2 = 52 \quad \Leftrightarrow$ $x = \pm 2$. Thus, the solutions are $(2, 1), (2, -1), (-2, 1)$, and $(-2, -1)$.

13. $\begin{cases} x^2 y = 16 \\ x^2 + 4y + 16 = 0 \end{cases}$ Now $x^2 y = 16 \quad \Leftrightarrow \quad x^2 = \dfrac{16}{y}$. Substituting for x^2 gives
$\dfrac{16}{y} + 4y + 16 = 0 \quad \Leftrightarrow \quad 4y^2 + 16y + 16 = 0 \quad \Leftrightarrow \quad y^2 + 4y + 4 = 0 \quad \Leftrightarrow \quad (y+2)^2 = 0$
$\Leftrightarrow \quad y = -2$. Therefore, $x^2 = \frac{16}{-2} = -8$ which has no real solution, and so the system has no solution.

15. $\begin{cases} x + \sqrt{y} = 0 \\ y^2 - 4x^2 = 12 \end{cases}$ Now $x + \sqrt{y} = 0 \quad \Leftrightarrow \quad x = -\sqrt{y}$. Substituting for x gives
$y^2 - 4(-\sqrt{y})^2 = 12 \quad \Leftrightarrow \quad y^2 - 4y - 12 = 0 \quad \Leftrightarrow \quad (y-6)(y+2) = 0 \quad \Rightarrow \quad y = 6, y = -2.$
Since $x = -\sqrt{-2}$ is not a real solution for $y = -2$, the only solution is $(-\sqrt{6}, 6)$.

17. $\begin{cases} x^2 + y^2 = 9 \\ x^2 - y^2 = 1 \end{cases}$ Adding the equations gives $2x^2 = 10 \quad \Leftrightarrow \quad x^2 = 5 \quad \Leftrightarrow \quad x = \pm\sqrt{5}$. For $x = \pm\sqrt{5} \quad \Rightarrow \quad y^2 = 9 - 5 = 4 \quad \Leftrightarrow \quad y = \pm 2$, and so the solutions are $(\sqrt{5}, 2)$, $(\sqrt{5}, -2)$, $(-\sqrt{5}, 2)$, and $(-\sqrt{5}, -2)$.

19. $\begin{cases} 2x^2 - 8y^3 = 19 \\ 4x^2 + 16y^3 = 34 \end{cases} \quad \Leftrightarrow \quad \begin{cases} 4x^2 - 16y^3 = 38 \\ 4x^2 + 16y^3 = 34 \end{cases}$ Adding the two equations gives $8x^2 = 72$ $\Leftrightarrow \quad x = \pm 3$, and then $2(9) - 8y^3 = 19 \quad \Leftrightarrow \quad y^3 = -\frac{1}{8} \quad \Leftrightarrow \quad y = -\frac{1}{2}$. Therefore, the solutions are $(3, -\frac{1}{2})$ and $(-3, -\frac{1}{2})$.

21. $\begin{cases} \dfrac{2}{x} - \dfrac{3}{y} = 1 \\ -\dfrac{4}{x} + \dfrac{7}{y} = 1 \end{cases}$ If we let $u = \dfrac{1}{x}$ and $v = \dfrac{1}{y}$, the system is equivalent to

$\begin{cases} 2u - 3v = 1 \\ -4u + 7v = 1 \end{cases} \quad \Leftrightarrow \quad \begin{cases} 4u - 6v = 2 \\ -4u + 7v = 1 \end{cases}$. Adding the equations gives $v = 3$, and $2u - 9 = 1 \quad \Leftrightarrow \quad u = 5$. Thus, the solution is $(\frac{1}{5}, \frac{1}{3})$.

23. $\begin{cases} 3\sqrt{x} + 5\sqrt{y} = 19 \\ 2\sqrt{x} + 7\sqrt{y} = 20 \end{cases}$ If we let $u = \sqrt{x}$ and $v = \sqrt{y}$, the system is equivalent to

$\begin{cases} 3u + 5v = 19 \\ 2u + 7v = 20 \end{cases} \quad \Leftrightarrow \quad \begin{cases} 6u + 10v = 38 \\ 6u + 21v = 60 \end{cases}$. Subtracting the equations gives $11v = 22 \quad \Leftrightarrow$ $v = 2$, and $3u + 5(2) = 19 \quad \Leftrightarrow \quad u = 3$. Then, $x = 9$ and $y = 4$, and so the solution is $(9, 4)$.

25. $\begin{cases} x^2 - xy + 2y^2 = 8 \\ x^3 - xy^2 = 0 \end{cases}$ From the second equation, it follows that $x(x^2 - y^2) = 0$, so $x = 0$ or $x = \pm y$. Now substitute into the first equation. For $x = 0$: $(0)^2 - (0)y + 2y^2 = 8 \quad \Leftrightarrow$ $y = \pm 2$. For $x = y$: $(y)^2 - (y)y + 2y^2 = 8 \quad \Leftrightarrow \quad y = \pm 2$ and $x = \pm 2$. For $x = -y$: $(-y)^2 - (-y)y + 2y^2 = 8 \quad \Leftrightarrow \quad y = \pm\sqrt{2}$, $x = \mp\sqrt{2}$. Hence, the solutions are $(0, 2)$, $(0, -2)$, $(2, 2)$, $(-2, -2)$, $(\sqrt{2}, -\sqrt{2})$, and $(-\sqrt{2}, \sqrt{2})$.

27. $\begin{cases} x - y = 2 \\ y + z = 0 \\ x^2 + y^2 + z^2 = 4 \end{cases} \quad \Leftrightarrow \quad \begin{cases} x = 2 + y \\ y = -z \\ x^2 + y^2 + z^2 = 4 \end{cases}$. Substituting for y in the first equation gives $x = 2 - z$. Next substituting for x and y gives $(2 - z)^2 + (-z)^2 + z^2 = 4 \quad \Leftrightarrow$ $4 - 4z + 3z^2 = 4 \quad \Leftrightarrow \quad 3z^2 - 4z = 0 \quad \Leftrightarrow \quad z(3z - 4) = 0 \quad \Rightarrow \quad z = 0$ or $z = \frac{4}{3}$. When $z = 0$ then $y = 0$ and $x = 2$. When $z = \frac{4}{3}$ then $y = -\frac{4}{3}$ and $x = \frac{2}{3}$. Therefore, the solutions are $(2, 0, 0)$ and $(\frac{2}{3}, -\frac{4}{3}, \frac{4}{3})$.

29. $\begin{cases} x^2 + yz = 0 \\ y + xz = 2 \\ xyz = 1 \end{cases}$ From the first equation, $x^2 + yz = 0 \quad \Leftrightarrow \quad yz = -x^2$. Substituting into the third equation we have $x(-x^2) = 1 \quad \Leftrightarrow \quad x^3 = -1 \quad \Leftrightarrow \quad x = -1$. Thus the last second equation becomes $y - z = 2 \quad \Leftrightarrow \quad y = z + 2$. Substituting for x and y into the third equation gives $(-1)(z + 2)z = 1 \quad \Leftrightarrow \quad z^2 + 2z + 1 = 0 \quad \Leftrightarrow \quad (z + 1)^2 = 0 \quad \Leftrightarrow \quad z = -1$. Finally, $(-1)y(-1) = 1 \quad \Leftrightarrow \quad y = 1$ and the solution is $(-1, 1, -1)$.

31. $\begin{cases} x^2 + y + z = 0 \\ 2x^2 - y + 3z = -4 \\ y^2 + yz = 0 \end{cases}$ Now $y^2 + yz = 0 \;\Rightarrow\; y(y+z) = 0 \;\Leftrightarrow\; y = 0$ or $y = -z$.

If $y = 0$, then substituting into the first and second equation we have

$\begin{cases} x^2 + z = 0 \\ 2x^2 + 3z = -4 \end{cases} \;\Leftrightarrow\; \begin{cases} 2x^2 + 2z = 0 \\ 2x^2 + 3z = -4 \end{cases}$ Subtracting gives $z = -4$, thus $x^2 - 4 = 0 \;\Rightarrow$ $x = \pm 2$.

If $y = -z$, then the first equation becomes $x^2 = 0$ so $x = 0$. Then substituting for x and y in the second equation yields $-(-z) + 3z = -4 \;\Leftrightarrow\; z = -1$. Thus, the solutions are $(2, 0, -4)$, $(-2, 0, -4)$, and $(0, 1, -1)$.

33. $\begin{cases} x^2 + y^2 = 25 \\ x + 3y = 2 \end{cases}$

$\Leftrightarrow \begin{cases} y = \pm\sqrt{25 - x^2} \\ y = -\frac{1}{3}x + \frac{2}{3} \end{cases}$

The solutions are $(-4.51, 2.17)$ and $(4.91, -0.97)$.

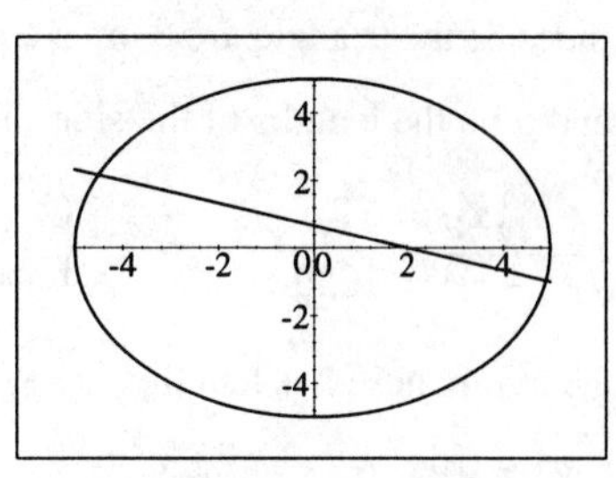

35. $\begin{cases} \frac{x^2}{9} + \frac{y^2}{18} = 1 \\ y = -x^2 + 6x - 2 \end{cases}$

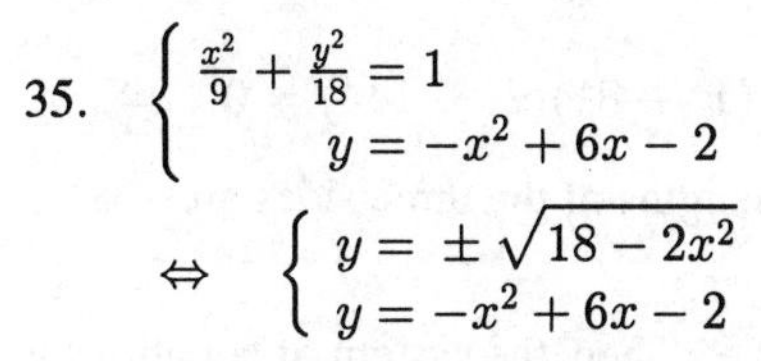

$\Leftrightarrow \begin{cases} y = \pm\sqrt{18 - 2x^2} \\ y = -x^2 + 6x - 2 \end{cases}$

The solutions are $(1.23, 3.87)$ and $(-0.35, -4.21)$.

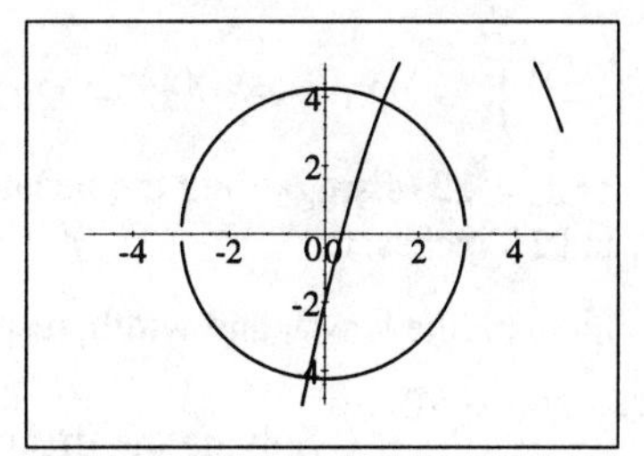

37. $\begin{cases} x^4 + 16y^4 = 32 \\ x^2 + 2x + y = 0 \end{cases}$

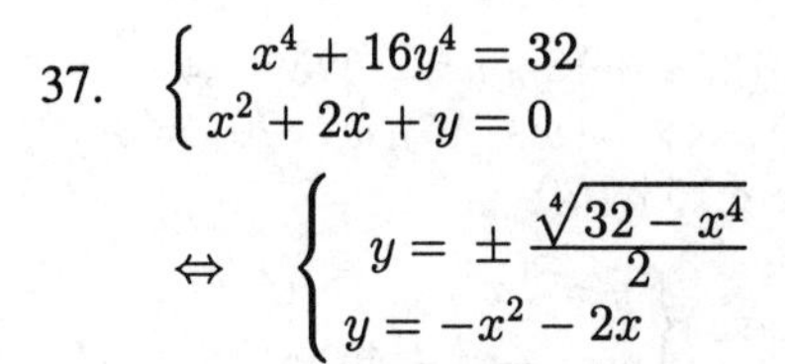

$\Leftrightarrow \begin{cases} y = \pm\dfrac{\sqrt[4]{32 - x^4}}{2} \\ y = -x^2 - 2x \end{cases}$

The solutions are $(-2.30, -0.70)$ and $(0.48, -1.19)$.

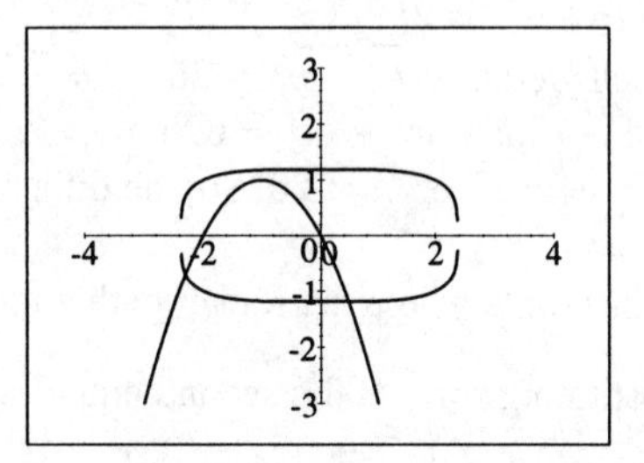

39. $\begin{cases} y = e^x + e^{-x} \\ y = 5 - x^2 \end{cases}$

The solutions are $(1.19, 3.59)$ and $(-1.19, 3.59)$.

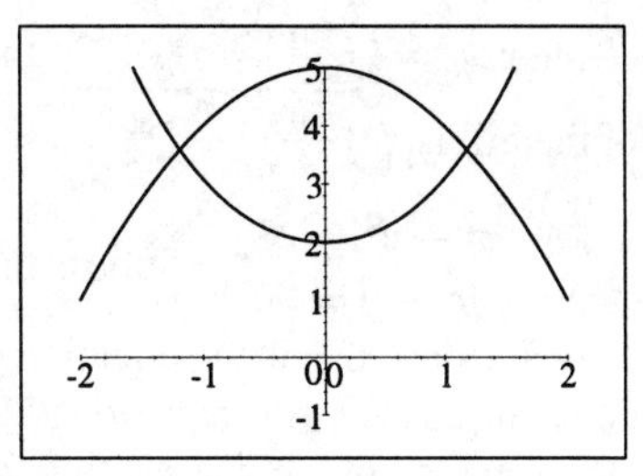

41. Let x and y be the lengths of the perpendicular sides in cm. Then $\sqrt{x^2+y^2}$ is the length of the hypotenuse. The system of equations is

$$\begin{cases} x+y+\sqrt{x^2+y^2}=40 \\ \frac{1}{2}xy=60 \end{cases}$$

From the first equation $\sqrt{x^2+y^2}=40-x-y \quad \Rightarrow \quad x^2+y^2=(40-x-y)^2$ $=1600+x^2+y^2+2xy-80x-80y \quad \Leftrightarrow \quad 0=1600+2xy-80x-80y$. Using the second equation we have $xy=120$ which we substitute for and get: $0=1600+2(120)-80(x+y) \quad \Leftrightarrow$ $x+y=23$. Again, using the second equation we have $x=\dfrac{120}{y}$, so $\frac{120}{y}+y=23 \quad \Rightarrow$ $y^2-23y+120=0 \quad \Leftrightarrow \quad (y-8)(y-15)=0 \quad \Rightarrow \quad y=8$ or $y=15$. Therefore, the lengths of the sides of the triangle are 8 cm, 15 cm, and 17 cm.

43. Let x and y be the lengths of the sides, in inches, that form the right angle. Then,

$$\begin{cases} \frac{1}{2}xy=54 \\ xy\sqrt{x^2+y^2}=1620 \end{cases} \quad \Leftrightarrow \quad \begin{cases} xy=108 \\ xy\sqrt{x^2+y^2}=1620 \end{cases}$$

since the hypotenuse has length $\sqrt{x^2+y^2}$. By substitution, $108\sqrt{x^2+y^2}=1620 \quad \Leftrightarrow$ $\sqrt{x^2+y^2}=15 \quad \Leftrightarrow \quad x^2+y^2=225$. From the first equation, $y=\dfrac{108}{x}$ and so $x^2+\left(\dfrac{108}{x}\right)^2=225 \quad \Leftrightarrow \quad x^4-225x^2+108^2=0 \quad \Leftrightarrow \quad (x^2-81)(x^2-144)=0 \quad \Rightarrow$ $x=9$ or $x=12$ (disregarding the negative values). Thus, the lengths of the three sides are $9, 12$, and 15 inches.

45. Let l and w be the length and width, respectively, of the rectangle. Then, the system of equations is $\begin{cases} 2l+2w=70 \\ \sqrt{l^2+w^2}=25 \end{cases}$ Solving the first equation for l we have $l=35-w$ and substituting into the second yields $\sqrt{l^2+w^2}=25 \quad \Leftrightarrow \quad l^2+w^2=625 \quad \Leftrightarrow \quad (35-w)^2+w^2=625 \quad \Leftrightarrow$ $1225-70w+w^2+w^2=625 \quad \Leftrightarrow \quad 2w^2-70w+600=0 \quad \Leftrightarrow \quad (w-15)(w-20)=0$ $\Rightarrow \quad w=15$ or $w=20$. So, the dimensions of the rectangle are 15 by 20.

47. At the points where the rocket path and the hillside meet, we have $\begin{cases} y=\frac{1}{2}x \\ y=-x^2+401x \end{cases}$ Substituting for y in the second equation gives $\frac{1}{2}x=-x^2+401x \quad \Leftrightarrow \quad x^2-\frac{801}{2}x=0 \quad \Leftrightarrow$ $x\left(x-\frac{801}{2}=0\right) \quad \Rightarrow \quad x=0, x=\frac{801}{2}$. When $x=0$, the rocket has not left the pad. When $x=\frac{801}{2}$, then $y=\frac{1}{2}\left(\frac{801}{2}\right)=\frac{801}{4}$. So the rocket lands at the point $\left(\frac{801}{2},\frac{801}{4}\right)$. The distance from the base of the hill is $\sqrt{\left(\frac{801}{2}\right)^2+\left(\frac{801}{4}\right)^2}\approx 447.77$ meters.

49. $\begin{cases} x^2+y^2=25 \\ x^2-2x+y^2+y=30 \end{cases}$ Subtracting the two equations, we get $3x-y=25-30 \quad \Leftrightarrow$ $y=3x+5$. Since this is the equation of a line, and since any point satisfying both the original equations must satisfy this equation, this must be the line that contains the points of intersection. Thus, the equation of the line passing through the intersection points is $y=3x+5$. (Notice, we did not have to find the points of intersection to determine the equation of the line that contains them.)

51. $\begin{cases} x - y = 3 \\ x^3 - y^3 = 387 \end{cases}$ Now $x - y = 3 \Leftrightarrow x = 3 + y$ and using the hint, $x^3 - y^3 = 387 \Leftrightarrow$ $(x - y)(x^2 + xy + y^2) = 387$. Next substituting for x we get $3[(3 + y)^2 + y(3 + y) + y^2] = 387$ $\Leftrightarrow 9 + 6y + y^2 + 3y + y^2 + y^2 = 129 \Leftrightarrow 3y^2 + 9y + 9 = 129 \Leftrightarrow (y + 8)(y - 5) = 0$ $\Rightarrow y = -8$ or $y = 5$. If $y = -8$ then $x = 3 + (-8) = -5$, while if $y = 5$ then $x = 3 + 5 = 8$. Hence, the solutions are $(-5, -8)$ and $(8, 5)$.

53. $\begin{cases} 2^x + 2^y = 10 \\ 4^x + 4^y = 68 \end{cases} \Leftrightarrow \begin{cases} 2^x + 2^y = 10 \\ 2^{2x} + 2^{2y} = 68 \end{cases}$

If we let $u = 2^x$ and $v = 2^y$, the system becomes $\begin{cases} u + v = 10 \\ u^2 + v^2 = 68 \end{cases}$

Solve the first equation for u, and substitute this into the second: $u + v = 10 \Leftrightarrow u = 10 - v$ so $(10 - v)^2 + v^2 = 68 \Leftrightarrow 100 - 20v + v^2 + v^2 = 68 \Leftrightarrow v^2 - 10v + 16 = 0 \Leftrightarrow$ $(v - 8)(v - 2) = 0 \Rightarrow v = 2$ or $v = 8$. If $v = 2$ then $u = 8$ and so $y = 1$ and $x = 3$. If $v = 8$ then $u = 2$ and so $y = 3$ and $x = 1$. Thus, the solutions are $(1, 3)$ and $(3, 1)$.

Exercises 6.8

1. $x \leq 2$

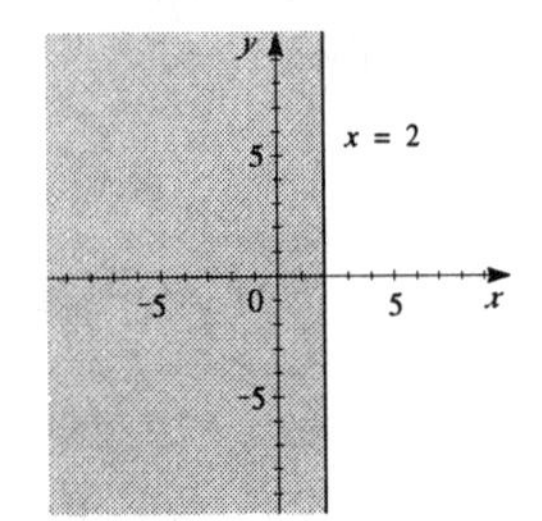

3. $y > x$

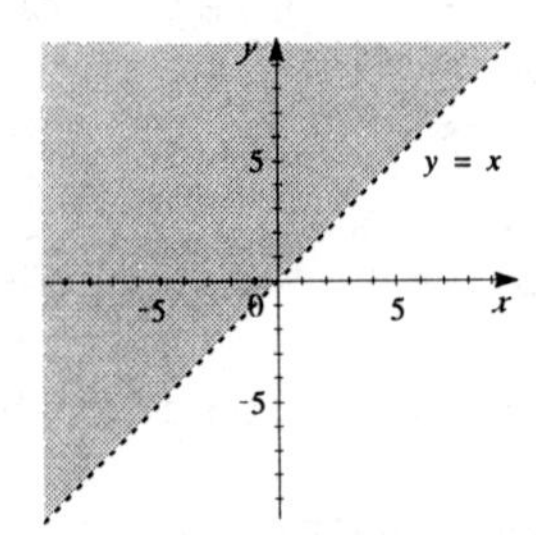

5. $y \geq 2x + 2$

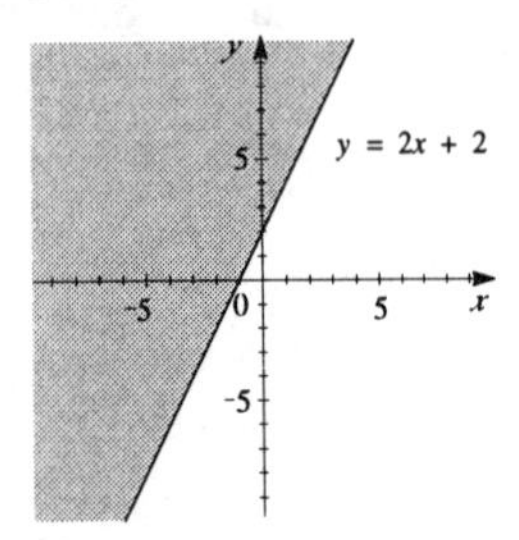

7. $2x - y \leq 8$

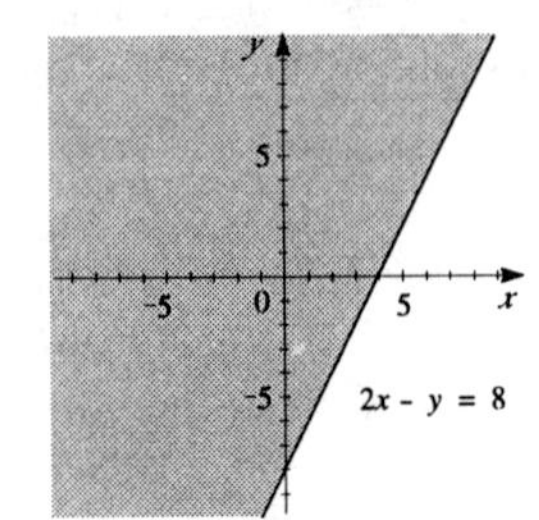

9. $4x + 5y < 25$

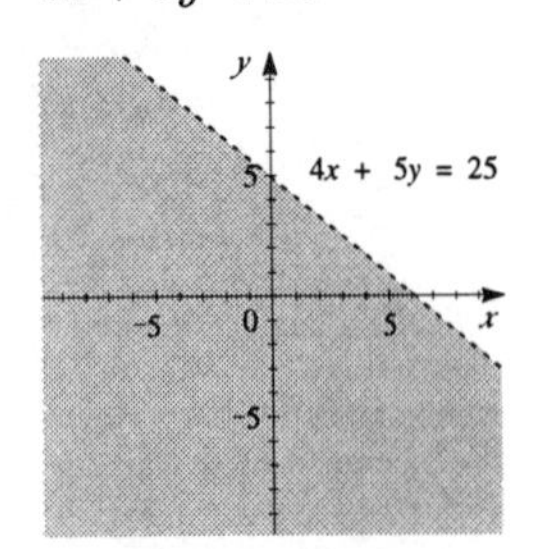

11. $y > x^3 + 1$

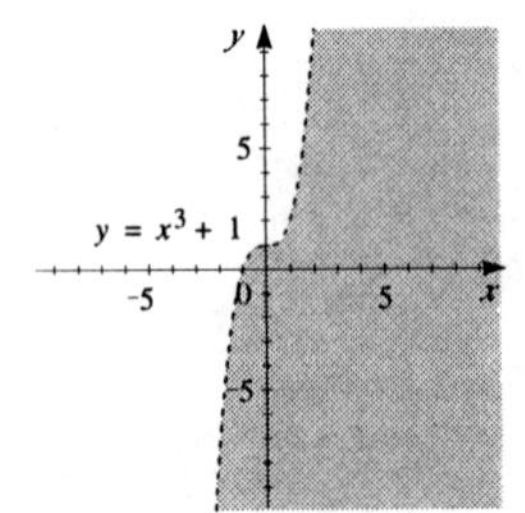

13. $x^2 + y^2 \geq 5$

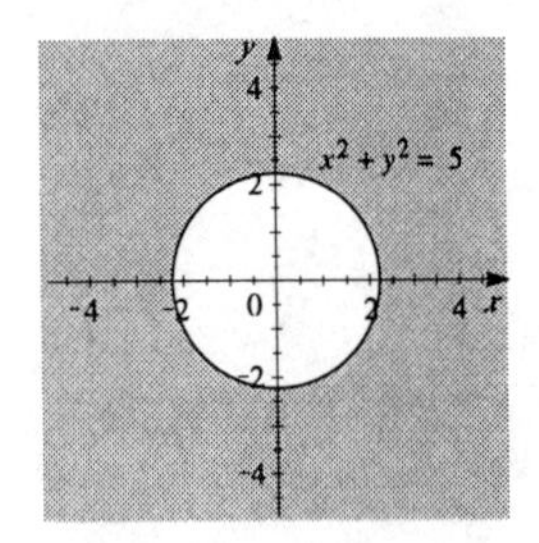

15. $\begin{cases} x + y \le 4 \\ y \ge x \end{cases}$ The vertices occur where $\begin{cases} x + y = 4 \\ y = x \end{cases}$
Substituting we have $2x = 4 \Leftrightarrow x = 2$. Since $y = x$, the vertex is at $(2, 2)$ and the solution set is not bounded.

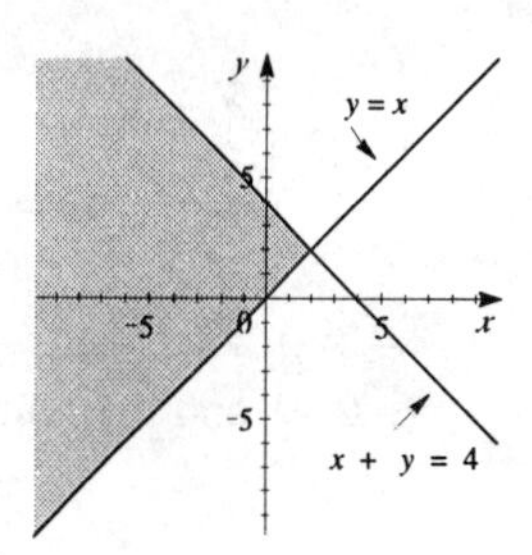

17. $\begin{cases} y < \frac{1}{4}x + 2 \\ y \ge 2x - 5 \end{cases}$ The vertices occur where $\begin{cases} y = \frac{1}{4}x + 2 \\ y = 2x - 5 \end{cases}$ and substituting for y gives $\frac{1}{4}x + 2 = 2x - 5 \Leftrightarrow \frac{7}{4}x = 7 \Leftrightarrow x = 4$, so $y = 3$. Hence, the vertex is at $(4, 3)$ and the solution is not bounded.

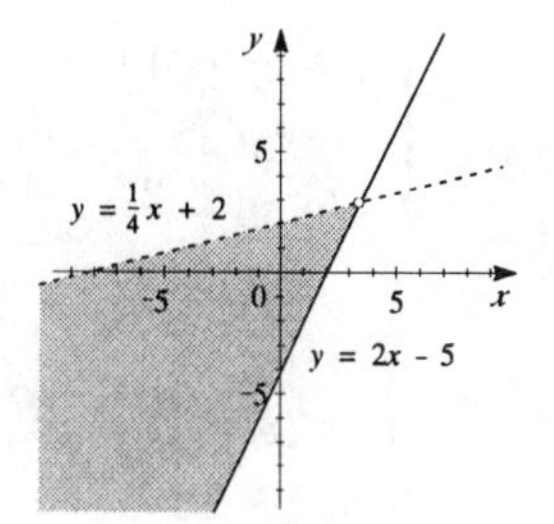

19. $\begin{cases} x \ge 0 \\ y \ge 0 \\ 3x + 5y \le 15 \\ 3x + 2y \le 9 \end{cases}$ From the graph, vertices occur at $(3, 0)$, $(0, 3)$ and $(0, 0)$, and the fourth vertex occurs where the lines $3x + 5y = 15$ and $3x + 2y = 9$ intersect. Subtracting these two equations gives $3y = 6 \Leftrightarrow y = 2$, and so $x = \frac{5}{3}$. Thus, the fourth vertex occurs at $(\frac{5}{3}, 2)$ and the solution set is bounded.

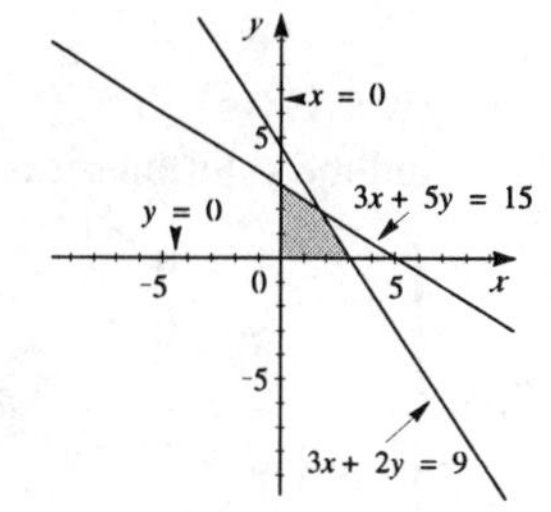

21. $\begin{cases} y < 9 - x^2 \\ y \ge x + 3 \end{cases}$ The vertices occur where $\begin{cases} y = 9 - x^2 \\ y = x + 3 \end{cases}$.
Substituting for y gives $9 - x^2 = x + 3 \Leftrightarrow x^2 + x - 6 = 0 \Leftrightarrow (x - 2)(x + 3) = 0 \Rightarrow x = -3, x = 2$. Therefore, the vertices occur at $(-3, 0)$ and $(2, 5)$ and the solution set is bounded.

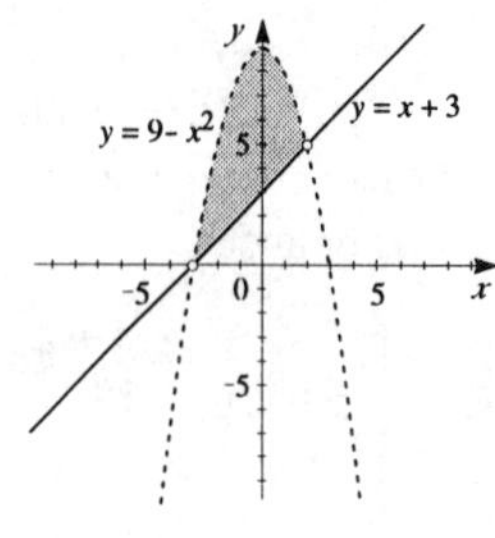

23. $\begin{cases} x^2 + y^2 \le 4 \\ x - y > 0 \end{cases}$ The vertices occur where $\begin{cases} x^2 + y^2 = 4 \\ x - y = 0 \end{cases}$.
Since $x - y = 0 \Leftrightarrow x = y$, substituting for x gives $y^2 + y^2 = 4 \Leftrightarrow y^2 = 2 \Rightarrow y = \pm\sqrt{2}$, and $x = \pm\sqrt{2}$. Therefore, the vertices occur at $(-\sqrt{2}, -\sqrt{2})$ and $(\sqrt{2}, \sqrt{2})$, and the solution set is bounded.

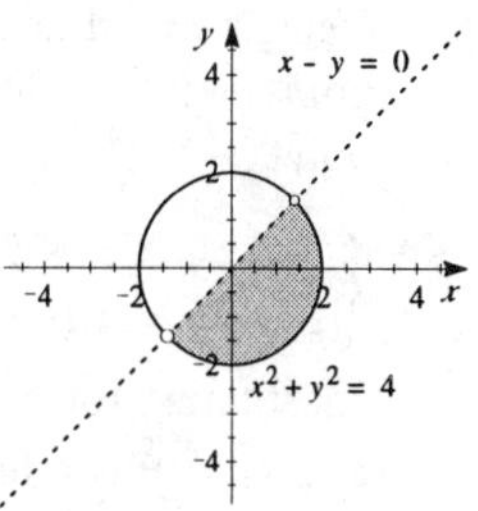

25. $\begin{cases} x^2 - y \le 0 \\ 2x^2 + y \le 12 \end{cases}$ The vertices occur where $\begin{cases} x^2 - y = 0 \\ 2x^2 + y = 12 \end{cases}$

$\Leftrightarrow \quad \begin{cases} 2x^2 - 2y = 0 \\ 2x^2 + y = 12 \end{cases}$. Subtracting the equations gives $3y = 12$

$\Leftrightarrow \quad y = 4$, and $x = \pm 2$. Thus, the vertices occur at $(2, 4)$ and $(-2, 4)$, and the solution set is bounded.

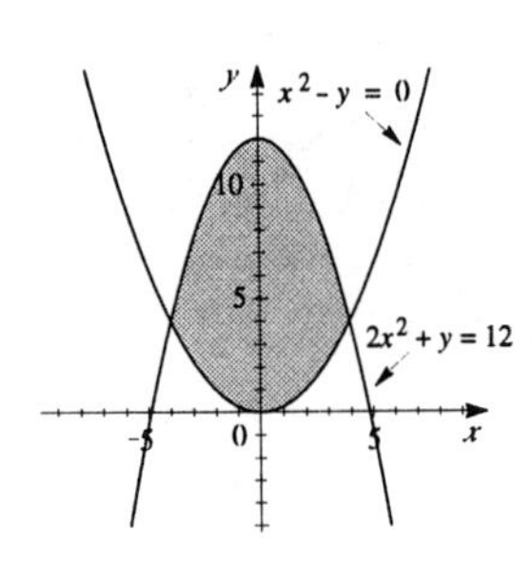

27. $\begin{cases} x + 2y \le 14 \\ 3x - y \ge 0 \\ x - y \ge 2 \end{cases}$ We now find the vertices of the region by solving pairs of the corresponding equations.

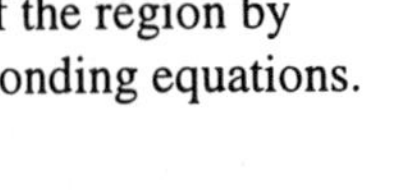

$\begin{cases} x + 2y = 14 \\ x - y = 2 \end{cases}$ Subtracting the second equation from the first gives $3y = 12 \quad \Leftrightarrow \quad y = 4$, and $x = 6$.

$\begin{cases} 3x - y = 0 \\ x - y = 2 \end{cases}$ Subtracting the second equation from the first

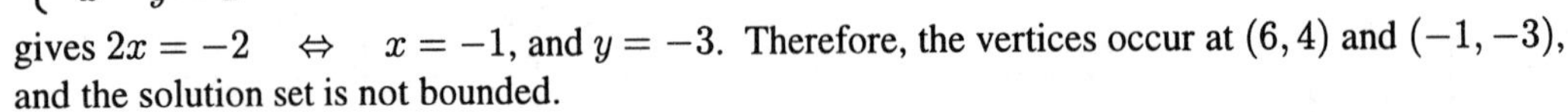

gives $2x = -2 \quad \Leftrightarrow \quad x = -1$, and $y = -3$. Therefore, the vertices occur at $(6, 4)$ and $(-1, -3)$, and the solution set is not bounded.

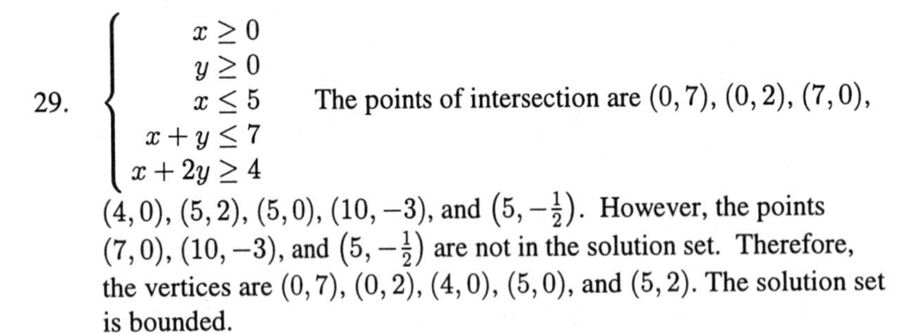

29. $\begin{cases} x \ge 0 \\ y \ge 0 \\ x \le 5 \\ x + y \le 7 \\ x + 2y \ge 4 \end{cases}$ The points of intersection are $(0, 7)$, $(0, 2)$, $(7, 0)$, $(4, 0)$, $(5, 2)$, $(5, 0)$, $(10, -3)$, and $\left(5, -\frac{1}{2}\right)$. However, the points $(7, 0)$, $(10, -3)$, and $\left(5, -\frac{1}{2}\right)$ are not in the solution set. Therefore, the vertices are $(0, 7)$, $(0, 2)$, $(4, 0)$, $(5, 0)$, and $(5, 2)$. The solution set is bounded.

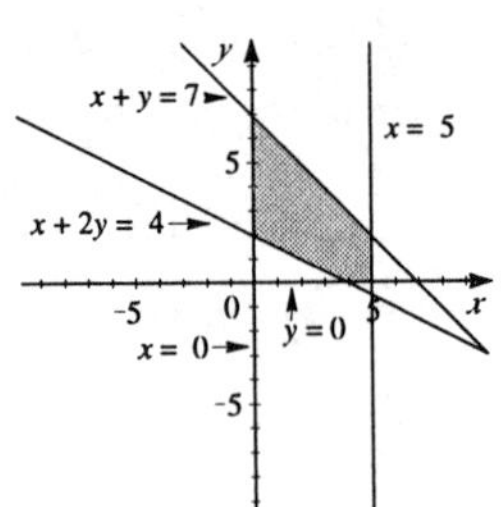

31. $\begin{cases} y > x + 1 \\ x + 2y \le 12 \\ x + 1 > 0 \end{cases}$ We now find the vertices of the region by solving pairs of the corresponding equations.

Using $x = -1$ and substituting for x in the line $y = x + 1$ gives the point $(-1, 0)$. Substituting for x in the line $x + 2y = 12$ gives the point $\left(-1, \frac{13}{2}\right)$.

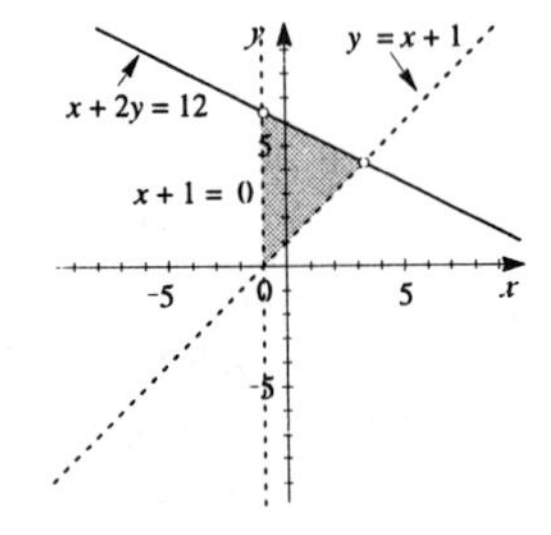

$\begin{cases} y = x + 1 \\ x + 2y = 12 \end{cases}$ From the first equation $x = y - 1$, and substituting for x into the second equation gives $y - 1 + 2y = 12 \quad \Leftrightarrow \quad 3y = 13 \quad \Leftrightarrow \quad y = \frac{13}{3}$, and $x = \frac{10}{3}$. So the vertices are $(-1, 0)$, $\left(-1, \frac{13}{2}\right)$, and $\left(\frac{10}{3}, \frac{13}{3}\right)$, and none of these vertices are in the solution set. The solution set is bounded.

33. $\begin{cases} x^2 + y^2 \le 8 \\ x \ge 2 \\ y \ge 0 \end{cases}$ The intersection points occur at $(2, \pm 2)$, $(2, 0)$, and $(2\sqrt{2}, 0)$. However, since $(2, -2)$ is not part of the solution set, the vertices are $(2, 2)$, $(2, 0)$, and $(2\sqrt{2}, 0)$. The solution set is bounded.

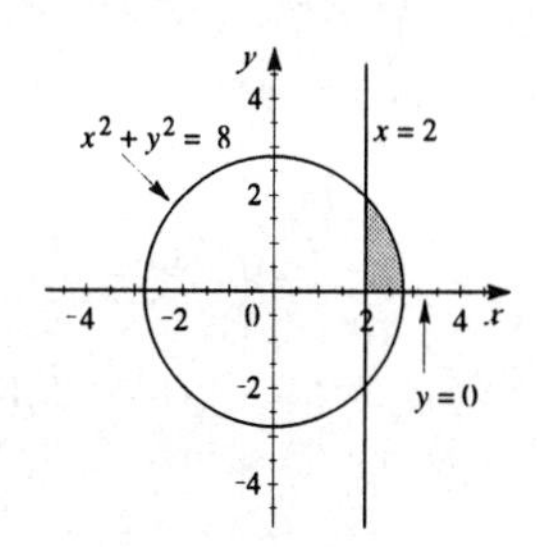

35. $\begin{cases} x^2 + y^2 < 9 \\ x + y > 0 \\ x \le 0 \end{cases}$ Substituting $x = 0$ into the equations $x^2 + y^2 = 9$ and $x + y = 0$ gives the vertices $(0, \pm 3)$ and $(0, 0)$.

To find the points of intersection for the equations $\begin{cases} x^2 + y^2 = 9 \\ x + y = 0 \end{cases}$ we solve for $x = -y$ and substitute into the first equation. This gives $(-y)^2 + y^2 = 9 \quad \Rightarrow \quad y = \pm \frac{3\sqrt{2}}{2}$. The points $(0, -3)$ and $\left(\frac{3\sqrt{2}}{2}, -\frac{3\sqrt{2}}{2}\right)$ lie away from the solution set, so the vertices are $(0, 0)$, $(0, 3)$, and $\left(-\frac{3\sqrt{2}}{2}, \frac{3\sqrt{2}}{2}\right)$. (Note that the vertices are not solutions in this case.) The solution set is bounded.

37. Let x be the number of fiction books published in a year, and y be the number of non-fiction books. Then, the following system of inequalities holds:

$$\begin{cases} x \ge 0 \\ y \ge 0 \\ x + y \le 100 \\ y \ge 20 \\ x \ge y \end{cases}$$

From the graph, we see that the vertices occur at $(50, 50)$, $(80, 20)$ and $(20, 20)$.

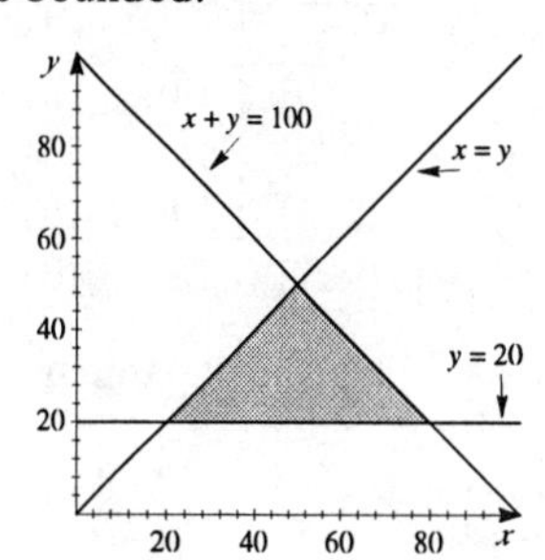

Exercises 6.9

1.

Vertex	$M = 2x + 3y$	
$(0, 0)$	0	$\leftarrow$ minimum value
$(3, 0)$	$2(3) + 3(0) = 6$	$\leftarrow$ maximum value
$(0, 2)$	$2(0) + 3(2) = 6$	$\leftarrow$ maximum value

3.

Vertex	$P = 200 - x - y$	
$(0, 2)$	$200 - (0) - (2) = 198$	$\leftarrow$ maximum value
$(0, 5)$	$200 - (0) - (5) = 195$	$\leftarrow$ minimum value
$(4, 0)$	$200 - (4) - (0) = 196$	

5. $\begin{cases} x \geq 0, y \geq 0 \\ 2x + y \leq 10 \\ 2x + 4y \leq 28 \end{cases}$

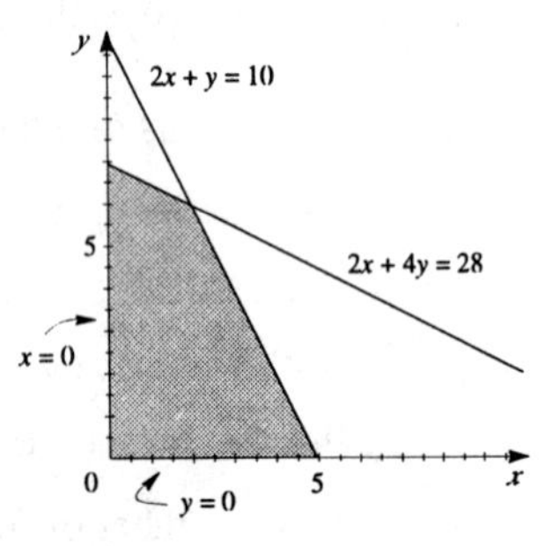

Objective function: $P = 140 - x + 3y$

From the graph, the vertices occur at $(0, 0)$, $(5, 0)$, $(2, 6)$, and $(0, 7)$.

Vertex	$P = 140 - x + 3y$	
$(0, 0)$	$140 - (0) + 3(0) = 140$	
$(5, 0)$	$140 - (5) + 3(0) = 135$	$\leftarrow$ minimum value
$(2, 6)$	$140 - (2) + 3(6) = 156$	
$(0, 7)$	$140 - (0) + 3(7) = 161$	$\leftarrow$ maximum value

Thus the maximum value is 161 and the minimum value is 135.

7. $\begin{cases} x \geq 3, y \geq 4 \\ 2x + y \leq 24 \\ 2x + 3y \leq 36 \end{cases}$

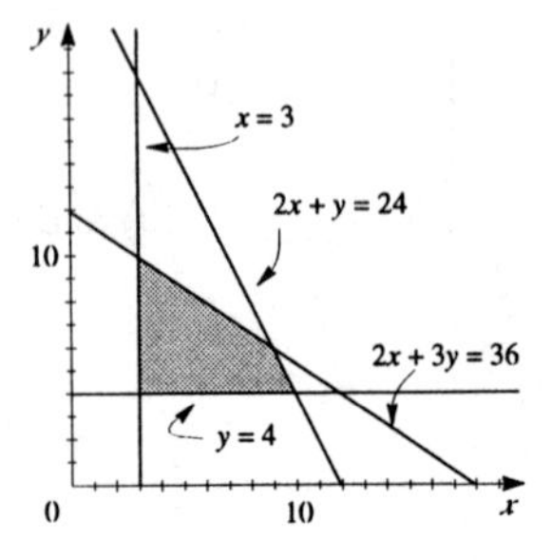

Objective function: $R = 12 + x + 4y$

From the graph, the vertices occur at $(3, 4)$, $(3, 10)$, $(9, 6)$, and $(10, 4)$.

Vertex	$R = 12 + x + 4y$	
$(3, 4)$	$12 + (3) + 4(4) = 31$	$\leftarrow$ minimum value
$(3, 10)$	$12 + (3) + 4(10) = 55$	$\leftarrow$ maximum value
$(9, 6)$	$12 + (9) + 4(6) = 45$	
$(10, 4)$	$12 + (10) + 4(4) = 38$	

Therefore, the maximum value of R is 55 and the minimum value is 31.

9. Let t be the number of tables made daily, c be the number of chairs made daily. Then the data given can be summarized by the table:

	Tables, t	Chairs, c	available time
carpentry	2 h	3 h	108 h
finishing	1 h	$\frac{1}{2}$ h	20 h
Profit	\$35	\$20	

Thus we wish to maximize the total profit, $P = 35t + 20c$, subject to the constraints:

$$\begin{cases} 2t + 3c \leq 108 \\ t + \frac{1}{2}c \leq 20 \\ t, c \geq 0 \end{cases}$$

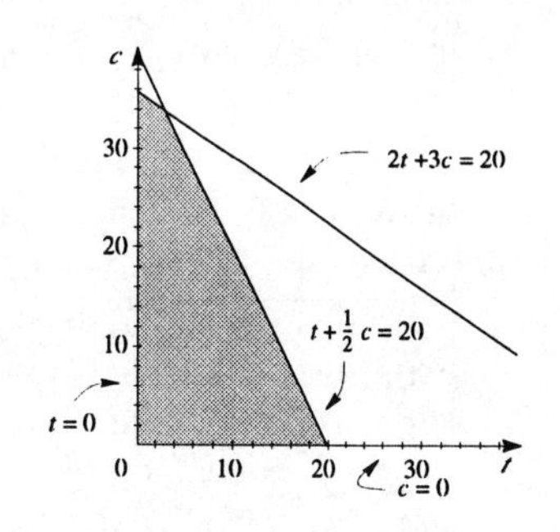

From the graph, the vertices occur at $(0, 0)$, $(20, 0)$, $(0, 36)$, and $(3, 34)$.

Vertex	$P = 35t + 20c$
$(0, 0)$	$35(0) + 20(0) = 0$
$(20, 0)$	$35(20) + 20(0) = 700$
$(0, 36)$	$35(0) + 20(36) = 720$
$(3, 34)$	$35(3) + 20(34) = 785$ ← maximum value

Hence, 3 tables and 34 chairs should be produced daily for a maximum profit of \$785.

11. Let x be the number of crates of oranges, y be the number of crates of grapefruit. Then the data given can be summarized by the table:

	oranges	grapefruit	available
volume	4 ft^3	6 ft^3	300 ft^3
weight	80 lb	100 lb	5600 lb
Profit	\$2.50	\$4.00	

In addition, $x \geq y$. Thus we wish to maximize the total profit $P = 2.5x + 4y$ subject to the constraints:

$$\begin{cases} x \geq 0, y \geq 0 \\ x \geq y \\ 4x + 6y \leq 300 \\ 80x + 100y \leq 5600 \end{cases}$$

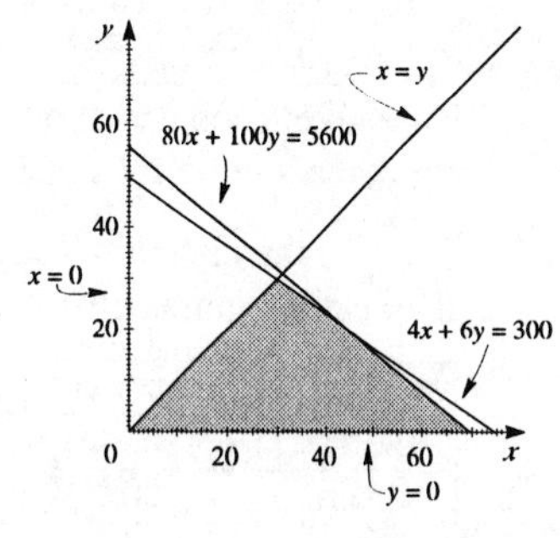

From the graph, the vertices occur at $(0, 0)$, $(30, 30)$, $(45, 20)$, and $(70, 0)$.

Vertex	$P = 2.5x + 4y$
$(0, 0)$	$2.5(0) + 4(0) = 0$
$(30, 30)$	$2.5(30) + 4(30) = 195$ ← maximum value
$(45, 20)$	$2.5(45) + 4(20) = 192.5$
$(70, 0)$	$2.5(70) + 4(0) = 175$

Thus, she should carry 30 crates of oranges and 30 crates of grapefruit for a maximum profit of \$195.

13. Let x be the number of stereo sets shipped from Long Beach to Santa Monica and y be the number of stereo sets shipped from Long Beach to El Toro. Thus, $15 - x$ sets must be shipped to Santa Monica from Pasadena, and $19 - y$ sets to El Toro from Pasadena. Thus $x \geq 0$, $y \geq 0$, $15 - x \geq 0$, $19 - y \geq 0$, $x + y \leq 24$, and $(15 - x) + (19 - y) \leq 18$. Simplifying, we get the inequalities (constraints):

$$\begin{cases} x \geq 0, y \geq 0 \\ x \leq 15, y \leq 19 \\ x + y \leq 24 \\ x + y \geq 16 \end{cases}$$

Thus the objective function is cost,
$C = 5x + 6y + 4(15 - x) + 5.5(19 - y) = x + 0.5y + 164.5$
which we wish to minimize. From the graph, the vertices occur at $(0,16)$, $(0,19)$, $(5,19)$, $(15,9)$, and $(15,1)$.

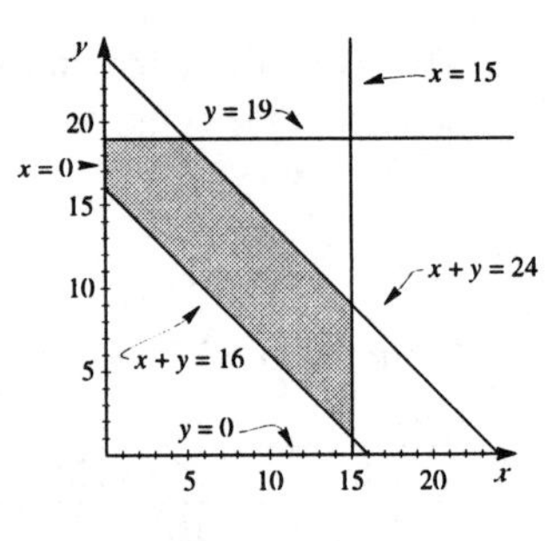

Vertex	$C = x + 0.5y + 164.5$
$(0,16)$	$(0) + 0.5(16) + 164.5 = 172.5$ ← minimum value
$(0,19)$	$(0) + 0.5(19) + 164.5 = 174$
$(5,19)$	$(5) + 0.5(19) + 164.5 = 179$
$(15,9)$	$(15) + 0.5(9) + 164.5 = 184$
$(15,1)$	$(15) + 0.5(1) + 164.5 = 180$

The minimum cost is \$172.50 and occurs when $x = 0$, $y = 16$. Hence, 0 sets should be shipped from Long Beach to Santa Monica, 16 from Long Beach to El Toro, 15 from Pasadena to Santa Monica, and 3 from Pasadena to El Toro.

15. From Example 4 of Section 6.8, the vertices of the feasible region occur at $(0,0)$, $(0,18)$, $(10.5,14.5)$, and $(25,0)$. If F is the number of cubic yards of Foamboard loaded and P is the number of cubic yards of Plastiflex loaded, then the objective function is profit, $P = 0.5(12F) + 0.7(10P) = 6F + 7P$ which we wish to maximize. (See Example 4 for graph and list of vertices.)

Vertex	$P = 6F + 7P$
$(0,0)$	$6(0) + 7(0) = 0$
$(0,18)$	$6(0) + 7(18) = 126$
$(10.5,14.5)$	$6(10.5) + 7(14.5) = 164.5$ ← maximum value
$(25,0)$	$6(25) + 7(0) = 150$

Therefore, each cart should have 10.5 cubic yards of Foamboard and 14.5 cubic yards of Plastiflex for a maximum profit of \$164.50 per minute.

17. Let x be the number of days that the Vancouver plant is operated and y be the number of days that the Seattle plant is operated. The data can then be summarized in the following table.

	Vancouver	Seattle	required
sofas	20	50	400
chairs	20	25	300
ottomans	35	25	375
cost	\$3000	\$4000	

Also, the order must be completed in 30 days. Thus we wish to minimize the cost $C = 3000x + 4000y$ subject to the constraints:

$$\begin{cases} x \geq 0, y \geq 0 \\ 20x + 50y \geq 400 \\ 20x + 25y \geq 300 \\ 35x + 25y \geq 375 \\ x \leq 30 \\ y \leq 30 \end{cases}$$

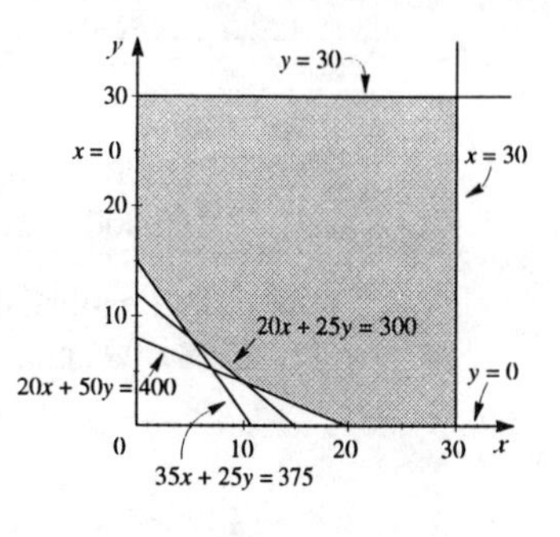

From the graph, the vertices occur at $(20,0)$, $(10,4)$, $(5,8)$, $(0,15)$, $(30,30)$, $(30,0)$, and $(0,30)$.

Vertex	$C = 3000x + 4000y$
$(20,0)$	$3000(20) + 4000(0) = 60,000$
$(10,4)$	$3000(10) + 4000(4) = 46,000$ ← minimum value
$(0,15)$	$3000(0) + 4000(15) = 60,000$
$(30,30)$	$3000(30) + 4000(30) = 210,000$
$(30,0)$	$3000(30) + 4000(0) = 90,000$
$(0,30)$	$3000(0) + 4000(30) = 120,000$
$(5,8)$	$3000(5) + 4000(8) = 47,000$

Thus, they should operate the Vancouver plant for 10 days and the Seattle plant for 4 days, for a minimum cost of \$46,000.

19. Let x be the amount in municipal bonds, y be the amount in bank certificates, both in dollars. Then, $12000 - x - y$ is the amount in high-risk bonds. So our constraints can be stated as:

$$\begin{cases} x \geq 0, y \geq 0 \\ 12{,}000 - x - y \geq 0 \\ x \geq 3y \\ 12000 - x - y \leq 2000 \end{cases} \Leftrightarrow \begin{cases} x \geq 0, y \geq 0 \\ x + y \leq 12{,}000 \\ x \geq 3y \\ x + y \geq 10{,}000 \end{cases}$$

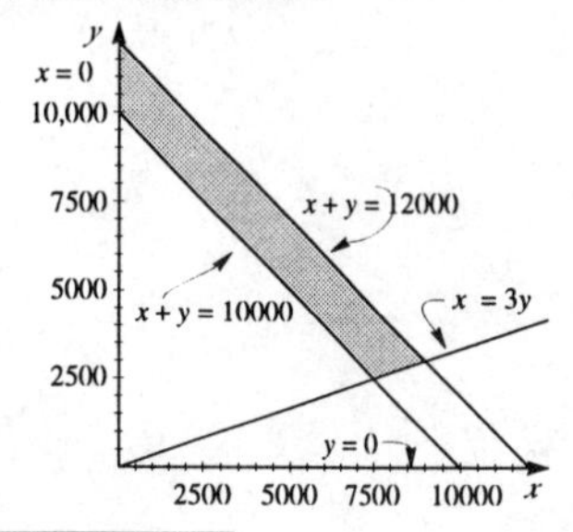

From the graph, the vertices occur at $(7500, 2500)$, $(10000, 0)$, $(12000,0)$, and $(9000, 3000)$. And the objective function is $P = 0.07x + 0.08y + 0.12(12000 - x - y)$ $= 1440 - 0.05x - 0.04y$, which we wish to maximize.

Vertex	$P = 1440 - 0.05x - 0.04y$
$(7500, 2500)$	$1440 - 0.05(7500) - 0.04(2500) = 965$ ← maximum value
$(10000, 0)$	$1440 - 0.05(10000) - 0.04(0) = 940$
$(12000, 0)$	$1440 - 0.05(12000) - 0.04(0) = 840$
$(9000, 3000)$	$1440 - 0.05(9000) - 0.04(3000) = 870$

Hence, she should invest \$7500 in municipal bonds, \$2500 in bank certificates, and \$2000 in high-risk bonds for a maximum yield of \$965.

21. Let g be the number of games published and e be the number of educational programs published. Then, the number of utility programs published is $36 - g - e$. Hence, we wish to maximize profit, $P = 5000g + 8000e + 6000(36 - g - e) = 216{,}000 - 1000g + 2000e$, subject to the constraints:

$$\begin{cases} g \geq 4, e \geq 0 \\ 36 - g - e \geq 0 \\ 36 - g - e \leq 2e \end{cases} \Leftrightarrow \begin{cases} g \geq 4, e \geq 0 \\ g + e \leq 36 \\ g + 3e \geq 36 \end{cases}$$

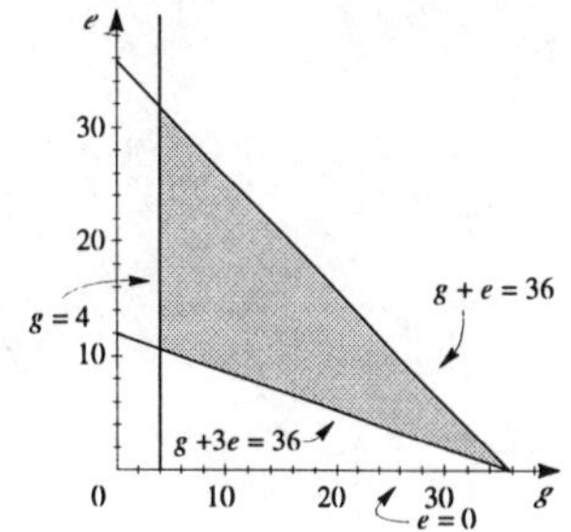

From the graph, the vertices occur at $\left(4, \frac{32}{3}\right)$, $(4, 32)$, and $(36, 0)$. The objective function is

Vertex	$P = 216{,}000 - 1000g + 2000e$	
$(4, \frac{32}{3})$	$216{,}000 - 1000(4) + 2000(\frac{32}{3}) = 233{,}333.33$	
$(4, 32)$	$216{,}000 - 1000(4) + 2000(32) = 276{,}000$	← maximum value
$(36, 0)$	$216{,}000 - 1000(36) + 2000(0) = 180{,}000$	

So, they should publish 4 games, 32 educational programs, and 0 utility programs for a maximum profit of $276,000 annually.

Exercises 6.10

1. $\dfrac{1}{(x-1)(x+2)} = \dfrac{A}{x-1} + \dfrac{B}{x+2}$

3. $\dfrac{x^2-3x+5}{(x-2)^2(x+4)} = \dfrac{A}{x-2} + \dfrac{B}{(x-2)^2} + \dfrac{C}{x+4}$

5. $\dfrac{x^2}{(x-3)(x^2+4)} = \dfrac{A}{x-3} + \dfrac{Bx+C}{x^2+4}$

7. $\dfrac{x^3-4x^2+2}{(x^2+1)(x^2+2)} = \dfrac{Ax+B}{x^2+1} + \dfrac{Cx+D}{x^2+2}$

9. $\dfrac{x^3+x+1}{x(2x-5)^3(x^2+2x+5)^2} = \dfrac{A}{x} + \dfrac{B}{2x-5} + \dfrac{C}{(2x-5)^2} + \dfrac{D}{(2x-5)^3} +$

$$\dfrac{Ex+F}{x^2+2x+5} + \dfrac{Gx+H}{(x^2+2x+5)^2}$$

11. $\dfrac{5}{(x-1)(x+4)} = \dfrac{A}{x-1} + \dfrac{B}{x+4}$. Multiplying by $(x-1)(x+4)$we get:

$5 = A(x+4) + B(x-1) \quad \Leftrightarrow \quad 5 = Ax + 4A + Bx - B$. Thus

$\begin{cases} A+B=0 \\ 4A-B=5 \end{cases}$. Now $A+B=0 \quad \Leftrightarrow \quad B=-A$, so substituting we get $4A-(-A)=5 \quad \Leftrightarrow$ $5A=5 \quad \Leftrightarrow \quad A=1$ and $B=-1$. The required partial fraction decomposition is $\dfrac{5}{(x-1)(x+4)} = \dfrac{1}{x-1} - \dfrac{1}{x+4}$.

13. $\dfrac{12}{x^2-9} = \dfrac{12}{(x-3)(x+3)} = \dfrac{A}{(x-3)} + \dfrac{B}{(x+3)}$. Multiplying by $(x-3)(x+3)$we get:

$12 = A(x+3) + B(x-3) \quad \Leftrightarrow \quad 12 = Ax + 3A + Bx - 3B$. Thus

$\begin{cases} A+B=0 \\ 3A-3B=12 \end{cases} \quad \Leftrightarrow \quad \begin{cases} A+B=0 \\ A-B=4 \end{cases}$. Adding we get $2A=4 \quad \Leftrightarrow \quad A=2$. So $2+B=2 \quad \Leftrightarrow$ and $B=-2$. The required partial fraction decomposition is $\dfrac{12}{x^2-9} = \dfrac{2}{x-3} - \dfrac{2}{x+3}$.

15. $\dfrac{4}{x^2-4} = \dfrac{4}{(x-2)(x+2)} = \dfrac{A}{x-2} + \dfrac{B}{x+2}$. Multiplying by x^2-4 we get:
$4 = A(x+2) + B(x-2) = (A+B)x + (2A-2B)$, and so

$\begin{cases} A+B=0 \\ 2A-2B=4 \end{cases} \quad \Leftrightarrow \quad \begin{cases} A+B=0 \\ A-B=2 \end{cases}$ Adding we get $2A=2 \quad \Leftrightarrow \quad A=1$, and $B=-1$. Therefore, $\dfrac{4}{x^2-4} = \dfrac{1}{x-2} - \dfrac{1}{x+2}$.

17. $\dfrac{x+14}{x^2-2x-8}=\dfrac{x+14}{(x-4)(x+2)}=\dfrac{A}{x-4}+\dfrac{B}{x+2}$. Hence,
$x+14=A(x+2)+B(x-4)=(A+B)x+(2A-4B)$, and so

$\begin{cases} A+B=1 \\ 2A-4B=14 \end{cases} \Leftrightarrow \begin{cases} 2A+2B=2 \\ A-2B=7 \end{cases}$. Adding we get $3A=9 \Leftrightarrow A=3$.

$(3)+B=1 \Leftrightarrow B=-2$. Therefore, $\dfrac{x+14}{x^2-2x-8}=\dfrac{3}{x-4}-\dfrac{2}{x+2}$.

19. $\dfrac{x}{8x^2-10x+3}=\dfrac{x}{(4x-3)(2x-1)}=\dfrac{A}{4x-3}+\dfrac{B}{2x-1}$. Hence,
$x=A(2x-1)+B(4x-3)=(2A+4B)x+(-A-3B)$, and so

$\begin{cases} 2A+4B=1 \\ -A-3B=0 \end{cases} \Leftrightarrow \begin{cases} 2A+4B=1 \\ -2A-6B=0 \end{cases}$. Adding we get $-2B=1 \Leftrightarrow B=-\frac{1}{2}$, and

$A=\frac{3}{2}$. Therefore, $\dfrac{x}{8x^2-10x+3}=\dfrac{\frac{3}{2}}{4x-3}-\dfrac{\frac{1}{2}}{2x-1}$.

21. $\dfrac{9x^2-9x+6}{2x^3-x^2-8x+4}=\dfrac{9x^2-9x+6}{(x-2)(x+2)(2x-1)}=\dfrac{A}{x-2}+\dfrac{B}{x+2}+\dfrac{C}{2x-1}$. Thus,

$$\begin{aligned} 9x^2-9x+6 &= A(x+2)(2x-1)+B(x-2)(2x-1)+C(x-2)(x+2) \\ &= A(2x^2+3x-2)+B(2x^2-5x+2)+C(x^2-4) \\ &= (2A+2B+C)x^2+(3A-5B)x+(-2A+2B-4C). \end{aligned}$$

This leads to a system with the following matrix representation:

$$\begin{bmatrix} 2 & 2 & 1 & 9 \\ 3 & -5 & 0 & -9 \\ -2 & 2 & -4 & 6 \end{bmatrix} \xrightarrow[R_1 \to R_1+R_3]{R_2 \to R_2+R_3} \begin{bmatrix} 0 & 4 & -3 & 15 \\ 1 & -3 & -4 & -3 \\ -2 & 2 & -4 & 6 \end{bmatrix} \xrightarrow{R_1 \leftrightarrow R_2}$$

$$\begin{bmatrix} 1 & -3 & -4 & -3 \\ 0 & 4 & -3 & 15 \\ -2 & 2 & -4 & 6 \end{bmatrix} \xrightarrow{R_3 \to R_3+2R_1} \begin{bmatrix} 1 & -3 & -4 & -3 \\ 0 & 4 & -3 & 15 \\ 0 & -4 & -12 & 0 \end{bmatrix} \xrightarrow{R_3 \to R_3+R_2}$$

$$\begin{bmatrix} 1 & -3 & -4 & -3 \\ 0 & 4 & -3 & 15 \\ 0 & 0 & -15 & 15 \end{bmatrix}$$

Hence, $-15C=15 \Leftrightarrow C=-1$, $4B+3=15 \Leftrightarrow B=3$, and $A-9+4=-3 \Leftrightarrow$
$A=2$. Therefore, $\dfrac{9x^2-9x+6}{2x^3-x^2-8x+4}=\dfrac{2}{x-2}+\dfrac{3}{x+2}-\dfrac{1}{2x-1}$.

23. $\dfrac{x^2+1}{x^3+x^2}=\dfrac{x^2+1}{x^2(x+1)}=\dfrac{A}{x}+\dfrac{B}{x^2}+\dfrac{C}{x+1}$. Hence,
$x^2+1=Ax(x+1)+B(x+1)+Cx^2=(A+C)x^2+(A+B)x+B$, and so $B=1$, $A+1=0$
$\Leftrightarrow A=-1$, and $-1+C=1 \Leftrightarrow C=2$. Therefore, $\dfrac{x^2+1}{x^3+x^2}=\dfrac{-1}{x}+\dfrac{1}{x^2}+\dfrac{2}{x+1}$.

25. $\dfrac{2x}{4x^2+12x+9} = \dfrac{2x}{(2x+3)^2} = \dfrac{A}{2x+3} + \dfrac{B}{(2x+3)^2}$. Hence,
$2x = A(2x+3) + B = 2Ax + (3A+B)$. So $2A = 2 \Leftrightarrow A = 1$, and $3(1) + B = 0 \Leftrightarrow$ $B = -3$. Therefore, $\dfrac{2x}{4x^2+12x+9} = \dfrac{1}{2x+3} - \dfrac{3}{(2x+3)^2}$.

27. $\dfrac{4x^2-x-2}{x^4+2x^3} = \dfrac{4x^2-x-2}{x^3(x+2)} = \dfrac{A}{x} + \dfrac{B}{x^2} + \dfrac{C}{x^3} + \dfrac{D}{x+2}$. Hence,
$4x^2 - x - 2 = Ax^2(x+2) + Bx(x+2) + C(x+2) + Dx^3$
$= (A+D)x^3 + (2A+B)x^2 + (2B+C)x + 2C$.
So $2C = -2 \Leftrightarrow C = -1$; $2B - 1 = -1 \Leftrightarrow B = 0$; $2A + 0 = 4 \Leftrightarrow A = 2$; $2 + D = 0 \Leftrightarrow D = -2$. Therefore, $\dfrac{4x^2-x-2}{x^4+2x^3} = \dfrac{2}{x} - \dfrac{1}{x^3} - \dfrac{2}{x+2}$.

29. $\dfrac{-10x^2+27x-14}{(x-1)^3(x+2)} = \dfrac{A}{x+2} + \dfrac{B}{x-1} + \dfrac{C}{(x-1)^2} + \dfrac{D}{(x-1)^3}$. Thus,
$-10x^2 + 27x - 14 = A(x-1)^3 + B(x+2)(x-1)^2 + C(x+2)(x-1) + D(x+2)$
$= A(x^3 - 3x^2 + 3x - 1) + B(x+2)(x^2 - 2x + 1) + C(x^2 + x - 2) + D(x+2)$
$= A(x^3 - 3x^2 + 3x - 1) + B(x^3 - 3x + 2) + C(x^2 + x - 2) + D(x+2)$
$= (A+B)x^3 + (-3A+C)x^2 + (3A - 3B + C + D)x + (-A + 2B - 2C + 2D)$, which leads to the following system of equations:

$$\begin{bmatrix} 1 & 1 & 0 & 0 & 0 \\ -3 & 0 & 1 & 0 & -10 \\ 3 & -3 & 1 & 1 & 27 \\ -1 & 2 & -2 & 2 & -14 \end{bmatrix} \xrightarrow[R_4 \to R_4+R_1]{R_2 \to R_2+3R_1,\ R_3 \to R_3-3R_1} \begin{bmatrix} 1 & 1 & 0 & 0 & 0 \\ 0 & 3 & 1 & 0 & -10 \\ 0 & -6 & 1 & 1 & 27 \\ 0 & 3 & -2 & 2 & -14 \end{bmatrix}$$

$$\xrightarrow[R_4 \to R_4-R_2]{R_3 \to R_3+2R_2} \begin{bmatrix} 1 & 1 & 0 & 0 & 0 \\ 0 & 3 & 1 & 0 & -10 \\ 0 & 0 & 3 & 1 & 7 \\ 0 & 0 & -3 & 2 & -4 \end{bmatrix} \xrightarrow{R_4 \to R_4+R_3} \begin{bmatrix} 1 & 1 & 0 & 0 & 0 \\ 0 & 3 & 1 & 0 & -10 \\ 0 & 0 & 3 & 1 & 7 \\ 0 & 0 & 0 & 3 & 3 \end{bmatrix}$$

Hence, $3D = 3 \Leftrightarrow D = 1$; $3C + 1 = 7 \Leftrightarrow C = 2$; $3B + 2 = -10 \Leftrightarrow B = -4$, and $A - 4 = 0 \Leftrightarrow A = 4$. Therefore,
$\dfrac{-10x^2+27x-14}{(x-1)^3(x+2)} = \dfrac{4}{x+2} - \dfrac{4}{x-1} + \dfrac{2}{(x-1)^2} + \dfrac{1}{(x-1)^3}$.

31. $\dfrac{3x^3+22x^2+53x+41}{(x+2)^2(x+3)^2} = \dfrac{A}{x+2} + \dfrac{B}{(x+2)^2} + \dfrac{C}{x+3} + \dfrac{D}{(x+3)^2}$. Thus,
$3x^3 + 22x^2 + 53x + 41 = A(x+2)(x+3)^2 + B(x+3)^2 + C(x+2)^2(x+3) + D(x+2)^2$
$= A(x^3 + 8x^2 + 21x + 18) + B(x^2 + 6x + 9) + C(x^3 + 7x^2 + 16x + 12) + D(x^2 + 4x + 4)$
$= (A+C)x^3 + (8A + B + 7C + D)x^2 + (21A + 6B + 16C + 4D)x +$
$(18A + 9B + 12C + 4D)$, so we must solve the system:

$$\begin{bmatrix} 1 & 0 & 1 & 0 & 3 \\ 8 & 1 & 7 & 1 & 22 \\ 21 & 6 & 16 & 4 & 53 \\ 18 & 9 & 12 & 4 & 41 \end{bmatrix} \xrightarrow[R_4 \to R_4 - 18R_1]{R_2 \to R_2 - 8R_1,\ R_3 \to R_3 - 21R_1} \begin{bmatrix} 1 & 0 & 1 & 0 & 3 \\ 0 & 1 & -1 & 1 & -2 \\ 0 & 6 & -5 & 4 & -10 \\ 0 & 9 & -6 & 4 & -13 \end{bmatrix}$$

$$\xrightarrow[R_4 \to R_4 - 9R_2]{R_3 \to R_3 - 6R_2} \begin{bmatrix} 1 & 0 & 1 & 0 & 3 \\ 0 & 1 & -1 & 1 & -2 \\ 0 & 0 & 1 & -2 & 2 \\ 0 & 0 & 3 & -5 & 5 \end{bmatrix} \xrightarrow{R_4 \to R_4 - 3R_3} \begin{bmatrix} 1 & 0 & 1 & 0 & 3 \\ 0 & 1 & -1 & 1 & -2 \\ 0 & 0 & 1 & -2 & 2 \\ 0 & 0 & 0 & 1 & -1 \end{bmatrix}$$

Hence, $D = -1$, $C + 2 = 2 \quad \Leftrightarrow \quad C = 0$, $B - 0 - 1 = -2 \quad \Leftrightarrow \quad B = -1$, and $A + 0 = 3$ $\Leftrightarrow \quad A = 3$. Therefore, $\dfrac{3x^3 + 22x^2 + 53x + 41}{(x+2)^2(x+3)^2} = \dfrac{3}{x+2} - \dfrac{1}{(x+2)^2} - \dfrac{1}{(x+3)^2}$.

33. $\dfrac{x-3}{x^3+3x} = \dfrac{x-3}{x(x^2+3)} = \dfrac{A}{x} + \dfrac{Bx+C}{x^2+3}$. Hence, $x - 3 = A(x^2+3) + Bx^2 + Cx$ $= (A+B)x^2 + Cx + 3A$. So $3A = -3 \quad \Leftrightarrow \quad A = -1$, $C = 1$, and $-1 + B = 0 \quad \Leftrightarrow$ $B = 1$. Therefore, $\dfrac{x-3}{x^3+3x} = -\dfrac{1}{x} + \dfrac{x+1}{x^2+3}$.

35. $\dfrac{2x^3+7x+5}{(x^2+x+2)(x^2+1)} = \dfrac{Ax+B}{x^2+x+2} + \dfrac{Cx+D}{x^2+1}$. Thus,
$2x^3 + 7x + 5 = (Ax+B)(x^2+1) + (Cx+D)(x^2+x+2)$
$= Ax^3 + Ax + Bx^2 + B + Cx^3 + Cx^2 + 2Cx + Dx^2 + Dx + 2D$
$= (A+C)x^3 + (B+C+D)x^2 + (A+2C+D)x + (B+2D)$. We must solve the system:

$$\begin{bmatrix} 1 & 0 & 1 & 0 & 2 \\ 0 & 1 & 1 & 1 & 0 \\ 1 & 0 & 2 & 1 & 7 \\ 0 & 1 & 0 & 2 & 5 \end{bmatrix} \xrightarrow[R_4 \to R_4 - R_2]{R_3 \to R_3 - R_1} \begin{bmatrix} 1 & 0 & 1 & 0 & 2 \\ 0 & 1 & 1 & 1 & 0 \\ 0 & 0 & 1 & 1 & 5 \\ 0 & 0 & -1 & 1 & 5 \end{bmatrix} \xrightarrow{R_4 \to R_4 + R_3} \begin{bmatrix} 1 & 0 & 1 & 0 & 2 \\ 0 & 1 & 1 & 1 & 0 \\ 0 & 0 & 1 & 1 & 5 \\ 0 & 0 & 0 & 2 & 10 \end{bmatrix}$$

Hence, $2D = 10 \quad \Leftrightarrow \quad D = 5$, $C + 5 = 5 \quad \Leftrightarrow \quad C = 0$, $B + 0 + 5 = 0 \quad \Leftrightarrow \quad B = -5$, and $A + 0 = 2 \quad \Leftrightarrow \quad A = 2$. Therefore, $\dfrac{2x^3+7x+5}{(x^2+x+2)(x^2+1)} = \dfrac{2x-5}{x^2+x+2} + \dfrac{5}{x^2+1}$.

37. $\dfrac{x^4+x^3+x^2-x+1}{x(x^2+1)^2} = \dfrac{A}{x} + \dfrac{Bx+C}{x^2+1} + \dfrac{Dx+E}{(x^2+1)^2}$. Hence,
$x^4 + x^3 + x^2 - x + 1 = A(x^2+1)^2 + (Bx+C)x(x^2+1) + x(Dx+E)$
$= A(x^4 + 2x^2 + 1) + (Bx^2 + Cx)(x^2+1) + Dx^2 + Ex$
$= A(x^4 + 2x^2 + 1) + Bx^4 + Bx^2 + Cx^3 + Cx + Dx^2 + Ex$
$= (A+B)x^4 + Cx^3 + (2A+B+D)x^2 + (C+E)x + A$, and so $A = 1$, $1 + B = 1 \quad \Leftrightarrow$ $B = 0$, $C = 1$, $2 + 0 + D = 1 \quad \Leftrightarrow \quad D = -1$, and $1 + E = -1 \quad \Leftrightarrow \quad E = -2$. Therefore, $\dfrac{x^4+x^3+x^2-x+1}{x(x^2+1)^2} = \dfrac{1}{x} + \dfrac{1}{x^2+1} - \dfrac{x+2}{(x^2+1)^2}$.

39. We must first get a proper rational function. Using long division we have:

$$\begin{array}{r} x^2 \\ x^3-2x^2+x-2\overline{\big)\,x^5-2x^4+x^3+0x^2+x+5} \\ \underline{x^5-2x^4+x^3-2x^2} \\ 2x^2+x+5 \end{array}$$

Therefore, $\dfrac{x^5-2x^4+x^3+x+5}{x^3-2x^2+x-2}=x^2+\dfrac{2x^2+x+5}{x^3-2x^2+x-2}=x^2+\dfrac{2x^2+x+5}{(x-2)(x^2+1)}$ $=x^2+\dfrac{A}{x-2}+\dfrac{Bx+C}{x^2+1}$. Hence, $2x^2+x+5=A(x^2+1)+(Bx+C)(x-2)$ $=Ax^2+A+Bx^2+Cx-2Bx-2C=(A+B)x^2+(C-2B)x+(A-2C)$. Equating coefficients, we get the system:

$$\begin{bmatrix} 1 & 1 & 0 & 2 \\ 0 & -2 & 1 & 1 \\ 1 & 0 & -2 & 5 \end{bmatrix} \xrightarrow{R_3 \to R_3 - R_1} \begin{bmatrix} 1 & 1 & 0 & 2 \\ 0 & -2 & 1 & 1 \\ 0 & -1 & -2 & 3 \end{bmatrix} \xrightarrow[-R_3 \leftrightarrow R_2]{R_2 \to R_2 - 2R_3} \begin{bmatrix} 1 & 1 & 0 & 2 \\ 0 & 1 & 2 & -3 \\ 0 & 0 & 5 & -5 \end{bmatrix}$$

Therefore, $5C=-5 \Leftrightarrow C=-1$, $B-2=-3 \Leftrightarrow B=-1$, and $A-1=2 \Leftrightarrow$ $A=3$. Therefore, $\dfrac{x^5-2x^4+x^3+x+5}{x^3-2x^2+x-2}=x^2+\dfrac{3}{x-2}-\dfrac{x+1}{x^2+1}$.

41. $\dfrac{ax+b}{x^2-1}=\dfrac{A}{x-1}+\dfrac{B}{x+1}$. Hence, $ax+b=A(x+1)+B(x-1)=(A+B)x+(A-B)$. So $\begin{cases} A+B=a \\ A-B=b \end{cases}$ Adding we get $2A=a+b \Leftrightarrow A=\dfrac{a+b}{2}$. Substituting we get $B=a-A=\dfrac{2a}{2}-\dfrac{a+b}{2}=\dfrac{a-b}{2}$. Therefore, $A=\dfrac{a+b}{2}$ and $B=\dfrac{a-b}{2}$.

Review Exercises for Chapter 6

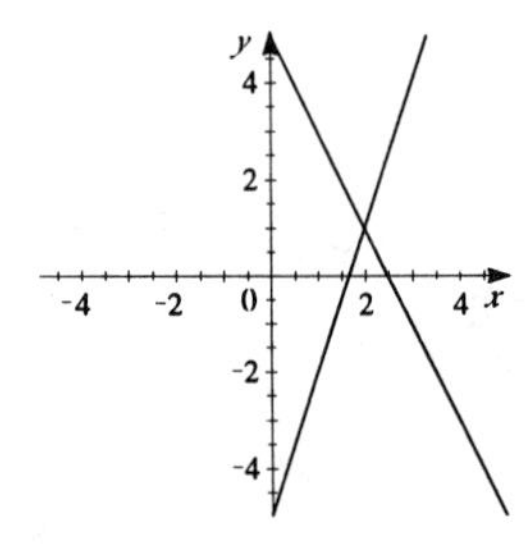

1. $\begin{cases} 3x - y = 5 \\ 2x + y = 5 \end{cases}$ Adding we get $5x = 10 \quad \Leftrightarrow \quad x = 2$. So $2(2) + y = 5 \quad \Leftrightarrow \quad y = 1$. Thus the solution is $(2, 1)$.

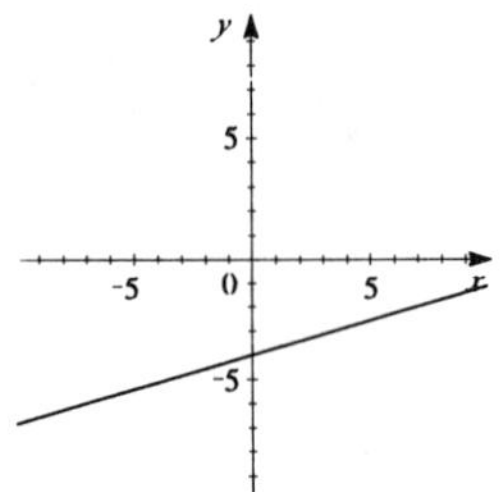

3. $\begin{cases} 2x - 7y = 28 \\ y = \frac{2}{7}x - 4 \end{cases} \quad \Leftrightarrow \quad \begin{cases} 2x - 7y = 28 \\ 2x - 7y = 28 \end{cases}$ Since these equations represent the same line, any point on this line will satisfy the system. Thus the solution is $\left(x, \frac{2}{7}x - 4\right)$, where x is any real number.

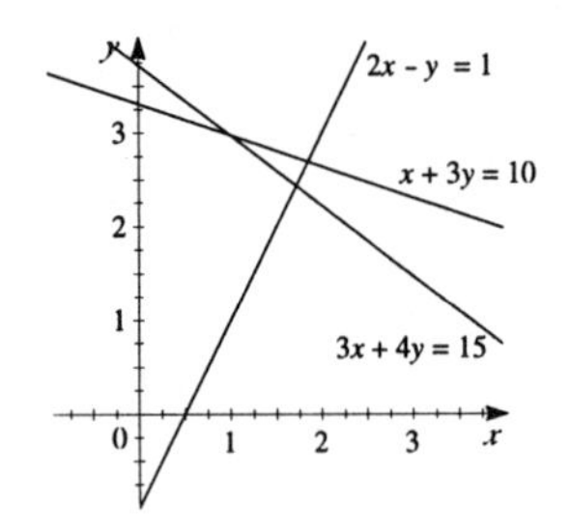

5. $\begin{cases} 2x - y = 1 \\ x + 3y = 10 \\ 3x + 4y = 15 \end{cases}$ Solving the first equation for y we get $y = -2x + 1$. Substituting into the second equation we have: $x + 3(-2x + 1) = 10 \quad \Leftrightarrow \quad -5x = 7 \quad \Leftrightarrow \quad x = -\frac{7}{5}$. So $y = -\left(-\frac{7}{5}\right) + 1 = \frac{12}{5}$. Checking the point $\left(-\frac{7}{5}, \frac{12}{5}\right)$ in the third equation we have $3\left(-\frac{7}{5}\right) + 4\left(\frac{12}{5}\right) \stackrel{?}{=} 15$ but $-\frac{21}{5} + \frac{48}{5} \neq 15$. Thus, there is no solution, and the lines do not intersect at one point.

7. $\begin{bmatrix} 1 & 1 & 2 & 6 \\ 2 & 0 & 5 & 12 \\ 1 & 2 & 3 & 9 \end{bmatrix} \xrightarrow[\text{R}_3 \to \text{R}_3 - \text{R}_1]{\text{R}_2 \to \text{R}_2 - 2\text{R}_1} \begin{bmatrix} 1 & 1 & 2 & 6 \\ 0 & -2 & 1 & 0 \\ 0 & 1 & 1 & 3 \end{bmatrix} \xrightarrow{\text{R}_3 \leftrightarrow \text{R}_2}$

$\begin{bmatrix} 1 & 1 & 2 & 6 \\ 0 & 1 & 1 & 3 \\ 0 & -2 & 1 & 0 \end{bmatrix} \xrightarrow{\text{R}_3 \to \text{R}_3 + 2\text{R}_2} \begin{bmatrix} 1 & 1 & 2 & 6 \\ 0 & 1 & 1 & 3 \\ 0 & 0 & 3 & 6 \end{bmatrix}$

Therefore, $3z = 6 \quad \Leftrightarrow \quad z = 2$; $y + 2 = 3 \quad \Leftrightarrow \quad y = 1$; and $x + 1 + 4 = 6 \quad \Leftrightarrow \quad x = 1$. Hence, the solution is $(1, 1, 2)$.

9. $\begin{bmatrix} 1 & -2 & 3 & 1 \\ 2 & -1 & 1 & 3 \\ 2 & -7 & 11 & 2 \end{bmatrix} \xrightarrow[\text{R}_3 \to \text{R}_3 - 2\text{R}_1]{\text{R}_2 \to \text{R}_2 - 2\text{R}_1} \begin{bmatrix} 1 & -2 & 3 & 1 \\ 0 & 3 & -5 & 1 \\ 0 & -3 & 5 & 0 \end{bmatrix} \xrightarrow{\text{R}_3 \to \text{R}_3 + \text{R}_2}$

$\begin{bmatrix} 1 & -2 & 3 & 1 \\ 0 & 3 & -5 & 1 \\ 0 & 0 & 0 & 1 \end{bmatrix}$

Therefore, the system is inconsistent and has no solution.

11. $\begin{bmatrix} 1 & -3 & 1 & 4 \\ 4 & -1 & 15 & 5 \end{bmatrix} \xrightarrow{R_2 \to R_2 - 4R_1} \begin{bmatrix} 1 & -3 & 1 & 4 \\ 0 & 11 & 11 & -11 \end{bmatrix} \xrightarrow{R_2 \to \frac{1}{11} R_2} \begin{bmatrix} 1 & -3 & 1 & 4 \\ 0 & 1 & 1 & -1 \end{bmatrix}$

Thus, the system has infinitely many solutions: $y + z = -1 \quad \Leftrightarrow \quad y = -1 - z$, and $x + 3(1 + z) + z = 4 \quad \Leftrightarrow \quad x = 1 - 4z$. Therefore, the solutions are $(1 - 4z, -1 - z, z)$, where z is any real number.

13. $\begin{bmatrix} -1 & 4 & 1 & 8 \\ 2 & -6 & 1 & -9 \\ 1 & -6 & -4 & -15 \end{bmatrix} \xrightarrow{R_1 \leftrightarrow R_3} \begin{bmatrix} 1 & -6 & -4 & -15 \\ 2 & -6 & 1 & -9 \\ -1 & 4 & 1 & 8 \end{bmatrix} \xrightarrow[R_3 \to R_3 + R_1]{R_2 \to R_2 - 2R_1}$

$\begin{bmatrix} 1 & -6 & -4 & -15 \\ 0 & 6 & 9 & 21 \\ 0 & -2 & -3 & -7 \end{bmatrix} \xrightarrow[R_3 \to \frac{1}{6} R_2]{R_3 \to 3R_3 + R_2} \begin{bmatrix} 1 & -6 & -4 & -15 \\ 0 & 1 & \frac{3}{2} & \frac{7}{2} \\ 0 & 0 & 0 & 0 \end{bmatrix}$

Thus, the system has infinitely many solutions: $y + \frac{3}{2} z = \frac{7}{2} \quad \Leftrightarrow \quad y = \frac{7}{2} - \frac{3}{2} z$, and $x - 6(\frac{7}{2} - \frac{3}{2} z) - 4z = -15 \quad \Leftrightarrow \quad x = 6 - 5z$. Therefore, the solution is $(6 - 5z, \frac{7}{2} - \frac{3}{2} z, z)$, where z is any real number.

15. Let x be the amount in the 6% account, and y be the amount in the 7% account. Thus, the system is:

$$\begin{cases} y = 2x \\ 0.06x + 0.07y = 600 \end{cases}$$

Substituting we have $0.06x + 0.07(2x) = 600 \quad \Leftrightarrow \quad 0.2x = 600 \quad \Leftrightarrow \quad x = 3000$. So $y = 2(3000) = 6000$. Hence, the man has \$3,000 invested at 6% and \$6,000 invested at 7%.

In Exercises 17 – 27,

$A = \begin{bmatrix} 2 & 0 & -1 \end{bmatrix}$ $\quad D = \begin{bmatrix} 1 & 4 \\ 0 & -1 \\ 2 & 0 \end{bmatrix}$ $\quad F = \begin{bmatrix} 4 & 0 & 2 \\ -1 & 1 & 0 \\ 7 & 5 & 0 \end{bmatrix}$

$B = \begin{bmatrix} 1 & 2 & 4 \\ -2 & 1 & 0 \end{bmatrix}$ $\quad E = \begin{bmatrix} 2 & -1 \\ -\frac{1}{2} & 1 \end{bmatrix}$ $\quad G = \begin{bmatrix} 5 \end{bmatrix}$

$C = \begin{bmatrix} \frac{1}{2} & 3 \\ 2 & \frac{3}{2} \\ -2 & 1 \end{bmatrix}$

17. $A + B$ cannot be performed because the matrix dimensions (1×3 and 2×3) are not compatible.

19. $2C + 3D = 2 \begin{bmatrix} \frac{1}{2} & 3 \\ 2 & \frac{3}{2} \\ -2 & 1 \end{bmatrix} + 3 \begin{bmatrix} 1 & 4 \\ 0 & -1 \\ 2 & 0 \end{bmatrix} = \begin{bmatrix} 1 & 6 \\ 4 & 3 \\ -4 & 2 \end{bmatrix} + \begin{bmatrix} 3 & 12 \\ 0 & -3 \\ 6 & 0 \end{bmatrix} = \begin{bmatrix} 4 & 18 \\ 4 & 0 \\ 2 & 2 \end{bmatrix}$

21. $GA = \begin{bmatrix} 5 \end{bmatrix} \begin{bmatrix} 2 & 0 & -1 \end{bmatrix} = \begin{bmatrix} 10 & 0 & -5 \end{bmatrix}$

23. $BC = \begin{bmatrix} 1 & 2 & 4 \\ -2 & 1 & 0 \end{bmatrix} \begin{bmatrix} \frac{1}{2} & 3 \\ 2 & \frac{3}{2} \\ -2 & 1 \end{bmatrix} = \begin{bmatrix} -\frac{7}{2} & 10 \\ 1 & -\frac{9}{2} \end{bmatrix}$

25. $BF = \begin{bmatrix} 1 & 2 & 4 \\ -2 & 1 & 0 \end{bmatrix} \begin{bmatrix} 4 & 0 & 2 \\ -1 & 1 & 0 \\ 7 & 5 & 0 \end{bmatrix} = \begin{bmatrix} 30 & 22 & 2 \\ -9 & 1 & -4 \end{bmatrix}$

27. $(C+D)E = \left(\begin{bmatrix} \frac{1}{2} & 3 \\ 2 & \frac{3}{2} \\ -2 & 1 \end{bmatrix} + \begin{bmatrix} 1 & 4 \\ 0 & -1 \\ 2 & 0 \end{bmatrix} \right) \begin{bmatrix} 2 & -1 \\ -\frac{1}{2} & 1 \end{bmatrix}$

$= \begin{bmatrix} \frac{3}{2} & 7 \\ 2 & \frac{1}{2} \\ 0 & 1 \end{bmatrix} \begin{bmatrix} 2 & -1 \\ -\frac{1}{2} & 1 \end{bmatrix} = \begin{bmatrix} -\frac{1}{2} & \frac{11}{2} \\ \frac{15}{4} & -\frac{3}{2} \\ -\frac{1}{2} & 1 \end{bmatrix}$

29. $D = \begin{bmatrix} 1 & 4 \\ 2 & 9 \end{bmatrix}$. Then, $|D| = 1(9) - 2(4) = 1$, and so $D^{-1} = \begin{bmatrix} 9 & -4 \\ -2 & 1 \end{bmatrix}$

31. $D = \begin{bmatrix} 4 & -12 \\ -2 & 6 \end{bmatrix}$. Then, $|D| = 4(6) - 2(12) = 0$, and so D has no inverse.

33. $D = \begin{bmatrix} 3 & 0 & 1 \\ 2 & -3 & 0 \\ 4 & -2 & 1 \end{bmatrix}$. Then, $|D| = 1 \begin{vmatrix} 2 & -3 \\ 4 & -2 \end{vmatrix} + 1 \begin{vmatrix} 3 & 0 \\ 2 & -3 \end{vmatrix} = -4 + 12 - 9 = -1$. So D has an inverse.

$$\left[\begin{array}{ccc|ccc} 3 & 0 & 1 & 1 & 0 & 0 \\ 2 & -3 & 0 & 0 & 1 & 0 \\ 4 & -2 & 1 & 0 & 0 & 1 \end{array}\right] \xrightarrow{R_1 \to R_1 - R_2} \left[\begin{array}{ccc|ccc} 1 & 3 & 1 & 1 & -1 & 0 \\ 2 & -3 & 0 & 0 & 1 & 0 \\ 4 & -2 & 1 & 0 & 0 & 1 \end{array}\right] \xrightarrow[R_3 \to R_3 - 4R_1]{R_2 \to R_2 - 2R_1}$$

$$\left[\begin{array}{ccc|ccc} 1 & 3 & 1 & 1 & -1 & 0 \\ 0 & -9 & -2 & -2 & 3 & 0 \\ 0 & -14 & -3 & -4 & 4 & 1 \end{array}\right] \xrightarrow[R_3 \to -2R_3]{R_2 \to -3R_2} \left[\begin{array}{ccc|ccc} 1 & 3 & 1 & 1 & -1 & 0 \\ 0 & 27 & 6 & 6 & -9 & 0 \\ 0 & 28 & 6 & 8 & -8 & -2 \end{array}\right] \xrightarrow{R_3 \to R_3 - R_2}$$

$$\left[\begin{array}{ccc|ccc} 1 & 3 & 1 & 1 & -1 & 0 \\ 0 & 27 & 6 & 6 & -9 & 0 \\ 0 & 1 & 0 & 2 & 1 & -2 \end{array}\right] \xrightarrow[R_3 \to \frac{1}{3}R_3]{R_3 \leftrightarrow R_2} \left[\begin{array}{ccc|ccc} 1 & 3 & 1 & 1 & -1 & 0 \\ 0 & 1 & 0 & 2 & 1 & -2 \\ 0 & 9 & 2 & 2 & -3 & 0 \end{array}\right] \xrightarrow[R_1 \to R_1 - 3R_2]{R_3 \to R_3 - 9R_2}$$

$$\left[\begin{array}{ccc|ccc} 1 & 0 & 1 & -5 & -4 & 6 \\ 0 & 1 & 0 & 2 & 1 & -2 \\ 0 & 0 & 2 & -16 & -12 & 18 \end{array}\right] \xrightarrow[R_1 \to R_1 - R_3]{R_3 \to \frac{1}{2}R_3} \left[\begin{array}{ccc|ccc} 1 & 0 & 0 & 3 & 2 & -3 \\ 0 & 1 & 0 & 2 & 1 & -2 \\ 0 & 0 & 1 & -8 & -6 & 9 \end{array}\right]$$

Thus, $D^{-1} = \begin{bmatrix} 3 & 2 & -3 \\ 2 & 1 & -2 \\ -8 & -6 & 9 \end{bmatrix}$.

35. $\begin{bmatrix} 12 & -5 \\ 5 & -2 \end{bmatrix} \begin{bmatrix} x \\ y \end{bmatrix} = \begin{bmatrix} 10 \\ 17 \end{bmatrix}$. If we let $A = \begin{bmatrix} 12 & -5 \\ 5 & -2 \end{bmatrix}$, then $A^{-1} = \frac{1}{-24+25} \begin{bmatrix} -2 & 5 \\ -5 & 12 \end{bmatrix}$

$= \begin{bmatrix} -2 & 5 \\ -5 & 12 \end{bmatrix}$, and so $\begin{bmatrix} x \\ y \end{bmatrix} = \begin{bmatrix} -2 & 5 \\ -5 & 12 \end{bmatrix} \begin{bmatrix} 10 \\ 17 \end{bmatrix} = \begin{bmatrix} 65 \\ 154 \end{bmatrix}$. Therefore, the solution is $(65, 154)$.

37. $|D| = \begin{vmatrix} 2 & 7 \\ 6 & 16 \end{vmatrix} = 32 - 42 = -10$, $|D_x| = \begin{vmatrix} 13 & 7 \\ 30 & 16 \end{vmatrix} = 208 - 210 = -2$, and

$|D_y| = \begin{vmatrix} 2 & 13 \\ 6 & 30 \end{vmatrix} = 60 - 78 = -18$. Therefore, $x = \frac{-2}{-10} = \frac{1}{5}$ and $y = \frac{-18}{-10} = \frac{9}{5}$, and so the solution is $\left(\frac{1}{5}, \frac{9}{5}\right)$.

39. $|D| = \begin{vmatrix} 2 & -1 & 5 \\ -1 & 7 & 0 \\ 5 & 4 & 3 \end{vmatrix} = 5\begin{vmatrix} -1 & 7 \\ 5 & 4 \end{vmatrix} + 3\begin{vmatrix} 2 & -1 \\ -1 & 7 \end{vmatrix} = -195 + 39 = -156;$

$|D_x| = \begin{vmatrix} 0 & -1 & 5 \\ 9 & 7 & 0 \\ -9 & 4 & 3 \end{vmatrix} = 5\begin{vmatrix} 9 & 7 \\ -9 & 4 \end{vmatrix} + 3\begin{vmatrix} 0 & -1 \\ 9 & 7 \end{vmatrix} = 495 + 27 = 522;$

$|D_y| = \begin{vmatrix} 2 & 0 & 5 \\ -1 & 9 & 0 \\ 5 & -9 & 3 \end{vmatrix} = 5\begin{vmatrix} -1 & 9 \\ 5 & -9 \end{vmatrix} + 3\begin{vmatrix} 2 & 0 \\ -1 & 9 \end{vmatrix} = -180 + 54 = -126$; and

$|D_z| = \begin{vmatrix} 2 & -1 & 0 \\ -1 & 7 & 9 \\ 5 & 4 & -9 \end{vmatrix} = -9\begin{vmatrix} 2 & -1 \\ 5 & 4 \end{vmatrix} - 9\begin{vmatrix} 2 & -1 \\ -1 & 7 \end{vmatrix} = -117 - 117 = -234.$

Therefore, $x = \frac{522}{-156} = -\frac{87}{26}$, $y = \frac{-126}{-156} = \frac{21}{26}$, and $z = \frac{-234}{-156} = \frac{3}{2}$, and so the solution is $\left(-\frac{87}{26}, \frac{21}{26}, \frac{3}{2}\right)$.

41. $\begin{cases} x^2 + y^2 + 6y = 0 \\ x - 2y = 3 \end{cases}$ Solving the second equation for x we have $x = 2y + 3$. So by substitution $(2y+3)^2 + y^2 + 6y = 0 \Leftrightarrow 4y^2 + 12y + 9 + y^2 + 6y = 0 \Leftrightarrow (5y+3)(y+3) = 0 \Leftrightarrow$ $y = -\frac{3}{5}$ or $y = -3$. If $y = -\frac{3}{5}$, $x = \frac{9}{5}$, while if $y = -3$, $x = -3$. Thus, the solutions are $(-3, -3)$ and $\left(\frac{9}{5}, -\frac{3}{5}\right)$.

43. $\begin{cases} 3x^4 + \dfrac{4}{y} = 50 \\ x^4 - \dfrac{8}{y} = 12 \end{cases}$ If we let $u = x^4$ and $v = \frac{1}{y}$, then the system can be represented as

$\begin{cases} 3u + 4v = 50 \\ u - 8v = 12 \end{cases} \Leftrightarrow \begin{cases} 3u + 4v = 50 \\ 3u - 24v = 36 \end{cases}$. Subtracting we have $28v = 14 \Leftrightarrow v = \frac{1}{2}$, so $u = 16$. Then, $x^4 = 16 \Leftrightarrow x = \pm 2$ and $y = 2$ and the solutions are $(\pm 2, 2)$.

45. $\begin{cases} x^2 + y^2 < 9 \\ x + y < 0 \end{cases}$ The vertices occur where $y = -x$. By substitution, $x^2 + x^2 = 9 \Leftrightarrow x = \pm\frac{3}{\sqrt{2}}$, and so $y = \mp\frac{3}{\sqrt{2}}$. Therefore, the vertices occur at $\left(\frac{3}{\sqrt{2}}, -\frac{3}{\sqrt{2}}\right)$ and $\left(-\frac{3}{\sqrt{2}}, \frac{3}{\sqrt{2}}\right)$ and the solution set is bounded.

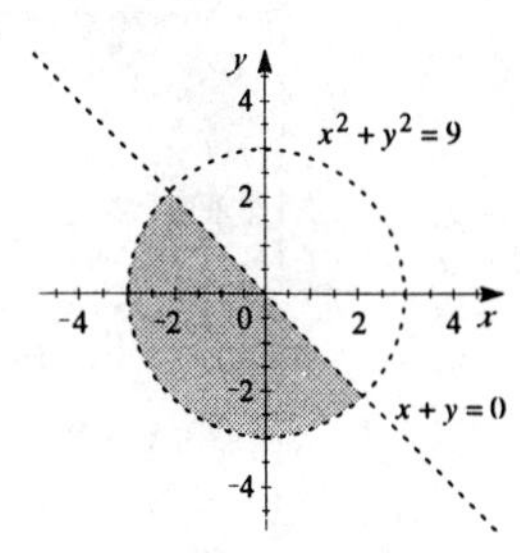

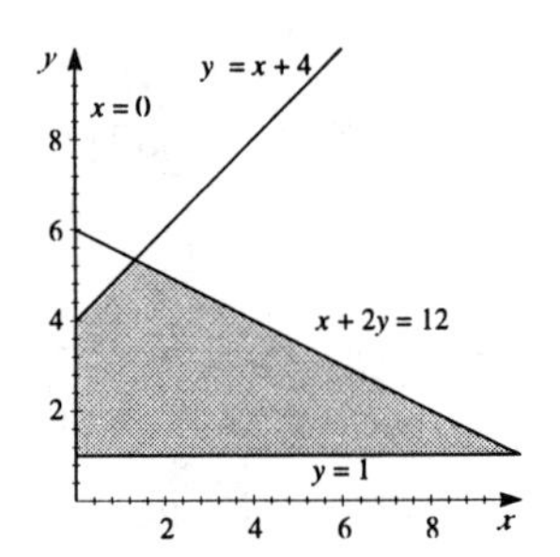

47. $\begin{cases} x \geq 0, y \geq 1 \\ x + 2y \leq 12 \\ y \leq x + 4 \end{cases}$ The intersection points occur at $(-3, 1)$, $(0, 4)$, $\left(\frac{4}{3}, \frac{16}{3}\right)$, $(0, 6)$, $(0, 1)$, and $(10, 1)$. Since the points $(-3, 1)$ and $(0, 6)$ are not in the solution set, the vertices are $(0, 4)$, $\left(\frac{4}{3}, \frac{16}{3}\right)$, $(10, 1)$, and $(0, 1)$. The solution set is bounded.

49.

Vertex	$P = 3x + 4y$
$(0, 4)$	$3(0) + 4(4) = 16$
$\left(\frac{4}{3}, \frac{16}{3}\right)$	$3\left(\frac{4}{3}\right) + 4\left(\frac{16}{3}\right) = \frac{76}{3}$
$(10, 1)$	$3(10) + 4(1) = 34$ ← maximum value
$(0, 1)$	$3(0) + 4(1) = 4$ ← minimum value

Therefore, the maximum value of the function is 34 and the minimum value is 4.

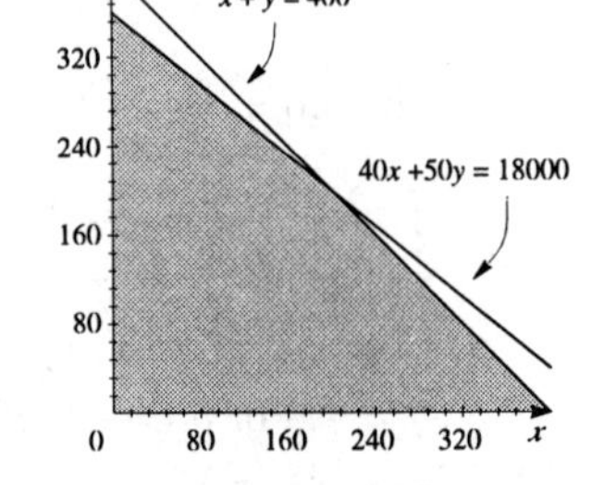

51. Let x be the number of acres of oats planted and y be the number of acres of barley planted. Then the given conditions lead to the inequalities:

$$\begin{cases} x + y \leq 400 \\ 40x + 50y \leq 18{,}000 \\ x \geq 0, y \geq 0 \end{cases}$$

From the graph the vertices occur at $(0, 0)$, $(400, 0)$, $(200, 200)$, and $(0, 360)$.

(a) The objective function is $P = 2.05(40x) + 1.80(50y) = 82x + 90y$.

Vertex	$P = 82x + 90y$
$(0, 0)$	$82(0) + 90(0) = 0$
$(400, 0)$	$82(400) + 90(0) = 32,800$
$(200, 200)$	$82(200) + 90(200) = 34,400$ ← maximum value
$(0, 360)$	$82(0) + 90(360) = 32,400$

Hence, he should plant 200 acres each of oats and barley for a profit of $34,400.

(b) The objective function is $P = 1.20(40x) + 1.60(50y) = 48x + 80y$.

Vertex	$P = 48x + 80y$
$(0, 0)$	$48(0) + 80(0) = 0$
$(400, 0)$	$48(400) + 80(0) = 19,200$
$(200, 200)$	$48(200) + 80(200) = 25,600$
$(0, 360)$	$48(0) + 80(360) = 28,800$ ← maximum value

Hence, he should plant 360 acres of barley and no oats for a profit of $28,800.

53. $$\begin{bmatrix} -1 & 1 & 1 & a \\ 1 & -1 & 1 & b \\ 1 & 1 & -1 & c \end{bmatrix} \xrightarrow{R_1 \to -R_1} \begin{bmatrix} 1 & -1 & -1 & -a \\ 1 & -1 & 1 & b \\ 1 & 1 & -1 & c \end{bmatrix} \xrightarrow[R_3 \to R_3 - R_1]{R_2 \to R_2 - R_1}$$

$$\begin{bmatrix} 1 & -1 & -1 & -a \\ 0 & 0 & 2 & a+b \\ 0 & 2 & 0 & a+c \end{bmatrix} \xrightarrow{R_2 \leftrightarrow R_3} \begin{bmatrix} 1 & -1 & -1 & -a \\ 0 & 2 & 0 & a+c \\ 0 & 0 & 2 & a+b \end{bmatrix}$$

Thus, $z = \frac{a+b}{2}$, $y = \frac{a+c}{2}$, and $x - \frac{a+c}{2} - \frac{a+b}{2} = -a \quad \Leftrightarrow \quad x = \frac{b+c}{2}$. The solution is $\left(\frac{b+c}{2}, \frac{a+c}{2}, \frac{a+b}{2}\right)$.

55. $$\begin{bmatrix} 1 & 1 & 12 \\ k & -1 & 0 \\ -1 & 1 & 2k \end{bmatrix} \xrightarrow{R_3 \to R_3 + R_1} \begin{bmatrix} 1 & 1 & 12 \\ k & -1 & 0 \\ 0 & 2 & 12+2k \end{bmatrix}$$

Thus, $2y = 12 + 2k \quad \Leftrightarrow \quad y = 6 + k$ and $x = \frac{6+k}{k}$, and so $\frac{6+k}{k} + 6 + k = 12 \Rightarrow$ $6 + k + 6k + k^2 = 12k \quad \Leftrightarrow \quad k^2 - 5k + 6 = 0 \quad \Leftrightarrow \quad (k-3)(k-2) = 0 \Rightarrow k = 2$ or $k = 3$.

57. $$\begin{cases} 0.32x + 0.43y = 0 \\ 7x - 12y = 341 \end{cases} \quad \Leftrightarrow \quad \begin{cases} y = -\dfrac{32}{43}x \\ y = \dfrac{7x - 341}{12} \end{cases}$$

The solution is $(21.41, -15.93)$.

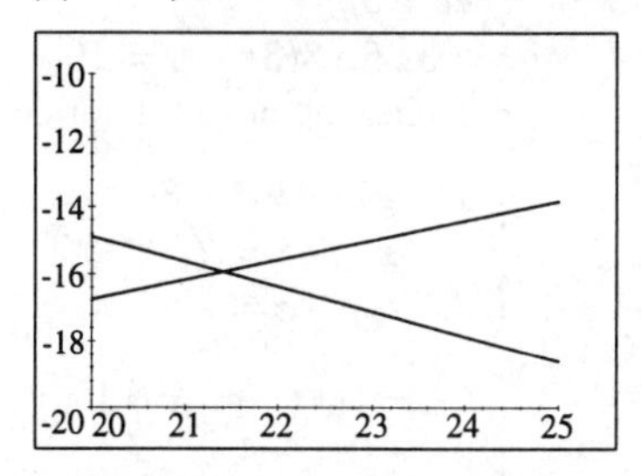

59. $$\begin{cases} x - y^2 = 10 \\ x = \frac{1}{22}y + 12 \end{cases} \quad \Leftrightarrow \quad \begin{cases} y = \pm\sqrt{x - 10} \\ y = 22(x - 12) \end{cases}$$

The solutions are $(11.94, -1.39)$ and $(12.07, 1.44)$.

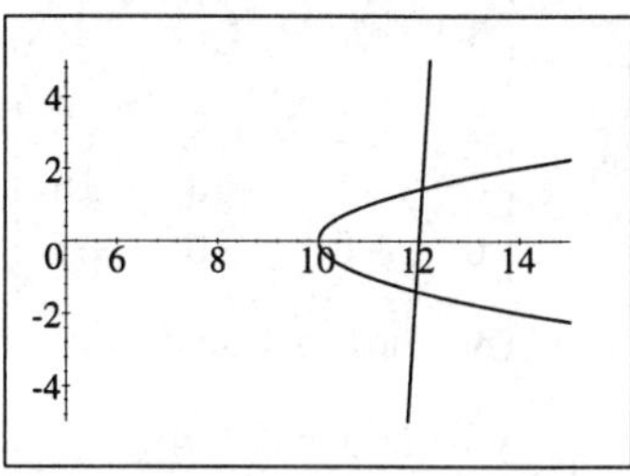

61. $\dfrac{3x+1}{x^2 - 2x - 15} = \dfrac{3x+1}{(x-5)(x+3)} = \dfrac{A}{x-5} + \dfrac{B}{x+3}$. Thus, $3x + 1 = A(x+3) + B(x-5)$ $= x(A+B) + (3A - 5B)$, and so

$$\begin{cases} A + B = 3 \\ 3A - 5B = 1 \end{cases} \quad \Leftrightarrow \quad \begin{cases} -3A - 3B = -9 \\ 3A - 5B = 1 \end{cases}$$ Adding we have $-8B = -8 \quad \Leftrightarrow \quad B = 1$,

and $A = 2$. Hence, $\dfrac{3x+1}{x^2 - 2x - 15} = \dfrac{2}{x-5} + \dfrac{1}{x+3}$.

63. $\dfrac{2x-4}{x(x-1)^2} = \dfrac{A}{x} + \dfrac{B}{x-1} + \dfrac{C}{(x-1)^2}$. Then, $2x - 4 = A(x-1)^2 + Bx(x-1) + Cx$ $= Ax^2 - 2Ax + A + Bx^2 - Bx + Cx = x^2(A+B) + x(-2A - B + C) + A$. So $A = -4$, $-4 + B = 0 \quad \Leftrightarrow \quad B = 4$, and $8 - 4 + C = 2 \quad \Leftrightarrow \quad C = -2$. So,

$$\frac{2x-4}{x(x-1)^2} = -\frac{4}{x} + \frac{4}{x-1} - \frac{2}{(x-1)^2}$$

Chapter 6 Test

1. Let w be the speed of the wind and a be the speed of the airplane in still air, in km per hour. Then the speed of the of the plane flying against the wind is $a - w$ and the speed of the plane flying with the wind is $a + w$. Then using *distance* = *time* × *rate* we get the system

$$\begin{cases} 600 = 2.5(a-w) \\ 300 = \frac{50}{60}(a+w) \end{cases} \quad \Leftrightarrow \quad \begin{cases} 240 = a - w \\ 360 = a + w \end{cases}$$

Adding the two equations we get $600 = 2a \quad \Leftrightarrow \quad a = 300$. Thus $360 = 300 + w \quad \Leftrightarrow \quad w = 60$. Thus the speed of the airplane in still air is 300 km/h and the speed of the wind is 60 km/h.

2. $\begin{cases} 3x - y = 10 \\ 2x + 5y = 1 \end{cases}$ Multiplying the first equation by 5 and then adding gives $17x = 51 \quad \Leftrightarrow$ $x = 3$. So $3(3) - y = 10 \quad \Leftrightarrow \quad y = -1$. Thus, the solution is $(3, -1)$, the system is linear, and it is neither inconsistent nor dependent.

3. $\begin{cases} x - y + 9z = -8 \\ x - 4z = 7 \\ 3x - y + z = 5 \end{cases}$ This system is linear and has the following matrix representation.

$$\begin{bmatrix} 1 & -1 & 9 & -8 \\ 1 & 0 & -4 & 7 \\ 3 & -1 & 1 & 5 \end{bmatrix} \xrightarrow[R_3 \to R_3 - 3R_1]{R_2 \to R_2 - R_1} \begin{bmatrix} 1 & -1 & 9 & -8 \\ 0 & 1 & -13 & 15 \\ 0 & 2 & -26 & 29 \end{bmatrix} \xrightarrow{R_3 \to R_3 - 2R_2}$$

$$\begin{bmatrix} 1 & -1 & 9 & -8 \\ 0 & 1 & -13 & 15 \\ 0 & 0 & 0 & -1 \end{bmatrix}$$

The third row states $0 = -1$, which is false. Hence the system is inconsistent.

4. $\begin{cases} 2x - y + z = 0 \\ 3x + 2y - 3z = 1 \\ x - 4y + 5z = -1 \end{cases}$ This system is linear and has the following matrix representation.

$$\begin{bmatrix} 2 & -1 & 1 & 0 \\ 3 & 2 & -3 & 1 \\ 1 & -4 & 5 & -1 \end{bmatrix} \xrightarrow{R_1 \leftrightarrow R_3} \begin{bmatrix} 1 & -4 & 5 & -1 \\ 3 & 2 & -3 & 1 \\ 2 & -1 & 1 & 0 \end{bmatrix} \xrightarrow[R_3 \to R_3 - 2R_1]{R_2 \to R_2 - 3R_1}$$

$$\begin{bmatrix} 1 & -4 & 5 & -1 \\ 0 & 14 & -18 & 4 \\ 0 & 7 & -9 & 2 \end{bmatrix} \xrightarrow{R_3 \to 2R_3 - R_2} \begin{bmatrix} 1 & -4 & 5 & -1 \\ 0 & 14 & -18 & 4 \\ 0 & 0 & 0 & 0 \end{bmatrix}$$

Then, $14\,y - 18z = 4 \quad \Leftrightarrow \quad y = \frac{2}{7} + \frac{9}{7}z$; $x - 4\left(\frac{2}{7} + \frac{9}{7}z\right) + 5z = -1 \quad \Leftrightarrow \quad x = \frac{1}{7} + \frac{1}{7}z$; and so the solutions are $\left(\frac{1}{7} + \frac{1}{7}z, \frac{2}{7} + \frac{9}{7}z, z\right)$ where z is any real number. The system is dependent.

5. $\begin{cases} 2x^2 + y^2 = 6 \\ 3x^2 - 4\,y = 11 \end{cases}$ Solving the second equation for x^2 we get $x^2 = \dfrac{11 + 4\,y}{3}$. Substituting into the first equation gives $2\left(\dfrac{11 + 4\,y}{3}\right) + y^2 = 6 \quad \Leftrightarrow \quad 22 + 8\,y + 3\,y^2 = 18 \quad \Leftrightarrow$ $3y^2 + 8y + 4 = 0 \quad \Leftrightarrow \quad (3y + 2)(y + 2) = 0 \quad \Rightarrow \quad y = -\frac{2}{3},\ y = -2$. Then $y = -\frac{2}{3} \quad \Rightarrow$

$x^2 = \frac{11-\frac{8}{3}}{3} = \frac{25}{9} \quad\Rightarrow\quad x = \pm\frac{5}{3}$, and $y = -2 \quad\Rightarrow\quad x^2 = \frac{11-8}{3} = 1 \quad\Rightarrow\quad x = \pm 1$.
Therefore the solutions are $\left(\frac{5}{3}, -\frac{2}{3}\right)$, $\left(-\frac{5}{3}, -\frac{2}{3}\right)$, $(1, -2)$ and $(-1, -2)$. The system is nonlinear.

In problems 6 – 13, the matrices A, B, and C are defined as follows:

$$A = \begin{bmatrix} 2 & 3 \\ 2 & 4 \end{bmatrix} \qquad B = \begin{bmatrix} 2 & 4 \\ -1 & 1 \\ 3 & 0 \end{bmatrix} \qquad C = \begin{bmatrix} 1 & 0 & 4 \\ -1 & 1 & 2 \\ 0 & 1 & 3 \end{bmatrix}$$

6. $A + B$ is undefined because A is 2×2 and B is 3×2, so they have incompatible dimensions.

7. AB is undefined because A is 2×2 and B is 3×2, so they have incompatible dimensions.

8. $BA - 3B = \begin{bmatrix} 2 & 4 \\ -1 & 1 \\ 3 & 0 \end{bmatrix} \begin{bmatrix} 2 & 3 \\ 2 & 4 \end{bmatrix} - 3 \begin{bmatrix} 2 & 4 \\ -1 & 1 \\ 3 & 0 \end{bmatrix} = \begin{bmatrix} 12 & 22 \\ 0 & 1 \\ 6 & 9 \end{bmatrix} - \begin{bmatrix} 6 & 12 \\ -3 & 3 \\ 9 & 0 \end{bmatrix} = \begin{bmatrix} 6 & 10 \\ 3 & -2 \\ -3 & 9 \end{bmatrix}$

9. $CBA = \begin{bmatrix} 1 & 0 & 4 \\ -1 & 1 & 2 \\ 0 & 1 & 3 \end{bmatrix} \begin{bmatrix} 2 & 4 \\ -1 & 1 \\ 3 & 0 \end{bmatrix} \begin{bmatrix} 2 & 3 \\ 2 & 4 \end{bmatrix} = \begin{bmatrix} 14 & 4 \\ 3 & -3 \\ 8 & 1 \end{bmatrix} \begin{bmatrix} 2 & 3 \\ 2 & 4 \end{bmatrix} = \begin{bmatrix} 36 & 58 \\ 0 & -3 \\ 18 & 28 \end{bmatrix}$

10. $A = \begin{bmatrix} 2 & 3 \\ 2 & 4 \end{bmatrix} \quad\Leftrightarrow\quad A^{-1} = \frac{1}{8-6} \begin{bmatrix} 4 & -3 \\ -2 & 2 \end{bmatrix} = \begin{bmatrix} 2 & -\frac{3}{2} \\ -1 & 1 \end{bmatrix}$

11. B^{-1} does not exist since B is not a square matrix.

12. $|B|$ is not defined since B is not a square matrix.

13. $|C| = \begin{vmatrix} 1 & 0 & 4 \\ -1 & 1 & 2 \\ 0 & 1 & 3 \end{vmatrix} = 1 \begin{vmatrix} 1 & 2 \\ 1 & 3 \end{vmatrix} + 4 \begin{vmatrix} -1 & 1 \\ 0 & 1 \end{vmatrix} = 1 - 4 = -3$

14. $\begin{cases} 4x - 3y = 10 \\ 3x - 2y = 30 \end{cases}$ is equivalent to the matrix equation $\begin{bmatrix} 4 & -3 \\ 3 & -2 \end{bmatrix} \begin{bmatrix} x \\ y \end{bmatrix} = \begin{bmatrix} 10 \\ 30 \end{bmatrix}$

$|D| = \begin{vmatrix} 4 & -3 \\ 3 & -2 \end{vmatrix} = 4(-2) - 3(-3) = 1$. So $D^{-1} = \begin{bmatrix} -2 & 3 \\ -3 & 4 \end{bmatrix}$ and

$\begin{bmatrix} x \\ y \end{bmatrix} = \begin{bmatrix} -2 & 3 \\ -3 & 4 \end{bmatrix} \begin{bmatrix} 10 \\ 30 \end{bmatrix} = \begin{bmatrix} 70 \\ 90 \end{bmatrix}$. Therefore, $x = 70$ and $y = 90$.

15. $\begin{cases} 2x - z = 14 \\ 3x - y + 5z = 0 \\ 4x + 2y + 3z = -2 \end{cases}$

Then, $|D| = \begin{vmatrix} 2 & 0 & -1 \\ 3 & -1 & 5 \\ 4 & 2 & 3 \end{vmatrix} = 2 \begin{vmatrix} -1 & 5 \\ 2 & 3 \end{vmatrix} - 1 \begin{vmatrix} 3 & -1 \\ 4 & 2 \end{vmatrix} = -26 - 10 = -36;$

$|D_x| = \begin{vmatrix} 14 & 0 & -1 \\ 0 & -1 & 5 \\ -2 & 2 & 3 \end{vmatrix} = 14 \begin{vmatrix} -1 & 5 \\ 2 & 3 \end{vmatrix} - 1 \begin{vmatrix} 0 & -1 \\ 4 & 2 \end{vmatrix} = -182 + 2 = -180;$

$$|D_y| = \begin{vmatrix} 2 & 14 & -1 \\ 3 & 0 & 5 \\ 4 & -2 & 3 \end{vmatrix} = -3\begin{vmatrix} 14 & -1 \\ -2 & 3 \end{vmatrix} - 5\begin{vmatrix} 2 & 14 \\ 4 & -2 \end{vmatrix} = -120 + 300 = 180; \text{ and}$$

$$|D_z| = \begin{vmatrix} 2 & 0 & 14 \\ 3 & -1 & 0 \\ 4 & 2 & -2 \end{vmatrix} = 2\begin{vmatrix} -1 & 0 \\ 2 & -2 \end{vmatrix} + 14\begin{vmatrix} 3 & -1 \\ 4 & 2 \end{vmatrix} = 4 + 140 = 144.$$

Therefore, $x = \frac{-180}{-36} = 5$, $y = \frac{180}{-36} = -5$, $z = \frac{144}{-36} = -4$, and so the solution is $(5, -5, -4)$.

16. $|A| = \begin{vmatrix} 1 & 4 & 1 \\ 0 & 2 & 0 \\ 1 & 0 & 1 \end{vmatrix} = 2\begin{vmatrix} 1 & 1 \\ 1 & 1 \end{vmatrix} = 0$; $|B| = \begin{vmatrix} 1 & 4 & 0 \\ 0 & 2 & 0 \\ -3 & 0 & 1 \end{vmatrix} = 2\begin{vmatrix} 1 & 0 \\ -3 & 1 \end{vmatrix} = 2.$

Since $|A| = 0$, A does not have an inverse, and since $|B| \neq 0$, B does have an inverse.

$$\begin{bmatrix} 1 & 4 & 0 & 1 & 0 & 0 \\ 0 & 2 & 0 & 0 & 1 & 0 \\ -3 & 0 & 1 & 0 & 0 & 1 \end{bmatrix} \xrightarrow{R_3 \to R_3 + 3R_1} \begin{bmatrix} 1 & 4 & 0 & 1 & 0 & 0 \\ 0 & 2 & 0 & 0 & 1 & 0 \\ 0 & 12 & 1 & 3 & 0 & 1 \end{bmatrix} \xrightarrow[R_3 \to R_3 - 6R_2]{R_1 \to R_1 - 2R_2}$$

$$\begin{bmatrix} 1 & 0 & 0 & 1 & -2 & 0 \\ 0 & 2 & 0 & 0 & 1 & 0 \\ 0 & 0 & 1 & 3 & -6 & 1 \end{bmatrix} \xrightarrow{R_2 \to \frac{1}{2}R_2} \begin{bmatrix} 1 & 0 & 0 & 1 & -2 & 0 \\ 0 & 1 & 0 & 0 & \frac{1}{2} & 0 \\ 0 & 0 & 1 & 3 & -6 & 1 \end{bmatrix}$$

Therefore, $B^{-1} = \begin{bmatrix} 1 & -2 & 0 \\ 0 & \frac{1}{2} & 0 \\ 3 & -6 & 1 \end{bmatrix}$.

17. $\begin{cases} x^2 - 2x - y + 5 \le 0 \\ y \le 5 + 2x \end{cases}$

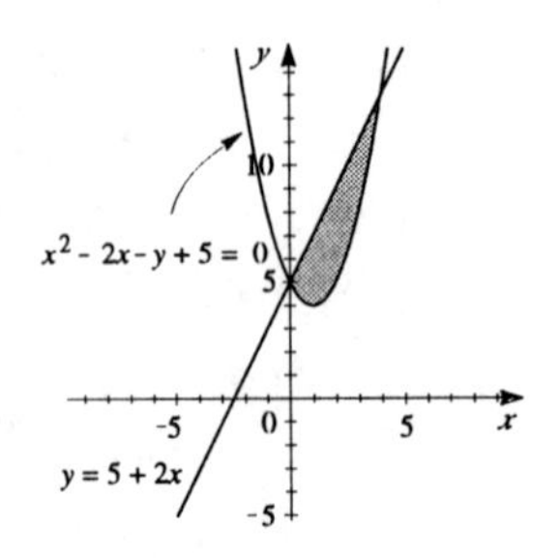

Substituting $y = 5 + 2x$ into the first equation gives
$x^2 - 2x - (5 + 2x) + 5 = 0 \quad \Leftrightarrow \quad x^2 - 4x = 0 \quad \Leftrightarrow$
$x(x - 4) = 0 \quad \Rightarrow \quad x = 0$ or $x = 4$. If $x = 0$ then
$y = 5 + 2(0) = 5$ and if $x = 4$ then $y = 5 + 2(4) = 13$. So the vertices are $(0, 5)$ and $(4, 13)$.

18. $\dfrac{4x - 1}{(x - 1)^2(x + 2)} = \dfrac{A}{x - 1} + \dfrac{B}{(x - 1)^2} + \dfrac{C}{x + 2}$. Thus,

$4x - 1 = A(x - 1)(x + 2) + B(x + 2) + C(x - 1)^2$
$= A(x^2 + x - 2) + B(x + 2) + C(x^2 - 2x + 1)$
$= (A + C)x^2 + (A + B - 2C)x + (-2A + 2B + C)$, which leads to the following system of equations:

$$\begin{cases} A + C = 0 \\ A + B - 2C = 4 \\ -2A + 2B + C = -1 \end{cases} \quad \text{So} \begin{bmatrix} 1 & 0 & 1 & 0 \\ 1 & 1 & -2 & 4 \\ -2 & 2 & 1 & -1 \end{bmatrix} \xrightarrow{R_1 \leftrightarrow R_2} \begin{bmatrix} 1 & 1 & -2 & 4 \\ 1 & 0 & 1 & 0 \\ -2 & 2 & 1 & -1 \end{bmatrix}$$

$$\xrightarrow[\text{R}_3 \to \text{R}_3+2\text{R}_1]{\text{R}_2 \to \text{R}_2-\text{R}_1} \begin{bmatrix} 1 & 1 & -2 & 4 \\ 0 & -1 & 3 & -4 \\ 0 & 4 & -3 & 7 \end{bmatrix} \xrightarrow{\text{R}_3 \to \text{R}_3+4\text{R}_1} \begin{bmatrix} 1 & 1 & -2 & 4 \\ 0 & -1 & 3 & -4 \\ 0 & 0 & 9 & -9 \end{bmatrix}$$

Therefore, $9C = -9 \Leftrightarrow C = -1$; $-B - 3 = -4 \Leftrightarrow B = 1$, and $A + 1 + 2 = 4 \Leftrightarrow A = 1$. Therefore, $\dfrac{4x-1}{(x-1)^2(x+2)} = \dfrac{1}{x-1} + \dfrac{1}{(x-1)^2} - \dfrac{1}{x+2}$.

19. Let x be the number of acres of wheat and let y be the number of acres of barley. Thus we wish to maximize the total profit, $P = 2.50(50x) + 2(40y) = 125x + 80y$, subject to the constraints:

$$\begin{cases} x + y \le 200 \\ 60x + 40y \le 10000 \\ x \ge 0, y \ge 0 \end{cases}$$

. From the graph the vertices occur at $(0,0)$, $(\frac{500}{3}, 0)$, $(100,100)$, and $(0,200)$.

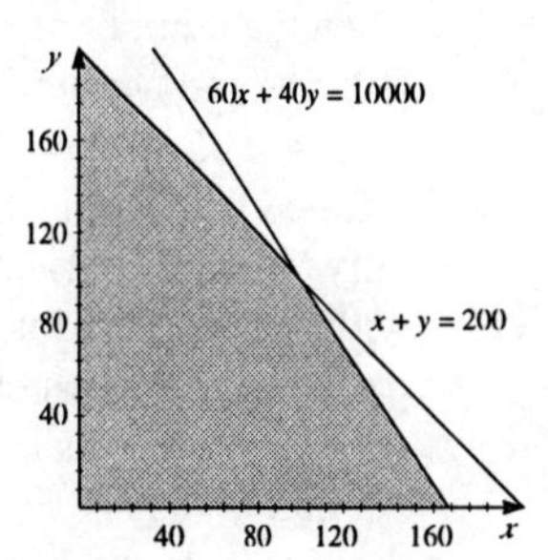

Vertex	$P = 125x + 80y$
$(0,0)$	$125(0) + 80(0) = 0$
$(\frac{500}{3}, 0)$	$125(\frac{500}{3}) + 80(0) = \frac{62500}{3} \approx 20833.33$ ← maximum value
$(100,100)$	$125(100) + 80(100) = 20,500$
$(0,200)$	$125(0) + 80(200) = 16,000$

Hence, $\frac{500}{3}$ acres of wheat and 0 acres of barley should be planted for a maximum profit of \$20,833.33.

20. $\begin{cases} 2x^2 + y^2 = 16 \\ y = x^4 - 4x^3 + 6x^2 - 4x \end{cases}$ We graph the system: $\begin{cases} y = \pm\sqrt{16 - 2x^2} \\ y = x^4 - 4x^3 + 6x^2 - 4x \end{cases}$

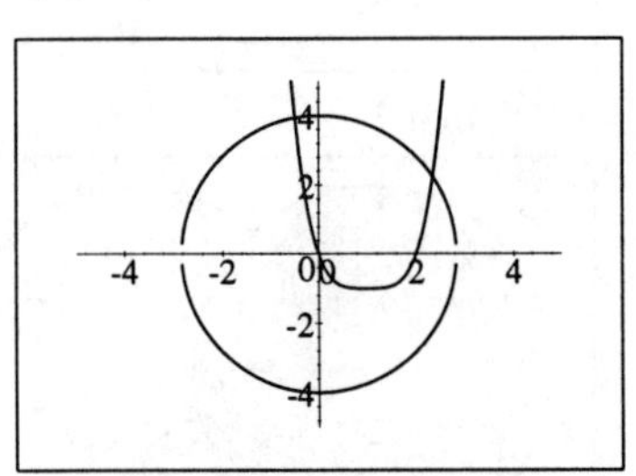

The solutions are approximately $(-0.49, 3.93)$ and $(2.34, 2.24)$.

Focus on Problem Solving

1. Solution #1: Drawing a Venn diagram, we may reason as follows: 30 business executives have all three cards, so the area common to all three circles has 30 people. Then the number of executives with *only* American Express and Mastercard is $33 - 30 = 3$. Similarly, there are $35 - 30 = 5$ people with only American Express and Visa, while $37 - 30 = 7$ executives have only Visa and Mastercard. Now, the number of people with *only* one card is found by subtracting away those parts of a circle which overlap other circles. Thus, American Express is $68 - 30 - 5 - 3 = 30$, and similarly, Visa is 10 and Mastercard is 12. Adding these parts together gives $30 + 10 + 12 + 3 + 5 + 7 + 30 = 97$ executives who have at least one card. But there are 100 executives in total; thus, 3 executives have none of these cards.

Solution 1

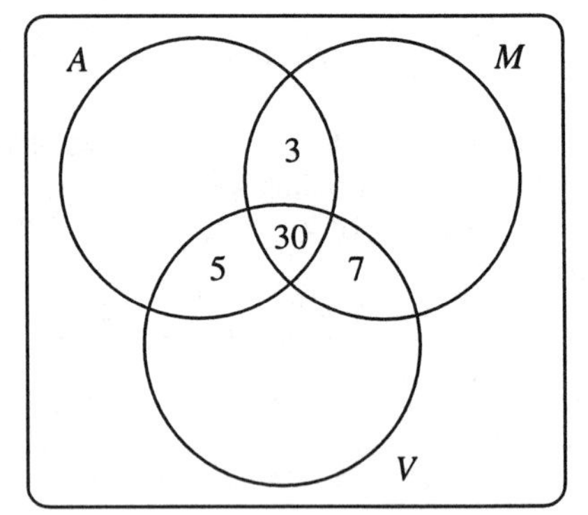

A = American Express
M = Mastercard
V = Visa

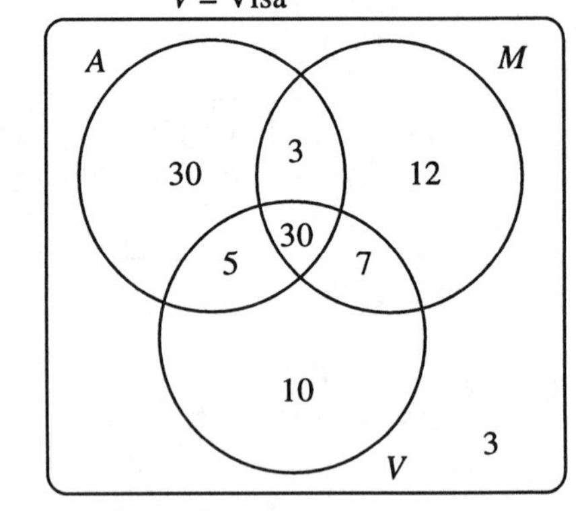

Solution #2: Suppose that for an arbitrary Venn diagram such as the one on the right, we let the number of things in circle A to be N_A, the number of things common to A and B to be N_{AB}, the number of things common to A, B, and C to be N_{ABC}, and so on. Then we can see that the number of things inside the union of the three circles is: $N = N_A + N_B + N_C - N_{AB} - N_{AB} - N_{BC} + N_{ABC}$. Furthermore, if we denote the number of things outside the circles (but inside the rectangle) to be x, then the total number of things in the rectangle is $A + x$. In our problem, the total number of business executives is 100. Therefore, $100 = N + x$

$= N_A + N_B + N_C - N_{AB} - N_{AB} - N_{BC} + N_{ABC} + x$
$\Leftrightarrow \quad 100 = 68 + 52 + 52 - 33 - 35 - 37 + 30 + x$
$= 97 + x \quad \Leftrightarrow \quad x = 100 - 97 = 3$. Thus, three executives do not own any of these cards.

Solution 2

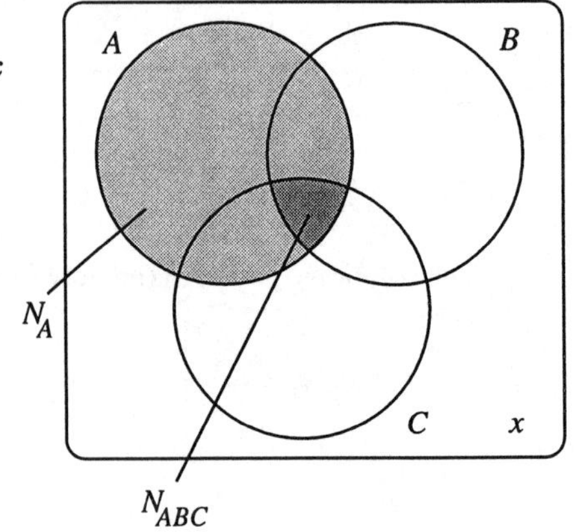

3. Drawing a Venn diagram, we can immediately fill in the region for birds but no cats or dogs since we are told directly that there are 10 such pet owners. But the total number of owners with birds is 30. Using the notation from Problem 1, solution 2, we see that $N_{BC} + N_{BD} - N_{BCD} + 10 = 30$. Thus, $8 + 15 - N_{BCD} + 10 = 30 \quad \Leftrightarrow$ $N_{BCD} = 33 - 30 = 3$. Having found the number in the region common to all three circles, we use reasoning similar to that in Problem 1 (above) to solve the rest of the Venn diagram:

   (a) 0

   (b) 20

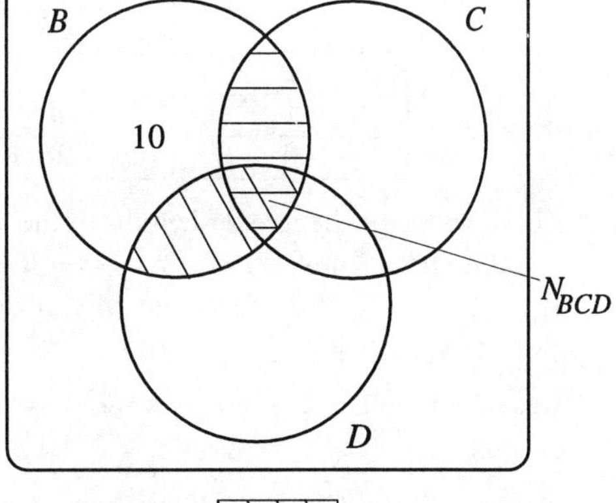

B = birds $\quad$ N_{BD}
C = cats
D = dogs $\quad$ N_{BC}

5. 1,000,000 is both a perfect square, 1000^2, and a perfect cube, 100^3. Therefore the positive integers less than or equal to 1000 are all roots of perfect squares that are less than or equal to 1,000,000 and so there are 1000 integers between 1 and 1,000,000 that are perfect squares. Similarly, the positive integers less than or equal to 100 are all roots of perfect cubes that are less than or equal to 1,000,000 and so there are 100 integers that are perfect cubes. Also, 10 of these numbers (1^6, 2^6, ..., 10^6) are both perfect squares and perfect cubes. So there are $1000 + 100 - 10 = 1090$ integers from one to a million that are either perfect squares or perfect cubes (or both).

7. Solution #1: Let J, S, and V represent the number of marbles that Justin, Sasha, and Vanessa have, respectively, to begin with. The number that they have after each step of the described procedure is given in the following chart:

	Justin	Sasha	Vanessa
Start	J	S	V
After 1st step	$J - S - V$	$2S$	$2V$
After 2nd step	$2J - 2S - 2V$	$2S - (J - S - V)$ $= 3S - J - V$	$4V$
After 3rd step	$4J - 4S - 4V$	$2(3S - J - V)$	$4V - 2(J - S - V) - (3S - J - V)$ $= 7V - J - S$

Each of these final quantities is 16. So we have the system of equations

$4J - 4S - 4V = 16$

$-2J + 6S - 2V = 16$

$-J - S + 7V = 16$

Solving this system using the methods of this chapter, we get $J = 26$, $S = 14$, and $V = 8$.

Solution #2: If we work backwards, we don't have to solve a system of equations. Remembering that at each stage the total number of marbles must be 48, we construct the following chart:

Stage	Justin	Sasha	Vanessa
3	16	16	16
2	8	8	32
1	4	28	16
0	26	14	8

9. Let m and n be the integer lengths of the sides of the rectangle. Since the perimeter and area are equal we have $2m + 2n = mn \Leftrightarrow mn - 2m - 2n = 0 \Leftrightarrow mn - 2m - 2n + 4 = 4 \Leftrightarrow (m-2)(n-2) = 4$. Since the only integer factors of 4 are $1 \cdot 4$ and $2 \cdot 2$, we consider two cases.

<u>Case 1:</u> $m - 2 = 1$ and $n - 2 = 4 \Rightarrow m = 3, n = 6$.

<u>Case 2:</u> $m - 2 = 2$ and $n - 2 = 2 \Rightarrow m = 4, n = 4$.

11. $3x + 5y + z + 7t = -8$, $4x + 8y + 2z + 6t = -8$, $6x + 2y + 8z + 4t = 8$, $7x + y + 5z + 3t = 8$. Adding the first and last equations, we get $10x + 6y + 6z + 10t = 0$. Adding the second and third equations, we get $10x + 10y + 10z + 10t = 0$. These equations can be written as $5(x+t) + 3(y+z) = 0$ and $(x+t) + (y+z) = 0$. If we let $u = x + t$ and $v = y + z$, then we have the system $5u + 3v = 0$, $u + v = 0$, whose only solution is $u = v = 0$. Thus $t = -x$ and $z = -y$, so the first two equations give $3x + 5y - y - 7x = -8$ and $4x + 8y - 2y - 6x = -8$ or $x - y = 2$ and $x - 3y = 4$, whose solution is $x = 1$, $y = -1$. Then $z = 1$ and $t = -1$.

13. (a) Since $x^3 + y^3 = (x+y)(x^2 - xy + y^2)$ we need to determine the value of xy

$$\begin{array}{lcr} x + y = 1 & \Leftrightarrow & x^2 + xy \quad = x \\ x + y = 1 & \Leftrightarrow & xy + y^2 = y \\ \hline & & x^2 + 2xy + y^2 = x + y \end{array}$$

So $4 + 2xy = 1 \Leftrightarrow xy = -\frac{3}{2}$. Thus
$x^3 + y^3 = (x+y)(x^2 - xy + y^2) = (1)(4 + \frac{3}{2}) = \frac{11}{2}$

(b) Again since $x^3 + y^3 = (x+y)(x^2 - xy + y^2)$ we need to determine the value of $x^2 + y^2$. $x + y = 4 \Leftrightarrow x^2 + 2xy + y^2 = 16 \Leftrightarrow x^2 + y^2 = 16 - 2xy = 16 - 2(1) = 14$. Thus $x^3 + y^3 = (x+y)(x^2 - xy + y^2) = (4)(14 - 1) = 52$.

15. We find the center of the circle by intersecting two of the perpendicular bisectors of two of the chords. For the points $(-19, 62)$ and $(45, 70)$, the slope of the chord is $\frac{70-62}{45+19} = \frac{1}{8}$ and the midpoint is $\left(\frac{-19+45}{2}, \frac{62+70}{2}\right) = (13, 66)$. Thus the equation of the perpendicular bisector is $y - 66 = -8(x - 13) \Leftrightarrow y = -8x + 170$. For the points $(20, -55)$ and $(45, 70)$, the slope of the chord is $\frac{70+55}{45-20} = 5$ and the midpoint is $\left(\frac{20+45}{2}, \frac{-55+70}{2}\right) = \left(\frac{65}{2}, \frac{15}{2}\right)$. Thus the equation of the perpendicular bisector is $y - \frac{15}{2} = -\frac{1}{5}\left(x - \frac{65}{2}\right) \Leftrightarrow y - \frac{15}{2} = -\frac{1}{5}x + \frac{13}{2} \Leftrightarrow x + 5y = 70$. Substituting the value of y from the first equation into the second equation, we have $x + 5(-8x + 170) = 70 \Leftrightarrow -39x = -780 \Leftrightarrow x = 20$. So $y = -8(20) + 170 = 10$. Thus the center is $(20, 10)$ and the equation of the circle is of the form $(x - 20)^2 + (y - 10)^2 = r^2$. We find r^2 by substituting for x and y, thus $(20 - 20)^2 + (-55 - 10)^2 = 65^2 = r^2$ and the equation we seek is $(x - 20)^2 + (y - 10)^2 = 65^2$.

17. Expanding about the first column we get $\begin{vmatrix} 1 & a & a^2 \\ 1 & b & b^2 \\ 1 & c & c^2 \end{vmatrix} = \begin{vmatrix} b & b^2 \\ c & c^2 \end{vmatrix} - \begin{vmatrix} a & a^2 \\ c & c^2 \end{vmatrix} + \begin{vmatrix} a & a^2 \\ b & b^2 \end{vmatrix}$

$= bc^2 - b^2c - (ac^2 - a^2c) + ab^2 - a^2b = a^2c + ab^2 + bc^2 - b^2c - ac^2 - a^2b$.

Now $(a-b)(b-c)(c-a) = (ab - ac - b^2 + bc)(c-a)$
$= abc - a^2b - ac^2 + a^2c - b^2c + ab^2 + bc^2 - abc = a^2c + ab^2 + bc^2 - b^2c - ac^2 - a^2b$. Thus

$$\begin{vmatrix} 1 & a & a^2 \\ 1 & b & b^2 \\ 1 & c & c^2 \end{vmatrix} = (a-b)(b-c)(c-a) = (ab - ac - b^2 + bc)(c-a)$$

19. $P(x) = |A - xI| = \begin{vmatrix} 3-x & 5 \\ -2 & 1-x \end{vmatrix} = (3-x)(1-x) + 10 = x^2 - 4x + 13$. Thus

$P(x) = x^2 - 4x + 13$. So

$$\begin{aligned} P(A) &= \begin{bmatrix} 3 & 5 \\ -2 & 1 \end{bmatrix}^2 - 4\begin{bmatrix} 3 & 5 \\ -2 & 1 \end{bmatrix} + 13\begin{bmatrix} 1 & 0 \\ 0 & 1 \end{bmatrix} \\ &= \begin{bmatrix} -1 & 20 \\ -8 & -9 \end{bmatrix} + \begin{bmatrix} -12 & -20 \\ 8 & -4 \end{bmatrix} + \begin{bmatrix} 13 & 0 \\ 0 & 13 \end{bmatrix} \\ &= \begin{bmatrix} 0 & 0 \\ 0 & 0 \end{bmatrix} \end{aligned}$$

Chapter Seven

Exercises 7.1

1. $y^2 = 4x$. Then $4p = 4 \quad \Leftrightarrow \quad p = 1$

   Focus: $(1, 0)$

   Directrix: $x = -1$

   Focal diameter: 4

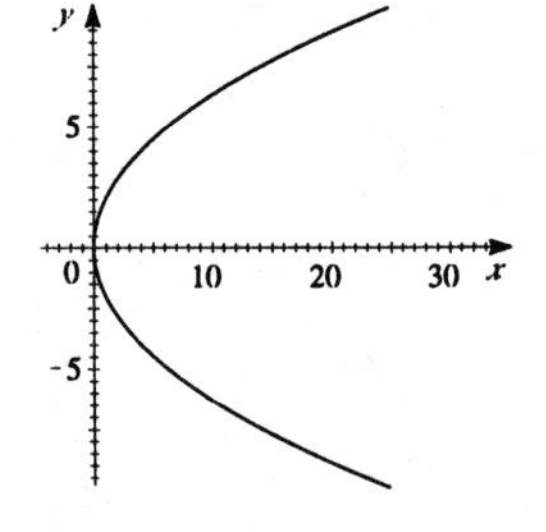

3. $x^2 = 9y$. Then $4p = 9 \quad \Leftrightarrow \quad p = \frac{9}{4}$

   Focus: $\left(0, \frac{9}{4}\right)$

   Directrix: $y = -\frac{9}{4}$

   Focal diameter: 9

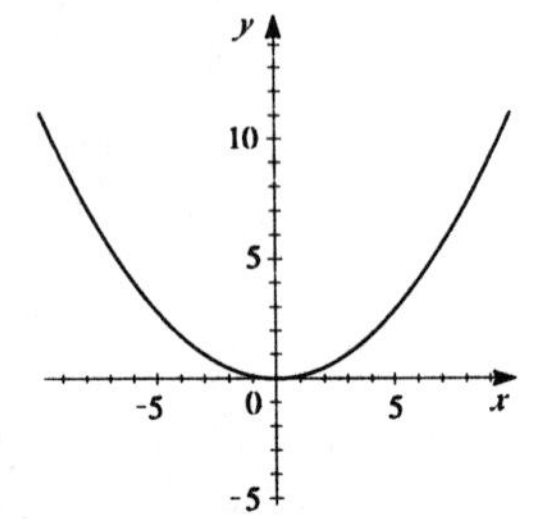

5. $y = 5x^2 \quad \Leftrightarrow \quad x^2 = \frac{1}{5}y$. Then $4p = \frac{1}{5} \quad \Leftrightarrow \quad p = \frac{1}{20}$

   Focus: $\left(0, \frac{1}{20}\right)$

   Directrix: $y = -\frac{1}{20}$

   Focal diameter: $\frac{1}{5}$

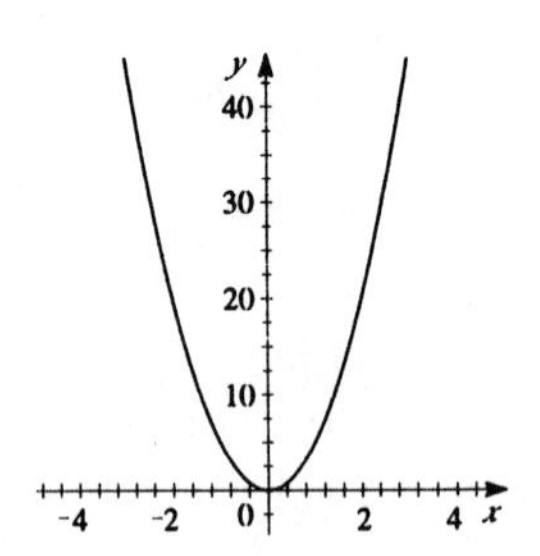

7. $x = -8y^2 \quad \Leftrightarrow \quad y^2 = -\frac{1}{8}x$. Then $4p = -\frac{1}{8} \quad \Leftrightarrow \quad p = -\frac{1}{32}$

   Focus: $\left(-\frac{1}{32}, 0\right)$

   Directrix: $x = \frac{1}{32}$

   Focal diameter: $\frac{1}{8}$

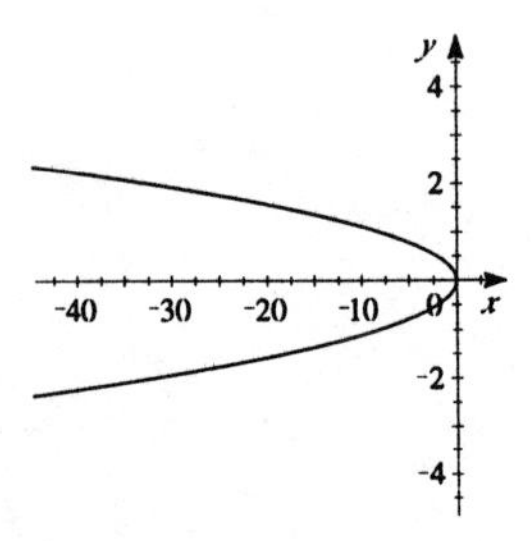

9. $x^2 + 6y = 0 \quad \Leftrightarrow \quad x^2 = -6y$. Then $4p = -6 \quad \Leftrightarrow \quad p = -\frac{3}{2}$

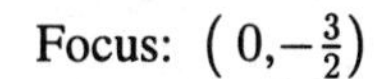
Focus: $\left(0, -\frac{3}{2}\right)$

Directrix: $y = \frac{3}{2}$

Focal diameter: 6

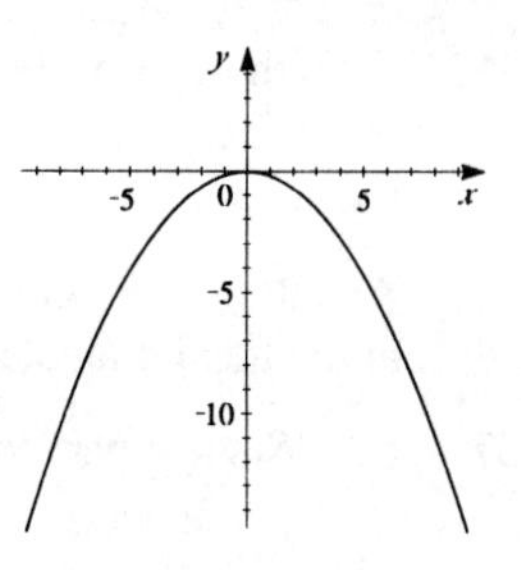

11. $5x + 3y^2 = 0 \quad \Leftrightarrow \quad y^2 = -\frac{5}{3}x$. Then $4p = -\frac{5}{3} \quad \Leftrightarrow \quad p = -\frac{5}{12}$

Focus: $\left(-\frac{5}{12}, 0\right)$

Directrix: $x = \frac{5}{12}$

Focal diameter: $\frac{5}{3}$

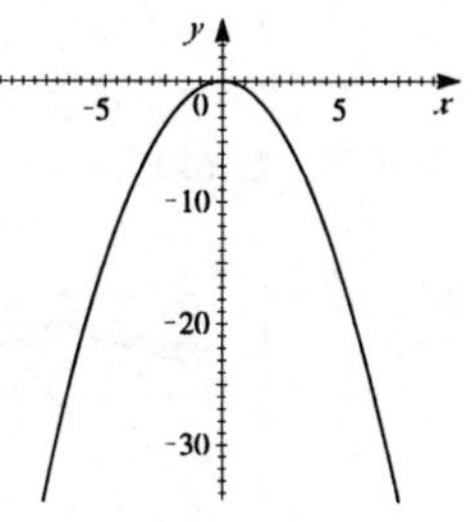

13. Since the focus is $(0,2)$, $p = 2 \quad \Leftrightarrow \quad 4p = 8$. Hence, the equation of the parabola is $x^2 = 8y$.

15. Since the focus is $(-8, 0)$, $p = -8 \quad \Leftrightarrow \quad 4p = -32$. Hence, the equation of the parabola is $y^2 = -32x$.

17. Since the directrix is $x = 2$, $p = -2 \quad \Leftrightarrow \quad 4p = -8$. Hence, the equation of the parabola is $y^2 = -8x$.

19. Since the directrix is $y = -10$, $p = 10 \quad \Leftrightarrow \quad 4p = 40$. Hence, the equation of the parabola is $x^2 = 40y$.

21. The focus is on the positive x-axis so the parabola opens horizontally with $2p = 2 \quad \Leftrightarrow \quad 4p = 4$. So the equation of the parabola is $y^2 = 4x$.

23. Since the parabola opens upward with focus 5 units from the vertex, the focus is at $(5, 0)$. So $p = 5 \quad \Leftrightarrow \quad 4p = 20$. Thus the equation of the parabola is $x^2 = 20y$.

25. $p = -3 \quad \Leftrightarrow \quad 4p = -12$. Since the parabola opens downward, its equation is $x^2 = -12y$.

27. The focal diameter is $4p = \frac{3}{2} + \frac{3}{2} = 3$. Since the parabola opens to the left, its equation is $y^2 = -3x$.

29. The equation of the parabola has the form $y^2 = 4px$. Since the parabola passes through the point $(4, -2)$, $(-2)^2 = 4p(4) \quad \Leftrightarrow \quad 4p = 1$, and so the equation is $y^2 = x$.

31. The area of the shades region $= \textit{width} \times \textit{height} = 4p \times p = 8$, and so $p^2 = 2 \quad \Leftrightarrow \quad p = -\sqrt{2}$ (because the parabola opens downward). Therefore, the equation is $x^2 = 4py = -4\sqrt{2}\,y \quad \Leftrightarrow$ $x^2 = -4\sqrt{2}\,y$.

33. (a) Since the focal diameter is 12 cm, then $4p = 12$. Hence, the parabola has equation $y^2 = 12x$.

    (b) At a point 20cm horizontally from the vertex, the parabola passes through the point $(20,y)$ and hence from part (a) $y^2 = 12(20) \quad \Leftrightarrow \quad y^2 = 240 \quad \Leftrightarrow \quad y = \pm 4\sqrt{15}$. Thus, $|\text{CD}| = 8\sqrt{15} \approx 31$ cm.

35. With the vertex at the origin, the top of one tower will be at the point $(300, 150)$. Inserting this point into the equation $x^2 = 4py$ gives: $(300)^2 = 4p(150) \quad \Leftrightarrow \quad 90000 = 600p \quad \Leftrightarrow \quad p = 150$. So the equation of the parabolic part of the cables is $x^2 = 4(150)y \quad \Leftrightarrow \quad x^2 = 600y$.

An alternative solution: since distance between towers $= 4 \times$ height, $p = 150$. This only happens at the focal diameter.

37. (a) Since a parabola with directrix $x = -p$ has equation $x^2 = 4py$ we have:
Directrix $y = \frac{1}{2} \Rightarrow p = -\frac{1}{2}$ so equation is $x^2 = 4\left(-\frac{1}{2}\right)y \Leftrightarrow x^2 = -2y$.
Directrix $y = 1 \Rightarrow p = -1$ so equation is $x^2 = 4(-1)y \Leftrightarrow x^2 = -4y$.
Directrix $y = 4 \Rightarrow p = -4$ so equation is $x^2 = 4(-4)y \Leftrightarrow x^2 = -16y$.
Directrix $y = 8 \Rightarrow p = -8$ so equation is $x^2 = 4(-8)y \Leftrightarrow x^2 = -32y$.

(b) 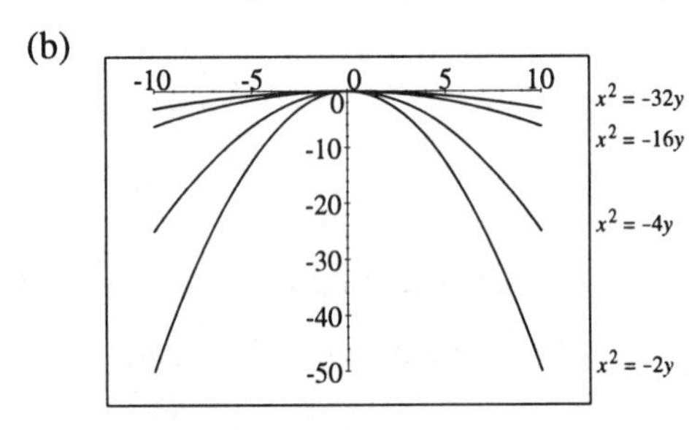

As the directrix moves further from the vertex, the parabola gets flatter.

Exercises 7.2

1. $\dfrac{x^2}{25}+\dfrac{y^2}{9}=1$. This ellipse has $a=5$, $b=3$, and so $c^2=a^2-b^2=16 \quad\Leftrightarrow\quad c=4$.

   Vertices: $(\pm 5,0)$; foci: $(\pm 4,0)$; eccentricity: $e=\frac{c}{a}=\frac{4}{5}=0.8$; length of the major axis: $2a=10$; length of the minor axis: $2b=6$.

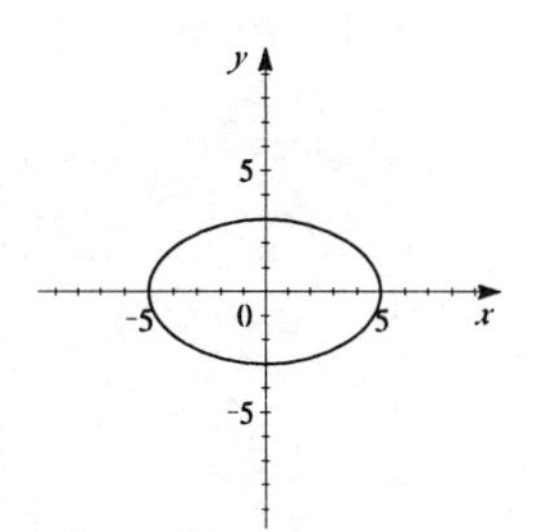

3. $9x^2+4y^2=36 \quad\Leftrightarrow\quad \dfrac{x^2}{4}+\dfrac{y^2}{9}=1$. This ellipse has $a=3$, $b=2$, and so $c^2=9-4=5 \quad\Leftrightarrow\quad c=\sqrt{5}$.

   Vertices: $(0,\pm 3)$; foci: $(0,\pm\sqrt{5})$; eccentricity: $e=\frac{c}{a}=\frac{\sqrt{5}}{3}$; length of the major axis: $2a=6$; length of the minor axis: $2b=4$.

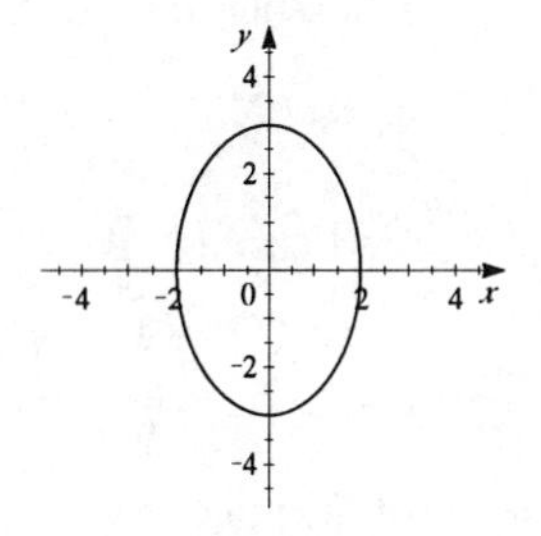

5. $x^2+4y^2=16 \quad\Leftrightarrow\quad \dfrac{x^2}{16}+\dfrac{y^2}{4}=1$. This ellipse has $a=4$, $b=2$, and so $c^2=16-4=12 \quad\Leftrightarrow\quad c=2\sqrt{3}$.

   Vertices: $(\pm 4,0)$; foci: $\left(\pm 2\sqrt{3},0\right)$; eccentricity: $e=\frac{c}{a}=\frac{2\sqrt{3}}{4}=\frac{\sqrt{3}}{2}$; length of the major axis: $2a=8$; length of the minor axis: $2b=4$.

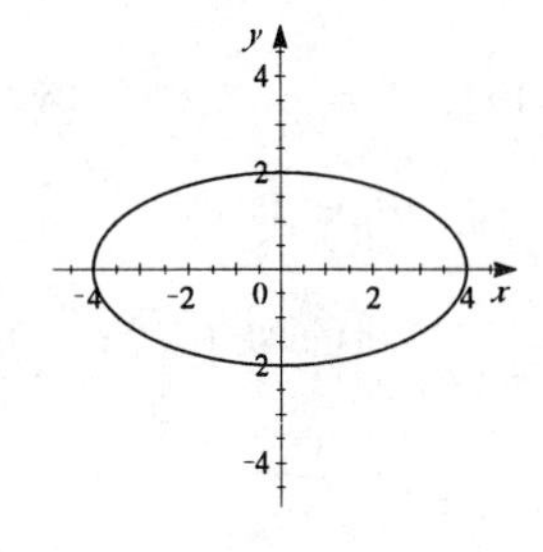

7. $2x^2+y^2=3 \quad\Leftrightarrow\quad \dfrac{x^2}{\frac{3}{2}}+\dfrac{y^2}{3}=1$. This ellipse has $a=\sqrt{3}$, $b=\sqrt{\frac{3}{2}}$, and so $c^2=3-\frac{3}{2}=\frac{3}{2} \quad\Leftrightarrow\quad c=\sqrt{\frac{3}{2}}$.

   Vertices: $\left(0,\pm\sqrt{3}\right)$; foci: $\left(0,\pm\sqrt{\frac{3}{2}}\right)$; eccentricity: $e=\frac{c}{a}=\dfrac{\sqrt{\frac{3}{2}}}{\sqrt{3}}=\dfrac{1}{\sqrt{2}}=\dfrac{\sqrt{2}}{2}$; length of the major axis: $2a=2\sqrt{3}$; length of the minor axis: $2b=2\sqrt{\frac{3}{2}}=\sqrt{6}$.

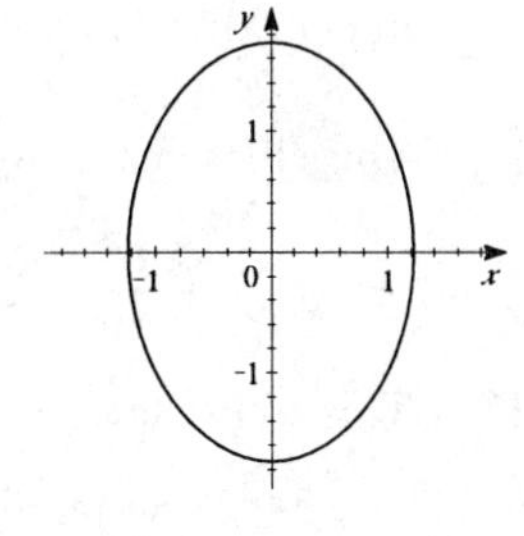

9. $x^2 + 4y^2 = 1 \quad \Leftrightarrow \quad \dfrac{x^2}{1} + \dfrac{y^2}{\frac{1}{4}} = 1$. This ellipse has $a = 1$, $b = \dfrac{1}{2}$, and so $c^2 = 1 - \frac{1}{4} = \frac{3}{4} \quad \Leftrightarrow \quad c = \frac{1}{2}\sqrt{3}$.

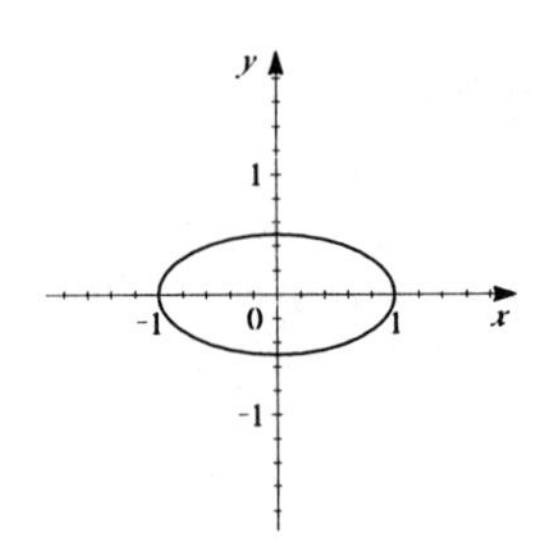

Vertices: $(\pm 1, 0)$; foci: $\left(\pm \frac{1}{2}\sqrt{3}, 0\right)$; eccentricity: $e = \frac{c}{a} = \frac{\frac{1}{2}\sqrt{3}}{1} = \frac{1}{2}\sqrt{3}$; length of the major axis: $2a = 2$; length of the minor axis: $2b = 1$.

11. $\dfrac{1}{2}x^2 + \dfrac{1}{8}y^2 = \dfrac{1}{4} \quad \Leftrightarrow \quad 2x^2 + \dfrac{1}{2}y^2 = 1 \quad \Leftrightarrow \quad \dfrac{x^2}{\frac{1}{2}} + \dfrac{y^2}{2} = 1$. This ellipse has $a = \sqrt{2}$, $b = \frac{1}{\sqrt{2}}$, and so $c^2 = 2 - \frac{1}{2} = \frac{3}{2} \quad \Leftrightarrow$ $c = \sqrt{\frac{3}{2}}$.

Vertices: $(0, \pm\sqrt{2})$; foci: $\left(0, \pm\sqrt{\frac{3}{2}}\right)$; eccentricity: $e = \dfrac{c}{a} = \dfrac{\sqrt{\frac{3}{2}}}{\sqrt{2}} = \dfrac{1}{2}\sqrt{3}$; length of the major axis: $2a = 2\sqrt{2}$; length of the minor axis: $2b = \sqrt{2}$.

13. $y^2 = 1 - 2x^2 \quad \Leftrightarrow \quad 2x^2 + y^2 = 1 \quad \Leftrightarrow \quad \dfrac{x^2}{\frac{1}{2}} + \dfrac{y^2}{1} = 1$. This ellipse has $a = 1$, $b = \frac{1}{\sqrt{2}}$, and so $c^2 = 1 - \frac{1}{2} = \frac{1}{2} \quad \Leftrightarrow \quad c = \frac{1}{\sqrt{2}}$.

Vertices: $(0, \pm 1)$; foci: $\left(0, \pm\frac{1}{\sqrt{2}}\right)$; eccentricity: $e = \dfrac{c}{a} = \dfrac{\frac{1}{\sqrt{2}}}{1} = \dfrac{1}{\sqrt{2}}$; length of the major axis: $2a = 2$; length of the minor axis: $2b = \sqrt{2}$.

15. This ellipse has a horizontal major axis with $a = 5$ and $b = 4$, so the equation is $\dfrac{x^2}{(5)^2} + \dfrac{y^2}{(4)^2} = 1 \quad \Leftrightarrow \quad \dfrac{x^2}{25} + \dfrac{y^2}{16} = 1$.

17. This ellipse has a vertical major axis with $c = 2$ and $b = 2$. $a^2 = c^2 + b^2 = 2^2 + 2^2 = 8 \quad \Leftrightarrow \quad a = 2\sqrt{2}$. So the equation is $\dfrac{x^2}{(2)^2} + \dfrac{y^2}{\left(2\sqrt{2}\right)^2} = 1 \quad \Leftrightarrow \quad \dfrac{x^2}{4} + \dfrac{y^2}{8} = 1$.

19. The foci are $(\pm 4, 0)$ and the vertices are $(\pm 5, 0)$. Thus, $c = 4$ and $a = 5$, and so $b^2 = 25 - 16 = 9$. Therefore, the equation of the ellipse is $\dfrac{x^2}{25} + \dfrac{y^2}{9} = 1$.

21. The length of the major axis is $2a = 4 \quad \Leftrightarrow \quad a = 2$, the length of the minor axis is $2b = 2 \quad \Leftrightarrow \quad b = 1$, and the foci are on the y-axis. Therefore, the equation of the ellipse is $x^2 + \dfrac{y^2}{4} = 1$.

23. The foci are $(0, \pm 2)$ and the length of the minor axis is $2b = 6 \quad \Leftrightarrow \quad b = 3$. Thus, $a^2 = 4 + 9 = 13$. Since the foci are on the y-axis, the equation is $\dfrac{x^2}{9} + \dfrac{y^2}{13} = 1$.

25. The endpoints of the major axis are $(\pm 10, 0) \quad \Leftrightarrow \quad a = 10$, and the distance between the foci is $2c = 6 \quad \Leftrightarrow \quad c = 3$. Therefore, $b^2 = 100 - 9 = 91$, and so the equation of the ellipse is $\dfrac{x^2}{100} + \dfrac{y^2}{91} = 1$.

27. The length of the major axis is 10 so $2a = 10 \quad \Leftrightarrow \quad a = 5$, and the foci are on the x-axis so the form of the equation is $\dfrac{x^2}{25} + \dfrac{y^2}{b^2} = 1$. Since the ellipse passes through $\left(\sqrt{5}, 2\right)$, we know that $\dfrac{\left(\sqrt{5}\right)^2}{25} + \dfrac{(2)^2}{b^2} = 1 \quad \Leftrightarrow \quad \dfrac{5}{25} + \dfrac{4}{b^2} = 1 \quad \Leftrightarrow \quad \dfrac{4}{b^2} = \dfrac{4}{5} \quad \Leftrightarrow \quad b^2 = 5$, and so the equation is $\dfrac{x^2}{25} + \dfrac{y^2}{5} = 1$.

29. Since the foci are $(\pm 1.5, 0)$, then $c = \frac{3}{2}$. Since the eccentricity is $0.8 = \frac{c}{a} \quad \Leftrightarrow \quad a = \dfrac{\frac{3}{2}}{\frac{4}{5}} = \frac{15}{8}$ and so $b^2 = \frac{225}{64} - \frac{9}{4} = \frac{225-16\cdot 9}{64} = \frac{81}{64}$. Therefore, the equation of the ellipse is $\dfrac{x^2}{\left({}^{15}/_{8}\right)^2} + \dfrac{y^2}{{}^{81}/_{64}} = 1 \quad \Leftrightarrow$ $\dfrac{64x^2}{225} + \dfrac{64y^2}{81} = 1$.

31. $\begin{cases} 4x^2 + y^2 = 4 \\ 4x^2 + 9y^2 = 36 \end{cases}$ Subtracting the first equation from the second equation gives: $8y^2 = 32 \quad \Leftrightarrow \quad y^2 = 4 \quad \Leftrightarrow \quad y = \pm 2$. Substituting $y = \pm 2$ in the first equation gives: $4x^2 + (\pm 2)^2 = 4 \quad \Leftrightarrow \quad x = 0$, and so the points of intersection are $(0, \pm 2)$.

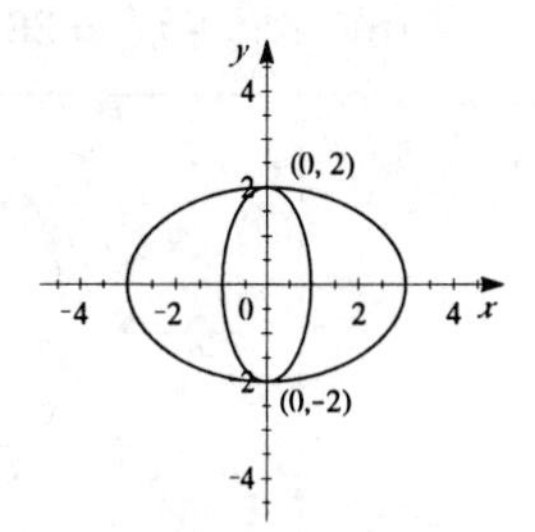

33. Using the perihelion, $a - c = 147{,}000{,}000$, while using the aphelion $a + c = 153{,}000{,}000$. Adding we have $2a = 300{,}000{,}000 \quad \Leftrightarrow \quad a = 150{,}000{,}000 = 1.5 \times 10^8$. Since $c = 153{,}000{,}000 - a$ we have $c = 3{,}000{,}000$. So, $b^2 = a^2 - c^2 = (15 \times 10^7)^2 - (3 \times 10^6)^2$ $= 2249 \times 10^{12} = 2.249 \times 10^{16}$ and so the equation of the ellipse is $\dfrac{x^2}{2.25 \times 10^{16}} + \dfrac{y^2}{2.249 \times 10^{16}} = 1$.

35. Using the perilune, $a - c = 1075 + 68 = 1143$ and using the apolune, $a + c = 1075 + 195 = 1270$. Adding we get $2a = 2413 \quad \Leftrightarrow \quad a = 1206.5$. So $c = 1270 - 1206.5 \quad \Leftrightarrow \quad c = 63.5$. Therefore, $b^2 = (1206.5)^2 - (63.5)^2 = 1{,}451{,}610$. Since $a^2 \approx 1{,}455{,}642$, the equation of Apollo 11's orbit is $\dfrac{x^2}{1{,}455{,}642} + \dfrac{y^2}{1{,}451{,}610} = 1$.

37. From the diagram, $a = 40$ and $b = 20$, and so the equation of the ellipse whose top half is the window is $\dfrac{x^2}{1600} + \dfrac{y^2}{400} = 1$. Since the ellipse passes through the point $(25, h)$ by substituting we

have $\dfrac{25^2}{1600} + \dfrac{h^2}{400} = 1 \quad \Leftrightarrow \quad 625 + 4y^2 = 1600 \quad \Leftrightarrow \quad y = \dfrac{1}{2}\sqrt{975} \approx 15.61$. Therefore, the window is approximately 15.6 inches high at the specified point.

39. The foci are at $(\pm c, 0)$ where $c^2 = a^2 - b^2$. The endpoints of one latus rectum are the points $(c, \pm k)$ and the length is $2k$. Substituting this point into the equation we get $\dfrac{c^2}{a^2} + \dfrac{k^2}{b^2} = 1 \quad \Leftrightarrow$ $\dfrac{k^2}{b^2} = 1 - \dfrac{c^2}{a^2} = \dfrac{a^2 - c^2}{a^2} \quad \Leftrightarrow \quad k^2 = \dfrac{b^2(a^2 - c^2)}{a^2}$. Since $b^2 = a^2 - c^2$, the last equation becomes $k^2 = \dfrac{b^4}{a^2} \quad \Rightarrow \quad k = \dfrac{b^2}{a}$. Thus the length of the latus rectum is $2k = 2\left(\dfrac{b^2}{a}\right) = \dfrac{2b^2}{a}$

41. (a) $\dfrac{x^2}{25} + \dfrac{y^2}{20} = 1 \quad \Leftrightarrow \quad \dfrac{y^2}{20} = 1 - \dfrac{x^2}{25} \quad \Leftrightarrow \quad y^2 = 20 - \dfrac{4x^2}{5} \quad \Rightarrow \quad y = \pm\sqrt{20 - \dfrac{4x^2}{5}}$

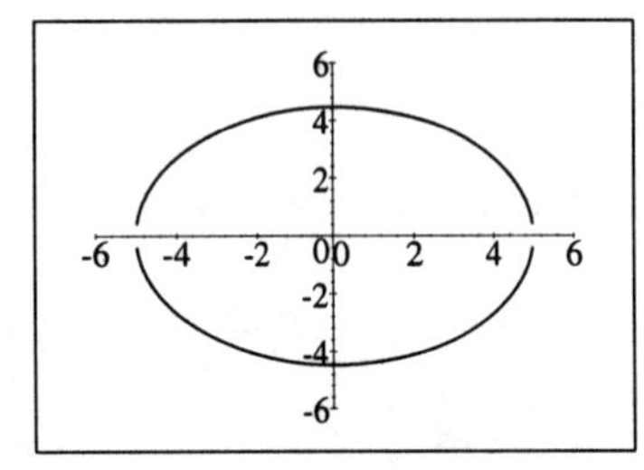

(b) $6x^2 + y^2 = 36 \quad \Leftrightarrow \quad y^2 = 36 - 6x^2 \quad \Rightarrow \quad y = \pm\sqrt{36 - 6x^2}$

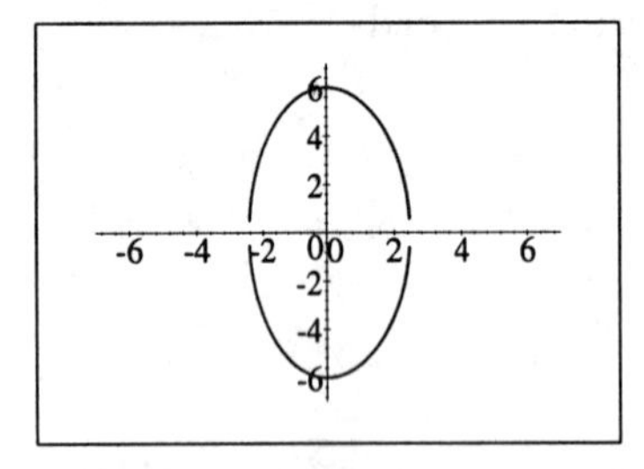

Exercises 7.3

1. The hyperbola $\dfrac{x^2}{4} - \dfrac{y^2}{16} = 1$ has $a = 2$, $b = 4$, and $c^2 = 16 + 4 \quad \Rightarrow$ $c = 2\sqrt{5}$.

   Vertices: $(\pm 2, 0)$; foci: $(\pm 2\sqrt{5}, 0)$; asymptotes: $y = \pm \frac{4}{2}x \quad \Leftrightarrow$ $y = \pm 2x$.

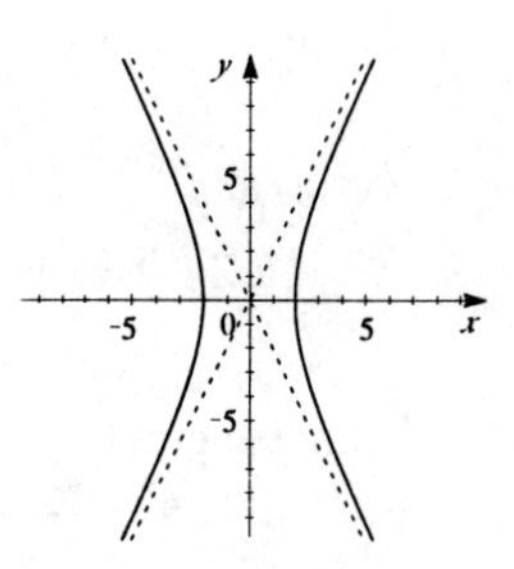

3. The hyperbola $\dfrac{y^2}{1} - \dfrac{x^2}{25} = 1$ has $a = 1$, $b = 5$, and $c^2 = 1 + 25 = 26$ $\Rightarrow \quad c = \sqrt{26}$.

   Vertices: $(0, \pm 1)$; foci: $(0, \pm \sqrt{26})$; asymptotes: $y = \pm \frac{1}{5}x$.

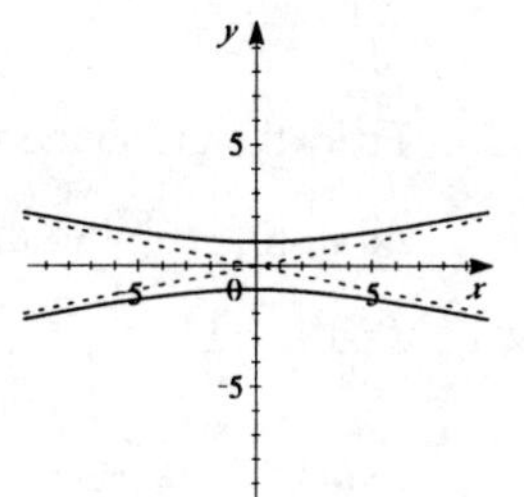

5. The hyperbola $x^2 - y^2 = 1$ has $a = 1$, $b = 1$, and $c^2 = 1 + 1 = 2$ $\Rightarrow \quad c = \sqrt{2}$.

   Vertices: $(\pm 1, 0)$; foci: $(\pm \sqrt{2}, 0)$; asymptotes: $y = \pm x$.

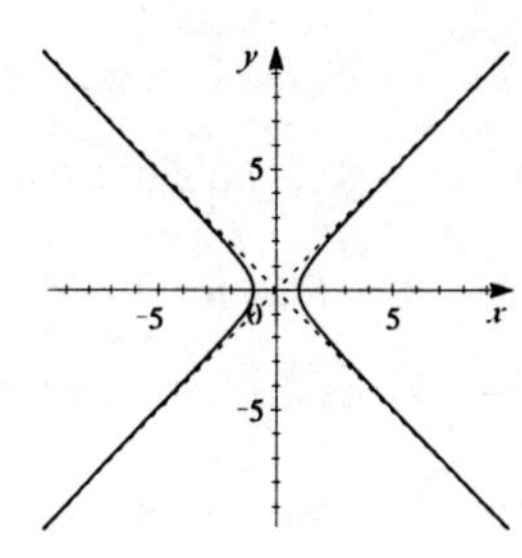

7. The hyperbola $25y^2 - 9x^2 = 225 \quad \Leftrightarrow \quad \dfrac{y^2}{9} - \dfrac{x^2}{25} = 1$ has $a = 3$, $b = 5$, and $c^2 = 25 + 9 = 34 \quad \Rightarrow \quad c = \sqrt{34}$.

   Vertices: $(0, \pm 3)$; foci: $(0, \pm \sqrt{34})$; asymptotes: $y = \pm \frac{3}{5}x$.

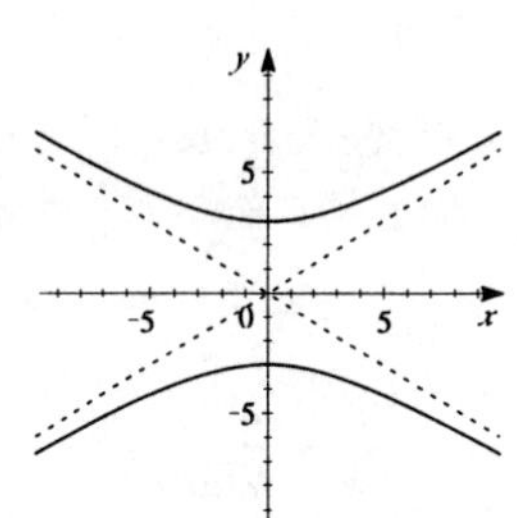

9. The hyperbola $x^2 - 4y^2 - 8 = 0 \quad \Leftrightarrow \quad \dfrac{x^2}{8} - \dfrac{y^2}{2} = 1$ has $a = \sqrt{8}$, $b = \sqrt{2}$, and $c^2 = 8 + 2 = 10 \quad \Rightarrow \quad c = \sqrt{10}$.

   Vertices: $(\pm 2\sqrt{2}, 0)$; foci: $(\pm \sqrt{10}, 0)$; asymptotes: $y = \pm \frac{\sqrt{2}}{\sqrt{8}}\,x$ $\Leftrightarrow \quad y = \pm \frac{1}{2}x$.

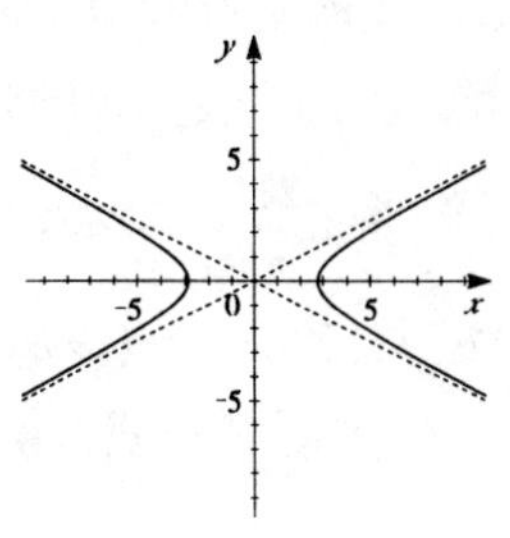

11. The hyperbola $4y^2 - x^2 = 1 \quad \Leftrightarrow \quad \dfrac{y^2}{\frac{1}{4}} - x^2 = 1$ has $a = \dfrac{1}{2}$, $b = 1$, and $c^2 = \frac{1}{4} + 1 = \frac{5}{4} \Rightarrow \quad c = \frac{1}{2}\sqrt{5}$.

Vertices: $(0, \pm\frac{1}{2})$; foci: $(0, \pm\frac{1}{2}\sqrt{5})$; asymptotes: $y = \pm\dfrac{\frac{1}{2}}{1}x$ $\Leftrightarrow \quad y = \pm\frac{1}{2}x$.

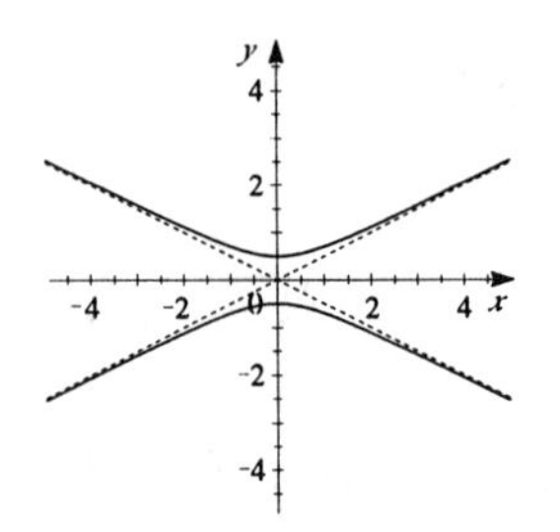

13. From the graph the foci are $(\pm 4, 0)$ and the vertices are $(\pm 2, 0)$, so $c = 4$ and $a = 2$. Thus $b^2 = 16 - 4 = 12$, and since the vertices are on the x-axis, the equation of the hyperbola is $\dfrac{x^2}{4} - \dfrac{y^2}{12} = 1$.

15. From the graph the vertices are $(0, \pm 4)$, foci are on the y-axis, and the hyperbola passes through the point $(3, -5)$. So the equation is of the form $\dfrac{y^2}{16} - \dfrac{x^2}{b^2} = 1$. Substituting the point $(3, -5)$ we have $\frac{(-5)^2}{16} - \frac{(3)^2}{b^2} = 1 \quad \Leftrightarrow \quad \dfrac{25}{16} - 1 = \dfrac{9}{b^2} \quad \Leftrightarrow \quad \dfrac{9}{16} = \dfrac{9}{b^2} \quad \Leftrightarrow \quad b^2 = 16$. Thus the equation of the hyperbola is $\dfrac{y^2}{16} - \dfrac{x^2}{16} = 1$

17. Since the foci are $(\pm 5, 0)$ and the vertices are $(\pm 3, 0)$, so $c = 5$ and $a = 3$. Then $b^2 = 25 - 9 = 16$, and since the vertices are on the x-axis, the equation of the hyperbola is $\dfrac{x^2}{9} - \dfrac{y^2}{16} = 1$.

19. The foci are $(0, \pm 2)$and the vertices are $(0, \pm 1)$, so $c = 2$ and $a = 1$. Then $b^2 = 4 - 1 = 3$, and since the vertices are on the y-axis, the equation is $y^2 - \dfrac{x^2}{3} = 1$.

21. The vertices are $(\pm 1, 0)$ and the asymptotes are $y = \pm 5x$, so $a = 1$ and using the asymptotes $y = \pm\frac{b}{a}x \quad \Leftrightarrow \quad \frac{b}{1} = 5 \quad \Leftrightarrow \quad b = 5$. Therefore, the equation of the hyperbola is $x^2 - \frac{y^2}{25} = 1$.

23. The foci are $(0, \pm 8)$ and the asymptotes are $y = \pm\frac{1}{2}x$, so $c = 8$ and using the asymptotes $y = \pm\frac{a}{b}x$, so $\frac{a}{b} = \frac{1}{2}$ and $b = 2a$. Since $a^2 + b^2 = c^2 = 64$, we have $a^2 + 4a^2 = 64 \quad \Leftrightarrow$ $a^2 = \frac{64}{5}$ and $b^2 = 4a^2 = \frac{256}{5}$. Thus, the equation of the hyperbola is $\dfrac{y^2}{64/5} - \dfrac{x^2}{256/5} = 1 \quad \Leftrightarrow$ $\dfrac{5y^2}{64} - \dfrac{5x^2}{256} = 1$.

25. The asymptotes of the hyperbola are $y = \pm x$ so $b = a$. Since the hyperbola passes through the point $(5, 3)$, its foci are on the x-axis, and the equation is of the form, $\dfrac{x^2}{a^2} - \dfrac{y^2}{a^2} = 1$, so it follows that $\dfrac{25}{a^2} - \dfrac{9}{a^2} = 1 \quad \Leftrightarrow \quad a^2 = 16 = b^2$. Therefore, the equation of the hyperbola is $\dfrac{x^2}{16} - \dfrac{y^2}{16} = 1$.

27. The foci are $(\pm 5, 0)$and the length of the transverse axis is 6, so $c = 5$ and $2a = 6 \quad \Leftrightarrow \quad a = 3$. Thus, $b^2 = 25 - 9 = 16$, and the equation is $\dfrac{x^2}{9} - \dfrac{y^2}{16} = 1$.

29. (a) The hyperbola $x^2 - y^2 = 5 \quad \Leftrightarrow \quad \dfrac{x^2}{5} - \dfrac{y^2}{5} = 1$ has $a = \sqrt{5}$ and $b = \sqrt{5}$. Then, the asymptotes are $y = \pm x$ and their slopes are $m_1 = 1$ and $m_2 = -1$. Since $m_1 \times m_2 = -1$, the asymptotes are perpendicular.

(b) Since the asymptotes are perpendicular to each other, $a = b$. Therefore, $c^2 = 2a^2 \quad \Leftrightarrow$ $a^2 = \frac{c^2}{2}$, and since the vertices are on the x-axis, the equation is $\dfrac{x^2}{\frac{1}{2}c^2} - \dfrac{y^2}{\frac{1}{2}c^2} = 1 \quad \Leftrightarrow$ $x^2 - y^2 = \frac{c^2}{2}$.

31. $\sqrt{(x+c)^2 + y^2} - \sqrt{(x-c)^2 + y^2} = \pm 2a$. Let us consider the positive case only. Then $\sqrt{(x+c)^2 + y^2} = 2a + \sqrt{(x-c)^2 + y^2}$, and squaring both sides gives
$x^2 + 2cx + c^2 + y^2 = 4a^2 + 4a\sqrt{(x-c)^2 + y^2} + x^2 - 2cx + c^2 + y^2 \quad \Leftrightarrow$
$4a\sqrt{(x-c)^2 + y^2} = 4cx - 4a^2$. Dividing by 4 and squaring both sides gives
$a^2(x^2 - 2cx + c^2 + y^2) = c^2x^2 - 2a^2cx + a^4 \quad \Leftrightarrow$
$a^2x^2 - 2a^2cx + a^2c^2 + a^2y^2 = c^2x^2 - 2a^2cx + a^4 \quad \Leftrightarrow \quad (c^2 - a^2)x^2 - a^2y^2 = a^2(c^2 - a^2)$.
The negative case gives the same result.

33. $d(AB) = 500 = 2c \quad \Leftrightarrow \quad c = 250$.

(a) Since, $\Delta t = 2640$, and $v = 980$, then $\Delta d = d(PA) - d(PB) = v\,\Delta t$

$= (980 \text{ ft}/\mu\text{s}) \cdot (2640\ \mu\text{s}) = 2{,}587{,}200 \text{ ft} = 490 \text{ mi}.$

(b) $c = 250$, $2a = 490 \quad \Leftrightarrow \quad a = 245$, and the foci are on the y-axis. Then, $b^2 = 250^2 - 245^2 = 2475$. Hence, the equation is $\dfrac{y^2}{60{,}025} - \dfrac{x^2}{2475} = 1$.

(c) Since P is due east of A, $c = 250$ is the y-coordinate of P. Therefore, P is at $(x, 250)$ and so $\dfrac{240^2}{245^2} - \dfrac{x^2}{2475} = 1 \quad \Leftrightarrow \quad x^2 = 2475\left(\dfrac{250^2}{245^2} - 1\right) \approx 102.05$. Then, $x \approx 10.1$ and so P is approximately 10.1 miles from A.

35. (a) From the equation $a^2 = k$ and $b^2 = 16 - k$. Thus $c^2 = a^2 + b^2 = k + 16 - k = 16 \quad \Rightarrow$ $c = \pm 4$. Thus the foci of the family of hyperbolas are $(0, \pm 4)$.

(b) $\dfrac{y^2}{k} - \dfrac{x^2}{16-k} = 1 \quad \Leftrightarrow \quad y^2 = k\left(1 + \dfrac{x^2}{16-k}\right) \quad \Rightarrow \quad y = \pm\sqrt{k + \dfrac{kx^2}{16-k}}$. For the top branch we graph $y = \sqrt{k + \dfrac{kx^2}{16-k}}$, $k = 1, 4, 8, 12$.

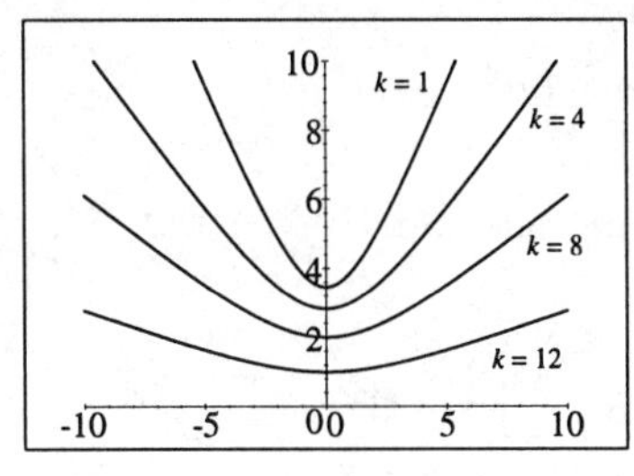

As k increases the shape gets wider and flatter.

Exercises 7.4

1. The ellipse $\dfrac{(x-2)^2}{9}+\dfrac{(y-1)^2}{4}=1$ is obtained from the ellipse $\dfrac{x^2}{9}+\dfrac{y^2}{4}=1$ by shifting it 2 units right and 1 unit upward. So $a=3$, $b=2$, and $c=\sqrt{9-4}=\sqrt{5}$.
   Center: $(2,1)$; foci: $(2\pm\sqrt{5},1)$; vertices: $(2\pm 3,1)$, so the vertices are $(-1,1)$ and $(5,1)$; length of the major axis: $2a=6$; length of the minor axis: $2b=4$.

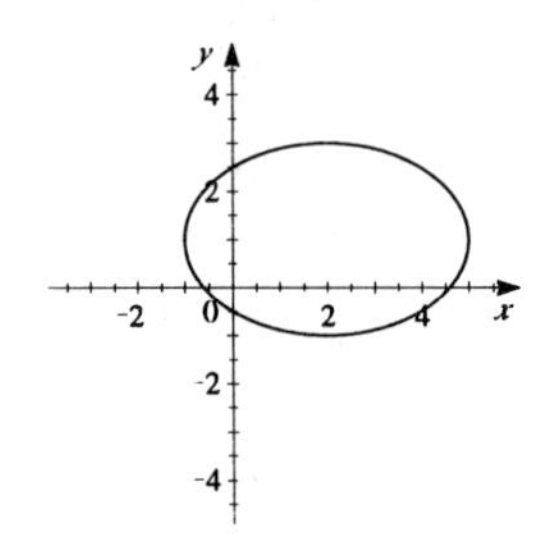

3. The ellipse $\dfrac{x^2}{9}+\dfrac{(y+5)^2}{25}=1$ is obtained from the ellipse $\dfrac{x^2}{9}+\dfrac{y^2}{25}=1$ by shifting it 5 units downward. So $a=5$, $b=3$, and $c=\sqrt{25-9}=4$.
   Center: $(0,-5)$; foci: $(0,-5\pm 4)=(0,-9)$ and $(0,-1)$; vertices: $(0,-5\pm 5)$, so the vertices are $(0,-10)$ and $(0,0)$; length of the major axis: $2a=10$; length of the minor axis: $2b=6$.

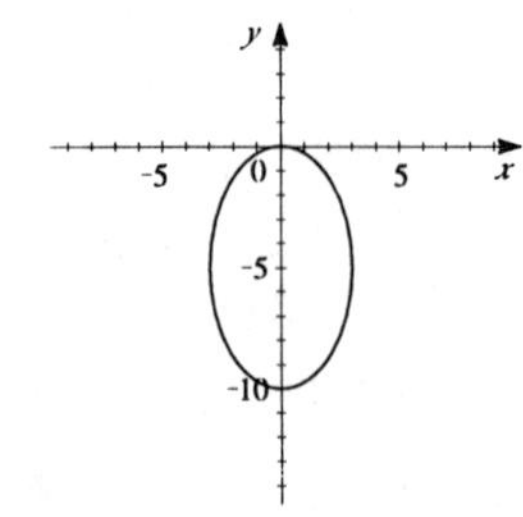

5. The parabola $(x-3)^2=8(y+1)$ is obtained from the parabola $x^2=8y$ by shifting it 3 units to the right and 1 unit down. So $4p=8$ $\Leftrightarrow$ $p=2$.
   Vertex: $(3,-1)$; focus: $(3,-1+2)=(3,1)$; directrix: $y=-1-2=-3$.

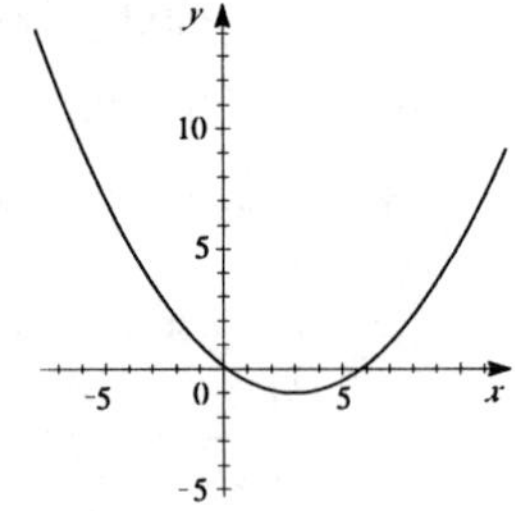

7. The parabola $-4\left(x+\frac{1}{2}\right)^2=y$ $\Leftrightarrow$ $\left(x+\frac{1}{2}\right)^2=-\frac{1}{4}y$ is obtained from the parabola $x^2=-\frac{1}{4}y$ by shifting it $\frac{1}{2}$ to the left.
   So $4p=-\frac{1}{4}$ $\Leftrightarrow$ $p=-\frac{1}{16}$.
   Vertex: $\left(-\frac{1}{2},0\right)$; focus: $\left(-\frac{1}{2},0-\frac{1}{16}\right)=\left(-\frac{1}{2},-\frac{1}{16}\right)$; directrix: $y=0+\frac{1}{16}=\frac{1}{16}$.

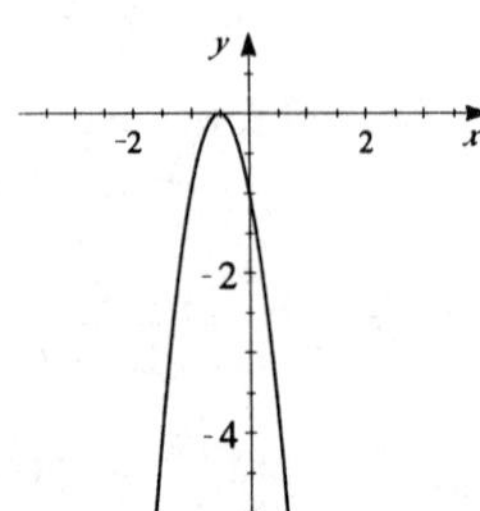

9. The hyperbola $\dfrac{(x+1)^2}{9}-\dfrac{(y-3)^2}{16}=1$ is obtained from the hyperbola $\dfrac{x^2}{9}-\dfrac{y^2}{16}=1$ by shifting it 1 unit left and 3 units up. So $a=3$, $b=4$, and $c=\sqrt{9+16}=5$.
   Center: $(-1,3)$; foci: $(-1\pm 5,3)$, so the foci are $(-6,3)$ and $(4,3)$; vertices: $(-1\pm 3,3)$, so the vertices are $(-4,3)$ and $(2,3)$; asymptotes: $(y-3)=\pm\frac{4}{3}(x+1)$ $\Leftrightarrow$ $3y=4x+13$ and $3y=-4x+5$.

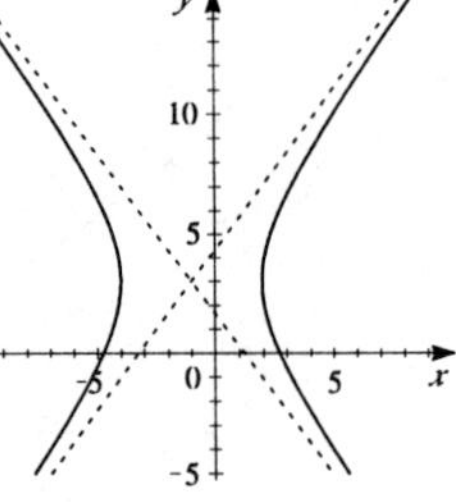

11. The hyperbola $y^2 - \dfrac{(x+1)^2}{4} = 1$ is obtained from hyperbola $y^2 - \dfrac{x^2}{4} = 1$ by shifting it 1 unit left. So $a = 1$, $b = 2$, and $c = \sqrt{1+4} = \sqrt{5}$.

Center: $(-1, 0)$; foci: $(-1, \pm\sqrt{5})$, so the foci are $(-1, -\sqrt{5})$ and $(-1, \sqrt{5})$; vertices: $(-1, \pm 1)$, so the vertices are $(-1, -1)$ and $(-1, 1)$; asymptotes: $y = \pm\frac{1}{2}(x+1) \quad\Leftrightarrow\quad y = \frac{1}{2}(x+1)$ and $y = -\frac{1}{2}(x+1)$.

13. This is a parabola that opens down with its vertex at $(0, 4)$ so its equation is of the form $x^2 = a(y-4)$. Since $(1, 0)$ is a point on this parabola, we have $(1)^2 = a(0-4) \quad\Leftrightarrow\quad 1 = -4a \quad\Leftrightarrow\quad a = -\frac{1}{4}$. Thus the equation is $x^2 = -\frac{1}{4}(y-4)$.

15. This is an ellipse with the major axis parallel to the x-axis with one vertex at $(0, 0)$, the other vertex at $(10, 0)$ and one foci at $(8, 0)$. The center is at $\left(\frac{0+10}{2}, 0\right) = (5, 0)$, $a = 5$, and $c = 3$ (the distance from one foci to the center). So $b^2 = a^2 - c^2 = 25 - 9 = 16$. Thus the equation is $\dfrac{(x-5)^2}{25} + \dfrac{y^2}{16} = 1$.

17. This is a hyperbola with center $(0, 1)$ and vertices $(0, 0)$ and $(0, 2)$. Since a is the distance form the center to a vertex, so $a = 1$. Because the transverse axis is vertical, the slope of an asymptote is $\frac{a}{b} = 1 \quad\Leftrightarrow\quad b = 1$. Thus the equation of the hyperbola is $(y-1)^2 - x^2 = 1$.

19. $9x^2 - 36x + 4y^2 = 0 \quad\Leftrightarrow\quad 9(x^2 - 4x + 4) - 36 + 4y^2 = 0 \quad\Leftrightarrow$ $9(x-2)^2 + 4y^2 = 36 \quad\Leftrightarrow\quad \dfrac{(x-2)^2}{4} + \dfrac{y^2}{9} = 1$. This is an ellipse that has $a = 3$, $b = 2$, $c = \sqrt{9-4} = \sqrt{5}$

Center: $(2, 0)$; foci: $(2, \pm\sqrt{5})$; vertices: $(2, \pm 3)$; length of major axis: $2a = 6$; length of minor axis: $2b = 4$.

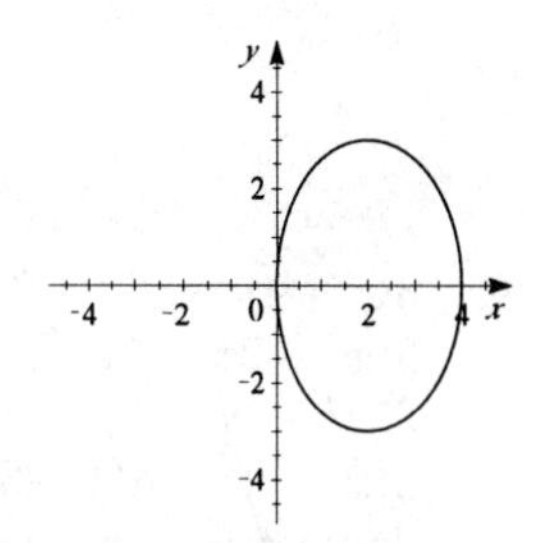

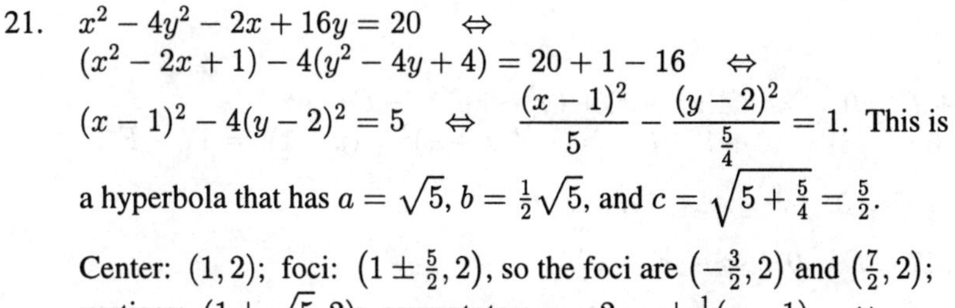

21. $x^2 - 4y^2 - 2x + 16y = 20 \quad\Leftrightarrow$
$(x^2 - 2x + 1) - 4(y^2 - 4y + 4) = 20 + 1 - 16 \quad\Leftrightarrow$
$(x-1)^2 - 4(y-2)^2 = 5 \quad\Leftrightarrow\quad \dfrac{(x-1)^2}{5} - \dfrac{(y-2)^2}{\frac{5}{4}} = 1$. This is a hyperbola that has $a = \sqrt{5}$, $b = \frac{1}{2}\sqrt{5}$, and $c = \sqrt{5 + \frac{5}{4}} = \frac{5}{2}$.

Center: $(1, 2)$; foci: $\left(1 \pm \frac{5}{2}, 2\right)$, so the foci are $\left(-\frac{3}{2}, 2\right)$ and $\left(\frac{7}{2}, 2\right)$; vertices: $(1 \pm \sqrt{5}, 2)$; asymptotes: $y - 2 = \pm\frac{1}{2}(x-1) \quad\Leftrightarrow$ $y = \frac{1}{2}x + \frac{3}{2}$ and $y = -\frac{1}{2}x + \frac{5}{2}$.

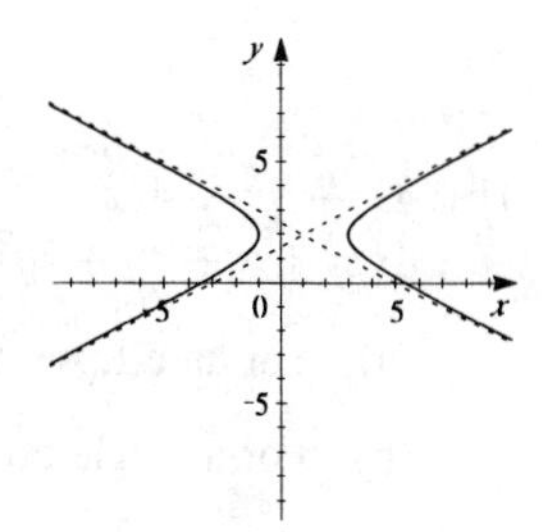

23. $4x^2 + 25y^2 - 24x + 250y + 561 = 0 \quad \Leftrightarrow$
$4(x^2 - 6x + 9) + 25(y^2 + 10y + 25) = -561 + 36 + 625 \quad \Leftrightarrow$
$4(x-3)^2 + 25(y+5)^2 = 100 \quad \Leftrightarrow \quad \dfrac{(x-3)^2}{25} + \dfrac{(y+5)^2}{4} = 1.$
This is an ellipse that has $a = 5$, $b = 2$, and $c = \sqrt{25 - 4} = \sqrt{21}$.

Center: $(3, -5)$; foci: $(3 \pm \sqrt{21}, -5)$; vertices: $(3 \pm 5, -5)$, so the vertices are $(-2, -5)$ and $(8, -5)$; length of the major axis: $2a = 10$; length of the minor axis: $2b = 4$.

25. $16x^2 - 9y^2 - 96x + 288 = 0 \quad \Leftrightarrow \quad 16(x^2 - 6x) - 9y^2 + 288 = 0$
$\Leftrightarrow \quad 16(x^2 - 6x + 9) - 9y^2 = 144 - 288 \quad \Leftrightarrow$
$16(x-3)^2 - 9y^2 = -144 \quad \Leftrightarrow \quad \dfrac{y^2}{16} - \dfrac{(x-3)^2}{9} = 1$. This is a hyperbola that has $a = 4$, $b = 3$, and $c = \sqrt{16 + 9} = 5$.

Center: $(3, 0)$; foci: $(3, \pm 5)$; vertices: $(3, \pm 4)$; asymptotes: $y = \pm \frac{4}{3}(x - 3) \quad \Leftrightarrow \quad y = \frac{4}{3}x - 4$ and $y = 4 - \frac{4}{3}x$.

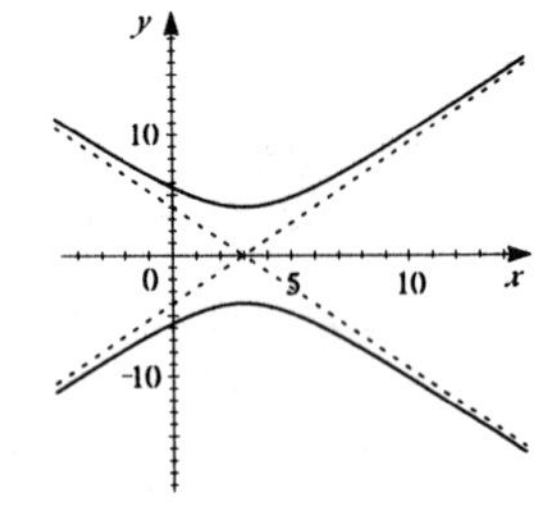

27. $x^2 + 16 = 4(y^2 + 2x) \quad \Leftrightarrow \quad x^2 - 8x - 4y^2 + 16 = 0 \quad \Leftrightarrow$
$(x^2 - 8x + 16) - 4y^2 = -16 + 16 \quad \Leftrightarrow \quad 4y^2 = (x-4)^2 \quad \Leftrightarrow$
$y = \pm \frac{1}{2}(x - 4)$. Thus, the conic is degenerate and is a pair of lines $y = \frac{1}{2}(x - 4)$ and $y = -\frac{1}{2}(x - 4)$.

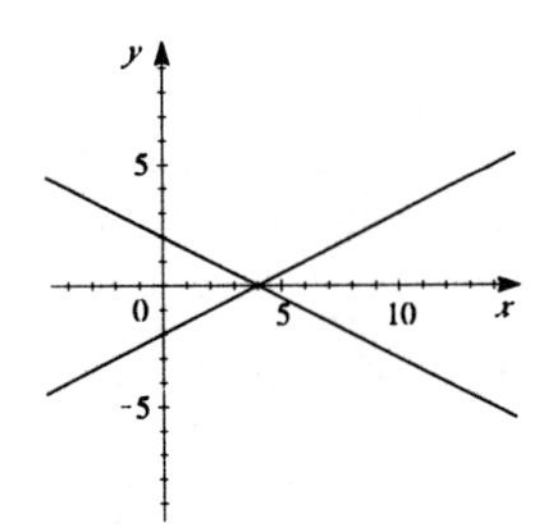

29. $3x^2 + 4y^2 - 6x - 24y + 39 = 0 \quad \Leftrightarrow$
$3(x^2 - 2x) + 4(y^2 - 6y) = -39 \quad \Leftrightarrow$
$3(x^2 - 2x + 1) + 4(y^2 - 6y + 9) = -39 + 3 + 36 \quad \Leftrightarrow$
$3(x-1)^2 + 4(y-3)^2 = 0 \quad \Leftrightarrow \quad x = 1$ and $y = 3$. This is a degenerate conic that is the point $(1, 3)$.

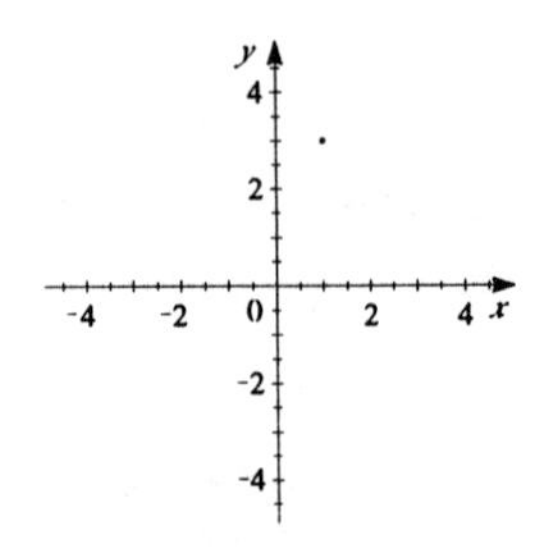

31 $4x^2 + y^2 + 4(x - 2y) + F = 0 \quad \Leftrightarrow \quad 4(x^2 + x) + (y^2 - 8y) = -F \quad \Leftrightarrow$
$4(x^2 + x + \frac{1}{4}) + (y^2 - 8y + 16) = 16 + 1 - F \quad \Leftrightarrow \quad 4(x + \frac{1}{2})^2 + (y - 1)^2 = 17 - F$

(a) For an ellipse, $17 - F > 0 \quad \Leftrightarrow \quad F < 17$.

(b) For a single point, $17 - F = 0 \quad \Leftrightarrow \quad F = 17$.

(c) For the empty set, $17 - F < 0 \quad \Leftrightarrow \quad F > 17$.

33. (a) $x^2 = 4p(y+p)$, $p = -2, -\frac{3}{2}, -1, -\frac{1}{2}, \frac{1}{2}, 1, \frac{3}{2}, 2$.

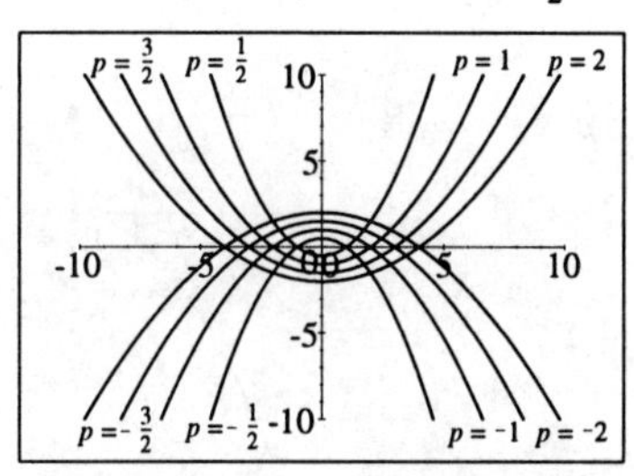

(b) $x^2 = 4p(y+p)$ is obtained by shifting the graph of $x^2 = 4py$ vertically $-p$ units so that the vertex is at $(0,-p)$. The focus of $x^2 = 4py$ is at $(0,p)$ so this point is also shifted $-p$ units vertically to the point $(0, p-p) = (0,0)$. Thus the focus is located at the origin.

(c) The parabolas become narrower.

Review Exercises for Chapter 7

1. $x^2 + 8y = 0 \quad \Leftrightarrow \quad x^2 = -8y$. This is a parabola that has $4p = -8 \quad \Leftrightarrow \quad p = -2$.

   Vertex: $(0, 0)$; focus: $(0, -2)$; directrix: $y = 2$.

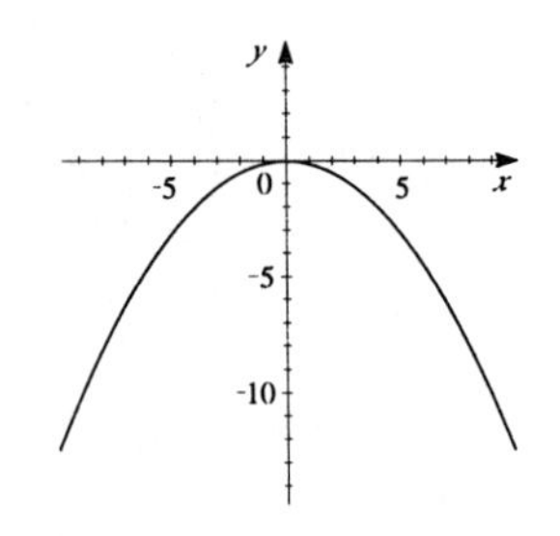

3. $x - y^2 + 4y - 2 = 0 \quad \Leftrightarrow \quad x - (y^2 - 4y + 4) - 2 = -4 \quad \Leftrightarrow \quad x - (y - 2)^2 = -2 \quad \Leftrightarrow \quad (y - 2)^2 = x + 2$. This is a parabola that has $4p = 1 \quad \Leftrightarrow \quad p = \frac{1}{4}$.

   Vertex: $(-2, 2)$; focus: $\left(-2 + \frac{1}{4}, 2\right) = \left(-\frac{7}{4}, 2\right)$; directrix: $x = -2 - \frac{1}{4} = -\frac{9}{4}$.

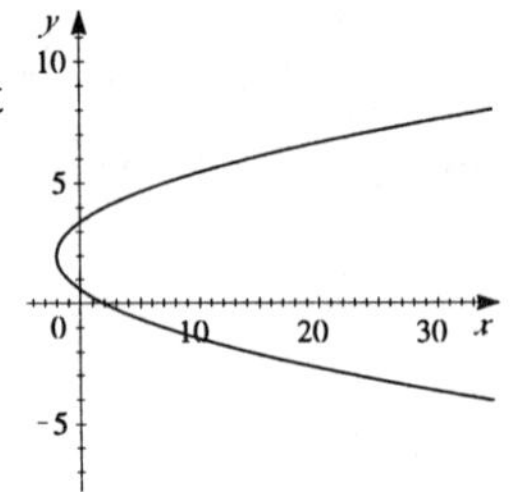

5. $x^2 + 4y^2 = 16 \quad \Leftrightarrow \quad \dfrac{x^2}{16} + \dfrac{y^2}{4} = 1$. This is an ellipse with $a = 4$, $b = 2$, and $c = \sqrt{16 - 4} = 2\sqrt{3}$.

   Center: $(0, 0)$; vertices: $(\pm 4, 0)$; foci: $(\pm 2\sqrt{3}, 0)$; length of the major axis: $2a = 8$; length of the minor axis: $2b = 4$.

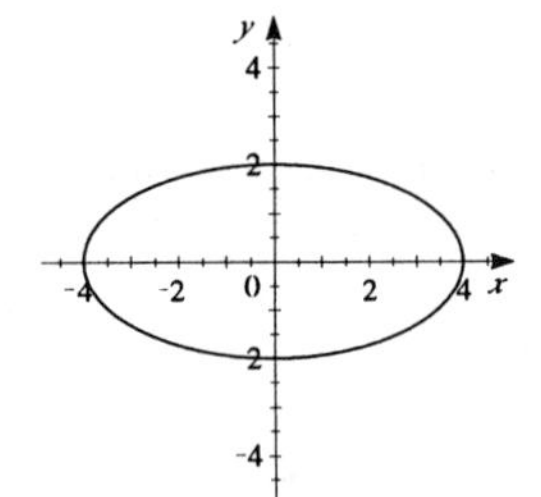

7. $4x^2 + 9y^2 = 36y \quad \Leftrightarrow \quad 4x^2 + 9(y^2 - 4y + 4) = 36 \quad \Leftrightarrow \quad 4x^2 + 9(y - 2)^2 = 36 \quad \Leftrightarrow \quad \dfrac{x^2}{9} + \dfrac{(y - 2)^2}{4} = 1$. This is an ellipse with $a = 3$, $b = 2$, and $c = \sqrt{9 - 4} = \sqrt{5}$.

   Center: $(0, 2)$; vertices: $(\pm 3, 2)$; foci: $(\pm \sqrt{5}, 2)$; length of the major axis: $2a = 6$; length of the minor axis: $2b = 4$.

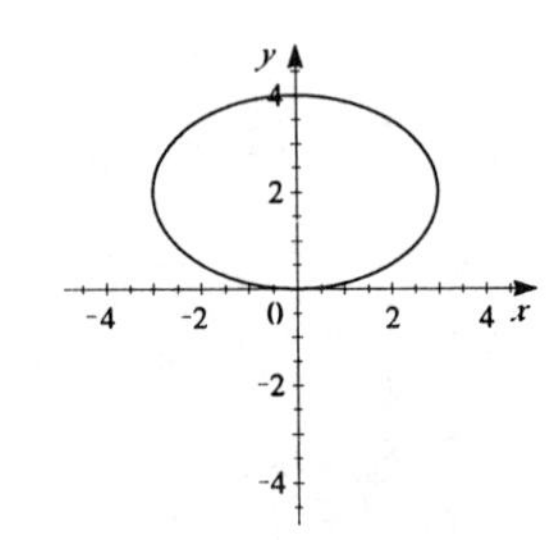

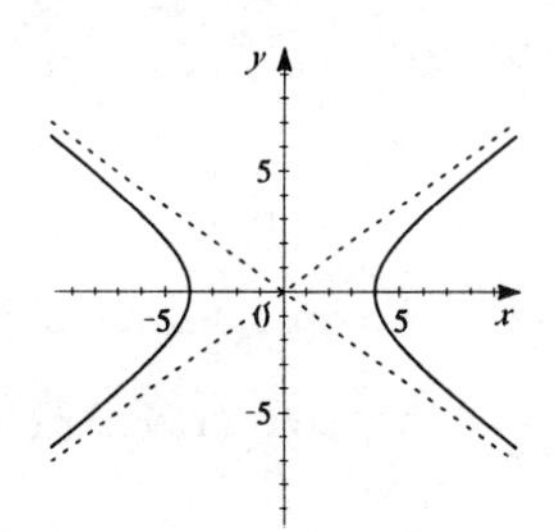

9. $x^2 - 2y^2 = 16 \quad \Leftrightarrow \quad \dfrac{x^2}{16} - \dfrac{y^2}{8} = 1$. This is a hyperbola that has $a = 4$, $b = 2\sqrt{2}$, and $c = \sqrt{16+8} = \sqrt{24} = 2\sqrt{6}$.

   Center: $(0,0)$; vertices: $(\pm 4, 0)$; foci: $(\pm 2\sqrt{6}, 0)$; asymptotes: $y = \pm \frac{2\sqrt{2}}{4}x \quad \Leftrightarrow \quad y = \pm \frac{1}{\sqrt{2}}x$.

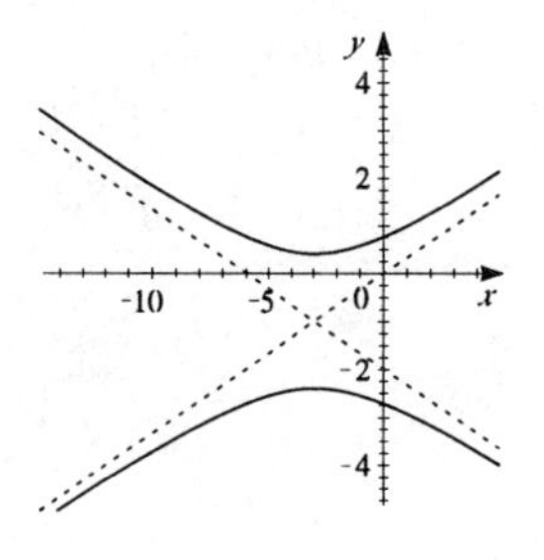

11. $9y^2 + 18y = x^2 + 6x + 18 \quad \Leftrightarrow$
    $9(y^2 + 2y + 1) = (x^2 + 6x + 9) + 9 - 9 + 18 \quad \Leftrightarrow$
    $9(y+1)^2 - (x+3)^2 = 18 \quad \Leftrightarrow \quad \dfrac{(y+1)^2}{2} - \dfrac{(x+3)^2}{18} = 1$. This is a hyperbola that has $a = \sqrt{2}$, $b = 3\sqrt{2}$, and $c = \sqrt{2+18} = 2\sqrt{5}$.

    Center: $(-3,-1)$; vertices: $(-3, -1 \pm \sqrt{2})$; foci: $(-3, -1 \pm 2\sqrt{5})$; asymptotes: $y + 1 = \pm\frac{1}{3}(x+3) \quad \Leftrightarrow \quad y = \frac{1}{3}x$ and $y = -\frac{1}{3}x - 2$.

13. This is a parabola that opens to the right with its vertex at $(0,0)$ and the focus at $(2,0)$. So $p = 2$ equation is $y^2 = 4(2)x \quad \Leftrightarrow \quad y^2 = 8x$.

15. From the graph, the center is at $(0,0)$ and the vertices are at $(0,-4)$ and $(0,4)$. Since a is the distance form the center to a vertex, so $a = 4$. Because one focus is at $(0,5)$ we have $c = 5$ and since $c^2 = a^2 + b^2$ we have $25 = 16 + b^2 \quad \Leftrightarrow \quad b^2 = 9$. Thus the equation of the hyperbola is $\dfrac{y^2}{16} - \dfrac{x^2}{9} = 1$.

17. From the graph the center of the ellipse is $(4,2)$, $a = 4$ and $b = 2$. Since the major axis is horizontal, the equation is $\dfrac{(x-4)^2}{4^2} + \dfrac{(y-2)^2}{2^2} = 1 \quad \Leftrightarrow \quad \dfrac{(x-4)^2}{16} + \dfrac{(y-2)^2}{4} = 1$.

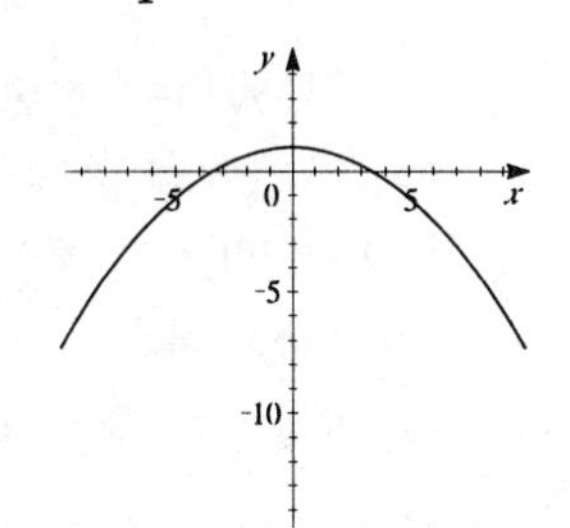

19. $\dfrac{x^2}{12} + y = 1 \quad \Leftrightarrow \quad \dfrac{x^2}{12} = -(y-1) \quad \Leftrightarrow \quad x^2 = -12(y-1)$. This is a parabola that has $4p = -12 \quad \Leftrightarrow \quad p = -3$.

    Vertex: $(0,1)$; focus: $(0, 1-3) = (0,-2)$.

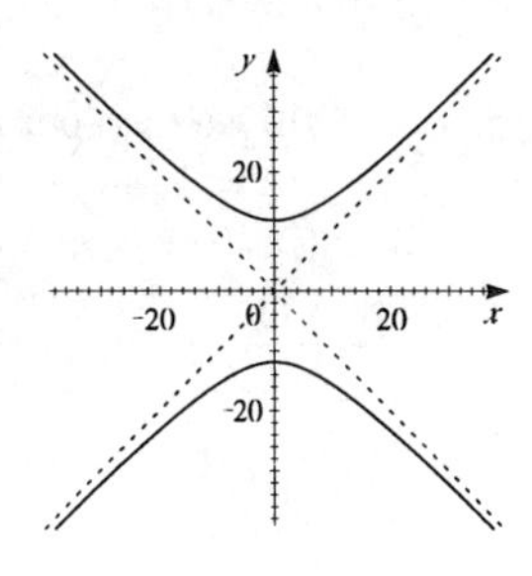

21. $x^2 - y^2 + 144 = 0 \quad \Leftrightarrow \quad \dfrac{y^2}{144} - \dfrac{x^2}{144} = 1$. This is a hyperbola that has $a = 12$, $b = 12$, and $c = \sqrt{144+144} = 12\sqrt{2}$.

    Vertices: $(0, \pm 12)$; foci: $(0, \pm 12\sqrt{2})$.

23. $4x^2 + y^2 = 8(x + y) \Leftrightarrow 4(x^2 - 2x) + (y^2 - 8y) = 0 \Leftrightarrow$
$4(x^2 - 2x + 1) + (y^2 - 8y + 16) = 4 + 16 \Leftrightarrow$
$4(x-1)^2 + (y-4)^2 = 20 \Leftrightarrow \dfrac{(x-1)^2}{5} + \dfrac{(y-4)^2}{20} = 1$. This is an ellipse that has $a = 2\sqrt{5}$, $b = \sqrt{5}$, and $c = \sqrt{20-5} = \sqrt{15}$.
Vertices: $(1, 4 \pm 2\sqrt{5})$; foci: $(1, 4 \pm \sqrt{15})$.

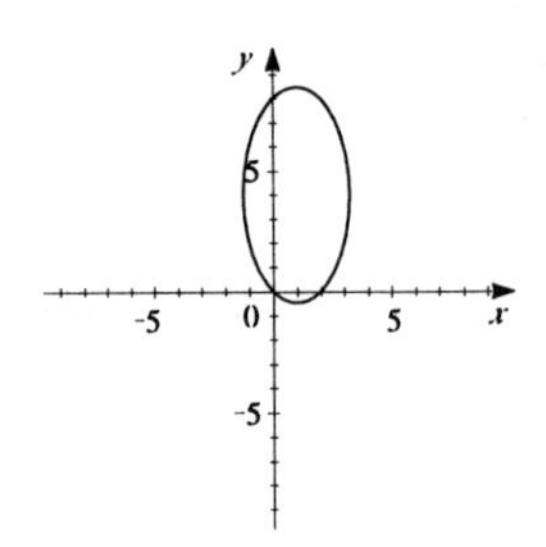

25. $x = y^2 - 16y \Leftrightarrow x + 64 = y^2 - 16y + 64 \Leftrightarrow$
$(y-8)^2 = x + 64$. This is a parabola that has $4p = 1 \Leftrightarrow p = \frac{1}{4}$.
Vertex: $(-64, 8)$; focus: $\left(-64 + \frac{1}{4}, 8\right) = \left(-\frac{255}{4}, 8\right)$.

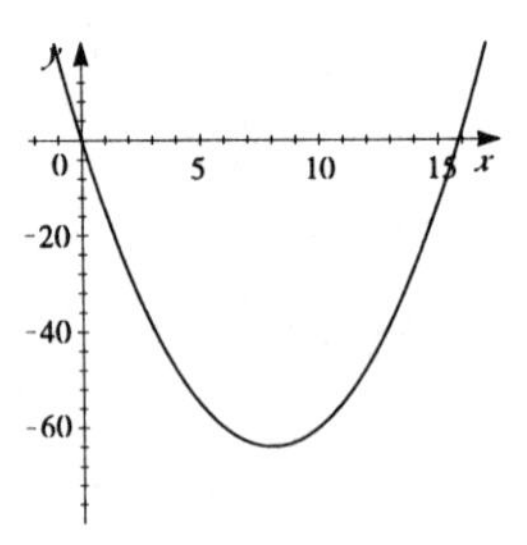

27. $2x^2 - 12x + y^2 + 6y + 26 = 0 \Leftrightarrow$
$2(x^2 - 6x) + (y^2 + 6y) = -26 \Leftrightarrow$
$2(x^2 - 6x + 9) + (y^2 + 6y + 9) = -26 + 18 + 9 \Leftrightarrow$
$2(x-3)^2 + (y+3)^2 = 1 \Leftrightarrow \dfrac{(x-3)^2}{\frac{1}{2}} + (y+3)^2 = 1$. This is an ellipse that has $a = 1$, $b = \frac{1}{\sqrt{2}}$, and $c = \sqrt{1 - \frac{1}{2}} = \frac{1}{\sqrt{2}}$.
Vertices: $(3, -3 \pm 1)$, so the vertices are $(3, -4)$ and $(3, -2)$; foci: $\left(3, -3 \pm \frac{1}{\sqrt{2}}\right)$.

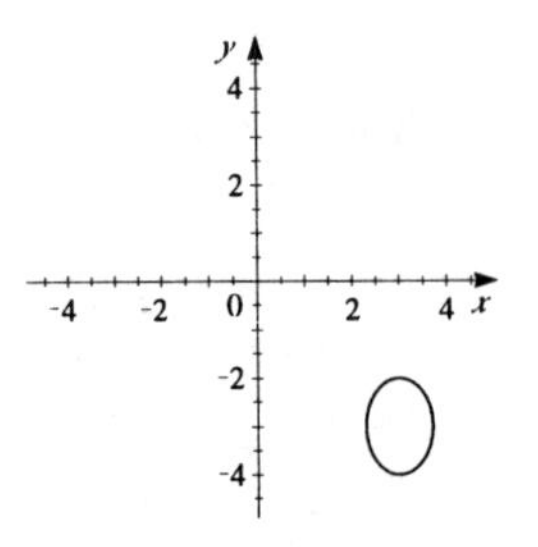

29. $9x^2 + 8y^2 - 15x + 8y + 27 = 0 \Leftrightarrow 9\left(x^2 - \frac{5}{3}x + \frac{25}{36}\right) + 8\left(y^2 + y + \frac{1}{4}\right) = -27 + \frac{25}{4} + 2 \Leftrightarrow$
$9(x - \frac{5}{6})^2 + 8(y + \frac{1}{2})^2 = -\frac{75}{4}$. However, since the left-hand side of the equation is greater than or equal to 0, there is no point that satisfies this equation. The graph is empty.

31. The parabola has focus $(0, 1)$ and directrix $y = -1$. Therefore, $p = 1$ and so $4p = 4$. Since the focus is on the y-axis and the center is at $(0, 0)$, the equation of the parabola is $x^2 = 4y$.

33. The hyperbola has vertices $(0, \pm 2)$ and asymptotes $y = \pm \frac{1}{2}x$. Therefore, $a = 2$ and the foci are on the y-axis. Since the slope of the asymptotes are $\pm \frac{1}{2} = \pm \dfrac{a}{b} \Leftrightarrow b = 2a = 4$. The equation of the hyperbola is $\dfrac{y^2}{4} - \dfrac{x^2}{16} = 1$.

35. The ellipse has foci $F_1(1, 1)$ and $F_2(1, 3)$ and one vertex is on the x-axis. Thus, $2c = 3 - 1 = 2 \Leftrightarrow c = 1$, and so the center of the ellipse is at $C(1, 2)$. Also, since one vertex is on the x-axis, $a = 2 - 0 = 2$, and thus $b^2 = 4 - 1 = 3$. So the equation of the ellipse is
$\dfrac{(x-1)^2}{3} + \dfrac{(y-2)^2}{4} = 1$.

37. The ellipse has vertices $V_1(7, 12)$ and $V_2(7, -8)$ and passes through the point $P(1, 8)$. Thus, $2a = 12 + 8 = 20 \Leftrightarrow a = 10$ and the center is $\left(7, \frac{-8+12}{2}\right) = (7, 2)$. Since the major axis of the

ellipse is vertical, its equation is $\dfrac{(x-7)^2}{b^2}+\dfrac{(y-2)^2}{100}=1$. Since the point $P(1,8)$ is on the ellipse, $\dfrac{(1-7)^2}{b^2}+\dfrac{(8-2)^2}{100}=1 \quad\Leftrightarrow\quad 3600+36b^2=100b^2 \quad\Leftrightarrow\quad 64b^2=3600 \quad\Leftrightarrow\quad b^2=\dfrac{225}{4}$.

Therefore, the equation of the ellipse is $\dfrac{(x-7)^2}{225/4}+\dfrac{(y-2)^2}{100}=1 \quad\Leftrightarrow$

$\dfrac{4(x-7)^2}{225}+\dfrac{(y-2)^2}{100}=1$.

39. Since $(0,0)$ and $(1600,0)$ are both points on the parabola, the x coordinate of the vertex is 800. And since the highest point it reaches is 3200, the y coordinate of the vertex is 3200. Thus the vertex is $(800,3200)$ and the equation is of the form $(x-800)^2=4p(y-3200)$. Substituting the point $(0,0)$ we get $(0-800)^2=4p(0-3200) \quad\Leftrightarrow\quad 640000=-12800p \quad\Leftrightarrow\quad p=-50$. So the equation is $(x-800)^2=4(-5)(y-3200) \quad\Leftrightarrow\quad (x-800)^2=-200(y-3200)$.

41. The length of the major axis is $2a=186,000,000 \quad\Leftrightarrow\quad a=93,000,000$. Next, the eccentricity is $e=\frac{c}{a}=0.017 \quad\Leftrightarrow\quad c=0.017(93,000,000)=1,581,000$.

(a) The earth is closes to the sun when the distance is $a-c=93,000,000-1,581,000$ $=91,419,000$.

(b) The earth is furthest from the sun when the distance is $a+c=93,000,000+1,581,000$ $=94,581,000$

43. (a) The graphs of $y=kx^2$ for $k=\frac{1}{2}$, 1, 2, and 4.

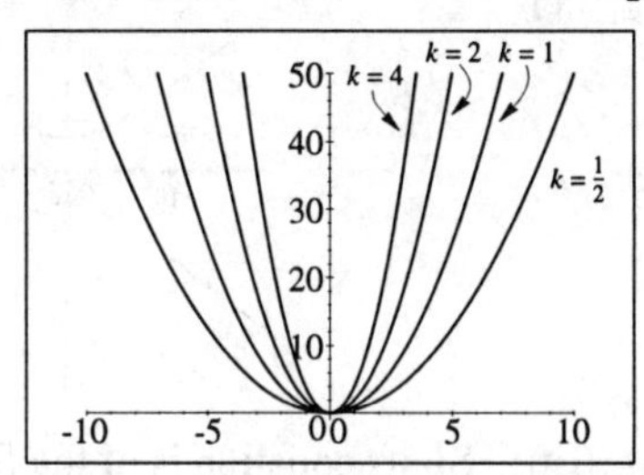

(b) $y=kx^2 \quad\Leftrightarrow\quad x^2=\frac{1}{k}y=4\left(\frac{1}{4k}\right)y$. Thus the focus is at $\left(0,\frac{1}{4k}\right)$.

(c) As k increases, the focus gets closer to the vertex.

Chapter 7 Test

1. $x^2 = -12y$. This is a parabola that has $4p = -12 \quad \Leftrightarrow \quad p = -3$.

   Focus: $(0, -3)$; directrix: $y = 3$.

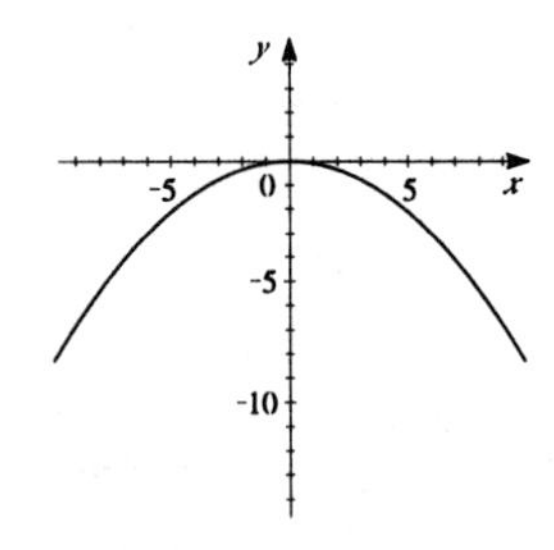

2. $\dfrac{x^2}{16} + \dfrac{y^2}{4} = 1$. This is an ellipse that has $a = 4$, $b = 2$, and $c = \sqrt{16 - 4} = 2\sqrt{3}$.

   Vertices: $(\pm 4, 0)$; foci: $(\pm 2\sqrt{3}, 0)$; length of the major axis: $2a = 8$; length of the minor axis: $2b = 4$.

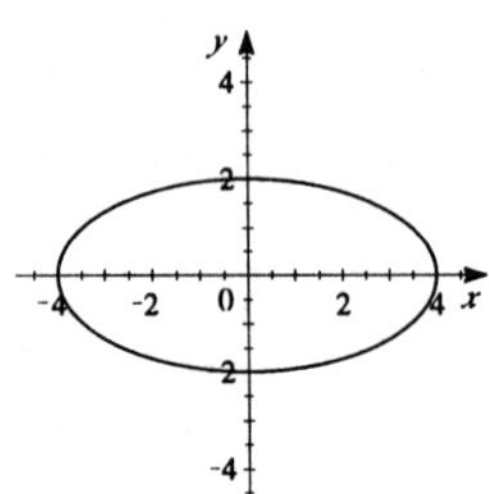

3. $\dfrac{y^2}{9} - \dfrac{x^2}{16} = 1$. This is a hyperbola that has $a = 3$, $b = 4$, and $c = \sqrt{9 + 16} = 5$.

   Vertices: $(0, \pm 3)$; foci: $(0, \pm 5)$; asymptotes: $y = \pm \frac{3}{4}x$.

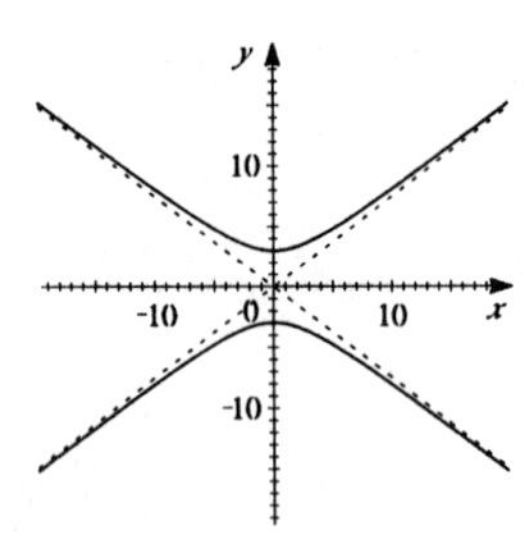

4. This is a parabola that opens to the left with its vertex at $(0, 0)$. So its equation is of the form $y^2 = 4px$ with $p < 0$. Substituting the point $(-4, 2)$ we have $2^2 = 4p(-4) \quad \Leftrightarrow \quad 4 = -16p \quad \Leftrightarrow \quad p = -\frac{1}{4}$. So the equation is $y^2 = 4\left(-\frac{1}{4}\right)x \quad \Leftrightarrow \quad y^2 = -x$.

5. This is an ellipse tangent to the x-axis at $(0, 0)$ and with one vertex at the point $(4, 3)$. The center is at $(0, 3)$, $a = 4$, and $b = 3$. Thus the equation is $\dfrac{x^2}{16} + \dfrac{(y-3)^2}{9} = 1$.

6. This a hyperbola with a horizontal transverse axis, vertices at $(1, 0)$ and $(3, 0)$, and foci at $(0, 0)$ and $(4, 0)$. Thus the center is $(2, 0)$, $a = 3 - 2 = 1$ and $c = 4 - 2 = 2$. Thus $b^2 = 2^2 - 1^2 = 3$. So the equation is $\dfrac{(x-2)^2}{1^2} - \dfrac{y^2}{3} = 1 \quad \Leftrightarrow \quad (x-2)^2 - \dfrac{y^2}{3} = 1$

7. $16x^2 + 36y^2 - 96x + 36y + 9 = 0 \quad \Leftrightarrow$
$16(x^2 - 6x) + 36(y^2 + y) = -9 \quad \Leftrightarrow$
$16(x^2 - 6x + 9) + 36\left(y^2 + y + \frac{1}{4}\right) = -9 + 144 + 9 \quad \Leftrightarrow$
$16(x-3)^2 + 36\left(y + \frac{1}{2}\right)^2 = 144 \quad \Leftrightarrow \quad \dfrac{(x-3)^2}{9} + \dfrac{\left(y+\frac{1}{2}\right)^2}{4} = 1.$
This is an ellipse that has $a = 3$, $b = 2$, and $c = \sqrt{9-4} = \sqrt{5}$.
Center: $\left(3, -\frac{1}{2}\right)$; vertices: $\left(3 \pm 3, -\frac{1}{2}\right)$, so vertices are $\left(0, -\frac{1}{2}\right)$ and $\left(6, -\frac{1}{2}\right)$.

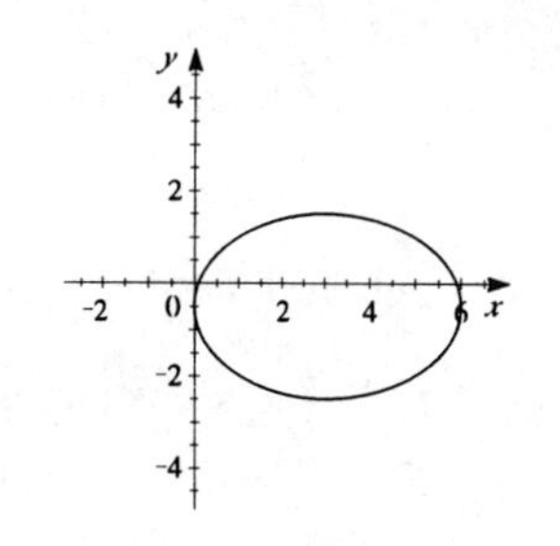

8. $9x^2 - 8y^2 + 36x + 64y = 92 \quad \Leftrightarrow \quad 9(x^2 + 4x) - 8(y^2 - 8y) = 92$
$\Leftrightarrow \quad 9(x^2 + 4x + 4) - 8(y^2 - 8y + 16) = 92 + 36 - 128 \quad \Leftrightarrow$
$9(x+2)^2 - 8(y-4)^2 = 0 \quad \Leftrightarrow \quad 9(x+2)^2 = 8(y-4)^2 \quad \Rightarrow$
$3(x+2) = \pm 2\sqrt{2}(y-4) \quad \Leftrightarrow \quad y = 4 \pm \frac{3}{2\sqrt{2}}(x+2)$. This is a degenerate hyperbola that consist of two intersecting lines.

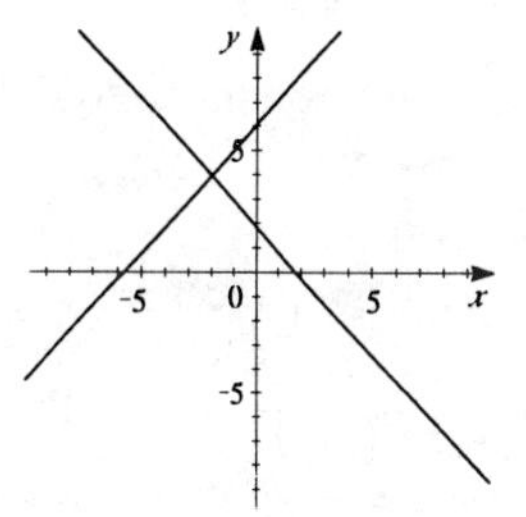

9. $2x + y^2 + 8y + 8 = 0 \quad \Leftrightarrow \quad y^2 + 8y + 16 = -2x - 8 + 16 \quad \Leftrightarrow$
$(y+4)^2 = -2(x-4)$. This is a parabola that has $4p = -2 \quad \Leftrightarrow$
$p = -\frac{1}{2}$.
Vertex: $(4, -4)$; focus: $\left(4 - \frac{1}{2}, -4\right) = \left(\frac{7}{2}, -4\right)$.

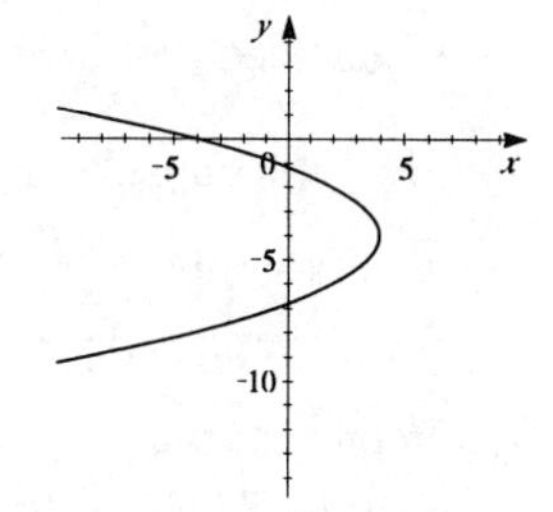

10. The hyperbola has foci at $(0, \pm 5)$ and asymptotes $y = \pm \frac{3}{4}x$. Since the foci are $(0, \pm 5)$, $c = 5$, the foci are on the y-axis, and the center is $(0,0)$. Also, since $y = \pm \frac{3}{4}x = \pm \frac{a}{b}x$, it follows that $\frac{a}{b} = \frac{3}{4} \quad \Leftrightarrow \quad a = \frac{3}{4}b$. Then $c^2 = 5^2 = 25 = a^2 + b^2 = \left(\frac{3}{4}b\right)^2 + b^2 = \frac{25}{16}b^2 \quad \Leftrightarrow \quad b^2 = 16$, and by substitution, $a = \frac{3}{4}(16) = 9$. Therefore, the equation of the hyperbola is $\dfrac{y^2}{9} - \dfrac{x^2}{16} = 1$.

11. The parabola has focus $(2, 4)$ and directrix the x-axis $(y = 0)$. Therefore, $2p = 4 - 0 = 4 \quad \Leftrightarrow$ $p = 2 \quad \Leftrightarrow \quad 4p = 8$, and the vertex is at $(2, 4 - p) = (2, 2)$. Hence, the equation of the parabola is $(x-2)^2 = 8(y-2) \quad \Leftrightarrow \quad x^2 - 4x + 4 = 8y - 16 \quad \Leftrightarrow \quad x^2 - 4x - 8y + 20 = 0$.

12. We place the vertex of the parabola at the origin, so the parabola contains the points $(3, \pm 3)$ and the equation is of the form $y^2 = 4px$. Substituting the point $(3,3)$ we get $3^2 = 4p(3) \quad \Leftrightarrow$ $9 = 12p \quad \Leftrightarrow \quad p = \frac{3}{4}$. So the focus is at $\left(\frac{3}{4}, 0\right)$ and we should place the light bulb $\frac{3}{4}$ inches from the vertex.

Focus on Problem Solving

1. Tetrahedron: faces are all triangles and three faces meet at each vertex. Thus $V = \dfrac{3F}{3} = F$ and $E = \dfrac{3F}{2}$. Substituting into Euler's formula we have $F - \dfrac{3F}{2} + F = 2 \quad \Leftrightarrow \quad \dfrac{1}{2}F = 2 \quad \Leftrightarrow$ $F = 4$. So $V = 4$ and $E = \frac{12}{2} = 6$.

   Octahedron: faces are all triangles and four faces meet at each vertex. Thus $V = \dfrac{3F}{4}$ and $E = \dfrac{3F}{2}$. Substituting into Euler's formula we have $F - \dfrac{3F}{2} + \dfrac{3F}{4} = 2 \quad \Leftrightarrow \quad \dfrac{1}{4}F = 2 \quad \Leftrightarrow \quad F = 8$. So $V = \frac{24}{4} = 6$ and $E = \frac{24}{2} = 12$.

   Cube: faces are all rectangles and three faces meet at each vertex. Thus $V = \dfrac{4F}{3}$ and $E = \frac{4F}{2} = 2F$. Substituting into Euler's formula we have $F - 2F + \dfrac{4F}{3} = 2 \quad \Leftrightarrow \quad \dfrac{1}{3}F = 2$ $\Leftrightarrow \quad F = 6$. So $V = \frac{24}{3} = 8$ and $E = 12$.

   Dodecahedron: faces are all pentagon and three faces meet at each vertex. Thus $V = \dfrac{5F}{3}$ and $E = \frac{5F}{2}$. Substituting into Euler's formula we have $F - \dfrac{5F}{2} + \dfrac{5F}{3} = 2 \quad \Leftrightarrow \quad \dfrac{1}{6}F = 2 \quad \Leftrightarrow$ $F = 12$. So $V = \frac{60}{3} = 20$ and $E = \frac{60}{2} = 30$.

   Icosahedron is done in the text.

3. In order to tile the plane with regular polygons, a whole number of polygons must meet at each corner point. Since the interior angle of a regular pentagon is 108°, it is impossible for a whole number of pentagons to meet at a corner. The plane can be tiled with regular hexagons, as shown is the figure on the right. For $n \geq 7$, the interior angle of a regular n-gon is strictly between 120° and 180°, so it is impossible for a whole number of such n-gons to meet at a corner. Thus it is only possible to tile the plane with equilateral triangles, squares, and regular hexagons.

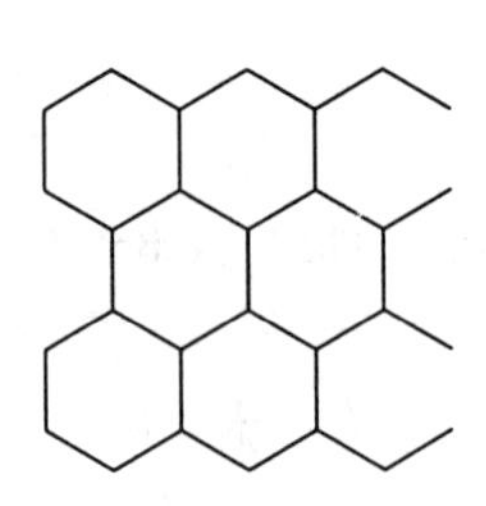

5. Imagine that the walls of the room are folded down as in the diagram. Notice that the diagram shows two points labeled B. So one of the straight lines from A to B gives the shortest path. If $a \leq b$ then the length of the shortest path is $\sqrt{(a+c)^2 + b^2}$ and if $a \geq b$, then the length of the shortest path is $\sqrt{a^2 + (b+c)^2}$.

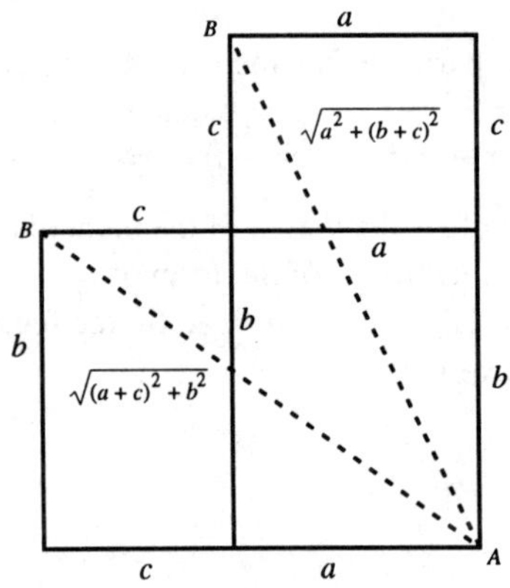

7. Let x represent the length of the third side of the triangle. By the Pythagorean Theorem, $3^2 + x^2 = 5^2$, so $x^2 = 16$ and hence $x = 4$. Thus the area of the triangle is $\frac{1}{2} \cdot 4 \cdot 3 = 6$. But if h is the length of the altitude perpendicular to the hypotenuse, then the area of the triangle is also $\frac{1}{2} \cdot 5 \cdot h$. Thus $\frac{1}{2} \cdot 5 \cdot h = 6 \quad \Leftrightarrow \quad h = \frac{12}{5} = 2.4$. Thus the length of the altitude is 2.4 cm.

9. Let x be the distance from the fourth corner to the point P. Then the Pythagorean Theorem gives

   (1) $a^2 + b^2 = x^2$

   (2) $b^2 + c^2 = 196$

   (3) $c^2 + d^2 = 100$

   (4) $d^2 + a^2 = 25$

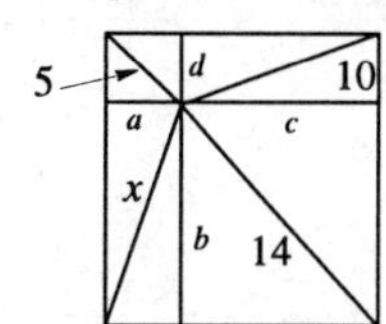

   Subtracting equation (1) from equation (2), we have $a^2 - c^2 = x^2 - 196$. And subtracting equation (1) from equation (4), we have $a^2 - c^2 = -75$. Thus $x^2 - 196 = -75 \quad \Leftrightarrow \quad x^2 = 121 \quad \Leftrightarrow$ $x = 11$ cm.

11. Let p be an odd prime and label the other integer sides of the triangle as shown. Then $p^2 + b^2 = c^2 \quad \Leftrightarrow \quad p^2 = c^2 - b^2 \quad \Leftrightarrow \quad p^2 = (c-b)(c+b)$. Now, $c - b$ and $c + b$ are integer factors of p^2 where p is a prime.

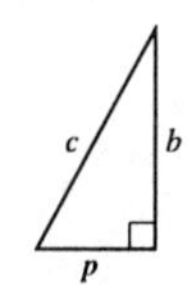

   Case 1: $c - b = p$ and $c + b = p$. Subtracting gives $b = 0$ which contradicts that b is the length of a side of a triangle.

   Case 2: $c - b = 1$ and $c + b = p^2$. Adding gives $2c = 1 + p^2 \quad \Leftrightarrow \quad c = \dfrac{1+p^2}{2}$ and $b = \dfrac{p^2-1}{2}$.

   Since p is odd, p^2 is odd. Thus $\dfrac{1+p^2}{2}$ and $\dfrac{p^2-1}{2}$ both define integers. For example, if we choose $p = 3$, then $c = 5$ and $b = 4$.

13. (a) If $2p + 1$ is a perfect square, then $2p + 1 = n^2 \quad \Leftrightarrow \quad 2p = n^2 - 1 \quad \Leftrightarrow$ $2p = (n+1)(n-1)$.

   Case 1: $n + 1 = 2$ and $n - 1 = p$. Thus $n = 1 \quad \Rightarrow \quad p = 0$ which contradicts the fact that p is prime.

   Case 2: $n + 1 = p$ and $n - 1 = 2$. Thus $n = 3 \quad \Rightarrow \quad p = 4$ which contradicts the fact that p is prime.

   Thus there are no such primes.

   (b) If $2p + 1$ is a perfect cube, then $2p + 1 = n^3 \quad \Leftrightarrow \quad 2p = n^3 - 1 \quad \Leftrightarrow$ $2p = (n-1)(n^2 + n + 1)$. Since p is prime we have 2 cases.

   Case 1: $n - 1 = 2$ and $n^2 + n + 1 = p$. Thus $n = 3 \quad \Rightarrow \quad p = n^2 + n + 1 = 9 + 3 + 1$ $= 13$. So $p = 13$ is a solution.

   Case 2: $n - 1 = p$ and $n^2 + n + 1 = 2$. In this case $n^2 + n - 1 = 0 \quad \Rightarrow$ $n = \frac{-1 \pm \sqrt{1+4}}{2} = \frac{-1 \pm \sqrt{5}}{2}$ which is not an integer..

   Thus the only solution is $p = 13$.

15. The belt is tangent to the pulley at the point where the belt leaves the pulley. Thus the length of the belt between adjacent tangent points is the same as the length between the centers of the corresponding pulleys. The length of the belt around the pulleys can be pieced together to form a circle of radius 1. Thus the length of the belt $L = P + 2\pi(1) = P + 2\pi$. (See the following figure.)

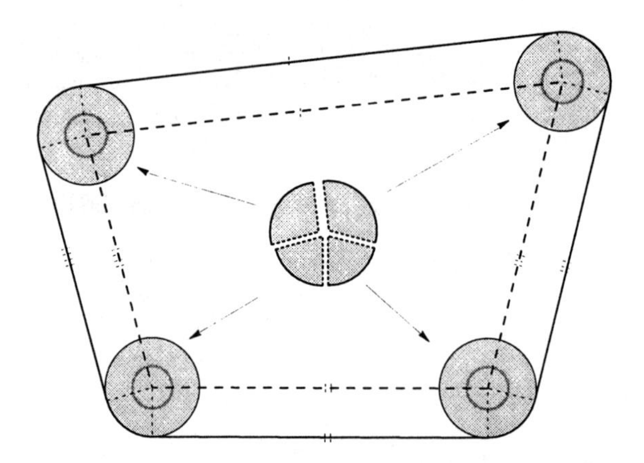

Chapter Eight

Exercises 8.1

1. $8! = 8 \cdot 7 \cdot 6 \cdot 5 \cdot 4 \cdot 3 \cdot 2 \cdot 1 = 40{,}320$

3. $\dfrac{12!}{10!} = \dfrac{12 \cdot 11 \cdot 10!}{10!} = 12 \cdot 11 = 132$

5. $\dfrac{100!}{98!} = \dfrac{100 \cdot 99 \cdot 98!}{98!} = 9900$

7. $\dfrac{8! + 9!}{8!} = \dfrac{8! + 9 \cdot 8!}{8!} = \dfrac{8!(1+9)}{8!} = 1 + 9 = 10$

9. $\dfrac{100! - 99!}{98!} = \dfrac{100 \cdot 99 \cdot 98! - 99 \cdot 98!}{98!} = \dfrac{98!(100 \cdot 99 - 99)}{98!} = 100 \cdot 99 - 99 = 9801$

11. $\dfrac{(n-1)!}{n!} = \dfrac{(n-1)!}{n \cdot (n-1)!} = \dfrac{1}{n}$

13. $\dfrac{(n+1)!}{(n-1)!\,n} = \dfrac{(n+1)n!}{n!} = n + 1$

15. By the Fundamental Counting Principle, the possible number of 3-letter words is $\binom{\text{number of ways to}}{\text{choose the 1st letter}} \cdot \binom{\text{number of ways to}}{\text{choose the 2nd letter}} \cdot \binom{\text{number of ways to}}{\text{choose the 3rd letter}}$
   (a) Since repetitions are allowed, we have 4 choices for each letter. Thus there are $4 \cdot 4 \cdot 4 = 64$ words.
   (b) Since repetitions are *not* allowed, we have 4 choices for the 1st letter, 3 choices for the 2nd letter, and 2 choices for the 3rd letter. Thus there are $4 \cdot 3 \cdot 2 = 24$ words.

17. Since there are four choices for each of the five questions, by the Fundamental Counting Principle there are $4 \cdot 4 \cdot 4 \cdot 4 \cdot 4 = 1024$ different ways the test can be completed.

19. Since a runner can only finish once, there are no repetitions. And since we are assuming that there is no tie, the number of different finishes is $\binom{\text{number of ways to}}{\text{choose the 1st runner}} \cdot \binom{\text{number of ways to}}{\text{choose the 2nd runner}} \cdot \binom{\text{number of ways to}}{\text{choose the 3rd runner}} \cdot \binom{\text{number of ways to}}{\text{choose the 4th runner}} \cdot \binom{\text{number of ways to}}{\text{choose the 5th runner}} = 5 \cdot 4 \cdot 3 \cdot 2 \cdot 1 = 120.$

21. Since there are 6 main courses, there are 6 ways to choose a main course. Likewise, there are 8 drinks and 3 desserts so there are 8 ways to choose a drink and 3 ways to choose a dessert. So the possible number of different meals consisting of a main course, a drink, and a dessert is $\binom{\text{number of ways to}}{\text{choose the main course}} \cdot \binom{\text{number of ways to}}{\text{choose a drink}} \cdot \binom{\text{number of ways to}}{\text{choose a dessert}} = (6)(8)(3) = 144.$

23. By the Fundamental Counting Principle, the number of different routes from town A to town D via towns B and C is $\binom{\text{number of routes}}{\text{from A to B}} \cdot \binom{\text{number of routes}}{\text{from B to C}} \cdot \binom{\text{number of routes}}{\text{from C to D}} = (4)(5)(6) = 120.$

25. The number of possible sequences of heads and tails when a coin is flipped 5 times is $\binom{\text{number of possible}}{\text{outcomes on the 1st flip}} \cdot \binom{\text{number of possible}}{\text{outcomes on the 2nd flip}} \cdot \binom{\text{number of possible}}{\text{outcomes on the 3rd flip}} \cdot \binom{\text{number of possible}}{\text{outcomes on the 4th flip}} \cdot \binom{\text{number of possible}}{\text{outcomes on the 5th flip}} = (2)(2)(2)(2)(2) = 2^5 = 32.$ (Here there are only two choices, head or tails, for each flip.)

27. Since there are six different faces on each die, the number of possible outcomes when a red die and a blue die and a white die are tossed is
$\binom{\text{number of possible}}{\text{outcomes on the red die}} \cdot \binom{\text{number of possible}}{\text{outcomes on the blue die}} \cdot \binom{\text{number of possible}}{\text{outcomes on the white die}} = (6)(6)(6) = 6^3 = 216.$

29. The number of possible skirt-blouse-shoe outfits is
$\binom{\text{number of ways}}{\text{to choose a skirt}} \cdot \binom{\text{number of ways}}{\text{to choose a blouse}} \cdot \binom{\text{number of ways}}{\text{to choose shoes}} = (5)(8)(12) = 480.$

31. The number of possible ID numbers of 1-letter followed by 2 digits is
$\binom{\text{number of ways}}{\text{to choose a letter}} \cdot \binom{\text{number of ways}}{\text{to choose a digit}} \cdot \binom{\text{number of ways}}{\text{to choose a digit}} = (26)(10)(10) = 2600.$ Since $2600 < 2844$, it is not possible to give each of the company's employees a different ID number using this scheme.

33. The number of different California license plates possible is
$\binom{\text{number of ways to}}{\text{choose a nonzero digit}} \cdot \binom{\text{number of ways}}{\text{to choose 3 letters}} \cdot \binom{\text{number of ways}}{\text{to choose 3 digits}} = (9)(26^3)(10^3) = 158{,}184{,}000.$

35. There are two possible ways to answer each question on a true-false test. Thus the possible number of different ways to complete a true-false test containing 10 questions is
$\binom{\text{number of ways}}{\text{to answer question 1}} \cdot \binom{\text{number of ways}}{\text{to answer question 2}} \cdot \cdots \cdot \binom{\text{number of ways}}{\text{to answer question 10}} = 2^{10} = 1024.$

37. The number of different classifications possible is
$\binom{\text{number of ways}}{\text{to choose a major}} \cdot \binom{\text{number of ways to}}{\text{choose a minor}} \cdot \binom{\text{number of ways}}{\text{to choose a year}} \cdot \binom{\text{number of ways}}{\text{to choose a sex}} = (32)(32)(4)(2) = 8192.$ To see that there are 32 ways to choose a minor, we start with 32 fields, subtract the major (you can not major and minor in the same subject), and then add one for the possibility of *NO MINOR*.

39. The number of possible license plates of two letters followed by three digits is
$\binom{\text{number of ways to}}{\text{choose 2 letters}} \cdot \binom{\text{number of ways}}{\text{to choose 3 digits}} = (26^2)(10^3) = 676{,}000.$ Since $676,000 < 8,000,000$, there will not be enough different license plates for the state's 8 million registered cars.

41. Since a student can hold only one office, the number of ways that a president, a vice-president and a secretary can be chosen from a class of 30 students is
$\binom{\text{number of ways}}{\text{to choose a president}} \cdot \binom{\text{number of ways to}}{\text{choose a vice-president}} \cdot \binom{\text{number of ways}}{\text{to choose a secretary}} = (30)(29)(28) = 24{,}360.$

43. The number of ways a chairman, vice-chairman and a secretary can be chosen if the chairman must be a Democrat and the vice-chairman must be a Republican is
$\left(\begin{array}{c}\text{number of ways to}\\ \text{choose a Dem. chairman}\\ \text{from the 10 Dem.}\end{array}\right) \cdot \left(\begin{array}{c}\text{number of ways to choose}\\ \text{a Rep. vice-chairman}\\ \text{from the 7 Rep}\end{array}\right) \cdot \left(\begin{array}{c}\text{number of ways to choose}\\ \text{a secretary from the}\\ \text{remaining 15 members}\end{array}\right) = (10)(7)(15) = 1050.$

45. The possible number of 5-letter words formed using the letters A, B, C, D, E, F, and G is calculated as follows:

(a) If repetition of letters is allowed then there are 7 ways to pick each letter in the word. Thus the number of 5-letter words is
$\binom{\text{number of ways to}}{\text{choose the 1st letter}} \cdot \binom{\text{number of ways to}}{\text{choose the 2nd letter}} \cdot \cdots \cdot \binom{\text{number of ways to}}{\text{choose the 5th letter}} = 7^5 = 16{,}807.$

(b) If no letter can be repeated in a word, then the number of 5-letter words is
$\binom{\text{number of ways to}}{\text{choose the 1st letter}} \cdot \binom{\text{number of ways to}}{\text{choose the 2nd letter}} \cdot \cdots \cdot \binom{\text{number of ways to}}{\text{choose the 5th letter}} = (7)(6)(5)(4)(3) = 2520.$

(c) If each word must begin with the letter A, then the number of 5-letter words (letters can be repeated) is

$\binom{\text{number of ways to}}{\text{choose the letter A}} \cdot \binom{\text{number of ways to}}{\text{choose the 2nd letter}} \cdot \cdots \cdot \binom{\text{number of ways to}}{\text{choose the 5th letter}} = (1)(7)(7)(7)(7) = 7^4 = 2401.$

(d) If the letter C must be in the middle , then the number of 5-letter words (letters can be repeated) is

$\binom{\text{number of ways to}}{\text{choose the 1st letter}} \cdot \binom{\text{number of ways to}}{\text{choose the 2nd letter}} \cdot \binom{\text{number of ways to}}{\text{choose the letter C}} \cdot \binom{\text{number of ways to}}{\text{choose the 4th letter}} \cdot \binom{\text{number of ways to}}{\text{choose the 5th letter}}$
$= (7)(7)(1)(7)(7) = 7^4 = 2401.$

(e) If the middle letter must be a vowel, then the number of 5-letter words is $(7)(7)(2)(7)(7)$ $= 2 \cdot 7^4 = 4802.$

47. The number of possible variable names is
$\binom{\text{number of ways to}}{\text{choose a letter}} \cdot \binom{\text{number of ways to choose}}{\text{a letter or a digit}} = (26)(36) = 936.$

49. The number of ways 4 men and 4 women may be seated in a row of 8 seats is calculated as follows:

(a) If the women are to be seated together and the men together, then the number of ways they can be seated is $\binom{\text{number of ways to seat the}}{\text{women first and the men second}} + \binom{\text{number of ways to seat the}}{\text{the men first and the women second}}$
$= (4)(3)(2)(1)(4)(3)(2)(1) + (4)(3)(2)(1)(4)(3)(2)(1) = (4!)(4!) + (4!)(4!) = 576 + 576$
$= 1152.$

(b) If they are to be seated alternately by gender , then the number of ways they can be seated is
$\binom{\text{number of ways to seat them}}{\text{alternately if a woman is first}} + \binom{\text{number of ways to seat them}}{\text{alternately if a man is first}}$
$= (4)(4)(3)(3)(2)(2)(1)(1) + (4)(4)(3)(3)(2)(2)(1)(1) = 1152.$

51. The number of ways that 8 mathematics books and 3 chemistry books may be placed on a shelf if the math books are to be next to each other and the chemistry books next to each other is
$\binom{\text{number of ways to place them}}{\text{if the math books are first}} + \binom{\text{number of ways to place them}}{\text{if the chemistry books are first}} = (8!)(3!) + (3!)(8!) = 483{,}840.$

53. In order for a number to be odd, the last digit must be odd. In this case, it must be a 1. Thus, the other two digits must be chosen from the three digits 2, 4, and 6. So, the number of 3-digit odd numbers that can be formed using the digits 1, 2, 4, and 6 if no digit may be used more than once is
$\left(\begin{array}{c}\text{number of ways}\\ \text{to choose a digit}\\ \text{from the 3 digits}\end{array}\right) \cdot \left(\begin{array}{c}\text{number of ways}\\ \text{to choose from the}\\ \text{2 remaining digits}\end{array}\right) \cdot \left(\begin{array}{c}\text{number of ways}\\ \text{to choose}\\ \text{an odd digit}\end{array}\right) = (3)(2)(1) = 6.$

Exercises 8.2

1. $P(8,3) = \frac{8!}{(8-3)!} = \frac{8!}{5!} = 8 \cdot 7 \cdot 6 = 336$

3. $P(11,4) = \frac{11!}{(11-4)!} = \frac{11!}{7!} = 11 \cdot 10 \cdot 9 \cdot 8 = 7920$

5. $P(100,1) = \frac{100!}{(100-1)!} = \frac{100!}{99!} = 100$

7. $P(15,5) = \frac{15!}{(15-5)!} = \frac{15!}{10!} = 15 \cdot 14 \cdot 13 \cdot 12 \cdot 11 = 360,360$

9. $P(n,n) = \frac{n!}{(n-n)!} = \frac{n!}{0!} = n!$

11. Since the order of finish is important, we want the number of permutations of 8 objects (the contestants) taken three at a time, which is $P(8,3) = 336$. See Exercise 1 for the calculation.

13. The number of ways of ordering 6 distinct objects (the 6 people) is $P(6,6) = 6! = 720$.

15. The number of ways of selecting 3 objects in order (a 3-letter word) from 6 distinct objects (the 6 letters) is $6^3 = 216$. Notice that this exercise does *not* say letters cannot be repeated. If we assume that the letters cannot be repeated, then the solution would be $P(6,3) = 6 \cdot 5 \cdot 4 = 120$.

17. The number of ways of selecting 3 objects in order (a 3-digit number) from 4 distinct objects (the 4 digits) with no repetition of the digits is $P(4,3) = 4 \cdot 3 \cdot 2 = 24$.

19. The number of ways of ordering 9 distinct objects (the contestants) is $P(9,9) = 9! = 362{,}880$. Here a runner cannot finish more than once, so no repetitions are allowed and order is important.

21. The number of ways of ordering 1000 distinct objects (the contestants) taking 3 at a time is $P(1000,3) = 1000 \cdot 999 \cdot 998 = 997,002,000$. We are assuming that a person cannot win more than once, that is, no repetitions.

23. We first place Jack in the first seat and then seat the remaining 4 students. Thus the number of these arrangements is $\left(\substack{\text{number of ways to} \\ \text{seat Jack in the 1st seat}}\right) \cdot \left(\substack{\text{number of ways to seat} \\ \text{the remaining 4 students}}\right) = P(1,1) \cdot P(4,4) = 1!\,4!\ 24$

25. <u>Begin with the letter L:</u> We first place the letter L in the first position (slot), then place the remaining 3 distinct letters in the next 3 slots. So this number is
$\left(\substack{\text{number of ways to} \\ \text{place L in the first slot}}\right) \cdot \left(\substack{\text{number of ways to place} \\ \text{the remaining 3 letters}}\right) = P(1,1) \cdot P(3,3) = 1!\,3! = 6.$

    <u>End with the letter L:</u> We first place the letter L in the last position (slot), then place the remaining 3 distinct letters in the next 3 slots. So this number is
$\left(\substack{\text{number of ways to} \\ \text{place L in the last slot}}\right) \cdot \left(\substack{\text{number of ways to place} \\ \text{the remaining 3 letters}}\right) = P(1,1) \cdot P(3,3) = 1!\,3! = 6$

27. We can have a 1-digit whole number *or* a 2-digit whole number *or* a 3-digit whole number *or* a 4-digit whole number. Since each digit must be distinct, the number of possible different whole

numbers made from the 4 distinct digits is $\left(\substack{\text{number of}\\ \text{1-digit numbers}}\right) + \left(\substack{\text{number of}\\ \text{2-digit numbers}}\right) + \left(\substack{\text{number of}\\ \text{3-digit numbers}}\right)$ $+ \left(\substack{\text{number of}\\ \text{4-digit numbers}}\right) = P(4,1) + P(4,2) + P(4,3) + P(4,4) = 4 + 12 + 24 + 24 = 64.$

29. Here we have 6 letters, of which 3 are A's, 2 are B's, and 1 is a C. Thus the number of distinguishable permutations is $\dfrac{6!}{3!\,2!\,1!} = \dfrac{(6)(5)(4)(3!)}{(2)(3!)} = 60.$

31. Here we have 5 letters, of which 2 are A's, 1 is a B, 1 is a C, and 1 is a D. Thus the number of distinguishable permutations is $\dfrac{5!}{2!\,1!\,1!\,1!} = \dfrac{(5)(4)(3)(2!)}{2!} = 60.$

33. Here we have 6 object, of which 2 are blue marbles and 4 are red marbles. Thus the number of distinguishable permutations is $\dfrac{6!}{2!\,4!} = \dfrac{(6)(5)(4!)}{4!} = 15.$

35. The number of distinguishable permutations of 12 objects (the 12 coins), from like groups of size 4 (the pennies), of size 3 (the nickels), of size 2 (the dimes) and of size 3 (the quarters) is $\dfrac{12!}{4!\,3!\,2!\,3!} = 277,200.$

37. The word ELEEMOSYNARY has 12 letters of which 3 are E's, 2 are Y's, and the remaining letters are distinct. So this is an exercise of distinguishable permutations of 12 objects (the 12 letters) from like groups of size 3, 2, and seven of size 1 and the number is $\dfrac{12!}{3!\,2!\,1!\,1!\,1!\,1!\,1!\,1!\,1!} = 39{,}916{,}800.$

39. The number of distinguishable permutations of 12 objects (the 12 factors, $4+5+2+1$) from like groups of size 4 (the a's), of size 5 (the b's), of size 2 (the c's) and of size 1 (the d) is $\dfrac{12!}{4!\,5!\,2!\,1!}$ $= 83{,}160.$

41. This is the number of distinguishable permutations of 7 objects (the students) from like groups of size 3 (the ones who stay in the 3-person room, size 2 (the ones who stay in the 2-person room), size 1 (the one who stays in the 1-person room) and size 1 (the one who sleeps in the car). This number is $\dfrac{7!}{3!\,2!\,1!\,1!} = 420.$

43. The number of distinguishable permutations of 13 objects (the total number of blocks he must travel) which can be partitioned into like groups of size 8 (the east blocks) and of size 5 (the north blocks) is $\dfrac{13!}{8!\,5!} = 1{,}287.$

Exercises 8.3

1. $C(8,3) = \dfrac{8!}{3!\,(8-3)!} = \dfrac{8!}{3!\,5!} = \dfrac{8\cdot 7\cdot 6}{3\cdot 2\cdot 1} = 56$

3. $C(11,4) = \dfrac{11!}{4!\,7!} = \dfrac{11\cdot 10\cdot 9\cdot 8}{4\cdot 3\cdot 2\cdot 1} = 330$

5. $C(100,1) = \dfrac{100!}{1!\,99!} = \dfrac{100}{1} = 100$

7. $C(15,5) = \dfrac{15!}{5!\,10!} = \dfrac{15\cdot 14\cdot 13\cdot 12\cdot 11}{5\cdot 4\cdot 3\cdot 2\cdot 1} = 3003$

9. $C(n,n) = \dfrac{n!}{n!\,(n-n)!} = \dfrac{n!}{n!\,0!} = 1$

11. We want the number of ways of choosing a group of three from a group of six. This number is $C(6,3) = \dfrac{6!}{3!\,3!} = 20$

13. We want the number of ways of choosing a group of six people from a group of ten people. The number of combinations of 10 objects (10 people) taken 6 at a time is $C(10,6) = \dfrac{10!}{6!\,4!} = 210$.

15. We want the number of ways of choosing a group (the 5-card hand) where order of selection is not important. The number of combinations of 52 objects (the 52 cards) taken 5 at a time is $C(52,5) = \dfrac{52!}{5!\,47!} = 2{,}598{,}960$.

17. The order of selection is not important, hence the number of combinations of 10 objects (the 10 questions) taken 7 at a time is $C(10,7) = \dfrac{10!}{7!\;3!} = 120$.

19. We assume that the order in which he plays the pieces in the recital is *not* important, so the number of combinations of 12 objects (the 12 pieces) taken 8 at a time is $C(12,8) = \dfrac{12!}{8!\,4!} = 495$.

21. We are just interest in the group of seven students taken from the class of 30 students. Thus the number is $C(30,7) = \dfrac{30!}{7!\,23!} = 2{,}035{,}800$.

23. We first take Jack out of the class of 30 students and select the 7students from the remaining 29 students. Thus there are $C(29,7) = \dfrac{29!}{7!\,22!} = 1{,}560{,}780$ ways to pick the 7 students for the field trip.

25. We count the number of different ways to choose the group of 6 numbers from the 53 numbers, thus we get that there is $C(53,6) = \dfrac{53!}{6!\,47!} = 22{,}957{,}480$ tickets. So it would cost $22,957,480.

27. (a) The number of ways of choosing 5 students from the 20 students is $C(20{,}5) = \dfrac{20!}{5!\,15!}$ $= 15{,}504$.

(b) The number of ways of choosing 5 students for the committee from the 12 females is $C(12,5) = \dfrac{12!}{5!\,7!} = 792.$

(c) We use the Fundamental Counting Principle to count the number of possible committees with 3 females and 2 males. Thus we get is $\left(\substack{\text{number of ways to choose the}\\ \text{3 females from the 12 females}}\right) \cdot \left(\substack{\text{number of ways to choose the}\\ \text{2 males from the 8 males}}\right) = C(12,3) \cdot C(8,2) = (220)(28) = 6160.$

29. We are interested in the number of distinguishable combinations taken from the set of six objects consisting of 3 A's, 2 B's, and 1 C. This number is $(3+1)(2+1)(1+1) = (4)(3)(2) = 24.$

31. We are interested in the number of distinguishable combinations taken from the set of five objects consisting of 2 A's, 1 B, 1 C, and 1 D. This number is $(2+1)(1+1)(1+1)(1+1) = (3)(2)(2)(2) = 24.$

33. (a) The number of ways to select 5 of the 8 objects is $C(8,5) = \dfrac{8!}{5!\,3!} = 56.$

(b) A set with 8 elements has $2^8 = 256$ subsets.

35. The number of distinguishable combinations taken from a set of 7 objects (the 7 coins) consisting of like groups of size 4 (the pennies) and 3 like groups of size 1 (the nickel, dime and quarter) is $(4+1)(1+1)(1+1)(1+1) = (5)(2)(2)(2) = 40.$ If Manual had 14 pennies then several combinations would give the same sum of money. For example, 7 pennies is the same *sum* as 2 pennies and 1 nickel.

37. Each divisor of 540 corresponds to a subset (distinguishable combination) of the prime factors of 540. Since $540 = 2^2 \cdot 3^3 \cdot 5^1$, the number of distinguishable combinations of 6 objects (the 6 prime factors) taken from like groups of size 2 (the 2's), of size 3 (the 3's) and of size 1 (the 5) is $(2+1)(3+1)(1+1) = (3)(4)(2) = 24.$

39. The number of distinguishable combinations of 8 objects (the 8 brochures) where 0, 1 or 2 may be chosen of each kind is $(3)(3)(3)(3)(3)(3)(3)(3) = 3^8 = 6561.$

41. We consider a set of 20 objects (the shoppers in the mall) and a subset corresponds to those shoppers that enter the store. Since a set of 20 objects has $2^{20} = 1{,}048{,}576$ subsets, there are $1,048,576$ outcomes to their decisions.

43. (a) $C(100,3) = \dfrac{100!}{3!\,97!} = \dfrac{100 \cdot 99 \cdot 98}{3 \cdot 2 \cdot 1} = 161{,}700$ and

$$C(100{,}97) = \frac{100!}{97!\,3!} = \frac{100 \cdot 99 \cdot 98}{3 \cdot 2 \cdot 1} = 161{,}700$$

(b) $C(n,r) = \dfrac{n!}{r!\,(n-r)!} = \dfrac{n!}{(n-r)!\,r!} = \dfrac{n!}{(n-r)!\,[n-(n-r)]!} = C(n, n-r)$

(c) Every subset of r objects chosen from a set of n objects determines a corresponding set of $(n-r)$ objects, namely, those not chosen. Therefore, the total number of combinations for each type are equal.

Exercises 8.4

1. If we consider the 2 people who insist on standing together as one object, then we need the number of ways of permuting 9 objects. We then multiply this by the number of ways of arranging the two people that insist on standing together. Thus the number of ways the ten people can be arranged in a row (with the two standing next to each other) is

$$\begin{pmatrix}\text{number of ways}\\ \text{the 10 people}\\ \text{can be arranged}\end{pmatrix} = \begin{pmatrix}\text{number of ways}\\ \text{that 9 objects}\\ \text{can be permuted}\end{pmatrix} \cdot \begin{pmatrix}\text{number of ways}\\ \text{the 2 picky people}\\ \text{can be arranged}\end{pmatrix} = P(9,9) \cdot P(2,2) = 362880 \times 2$$

$= 725{,}760$

3. If we consider the 4 math books as one object, then we need the number of ways of permuting 5 objects. We then multiply this by the number of ways of arranging the 4 math books. So the number of ways the 8 books can be arranged is $\begin{pmatrix}\text{number of ways}\\ \text{that 5 objects}\\ \text{can be permuted}\end{pmatrix} \cdot \begin{pmatrix}\text{number of ways}\\ \text{the 4 math books}\\ \text{can be arranged}\end{pmatrix} = P(5,5) \cdot P(4,4)$

$= 5! \times 4! = 2880$

5. It is easier to count the complement, where these two runners are placed next to each other, then subtract this number from the number of ways to arrange all 6 runners. As in previous exercises, we place the 2 runners together and treat them as one object and then arrange 5 objects. So the number of ways where these 2 runners are next to each other is $P(5,5) \cdot P(2,2) = 5! \cdot 2 = 240$. The number of ways to arrange 6 objects is $P(6,6) = 6! = 720$. So the number of ways to arrange the 6 runners so that these 2 runners are not placed next to each other is $720 - 240 = 480$.

7. (a) Because this necklace has no clasp, the necklace has no orientation. That is, when you turn the necklace over this does not yield a new arrangement. Thus each arrangement is counted twice when we just count the number of ways to arrange the 12 stones around the necklace. So the number of different necklaces is the number of ways 12 objects can be arranged (in a non-orientable circle) which is $\frac{11!}{2} = 19{,}958{,}400$.

   (b) The clasp distinguishes a beginning and an end and it also distinguishes an orientation. So this is different from part (a) where the stones are in a circle without orientation. So the number of different necklaces is the number of ways 12 objects can be arranged in a row, which is $12! = 479{,}001{,}600$. Note: we are assuming that there is a difference between the different sides of the clasp.

9. (a) We select the 3 welders from the group of 7 welders and the 2 machinists from the group of 6 machinists. Since order of selection is not important we use combinations, so the number of ways the workers can be assigned is $C(7,3) \cdot C(6,2) = 35 \times 15 = 525$

   (b) The complement of *at least one welder* is NO welder. So we count the number of ways to select 5 machinist from the group of 6 machinists and subtract this number from the number of ways to select 5 workers from the group of 13 workers. Thus the number of ways that the workers can be assigned with at least one welder is $C(13,5) - C(6,5) = 1287 - 6 = 1281$.

   (c) *At most two welders* means we can have 0 welders and 5 machinist or 1 welders and 4 machinist or 2 welders and 4 machinist. So the number of ways to select 5 worker with at most 2 welders is

$C(7,0)\cdot C(6,5)+C(7,1)\cdot C(6,4)+C(7,2)\cdot C(6,3)=(1)(6)+(7)(15)+(21)(20)$
$=6+105+420=531.$

11. The number of ways the committee can be chosen is
$\left(\begin{smallmatrix}\text{number of ways to}\\ \text{choose 2 of 6 freshmen}\end{smallmatrix}\right)\cdot\left(\begin{smallmatrix}\text{number of ways to}\\ \text{choose 3 of 8 sophomores}\end{smallmatrix}\right)\cdot\left(\begin{smallmatrix}\text{number of ways to}\\ \text{choose 4 of 12 juniors}\end{smallmatrix}\right)\cdot\left(\begin{smallmatrix}\text{number of ways to}\\ \text{choose 5 of 10 seniors}\end{smallmatrix}\right)$
$=C(6,2)\cdot C(8,3)\cdot C(12,4)\cdot C(10,5)=15\cdot 56\cdot 495\cdot 252=104{,}781{,}600$

13. We count the number of arrangements where Mike and Molly are standing together and subtract the arrangements in which John and Jane are also standing together. If we consider Mike and Molly as one object, then we need the number of ways of arranging 9 objects. We then need to arrange Mike and Molly. Thus the number of arrangements where Mike and Molly are standing together is $P(9,9)\cdot P(2,2)=9!\cdot 2!=725,760$. To count the number of arrangements where Mike and Molly stand together and John and Jane also stand together, we treat John and Jane as one object and Mike and Molly as one object and arrange the 8 objects. So there are $P(8,8)\cdot P(2,2)\cdot P(2,2)=8!\cdot 2!\cdot 2!=161,280$ ways to do this. (Remember, we still needed to arrange John and Jane within their group and Mike and Molly within their group.) Hence the number of ways to arrange the students is $725,760-161,280=564,480$.

15. We find the complement where these two children must hold hands. Again, we treat these two children as one object and arrange the 6 objects in a circle in $P(2,2)\cdot 5!=240$ ways. The number of ways to arrange the 7 children in a circle is $6!=720$. So the number of circles that can be formed where these children do not hold hands is $720-240=480$.

17. Here we are only concerned with dividing the 12 children into two groups, each with 3 boys and 3 girls. Each time we make a group of 3 boys and 3 girls we make a corresponding group of 3 boys and 3 girls. So when we count the number of ways to make the first group, we count the partition into two groups twice, once when we count the first group of 3 boys and 3 girls and once when we count the second group of 3 boys and 3 girls. As a result, we must divide this count by 2. So the number of ways to divide the 12 children into two groups, each with 3 boys and 3 girls, is $\frac{1}{2}\cdot C(6,3)\cdot C(6,3)=\left(\frac{1}{2}\right)(20)(20)=200$.

19. The complement of *at least one is played out of order* is *all eight sides are played in order*. The number of ways to play all eight sides in order is 1. The number of ways to play the 8 sides is $P(8,8)=8!$. Hence the number of ways to play the four album set so that at least one is played out of order is $8!-1=40,319$.

21. We find the number of ways to select the ice cream and the number of ways to select the toppings then use the Fundamental Counting Principle to find the number of banana splits that are possible. For determining the number of ways to select the three scoops of ice cream, we consider three cases:
$\left(\begin{smallmatrix}\text{all three scoops}\\ \text{are different}\end{smallmatrix}\right)+\left(\begin{smallmatrix}\text{two scoops from one flavor}\\ \text{one scoop from a different flavor}\end{smallmatrix}\right)+\left(\begin{smallmatrix}\text{all three scoops}\\ \text{from the same flavor}\end{smallmatrix}\right)$
$=\left(\begin{smallmatrix}\text{pick 3 of}\\ \text{the 12 flavors}\end{smallmatrix}\right)+\left(\begin{smallmatrix}\text{pick the flavor}\\ \text{for two of the scoops}\end{smallmatrix}\right)\cdot\left(\begin{smallmatrix}\text{pick the flavor}\\ \text{for the third scoop}\end{smallmatrix}\right)+\left(\begin{smallmatrix}\text{pick the flavor}\\ \text{for the three scoops}\end{smallmatrix}\right)$
$=C(12,3)+C(12,1)\cdot C(11,1)+C(12,1)=220+(12)(11)+12=364.$
Since the toppings can also be repeated the reasoning is similar in determining the number of ways to select the toppings. So the number of ways is
$C(8,3)+C(8,1)\cdot C(7,1)+C(8,1)=56+(8)(7)+8=120.$
Thus there are $364\cdot 120=43,680$ possible banana splits.

23. (a) To find the number of ways the men and women can be seated we first select and place a man in the first seat and then arrange the other 7 people. Thus we get
$\binom{\text{select 1 of}}{\text{the 4 men}} \cdot \binom{\text{arrange the}}{\text{remaining 7 people}} = C(4,1) \cdot P(7,7) = 4 \cdot 7! = 20,160.$

(b) To find the number of ways the men and women can be seated we first select and place a woman in the first and last seats and then arrange the other 6 people. Thus we get
$\binom{\text{arrange 2 of}}{\text{the 4 women}} \cdot \binom{\text{arrange the}}{\text{remaining 6 people}} = P(4,2) \cdot P(6,6) = 6 \cdot 6! = 8.640.$

25. (a) There are two ways to view this exercise, both yield the same number. In the first method we first think of the women as one object and arrange the 5 objects in a circle, then arrange the 4 women. This gives $4! \cdot P(4,4) = 4! \cdot 4! = 576$ ways. In the second method we arrange the women and anchor them down in the circle, then arrange the men. Again this yields $P(4,4) \cdot P(4,4) = 4! \cdot 4! = 576$ ways.

(b) Using the idea from Example 3, we first pick a woman as an anchor in the circle and arrange the remaining women in every other seat, then arrange the 4 men in the remaining seats. Thus the number of ways to alternately seat these 4 women and 4 men around the circle is $P(3,3) \cdot P(4,4) = 3! \cdot 4! = (6)(24) = 144$.

27. We partition the group into pairs by selecting groups of 2 without replacement, then we divide by the number of ways to arrange the 4 pairs, because the order in which the pairs are select is not important.. Thus we obtain $\dfrac{C(8,2) \cdot C(6,2) \cdot C(4,2) \cdot C(2,2)}{P(4,4)} = 105$ ways.

29. There are many different possibilities here, so we consider the complement where NO professors are chosen for the delegates and subtract this number form the way to select 3 people from the group of 8 people which is $C(8,3)$. If the professor cannot not be select, then we must select 3 people from a group of 5 and this can be done in $C(5,3)$ ways. Thus the number of delegations that contain a professor is $C(8,3) - C(5,3) = 56 - 10 = 46$.

31. Since two points determine a line and no three points are collinear, the exercise is how many ways can you select 2 of the 12 dots. Thus the number of possible lines is $C(12,2) = 66$.

33. Since a diagonal connects two *non-adjacent* vertices of the dodecagon, we count all the ways to chose 2 of the 12 vertices and subtract those that correspond to the sides of the dodecagon. This the are 12 sides we have $C(12,2) - 12 = 66 - 12 = 54$ diagonals.

35. (a) We assume that at least one flag must be hoisted to make a signal. Thus the number of signals using at most 3 flags is $\binom{\text{number of signals}}{\text{using 3 flags}} + \binom{\text{number of signals}}{\text{using 2 flags}} + \binom{\text{number of signals}}{\text{using 1 flag}}$
$= P(5,3) + P(5,2) + P(5,1) = 85$

(b) The number of signals using at least 4 flags is $\binom{\text{number of signals}}{\text{using 4 flags}} + \binom{\text{number of signals}}{\text{using 5 flags}}$
$= P(5,4) + P(5,5) = 5! + 5! = 240$

37. We arrange the two who must sit in the first row in the 4 seats available, next arrange the person who must sit in the second row in one of the three seats, then place the three in the available seats in the last row. Finally, we arrange the remaining 7 people in the remaining seats. Hence we obtain that there are $P(4,2) \cdot P(3,1) \cdot P(6,3) \cdot P(7,7) = 21,772,800$ ways to seat these people.

39. We consider both cases. If Jane chooses a mystery, then John has $C(6{,}2) \cdot C(8{,}3) = 840$ choices. If Jane chooses a biography, then John has $C(7{,}2) \cdot C(7{,}3) = 735$ choices. So John has more choices if Jane chooses a mystery.

41. We count the number of ways to get 5 cards from the same suit and remove those that contain a sequence (straight-flushes). To get 5 cards from the same suit we first select a suit and then 5 of the 13 cards from the suit. This can be done in $C(4,1) \cdot C(13,5) = 5,148$ ways. To find the number of straights, 5 card sequences, we look at the number of ways to select the low card in the sequence. The lowest straight is Ace-2-3-4-5 and the highest straight is 10-Jack-Queen-King-Ace. So there are 10 sequences of numbers and 4 ways to select the suit, so 40 straight-flushes. Thus the number of regular flushes is $5,148 - 40 = 5,108$.

Exercises 8.5

1. Let H stand for head and T for tails.
   (a) $S = \{HH, HT, TH, TT\}$
   (b) Let E be the event of getting exactly two heads. So $E = \{HH\}$. Then $P(E) = \dfrac{n(E)}{n(S)} = \dfrac{1}{4}$.
   (c) Let F be the event of getting at least one head. Then $F = \{HH, HT, TH\}$ and $P(F) = \dfrac{n(F)}{n(S)} = \dfrac{3}{4}$.
   (d) Let G be the event of getting exactly one head, $G = \{HT, TH\}$. So $P(G) = \dfrac{n(G)}{n(S)} = \dfrac{2}{4} = \dfrac{1}{2}$.

3. (a) Let E be the event of rolling a six. Then $P(E) = \dfrac{n(E)}{n(S)} = \dfrac{1}{6}$.
   (b) Let F be the event of rolling an even number, $F = \{2, 4, 6\}$. Then $P(F) = \dfrac{n(F)}{n(S)} = \dfrac{3}{6} = \dfrac{1}{2}$.
   (c) Let G be the event of rolling a number greater than 5. Since 6 is the only face greater than 5, $P(G) = \dfrac{n(G)}{n(S)} = \dfrac{1}{6}$.

5. (a) Let E be the event of choosing a king. Since a deck has 4 kings, $P(E) = \dfrac{n(E)}{n(S)} = \dfrac{4}{52} = \dfrac{1}{13}$.
   (b) Let F be the event of choosing a face card. Since there are 3 face cards per suit and 4 suits, then $P(F) = \dfrac{n(F)}{n(S)} = \dfrac{12}{52} = \dfrac{3}{13}$.
   (c) Let F be the event of choosing a face card. Then $P(F') = 1 - P(F) = 1 - \frac{3}{13} = \frac{10}{13}$.

7. (a) Let E be the event of selecting a red ball. Since the jar contains 5 red balls, $P(E) = \dfrac{n(E)}{n(S)} = \dfrac{5}{8}$.
   (b) Let F be the event of selecting a yellow ball. Since there is only one yellow ball, $P(F') = 1 - P(F) = 1 - \dfrac{n(F)}{n(S)} = 1 - \dfrac{1}{8} = \dfrac{7}{8}$.
   (c) Let G be the event of selecting a black ball. Since there are no black balls in the jar, $P(G) = \dfrac{n(G)}{n(S)} = \dfrac{0}{8} = 0$.

9. Let E be the event of rolling a number greater than 8 with a dodecahedral die. So $E = \{9, 10, 11, 12\}$ and $P(E) = \dfrac{n(E)}{n(S)} = \dfrac{4}{12} = \dfrac{1}{3}$.

11. (a) Let E be the event of choosing a number divisible by 3. Then $E = \{3, 6, 9, \ldots, 99\}$ and $P(E) = \frac{33}{100} = 0.33$.

(b) Let E be the event of choosing a number divisible by 3. Then
$P(E') = 1 - P(E) = 1 - \frac{33}{100} = \frac{67}{100} = 0.67$.

(c) Let F be the event of choosing a number divisible by 3 and by 5, or equivalently of choosing a number divisible by 15. Then $F = \{15, 30, 45, 60, 75, 90\}$ and $P(F) = \frac{6}{100} = 0.06$.

13. (a) Let E be the event of choosing a "T". Since 3 of the 16 letters are T's, $P(E) = \frac{3}{16}$.

(b) Let F be the event of choosing a vowel. Since there are 6 vowels, $P(F) = \frac{6}{16} = \frac{3}{8}$.

(c) Let F be the event of choosing a vowel. Then $P(F') = 1 - \frac{3}{8} = \frac{5}{8}$.

15. Let E be the event of choosing 5 cards of the same suit. Since there are 4 suits and 13 cards in each suit, then $n(E) = 4 \cdot C(13,5)$. Also by Exercise 15, Section 8.3, $n(S) = C(52,5)$. Therefore
$$P(E) = \frac{4 \cdot C(13,5)}{C(52,5)} = \frac{5,148}{2,598,960} \approx 0.00198.$$

17. Let E be the event of choosing four of a kind. To determine $n(E)$ we first choose a kind, then 4 cards from that kind, and finally one additional card. Thus $n(E) = C(13,1) \cdot C(4,4) \cdot C(48,1)$ $= 13 \cdot 1 \cdot 48 = 624$. Therefore $P(E) = \dfrac{624}{C(52,5)} = \dfrac{1}{4165} \approx 0.00024$.

19. Let E be the event of choosing 3 hearts and 2 diamonds. Since there are 13 hearts and 13 diamonds, then $n(\text{E}) = C(13,3) \cdot C(13,2)$. Therefore
$$P(E) = \frac{C(13,3) \cdot C(13,2)}{C(52,5)} = \frac{286 \cdot 78}{2,598,960} = \frac{22,308}{2,598,960} \approx 0.008583.$$

21. Let E be the event of choosing at most 4 hearts. Then E' is the event of choosing 5 hearts. Thus
$$P(E) = 1 - P(E') = 1 - \frac{C(13,5)}{C(52,5)} = 1 - \frac{1287}{2,598,960} \approx 1 - 0.000495 \approx 0.999505.$$

23. (a) Let B stand for "boy" and G stand for "girl". Then $S = \{$BBBB, GBBB, BGBB, BBGB, BBBG, GGBB, GBGB, GBBG, BGGB, BGBG, BBGG, BGGG, GBGG, GGBG, GGGB, GGGG$\}$.

(b) Let E be the event that the couple has only boys. Then $E = \{$BBBB$\}$ and $P(E) = \frac{1}{16}$.

(c) Let F be the event that the couple has 2 boys and 2 girls. Then $F = \{$GGBB, GBGB, GBBG, BGGB, BGBG, BBGG$\}$ so $P(F) = \frac{6}{16} = \frac{3}{8}$.

(d) Let G be the event that the couple has 4 children of the same sex. Then $G = \{$BBBB, GGGG$\}$ and $P(G) = \frac{2}{16} = \frac{1}{8}$.

(e) Let H be the event that the couple has at least 2 girls. Then H' is the event that the couple has fewer than two girls. Thus $H' = \{$BBBB, GBBB, BGBB, BBGB, BBBG$\}$ so $n(H') = 5$ and $P(H) = 1 - P(H') = 1 - \frac{5}{16} = \frac{11}{16}$.

25. Let E be the event that the ball lands in an odd numbered slot. Since there are 18 odd numbers between 1 and 36, $P(E) = \frac{18}{38} = \frac{9}{19}$.

27. Let E be the event of picking the 6 winning numbers. Since there is only one winning ticket,
$$P(E) = \frac{1}{C(49,6)} = \frac{1}{13,983,816} \approx 7.15 \times 10^{-8}.$$

29. Let E be the event that she opens the lock within an hour. The number of combinations she can try in one hour is $10 \cdot 60 = 600$. The number of possible combinations is $P(40,3)$ assuming that no number can be repeated. Thus $P(E) = \dfrac{600}{P(40,3)} = \dfrac{600}{59,280} = \dfrac{5}{494} \approx 0.010$.

31. The sample space consist of all possible True-False combinations, so $n(S) = 2^{10}$.

(a) Let E be the event that the student answers all 10 questions correctly. Since there is only one way to answer all 10 questions correctly, $P(E) = \frac{1}{2^{10}} = \frac{1}{1024}$.

(b) Let F be the event that the student answers exactly 7 questions correctly. The number of ways to answer exactly 7 of the 10 questions correctly is the number of ways to choose 7 of the 10 questions, so $n(F) = C(10,7)$. Therefore $P(E) = \dfrac{C(10,7)}{2^{10}} = \dfrac{120}{1024} = \dfrac{15}{128}$.

33. Let E be the event that Paul stands next to Phyllis. To find $n(E)$ we treat Paul and Phyllis as one object and find the number of ways to arrange the 19 objects and then multiply the result by the number of ways to arrange Paul and Phyllis. So $n(E) = 19! \cdot 2!$. The sample space is all the ways that 20 people can be arranged. Thus $P(E) = \dfrac{19! \cdot 2!}{20!} = \dfrac{2}{20} = 0.10$.

35. (a) Let E be the event that the monkey types "Hamlet" as his first word. Since "Hamlet" contains 6 letters and there are 48 typewriter keys, then $P(E) = \dfrac{1}{48^6} \approx 8.18 \times 10^{-11}$.

(b) Let F be the event that the monkey types "to be or not to be" as his first words. Since this phrase has 18 characters (including the blanks), then $P(F) = \dfrac{1}{48^{18}} \approx 5.47 \times 10^{-31}$.

37. Let E be the event that the monkey will arrange the 11 blocks to spell "PROBABILITY" as his first word. The number of ways of arranging these blocks is the number of distinguishable permutations of 11 blocks. Since there are two blocks labeled `B' and two blocks labeled `I', the number of distinguishable permutations is $\frac{11!}{2!\,2!}$. Only one of these arrangements spells the word "PROBABILITY', thus $P(E) = \dfrac{1}{\frac{11!}{2!\,2!}} = \dfrac{2!\,2!}{11!} \approx 1.00 \times 10^{-7}$.

39. (a) Let E be the event that the pea is tall. Since tall is dominant, $E = \{\text{TT, Tt, tT}\}$. So $P(E) = \frac{3}{4}$.

(b) Then E' is the event that the pea is short. So $P(E') = 1 - P(E) = 1 - \frac{3}{4} = \frac{1}{4}$.

Exercises 8.6

1. (a) YES, the events are mutually exclusive since a person cannot be both male and female.
   (b) NO, the events are not mutually exclusive since a person can be both tall and blond.

3. (a) YES, the events are mutually exclusive since the number cannot be both even and odd. So $P(E \cup F) = P(E) + P(F) = \frac{3}{6} + \frac{3}{6} = 1$.
   (b) NO, the events are not mutually exclusive since 6 is both even and greater than 4. So $P(E \cup F) = P(E) + P(F) - P(E \cap F) = \frac{3}{6} + \frac{2}{6} - \frac{1}{6} = \frac{2}{3}$.

5. (a) NO, the events E and F are not mutually exclusive since the Jack, Queen, and King of spades are both face cards and spades. So $P(E \cup F) = P(E) + P(F) - P(E \cap F) = \frac{13}{52} + \frac{12}{52} - \frac{3}{52} = \frac{11}{26}$.
   (b) YES, the events E and F are mutually exclusive since the card cannot be both a heart and a spade. So $P(E \cup F) = P(E) + P(F) = \frac{13}{52} + \frac{13}{52} = \frac{1}{2}$.

7. (a) Let E be the event that the spinner stops on red. Since 12 of the regions are red, $P(E) = \frac{12}{16} = \frac{3}{4}$.
   (b) Let F be the event that the spinner stops on an even number. Since 8 of the regions are even-numbered, $P(F) = \frac{8}{16} = \frac{1}{2}$.
   (c) Since 4 of the even-numbered regions are red $P(E \cup F) = P(E) + P(F) - P(E \cap F)$ $= \frac{3}{4} + \frac{1}{2} - \frac{4}{16} = 1$.

9. Let E be the event that the ball lands in an odd numbered slot and F be the event that it lands in a slot with a number higher than 31. Then $P(E \cup F) = P(E) + P(F) - P(E \cap F)$ $= \frac{18}{38} + \frac{5}{38} - \frac{2}{38} = \frac{21}{38}$.

11. Let E be the event of arranging the letters to spell "TRIANGLE" and F be the event of arranging them to spell "INTEGRAL". There are P(8, 8) possible ways of arranging the blocks and only one way to spell each of these words. Also, these events are mutually exclusive since the blocks cannot spell both words at the same time, so
$$P(E \cup F) = P(E) + P(F) = \frac{1}{P(8,8)} + \frac{1}{P(8,8)} = \frac{2}{8!} = \frac{1}{20160} \approx 4.96 \times 10^{-5}.$$

13. Let E be the event that the committee is all male and F the event it is all female. The sample space is the set of all ways that 5 people can be chosen from the group of 14. These events are mutually exclusive, so $P(E \cup F) = P(E) + P(F) = \dfrac{C(6,5)}{C(14,5)} + \dfrac{C(8,5)}{C(14,5)} = \dfrac{6+56}{2002} = \dfrac{31}{1001}$.

15. Let E be the event that the couple has 3 girls and F the event they have 4 girls. These events are mutually exclusive, so $P(E \cup F) = P(E) + P(F) = \dfrac{C(4,3)}{2^4} + \dfrac{1}{2^4} = \dfrac{5}{16}$. (See Exercise 23, Section 8.5 for the sample space and events.)

17. Let E be the event that the marble is red and F be the event that the number is odd-numbered. Then E' is the event that the marble is blue and F' is the event that the marble is even-numbered.
   (a) $P(E) = \frac{6}{16} = \frac{3}{8}$
   (b) $P(F) = \frac{8}{16} = \frac{1}{2}$

(c) $P(E \cup F) = P(E) + P(F) - P(E \cap F) = \frac{6}{16} + \frac{8}{16} - \frac{3}{16} = \frac{11}{16}$

(d) $P(E' \cup F') = P(E') + P(F') - P(E' \cap F') = \frac{10}{16} + \frac{8}{16} - \frac{5}{16} = \frac{13}{16}$.

Exercises 8.7

1. (a) YES, because the first flip does not influence the outcome of the second flip.
   (b) $P(E \cap F) = P(E) \cdot P(F) = \left(\frac{1}{2}\right)\left(\frac{1}{2}\right) = \frac{1}{4}$

3. Let E be the event that the die shows a 5 and let F be the event that it shows a number greater than 3. Since there are 3 faces greater than 3, $P(E \mid F) = \dfrac{n(E \cap F)}{n(F)} = \dfrac{1}{3}$.

5. Let E be the event that the card is a queen and let F be the event that it is a face card. Since 4 of the 12 face cards are queens, $P(E \mid F) = \dfrac{n(E \cap F)}{n(F)} = \dfrac{4}{12} = \dfrac{1}{3}$.

7. Let E be the event that the card is a spade and let F be the event that it is a king. Since only one of the 4 kings is also a spade, $P(E \mid F) = \dfrac{n(E \cap F)}{n(F)} = \dfrac{1}{4}$.

9. Let E be the event that the spinner stops on an even-numbered region and let F be the event that the spinner stops on red region. So $P(E \mid F) = \dfrac{n(E \cap F)}{n(F)} = \dfrac{4}{12} = \dfrac{1}{3}$.

11. (a) YES. What happens on spinner A does not influence what happens on spinner B.
    (b) $P(E \cap F) = P(E) \cdot P(F) = \left(\frac{2}{4}\right)\left(\frac{2}{8}\right) = \frac{1}{8}$

13. (a) After the red ball is removed from the jar, the jar then contains 4 red balls and 7 green balls. Thus the probability that the second is red given that the first is red is $\frac{4}{11}$.
    (b) After the green ball is removed from the jar, the jar then contains 5 red balls and 6 green balls. Thus the probability that the second is red given that the first is green is $\frac{5}{11}$.
    (c) The jar starts with 7 odd-numbered balls and 5 even-numbered balls. Since the first chosen ball is odd-numbered, after it is chosen the jar contains 6 odd-numbered balls and 5 even-numbered balls. So the probability that the second has an even number given that the first has an odd number is $\frac{5}{11}$.
    (d) Using the information from part (c), after the even-numbered ball is removed the jar contains 7 odd-numbered balls and 4 even-numbered balls. So the probability that the second has an even number given that the first has an even number is $\frac{4}{11}$.

15. Let E be the event that the sum of the tosses is less than 6 and let F be the event that the first toss is a 1. Since the first toss is a 1, the second toss must be less than 5. So $P(E \mid F) = P(2^{\text{nd}} \text{ toss is less than 5}) = \frac{4}{6} = \frac{2}{3}$.

17. Let E be the event of getting a 1 on the first toss and let F be the event of getting an even number on the second toss. Since these events are independent, $P(E \cap F) = P(E) \cdot P(F) = \frac{1}{6} \cdot \frac{3}{6} = \frac{1}{12}$.

19. (a) Let E be the event that the first card is an ace and let F be the event that the second card is an ace. Now $P(E) = \frac{4}{52}$ and $P(F \mid E) = \frac{3}{51}$ since after the first ace is removed there are 51 cards left of which 3 are aces. Thus $P(E \cap F) = P(E) \cdot P(F \mid E) = \frac{4}{52} \cdot \frac{3}{51} = \frac{1}{221}$.
    (b) Let E be the event that the first card is the ace of spades and let F be the event that the second card is a spade. Now $P(E) = \frac{1}{52}$ and $P(F \mid E) = \frac{12}{51}$ since after the ace of spades is removed

there are 51 cards left of which 12 are spades. SO $P(E \cap F) = P(E) \cdot P(F \mid E) = \frac{1}{52} \cdot \frac{12}{51} = \frac{1}{221}$.

21. Let E be the event that the player wins on spin 1 and let F be the event that the player wins on spin 2. What happens on the first spin does not influence what happens on the second spin, so the events are independent. Thus $P(E \cap F) = P(E) \cdot P(F) = \frac{1}{38} \cdot \frac{1}{38} = \frac{1}{1444}$.

23. Let E, F and G denote the events of rolling two ones on the first, second, and third rolls, respectively, of a pair of dice. The events are independent, so $P(E \cap F \cap \mathrm{G}) = P(E) \cdot P(F) \cdot P(\mathrm{G}) = \frac{1}{36} \cdot \frac{1}{36} \cdot \frac{1}{36} = \frac{1}{36^3} \approx 2.14 \times 10^{-5}$.

25. (a) Let E and F be the events that the first child is a boy and that the second child is a boy, respectively. The events are independent, so $P(F \mid E) = P(F) = \frac{1}{2}$.

(b) Let F be the event that the couple has at least one boy and let E be the event that the couple has two boys. There are 3 possibilities in which F can occur, BB, BG, and GB, and of these 3, only 1 in which E can occur. Therefore, $P(E \mid F) = \dfrac{n(E \cap F)}{n(F)} = \dfrac{1}{3}$.

27. The probability that the first is black is $\frac{4}{7}$, the probability that the second is white is $\frac{3}{6}$, the probability that the third is black is $\frac{3}{5}$, and so on. Thus the probability that the first is black and that the second is white and that the third is black, and so on, is $\frac{4}{7} \cdot \frac{3}{6} \cdot \frac{3}{5} \cdot \frac{2}{4} \cdot \frac{2}{3} \cdot \frac{1}{2} \cdot \frac{1}{1} = \frac{1}{35}$.

29. (a) The probability that the first wheel has a bar is $\frac{1}{11}$, and similarly for the second and third wheels. The events are independent, and so the probability of getting 3 bars is $\frac{1}{11} \cdot \frac{1}{11} \cdot \frac{1}{11} = \frac{1}{1331}$.

(b) The probability of getting a number on the first wheel is $\frac{10}{11}$, the probability of getting the same number on the second wheel is $\frac{1}{11}$, and the probability of getting the same number on the third wheel is $\frac{1}{11}$. Thus the probability of getting the same number on each wheel is $\frac{10}{11} \cdot \frac{1}{11} \cdot \frac{1}{11} = \frac{10}{1331}$.

(c) We use the complement, NO BARS, to determine the probability of at least one bar. The probability that the first wheel does not have a bar is $\frac{10}{11}$, and similarly for the second and third wheels. Since the events are independent, the probability of getting NO BARS is $\frac{10}{11} \cdot \frac{10}{11} \cdot \frac{10}{11} = \frac{10^3}{11^3} = \frac{1000}{1331}$ and so $P(\text{at least one BAR}) = 1 - \frac{1000}{1331} = \frac{331}{1331}$.

31. Since we are interested in "at least" 2 having the same birth month, we use the complement, no two having the same birth month. Let E be the event that at least two have a birthday in the same month, so E' is the event that no two have a birthday in the same month.

$P(E') = \dfrac{\text{number of ways to arrange 6 distinct birth months}}{\text{number of ways to arrange 6 birth months}} = \dfrac{P(12,6)}{12^6} = \dfrac{12 \cdot 11 \cdot 10 \cdot 9 \cdot 8 \cdot 7}{12^6}$
$= \frac{385}{1728}$. So $P(E) = 1 - P(E') = 1 - \frac{385}{1728} = \frac{1343}{1728}$.

Exercises 8.8

1. Mike gets \$2 with probability $\frac{1}{2}$ and \$1 with probability $\frac{1}{2}$. Thus $E = (2)\left(\frac{1}{2}\right) + (1)\left(\frac{1}{2}\right) = 1.5$, and so his expected winnings are \$1.50 per game.

3. Since the probability of drawing the ace of spades is $\frac{1}{52}$, the expected value of this game is $E = (100)\left(\frac{1}{52}\right) + (-1)\left(\frac{51}{52}\right) = \frac{49}{52} \approx 0.94$. So your expected winnings are \$0.94 per game.

5. Since the probability that Carol rolls a six is $\frac{1}{6}$, the expected value of this game is $E = (3)\left(\frac{1}{6}\right) + (0.50)\left(\frac{5}{6}\right) = \frac{5.5}{6} \approx 0.9167$. So Carol expects to win \$0.92 per game.

7. Since the probability that the die shows an even number equals the probability that that die shows an odd number, the expected value of this game is $E = (2)\left(\frac{1}{2}\right) + (-2)\left(\frac{1}{2}\right) = 0$. So Tom should expect to break even after playing this game many times.

9. Since it cost 50¢ to play, if you get a silver dollar, you only win $1 - 0.50 = \$.50$. Thus the expected value of this game is $E = (0.50)\left(\frac{2}{10}\right) + (-0.50)\left(\frac{8}{10}\right) = -0.30$. So your expected winnings are $-\$0.30$ per game. In other words, you should expect to lose \$0.30 per game.

11. You can either win \$35 or lose \$1, so the expected value of this game is $E = (35)\left(\frac{1}{38}\right) + (-1)\left(\frac{37}{38}\right) = -\frac{2}{38} = -0.0526$. So the expect value is $-\$0.0526$ per game.

13. By the rules of the game, a player can win \$10, \$5, \$0 or lose \$100. The expected value of this game is $E = (10)\left(\frac{10}{100}\right) + (5)\left(\frac{10}{100}\right) + (-100)\left(\frac{2}{100}\right) + (0)\left(\frac{78}{100}\right) = -0.50$. So the expected winnings per game are $-\$0.50$.

15. If the stock goes up to \$20, she expects to make $\$20 - \$5 = \$15$. And if the stock falls to \$1, then she is out the cost of the stock plus the value of the stock, that is $-\$5 + \$1 = -\$4$. So the expected value of her profit is $E = (15)(0.1) + (-4)(0.9) = -2.1$. Thus, her expected profit per share is $-\$2.10$, that is, she should expect to loose \$2.10 per share. She did not make a wise investment.

17. Each of the 3 wheels has 11 possible outcomes the ten digits or a watermelon, and is independent of the other wheels. The probability of getting n watermelons is the number of ways to get n of 3 wheels to show watermelons, times the probability that the wheels show n watermelons. Thus $P(n \text{ watermelons}) = C(3,n)\left(\frac{1}{11}\right)^n\left(\frac{10}{11}\right)^{3-n}$. That is, $P(3 \text{ watermelons}) = C(3,3)\left(\frac{1}{11}\right)^3\left(\frac{10}{11}\right)^0 = \frac{1}{11^3}$; $P(2 \text{ watermelons}) = C(3,2)\left(\frac{1}{11}\right)^2\left(\frac{10}{11}\right)^1 = \frac{30}{11^3}$; $P(1 \text{ watermelon}) = C(3,1)\left(\frac{1}{11}\right)^1\left(\frac{10}{11}\right)^2 = \frac{300}{11^3}$; $P(\text{no watermelon}) = C(3,0)\left(\frac{1}{11}\right)^0\left(\frac{10}{11}\right)^3 = \frac{10^3}{11^3}$. Therefore, the expected value of this game is $E = (4.75)\left(\frac{1}{11^3}\right) + (0.75)\left(\frac{30}{11^3}\right) + (0.25)\left(\frac{300}{11^3}\right) + (-0.25)\left(\frac{10^3}{11^3}\right) = -0.11$ or $-\$0.11$ per game.

19. Let x be the fair price to pay to play this game. Then the winnings will be $1 - x$ and the losing will be $-x$. So $E = 0 \Leftrightarrow (1-x)\left(\frac{2}{8}\right) + (-x)\left(\frac{6}{8}\right) = 0 \Leftrightarrow 1 - 4x = 0 \Leftrightarrow x = \frac{1}{4} = 0.25$. Thus, a fair price to play this game is \$0.25.

21. He can win between \$0 and \$5. So the expected value of this game is $E = 5 \cdot \left(\frac{5\cdot4\cdot3\cdot2\cdot1}{7\cdot6\cdot5\cdot4\cdot3}\right) + 4 \cdot \left(\frac{5\cdot4\cdot3\cdot2\cdot2}{7\cdot6\cdot5\cdot4\cdot3}\right) + 3 \cdot \left(\frac{5\cdot4\cdot3\cdot2}{7\cdot6\cdot5\cdot4}\right) + 2 \cdot \left(\frac{5\cdot4\cdot2}{7\cdot6\cdot5}\right) + 1 \cdot \left(\frac{5\cdot2}{7\cdot6}\right) + 0 \cdot \left(\frac{2}{7}\right) \approx 1.67$ or \$1.67.

Review Exercises for Chapter 8

1. The number of possible outcomes $= \binom{\text{number of outcomes}}{\text{when a coin is tossed}} \cdot \binom{\text{number of outcomes}}{\text{a die is rolled}} \cdot \binom{\text{number of ways}}{\text{to draw a card}}$
$= (2)(6)(52) = 624$

3. Let x be the number of the people in the group. Then $C(x,2) = 10 \quad \Leftrightarrow \quad \dfrac{x!}{2!(x-2)!} = 10 \quad \Leftrightarrow$
$\dfrac{x!}{(x-2)!} = 20 \quad \Leftrightarrow \quad x(x-1) = 20 \quad \Leftrightarrow \quad x^2 - x - 20 = 0 \quad \Leftrightarrow \quad (x-5)(x+4) = 0 \quad \Leftrightarrow$
$x = 5$ or $x = -4$. So there are 5 people in this group.

5. (a) Order is not important and there are no repetitions, so the number of different two-element subsets is $C(5,2) = \dfrac{5!}{2!\,3!} = \dfrac{5 \cdot 4}{2} = 10$.

   (b) Order is important and there are no repetitions, so the number of different two-letter words is $P(5,2) = \frac{5!}{3!} = 20$.

7. We pick the flips that result in a HEAD, the others will result in a TAIL. So the number of different ways is this can occur is $C(10,7) = C(10,3) = \dfrac{10!}{3!\,7!} = 120$.

9. You earn a score of 70% by answering exactly 7 of the 10 questions correctly. The number of different ways to answer the questions correctly is $C(10,7) = \dfrac{10!}{7!\,3!} = 120$.

11. You must choose two of the ten questions to omit, and the number of ways of choosing these two questions is $C(10,2) = \dfrac{10!}{2!\,8!} = 45$.

13. The maximum number of employees using this security system is
$\binom{\text{number of choices}}{\text{for the first letter}} \cdot \binom{\text{number of choices}}{\text{for the second letter}} \cdot \binom{\text{number of choices}}{\text{for the third letter}} = (26)(26)(26) = 17,576$

15. (a) Since we cannot choose a major and a minor in the same subject, the number of ways a student can select a major and a minor is $P(16,2) = 16 \cdot 15 = 240$.

    (b) Again, since we cannot have repetitions and the order of selection is important, the number of ways to select a major, a first minor and a second minor is $P(16,3) = 16 \cdot 15 \cdot 14 = 3360$.

    (c) When we select a major and 2 minors the order in which we chose the minors is not important. Thus the number of ways to select a major and 2 minors is $\binom{\text{number of ways}}{\text{to select a major}} \cdot \binom{\text{number of ways to}}{\text{select two minors}}$
    $= 16 \cdot C(15,2) = 16 \cdot 105 = 1680$

17. Since the letters are distinct, the number of anagrams of the word TRIANGLE is $8! = 40,320$.

19. Since MISSISSIPPI has 4 I's, 4 S's, 2 P's, the number of distinguishable anagrams of the word MISSISSIPPI is $\dfrac{11!}{4!\,4!\,2!\,1!} = 34,650$.

21. A letter can be represented by a sequence of length 1, or a sequence of length 2 or a sequence of length 3. Since each symbol is either a dot or a dash, the possible number of letters is
$\binom{\text{number of letters}}{\text{using 3 symbols}} + \binom{\text{number of letters}}{\text{using 2 symbols}} + \binom{\text{number of letters}}{\text{using 1 symbol}} = 2^3 + 2^2 + 2 = 14$.

23. (a) The possible number of committees is $C(18,7) = 31,824$.

(b) Since we must select the 4 men from the group 10 men and the 3 women from the group of 8 women, the possible number of committees is $\left(\substack{\text{number of ways to}\\ \text{choose 4 of 10 men}}\right) \cdot \left(\substack{\text{number of ways to}\\ \text{choose 3 of 8 women}}\right)$ $= C(10,4) \cdot C(8,3) = 210 \cdot 56 = 11,760$

(c) We remove Susie from the group of 18, so the possible number of committees is $C(17,7) = 19,448$.

(d) The possible number of committees is $\left(\substack{\text{possible number of}\\ \text{committees with 5 women}}\right) + \left(\substack{\text{possible number of}\\ \text{committees with 6 women}}\right) + \left(\substack{\text{possible number of}\\ \text{committees with 7 women}}\right)$ $= C(8,5) \cdot C(10,2) + C(8,6) \cdot C(10,1) + C(8,7) \cdot C(10,0) = 56 \cdot 45 + 28 \cdot 10 + 8 \cdot 1$ $= 2808$.

(e) Since the committee is to have 7 members, "at most two men" is the same as "at least five women" which we found in part (d). So the number is also 2808.

(f) We select the specific offices first, then complete the committee from the remaining members of the group. So the number of possible committees is $\left(\substack{\text{number of ways to choose}\\ \text{a chairman, a vice-chairman, a secretary}}\right) \cdot \left(\substack{\text{number of ways to}\\ \text{choose 4 other members}}\right) = P(18,3) \cdot C(15,4)$ $= 4896 \cdot 1365 = 6,683,040$.

25. Let R_n denote the event that the nth ball is red and let W_n denote the event that the nth ball is white.

(a) $P(\text{both balls are red}) = P(R_1 \cap R_2) = P(R_1) \cdot P(R_2 \mid R_1) = \frac{10}{15} \cdot \frac{9}{14} = \frac{3}{7}$.

(b) Solution 1: The probability that one is white and that the other is red is
$$\frac{\text{number of ways to select one white \& one red}}{\text{number of ways to select two balls}} = \frac{C(10,1) \cdot C(5,1)}{C(15,2)} = \frac{10}{21}.$$
Solution 2: $P(\text{one white and one red}) = P(W_1 \cap R_2) + P(R_1 \cap W_2)$ $= P(W_1) \cdot P(R_2 \mid W_1) + P(R_1) \cdot P(W_2 \mid R_1) = \frac{5}{15} \cdot \frac{10}{14} + \frac{10}{15} \cdot \frac{5}{14} = \frac{10}{21}$.

(c) Solution 1: Let E be the event "at least one is red". Then E' is the event "both are white". $P(E') = P(W_1 \cap W_2) = P(W_1) \cdot P(W_2 \mid W_1) = \frac{5}{15} \cdot \frac{4}{14} = \frac{2}{21}$. Thus $P(E) = 1 - \frac{2}{21} = \frac{19}{21}$.
Solution 2: $P(\text{at least one is red}) = P(\text{one red and one white}) + P(\text{both red})$ $= \frac{10}{21} + \frac{9}{21} = \frac{19}{21}$ (from (a) and (b)).

(d) Since 5 of the 15 balls are both red and even-numbered, the probability that both balls are red and even-numbered is $\frac{5}{15} \cdot \frac{4}{14} = \frac{2}{21}$.

(e) Since 2 of the 15 balls are both white and odd-numbered, the probability that both are white and odd-numbered is $\frac{2}{15} \cdot \frac{1}{14} = \frac{1}{105}$.

27. (a) $P(\text{ace}) = \frac{4}{52} = \frac{1}{13}$.

(b) Let E be the event the card chosen is an ace and let F be the event the card chosen is a jack. Then $P(E \cup F) = P(E) + P(F) = \frac{4}{52} + \frac{4}{52} = \frac{2}{13}$.

(c) Let E be the event the card chosen is an ace and let F be the event the card chosen is a spade. Then $P(E \cup F) = P(E) + P(F) - P(E \cap F) = \frac{4}{52} + \frac{13}{52} - \frac{1}{52} = \frac{4}{13}$.

(d) Let E be the event the card chosen is an ace and let F be the event the card chosen is a red card. Then $P(E \cap F) = \dfrac{n(E \cap F)}{n(S)} = \dfrac{2}{52} = \dfrac{1}{26}$.

29. (a) The probability the first die shows some number is 1, and the probability the second die shows the same number is $\frac{1}{6}$. So the probability each die shows the same number is $1 \cdot \frac{1}{6} = \frac{1}{6}$.

(b) By part (a) the event of showing the same number has probability of $\frac{1}{6}$ and the complement of this event is that the dice show different numbers. Thus the probability that the dice show different numbers is $1 - \frac{1}{6} = \frac{5}{6}$.

31. Using the same logic as in Exercise 29 (a), the probability that all three die show the same number is $1 \cdot \frac{1}{6} \cdot \frac{1}{6} = \frac{1}{36}$ while the probability they are not all the same is $1 - \frac{1}{36} = \frac{35}{36}$. Thus the expected value of this game is $E = (5)\left(\frac{1}{36}\right) + (-1)\left(\frac{35}{36}\right) = -\frac{30}{36} = -0.83$. So John's expected winnings per game is $-\$0.83$, that is, he expects to loose \$0.83 per game.

33. Since Mary makes a guess as to the order of ratification of the 13 original states, the number of such guesses is $P(13, 13) = 13!$, while the probability that she guesses the correct order is $\frac{1}{13!}$. Thus the expected value is $E = (1,000,000)\left(\frac{1}{13!}\right) + (0)\left(\frac{13!-1}{13!}\right) = 0.00016$. So Mary's expected winnings are \$0.00016.

35. (a) Since there are only two colors of socks, any 3 socks must have a matching pair.

(b) Solution 1: If the two socks drawn form a matching pair then they are either both red or both blue. So $P(\text{choosing a matching pair}) = P(\text{both red or both blue})$

$$= P(\text{both red}) + P(\text{both blue}) = \frac{C(20,2)}{C(50,2)} + \frac{C(30,2)}{C(50,2)} \approx 0.51.$$

Solution 2: The complement of choosing a matching pair is choosing one sock of each color. So $P(\text{choosing a matching pair}) = 1 - P(\text{different colors}) = 1 - \dfrac{C(20,1) \cdot C(30,1)}{C(50,2)}$

$\approx 1 - 0.49 = 0.51$.

37. (a) $\left(\begin{array}{c}\text{number of different}\\ \text{zip codes}\end{array}\right) = \left(\begin{array}{c}\text{number of ways to}\\ \text{choose the 1st digit}\end{array}\right) \cdot \left(\begin{array}{c}\text{number of ways to}\\ \text{choose the 2nd digit}\end{array}\right) \cdot \cdots \cdot \left(\begin{array}{c}\text{number of ways to}\\ \text{choose the 5th digit}\end{array}\right)$

$= 10 \cdot 10 \cdot 10 \cdot 10 \cdot 10 = 10^5$

(b) Since there are five numbers (0, 1, 6, 8 and 9) that can be read upside down, we have $\left(\begin{array}{c}\text{number of different}\\ \text{zip codes}\end{array}\right) = \left(\begin{array}{c}\text{number of ways to}\\ \text{choose the 1st digit}\end{array}\right) \cdot \left(\begin{array}{c}\text{number of ways to}\\ \text{choose the 2nd digit}\end{array}\right) \cdot \cdots \cdot \left(\begin{array}{c}\text{number of ways to}\\ \text{choose the 5th digit}\end{array}\right) = 5^5$

(c) Let E be the event that a zip code can be read upside down. Then by parts (a) and (b), $P(E) = \dfrac{n(E)}{n(S)} = \dfrac{5^5}{10^5} = \dfrac{1}{32}$.

(d) Suppose a zip code is turned upside down. Then the middle digit remains the middle digit so it must be a digit that reads the same when turned upside down, that is, a 0, 1 or 8. Also, the last digit becomes the first digit and the next to last digit becomes the second digit. Thus, once the first two digits are chosen, the last two are determined. Therefore, the number of zip codes that read the same upside down as right side up is

$\left(\begin{array}{c}\text{number of ways to}\\ \text{choose the 1st digit}\end{array}\right) \cdot \left(\begin{array}{c}\text{number of ways to}\\ \text{choose the 2nd digit}\end{array}\right) \cdot \left(\begin{array}{c}\text{number of ways to}\\ \text{choose the 3rd digit}\end{array}\right) \cdot \left(\begin{array}{c}\text{number of ways to}\\ \text{choose the 4th digit}\end{array}\right) \cdot \left(\begin{array}{c}\text{number of ways to}\\ \text{choose the 5th digit}\end{array}\right)$

$= 5 \cdot 5 \cdot 3 \cdot 1 \cdot 1 = 75$.

39. (a) Using the rule for the number of distinguishable combinations, the number of divisors of N is $(7 + 1)(2 + 1)(5 + 1) = 144$.

(b) An even divisor of N must contain 2 as a factor. Thus we place a 2 as one of the factors and count the number of distinguishable combinations of $M = 2^6 3^2 5^5$. So using the rule for the number of distinguishable combinations, the number of even divisors of N is $(6+1)(2+1)(5+1) = 126$.

(c) A divisor is a multiple of 6 if 2 is a factor and 3 is a factor. Thus we place a 2 as one of the factors and a 3 as one of the factors. Then count the number of distinguishable combinations of $K = 2^6 3^1 5^5$. So using the rule for the number of distinguishable combinations, the number of even divisors of N is $(6+1)(1+1)(5+1) = 84$.

(d) Let E be the event that the divisor is even. Then using parts (a) and (b), $P(E) = \dfrac{n(E)}{n(S)}$ $= \frac{126}{144} = \frac{7}{8}$.

41. This exercise is about the number of ways of partitioning the 10 men and 8 women into groups of 5 men and 4 women (not the number of groups). Since each choice of 5 men and 4 women determines two different groups (the other being the 5 men and 4 women not chosen), the number of ways of dividing the group in the desired way is $\frac{1}{2} \cdot C(10,5) \cdot C(8,4) = 8820$.

Chapter 8 Test

1. (a) Repetition is allowed. Then each letter of the word can be chosen in 10 ways since there are 10 letters. Thus the number of possible 5 letter words if repetition is allowed is $10 \cdot 10 \cdot 10 \cdot 10 \cdot 10 = 10^5 = 100,000$.

   (b) Repetition is not allowed. Then since order is important, we need the number of permutations of 10 objects (the 10 letters) taken 5 at a time. Therefore, the number of possible 5 letter words if repetition is not allowed is $P(10,5) = 10 \cdot 9 \cdot 8 \cdot 7 \cdot 6 = 30,240$.

2. There are three choices to be made: one choice each of a main course, a dessert, and a drink. Since a main course can be chosen in one of five ways, a dessert in one of three ways, and a drink in one of four ways, so there are $5 \cdot 3 \cdot 4 = 60$ possible ways that a customer could order a meal.

3. We choose the offices first. Here order is important, because the offices are different; thus there are $P(30,3)$ ways to do this. Next we choose the other 5 members from the remaining 27 members. Here order is not important, so there are $C(27,5)$ ways to do this. Therefore the number of ways that the board of directors can be chosen is $P(30,3) \cdot C(27,5) = 30 \cdot 29 \cdot 28 \cdot \frac{27!}{5!\,22!} = 1,966,582,800$.

4. There are two choices to be made: choose a road to travel from Ajax to Berry, and then choose a different road from Berry to Ajax. Since there are 4 roads joining the two cities, we need the number of permutations of 4 objects (the roads) taken 2 at a time (the road there and the road back). This number is $P(4,2) = 4 \cdot 3 = 12$.

5. A customer must choose a size of pizza and must make a choice of toppings. There are 4 sizes of pizza, and each choice of toppings from the 14 available corresponds to a subset of 14 objects. Since a set with 14 objects has 2^{14} subsets, the number of different pizzas this parlor offers is $4 \cdot 2^{14} = 65{,}536$.

6. (a) We need the number of ways of arranging the 10 students in order. Thus the number of permutations of 10 objects taken 10 at a time which is $P(10,10) = 10! = 3{,}628{,}800$.

   (b) The two tallest students are to stand next to each other. We arrange these students and then consider them as one object and arrange 9 objects (treat the two tallest as one object). Thus the number of ways the students can stand in a row is $P(2,2) \cdot P(9,9) = 2! \cdot 9! = 725,760$.

   (c) The two tallest students refuse to stand next to each other. There are 10! ways in which these 10 students can stand in a row. By part (b), the number of arrangements where the two tallest stand together is $2 \cdot 9!$. Therefore, the number of arrangements where the two tallest stand apart is $10! - 2 \cdot 9! = 9!(10-2) = 8 \cdot 9! = 2,903,040$.

7. (a) We want the number of ways of arranging 4 distinct objects (the letters L, O, V, E). This is the number of permutations of 4 objects taken 4 at a time. Therefore, the number of anagrams of the word LOVE is $P(4,4) = 4! = 24$.

   (b) We want the number of distinguishable permutations of 6 objects (the letters K, I, S, S, E, S) consisting of three like groups of size 1 and a like group of size 3 (the S's). Therefore, the number of different anagrams of the word KISSES is $\dfrac{6!}{1!\,1!\,1!\,3!} = \dfrac{6!}{3!} = 120$.

8. Let E be the event of choosing a number between 1 and 100 that is divisible by 3.

(a) Since there are 33 numbers between 1 and 100 that are divisible by 3 (3, 6, 9, … , 99), $P(E) = \dfrac{n(E)}{n(S)} = \dfrac{33}{100} = 0.33$.

(b) E' is the event of choosing a number between 1 and 100 that is not divisible by 3. Therefore, by using part (a), $P(E') = 1 - P(E) = 1 - \dfrac{33}{100} = \dfrac{67}{100} = 0.67$.

9. Let E be the event of choosing 3 men. Then $P(E) = \frac{n(E)}{n(S)} = \dfrac{\text{number of ways to choose 3 men}}{\text{number of ways to choose 3 people}} = \dfrac{C(5,3)}{C(15,3)} \approx 0.022$.

10. Two dice are rolled. Let E be the event of getting doubles. Since a double may occur in 6 ways, $P(E) = \dfrac{n(E)}{n(S)} = \dfrac{6}{36} = \dfrac{1}{6}$.

11. Two cards are drawn from a deck.

(a) Let E be the event of choosing 2 face cards. Since there are 12 face cards, $P(E) = \frac{C(12,2)}{C(52,2)} \approx 0.0498$.

(b) Let E be the event that the first card is an ace and F be the event that the second card is an ace, then we want $P(F \mid E)$. But $P(F \mid E) = \frac{3}{51} \approx 0.0588$, since there are 51 cards left of which 3 are aces.

12. Let R be the event that the ball chosen is red. Let E be the event that the ball chosen is even-numbered.

(a) Since 5 of the 13 balls are red, $P(R) = \frac{5}{13} \approx 0.3846$.

(b) Since 6 of the 13 balls are even-numbered, $P(E) = \frac{6}{13} \approx 0.4615$.

(c) $P(R \text{ or } E) = P(R) + P(E) - P(R \cap E) = \frac{5}{13} + \frac{6}{13} - \frac{2}{13} = \frac{9}{13} \approx 0.6923$.

(d) $P(E \mid R) = \dfrac{P(R \cap E)}{P(R)} = \dfrac{2/13}{5/13} = \dfrac{2}{5} = 0.4$.

(e) $P(R \mid E) = \dfrac{P(R \cap E)}{P(E)} = \dfrac{2/13}{6/13} = \dfrac{2}{6} \approx 0.333$.

(f) $P(R \cap E) = \frac{2}{13} \approx 0.1538$.

13. A deck of cards contains 4 aces, 12 face cards, and 36 other cards. Thus the probability of an ace is $\frac{4}{52} = \frac{1}{13}$, the probability of a face card is $\frac{12}{52} = \frac{3}{13}$, and the probability of a non-ace, non-face card is $\frac{36}{52} = \frac{9}{13}$. Thus the expected value of this game is $E = (10)\left(\frac{1}{13}\right) + (1)\left(\frac{3}{13}\right) + (-.5)\left(\frac{9}{13}\right) = \frac{8.5}{13} \approx 0.654$.

14. There are 4 students and 12 astrological signs. Let E be the event that at least 2 have the same astrological sign. Then E' is the event that no 2 have the same astrological sign. It is easier to find E'. So $P(E') = \dfrac{\text{number of ways to assign 4 different astrological signs}}{\text{number of ways to assign 4 astrological signs}} = \dfrac{P(12,4)}{12^4}$ $= \dfrac{12 \cdot 11 \cdot 10 \cdot 9}{12 \cdot 12 \cdot 12 \cdot 12} = \dfrac{55}{96}$. Therefore, $P(E) = 1 - P(E') = 1 - \dfrac{55}{96} = \dfrac{41}{96} \approx 0.427$.

15. The probability of drawing an ace is $\frac{4}{52}$, of drawing a two is $\frac{4}{52}$, … , of drawing a ten is $\frac{4}{52}$, and of drawing a face card is $\frac{12}{52}$. Therefore the expected number of problems is

$E = (1)\left(\frac{4}{52}\right) + (2)\left(\frac{4}{52}\right) + (3)\left(\frac{4}{52}\right) + (4)\left(\frac{4}{52}\right) + (5)\left(\frac{4}{52}\right) + (6)\left(\frac{4}{52}\right) + (7)\left(\frac{4}{52}\right) + (8)\left(\frac{4}{52}\right) + (9)\left(\frac{4}{52}\right)$

$+ (10)\left(\frac{4}{52}\right) + (12)\left(\frac{12}{52}\right) = \frac{364}{52} = 7$. So the students should expect to be assigned 7 problems each day.

Focus on Problem Solving

1. (a) When the contestant first made her choice, the probability of winning was $\frac{1}{10}$. If the contestant does not switch, the probability of winning remains $\frac{1}{10}$.

   (b) If the contestant switches the probability is $1 - \frac{1}{10} = \frac{9}{10}$.

3. (a) The game must end in at most 3 tosses. Hence the possible equally likely ways of completing the game are HHH, HHT, HTH, HTT, THH, THT, TTH, TTT. Player A wins in 7 of these 8 ways, so the probability of A winning is $\frac{7}{8}$. Thus the jackpot should be divided as follows: \$7000 for player A and \$1000 for player B. (This may seem paradoxical at first, but part (b) should convince you.)

   (b) Continue the game from the point it is interrupted by tossing a coin and keeping track of the number of times A wins the game and the number of times B wins. We have performed this experiment ten times and obtained the following for the number of times A wins in 40 trials: 37, 35, 36, 38, 34, 36, 36, 35, 32, 38.

5. Since there are 4 customers there are 4! ways to return their hats. To find the number of ways in which no customer get his own hat back, we first count the number of ways that the complement of this event can happen—that is, the number of ways in which *at least one* customer gets his own hat back. To do this, we will count the number of ways in which exactly one customer gets his hat back, plus the number of ways in which exactly two customers get their hats back, plus the number of ways that all four customers get their hats back. (Note that it is impossible for exactly three customers to get their own hats back, since then the hat left over will belong to the customer left over.) To count the number of ways exactly one person gets his hat back, we reason as follows. First we choose the one lucky person: there are $C(4,1)$ ways of doing that. Then we distribute the remaining three hats to the other three people in such a way that none of these gets his own hat back. A moment's reflection will convince you that there are just two ways for this to happen. (Any fixed individual can be given one of two wrong hats, and then it is determined where the remaining two hats must go.) Thus there are $C(4,1) \cdot 2 = 8$ ways for exactly one person to get his hat back. For exactly two people to get their hats back, we can first choose the two lucky people, and then switch around the remaining two hats. Thus there are $C(4,2) \cdot 1 = 6$ ways of doing this. Finally, there is of course only one way to give all hats back to the right people. Thus the number of ways in which no one gets his own hat back is $4! - (8 + 6 + 1) = 24 - 15 = 9$, and so the probability that this happens is $\frac{9}{24} = \frac{3}{8}$.

7. Since Ontario or Quebec must vote YES we consider four cases:

   <u>Case 1:</u> Exactly 7 provinces vote yes.

   (i) Ontario or Quebec and 6 of the 8 remaining provinces vote yes. Since each of the remaining 3 provinces can either vote no or abstain, the number of ways is
   $C(2,1) \cdot C(8,6) \cdot 2^3 = 2 \cdot 28 \cdot 8 = 448$

   (ii) Ontario and Quebec and 5 of the 8 remaining provinces vote yes. Number of ways:
   $C(2,2) \cdot C(8,5) \cdot 2^3 = 1 \cdot 56 \cdot 8 = 448$

   So in Case 1 the number of ways is $448 + 448 = 896$.

   <u>Case 2:</u> Exactly 8 provinces vote yes.

(i) Ontario or Quebec and 7 of the 8 remaining provinces vote yes. Number of ways: $C(2,1) \cdot C(8,7) \cdot 2^2 = 2 \cdot 8 \cdot 4 = 64$

(ii) Ontario and Quebec and 6 of the 8 remaining provinces vote yes. Number of ways: $C(2,2) \cdot C(8,6) \cdot 2^2 = 1 \cdot 28 \cdot 4 = 112$

So in Case 2 the number of ways is $64 + 112 = 176$.

Case 3: Exactly 9 provinces vote yes. Number of ways: $C(10,9) \cdot 2^1 = 10 \cdot 2 = 20$

Case 4: All 10 provinces vote yes. Number of ways: $C(10,10) = 1$

Thus the total number of ways is $896 + 176 + 20 + 1 = 1093$.

9. We see that 1 can be written in 1 way, 2 can be written in 2 ways $(2, 1+1)$, 3 can be written in 4 ways $(3, 1+2, 2+1, 1+1+1)$, 4 can be written in 8 ways (as in the statement of the problem), and 5 can be written in 16 ways. These are all powers of 2, so we make the guess that the number n can be written in 2^{n-1} ways.

To prove this, notice that the number of ways of writing 5 as a sum of natural numbers is the same as the number of ways of partitioning five 1's by inserting plus signs into the spaces between the 1's:

$1\,1\,1\,1\,1\,(5)$ $\quad 1+1\,1\,1\,1\,(1+4)$ $\quad 1\,1+1+1\,1\,(2+1+2)$ $\quad 1+1\,1\,1+1\,(1+3+1)$ and so on. For the number n, we want the number of ways of partitioning n 1's by inserting plus signs into the $n-1$ spaces between the 1's. For each of the $n-1$ spaces, we can either insert a plus sign or not insert it, so the number of ways is $2 \cdot 2 \cdot \cdots \cdot 2 = 2^{n-1}$.

11. We draw 1, 2, 3, 4, and 5 lines and obtain the following results for the resulting number of intersection points and regions created. ($n = 3$ is shown in the text.)

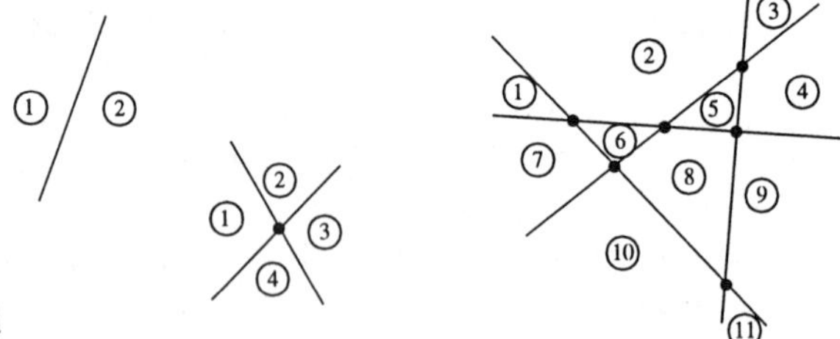

Lines	1	2	3	4	5
Points	0	1	3	6	10
Regions	2	4	7	11	16

It appears that the number of points created after placing n lines (assuming $n \geq 2$) is $1 + 2 + 3 + \cdots + (n-1)$; and the number of regions is $1 + (1 + 2 + 3 + \cdots + n)$, since the numbers in the last row of the above table are each 1 more than the number that appears above it to the right, and we start with 1 region when there are no lines at all.

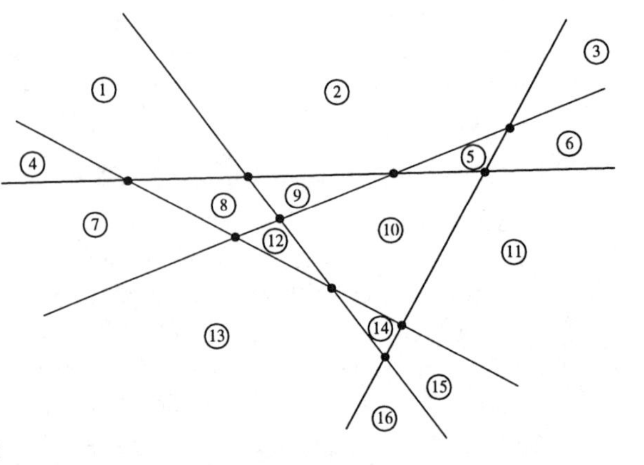

To confirm this, note that when we add the n-th line, to have as many intersections and regions as possible, the new line should have a different slope than the existing lines, and should not cross any of the existing intersection points. Thus each of the old lines is crossed exactly once, so that $n-1$ new intersection points are added when adding the n-th line. Moreover, as we draw this new line, each time we make a new intersection point, an existing region has been divided in two; then after all new intersection points have been added, we complete the picture by dividing one final region into two pieces when we complete the line. Thus we add n regions when inserting the n-th line. This means that adding the n-th line gives the following total number of points and regions:

Points: $1 + 2 + 3 + \cdots + (n - 1)$

Regions: $1 + (1 + 2 + 3 + \cdots + n)$

(a) After drawing 10 lines, the maximum number of points of intersection is therefore $1 + 2 + 3 + 4 + 5 + 6 + 7 + 8 + 9 = 45$.

(b) The maximum number of lines is $1 + (1 + 2 + 3 + 4 + 5 + 6 + 7 + 8 + 9 + 10) = 56$.

13. There are 121 points shown in the figure.

(a) There are 11 points whose x-coordinate is 2. Thus the probability is $\frac{11}{121} = \frac{1}{11} \approx 0.091$.

(b) There are 11 points in the figure that lie on the line $y = x$. Thus the probability is $\frac{11}{121} = \frac{1}{11} \approx 0.091$.

(c) There are 12 points in the figure that lie on the circle $x^2 + y^2 = 25$. (The points that touch the circle but are not darkened are not really on the circle.) Thus the probability is $\frac{12}{121} \approx 0.099$.

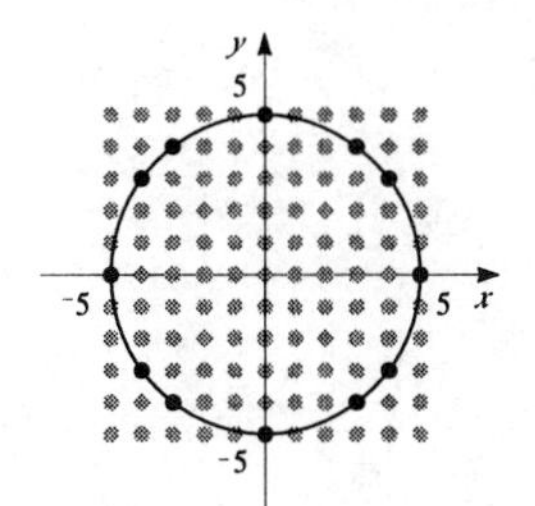

(d) There are 69 points in the figure that lie inside (but not on) the circle $x^2 + y^2 = 25$. Thus the probability is $\frac{69}{121} \approx 0.570$.

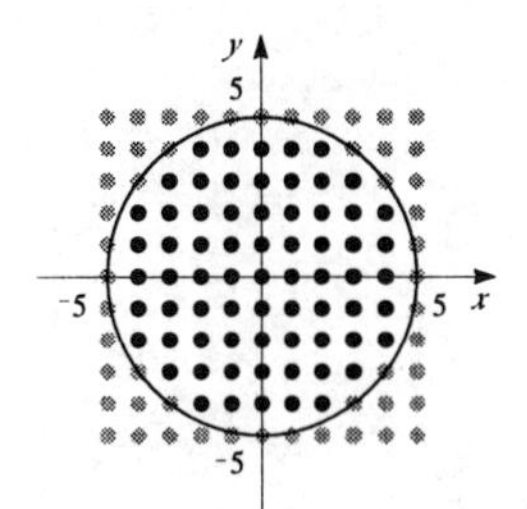

(e) There are 63 points in the figure that lie above but not on the line $y = 2x - 1$. Thus the probability is $\frac{63}{121} \approx 0.521$.

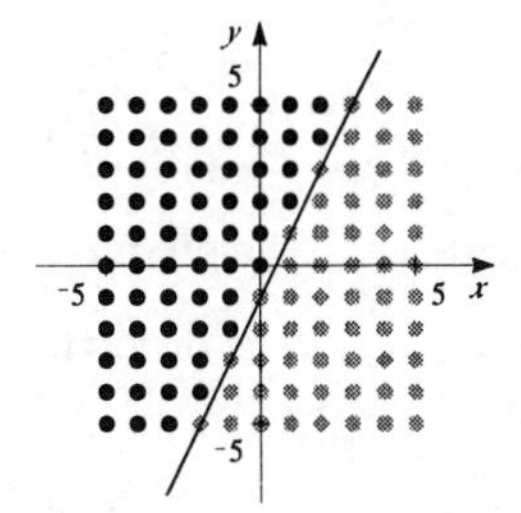

Chapter Nine

Exercises 9.1

1. $a_n = n + 1$. Then $a_1 = 1 + 1 = 2$; $a_2 = 2 + 1 = 3$; $a_3 = 3 + 1 = 4$; $a_4 = 4 + 1$; $a_{1000} = 1000 + 1 = 1001$.

3. $a_n = \frac{1}{n+1}$. Then $a_1 = \frac{1}{1+1} = \frac{1}{2}$; $a_2 = \frac{1}{2+1} = \frac{1}{3}$; $a_3 = \frac{1}{3+1} = \frac{1}{4}$; $a_4 = \frac{1}{4+1} = \frac{1}{5}$; $a_{1000} = \frac{1}{1000+1} = \frac{1}{1001}$.

5. $a_n = \frac{(-1)^n}{n^2}$. Then $a_1 = \frac{(-1)^1}{1^2} = -1$; $a_2 = \frac{(-1)^2}{2^2} = \frac{1}{4}$; $a_3 = \frac{(-1)^3}{3^2} = -\frac{1}{9}$; $a_4 = \frac{(-1)^4}{4^2} = \frac{1}{16}$; $a_{1000} = \frac{(-1)^{1000}}{1000^2} = \frac{1}{1{,}000{,}000}$.

7. $a_n = 1 + (-1)^n$. Then $a_1 = 1 + (-1)^1 = 0$; $a_2 = 1 + (-1)^2 = 2$; $a_3 = 1 + (-1)^3 = 0$; $a_4 = 1 + (-1)^4 = 2$; $a_{1000} = 1 + (-1)^{1000} = 2$.

9. $a_n = (-2)^n$. Then $a_1 = (-2)^1 = -2$; $a_2 = (-2)^2 = 4$; $a_3 = (-2)^3 = -8$; $a_4 = (-2)^4 = 16$; $a_{1000} = (-2)^{1000} = 2^{1000}$.

11. $a_n = \frac{(-1)^{n+1}n}{n+1}$. Then $a_1 = \frac{(-1)^2 \cdot 1}{1+1} = \frac{1}{2}$; $a_2 = \frac{(-1)^3 \cdot 2}{2+1} = -\frac{2}{3}$; $a_3 = \frac{(-1)^4 \cdot 3}{3+1} = \frac{3}{4}$; $a_4 = \frac{(-1)^5 \cdot 4}{4+1} = -\frac{4}{5}$; $a_{1000} = \frac{(-1)^{1001} \cdot 1000}{1001} = -\frac{1000}{1001}$.

13. $a_n = n^n$. Then $a_1 = 1^1 = 1$; $a_2 = 2^2 = 4$; $a_3 = 3^3 = 27$; $a_4 = 4^4 = 256$; $a_{1000} = 1000^{1000} = 10^{3000}$.

15. $a_n = 2(a_{n-1} - 2)$ and $a_1 = 3$. Then $a_2 = 2[(3) - 2] = 2$; $a_3 = 2[(2) - 2] = 0$; $a_4 = 2[(0) - 2] = -4$; $a_5 = 2[(-4) - 2] = -12$.

17. $a_n = 2a_{n-1} + 1$ and $a_1 = 1$. Then $a_2 = 2(1) + 1 = 3$; $a_3 = 2(3) + 1 = 7$; $a_4 = 2(7) + 1 = 15$; $a_5 = 2(15) + 1 = 31$.

19. $a_n = a_{n-1} - a_{n-2}$ and $a_1 = 0$; $a_2 = 1$. Then $a_3 = 1 - 0 = 1$; $a_4 = 1 - 1 = 0$; $a_5 = 0 - 1 = -1$.

21. $a_n = a_{n-1} + a_{n-2}$ and $a_1 = 1$; $a_2 = 2$. Then $a_3 = 2 + 1 = 3$; $a_4 = 3 + 2 = 5$; $a_5 = 5 + 3 = 8$.

23. $2, 4, 8, 16, \ldots$. All are multiples of 2, so $a_1 = 2$, $a_2 = 2^2$, $a_3 = 2^3$, $a_4 = 2^4, \ldots$ Thus $a_n = 2^n$.

25. $1, 4, 7, 10, \ldots$. The difference between to consecutive terms is 3, so $a_1 = 3(1) - 2$, $a_2 = 3(2) - 2$, $a_3 = 3(3) - 2$, $a_4 = 3(4) - 2, \ldots$ Thus $a_n = 3n - 2$.

27. $1, \frac{3}{4}, \frac{5}{9}, \frac{7}{16}, \frac{9}{25}, \ldots$. We consider the numerator separate from the denominator. The numerators of the terms differ by 2 and the denominators are perfect squares. So $a_1 = \frac{2(1)-1}{1^2}$, $a_2 = \frac{2(2)-1}{2^2}$, $a_3 = \frac{2(3)-1}{3^2}$, $a_4 = \frac{2(4)-1}{4^2}$, $a_5 = \frac{2(5)-1}{5^2}, \ldots$. Thus $a_n = \frac{2n-1}{n^2}$.

29. $r, r^2, r^3, r^4, \ldots$. The exponent corresponds to the number of the term. So $a_n = r^n$.

31. 0, 2, 0, 2, 0, 2, … . These terms alternate between 0 and 2. So $a_1 = 1 - 1$, $a_2 = 1 + 1$, $a_3 = 1 - 1$, $a_4 = 1 + 1$, $a_5 = 1 - 1$, $a_6 = 1 + 1$, … Thus $a_n = 1 + (-1)^n$.

33. (a) $a_n = n^2$. Then $a_1 = 1^2 = 1$, $a_2 = 2^2 = 4$, $a_3 = 3^2 = 9$, $a_4 = 4^2 = 16$.

(b) $a_n = n^2 + (n-1)(n-2)(n-3)(n-4)$

$a_1 = 1^2 + (1-1)(1-2)(1-3)(1-4) = 1 + 0(-1)(-2)(-3) = 1$

$a_2 = 2^2 + (2-1)(2-2)(2-3)(2-4) = 4 + 1 \cdot 0(-1)(-2) = 4$

$a_3 = 3^2 + (3-1)(3-2)(3-3)(3-4) = 9 + 2 \cdot 1 \cdot 0(-1) = 9$

$a_4 = 4^2 + (4-1)(4-2)(4-3)(4-4) = 16 + 3 \cdot 2 \cdot 1 \cdot 0 = 16$

Hence, the sequences agree in the first four terms. However, for the second sequence $a_5 = 5^2 + (5-1)(5-2)(5-3)(5-4) = 25 + 4 \cdot 3 \cdot 2 \cdot 1 = 49$, and for the first sequence, $a_5 = 5^2 = 25$, and thus the sequences disagree from the fifth term on.

(c) $a_n = n^2 + (n-1)(n-2)(n-3)(n-4)(n-5)(n-6)$ agrees with $a_n = n^2$ in the first six terms only.

35. $\sqrt{2}, \sqrt{2\sqrt{2}}, \sqrt{2\sqrt{2\sqrt{2}}}, \sqrt{2\sqrt{2\sqrt{2\sqrt{2}}}}, \ldots$. We simplify each term in an attempt to determine a formula for a_n. So $a_1 = 2^{1/2}$, $a_2 = \sqrt{2 \cdot 2^{1/2}} = \sqrt{2^{3/2}} = 2^{3/4}$, $a_3 = \sqrt{2 \cdot 2^{3/4}} = \sqrt{2^{7/4}} = 2^{7/8}$, $a_4 = \sqrt{2 \cdot 2^{7/8}} = \sqrt{2^{15/8}} = 2^{15/16}, \ldots$. Thus $a_n = 2^{(2^n - 1)/2^n}$.

37. $$a_{n+1} = \begin{cases} \dfrac{a_n}{2} & \text{if } a_n \text{ is an even number} \\ 3a_n + 1 & \text{if } a_n \text{ is an odd number} \end{cases}$$

Since $a_1 = 25$, we have $a_2 = 76$, $a_3 = 38$, $a_4 = 19$, $a_5 = 58$, $a_6 = 29$, $a_7 = 88$, $a_8 = 44$, $a_9 = 22$, $a_{10} = 11$, $a_{11} = 34$, $a_{12} = 17$, $a_{13} = 52$, $a_{14} = 26$, $a_{15} = 13$, $a_{16} = 40$, $a_{17} = 20$, $a_{18} = 10$, $a_{19} = 5$, $a_{20} = 16$, $a_{21} = 8$, $a_{22} = 4$, $a_{23} = 2$, $a_{24} = 1$, $a_{25} = 4$, $a_{26} = 2$, $a_{27} = 1$, … (with 4, 2, 1 repeating).

39. Let F_n be the number of rabbits in the nth month. Clearly $F_1 = F_2 = 1$. In the nth month each pair that is two or more months old (that is, F_{n-2} pairs) will add a pair of offspring to the F_{n-1} pairs already present. Thus $F_n = F_{n-1} + F_{n-2}$. So F_n is the Fibonacci sequence.

Exercises 9.2

1. $2, 5, 8, 11, \ldots$. Then $d = a_2 - a_1 = 5 - 2 = 3$; $a_5 = a_4 + 3 = 11 + 3 = 14$; $a_n = 2 + 3(n-1)$; $a_{100} = 2 + 3(99) = 299$.

3. $4, 9, 14, 19, \ldots$. Then $d = a_2 - a_1 = 9 - 4 = 5$; $a_5 = a_4 + 5 = 19 + 5 = 24$; $a_n = 4 + 5(n-1)$; $a_{100} = 4 + 5(99) = 499$.

5. $-12, -8, -4, 0, \ldots$. Then $d = a_2 - a_1 = -8 - (-12) = 4$; $a_5 = a_4 + 4 = 0 + 4 = 4$; $a_n = -12 + 4(n-1)$; $a_{100} = -12 + 4(99) = 384$.

7. $25, 26.5, 28, 29.5, \ldots$. Then $d = a_2 - a_1 = 26.5 - 25 = 1.5$; $a_5 = 29.5 + 1.5 = 31$; $a_n = 25 + 1.5(n-1)$; $a_{100} = 25 + 1.5(99) = 173.5$.

9. $2, 2+s, 2+2s, 2+3s, \ldots$. Then $d = a_2 - a_1 = 2 + s - 2 = s$, $a_5 = 2 + 3s + s = 2 + 4s$; $a_n = 2 + (n-1)s$, $a_{100} = 2 + 99s$.

11. $2, 6, 18, 54, \ldots$. Then $r = \dfrac{a_2}{a_1} = \dfrac{6}{2} = 3$; $a_5 = 54(3) = 162$; $a_n = 2 \cdot 3^{n-1}$.

13. $0.3, -0.09, 0.027, -0.0081, \ldots$. Then $r = \dfrac{a_2}{a_1} = \dfrac{-0.09}{0.3} = -0.3$; $a_5 = -0.0081\,(-0.3) = 0.00243$; $a_n = 0.3(-0.3)^{n-1}$.

15. $144, -12, 1, -\frac{1}{12}, \ldots$. Then $r = \dfrac{a_2}{a_1} = \dfrac{-12}{144} = -\dfrac{1}{12}$; $a_5 = -\dfrac{1}{12}\left(-\dfrac{1}{12}\right) = \dfrac{1}{144}$; $a_n = 144\left(-\frac{1}{12}\right)^{n-1}$.

17. $3, 3^{5/3}, 3^{7/3}, 27, \ldots$. Then $r = \dfrac{a_2}{a_1} = \dfrac{3^{5/3}}{3} = 3^{2/3}$; $a_5 = 27 \cdot 3^{2/3} = 3^{11/3}$; $a_n = 3\left(3^{2/3}\right)^{n-1} = 3 \cdot 3^{(2n-2)/3} = 3^{(2n+1)/3}$.

19. $1, s^{2/7}, s^{4/7}, s^{6/7}, \ldots$. Then $r = \dfrac{a_2}{a_1} = \dfrac{s^{2/7}}{1} = s^{2/7}$; $a_5 = s^{6/7} \cdot s^{2/7} = s^{8/7}$; $a_n = \left(s^{2/7}\right)^{n-1} = s^{(2n-2)/7}$.

In Exercises 21 – 31, let $d_{i,j}$ represent the difference between the jth and i^{th} terms and let $r_{i,j}$ represent the corresponding ratio.

21. $5, -3, 5, -3, \ldots$. $d_{1,2} = -3 - 5 = -8$, but $d_{2,3} = 5 - (-3) = 8$, and $r_{1,2} = \frac{-3}{5}$, but $r_{2,3} = \frac{5}{-3}$. Thus, the sequence is neither arithmetic nor geometric.

23. $\sqrt{3}, 3, 3\sqrt{3}, 9, \ldots$. $r_{1,2} = \frac{3}{\sqrt{3}} = \sqrt{3}$, $r_{2,3} = \frac{3\sqrt{3}}{3} = \sqrt{3}$, $r_{3,4} = \frac{9}{3\sqrt{3}} = \sqrt{3}$. Hence, the sequence is geometric with $r = \sqrt{3}$, and $a_5 = 9\sqrt{3}$.

25. $2, -1, \frac{1}{2}, 2, \ldots$. $d_{1,2} = -1 - 2 = -3$, but $d_{2,3} = \frac{1}{2} + 1 = \frac{3}{2}$ so the sequence is not arithmetic. $r_{1,2} = \frac{-1}{2}$, but $r_{3,4} = \frac{2}{\frac{1}{2}} = 4$ so the sequence is not geometric. Thus, the sequence is neither arithmetic nor geometric.

27. $x-1, x, x+1, x+2, \ldots$. $d_{1,2} = x-(x-1) = 1$, $d_{2,3} = (x+1)-x = 1$, $d_{3,4} = (x+2)-(x+1) = 1$. So the sequence is arithmetic with $d = 1$ and $a_5 = (x+2)+1 = x+3$.

29. $16, 8, 4, 1, \ldots$. $d_{1,2} = 8-16 = -8$, but $d_{2,3} = 4-8 = -4$ so the sequence is not arithmetic. $r_{1,2} = \frac{8}{16} = \frac{1}{2}$, but $r_{3,4} = \frac{1}{4}$. Therefore, the sequence is neither arithmetic. nor geometric.

31. $1, \frac{3}{2}, 2, \frac{5}{2}, \ldots$. $d_{1,2} = \frac{3}{2} - 1 = \frac{1}{2}$, $d_{2,3} = 2 - \frac{3}{2} = \frac{1}{2}$, $d_{3,4} = \frac{5}{2} - 2 = \frac{1}{2}$. Hence, the sequence is arithmetic. with $d = \frac{1}{2}$, and $a_5 = \frac{5}{2} + \frac{1}{2} = 3$.

33. $a_{10} = \frac{55}{2}$, $a_2 = \frac{7}{2}$, and $a_n = a + d(n-1)$. Then $a_2 = a + d = \frac{7}{2} \Leftrightarrow d = \frac{7}{2} - a$. Substituting into $a_{10} = a + 9d = \frac{55}{2}$ gives $a + 9\left(\frac{7}{2} - a\right) = \frac{55}{2} \Leftrightarrow a = \frac{1}{2}$. Thus, the first term is $a_1 = \frac{1}{2}$.

35. $a_{100} = 98$, and $d = 2$. Since $a_{100} = a + 99d = a + 99(2) = a + 198 = 98 \Leftrightarrow a = -100$. Hence, $a_1 = -100$, $a_2 = -100 + 2 = -98$, and $a_3 = -100 + 4 = -96$.

37. $a_1 = 3$, $a_3 = \frac{4}{3}$. Thus $r^2 = \dfrac{\frac{4}{3}}{3} = \dfrac{4}{9} \Leftrightarrow r = \pm\dfrac{2}{3}$. If $r = \dfrac{2}{3}$, then $a_5 = 3\left(\dfrac{2}{3}\right)^4 = \dfrac{16}{27}$, and if $r = -\frac{2}{3}$, $a_5 = 3\left(-\frac{2}{3}\right)^4 = \frac{16}{27}$. Therefore, $a_5 = \frac{16}{27}$.

39. $r = \frac{2}{5}$, $a_4 = \frac{5}{2}$. Since $r = \dfrac{a_4}{a_3}$, then $a_3 = \dfrac{a_4}{r} = \dfrac{\frac{5}{2}}{\frac{2}{5}} = \dfrac{25}{4}$.

41. The arithmetic sequence is $1, 4, 7, \ldots$. So $d = 4 - 1 = 3$ and $a_n = 1 + 3(n-1) = 88 \Leftrightarrow 3(n-1) = 87 \Leftrightarrow n - 1 = 29 \Leftrightarrow n = 30$. So, 88 is the 30th term.

43. We have an arithmetic sequence with $a = 1$, $d = 4$. Thus, $a_n = 1 + 4(n-1) = 11{,}937 \Leftrightarrow 4(n-1) = 11{,}936 \Leftrightarrow n - 1 = 2984 \Leftrightarrow n = 2985$. Thus, 11,937 is a term in the sequence and it is the 2985th term.

45. Since the ball starts at 80 feet, $a = 80$ and since it rebounds three-fourths of the distance fallen, $r = \frac{3}{4}$. So on the nth bounce, the ball attains a height of $a_n = 80\left(\frac{3}{4}\right)^n$. Hence, on the 5th bounce the ball goes $a_5 = 80\left(\frac{3}{4}\right)^5 = \dfrac{80 \cdot 243}{1024} \approx 19$ ft high.

47. Since the original value of the machine is 12,500 and machine depreciates 15% each year, the value of the machine is 85% of the value of the machine the previous year. Thus $a = 12{,}500$ and $r = 0.85$. So after n years, the value of the machine is $a_n = 12{,}500 \cdot (0.85)^n$. Thus after 6 years, the value of the machine is $a_6 = 12{,}500 \cdot (0.85)^6 = \$4714.37$.

49. We start with 5 gallons, so $a = 5$. Each time 1 gallon or $\frac{1}{5}$ of the liquid is removed, thus $\frac{4}{5}$ remains after every repetition of the process, so $r = \frac{4}{5}$. Thus, after n repetitions $a_n = 5\left(\frac{4}{5}\right)^n$ gallons of water remains. Since $a_3 = 5\left(\frac{4}{5}\right)^3 = 2.56$, the radiator contains 2.56 gallons of water after the process is repeated three times. And since $a_5 = 5\left(\frac{4}{5}\right)^5 = 1.6384$, the radiator contains 1.6384 gallons of water after the process is repeated five times.

51. $a_{n+1} = 3a_n$ and $a_1 = c$. Since $\dfrac{a_{n+1}}{a_n} = 3$ for all n, the sequence is geometric and the common ratio is $r = 3$.

53. Since $a_1, a_2, a_3, \ldots$ is a geometric sequence with common ratio r, the terms can be expressed as $a_2 = a_1 \cdot r$, $a_3 = a_1 \cdot r^2, \ldots, a_n = a_1 \cdot r^{n-1}$. Hence $a_2^2 = (a_1 \cdot r)^2 = a_1^2 \cdot r^2$, $a_3^2 = (a_1 \cdot r^2)^2 = a_1^2(r^2)^2, \ldots, a_n^2 = (a_1 \cdot r^{n-1})^2 = a_1^2(r^2)^{n-1}$ and so the sequence $a_1^2, a_2^2, a_3^2, \ldots$ is a geometric sequence with common ratio r^2.

55. Since $a_1, a_2, a_3, \ldots$ is an arithmetic. sequence with common difference d, the terms can be expressed as $a_2 = a_1 + d$, $a_3 = a_1 + 2d, \ldots, a_n = a_1 + (n-1)d$. So $10^{a_2} = 10^{a_1+d} = 10^{a_1} \cdot 10^d$, $10^{a_3} = 10^{a_1+2d} = 10^{a_1} \cdot (10^d)^2, \ldots, 10^{a_n} = 10^{a_1+(n-1)\,d} = 10^{a_1} \cdot (10^d)^{n-1}$ and so $10^{a_1}, 10^{a_2}, 10^{a_3}$, $\ldots$ is a geometric sequence with common ratio $r = 10^d$.

57. Using the hint given, let x denote the middle term of the sequence and d the common difference. Then, $(x-d) + x + (x+d) = 15 \quad \Leftrightarrow \quad 3x = 15 \quad \Leftrightarrow \quad x = 5$. Substituting for x in the second condition we have $(x-d) \cdot x \cdot (x+d) = 80 \quad \Rightarrow \quad (5-d) \cdot 5 \cdot (5+d) = 80 \quad \Leftrightarrow$ $25 - d^2 = 16 \quad \Leftrightarrow \quad d^2 = 9 \quad \Leftrightarrow \quad d = \pm 3$. Thus, the three terms of the sequence are 2, 5, and 8.

59. The first three terms of an arithmetic. sequence are $x-1, x+1, 3x+3$. Then, $a_1 = x-1$, $a_2 = a_1 + d = x+1$, and $a_3 = a_1 + 2d = 3x+3$ and so $d = a_2 - a_1 = (x+1) - (x-1) = 2$. Substituting for $a_1$and d into a_3 gives $x - 1 + 2 \cdot 2 = 3x + 3 \quad \Leftrightarrow \quad 0 = 2x \quad \Leftrightarrow \quad x = 0$.

61. a_n is a geometric sequence with common ratio r. Thus $a_{n+1} = r \cdot a_n$ and $a_{n+2} = r^2 \cdot a_n$. Since we are given $a_n = a_{n+1} + a_{n+2}$, we have $a_n = r \cdot a_n + r^2 \cdot a_n \quad \Leftrightarrow \quad 1 = r + r^2 \quad \Leftrightarrow$ $r^2 + r - 1 = 0 \quad \Leftrightarrow \quad r = \frac{-1 \pm \sqrt{1+4}}{2}$. So $r = \frac{-1-\sqrt{5}}{2}$ (which is inadmissible since the terms are positive) or $r = \frac{-1+\sqrt{5}}{2}$. Therefore, the common ratio is $\frac{\sqrt{5}-1}{2}$.

63. Since we have 5 terms, let us denote $a_1 = 5$ and $a_5 = 80$. Also, $\frac{a_5}{a_1} = r^4$ because the sequence is geometric and so $r^4 = \frac{80}{5} = 16 \quad \Leftrightarrow \quad r = \pm 2$. If $r = 2$, the three geometric means are $a_2 = 10$, $a_3 = 20$, and $a_4 = 40$. (If $r = -2$, the three geometric means are $a_2 = -10$, $a_3 = 20$, and $a_4 = -40$, but these are not between 5 and 80.)

65. The two original numbers are 3 and 5. Thus, the reciprocals are $\frac{1}{3}$ and $\frac{1}{5}$ and their arithmetic. mean is $\frac{1}{2}\left(\frac{1}{3} + \frac{1}{5}\right) = \frac{1}{2}\left(\frac{5}{15} + \frac{3}{15}\right) = \frac{4}{15}$. Therefore, the harmonic mean is $\frac{15}{4}$.

Exercises 9.3

1. $\sum_{k=1}^{4} k = 1 + 2 + 3 + 4 = 10$

3. $\sum_{k=1}^{3} \frac{1}{k} = 1 + \frac{1}{2} + \frac{1}{3} = \frac{6}{6} + \frac{3}{6} + \frac{2}{6} = \frac{11}{6}$

5. $\sum_{i=1}^{8} [1 + (-1)^i] = 0 + 2 + 0 + 2 + 0 + 2 + 0 + 2 = 8$

7. $\sum_{k=1}^{5} 2^{k-1} = 2^0 + 2^1 + 2^2 + 2^3 + 2^4 = 1 + 2 + 4 + 8 + 16 = 31$

9. $\sum_{m=0}^{4} (-3)^{m+2} = (-3)^2 + (-3)^3 + (-3)^4 + (-3)^5 + (-3)^6 = 9 - 27 + 81 - 243 + 729 = 549$

11. $\sum_{m=3}^{5} (2^m + m^2) = (2^3 + 3^2) + (2^4 + 4^2) + (2^5 + 5^2) = 8 + 9 + 16 + 16 + 32 + 25 = 106$

13. $\sum_{k=1}^{5} \sqrt{k} = \sqrt{1} + \sqrt{2} + \sqrt{3} + \sqrt{4} + \sqrt{5}$

15. $\sum_{k=0}^{6} \sqrt{k+4} = \sqrt{4} + \sqrt{5} + \sqrt{6} + \sqrt{7} + \sqrt{8} + \sqrt{9} + \sqrt{10}$

17. $\sum_{i=1}^{8} i\, x^{i+1} = x^2 + 2x^3 + 3x^4 + 4x^5 + 5x^6 + 6x^7 + 7x^8 + 8x^9$

19. $\sum_{j=1}^{n} (-1)^{j+1} x^j = (-1)^2 x + (-1)^3 x^2 + (-1)^4 x^3 + \cdots + (-1)^{n+1} x^n$
$= x - x^2 + x^3 - \cdots + (-1)^{n+1} x^n$

21. $1 + 2 + 3 + 4 + \cdots + 100 = \sum_{k=1}^{100} k$

23. $$\frac{1}{2\ln 2} - \frac{1}{3\ln 3} + \frac{1}{4\ln 4} - \frac{1}{5\ln 5} + \cdots + \frac{1}{100\ln 100} = \sum_{k=2}^{100} \frac{(-1)^k}{k\ln k}$$

25. $$1 - \frac{x}{3} + \frac{x^2}{9} - \frac{x^3}{27} + \frac{x^4}{81} - \frac{x^5}{243} = \sum_{k=0}^{5} \frac{(-1)^k x^k}{3^k}$$

27. $1 - 2x + 3x^2 - 4x^3 + 5x^4 + \cdots - 100x^{99} = \sum_{k=1}^{100} (-1)^{k+1} \cdot k \cdot x^{k-1}$

29. $1 \cdot 2 \cdot 3 + 2 \cdot 3 \cdot 4 + 3 \cdot 4 \cdot 5 + \cdots + 97 \cdot 98 \cdot 99 = \sum_{k=1}^{97} k(k+1)(k+2)$

31. $1 + 3 + 5 + 7 + \cdots + 10001$. Then $S_1 = 1$; $S_2 = 1 + 3 = 4$; $S_3 = 1 + 3 + 5 = 9$;
$S_4 = 1 + 3 + 5 + 7 = 16$; $S_5 = 1 + 3 + 5 + 7 + 9 = 25$; $S_6 = 1 + 3 + 5 + 7 + 9 + 11 = 36$.

33. $\sum_{k=1}^{100} \frac{1}{3^k}$. Then $S_1 = \frac{1}{3}$; $S_2 = \frac{1}{3} + \frac{1}{9} = \frac{4}{9}$; $S_3 = \frac{1}{3} + \frac{1}{9} + \frac{1}{27} = \frac{13}{27}$;

$S_4 = \frac{1}{3} + \frac{1}{9} + \frac{1}{27} + \frac{1}{81} = \frac{40}{81}$; $S_5 = \frac{1}{3} + \frac{1}{9} + \frac{1}{27} + \frac{1}{81} + \frac{1}{243} = \frac{121}{243}$;

$S_6 = \frac{1}{3} + \frac{1}{9} + \frac{1}{27} + \frac{1}{81} + \frac{1}{243} + \frac{1}{729} = \frac{364}{729}$.

35. $\sum_{k=1}^{1000} \left(\frac{1}{k+1} - \frac{1}{k+2} \right)$. Then $S_1 = \frac{1}{1+1} - \frac{1}{1+2} = \frac{1}{2} - \frac{1}{3}$,
$S_2 = \left(\frac{1}{2} - \frac{1}{3}\right) + \left(\frac{1}{3} - \frac{1}{4}\right) = \frac{1}{2} - \frac{1}{4}$, $S_3 = S_2 + a_3 = \left(\frac{1}{2} - \frac{1}{4}\right) + \left(\frac{1}{4} - \frac{1}{5}\right) = \frac{1}{2} - \frac{1}{5}$,
$S_4 = S_3 + a_4 = \left(\frac{1}{2} - \frac{1}{5}\right) + \left(\frac{1}{5} - \frac{1}{6}\right) = \frac{1}{2} - \frac{1}{6}, \ldots$. Therefore, $S_n = \frac{1}{2} - \frac{1}{n+2}$. Thus, the sum is $S_{1000} = \frac{1}{2} - \frac{1}{1002} = \frac{250}{501}$.

37. $\sum_{k=1}^{20} \frac{2}{3^k}$. Then $S_1 = \frac{2}{3}$, $S_2 = \frac{2}{3} + \frac{2}{9} = \frac{8}{9}$, $S_3 = \frac{8}{9} + \frac{2}{27} = \frac{26}{27}, \ldots$. Therefore, $S_n = \frac{3^n - 1}{3^n} = 1 - \frac{1}{3^n}$.
Thus, the sum is $S_{20} = 1 - \frac{1}{3^{20}}$.

39. $\sum_{i=1}^{99} (\sqrt{i} - \sqrt{i+1})$. Then $S_1 = 1 - \sqrt{2}$, $S_2 = (1 - \sqrt{2}) + (\sqrt{2} - \sqrt{3}) = 1 - \sqrt{3}$,
$S_3 = S_2 + a_3 = \left(1 - \sqrt{3}\right) + \left(\sqrt{3} - \sqrt{4}\right) = 1 - \sqrt{4}, \ldots, S_n = 1 - \sqrt{n+1}$. Therefore, the sum of the series is $S_{99} = 1 - \sqrt{100} = -9$.

41. $\sum_{k=1}^{999999} \log \frac{k}{k+1} = \sum_{k=1}^{999999} [\log k - \log(k+1)]$. Then $S_1 = \log 1 - \log 2 = -\log 2$,

$S_2 = (-\log 2) + (\log 2 - \log 3) = -\log 3$, $S_3 = (-\log 3) + (\log 3 - \log 4) = -\log 4, \ldots$.
Therefore, $S_n = -\log(n+1)$. The sum of the series is $S_{999999} = -\log(999999 + 1)$
$= -\log 1{,}000{,}000 = -6$.

Exercises 9.4

1. $a = 1, d = 2, n = 10$. Then $S_{10} = \frac{10}{2}[2a + (10-1)d] = \frac{10}{2}[2 \cdot 1 + 9 \cdot 2] = 100$

3. $a = 4, d = 2, n = 20$. Then $S_{20} = \frac{20}{2}[2a + (20-1)d] = \frac{20}{2}[2 \cdot 4 + 19 \cdot 2] = 460$

5. $a_1 = 55, d = 12, n = 10$. Then $S_{10} = \frac{10}{2}[2a + (10-1)d] = \frac{10}{2}[2 \cdot 55 + 9 \cdot 12] = 1090$

7. $a_3 = 980, a_{10} = 910, n = 5$. So, $980 = a + 2d$ and $910 = a + 9d$. Subtracting the second equation from the first gives $70 = -7d \quad \Leftrightarrow \quad d = -10$. Substituting for d in the first equation gives $980 = a + 2(-10) \quad \Leftrightarrow \quad a = 1000$. Thus $S_5 = \frac{5}{2}[2 \cdot 1000 + 4 \cdot (-10)] = 4900$.

9. $a = 5, r = 2, n = 6$. Then $S_6 = a\dfrac{1-r^n}{1-r} = 5\dfrac{1-2^6}{1-2} = (-5)(-63) = 315$.

11. $a_3 = 28, a_6 = 224, n = 6$. So, $\dfrac{a_6}{a_3} = \dfrac{224}{28} \quad \Leftrightarrow \quad r^3 = 8 \quad \Leftrightarrow \quad r = 2$. Since $a_3 = a \cdot r^2$, we get $a = \dfrac{a_3}{r^2} = \dfrac{28}{2^2} = 7$. Then $S_6 = 7\dfrac{1-2^6}{1-2} = (-7)(-63) = 441$.

13. $a_3 = 0.18, r = 0.3, n = 5$. So, $\dfrac{a_3}{a_1} = r^2 \quad \Leftrightarrow \quad a = a_1 = \dfrac{a_3}{r^2} = \dfrac{0.18}{0.09} = 2$. Thus $S_5 = 2 \cdot \dfrac{1-(0.3)^5}{1-0.3} = 2.8502$.

15. $1 + 5 + 9 + \cdots + 401$ is an arithmetic series. Then $a = 1, d = 5 - 1 = 4$, and $401 = a_n = 1 + 4(n-1) \quad \Leftrightarrow \quad n - 1 = 100 \quad \Leftrightarrow \quad n = 101$. So, $S_{101} = \frac{101}{2}(1 + 401) = 101 \cdot 201 = 20{,}301$.

17. $0.7 + 2.7 + 4.7 + \cdots + 56.7$ is an arithmetic series. Then $a = 0.7, d = 2$, and $56.7 = a_n = 0.7 + 2(n-1) \quad \Leftrightarrow \quad 28 = n - 1 \quad \Leftrightarrow \quad n = 29$. So, $S_{29} = \frac{29}{2}(0.7 + 56.7) = 832.3$.

19. $\sum_{k=0}^{10}(3 + 0.25k)$ is an arithmetic series. Then $a = 3 + 0.25 \cdot 0 = 3, d = 0.25$, and $a_{11} = 3 + 0.25 \cdot 10 = 5.5$. So, $S_{11} = \frac{11}{2}(3 + 5.5) = 46.75$.

21. $1 + 3 + 9 + \cdots + 2187$ is a geometric series. Then $a = 1, r = \dfrac{a_2}{a_1} = \dfrac{3}{1} = 3$, and $2187 = a_n = 1 \cdot 3^{n-1} \quad \Leftrightarrow \quad n - 1 = \log_3 2187 = 7 \quad \Leftrightarrow \quad n = 8$. So, $S_8 = (1)\dfrac{1-3^8}{1-3} = 3280$.

23. $0.7 + 0.49 + 0.343 + \cdots + 0.16807$ is a geometric series. Then $a = 0.7, r = \dfrac{a_2}{a_1} = \dfrac{0.49}{0.7} = 0.7$ and $0.16807 = a_n = 0.7^n \quad \Leftrightarrow \quad n = \log_{0.7} 0.16807 = 5$. So, $S_5 = (0.7)\dfrac{1-(0.7)^5}{1-0.7} = 1.94117$.

25. $\sum_{k=0}^{10} 3(\frac{1}{2})^k$ is a geometric series. Then $a = 3$, $r = \frac{1}{2}$, and $n = 11$. So,

$$S_{11} = (3)\frac{1-\left(\frac{1}{2}\right)^{11}}{1-\left(\frac{1}{2}\right)} = 6\left[1-\left(\frac{1}{2}\right)^{11}\right] = 5.997070313.$$

27. $4 + 2.4 + 1.44 + \cdots + 0.5184$ is a geometric series, where $a = 4$ and $r = \dfrac{a_2}{a_1} = \dfrac{2.4}{4} = 0.6$. Then $0.5184 = a_n = 4 \cdot 0.6^{n-1} \quad \Leftrightarrow \quad n - 1 = \log_{0.6} 0.1296 = 4 \quad \Leftrightarrow \quad n = 5$. So, $S_5 = (4)\dfrac{1-(0.6)^5}{1-0.6} = 9.2224$.

29. $1 - x + x^2 - x^3 + \cdots + x^{20}$ is a geometric series, where $a = 1$ and $r = \dfrac{-x}{1} = -x$. Then $x^{20} = a_n = 1(-x)^{n-1} \quad \Leftrightarrow \quad n = 21$. So, $S_{21} = (1)\dfrac{1-(-x)^{21}}{1-(-x)} = \dfrac{1+x^{21}}{1+x}$.

31. $2 + 4 + 6 + \cdots + 1000$ is an arithmetic series, where $a = 2$ and $d = a_2 - a_1 = 4 - 2 = 2$. Then $1000 = a_n = 2 + 2(n-1) \quad \Leftrightarrow \quad n = 500$. So, $S_{500} = \frac{500}{2}(2 + 1000) = 250{,}500$.

33. $\frac{1}{2} + 1 + \frac{3}{2} + \cdots + 64$ is an arithmetic series, where $a = \frac{1}{2}$ and $d = \frac{1}{2}$. Then $64 = a_n = \frac{1}{2} + \frac{1}{2}(n-1)$ $\Leftrightarrow \quad n = 128$. So, $S_{128} = \frac{128}{2}\left(\frac{1}{2} + 64\right) = 4128$.

35. $\sum_{i=0}^{8}(1 + \sqrt{2}\,i)$ is an arithmetic series, where $a = 1$ and $d = \sqrt{2}$. Then $a_9 = 1 + 8\sqrt{2}$. So, $S_9 = \frac{9}{2}(1 + 1 + 8\sqrt{2}) = 9(1 + 4\sqrt{2})$.

37. $\sum_{n=0}^{8}(5)^{n/3}$ is a geometric series with $a = 1$, $r = {}^3\sqrt{5}$, and $n = 9$. So,

$$S_9 = (1)\frac{1-\left(\sqrt[3]{5}\right)^9}{1-\sqrt[3]{5}} = \frac{1-5^3}{1-\sqrt[3]{5}} = \frac{-124}{1-\sqrt[3]{5}} = \frac{124}{\sqrt[3]{5}-1}.$$

39. We have an arithmetic sequence with $a = 5$ and $d = 2$. Since $2700 = S_n = \frac{n}{2}[2a + (n-1)d]$, we have $2700 = \frac{n}{2}[10 + 2(n-1)] \quad \Leftrightarrow \quad 5400 = 10n + 2n^2 - 2n \quad \Leftrightarrow \quad n^2 + 4n - 2700 = 0$ $\Leftrightarrow \quad (n-50)(n+54) = 0 \quad \Leftrightarrow \quad n = 50$ or $n = -54$. Since n is a positive integer, 50 terms of the sequence must be added.

41. $S_4 = 50$ and $r = \frac{1}{2}$ in a geometric series. Since $50 = S_4 = (a)\dfrac{1-\left(\frac{1}{2}\right)^4}{1-\left(\frac{1}{2}\right)} = (a)\dfrac{15}{8}$, we have $a = 50 \cdot \frac{8}{15} = \frac{80}{3}$. Thus, the first term is $\frac{80}{3}$.

43. We have an arithmetic series where $S_{20} = 155$ and $a_1 = 3$. Since $155 = S_{20} = \frac{20}{2}(2 \cdot 3 + 19d)$, we have $155 = 10(6 + 19d) \quad \Leftrightarrow \quad 95 = 190d \quad \Leftrightarrow \quad d = \frac{1}{2}$. So the common difference is $\frac{1}{2}$.

45. We have a geometric sequence with $a_2 = \dfrac{14}{3}$ and $a_5 = \dfrac{112}{81}$. Then $r^3 = \dfrac{a_5}{a_2} = \dfrac{\frac{112}{81}}{\frac{14}{3}} = \dfrac{8}{27} \quad \Leftrightarrow$

$r = \frac{2}{3}$. Also, $a = a_1 = \dfrac{a_2}{r} = \dfrac{\frac{14}{3}}{\frac{2}{3}} = 7$. So, $S_4 = (7)\dfrac{1 - \left(\frac{2}{3}\right)^4}{1 - \left(\frac{2}{3}\right)} = \dfrac{455}{27}$.

47. $P = 10^{1/10} \cdot 10^{2/10} \cdot 10^{3/10} \cdot \dots \cdot 10^{19/10} = 10^{(1+2+3+\ldots+19)/10}$. Now, $1 + 2 + 3 + \cdots + 19$ is an arithmetic series with $a = 1$, $d = 1$, and $n = 19$. Thus, $1 + 2 + 3 + \cdots + 19 = \frac{19}{2}(1 + 19) = 190$, and so $P = 10^{190/10} = 10^{19}$.

49. We have a geometric sequence with $a = 1$, $r = 2$. Then $S_n = (1)\dfrac{1 - 2^n}{1 - 2} = 2^n - 1$. At the end of 30 days she will have $S_{30} = 2^{30} - 1 = 1{,}073{,}741{,}823$ cents $=$ \$10,737,418.23. To become a billionaire we want $2^n - 1 = 10^{11}$ or approximately $2^n = 10^{11}$. So $\log 2^n = \log 10^{11} \quad \Leftrightarrow$ $n = \frac{11}{\log 2} \approx 36.5$. Thus it will take 37days.

51. The sequence is 16, 48, 80, … . This is an arithmetic sequence with $a = 16$ and $d = 32$.
   (a) The total distance after 6 seconds is $S_6 = \frac{6}{2}(32 + 5 \cdot 32) = 3 \cdot 192 = 576$ ft.
   (b) The total distance after n seconds is $S_n = \frac{n}{2}[32 + 32(n - 1)] = 16n^2$ ft.

53. Let $a_1 = 1$ be the man with 7 wives. Also let $a_2 = 7$ (the wives), $a_3 = 7a_2 = 7^2$ (the sacks), $a_4 = 7a_3 = 7^3$ (the cats), $a_5 = 7a_4 = 7^4$ (the kits). The total is $a_1 + a_2 + a_3 + a_4 + a_5 = 1 + 7 + 7^2 + 7^3 + 7^4$ which is a geometric series with $a = 1$ and $r = 7$. Thus, the number in the party is $S_5 = 1 \cdot \frac{1-7^5}{1-7} = 2801$.

Exercises 9.5

1. $n = 10$, $R = \$1000$, $i = 0.06$. So, $A_f = R\dfrac{(1+i)^n - 1}{i} = 1000\dfrac{(1+0.06)^{10} - 1}{0.06} = \$13,180.79$.

3. $n = 20$, $R = \$5000$, $i = 0.12$. So, $A_f = R\dfrac{(1+i)^n - 1}{i} = 5000\dfrac{(1+0.12)^{20} - 1}{0.12} = \$360{,}262.21$.

5. $n = 16$, $R = \$300$, $i = \dfrac{0.08}{4} = 0.02$. So, $A_f = R\dfrac{(1+i)^n - 1}{i} = 300\dfrac{(1+0.02)^{16} - 1}{0.02}$
$= \$5{,}591.79$.

7. $A_f = \$2000$, $i = \dfrac{0.06}{12} = 0.005$, $n = 8$. Then, $R = \dfrac{iA_f}{(1+i)^n - 1} = \dfrac{(0.005)(2000)}{(1+0.005)^8 - 1} = \$245.66$.

9. $R = \$200$, $n = 20$, $i = \dfrac{0.09}{2} = 0.045$. So, $A_p = R\dfrac{1-(1+i)^{-n}}{i} = (200)\dfrac{1-(1+0.045)^{-20}}{0.045}$
$= \$2601.59$.

11. $A_p = \$12,000$, $i = \dfrac{0.105}{12} = 0.00875$, $n = 48$. Then, $R = \dfrac{iA_p}{1-(1+i)^{-n}} = \dfrac{(0.00875)(12000)}{1-(1+0.00875)^{-48}}$
$= \$307.24$.

13. $A_p = \$100{,}000$, $i = \dfrac{0.08}{12} \approx 0.006667$, $n = 360$. Then, $R = \dfrac{iA_p}{1-(1+i)^{-n}}$
$= \dfrac{(0.006667)(100{,}000)}{1-(1+0.006667)^{-360}} = \$733.76$. Therefore, the total amount paid on this loan over the 30 year period is $(360)(733.76) = \$264{,}153.60$.

15. $A_p = 100{,}000$, $n = 360$, $i = \frac{0.0975}{12} = 0.008125$.
    (a) $R = \dfrac{i\,A_p}{1-(1+i)^{-n}} = \dfrac{(0.008125)(100{,}000)}{1-(1+0.008125)^{-360}} = \$859.15$.
    (b) The total amount that will be paid over the 30 year period is $(360)(859.15) = \$309{,}294.00$.
    (c) $R = \$859.15$, $i = \dfrac{0.0975}{12} = 0.008125$, $n = 360$. So, $A_f = 859.15\dfrac{(1+0.008125)^{360} - 1}{0.008125}$
    $= \$1{,}841{,}519.29$.

17. $R = \$30$, $i = \dfrac{0.10}{12} \approx 0.008333$, $n = 12$. Then, $A_p = R\dfrac{1-(1+i)^{-n}}{i}$
$= 30\dfrac{1-\left(1+0.008333\right)^{-12}}{0.008333} = \$341.24$.

19. $A_p = \$640$, $R = \$32$, $n = 24$. We want to solve the equation $R = \dfrac{iA_p}{1-(1+i)^n}$ for the interest rate i. Let x be the interest rate, then $i = \frac{x}{12}$. Substituting the values we know, we can express R as a function of x by: $R(x) = \dfrac{\frac{x}{12}\cdot 640}{1-\left(1+\frac{x}{12}\right)^{-24}}$. We graph $R(x)$ and $y = 32$ in the rectangle $[0.12, 0.22] \times [30, 34]$. The x-coordinate of the intersection is about 0.1816, which corresponds to an interest rate of 18.16%.

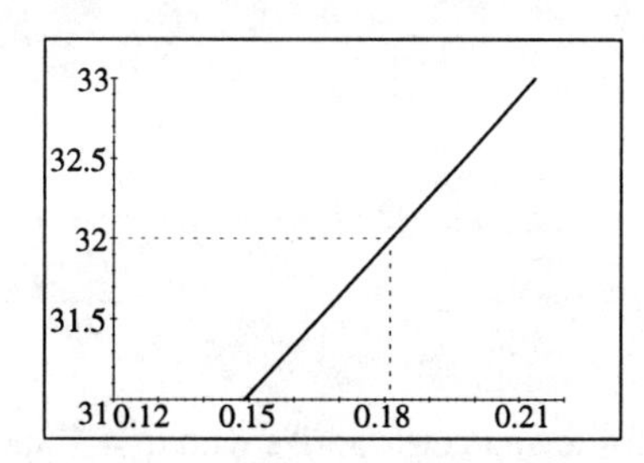

21. $A_p = \$189.99$, $R = \$10.50$, $n = 20$. We want to solve the equation $R = \dfrac{iA_p}{1-(1+i)^n}$ for the interest rate i. Let x be the interest rate, then $i = \frac{x}{12}$. Substituting the values we know, we can express R as a function of x by: $R(x) = \dfrac{\frac{x}{12} \cdot 189.99}{1-\left(1+\frac{x}{12}\right)^{-20}}$. We graph $R(x)$ and $y = 10.50$ in the rectangle $[0.10, 0.18] \times [10, 11]$. The x-coordinate of the intersection is about 0.1168, which corresponds to an interest rate of 11.68%.

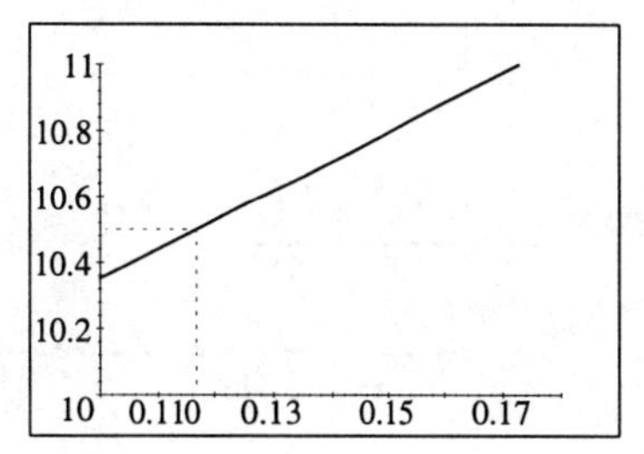

Exercises 9.6

1. $1+\frac{1}{3}+\frac{1}{9}+\frac{1}{27}+\cdots$ is an infinite geometric series with $a=1$ and $r=\frac{1}{3}$. Therefore,
$S=\dfrac{a}{1-r}=\dfrac{1}{1-\left(\frac{1}{3}\right)}=\dfrac{3}{2}$.

3. $1-\frac{1}{3}+\frac{1}{9}-\frac{1}{27}+\cdots$ is an infinite geometric series with $a=1$ and $r=-\frac{1}{3}$. Therefore,
$S=\dfrac{a}{1-r}=\dfrac{1}{1-\left(-\frac{1}{3}\right)}=\dfrac{3}{4}$.

5. $\dfrac{1}{3^6}+\dfrac{1}{3^8}+\dfrac{1}{3^{10}}+\dfrac{1}{3^{12}}+\cdots$ is an infinite geometric series with $a=\dfrac{1}{3^6}$ and $r=\dfrac{1}{3^2}=\dfrac{1}{9}$. Therefore,
$S=\dfrac{a}{1-r}=\dfrac{\frac{1}{3^6}}{1-\left(\frac{1}{9}\right)}=\dfrac{1}{3^6}\cdot\dfrac{9}{8}=\dfrac{1}{648}$.

7. $-\dfrac{100}{9}+\dfrac{10}{3}-1+\dfrac{3}{10}-\cdots$ is an infinite geometric series with $a=-\dfrac{100}{9}$ and $r=-\dfrac{3}{10}$.
Therefore, $S=\dfrac{a}{1-r}=\dfrac{-\frac{100}{9}}{1-\left(-\frac{3}{10}\right)}=\dfrac{-\frac{100}{9}}{\frac{13}{10}}=-\dfrac{100}{9}\cdot\dfrac{10}{13}=-\dfrac{1000}{117}$.

9. $5^{4/3}-5^{5/3}+5^{6/3}-5^{7/3}+\cdots$ is an infinite geometric series with $a=5^{4/3}$ and $r=-5^{1/3}=-\sqrt[3]{5}$.
Thus, $S=\dfrac{a}{1-r}=\dfrac{5^{4/3}}{1+\sqrt[3]{5}}=\dfrac{5^{4/3}}{1+5^{1/3}}$.

11. $0.777\ldots=\dfrac{7}{10}+\dfrac{7}{100}+\dfrac{7}{1000}+\cdots$ is a geometric series with $a=\dfrac{7}{10}$ and $r=\dfrac{1}{10}$. Thus
$0.777\ldots=\dfrac{a}{1-r}=\dfrac{\frac{7}{10}}{1-\frac{1}{10}}=\dfrac{7}{9}$.

13. $0.030303\ldots=\dfrac{3}{100}+\dfrac{3}{10{,}000}+\dfrac{3}{1{,}000{,}000}+\cdots$ is a geometric series with $a=\dfrac{3}{100}$ and
$r=\dfrac{1}{100}$. Thus $0.030303\ldots=\dfrac{a}{1-r}=\dfrac{\frac{3}{100}}{1-\frac{1}{100}}=\dfrac{3}{99}=\dfrac{1}{33}$.

15. $0.\overline{112}=0.112112112\ldots=\dfrac{112}{1000}+\dfrac{112}{1{,}000{,}000}+\dfrac{112}{1{,}000{,}000{,}000}+\cdots$ is a geometric series with
$a=\frac{112}{1000}$ and $r=\frac{1}{1000}$. Thus $0.112112112\ldots=\frac{a}{1-r}=\frac{\frac{112}{1000}}{1-\frac{1}{1000}}=\frac{112}{999}$.

17. Let n be the number of the bounce and a_n the height the ball reaches on the nth bounce. Then $a_0=12$ and $a_n=\frac{2}{3}a_{n-1}$. Since the total distance traveled D includes the bounce up as well and the distance down $D=a_0+2\cdot a_1+2\cdot a_2+\ldots=-a_0+2\cdot a_0+2\cdot a_1+2\cdot a_2+\ldots$
$=-a_0+2\sum\limits_{i=0}^{\infty}a_i=-12+2\sum\limits_{i=0}^{\infty}12\left(\frac{2}{3}\right)^i=-12+24\sum\limits_{i=0}^{\infty}\left(\frac{2}{3}\right)^i=-12+24\cdot\frac{1}{1-\frac{2}{3}}=-12+24\cdot 3$
$=60$. Thus, the total distance traveled is 60 ft.

19. The time required for the ball to stop bouncing is $t = 1 + \frac{1}{\sqrt{2}} + \left(\frac{1}{\sqrt{2}}\right)^2 + \cdots$ which is an infinite geometric series with $a = 1$ and $r = \frac{1}{\sqrt{2}}$. The sum of this series is $t = \dfrac{1}{1 - \frac{1}{\sqrt{2}}} = \dfrac{\sqrt{2}}{\sqrt{2} - 1}$ $= \dfrac{\sqrt{2}}{\sqrt{2} - 1} \cdot \dfrac{\sqrt{2} + 1}{\sqrt{2} + 1} = \dfrac{2 + \sqrt{2}}{2 - 1} = 2 + \sqrt{2}$. Thus the time required for the ball to stop is $2 + \sqrt{2} \approx 3.41$ s.

21. Let A_n be the area of the disks of paper placed at the nth stage. Then $A_1 = \pi R^2$, $A_2 = 2 \cdot \pi\left(\frac{1}{2}R\right)^2 = \frac{\pi}{2}R^2$, $A_3 = 4 \cdot \pi\left(\frac{1}{4}R\right)^2 = \frac{\pi}{4}R^2, \ldots$. We see from this pattern that the total area is $A = \pi R^2 + \frac{1}{2}\pi R^2 + \frac{1}{4}\pi R^2 + \cdots$. Thus, the total area, A, is an infinite geometric series with $a = \pi R^2$ and $r = \frac{1}{2}$. So, $A = \dfrac{\pi R^2}{1 - \frac{1}{2}} = 2\pi R^2$.

23. Using the result from Exercise 22, we have $R = 5000$ and $i = 0.10$. Then $A_p = \dfrac{R}{i} = \dfrac{5000}{0.10}$ $= \$50{,}000$.

25. Let a_n denote the area colored blue as the result of the nth subdivision. Since only the middle square is colored blue $a_n = \frac{1}{9}$(area colored yellow after the $n - 1^{\text{st}}$ subdivision). So $a_1 = \frac{1}{9}$, $a_2 = \frac{1}{9}\left(\frac{8}{9}\right)$, $a_3 = \frac{1}{9}\left(\frac{8}{9}\right)^2$, $a_4 = \frac{1}{9}\left(\frac{8}{9}\right)^3, \ldots$. Thus the total area colored blue $= \frac{1}{9} + \frac{1}{9}\left(\frac{8}{9}\right) + \frac{1}{9}\left(\frac{8}{9}\right)^2 + \frac{1}{9}\left(\frac{8}{9}\right)^3 + \ldots$ is an infinite geometric series with $a = \frac{1}{9}$ and $r = \frac{8}{9}$. So the total area is $\dfrac{\frac{1}{9}}{1 - \frac{8}{9}} = 1$.

Exercises 9.7

1. Let $P(n)$ denote the statement $2+4+6+\cdots+2n = n(n+1)$.

   Step 1 $P(1)$ is the statement that $2 = 1(1+1)$, which is true.

   Step 2 Assume that $P(k)$ is true; that is $2+4+6+\cdots+2k = k(k+1)$. We want to use this to show that $P(k+1)$ is true. Now

   $$2+4+6+\cdots+2k+2(k+1) =$$
   $$k(k+1)+2(k+1) = \qquad \text{induction hypothesis}$$
   $$(k+1)(k+2) = (k+1)[(k+1)+1]$$

   Thus, $P(k+1)$ follows from $P(k)$. So by the Principle of Mathematical Induction, $P(n)$ is true for all n.

3. Let $P(n)$ denote the statement $5+8+11+\cdots+(3n+2) = \dfrac{n(3n+7)}{2}$.

   Step 1 We need to show that $P(1)$ is true. But $P(1)$ says that $5 = \frac{1 \cdot 10}{2}$, which is true.

   Step 2 Assume that $P(k)$ is true; that is $5+8+11+\cdots+(3k+2) = \dfrac{k(3k+7)}{2}$. We want to use this to show that $P(k+1)$ is true. Then

   $$5+8+11+\cdots+(3k+2)+[3(k+1)+2] =$$
   $$\frac{k(3k+7)}{2}+3k+5 = \qquad \text{induction hypothesis}$$
   $$\frac{3k^2+7k}{2}+\frac{6k+10}{2} =$$
   $$\frac{3k^2+13k+10}{2} = \frac{(3k+10)(k+1)}{2} = \frac{(k+1)[3(k+1)+7]}{2}$$

   Thus, $P(k+1)$ follows from $P(k)$. So by the Principle of Mathematical Induction, $P(n)$ is true for all n

5. Let $P(n)$ denote the statement $1\cdot 2+2\cdot 3+3\cdot 4+\cdots+n(n+1) = \dfrac{n(n+1)(n+2)}{3}$.

   Step 1 $P(1)$ is the statement that $1 \cdot 2 = \frac{1\cdot 2\cdot 3}{3}$, which is true.

   Step 2 Assume that $P(k)$ is true; that is $1\cdot 2+2\cdot 3+3\cdot 4+\cdots+k(k+1) = \dfrac{k(k+1)(k+2)}{3}$. We want to use this to show that $P(k+1)$ is true. Then

   $$1\cdot 2+2\cdot 3+3\cdot 4+\cdots+k(k+1)+(k+1)[(k+1)+1] =$$
   $$\frac{k(k+1)(k+2)}{3}+(k+1)(k+2) = \qquad \text{induction hypothesis}$$
   $$\frac{k(k+1)(k+2)}{3}+\frac{3(k+1)(k+2)}{3} = \frac{(k+1)(k+2)(k+3)}{3}$$

   Thus, $P(k+1)$ follows from $P(k)$. So by the Principle of Mathematical Induction, $P(n)$ is true for all n.

7. Let $P(n)$ denote the statement $1^3+2^3+3^3+\cdots+n^3 = \dfrac{n^2(n+1)^2}{4}$.

Step 1 $P(1)$ is the statement that $1^3 = \frac{1^2 \cdot 2^2}{4}$, which is clearly true.

Step 2 Assume that $P(k)$ is true; that is $1^3 + 2^3 + 3^3 + \cdots + k^3 = \frac{k^2(k+1)^2}{4}$. We want to use this to show that $P(k+1)$ is true.

$1^3 + 2^3 + 3^3 + \cdots + k^3 + (k+1)^3 =$

$\frac{k^2(k+1)^2}{4} + (k+1)^3 =$ induction hypothesis

$\frac{(k+1)^2[k^2 + 4(k+1)]}{4} =$

$\frac{(k+1)^2[k^2+4k+4]}{4} = \frac{(k+1)^2(k+2)^2}{4} = \frac{(k+1)^2[(k+1)+1]^2}{4}$

Thus, $P(k+1)$ follows from $P(k)$. So by the Principle of Mathematical Induction, $P(n)$ is true for all n.

9. Let $P(n)$ denote the statement $2^3 + 4^3 + 6^3 + \cdots + (2n)^3 = 2n^2(n+1)^2$.

Step 1 $P(1)$ is true since $2^3 = 2(1)^2(1+1)^2 = 2 \cdot 4 = 8$.

Step 2 Assume that $P(k)$ is true; that is $2^3 + 4^3 + 6^3 + \cdots + (2k)^3 = 2k^2(k+1)^2$. We want to use this to show that $P(k+1)$ is true.

$2^3 + 4^3 + 6^3 + \cdots + (2k)^3 + [2(k+1)]^3 =$

$2k^2(k+1)^2 + [2(k+1)]^3 =$ induction hypothesis

$2k^2(k+1)^2 + 8(k+1)(k+1)^2 =$

$(k+1)^2(2k^2 + 8k + 8) = 2(k+1)^2(k+2)^2 = 2(k+1)^2[(k+1)+1]^2$

So $P(k+1)$ follows from $P(k)$. Thus by the Principle of Mathematical Induction, $P(n)$ is true for all n.

11. Let $P(n)$ denote the statement $1 \cdot 2 + 2 \cdot 2^2 + 3 \cdot 2^3 + 4 \cdot 2^4 + \cdots + n \cdot 2^n = 2[1 + (n-1)2^n]$.

Step 1 $P(1)$ is the statement that $1 \cdot 2 = 2[1+0]$, which is clearly true.

Step 2 Assume that $P(k)$ is true; that is $1 \cdot 2 + 2 \cdot 2^2 + 3 \cdot 2^3 + 4 \cdot 2^4 + \cdots + k \cdot 2^k = 2[1 + (k-1)2^k]$. We want to use this to show that $P(k+1)$ is true.

$1 \cdot 2 + 2 \cdot 2^2 + 3 \cdot 2^3 + 4 \cdot 2^4 + \cdots + k \cdot 2^k + (k+1) \cdot 2^{(k+1)} =$

$2[1 + (k-1)2^k] + (k+1) \cdot 2^{(k+1)} =$ induction hypothesis

$2[1 + (k-1) \cdot 2^k + (k+1) \cdot 2^k] =$

$2[1 + 2k \cdot 2^k] = 2[1 + k \cdot 2^{k+1}] = 2\{1 + [(k+1) - 1]2^{k+1}\}$

Thus, $P(k+1)$ follows from $P(k)$. So by the Principle of Mathematical Induction, $P(n)$ is true for all n.

13. Let $P(n)$ denote the statement $n^2 + n$ is divisible by 2.

Step 1 $P(1)$ is the statement that $1^2 + 1 = 2$ is divisible by 2, which is clearly true.

Step 2 Assume that $P(k)$ is true; that is $k^2 + k$ is divisible by 2. Now $(k+1)^2 + (k+1) = k^2 + 2k + 1 + k + 1 = (k^2 + k) + 2k + 2 = (k^2 + k) + 2(k+1$. By the induction hypothesis, $k^2 + k$ is divisible by 2, and clearly $2(k+1)$ is divisible by 2.

Thus the sum is divisible by 2, so $P(k+1)$ true. Since $P(k+1)$ follows from $P(k)$, by the Principle of Mathematical Induction, $P(n)$ is true for all n.

15. Let $P(n)$ denote the statement that $n^2 - n + 41$ is odd.
Step 1 $P(1)$ is the statement that $1^2 - 1 + 41 = 41$ is odd, which is clearly true.
Step 2 Assume that $P(k)$ is true; that is $k^2 - k + 41$ is odd. We want to use this to show that $P(k+1)$ is true. Now, $(k+1)^2 - (k+1) + 41 = k^2 + 2k + 1 - k - 1 + 41 = (k^2 - k + 41) + 2k$ which is also odd because $k^2 - k + 41$ is odd by the induction hypothesis, $2k$ is always even, and an odd number plus an even number is always odd. Therefore, $P(k+1)$ follows from $P(k)$ and by the Principle of Mathematical Induction, $P(n)$ is true for all n.

17. Let $P(n)$ denote the statement that $8^n - 3^n$ is divisible by 5.
Step 1 $P(1)$ is the statement that $8^1 - 3^1 = 5$ is divisible by 5, which is clearly true.
Step 2 Assume that $P(k)$ is true; that is, $8^k - 3^k$ is divisible by 5. We want to use this to show that $P(k+1)$ is true. Now, $8^{k+1} - 3^{k+1} = 8 \cdot 8^k - 3 \cdot 3^k = 8 \cdot 8^k - (8-5) \cdot 3^k$ $= 8 \cdot (8^k - 3^k) + 5 \cdot 3^k$, which is divisible by 5 because $8^k - 3^k$ is divisible by 5 by our induction hypothesis and $5 \cdot 3^k$ is divisible by 5. So $P(k+1)$ follows from $P(k)$ and by the Principle of Mathematical Induction, $P(n)$ is true for all n.

19. Let $P(n)$ denote the statement $n < 2^n$.
Step 1 $P(1)$ is the statement that $1 < 2^1 = 2$, which is clearly true.
Step 2 Assume that $P(k)$ is true; that is $k < 2^k$. We want to use this to show that $P(k+1)$ is true. Adding 1 to both sides of $P(k)$ we have $k+1 < 2^k + 1$. Since $1 < 2^k$ for $k \geq 1$ we have $2^k + 1 < 2^k + 2^k = 2 \cdot 2^k = 2^{k+1}$. Thus $k+1 < 2^{k+1}$, which is exactly $P(k+1)$. Therefore, $P(k+1)$ follows from $P(k)$. So by the Principle of Mathematical Induction, $P(n)$ is true for all n.

21. Let $P(n)$ denote the statement $(1+x)^n \geq 1 + nx$, if $x > -1$.
Step 1 $P(1)$ is the statement that $(1+x)^1 \geq 1 + 1x$, which is clearly true.
Step 2 Assume that $P(k)$ is true; that is $(1+x)^k \geq 1 + kx$. Now $(1+x)^{k+1} = (1+x)(1+x)^k \geq (1+x)(1+kx)$, by induction hypothesis. Since $(1+x)(1+kx) = 1 + (k+1)x + kx^2 \geq 1 + (k+1)x$ (since $kx^2 \geq 0$), we have $(1+x)^{k+1} \geq 1 + (k+1)x$, which is $P(k+1)$. Since $P(k+1)$ follows from $P(k)$, by the Principle of Mathematical Induction, $P(n)$ is true for all n.

23. Let $P(n)$ be the statement that $a_n = 5 \cdot 3^{n-1}$.
Step 1 $P(1)$ is the statement that $a_1 = 5 \cdot 3^0 = 5$, which is true.
Step 2 Assume that $P(k)$ is true; that is, $a_k = 5 \cdot 3^{k-1}$. We want to use this to show that $P(k+1)$ is true. Now, $a_{k+1} = 3a_k = 3 \cdot (5 \cdot 3^{k-1})$ by the induction hypothesis. Thus $a_{k+1} = 3 \cdot (5 \cdot 3^{k-1}) = 5 \cdot 3^k$, which is exactly $P(k+1)$. Thus, $P(k+1)$ follows from $P(k)$. So by the Principle of Mathematical Induction, $P(n)$ is true for all n.

25. Let $P(n)$ be the statement that $x - y$ is a factor of $x^n - y^n$ for all natural numbers n.
Step 1 $P(1)$ is the statement that $x - y$ is a factor of $x^1 - y^1$, which is clearly true.
Step 2 Assume that $P(k)$ is true; that is, $x - y$ is a factor of $x^k - y^k$. We want to use this to show that $P(k+1)$ is true. Now, $x^{k+1} - y^{k+1} = x^{k+1} - x^k y + x^k y - y^{k+1}$

$= x^k(x-y) + (x^k - y^k)y$ for which $x-y$ is a factor because $x-y$ is a factor of $x^k(x-y)$ and $x-y$ is a factor of $(x^k - y^k)y$ by the induction hypothesis. Thus, $P(k+1)$ follows from $P(k)$. So by the Principle of Mathematical Induction, $P(n)$ is true for all n.

27. (a) $P(n) = n^2 - n + 11$ is prime for all n. This is false as the case for $n = 11$ demonstrates: $P(11) = 11^2 - 11 + 11 = 121$, which is not prime since $11^2 = 121$.

(b) $n^2 > n$ for all $n \geq 2$. This is true. Let $P(n)$ denote the statement that $n^2 > n$.

Step 1 $P(2)$ is the statement that $2^2 = 4 > 2$, which is clearly true.

Step 2 Assume that $P(k)$ is true; that is, $k^2 > k$. We want to use this to show that $P(k+1)$ is true. Now $(k+1)^2 = k^2 + 2k + 1$. Using the induction hypothesis to replace k^2 we have $k^2 + 2k + 1 > k + 2k + 1 = 3k + 1 > k + 1$, since $k \geq 2$. Thus $(k+1)^2 > k + 1$, which is exactly $P(k+1)$. Thus, $P(k+1)$ follows from $P(k)$. So by the Principle of Mathematical Induction, $P(n)$ is true for all n.

(c) $2^{2n+1} + 1$ is divisible by 3 for all $n \geq 1$. This is true. Let $P(n)$ denote the statement that $2^{2n+1} + 1$ is divisible by 3.

Step 1 $P(1)$ is the statement that $2^3 + 1 = 9$ is divisible by 3, which is clearly true.

Step 2 Assume that $P(k)$ is true; that is, $2^{2k+1} + 1$ is divisible by 3. We want to use this to show that $P(k+1)$ is true. Now, $2^{2(k+1)+1} + 1 = 2^{2k+3} + 1 = 4 \cdot 2^{2k+1} + 1$ $= (3+1)2^{2k+1} + 1 = 3 \cdot 2^{2k+1} + (2^{2k+1} + 1)$, which is divisible by 3 since $2^{2k+1} + 1$ is divisible by 3 by the induction hypothesis and $3 \cdot 2^{2k+1}$ is divisible by 3. Thus, $P(k+1)$ follows from $P(k)$. So by the Principle of Mathematical Induction, $P(n)$ is true for all n.

(d) The statement $n^3 \geq (n+1)^2$ for all $n \geq 2$ is false. The statement fails when $n = 2$: $2^3 = 8 < (2+1)^2 = 9$.

(e) $n^3 - n$ is divisible by 3 for all $n \geq 2$. This is true. Let $P(n)$ denote the statement that $n^3 - n$ is divisible by 3.

Step 1 $P(2)$ is the statement that $2^3 - 2 = 6$ is divisible by 3, which is clearly true.

Step 2 Assume that $P(k)$ is true; that is, $k^3 - k$ is divisible by 3. We want to use this to show that $P(k+1)$ is true. Now, $(k+1)^3 - (k+1) = k^3 + 3k^2 + 3k + 1 - (k+1) = k^3 + 3k^2 + 2k = k^3 - k + 3k^2 + 2k + k = (k^3 - k) + 3(k^2 + k)$. The term $k^3 - k$ is divisible by 3 by our induction hypothesis and the term $3(k^2 + k)$ is clearly divisible by 3. Thus $(k+1)^3 - (k+1)$ is divisible by 3, which is exactly $P(k+1)$. So by the Principle of Mathematical Induction, $P(n)$ is true for all n.

(f) $n^3 - 6n^2 + 11n$ is divisible by 6 for all $n \geq 1$. This is true. Let $P(n)$ denote the statement that $n^3 - 6n^2 + 11n$ is divisible by 6.

Step 1 $P(1)$ is the statement that $(1)^3 - 6(1)^2 + 11(1) = 6$ is divisible by 6, which is clearly true.

Step 2 Assume that $P(k)$ is true; that is, $k^3 - 6k^2 + 11k$ is divisible by 6. We show that $P(k+1)$ is then also true. Now $(k+1)^3 - 6(k+1)^2 + 11(k+1)$
$= k^3 + 3k^2 + 3k + 1 - 6k^2 - 12k - 6 + 11k + 11 = k^3 - 3k^2 + 2k + 6$
$= k^3 - 6k^2 + 11k + (3k^2 - 9k + 6) = (k^3 - 6k^2 + 11k) + 3(k^2 - 3k + 2)$
$= (k^3 - 6k^2 + 11k) + 3(k-1)(k-2)$. In this last expression the first term is divisible by 6 by our induction hypothesis. The second term is also divisible by 6. To

see this notice that $(k-1)$ and $(k-2)$ are consecutive natural numbers and so one of them must be even (divisible by 2). Since 3 also appears in this second term, it follows that this term is divisible by 2 and 3 and so is divisible by 6. Thus $P(k+1)$ follows from $P(k)$. So by the Principle of Mathematical Induction, $P(n)$ is true for all n.

29. Let $P(n)$ denote the statement that $F_1 + F_2 + F_3 + \cdots + F_n = F_{n+2} - 1$.

Step 1 $P(1)$ is the statement that $F_1 = F_3 - 1$. But $F_1 = 1 = 2 - 1 = F_3 - 1$, which is true.

Step 2 Assume that $P(k)$ is true; that is $F_1 + F_2 + F_3 + \cdots + F_k = F_{k+2} - 1$. We want to use this to show that $P(k+1)$ is true. Now, $F_{(k+1)+2} - 1 = F_{k+3} - 1 = F_{k+2} + F_{k+1} - 1$ $= (F_{k+2} - 1) + F_{k+1} = F_1 + F_2 + F_3 + \cdots + F_k + F_{k+1}$ applying the induction hypothesis. Thus $P(k+1)$ follows from $P(k)$. So by the Principle of Mathematical Induction, $P(n)$ is true for all n.

31. Let $a_1 = a_2 = 2$ and $a_{n+2} = a_{n+1} \cdot a_n$ for $n \geq 1$. Let $P(n)$ denote the statement that $a_m = 2^{F_m}$ for $m = 1, \ldots, n$. (Notice that this is a slightly different statement, $P(n)$ is a statement that involves all the previous cases, not only n.)

Step 1 $P(1)$ is the statement that $a_1 = 2^{F_1}$, which is true since $a_1 = 2$ and $2^{F_1} = 2^1 = 2$.

Step 2 Assume that $P(k)$ is true; that is $a_m = 2^{F_m}$ for $m = 1, \ldots, k$. We want to use this to show that $P(k+1)$ is true; that is, $a_{m+1} = 2^{F_{m+1}}$ for $m = 1, \ldots, k+1$. Now, by definition of a_n, $a_{k+1} = a_k \cdot a_{k-1}$. And by applying the induction hypothesis for $m = k$ and $m = k-1$ we have $a_k \cdot a_{k-1} = 2^{F_k} \cdot 2^{F_{k-1}} = 2^{F_k + F_{k-1}} = 2^{F_{k+1}}$, the last equality is by the definition of the Fibonacci sequence. Therefore, $a_m = 2^{F_m}$ for $m = 1, 2, \ldots, k, k+1$, which is exactly $P(k+1)$. So by the Principle of Mathematical Induction, $P(n)$ is true for all n.

33. Let $a_1 = 1$ and $a_{n+1} = \dfrac{1}{1 + a_n}$ for $n \geq 1$. Let $P(n)$ be the statement that $a_n = \dfrac{F_n}{F_{n+1}}$ for all $n \geq 1$.

Step 1 $P(1)$ is the statement that $a_1 = \dfrac{F_1}{F_2}$, which is true since $a_1 = 1$ and $\dfrac{F_1}{F_2} = \dfrac{1}{1} = 1$.

Step 2 Assume that $P(k)$ is true; that is, $a_k = \dfrac{F_k}{F_{k+1}}$. We want to use this to show that $P(k+1)$ is true. Now, $a_{k+1} = \dfrac{1}{1 + a_k}$ by the definition of the a_n, and by using the induction hypothesis, $\dfrac{1}{1 + a_k} = \dfrac{1}{1 + \frac{F_k}{F_{k+1}}} = \dfrac{F_{k+1}}{F_k + F_{k+1}} = \dfrac{F_{k+1}}{F_{k+2}}$ (by definition of the Fibonacci sequence). So $P(k+1)$ follows from $P(k)$, and by the Principle of Mathematical Induction, $P(n)$ is true for all n.

35. Since $100 \cdot 10 = 10^3$, $100 \cdot 11 < 11^3$, $100 \cdot 12 < 12^3$, $\ldots$ our conjecture is that $100n \leq n^3$ for all natural numbers $n \geq 10$. Let $P(n)$ denote the statement that $100n \leq n^3$.

Step 1 $P(10)$ is the statement that $100 \cdot 10 = 1{,}000 \leq 10^3 = 1{,}000$, which is true.

Step 2 Assume that $P(k)$ is true; that is, $100k \leq k^3$. We want to use this to show that $P(k+1)$ is true. Now, $100(k+1) = 100k + 100 \leq k^3 + 100$ (by the induction hypothesis) $\leq k^3 + k^2$ (because $k \geq 10$, so $k^2 \geq 100$) $\leq k^3 + 3k^2 + 3k + 1 = (k+1)^3$. Thus $100(k+1) \leq (k+1)^3$, which is exactly $P(k+1)$. So $P(k+1)$ follows from $P(k)$, and by the Principle of Mathematical Induction, $P(n)$ is true for all $n \geq 10$.

Exercises 9.8

1. $(x+y)^6 = x^6 + 6x^5y + 15x^4y^2 + 20x^3y^3 + 15x^2y^4 + 6xy^5 + y^6$

3. $\left(x+\dfrac{1}{x}\right)^4 = x^4 + 4x^3 \cdot \dfrac{1}{x} + 6x^2\left(\dfrac{1}{x}\right)^2 + 4x\left(\dfrac{1}{x}\right)^3 + \left(\dfrac{1}{x}\right)^4 = x^4 + 4x^2 + 6 + \dfrac{4}{x^2} + \dfrac{1}{x^4}$

5. $(x-1)^5 = x^5 - 5x^4 + 10x^3 - 10x^2 + 5x - 1$

7. $(x^2y-1)^5 = (x^2y)^5 - 5(x^2y)^4 + 10(x^2y)^3 - 10(x^2y)^2 + 5x^2y - 1$
$= x^{10}y^5 - 5x^8y^4 + 10x^6y^3 - 10x^4y^2 + 5x^2y - 1$

9. $(2x-3y)^3 = (2x)^3 - 3(2x)^2 3y + 3 \cdot 2x(3y)^2 - (3y)^3 = 8x^3 - 36x^2y + 54xy^2 - 27y^3$

11. $\left(\dfrac{1}{x} - \sqrt{x}\right)^5 = \left(\dfrac{1}{x}\right)^5 - 5\left(\dfrac{1}{x}\right)^4 \sqrt{x} + 10\left(\dfrac{1}{x}\right)^3 x - 10\left(\dfrac{1}{x}\right)^2 x\sqrt{x} + 5\left(\dfrac{1}{x}\right)x^2 - x^2\sqrt{x}$
$= \dfrac{1}{x^5} - \dfrac{5}{x^{7/2}} + \dfrac{10}{x^2} - \dfrac{10}{x^{1/2}} + 5x - x^{5/2}$

13. $\dbinom{6}{4} = \dfrac{6!}{4!\,2!} = \dfrac{6 \cdot 5 \cdot 4!}{2 \cdot 1 \cdot 4!} = 15$

15. $\dbinom{100}{98} = \dfrac{100!}{98!\,2!} = \dfrac{100 \cdot 99 \cdot 98!}{98! \cdot 2 \cdot 1} = 4950$

17. $\dbinom{3}{1}\dbinom{4}{2} = \dfrac{3!}{1!\,2!}\dfrac{4!}{2!\,2!} = \dfrac{3 \cdot 2! \cdot 4 \cdot 3 \cdot 2!}{1 \cdot 2! \cdot 2 \cdot 1 \cdot 2!} = 18$

19. $\dbinom{5}{0} + \dbinom{5}{1} + \dbinom{5}{2} + \dbinom{5}{3} + \dbinom{5}{4} + \dbinom{5}{5} = (1+1)^5 = 2^5 = 32$

21. $(x+2y)^4 = \dbinom{4}{0}x^4 + \dbinom{4}{1}x^3 \cdot 2y + \dbinom{4}{2}x^2 \cdot 4y^2 + \dbinom{4}{3}x \cdot 8y^3 + \dbinom{4}{4}16y^4$
$= x^4 + 8x^3y + 24x^2y^2 + 32xy^3 + 16y^4$

23 $\left(1+\dfrac{1}{x}\right)^6 = \dbinom{6}{0}1^6 + \dbinom{6}{1}1^5\left(\dfrac{1}{x}\right) + \dbinom{6}{2}1^4\left(\dfrac{1}{x}\right)^2 + \dbinom{6}{3}1^3\left(\dfrac{1}{x}\right)^3 + \dbinom{6}{4}1^2\left(\dfrac{1}{x}\right)^4$
$+ \dbinom{6}{5}1\left(\dfrac{1}{x}\right)^5 + \dbinom{6}{6}\left(\dfrac{1}{x}\right)^6 = 1 + \dfrac{6}{x} + \dfrac{15}{x^2} + \dfrac{20}{x^3} + \dfrac{15}{x^4} + \dfrac{6}{x^5} + \dfrac{1}{x^6}$

25. The first three terms in the expansion of $(x+2y)^{20}$ are: $\dbinom{20}{0}x^{20} = x^{20}$, $\dbinom{20}{1}x^{19} \cdot 2y = 40x^{19}y$, and $\dbinom{20}{2}x^{18} \cdot 4y^2 = 760x^{18}y^2$.

27. The last two terms in the expansion of $(a^{2/3} + a^{1/3})^{25}$ are: $\dbinom{25}{24}a^{2/3} \cdot a^8 = 25a^{26/3}$, and $\dbinom{25}{25}a^{25/3} = a^{25/3}$.

29. The middle term in the expansion of $(x^2+1)^{18}$ occurs when both terms are raised to the 9th power. So, this term is $\binom{18}{9}(x^2)^9 1^9 = 48{,}620x^{18}$.

31. The 24th term in the expansion of $(a+b)^{25}$ is $\binom{25}{23}a^2b^{23} = 300a^2b^{23}$.

33. The 100th term in the expansion of $(1+y)^{100}$ is $\binom{100}{99}1^1 \cdot y^{99} = 100y^{99}$.

35. The term that contains x^4 in the expansion of $(x+2y)^{10}$ has exponent $r=4$. So, this term is $\binom{10}{4}x^4 \cdot (2y)^{10-4} = 13{,}440x^4y^6$.

37. The rth term is $\binom{12}{r}a^r(b^2)^{12-r} = \binom{12}{r}a^r b^{24-2r}$. Thus the term that contains b^8 occurs where $24-2r=8 \quad \Leftrightarrow \quad r=8$. Thus, the term is $\binom{12}{8}a^8b^8 = 495a^8b^8$.

39. The rth term is $\binom{8}{r}(8x)^r\left(\frac{1}{2x}\right)^{8-r} = \binom{8}{r}\frac{8^r}{2^{8-r}} \cdot \frac{x^r}{x^{8-r}} = \binom{8}{r}\frac{8^r}{2^{8-r}} \cdot x^{2r-8}$. So the term that does not contain x occurs when $2r-8=0 \quad \Leftrightarrow \quad r=4$. Thus, the term is $\binom{8}{4}(8x)^4\left(\frac{1}{2x}\right)^4 = 17{,}920$.

41. The rth term is $\binom{8}{r}(2c)^r(c^{1/2})^{8-r} = \binom{8}{r}2^r c^r c^{4-r/2} = \binom{8}{r}2^r c^{4+r/2}$. The term that contains c^7 occurs when $4+\frac{r}{2}=7 \quad \Leftrightarrow \quad \frac{r}{2}=3 \quad \Leftrightarrow \quad r=6$. So, the term is $\binom{8}{6}2^6c^7 = 1792c^7$.

43. $\left(x+\frac{1}{x}\right)^{20} = \binom{20}{0}x^{20} + \binom{20}{1}x^{19}\left(\frac{1}{x}\right) + \binom{20}{2}x^{18}\left(\frac{1}{x}\right)^2 + \cdots + \binom{20}{20}x^0\left(\frac{1}{x}\right)^{20}$ which therefore has 21 distinct terms.

45. $(a^2-2ab+b^2)^5 = [(a-b)^2]^5 = (a-b)^{10}$ which has 11 distinct terms.

47. $x^4+4x^3y+6x^2y^2+4xy^3+y^4 = (x+y)^4$

49. $8a^3+12a^2b+6ab^2+b^3 = \binom{3}{0}(2a)^3 + \binom{3}{1}(2a)^2b + \binom{3}{2}2ab^2 + \binom{3}{3}b^3 = (2a+b)^3$

51. $(a^2+a+1)^4 = [a^2+(a+1)]^4 = a^8+4a^6(a+1)+6a^4(a+1)^2+4a^2(a+1)^3+(a+1)^4$
$= a^8+4a^7+4a^6+6a^4(a^2+2a+1)+4a^2(a^3+3a^2+3a+1)+a^4+4a^3+6a^2+4a+1$
$= a^8+4a^7+10a^6+16a^5+19a^4+16a^3+10a^2+4a+1$

53. Notice that $(100!)^{101} = (100!)^{100} \cdot 100!$ and $(101!)^{100} = (101 \cdot 100!)^{100} = 101^{100} \cdot (100!)^{100}$. Now $100! = 1 \cdot 2 \cdot 3 \cdot 4 \cdot \cdots \cdot 99 \cdot 100$ and $101^{100} = 101 \cdot 101 \cdot 101 \cdot \cdots \cdot 101$. Thus each of these last two expressions consists of 100 factors multiplied together and since each factor in the product for 101^{100} is larger than each factor in the product for $100!$ it follows that $100! < 101^{100}$. Thus $(100!)^{100} \cdot 100! < (100!)^{100} \cdot 101^{100}$. So, $(100!)^{101} < (101!)^{100}$.

55. $\binom{n}{1} = \frac{n!}{1!\,(n-1)!} = \frac{n(n-1)!}{1(n-1)!} = \frac{n}{1} = n.$ $\binom{n}{n-1} = \frac{n!}{(n-1)!\,1!} = \frac{n(n-1)!}{(n-1)!\,1} = n.$
Therefore, $\binom{n}{1} = \binom{n}{n-1}$.

57. (a) $2^n = (1+1)^n = \binom{n}{0}1^0 \cdot 1^n + \binom{n}{1}1^1 \cdot 1^{n-1} + \binom{n}{2}1^2 \cdot 1^{n-2} + \cdots + \binom{n}{n}1^n \cdot 1^0 =$
$\binom{n}{0} + \binom{n}{1} + \binom{n}{2} + \cdots + \binom{n}{n}$.

(b) $\binom{n}{r}$ is the number of ways of choosing r objects from n objects. That is, $\binom{n}{r}$ is the number of r element subsets of the n objects. Thus $\binom{n}{0} + \binom{n}{1} + \binom{n}{2} + \cdots + \binom{n}{n-1} + \binom{n}{n}$ is the number of subsets with 0 elements, plus the number of subsets with 1 element, plus the number of subsets with 2 elements, and so on. But this is the total number of subsets of n objects and therefore should be equal to 2^n.

59. (a) $\binom{n}{r-1} + \binom{n}{r} = \frac{n!}{(r-1)!\,[n-(r-1)]!} + \frac{n!}{r!\,(n-r)!}$

(b) $\frac{n!}{(r-1)!\,[n-(r-1)]!} + \frac{n!}{r!\,(n-r)!} = \frac{r \cdot n!}{r \cdot (r-1)!\,(n-r+1)!} + \frac{(n-r+1)\,n!}{r!\,(n-r+1)(n-r)!}$
$= \frac{r \cdot n!}{r!\,(n-r+1)!} + \frac{(n-r+1)\,n!}{r!\,(n-r+1)!}$. Thus a common denominator is $r!\,(n-r+1)!$.

(c) Therefore, using the results of parts (a) and (b),
$$\binom{n}{r-1} + \binom{n}{r} = \frac{n!}{(r-1)!\,[n-(r-1)]!} + \frac{n!}{r!\,(n-r)!} = \frac{r \cdot n!}{r!\,(n-r+1)!} + \frac{(n-r+1)\,n!}{r!\,(n-r+1)!}$$
$$= \frac{r \cdot n! + (n-r+1)n!}{r!\,(n-r+1)!} = \frac{n! \cdot (r+n-r+1)}{r!\,(n-r+1)!} = \frac{n! \cdot (n+1)}{r!\,(n+1-r)!}$$
$$= \frac{(n+1)!}{r\,!(n+1-r)!} = \binom{n+1}{r}.$$

61. (a) $(x+y)^5 = (x+y)(x+y)(x+y)(x+y)(x+y)$
$= (x+y)(x+y)(x+y)(xx + xy + yx + yy)$
$= (x+y)(x+y)(xxx + xxy + xyx + xyy + yxx + yxy + yyx + yyy)$
$= (x+y)(xxxx + xxxy + xxyx + xxyy + xyxx + xyxy + xyyx + xyyy + yxxx + yxxy$
$+ yxyx + yxyy + yyxx + yyxy + yyyx + yyyy)$
$= xxxxx + xxxxy + xxxyx + xxxyy + xxyxx + xxyxy + xxyyx + xxyyy + xyxxx$
$+ xyxxy + xyxyx + xyxyy + xyyxx + xyyxy + xyyyx + xyyyy + yxxxx + yxxxy$
$+ yxxyx + yxxyy + yxyxx + yxyxy + yxyyx + yxyyy + yyxxx + yyxxy + yyxyx$
$+ yyxyy + yyyxx + yyyxy + yyyyx + yyyyy$

(b) There are ten terms:
$xxyyy + xyxyy + xyyxy + xyyyx + yxxyy + yxyxy + yxyyx + yyxxy + yyxyx + yyyxx$

(c) The x's in part (b) appear is all possible ways. Thus the number of ways to pick 2 of the 5 position to hold the x's (and we fill the remaining positions with y's) is $C(5, 2)$. Thus the coefficient of x^2y^3 is $C(5, 2)$.

Review Exercises for Chapter 9

1. $a_n = \dfrac{n^2}{n+1}$. Then $a_1 = \dfrac{1^2}{1+1} = \dfrac{1}{2}$; $a_2 = \dfrac{2^2}{2+1} = \dfrac{4}{3}$; $a_3 = \dfrac{3^2}{3+1} = \dfrac{9}{4}$; $a_4 = \dfrac{4^2}{4+1} = \dfrac{16}{5}$; and $a_{10} = \dfrac{10^2}{10+1} = \dfrac{100}{11}$.

3. $a_n = \dfrac{[(-1)^n + 1]}{n^3}$. Then $a_1 = \dfrac{(-1)^1+1}{1^3} = 0$; $a_2 = \dfrac{(-1)^2+1}{2^3} = \dfrac{2}{8} = \dfrac{1}{4}$; $a_3 = \dfrac{(-1)^3+1}{3^3} = 0$; $a_4 = \dfrac{(-1)^4+1}{4^3} = \dfrac{2}{64} = \dfrac{1}{32}$; and $a_{10} = \dfrac{(-1)^{10}+1}{10^3} = \dfrac{1}{500}$.

5. $a_n = \dfrac{(2n)!}{2^n n!}$. Then $a_1 = \dfrac{(2\cdot 1)!}{2^1 \cdot 1!} = 1$; $a_2 = \dfrac{(2\cdot 2)!}{2^2\cdot 2!} = 3$; $a_3 = \dfrac{(2\cdot 3)!}{2^3\cdot 3!} = \dfrac{6\cdot 5\cdot 4}{8} = 15$; $a_4 = \dfrac{(2\cdot 4)!}{2^4\cdot 4!} = \dfrac{8\cdot 7\cdot 6\cdot 5}{16} = 105$; and $a_{10} = \dfrac{(2\cdot 10)!}{2^{10}\cdot 10!} = 654{,}729{,}075$.

7. $a_n = a_{n-1} + 2n - 1$ and $a_1 = 1$. Then $a_2 = a_1 + 4 - 1 = 4$; $a_3 = a_2 + 6 - 1 = 9$; $a_4 = a_3 + 8 - 1 = 16$; $a_5 = a_4 + 10 - 1 = 25$; $a_6 = a_5 + 12 - 1 = 36$; and $a_7 = a_6 + 14 - 1 = 49$.

9. $a_n = a_{n-1} + 2a_{n-2}$, $a_1 = 1$ and $a_2 = 3$. Then $a_3 = a_2 + 2a_1 = 5$; $a_4 = a_3 + 2a_2 = 11$; $a_5 = a_4 + 2a_3 = 21$; $a_6 = a_5 + 2a_4 = 43$; and $a_7 = a_6 + 2a_5 = 85$.

11. $a_n = (a_{n-1} - 1)!$ and $a_1 = 3$. Then $a_2 = (a_1 - 1)! = (3-2)! = 2$; $a_3 = (a_2 - 1)! = (2-1)! = 1$; $a_4 = (a_3 - 1)! = (1-1)! = 1$; $a_5 = a_6 = a_7 = 1$.

13. $5, 5.5, 6, 6.5, \ldots$. Since $5.5 - 5 = 6 - 5.5 = 6.5 - 6 = 0.5$, this is an arithmetic sequence with $a_1 = 5$ and $d = 0.5$. Then $a_5 = a_4 + 0.5 = 7$.

15. $\sqrt{2}, 2\sqrt{2}, 3\sqrt{2}, 4\sqrt{2}, \ldots$. Since $2\sqrt{2} - \sqrt{2} = 3\sqrt{2} - 2\sqrt{2} = 4\sqrt{2} - 3\sqrt{2} = \sqrt{2}$, this is an arithmetic sequence with $a_1 = \sqrt{2}$ and $d = \sqrt{2}$. Then $a_5 = a_4 + \sqrt{2} = 4\sqrt{2} + \sqrt{2} = 5\sqrt{2}$.

17. $t-3, t-2, t-1, t, \ldots$. Since $(t-2)-(t-3) = (t-1)-(t-2) = t-(t-1) = 1$, this is an arithmetic sequence with $a_1 = t - 3$ and $d = 1$. Then $a_5 = a_4 + 1 = t + 1$.

19. $\dfrac{3}{4}, \dfrac{1}{2}, \dfrac{1}{3}, \dfrac{2}{9}, \ldots$. Since $\dfrac{\frac{1}{2}}{\frac{3}{4}} = \dfrac{\frac{1}{3}}{\frac{1}{2}} = \dfrac{\frac{2}{9}}{\frac{1}{3}} = \dfrac{2}{3}$, this is a geometric sequence with $a_1 = \dfrac{3}{4}$ and $r = \frac{2}{3}$. Then $a_5 = a_4 \cdot r = \frac{2}{9}\cdot\frac{2}{3} = \frac{4}{27}$.

21. $\ln a, \ln 2a, \ln 3a, \ln 4a, \ldots$. Since $\ln 2a - \ln a \neq \ln 3a - \ln 2a$ and $\dfrac{\ln 2a}{\ln a} \neq \dfrac{\ln 3a}{\ln 2a}$ this series is neither arithmetic nor geometric.

23. $a, abc^3, ab^2c^6, ab^3c^9, \ldots$. Since $\dfrac{abc^3}{a} = \dfrac{ab^2c^6}{abc^3} = \dfrac{ab^3c^9}{ab^2c^6} = bc^3$, this is a geometric sequence with $a_1 = a$ and $r = bc^3$. Then $a_5 = a_4 \cdot r = ab^3c^9 \cdot bc^3 = ab^4c^{12}$.

25. $3, 6i, -12, -24i, \ldots$. Since $\dfrac{6i}{3} = 2i$, $\dfrac{-12}{6i} = \dfrac{-2}{i} = \dfrac{-2i}{i^2} = 2i$, $\dfrac{-24i}{-12} = 2i$, this is a geometric sequence with common ratio $r = 2i$.

27. $a_6 = 17 = a + 5d$ and $a_4 = 11 = a + 3d$. Then, $a_6 - a_4 = 17 - 11 \Leftrightarrow (a + 5d) - (a + 3d) = 6 \Leftrightarrow 6 = 2d \Leftrightarrow d = 3$. Substituting into $11 = a + 3d$ gives $11 = a + 3 \cdot 3$ and so $a = 2$. So, $a_2 = a + (2 - 1)d = 2 + 3 = 5$.

29. $a_3 = 9$ and $r = \frac{3}{2}$. Then $a_5 = a_3 \cdot r^2 = 9 \cdot \left(\frac{3}{2}\right)^2 = \frac{81}{4}$.

31. Let a_{mc} denote the term of the geometric series which is the frequency of middle C. Then $a_{\text{mc}} = 256$ and $a_{\text{mc}+1} = 512$. Since this is a geometric sequence, $r = \frac{512}{256} = 2$, and so $a_{\text{mc}-2} = \dfrac{a_{\text{mc}}}{r^2} = \dfrac{256}{2^2} = 64$.

33. Let a_n be the number of bacteria in the dish at the end of $5n$ seconds. So, $a_0 = 3$, $a_1 = 3 \cdot 2$, $a_2 = 3 \cdot 2^2$, $a_3 = 3 \cdot 2^3$, $\ldots$. Then, clearly, a_n is a geometric sequence with $r = 2$ and $a = 3$. Thus at the end of $60 = 5(12)$ seconds, the number of bacteria is $a_{12} = 3 \cdot 2^{12} = 12{,}288$.

35. Suppose that the common ratio in the sequence $a_1, a_2, a_3, \ldots$ is r_1. Also suppose that the common ratio in the sequence $b_1, b_2, b_3, \ldots$ is r_2. Then $a_n = a_1 r_1^{n-1}$ and $\text{b}_n = b_1 r_2^{n-1}$, $n = 1, 2, 3, \ldots$. Thus $a_n b_n = a_1 r_1^{n-1} \cdot b_1 r_2^{n-1} = (a_1 b_1)(r_1 r_2)^{n-1}$. So the sequence $a_1 b_1, a_2 b_2, a_3 b_3, \ldots$ is geometric with common ratio $r_1 r_2$.

37. (a) $6, x, 12, \ldots$ is arithmetic if $x - 6 = 12 - x \Leftrightarrow 2x = 18 \Leftrightarrow x = 9$.

(b) $6, x, 12, \ldots$ is geometric if $\dfrac{x}{6} = \dfrac{12}{x} \Leftrightarrow x^2 = 72 \Leftrightarrow x = \pm 6\sqrt{2}$.

39. $\displaystyle\sum_{k=3}^{6} (k+1)^2 = (3+1)^2 + (4+1)^2 + (5+1)^2 + (6+1)^2 = 16 + 25 + 36 + 49 = 126$

41. $\displaystyle\sum_{k=1}^{6} (k+1)2^{k-1} = 2 \cdot 2^0 + 3 \cdot 2^1 + 4 \cdot 2^2 + 5 \cdot 2^3 + 6 \cdot 2^4 + 7 \cdot 2^5 =$ $2 + 6 + 16 + 40 + 96 + 224 = 384$

43. $\displaystyle\sum_{k=1}^{10} (k-1)^2 = 0^2 + 1^2 + 2^2 + 3^2 + 4^2 + 5^2 + 6^2 + 7^2 + 8^2 + 9^2$

45. $\displaystyle\sum_{k=1}^{50} \frac{3^k}{2^{k+1}} = \frac{3}{2^2} + \frac{3^2}{2^3} + \frac{3^3}{2^4} + \frac{3^4}{2^5} + \cdots + \frac{3^{49}}{2^{50}} + \frac{3^{50}}{2^{51}}$

47. $3 + 6 + 9 + 12 + \cdots + 99 = 3(1) + 3(2) + 3(3) + \cdots + 3(33) = \displaystyle\sum_{k=1}^{33} 3k$

49. $1 \cdot 2^3 + 2 \cdot 2^4 + 3 \cdot 2^5 + 4 \cdot 2^6 + \cdots + 100 \cdot 2^{102} = (1)2^{(1)+2} + (2)2^{(2)+2} + (3)2^{(3)+2} + (4)2^{(4)+2} + \cdots + (100)2^{(100)+2} = \displaystyle\sum_{k=1}^{100} k \cdot 2^{k+2}$

51. $1 + 0.9 + (0.9)^2 + \cdots + (0.9)^5$ is a geometric series with $a = 1$ and $r = \frac{0.9}{1} = 0.9$. Thus, the sum of the series is $S_6 = \dfrac{1 - (0.9)^6}{1 - 0.9} = \dfrac{1 - 0.531441}{0.1} = 4.68559$.

53. $1-\sqrt{5}+5-5\sqrt{5}+\cdots+625$ is a geometric series with $a=1$ and $r=-\sqrt{5}$. Then, $a_n=625=5^4=(-\sqrt{5})^{n-1} \quad\Leftrightarrow\quad n=9$, and so the sum of the series is $S_9=\dfrac{1-(-\sqrt{5})^9}{1-(-\sqrt{5})}$ $=\dfrac{1+5^{9/2}}{1+5^{1/2}}\approx 432.17$.

55. $\frac{1}{3}+\frac{2}{3}+1+\frac{4}{3}+\cdots+33$ is an arithmetic series with $a=\frac{1}{3}$ and $d=\frac{1}{3}$. Then, $a_n=33=\frac{1}{3}+\frac{1}{3}(n-1) \quad\Leftrightarrow\quad n=99$, and so the sum is $S_{99}=\frac{99}{2}\left(\frac{2}{3}+\frac{99-1}{3}\right)=\frac{99}{2}\cdot\frac{100}{3}=1650$.

57. $\sum\limits_{n=0}^{6}3\cdot(-4)^n$ is a geometric series with $a=3$, $r=-4$, and $n=7$. Therefore, the sum of the series is $S_7=3\cdot\dfrac{1-(-4)^7}{1-(-4)}=\dfrac{3}{5}(1+4^7)=9831$.

59. We have an arithmetic sequence with $a=7$ and $d=3$. Then $S_n=325=\frac{n}{2}[2a+(n-1)d]$ $=\frac{n}{2}[14+3(n-1)]=\frac{n}{2}(11+3n) \quad\Leftrightarrow\quad 650=3n^2+11n \quad\Leftrightarrow\quad (3n+50)(n-13)=0$ $\Leftrightarrow\quad n=13$ (because $n=-\frac{50}{3}$ is inadmissible). Thus, 13 terms must be added.

61. We have an arithmetic series with $S_8=100$ and $a=2$. Then $S_8=100=\frac{8}{2}(2+2+7d)$ $=16+28d \quad\Leftrightarrow\quad d=3$. Therefore, $a_{10}=2+9\cdot3=29$.

63. Initial population $=a=100{,}000$ and the rate of increase is 10%. After the 1st year the population will be $100{,}000\cdot1.1$, and the population after 10 years will be $100{,}000\cdot(1.1)^{10}=259{,}374$. In general, the population after n years will be $P_n=100{,}000\cdot(1.1)^n$. (This is a geometric sequence with $a=100{,}000$ and $r=1.1$.)

65. $R=1000$, $i=0.08$, and $n=16$. Thus, $A=1000\dfrac{1.08^{16}-1}{1.08-1}=12{,}500[1.08^{16}-1]=\$30{,}324.28$.

67. $A=60{,}000$ and $i=\dfrac{0.09}{12}=0.0075$.

(a) If the period is 30 years, $n=360$ and $R=\dfrac{60{,}000\cdot0.0075}{1-1.0075^{-360}}=\$482.77$.

(b) If the period is 15 years, $n=180$ and $R=\dfrac{60{,}000\cdot0.0075}{1-1.0075^{-180}}=\$608.56$.

69. $0.1+0.01+0.001+0.0001+\cdots$ is an infinite geometric series with $a=0.1$ and $r=0.1$. Therefore, the sum is $S=\frac{0.1}{1-0.1}=\frac{1}{9}$.

71. $a+ab^2+ab^4+ab^6+\cdots$ is an infinite geometric series with first term a and common ratio b^2. Thus, the sum is $S=\dfrac{a}{1-b^2}$.

73. Let $P(n)$ denote the statement that $1^4+2^4+3^4+\cdots+n^4=\dfrac{n(n+1)(2n+1)(3n^2+3n-1)}{30}$.

Step 1 $P(1)$ is the statement: $1^4=\dfrac{1\cdot2\cdot3\cdot5}{30}$, which is clearly true.

Step 2 Assume that $P(k)$ is true; that is

$1^4 + 2^4 + 3^4 + \cdots + k^4 = \dfrac{k(k+1)(2k+1)(3k^2+3k-1)}{30}$. We want to use this to show that $P(k+1)$ is true.

$$1^4 + 2^4 + 3^4 + \cdots + k^4 + (k+1)^4 =$$

$$\frac{k(k+1)(2k+1)(3k^2+3k-1)}{30} + (k+1)^4 = \qquad \text{induction hypothesis}$$

$$= \frac{k(k+1)(2k+1)(3k^2+3k-1) + 30(k+1)^4}{30}$$

$$= \frac{(k+1)[k(2k+1)(3k^2+3k-1) + 30(k+1)^3]}{30}$$

$$= \frac{(k+1)(6k^4 + 39k^3 + 91k^2 + 89k + 30)}{30}$$

$$= \frac{(k+1)(k+2)(2k+3)(3k^2+9k+5)}{30} \qquad \text{Factor using the methods of Chapter 4}$$

$$= \frac{(k+1)(k+2)(2k+3)(3k^2+6k+3+3k+3-1)}{30}$$

$= \dfrac{(k+1)(k+1+1)[2(k+1)+1][3(k+1)^2+3(k+1)-1]}{30}$, which is exactly $P(k+1)$. Thus, $P(k+1)$ follows from $P(k)$. So by the Principle of Mathematical Induction, $P(n)$ is true for all n.

75. Let $P(n)$ denote the statement that $\left(1+\frac{1}{1}\right)\left(1+\frac{1}{2}\right)\left(1+\frac{1}{3}\right)\cdot \cdots \cdot \left(1+\frac{1}{n}\right) = n+1$.

Step 1 $P(1)$ is the statement: $\left(1+\frac{1}{1}\right) = 1+1$ which is clearly true.

Step 2 Assume that $P(k)$ is true; that is, $\left(1+\frac{1}{1}\right)\left(1+\frac{1}{2}\right)\left(1+\frac{1}{3}\right)\cdot \cdots \cdot \left(1+\frac{1}{k}\right) = k+1$. We want to use this to show that $P(k+1)$ is true.

$$\left[\left(1+\tfrac{1}{1}\right)\left(1+\tfrac{1}{2}\right)\left(1+\tfrac{1}{3}\right)\cdot \cdots \cdot \left(1+\tfrac{1}{k}\right)\right]\left(1+\tfrac{1}{k+1}\right) =$$

$$(k+1)\left(1+\frac{1}{k+1}\right) = \qquad \text{induction hypothesis}$$

$$(k+1)+1 =$$

which is exactly $P(k+1)$. Thus, $P(k+1)$ follows from $P(k)$. So by the Principle of Mathematical Induction, $P(n)$ is true for all n.

77. Let $P(n)$ denote the statement that $11^{n+2} + 12^{2n+1}$ is divisible by 133.

Step 1 $P(1)$ is the statement is $11^3 + 12^3 = 3059$ and since $3059 = 23 \cdot 133$, the statement is true.

Step 2 Assume that $P(k)$ is true; that is, $11^{k+2} + 12^{2k+1}$ is divisible by 133. We want to use this to show that $P(k+1)$ is true. Now, $11^{k+1+2} + 12^{2(k+1)+1} = 11^{k+3} + 12^{2k+3}$
$= 11 \cdot 11^{k+2} + 144 \cdot 12^{2k+1} = (144-133)11^{k+2} + 144 \cdot 12^{2k+1}$
$= 144 \cdot 11^{k+2} + 144 \cdot 12^{2k+1} - 133 \cdot 11^{k+2} = 144(11^{k+2} + 12^{2k+1}) - 133 \cdot 11^{k+2}$ which is divisible by 133 because $11^{k+2} + 12^{2k+1}$ is divisible by 133 by our induction hypothesis and

clearly 133 is divisible by 133. Thus, $P(k+1)$ follows from $P(k)$, and so by the Principle of Mathematical Induction, $P(n)$ is true for all n.

79. Let $P(n)$ denote the statement that F_{4n} is divisible by 3.

Step 1 Show that $P(1)$ is true. But $P(1)$ is true since $F_4 = 3$ is divisible by 3.

Step 2 Assume that $P(k)$ is true; that is, F_{4k} is divisible by 3. We want to use this to show that $P(k+1)$ is true. Now, $F_{4(k+1)} = F_{4k+4} = F_{4k+2} + F_{4k+3}$
$= (F_{4k} + F_{4k+1}) + (F_{4k+1} + F_{4k+2}) = F_{4k} + F_{4k+1} + F_{4k+1} + (F_{4k} + F_{4k+1})$
$= 2 \cdot F_{4k} + 3 \cdot F_{4k+1}$ which is divisible by 3 because F_{4k} is divisible by 3 by our induction hypothesis and $3 \cdot F_{4k+1}$ is clearly divisible by 3. Thus, $P(k+1)$ follows from $P(k)$, and so by the Principle of Mathematical Induction, $P(n)$ is true for all n.

81. $\binom{5}{2}\binom{5}{3} = \frac{5!}{2!3!} \cdot \frac{5!}{3!2!} = \frac{5 \cdot 4}{2} \cdot \frac{5 \cdot 4}{2} = 10 \cdot 10 = 100$

83. $\sum_{k=0}^{5} \binom{5}{k} = \binom{5}{0} + \binom{5}{1} + \binom{5}{2} + \binom{5}{3} + \binom{5}{4} + \binom{5}{5} = 2\left(\frac{5!}{0!\,5!} + \frac{5!}{1!\,4!} + \frac{5!}{2!\,3!}\right)$
$= 2(1 + 5 + 10) = 32$

85. $(1 - x^2)^6 = \binom{6}{0}1^6 - \binom{6}{1}1^5x^2 + \binom{6}{2}1^4x^4 - \binom{6}{3}1^3x^6 + \binom{6}{4}1^2x^8 - \binom{6}{5}x^{10}$
$+ \binom{6}{6}x^{12} = 1 - 6x^2 + 15x^4 - 20x^6 + 15x^8 - 6x^{10} + x^{12}$

87. There are 23 terms in the expansion of $(a+b)^{22}$ and the 20th term occurs when $r = 19$. Thus, the 20th term is $\binom{22}{19}a^3b^{19} = 1540a^3b^{19}$.

89. The term that contains s^5 in the expansion of $\left(\frac{s^3}{2} - \frac{2}{s^2}\right)^5$ occurs when $\left(\frac{s^3}{2}\right)^{5-r}\left(\frac{2}{s^2}\right)^r = s^5 \Leftrightarrow$
$3(5) - 3r - 2r = 5 \Leftrightarrow 5r = 15 - 5 = 10 \Leftrightarrow r = 2$. Hence, this term is
$\binom{5}{2}\left(\frac{s^3}{2}\right)^3\left(-\frac{2}{s^2}\right)^2 = 10 \cdot \frac{s^9}{8} \cdot \frac{4}{s^4} = 5s^5$ and the coefficient is 5.

91. $(a+1)^3 - 3(a+1)^2 + 3(a+1) - 1 = [(a+1) - 1]^3 = a^3$

Chapter 9 Test

1. $a_n = \frac{n}{1-n^2}$. Then $a_{10} = \frac{10}{1-10^2} = -\frac{10}{99}$.

2. $a_{n+2} = (a_n)^2 - a_{n+1}$, $a_1 = 1$ and $a_2 = 1$. Then $a_3 = a_1^2 - a_2 = 1^2 - 1 = 0$, $a_4 = a_2^2 - a_3 = 1^2 - 0 = 1$, and $a_5 = a_3^2 - a_4 = 0^2 - 1 = -1$.

3. 80, 76, 72, ... is an arithmetic sequence with a = 80 and $r = -4$. Thus, $a_{30} = 80 + 29(-4) = -36$.

4. We have a geometric sequence with $a_2 = 125$ and $a_5 = 1$. Since $\frac{a_5}{a_2} = \dfrac{ar^4}{ar} = r^3$, then $r^3 = \dfrac{1}{125}$ $\Leftrightarrow$ $r = \frac{1}{5}$. Therefore, $a_6 = a_5 \cdot r = 1 \cdot \frac{1}{5} = \frac{1}{5}$ and so the 6th term of this sequence is $\frac{1}{5}$.

5. (a) If $a_1, a_2, a_3, \ldots$ is an arithmetic sequence, then the sequence $a_1^2, a_2^2, a_3^2, \ldots$ is also arithmetic. This statement is FALSE. For a counterexample, consider $a_1 = 1, a_2 = 2, a_3 = 3\ldots$. Then $a_1, a_2, a_3, \ldots$ is an arithmetic sequence with $a = 1$ and $d = 1$. However, the sequence $a_1^2, a_2^2, a_3^2, \ldots$ becomes 1, 4, 9, ... which is NOT arithmetic because there is no common difference.

   (b) If $a_1, a_2, a_3, \ldots$ is a geometric sequence, then the sequence $a_1^2, a_2^2, a_3^2, \ldots$ is also geometric. This statement is TRUE. Let the common ratio for the geometric series $a_1, a_2, a_3, \ldots$ be r so that $a_n = a_1 r^{n-1}$, $n = 1, 2, 3, \ldots$. Then $a_n^2 = (a_1 r^{n-1})^2 = (a_1^2)(r^2)^{n-1}$. Therefore the sequence $a_1^2, a_2^2, a_3^2, \ldots$ is geometric with common ratio r^2.

6. (a) The sum of a finite arithmetic series is $S_n = \dfrac{n}{2}(\text{a} + a_n)$ or $S_n = \dfrac{n}{2}[2\text{a} + (n-1)d]$.

   (b) $a_1 = 10$ and $a_{10} = 2$. Then $S_{10} = \dfrac{10}{2}(10 + 2) = 60$.

   (c) Since $S_{10} = \dfrac{10}{2}(2 \cdot 10 + 9d) = 60$, then $5(20 + 9d) = 60$ $\Leftrightarrow$ $d = -\dfrac{8}{9}$ and so the common difference is $-\frac{8}{9}$. Also, $a_{100} = a + 99d = 10 - \frac{8}{9} \cdot 99 = -78$.

7. (a) The sum of a finite geometric series is $S_n = a\dfrac{1 - r^n}{1 - r}$ where $r \neq 1$.

   (b) The geometric series $\dfrac{1}{3} + \dfrac{2}{3^2} + \dfrac{2^2}{3^3} + \dfrac{2^3}{3^4} + \cdots + \dfrac{2^9}{3^{10}}$ has a $= \dfrac{1}{3}$, $r = \dfrac{2}{3}$, and $n = 10$. So,

$$S_{10} = \frac{1}{3} \cdot \frac{1 - \left(\frac{2}{3}\right)^{10}}{1 - \left(\frac{2}{3}\right)} = \frac{1}{3} \cdot 3\left(1 - \frac{1024}{59{,}049}\right) = \frac{58{,}025}{59{,}049}.$$

8. The infinite geometric series $1 + \dfrac{1}{2^{1/2}} + \dfrac{1}{2} + \dfrac{1}{2^{3/2}} + \cdots$ has a $= 1$ and $r = 2^{-1/2} = \dfrac{1}{\sqrt{2}}$. Thus,

$$S = \frac{1}{1 - \left(\frac{1}{\sqrt{2}}\right)} = \frac{\sqrt{2}}{\sqrt{2}-1} = \frac{\sqrt{2}}{\sqrt{2}-1} \cdot \frac{\sqrt{2}+1}{\sqrt{2}+1} = 2 + \sqrt{2}.$$

9. Let $P(n)$ denote the statement that $1^2 + 2^2 + 3^2 + \cdots + n^2 = \frac{n(n+1)(2n+1)}{6}$.

   Step 1 Show that $P(1)$ is true. But $P(1)$ says that $1^2 = \frac{1 \cdot 2 \cdot 3}{6}$, which is true.

   Step 2 Assume that $P(k)$ is true; that is, $1^2 + 2^2 + 3^2 + \cdots + k^2 = \frac{k(k+1)(2k+1)}{6}$. We want to use this to show that $P(k + 1)$ is true.

$$\begin{aligned} 1^2+2^2+3^2+\cdots+k^2+(k+1)^2 &= \frac{k(k+1)(2k+1)}{6} \quad \text{induction hypothesis} \\ &= \frac{k(k+1)(2k+1)+6(k+1)^2}{6} \\ &= \frac{(k+1)(2k^2+1)+(6k+6)(k+1)}{6} \\ &= \frac{(k+1)(2k^2+k+6k+6)}{6} \\ &= \frac{(k+1)(2k^2+7k+6)}{6} \\ &= \frac{(k+1)(k+2)(2k+3)}{6} \\ &= \frac{(k+1)(k+2)(2k+3)}{6} \end{aligned}$$

which is exactly $P(k+1)$. Thus $P(k+1)$ follows from $P(k)$. So by the Principle of Mathematical Induction, $P(n)$ is true for all n.

10. (a) $\sum_{n=1}^{5}(1-n^2) = (1-1^2)+(1-2^2)+(1-3^2)+(1-4^2)+(1-5^2)$
$= 0-3-8-15-24 = -50$

(b) $\sum_{n=3}^{6}(-1)^n 2^{n-2} = (-1)^3 2^{3-2} + (-1)^4 2^{4-2} + (-1)^5 2^{5-2} + (-1)^6 2^{6-2}$
$= -2+4-8+16 = 10$

11. From the text, $(a+b)^n = \sum_{k=0}^{n}\binom{n}{k}a^{n-k}b^k$.

12. $(2x+y^2)^5 = \binom{5}{0}(2x)^5 + \binom{5}{1}(2x)^4y^2 + \binom{5}{2}(2x)^3\left(y^2\right)^2 + \binom{5}{3}(2x)^2\left(y^2\right)^3 + \binom{5}{4}(2x)\left(y^2\right)^4$
$+ \binom{5}{5}\left(y^2\right)^5 = 32x^5 + 80x^4y^2 + 80x^3y^4 + 40x^2y^6 + 10xy^8 + y^{10}$

13. The rth term in the expansion of $(2a+b)^{100}$ is $\binom{100}{r}(2a)^r b^{100-r} = \binom{100}{r}2^r a^r b^{100-r}$. The term that contains a^3 occurs when $r=3$. Thus, the term is $\binom{100}{3}(2a)^3b^{97} = 1{,}293{,}600a^3b^{97}$.

14. The rth term in the expansion of $\left(2x+\frac{1}{x}\right)^{10}$ is $\binom{10}{r}(2x)^r\left(\frac{1}{x}\right)^{10-r} = \binom{10}{r}2^r x^r\left(x^{-1}\right)^{10-r}$
$= \binom{10}{r}2^r x^{2r-10}$. Thus the term that does not contain x occurs when $2r-10=0 \quad \Leftrightarrow \quad r=5$.
This term is $\binom{10}{5}(2x)^5\left(\frac{1}{x}\right)^5 = \frac{10!}{5!5!}\cdot 32 = 8064$.

Focus on Problem Solving

1. By dividing the successive gnomons into squares with side 1, we see that the sum of their areas is $1 + 3 + 5 + \cdots + (2n - 1) = n^2$ because the total area is the area of an $n \times n$ square.

3. Using the formula from the page 649 we have:
$$1^3 + 2^3 + \cdots + (n-1)^3 + n^3 = (1 + 2 + \cdots + n)^2$$
$$1^3 + 2^3 + \cdots + (n-1)^3 = (1 + 2 + \cdots + (n-1))^2$$
Subtracting: $n^3 = (1 + 2 + \cdots + n)^2 - (1 + 2 + \cdots + (n-1))^2$

5. $S_1 = 1$
$S_2 = 1 + 4 = 5$
$S_3 = 5 + 18 = 23$
$S_4 = 23 + 96 = 119$
These numbers are all 1 less than factorials, so we conjecture that $S_n = (n+1)! - 1$. To prove this, we follow the hint and first note that $(k+1)! - k! = [(k+1) - 1]k! = k \cdot k!$. Thus $S_n = (2! - 1!) + (3! - 2!) + (4! - 3!) + \cdots + [(n+1)! - n!]$. We see that all the terms except $-1!$ and $(n+1)!$ cancel, and so $S_n = (n+1)! - 1$.

7. Let $P(n)$ be the statement that $\frac{n^5}{5} + \frac{n^4}{2} + \frac{n^3}{3} - \frac{n}{30}$ is an integer.

Step 1 $P(1)$ is true because $\frac{1}{5} + \frac{1}{2} + \frac{1}{3} - \frac{1}{30} = 1$.

Step 2 Suppose that $P(k)$ is true. Then $\frac{(k+1)^5}{5} + \frac{(k+1)^4}{2} + \frac{(k+1)^3}{3} - \frac{k+1}{30}$
$$= \frac{k^5 + 5k^4 + 10k^3 + 10k^2 + 5k + 1}{5} + \frac{k^4 + 4k^3 + 6k^2 + 4k + 1}{2} + \frac{k^3 + 3k^2 + 3k + 1}{3} - \frac{k+1}{30}$$
$$= \left[\frac{k^5}{5} + \frac{k^4}{2} + \frac{k^3}{3} - \frac{k}{30}\right] + \left[(k^4 + 2k^3 + 2k^2 + k) + (2k^3 + 3k^2 + 2k) + (k^2 + k)\right]$$
$+ \left[\frac{1}{5} + \frac{1}{2} + \frac{1}{3} - \frac{1}{30}\right]$. The first term is an integer by the induction hypothesis. The second term is an integer since k is an integer. The third term equals 1. So their sum is an integer. Therefore $P(k+1)$ is true. Thus $P(n)$ is true for all n by mathematical induction.

9. Since all rows have the same sum S, and there are n rows, the total sum is nS. But this total sum must be $1 + 2 + 3 + \cdots + n^2 = \frac{n^2(n^2+1)}{2}$ (using the formula for the sum of an arithmetic series or the formula in Example 2 in Section 9.7.) Therefore, $nS = \frac{n^2(n^2+1)}{2} \quad \Leftrightarrow \quad S = \frac{n(n^2+1)}{2}$.

11. (a) 16
(b) 2^{n-1} (one would think)
(c) No, we get 31 instead of 32.

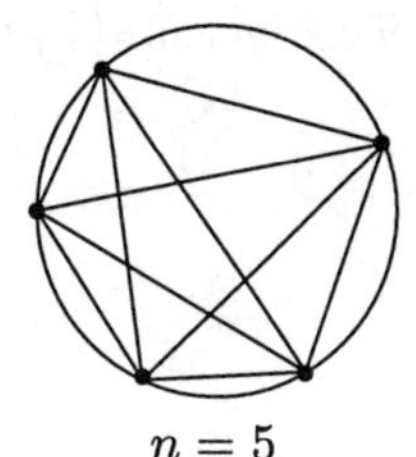

$n = 5$

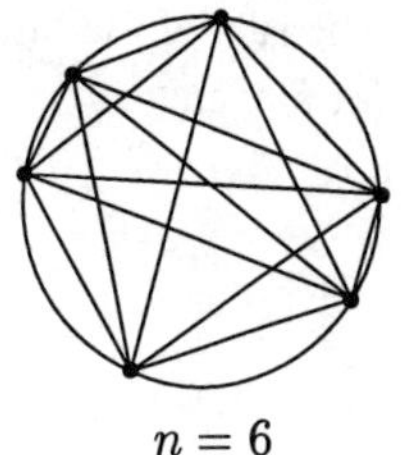

$n = 6$

(d)

n	1	2	3	4	5	6	7	8
Regions	1	2	4	8	16	31	57	99
First difference	1	2	4	8	15	26	42	
Second difference	1	2	4	7	11	16	22	
Third difference	1	2	3	4	5	6		
Fourth difference	1	1	1	1	1			

n	7	8	9	10
Regions	57	99	163	256
First difference	42	64	93	
Second difference	16	22	29	
Third difference	6	7	8	
Fourth difference	1	1	1	

From the difference table, we see that for $n = 7, 8, 9$, and 10 points, we get 57, 99, 163, and 256 regions, respectively.

13. (a) We assume $|r| < 1$. Following the hint, let S denote the sum. Thus

$$\begin{aligned} S &= 1 + 2r + 3r^2 + 4r^3 + \cdots \\ -rS &= \quad\; -r - 2r^2 - 3r^3 - \cdots \\ \hline S - rS &= 1 + r + r^2 + r^3 + \cdots \end{aligned}$$

The right hand side of this last equation is a geometric series whose sum is $\dfrac{1}{1-r}$, that is

$$S - rS = \frac{1}{1-r} \quad\Leftrightarrow\quad S(1-r) = \frac{1}{1-r} \quad\Leftrightarrow\quad S = \frac{1}{(1-r)^2}.$$

(b) $\dfrac{1}{6} + \dfrac{2}{6^2} + \dfrac{3}{6^3} + \dfrac{4}{6^4} + \cdots = \dfrac{1}{6}\left(1 + \dfrac{2}{6} + \dfrac{3}{6^2} + \dfrac{4}{6^3} + \cdots\right)$. So using $r = \dfrac{1}{6}$,

$$S = \frac{1}{6} \cdot \frac{1}{\left(1 - \frac{1}{6}\right)^2} = \frac{1}{6} \cdot \frac{1}{\frac{25}{36}} = \frac{6}{25}.$$

15. We expand $(1-1)^n$. We get

$0 = (1-1)^n = \binom{n}{0}(1)^n(-1)^0 + \binom{n}{1}(1)^{n-1}(-1)^1 + \binom{n}{2}(1)^{n-2}(-1)^2 + \cdots + \binom{n}{n}(1)^0(-1)^n \quad\Leftrightarrow$

$0 = \binom{n}{0} - \binom{n}{1} + \binom{n}{2} - \binom{n}{3} + \binom{n}{4} - \cdots$

Next adding the $\binom{n}{k}$'s for odd k to both sides yields

$\binom{n}{1} + \binom{n}{3} + \binom{n}{5} + \cdots = \binom{n}{0} + \binom{n}{2} + \binom{n}{4} + \cdots$

Thus the sum of the binomial coefficients $\binom{n}{k}$ for k odd is equal to the sum of binomial coefficients $\binom{n}{k}$ for k even.